2014 International SoC Design Conference

(ISOCC 2014)

Jeju, South Korea
3-6 November 2014

IEEE Catalog Number: CFP1469E-POD
ISBN: 978-1-4799-5128-4

Copyright © 2014 by the Institute of Electrical and Electronic Engineers, Inc
All Rights Reserved

Copyright and Reprint Permissions: Abstracting is permitted with credit to the source. Libraries are permitted to photocopy beyond the limit of U.S. copyright law for private use of patrons those articles in this volume that carry a code at the bottom of the first page, provided the per-copy fee indicated in the code is paid through Copyright Clearance Center, 222 Rosewood Drive, Danvers, MA 01923.

For other copying, reprint or republication permission, write to IEEE Copyrights Manager, IEEE Service Center, 445 Hoes Lane, Piscataway, NJ 08854. All rights reserved.

***This publication is a representation of what appears in the IEEE Digital Libraries. Some format issues inherent in the e-media version may also appear in this print version.**

IEEE Catalog Number: CFP1469E-POD
ISBN 13:   978-1-4799-5128-4

Additional Copies of This Publication Are Available From:

Curran Associates, Inc
57 Morehouse Lane
Red Hook, NY 12571 USA
Phone: (845) 758-0400
Fax: (845) 758-2633
E-mail: curran@proceedings.com
Web: www.proceedings.com

2014 International SoC Design Conference (ISOCC 2014)

Jeju, South Korea
3-6 November 2014

IEEE Catalog Number: CFP1469E-POD
ISBN: 978-1-47995-128-4

2014 International SoC Design Conference

Contents

Foreword ··· i

Message from TPC Chair ··· ii

Invitation to ISOCC 2015 ·· iii

Committee ·· iv

Time Table ·· vii

Tutorial ··· ix

Opening Ceremony ·· x ii

Keynote Speeches

Keynote Speech I
Flash Storage Innovations Reshape Cloud Data Centers
Duckhyun Chang(Senior Vice President / Samsung Electronics Co. Ltd., Korea) ··············· x iii

Keynote Speech II
Managing the Challenges of Embedded Vision Processing - VISIE Vision Processor Architecture
Wei-Jin Dai(CEO and President / Vivante Corporation, USA) ······················· x iv

Keynote Speech III
Designing Change - Leveraging Innovation and Collaboration
Joachim Kunkel(Senior Vice President and General Manager Solutions Group /
    Synopsys Inc., Germany) ·· x v

Keynote Speech IV
Is the Time-domain circuit a new paradigm for Analog?
- possibility and limit of the time-domain approach -
Kunihiro Asada (Professor / VLSI Design and Education Center,
    The University of Tokyo, Japan) ····································· x vi

Keynote Speech V
Modeling, Design, and EDA Research for Stacked-Die 3D IC at GTCAD Lab
Sung Kyu Lim (Dan Fielder Professor / School of Electrical and Computer Engineering,
    Georgia Institute of technology, USA) ······························· x vii

Keynote Speech VI
The Next PC: Personal Care Enabled by Internet of Medical Things
Liang-Gee Chen (Professor / National Taiwan University, Taiwan) ······················ x viii

2014 International SoC Design Conference

Foreword

"Welcome to 11ᵗʰ ISOCC 2014"

On behalf of the program committee, it is my great pleasure to invite you to the 11th International SoC Design Conference, ISOCC 2014. ISOCC is an annual conference providing an open forum for SoC designers to exchange information, to gain new knowledge, and to network with leading experts in this field. ISOCC 2014 will continue the tradition.

ISOCC 2014 will be held from November 3rd till 6th at Ramada Plaza, Jeju Island, Korea. The conference will have technical sessions of oral and poster presentation. The technical sessions will provide a timely chance to discuss the recent progress in diverse fields of SoC design. The keynote speeches will also be delivered by the leading experts to provide the deepest insights into SoC design. During your stay in Korea, I hope you will enjoy not only the technical diversity of topics but also the natural beauty of fantastic Jeju Island. I look forward to meeting you at ISOCC 2014.

Jun Rim Choi
General Chair
ISOCC 2014

2014 International SoC Design Conference

Message from TPC Chair

Welcome to ISOCC 2014, the 11th International SoC Design Conference at Ramada Plaza Jeju Hotel, Jeju, Korea. ISOCC 2014 is a feast for SoC engineers and contributors including authors, attendees, organizing committee members, and technical program committee members, and reviewers. This feast is especially enjoyable this year thanks to the efforts of the contributors.

ISOCC is an annual conference providing the world's premier SoC design forum for leading researchers from academia and industry. The theme of ISOCC 2014 is "SoC for Smart Connectivity" to make a vision together towards the impending technological era.

In 5 technical tracks, "Analog and Mixed signal Circuits", "Digital VLSI Circuits and Embedded Systems", "SoC Design Methodology", "Low Power & Power Management ICs", and "Application Specific SoCs and Emerging Technologies", 265 papers has been submitted from 14 countries for regular oral and poster sessions. The technical program committee has carefully chosen only a limited number of papers. As a result, 62 papers in the regular oral sessions and 55 papers in the poster session (total 117 papers) are selected to be presented, which makes the acceptance rate 44.2%. In addition to the contributed papers, 6 keynote speeches, 5 invited papers, 26 invited special session papers, 4 tutorials, industrial demos, and chip design contest demos (CDC) are all mixed harmoniously in ISOCC 2014. 6 keynote speeches are one of the highlights of ISOCC 2014 to hear from leaders of SoC industry and academia.

It is my great honor and privilege to serve as technical program committee chair with all the wonderful colleagues and contributors. On behalf of the technical program committee, I would like to express my sincere and heartful gratitude for your help and contribution to make ISOCC 2014 better.

Jinwook Burm
Technical Program Committee Chair
ISOCC 2014

2014 International SoC Design Conference

Invitation to ISOCC 2014

In 2015, we will continue ISOCC tradition in a very special city of Korea, Gyeongju. Being an ancient capital of Korea almost 1,000 years, Gyeongju possesses many artistic attractions and stories to go with state-of-the-art technical presentations. The recent inauguration of Korea Train eXpress (KTX) to Gyeongju allows easy access to ISOCC 2015. I wish to invite every one of you and colleagues to Gyeongju to enjoy your discussions and leisure in the beautiful resort, Gyeongju.

Jinwook Burm
General Chair, ISOCC 2015

2014 International SoC Design Conference

Organizing Committee

General Chair
Jun Rim Choi (Kyungpook National University, Korea)

General Co-Chairs
Jin-Ku Kang (Inha University, Korea)
Makoto Ikeda (The University of Tokyo, Japan)
Yeo Kiat Seng (Singapore Univ. of Technology and Design)
Shyh-Jye Jou (National Chiao Tung University, Taiwan)
Jun Jin Kong (Samsung Electronics, Korea)

General Vice Chairs
Shiho Kim (Yonsei University, Korea)
Chi Ho In (Semyung University, Korea)
Joongho Choi (University of Seoul, Korea)

Conference Secretary
Kyung Ki Kim (Daegu University, Korea)

Special Session Chairs
Chol Woo Kim (Korea University, Korea)
Woo Young Choi (Yonsei University, Korea)
Jongsun Park (Korea University, Korea)

Finance Chairs
Yoon Sik Lee (KETI, Korea)
Jin-Ku Kang (Inha University, Korea)
Min Kyu Song (Dongkuk University, Korea)
Kyung Ki Kim (Daegu University, Korea)

IEEE Liason Chairs
Myung Hun Sunwoo (Ajou University, Korea)
Yeon Mo Chung (Kyung Hee University, Korea)
Sang Bock Cho (University of Ulsan, Korea)
Jin Sang Kim (Kyung Hee University, Korea)
Publication Chairs
Yong Moon (Soongsil University, Korea)

2014 International SoC Design Conference

Committee

Kyeong-Sik Min (Kookmin University, Korea)
Yong Ho Song (Hanyang University, Korea)

Publicity Chairs
Changsik Yoo (Hanyang University, Korea)
Hyung Tak Kim (Hongik University, Korea)
Nak Woong Eum (ETRI, Korea)
Hyuk-Jae Lee (Seoul National University, Korea)

Local Arrangement Chairs
Seok Jun Ko (Jeju National University, Korea)
Dong Kyue Kim (Hanyang University, Korea)
Byungin Moon (Kyungpook National University, Korea)
Jaehoon Shim (Kyungpook National University , Korea)

Poster Session Chairs
Tae Wook Kim (Yonsei University, Korea)
Young-Chan Jang (Kumoh National Institute of Technology, Korea)

Chip Design Contest Chairs
Kyoung Rok Cho (Chungbuk National University, Korea)
Kwang Hyun Baek (Chung-Ang University, Korea)
Jin-Gyun Chung (Chonbuk National University, Korea)
Yong Chae Jeong (Chonbuk National University, Korea)

Honorary Chairs
Young Hwan Kim (POSTECH, Korea)
Hang Geun Jeong (Chonbuk National University, Korea)
Hi Seok Kim (Cheongju University, Korea)
Shin Il Lim (Seokyeong Unversity, Korea)
Kyeongsoon Cho (Hankuk University of Foreign Studies, Korea)
Kwang Sub Yoon (Inha University , Korea)

2014 International SoC Design Conference

Technical Program Committee

Technical Program Chair
Jinwook Burm (Sogang University, Korea)

Technical Program Co-Chairs
Ken Choi (Illinois Institute of Technology, USA)
Tony Tae Hyoung Kim (Nanyang Thchnological University, Singapore)
An-Yeu(Andy) Wu (National Taiwan University, Taiwan)

Technical Program Vice Chairs
Hyunchol Shin (Kwangwoon University, Korea)
Hanho Lee (Inha University, Korea)

Analog & Mixed Circuits
Min Kyu Song (Dongguk University, Korea)
Changsik Yoo (Hanyang University, Korea)
Kang-Yoon Lee (Sungkyunkwan University, Korea)

Low Power & PMICs
Jeongjin Roh (Hanyang University, Korea)
Kyeong-Sik Min (Kookmin University, Korea)

Digital Circuits & Embedded System
Tae Hee Han (Sungkyunkwan University, Korea)
Ji-Hoon Kim (Chungnam National University, Korea)

SoC Design Methodology
Youngmin Kim (UNIST, Korea)
Ikjun Chang (Kyunghee University, Korea)

Application Specific SoCs & Emerging Tech.
TaeWook Kim (Yonsei University, Korea)
Youngcheol Chae (Yonsei University, Korea)

2014 International SoC Design Conference

Time Table

Monday ~ Tuesday, November 3~4, 2014

Nov. 3 Monday			Time		Nov. 4 Tuesday							
Lobby	Ballroom 2	Ballroom 3	From	Till	Ballroom 1	Ballroom 2	Ballroom 3	Ballroom 4	Mara	Udo	Chuja	Lobby
			9:00	9:15		ETRI Aldebaran Demo	CDC-1	CDC-2	CDC-3	CDC-4	CDC-5	CDC & ETRI Demo & Panel 1
			9:15	9:30								
			9:30	9:45								
			9:45	10:00		Break						
			10:00	10:15		Opening Ceremony						
			10:15	11:00		Keynote - 1						
			11:00	11:45		Keynote - 2						
			11:45	12:30		Keynote - 3						
			12:30	13:30		Lunch						
			13:30	13:35		ETRI Aldebaran Demo	A1	DV	ET	LP	SS-A		CDC & ETRI Demo & Panel 2
			13:35	13:50			346	574	585	218	721		
			13:50	14:05			572	1191	189	1068	1155		
			14:05	14:20			369	364	655	486	1165		
			14:20	14:35			123	1003	145	192	1169		
			14:35	14:50			841	1052	391	967	1159		
			14:50	15:05				1060	742	225			
			15:05	15:20		CDC Tour Break						
			15:20	15:35								
Registration	Tutorial 1-1	Tutorial 2-1	15:35	15:50		ETRI Aldebaran Demo	COSAR Workshop	SS-B	748	SS-C	788	SS-D	1147
			15:50	16:05					755		558		1143
			16:05	16:20					569		1138		1086
			16:20	16:35					753		793		630
	Break		16:35	16:50					743		658		611
			16:50	17:05					640				
	Tutorial 1-2	Tutorial 2-2	17:05	17:20					1021				
			17:20	18:00		Break						
			18:00	18:20		Banquet						
			18:20	18:30								
	Welcome Reception		18:30	19:00								
			19:00	20:00								

A1 Analog and Mixed-Signal Techniques I
DV Digital Circuits and VLSI Architectures
ET Emerging technology
LP Power Electronics / Energy Harvesting Circuits
SS-A Invited Special Session: Near-Threshold Voltage Circuit Design
SS-B Invited Special Session: Image Signal Processing for Vision/Multimedia SoC
SS-C Invited Special Session: Analog/Digital Circuits for Mobile SoC
SS-D Invited Special Session: Design, Analysis and Tools for Integrated Circuits and Systems (DATICS)

vii

2014 International SoC Design Conference

Time Table

Wednesday ~ Thursday, November 5~6, 2014

From	Till	Ballroom 1	Ballroom 2	Ballroom 3	Ballroom 4	Mara	Udo	Lobby	Nov. 6 Thursday — Mara
8:30	8:45		1140	458	287	822	1099		
8:45	9:00		397	148	1186	998	1123		Discussion SoC for Smart Connectivity
9:00	9:15	MC	912	DC — 276	MP — 1136	IMMD — 360	SS-E — 854		
9:15	9:30		383	686	747		1054		
9:30	9:45			212	804		870		
9:45	10:00	Break							
10:00	10:15	Break							
10:15	11:00	Keynote - 4							
11:00	11:45	Keynote - 5							
11:45	12:30	Keynote - 6							
12:30	13:30	Lunch							
13:30	13:35								
13:35	13:50			1179	659	990	327		
13:50	14:05			641	470	424	859		
14:05	14:20			615	1026	209	619		
14:20	14:35		HI	995	A2 — 1104	ME — 163	SD — 716		
14:35	14:50			223	384	363	433		
14:50	15:05			196	825	259	430		
15:05	15:20			388	971		714		
15:20	15:35	Break							
15:35	15:50							Poster Session	
15:50	16:05							Poster Session	
16:05	16:20							Poster Session	
16:20	16:35							Poster Session	
16:35	16:50							Poster Session	
16:50	17:05							Poster Session	
17:05	17:15	Break							
17:15	18:00	Closing Ceremony							

MC Multimedia (A/V) Algorithm and SoCs/Communication SoCs

DC Data Converters

IMMD Image Sensors/Sensors and MEMS Circuits/Biomedical SoCs

MP Processors / Multi-Core Architectures & Software

HI High-Speed Signal Interfaces/Wireline and Wireless ICs (RF ICs)

A2 Analog and Mixed-Signal Techniques II

ME Memory Circuits and Embedded Memory/Embedded Systems

SD SoC Testing/Design Verification/Signal Integrity / Interconnect Modeling and Simulation

SS-E Invited Special Session:Design Challenges for Advanced Mobile Intelligence: Low Power, Object Detection, and Mobile Security

2014 International SoC Design Conference

Tutorials

15:05~18:20 Monday, November 3, 2014

[Tutorial 1-1(15:05~16:35 Ballroom 2)]
Design Technologies for Application Processor

Youngmin Shin
Master, Samsung Electronics Co., Ltd., Korea

Biography

Youngmin Shin is a Samsung Master of the Semiconductor Business, Samsung Electronics Co., Ltd.
He is a heading of Advanced Design & Technology Group in SoC Processor development team. His role in Samsung is to develop next generation ARM CPU using new process and work for technology leadership. After received BS degree from Yonsei University, Seoul, Korea in 1988, he has worked for Samsung only around 24 years. He worked for 0.35~0.13um Alpha CPU joint-project with DEC(1996~2001). He has developed high performance ARM CPUs after 2002 not only for mobile CPU but also ARM CPU for consumer SoC from 130nm to 14nm.

Abstract

Application Processor SoC market is growing very fast in volume and becoming more intelligent and powerful. From 2012, volume of Smartphone is higher than that of PC. In order to get more competitiveness, high performance and low power IPs must be needed in this Application Processor SoC. In this lecture, he will present how CPU and GPU in Application Processor has been developed for this purpose and what kind of design techniques are used for Application Processor. He also present industry trend of mobile CPU development.

[Tutorial 1-2(16:50~18:20 Ballroom 2)]
Current and Future Trends of the Specialized Non-mobile CMOS Image Sensor

Juil Lee
Dongbu HiTek, Co., Ltd., Korea

2014 International SoC Design Conference

Biography

Bachelor of Science in Metallurgical Engineering at Yonsei University (1984~1988)
Master of Science in Metallurgical Engineering at Yonsei University (1988~1990)
Hyundai Electronics Industry (1992~1999)
   - 0.5um~0.35um CIS Tech. Development (8.0um~5.6um Pixel)
Hynix (1999~2003)
   - 0.18um CIS Tech. Development (5.0um~3.2um Pixel)
Magnachip (2003~2008)
   - 0.13um~0.11um CIS Tech. Development (2.8um~1.75um Pixel)
Dongbu HiTek (2008~Present)
   - 0.13um~0.11um CIS Tech. Development (2.2um~1.4um Pixel)
   - Under 90nm CIS Tech. Development (1.1um Pixel)

Abstract

The sensor is an easy device to artificially reproduce, or even transcend a person's senses. Humans have five senses - sight, hearing, touch, smell, and taste. Among them, sensors related to sight and hearing can be easily reproduced using semiconductor material, but sensors related with touch, smell, and taste are relatively difficult to reproduce. In this lecture, I will introduce the CMOS Image Sensor, with a focus on the types and characteristics of the specialized non-mobile CMOS Image Sensor. This type of sensor is generally used for automotive sensors, security sensors, ambient light/proximity sensors, 3D TOF sensors, line/linear sensors, touch screen sensors, rain sensors, optical mouse sensors, fingerprint sensors, smoke sensors, encoder sensors, and machine vision sensors. I will explain the various technological developments of the specialized non-mobile CMOS Image Sensor, and will conclude with a forecast of future trends. A shift has occurred these days from a conventional mobile CMOS Image Sensor to a more specialized non-mobile CMOS Image Sensor. The key words of the shift are convergence, global leading technology, differentiation, value adding and many products with small volume.

[Tutorial 2-1(15:05~16:35 Ballroom 3)]
Device Noise-Aware Analog Design and Ultra-Low Power Analog Circuit Design in an Ubiquitous World

Jaejin Park
Master, Samsung Electronics Co., Ltd., Korea

Biography

Jaejin Park received the Ph. D degree at Carnegie Mellon University, USA. He is a Samsung Master (technical vice president) in System LSI division, Samsung Electronics Co., Ltd. His interests are the design of data converters (ADC/DAC), clock generators (PLL/DLL), and high-speed interfaces. His recent works include analog device engineering, analog/mixed signal design methodology, and ultra-low power circuits. He has coauthored internal and external publications and hold patents in the analog design group.

2014 International SoC Design Conference

Abstract

The first part of this talk covers fundamentals of thermal, Flicker, and RTS (Random Telegraph Signal) noises and the analog device noise analysis based on input referred noise. The device noise-aware analog design and methodology are also presented with an example. The second part shows requirements and challenges of ultra-low power dissipation in an ubiquitous world and approaches to address facing problems are presented with analog techniques of several sensor conditioning circuits and energy harvesting.

[Tutorial 2-2(16:50~18:20 Ballroom 3)]
Electrostatic Discharge (ESD) Protection of Automotive Electronics in High-Voltage Si BiCMOS/BCD Technologies

Juin J. Liou
Professor, University of Central Florida, USA

Biography

Juin J. Liou received the B.S. (honors), M.S., and Ph.D. degrees in electrical engineering from the University of Florida, Gainesville, in 1982, 1983, and 1987, respectively. In 1987, he joined the Department of Electrical and Computer Engineering at the University of Central Florida (UCF), Orlando, Florida where he is now the UCF Pegasus Distinguished Professor and Lockheed Martin St. Laurent Professor of Engineering. His current research interests are Micro/nanoelectronics computer-aided design, RF device modeling and simulation, and electrostatic discharge (ESD) protection design and simulation. Dr. Liou has served as the IEEE EDS Vice-President of Regions/Chapters, IEEE EDS Treasurer, IEEE EDS Finance Committee Chair, Member of IEEE EDS Board of Governors and Member of IEEE EDS Educational Activities Committee.

Abstract

Electrostatic discharge (ESD) is one of the most prevalent threats to electronic components. It is an event in which a finite amount of charge is transferred from one object (i.e., human body) the other (i.e., microchip). This process can result in a very high current passing through the microchip within a very short period of time, and more than 35% of chip damages can be attributed to such an event. As such, designing on-chip ESD structures to protect integrated circuits against ESD stresses is a high priority in the semiconductor industry. The continuing advancement of MOS technology makes the ESD-induced failures even more prominent, and one can predict with certainty that the availability of effective and robust ESD protection solutions will become a critical and essential component to the successful commercialization of the modern and future MOS-based electronics. An overview on the ESD sources, models, protection schemes, and testing will first be given in this tutorial. This is followed by presenting the approaches and challenges of designing and realizing ESD protection solutions for automotive electronics integrated circuits fabricated in the high-voltage BiCMOS/Bipolar-CMOS-DMOS (BCD) technologies.

Outline

1. Overview of ESD (i.e., ESD standards, ESD protection schemes, ESD testing)
2. Background and challenges of designing ESD protection solutions for high-voltage integrated circuits (i.e., automotive integrated circuits)
3. Development of effective ESD protection solutions for automotive electronics in high-voltage Si BiCMOS/BCD technologies

10:00~10:15 Tuesday, November 4, 2014
RAMADA Ballroom 1 (2F)

Chair : Jinwook Burm (Sogang University, Korea)

Welcome Address

Jun Rim Choi, General Chair (Kyungpook National University, Korea)

Conference Statistics

Jinwook Burm, TPC Chair (Sogang University, Korea)

Announcements

Jinwook Burm, TPC Chair (Sogang University, Korea)

2014 International SoC Design Conference

Keynote Speech I
10:15~11:00 Tuesday, November 4, 2014

Duckhyun Chang
Senior Vice President
Samsung Electronics Co. Ltd., Korea

Flash Storage Innovations Reshape Cloud Data Centers

Biography

Dr. Duck-Hyun Chang was appointed as the Senior Vice President of Solution Product & Development in December of 2013 since he joined Samsung in 2002. After working in Mentor Graphics and Motorola Semiconductor, he served as leader of High Speed Interface Core Development and Storage Controller Design in System LSI Division before moving to Memory Division in 2009. He received his B.S. in Electronic Engineering from Seoul National University in 1986 and Ph.D. in Electrical Engineering from University of Florida in 1997.

Abstract

The demand for quickly storing and retrieving a large amount of data is skyrocketing. Traditional rotating media (hard drives) have failed to offer adequate I/O capabilities under energy and space constraints. Inevitably, flash memory solid-state drives are rapidly replacing hard drives in modern cloud data centers.

In this keynote speech, he will first explain why this trend will accelerate in the future with continued innovations in the flash memory technology. Furthermore, he will discuss advances in today's flash memory solutions, including sophisticated SoC designs, complex media management software, and robust system design techniques.

Finally, he will touch upon "in-storage computing", a new big data processing paradigm where data are processed within the solid-state drive, for improved energy consumption and performance of the overall system.

2014 International SoC Design Conference

Keynote Speech II

11:00~11:45 Tuesday, November 4, 2014

Wei-Jin Dai
CEO and President
Vivante Corporation, USA

Managing the Challenges of Embedded Vision Processing - VISIE Vision Processor Architecture

Biography

Wei-Jin has 30 years of experience managing both business and product development. Vivante Corporation, a leader in multi-core GPU, OpenCLTM, CPC Composition Engine and Vector Graphics IP solutions, provide the highest performance and lowest power silicon characteristics across a range of KhronosTM Group API conformant standards based on the Vega architecture. Previously Corporate Vice President at Cadence Design Systems, Wei-Jin has also driven the success of multiple startup companies, and he co-founded Silicon Perspective Corporation. He has also held engineering and management positions at Hewlett Packard and Lucent Bell Labs. Wei-Jin holds M.S. and B.S. degrees in Electrical Engineering and Computer Science from the University of California, Berkeley.

Abstract

Embedded VISION is at a tipping point where widespread adoption of devices that "see and understand" the surrounding physical world will change many industries and applications. Vivante has been working with leading industry partners to deliver high performance, real-time, intelligent vision processing in mobile power budgets for smartphones, automotive ADAS (Advanced Driver Assistance System), and security/surveillance.

This keynote will highlight exciting innovations and disruptive technologies that enable GPU vision in Vivante GC7000 VX Series GPUs. GC7000 VX is the industry's first GPU product line that enables photorealistic 3D rendering and embedded vision acceleration on the same processing core. This innovation is achieved through Vivante's latest Dynamic VLIW vision instruction set and enhanced shader extensions that achieve single cycle efficiency for most Khronos® OpenVX™ (Vision Acceleration) API instructions, multi-threaded unified cache system, Stream Interface and fine-grained power control to keep power consumption and thermals at mobile levels for real time vision processing. Vision use cases highlighted in the presentation include examples of collaboration with Freescale/Automotive OEMs, and partners in the security market.

2014 International SoC Design Conference

Keynote Speech III

11:45~12:30 Tuesday, November 4, 2014

Joachim Kunkel
Senior Vice President and General Manager Solutions Group,
Synopsys Inc., Germany

Designing Change – Leveraging Innovation and Collaboration

Biography

Joachim Kunkel joined Synopsys in 1994 and is currently senior vice president and general manager of the Solutions Group. In that capacity, he manages the business unit responsible for Synopsys DesignWare intellectual property (IP), FPGA implementation tools, and prototyping solutions. Before coming to Synopsys, Mr. Kunkel was co-founder of CADIS GmbH in Aachen, Germany. There, he served as managing director and performed myriad duties in engineering, sales and marketing. Mr. Kunkel holds an MSEE degree, the Dipl.-Ing. der Nachrichtentechnik, from the Aachen University of Technology.

Abstract

In the semiconductor design community, the word "change" is, to say the least, an understatement. From a dizzying array of emerging "smart" niche end-products to major market trend shifts to ecosystem reconfigurations at every level, the world around us is changing at an unprecedented pace. The demand on designers to innovate ever faster is driving everything from increased IP usage to prototyping and deeper ecosystem collaboration. In his presentation, Joachim will provide an overview of many years of investment and innovation efforts, resulting in an exciting sweep of major new design productivity advancements in our design and verification platforms.

2014 International SoC Design Conference

Keynote Speech IV

10:15~11:00 Wednesday, November 5, 2014

Kunihiro Asada
Professor
VLSI Design and Education Center, The University of Tokyo, Japan

Is the Time-domain circuit a new paradigm for Analog?
- possibility and limit of the time-domain approach

Biography

Kunihiro Asada received the B. S., M. S., and Ph.D. from University of Tokyo in 1975, 1977, and 1980, respectively. In 1980 he joined the Faculty of Engineering, University of Tokyo. From 1985 to 1986 he stayed at Edinburgh University as a visiting scholar. From 1990 to 1992 he served as the Editor of IEICE Transactions on Electronics. In 1996 he established VDEC in University of Tokyo. He served as the Chair of IEEE/SSCS Japan Chapter in 2001-2002 and the Chair of IEEE Japan Chapter Operation Committee in 2007-2008. He is currently professor, director of VDEC. His research interest is design and evaluation of integrated systems and component devices. He is a member of IEEE, IEICE and IEEJ.

Abstract

One of serious issues in the conventional analog circuits for advanced CMOS technologies is degradation of dynamic range due to lower supply voltage. It is a good contrast with time-domain circuits, where sharper edge of voltage transition results in lower timing jitter with improved dynamic range. So, many researchers have proposed time-domain circuits as replacements of voltage domain circuits, such as TDAs (time difference amplifier), TDCs (time-to-digital-converter), PLLs and so on. However, it is not yet proved systematically, whether time-domain is really advantageous over voltage-domain in future CMOS technologies.

In this presentation, after discussing similarity and dissimilarity between voltage-domain circuits and time-domain circuits, a comparison in terms of dynamic range will be introduced for advanced CMOS technologies, by means of numerical simulation and theoretical analysis. It will clarify conditions for time-domain circuits to be advantageous over voltage domain. It will also show that there is no fundamental difference between voltage-domain and time-domain in specific conditions. Finally, some of our recent results of time-domain circuits will be introduced as concrete examples of time-domain circuits.

Keywords - dynamic range, voltage-domain circuit, time-domain circuit, TDA, TDC, PW-PLL, CDR

2014 International SoC Design Conference

Keynote Speech V

11:00~11:45 Wednesday, November 5, 2014

Sung-Kyu Lim
Dan Fielder Professor,
School of Electrical and Computer Engineering,
Georgia Institute of technology, USA

Modeling, Design, and EDA Research for Stacked-Die 3D IC at GTCAD Lab

Biography

Sung Kyu Lim holds the Dan Fielder Professorship at the School of Electrical and Computer Engineering, Georgia Institute of Technology. He received the Ph.D. degree from UCLA in 2000. His research focus is on the architecture, design, test, and EDA solutions for 3D ICs. His research on 3D IC reliability is featured as a Research Highlight in the Communication of the ACM in 2014.

Dr. Lim received the NSF CAREER Award in 2006. Some of his recent Best Paper Award nominations include DAC (2011, 2012, 2014), ISPD (2014), and ISLPED (2012). He is currently an Associate Editor of the IEEE Transactions on Computer-Aided Design of Integrated Circuits and Systems (TCAD).

Abstract

This talk presents an overview of stacked-die 3D IC research at the Georgia Tech Computer-Aided Design (GTCAD) laboratory. First, we present commercial-grade RTL-to-GDSII design and analysis tools for 3D ICs built with (1) through-silicon-via (TSV) and (2) monolithic inter-tier via (MIV). Second, we build large-scale 3D IC designs using the tools and discuss their power, performance, area, reliability (PPAR) tradeoffs.

Our designs are based on open source and foundry PDKs at 130nm, 45nm, and 28nm. We also study PPAR tradeoffs among various die stacking options including face-to-face, face-to-back, and monolithic integration. Lastly, we present our 2-tier test chip that features 64 general-purpose cores and SRAM presented at ISSCC 2012. This is arguably the first general-purpose many-core 3D processor ever developed in academia and fully tested using real applications.

2014 International SoC Design Conference

Keynote Speech Ⅵ

11:45~12:30 Wednesday, November 5, 2014

Liang-Gee Chen
Professor, Vice President
National Taiwan University, Taiwan

The Next PC: Personal Care Enabled by Internet of Medical Things

Biography

Prof. Liang-Gee Chen received the B.S., M.S., and Ph.D. degrees in electrical engineering from National Cheng Kung University, Tainan, Taiwan, R.O.C. in 1979, 1981, and 1986, respectively. In 1988, he joined the Department of Electrical Engineering, National Taiwan University. Currently, he serves as the Chair Professor and received the National Professorship of Taiwan. He is the IEEE Fellow from 2001. His research interests include DSP architecture design, video processor design, and video coding systems. He has over 420 publications and 15 US patents. He has conducted more than 100 technology transfers and helped two start-ups IPO successfully.
Dr. Chen has served as Editorial Board for many IEEE Transactions in VLSI, Video Technology and Circuits and Systems. He also have been invited as the Chair of technical program committee for 2009 IEEE ICASSP and ISCAS 2012. In 2011, he received the Best paper award of IEEE CICC, Custom Integrated Circuits Conference.
Dr. Chen is also an enthusiastic entrepreneur. He is the founder of Taiwan IC design Society, CIC (Chip Implementation Center), NTU SOC center and Graduate Institute of Electronics Engineering. From 2007, he established the Creativity and Entrepreneurship Program at National Taiwan University. The program built the curriculum of innovation, entrepreneurship and start-up. It changes the culture of NTU students and become one of the most welcome program at NTU.

Abstract

It is widely accepted that preventive care armed with ICT technologies will play an important role in increasing the quality of healthcare and decreasing the medical costs. However, very few evidences have been reported to prove this common belief. We have established a telehealth center at National Taiwan University Hospital to safe-guard the patients at home using internet of the things (IOT) technologies. It is found that the mean number of monthly emergency department visits, the length of hospitalization, intensive care unit admissions and all medical costs are much reduced for the patients in the telehealth group. In addition to traditional vital sign medical devices, biomedical SoCs for virus/DNA/protein detection, ECG/neural signal processing, drug delivery, nerve stimulating, body cruise, and silicon brain have also been developed in support of the telehealth center. We believe that the advance of these new biomedical SoCs combined with the IOT technologies will open up a new PC (Personal Care) era for the welfare of the whole human being.

2014 International SoC Design Conference

Oral Session

Tuesday, November 4, 2014

A1

Analog and Mixed-Signal Techniques I

13:35~14:50 Ballroom 3
Chair: Hung-Wen Lin*(YuanZe University, Taiwan)*

[A1-1] 12.5-Gb/s Monolithically Integrated Optical Receiver With CMOS Avalanche Photodetector
Hyun-Yong Jung[1], Jeong-Min Lee[1], Jin-Sung Youn[1], Woo-Young Choi[1], and Myung-Jae Lee [2]
[1] Yonsei University, Korea
[2] Delft University of Technology, Netherlands 1

[A1-2] An Area Saving Inductor Current Sensor with Load Transient Enhancement in DC-DC Converter
Ngan K. Hoang[1], Xuan-Dien Do[1], Young-Jin Woo[2], and Sang-Gug Lee[1]
[1] KAIST, Korea
[2] Silicon Works Co. Ltd., Korea 3

[A1-3] A Low-IF AGC Amplifier for DSRC Receiver
Hung-Wen Lin, Wu-Wei Lin, and Chun-Yen Lin
YuanZe University, Taiwan 5

[A1-4] A 10-bit Fast Lock Data Recovery Compensating Pulse-Width Distortion for Isolated Data Communications
Hironobu Akita, Takasuke Ito, Keita Hayakawa, Nobuaki Matsudaira, Hirofumi Yamamoto, Chao Chen, Shigeki Ohtsuka, and Shinichirou Taguchi
DENSO Corporation, Japan 7

[A1-5] Auto-delay offset cancellation technique for time difference repeating amplifier
In-Seok Kong, Eun-Ho Yang, Kyung-Sub Son, Young-Jin Kim, and Jin-Ku Kang
Inha University,Korea 9

DV

Digital Circuits and VLSI Architectures

13:30~15:05 Ballroom 4
Chair: Saleh Abdel-hafeez*(Jordan University of Science and Technology, Jordan)*

x ix

2014 International SoC Design Conference

[DV-1] Invited Paper : A CMOS Low Power Biohybrid Lamprey Robot Controller Design
Yong-Bin Kim
Northeastern University, USA 11

[DV-2] Asynchronous Circuit Design using New High Speed NCL Gates
Minsu Choi[1], Byung-Ho Kang[2], Yong-Bin Kim[3], and Kyung Ki Kim[2]
[1] Missouri University of Science & Technology, USA
[2] Daegu University, Korea
[3] Northeastern University, USA 13

[DV-3] AES Sbox GF(24) inversion functions based PUFs
Hyunmin Kim and Seokhie Hong
Korea University, Korea 15

[DV-4] Modulo 2n+1 Squarer Design for Efficient Hardware Implementation
Rajashekhar Modugu[1], Yong-Bin Kim[2], Kyung Ki Kim[3], and Minsu Choi[4]
[1] Qualcomm, USA
[2] Northeastern University, USA
[3] Daegu University, Korea
[4] Missouri University of Science and Technology, USA 17

[DV-5] Low Power Challenge and Solution for Advanced Mobile Device Design
Haiqing Nan[1], Dong Ke[1], and Ken Choi[2]
[1] Intel Mobile Communications Technology Ltd., China
[2] Illinois Institute of Technology, USA 19

[DV-6] Distributed Architecture of Touch Screen Controller SoC for Large Touch Screen Panels
Gyeongseop Choi[1], M.G.A. Mohamed[1,2], and Hyung Won Kim[1]
[1] Chungbuk National University, Korea
[2] MiniaUniversity,Egypt 22

ET
Emerging Technology

13:30~15:05 Mara
Chair: Hyunchol Shin *(Kwangwoon University, Korea)*

[ET-1] Invited Paper : Recent Advances in ASIC-compatible Circuit Techniques for a SOC in Newly Emerging Application Areas
Sang H. Dhong[1] and Wei Hwang[2]
[1] Taiwan Semiconductor Manufacturing Company, Taiwan
[2] National Chiao-Tung University, Taiwan 24

[ET-2] Exploring Hybrid SRAM/MRAM L2 NUCA Stacked on 3D Chip-Multiprocessors
Seunghan Lee[1], Kyungsu Kang[2], Jongpil Jung[1], and Chong-Min Kyung[1]
[1] KAIST, Korea
[2] Samsung Electronics Corporation, Korea 26

[ET-3] A New Multi-target Detection for Vechicle Radar
Zhenyi Liu and Franklin Bien
Ulsan National University of Science and Technology, Korea 28

2014 International SoC Design Conference

[ET-4] Designing with FinFET Technology
Andrew Marshall
The University of Texas at Dallas, USA 30

[ET-5] Design and Performance Benchmarking of Steep-Slope Tunnel Transistors for Low Voltage Digital and Analog Circuits Enabling Self-Powered SOCs
Gaurav Kaushal[1], K. Subramanyam[2], Siva Nageswar Rao[2], G.Vidya[2], Radhika Ramya[2], Sadulla Shaik[2], H. Jeong1, S. O. Jung[1], and Ramesh Vaddi[3]
[1] Yonsei University, Korea
[2] VFSTR University, India1
[3] Shiv NadarUniversity, India 32

[ET-6] 77GHz frequency synthesizer for invention of radar systems
HyonGi Yoo and Franklin Bien
Ulsan National Institute of Science and Technology, Korea 34

LP

Power Electronics / Energy Harvesting Circuits

13:30~15:05 Udo
Chair: Kyeongsik Min *(Kookmon University, Korea)*

[LP-1] Invited Paper : Low Power Cross-Domain High-Voltage Transmitters for Battery Management Systems
Chih-Lin Chen, Yu-Hsun Su, and Chua-Chin Wang
National Sun Yat-Sen University, Taiwan 36

[LP-2] A Single Chip Li-Ion Battery Protection IC with Low Standby Mode Auto Realease
Seunghyeong Lee[2], Yongjae Jeong[1], Yungwi Song[1], and Jongsun Kim[2]
[1] SANBUD, Korea
[2] HongikUniversity, Korea 38

[LP-3] A High Energy Extraction Self-controllable CMOS Resonant Rectifier Circuit for Piezoelectric Energy Scavenging System
Amadud Din, Doyoung Chung, Dasom Park, Hyunsik Lee, and Jong Wook Lee
Kyung Hee University, Korea 40

[LP-4] A High Efficiency Multi-Channel LED Driver Based on Converter-Free Technique and Load Adaptive Method
Si FU, Minjie Chen, Xutao Lee, and Tsutomu Yoshihara
Waseda University, Japan 42

[LP-5] A Fully Integrated AC-DC Regulator Over Wide Frequency Range for Implantable Bio-Medical Devices
Qi Cheng[1,2], Liuyan Chen[1,2], and Jianping Guo[1,2]
[1] Sun Yat-sen University, China
[2] SYSU-CMU Shunde International Joint Research Institute, China 44

[LP-6] Dynamic Performance Improvement of DC-DC Buck Converter by Slope Adjustable Triangular Wave Generator
Shu Wu, Yasunori Kobori, and Haruo Kobayashi
Gunma University, Japan 46

2014 International SoC Design Conference

Wednesday, November 5, 2014

MC

Multimedia (A/V) Algorithm and SoCs / Communication

08:30~09:45 Ballroom 2
Chair: Truong Nguyen *(University of Califonia, San Diego, USA)*

[MC-1] Transfer Function Design for Volume Rendering using K-Means Clustering
Jae Hwan Lim and Young Hwan Kim
POSTECH, Korea 48

[MC-2] Implementation of Pedestrian Detection using a CENTRIST-ROI in Embedded Environment
Yun-seop Hwang, Chang-min Jung, Tae-ryong Park, and Kwang-yeob Lee
SeoKyeong University, Korea 50

[MC-3] Anisotropic Diffusion-based Denoising Using Residual image for Preservation of Image Details
GyuJin Bae, Sung In Cho, and Young Hwan Kim
Pohang University of Science and Technology, Korea 52

[MC-4] Propagation Delay Detection based on I-Q Modulation for Indoor Positioing System
Soo-Yong Kim, Chaehag Yi, and Andrey Zobenko
Samsung Electronics Co., Ltd., Korea 54

DC

Data Converters

08:30~09:45 Ballroom 3
Chair: Youngcheol Chae *(Yonsei University, Korea)*

[DC-1] A 10-bit 20-MS/s Dual-Channel Algorithmic ADC With Improved Clocking Scheme
Joo-Won Oh, Yong-SikKwak, and Gil-Cho Ahn
Sogang University, Korea 56

[DC-2] A 12-bit 750KS/s 69dB-SNDR 0.48mW Dual-Sampling SAR ADC with Reduced C-DAC for Wireless Charging Receiver
Hamed Abbasizadeh, Behnam Samadpoor Rikan, Ji-Hun Kang, Hyung-Gu Park, and Kang-Yoon Lee
Sungkyunkwan University, Korea 58

[DC-3] Digital Calibration Algorithm for Half-Unary Current-Steering DAC for Linearity Improvement
ShaifulNizam Mohyar[1,2] and Haruo Kobayashi[1]
[1] Gunma University,Japan

[2] Universiti Malaysia Perlis (UniMAP) Pauh Putra Campus, Malaysia 60

[DC-4] Design of a low power 4th Order $\Sigma\Delta$ Modulator with the reused opamps

Su hun Yang, JeongHoon Choi, and Kwang sub Yoon

InhaUniversity, Korea 62

[DC-5] Design of a Full-Swing CMOS Current Steering D/A Converter with an Adaptive High-Impedance Current Cell

Dongjoo Kim[1], Woongtaek Lim[1], Jongyoon Hwang[1], Sangil Han[2], and Minkyu Song[2]

[1] Dongguk University, KOREA
[2] Exicon Co. Ltd., KOREA 64

MP

Processors / Multi-Core Architectures & Software

08:30~09:45 Ballroom 4
Chair: Hanho Lee*(Inha University, Korea)*

[MP-1] Energy-Optimal Algorithm for Dynamic Voltage Scaling with Non-Convex Power Functions

Juyeon Kim and Taewhan Kim

Seoul National University, Korea 66

[MP-2] On-chip Aging Prediction Circuit in Nanometer Digital Circuits

Byunghyun Jang[1], Jin Kyung Lee[2], Minsu Choi[3], and Kyung Ki Kim[2]

[1] The University of Mississippi, USA
[2] Daegu University, Korea
[3] Missouri University of Science & Technology, USA 68

[MP-3] A 0.5V 29pJ/Cycle Sensor Node Processor for Intelligent Sensing Applications

Jun Zhou[1], Xin Liu[1], Chao Wang[1], Kah-Hyong Chang[1], Jianwen Luo[1], Jingjing Lan[1], Lei Liao[1], Yat-Hei Lam[1], YongkuiYang[2], Bo Wang[2], Xin Zhang[1], Wang Ling Goh[2], Tony Tae-Hyoung Kim[2], and Minkyu Je[3]

[1] A*STAR (Agency for Science, Technology and Research), Singapore
[2] Nanyang Technological University, Singapore
[3] DGIST, Korea 70

[MP-4] Multi-Core Architecture for Real-Time and Energy-Efficient Bearing Fault Diagnosis

Myeongsu Kang, InkyuJeong, and Jong-Myon Kim

University of Ulsan, Korea 72

[MP-5] Accelerating Forex Trading System Through Transaction Log Compression

JiHoon Jang, Sang Muk Lee, Sang Don Kim, Oh SeongGwon, EunnuriKo, Seong Mo Lee, Jung Woo Shin, and SeungEun Lee

Seoul National University of Science and Technology, Korea 74

2014 International SoC Design Conference

IMMD
Image Sensors/Sensors and MEMS Circuits / Biomedical SoCs

08:30~09:15 Mara
Chair: Andrew Marshall*(University of Texas at Dallas, USA)*

[IMMD-1] FPN Correction for a Linear-logarithmic CMOS Image Sensor with a Tunable Linear Range Using Two-step Charge Transfer
ByeungseokYoo, InkyuBaek, and Kyounghoon Yang
KAIST, Korea 76

[IMMD-2] A RFID/NFC Based Programmable SOC for Biomedical Applications
Mayukh Bhattacharyya[1], Waldemar Gruenwald[1], Benjamin Dusch[1], Jasmin Aghassi-Hagmann[1], Dirk Jansen[1], and Leonhard Reindl[2]
[1] *University of Applied Sciences Offenburg, Germany*
[2] *University of Freiburg, Germany* 78

[IMMD-3] Multi Sensor Voltage Signal Conditioner with Adaptive Level Shift and DC Offset Calibration on a Single Chip
Ji-HoonSuh, Mauricio Velazquez Lopez, Jeong-Ho Park, and Hyung-JounYoo
Korea Advanced Institute of Science and Technology, Korea 80

HI
High-Speed Signal Interfaces/Wireline and Wireless ICs (RF ICs)

13:30~15:20 Ballroom 3
Chair: Kwan-woo Kim*(Qualcomm, USA)*

[HI-1] Invited Paper : A Transformed Radial Stub Low-pass Filter Using Defected Ground Structure for Stopband Extension
Shanshan Xu[1], Kaixue Ma[1,2], and Kiat Seng Yeo[1,3]
[1] *Nanyang Technological University, Singapore*
[2] *University of Electronic Science and Technology of China (UESTC), China*
[3] *Singapore University of Technology and Design, Singapore* 82

[HI-2] Design of 80 - 1150 MHz CMOS LNA-less Receiver for Long-Range Wireless Sensor Network Applications
Seunghyeon Kim and Hyunchol Shin
Kwangwoon University, Korea 84

[HI-3] A U-band VCO in 65nm CMOS with 0.44dBm output power
Jongsuk Lee[1], Sangho Shin[2], and Yong Moon[1]
[1] *Soongsil University, Korea*
[2] *University of California Santa Cruz, USA* 86

[HI-4] Multi objective Optimization of Input Low Noise Amplifier for Common GPS/Galileo/GLONASS/Compass Satellite Navigation System Receiver
Josef Dobes, Jan M´ıchal, Frantisek Vejrazka, Jakub Popp, and V´aclav Panko
Czech Technical University in Prague, Czech Republic 88

[HI-5] An 10-Gb/s Pulse-Mode I/O for On-Chip 5-mm interconnect

Hung-Wen Lin, Guan-Ru Wu, Zhi-Xiang Shao, and Yong-Hsin Huang
YuanZe University, Taiwan 90

[HI-6] A Low Power High Linearity Phase Interpolator Design for High Speed IO Interfaces
Siddharth Katare[1], Sitaraman V. Iyer[2], Guluke Tong[2], Lasya R Munagala[1], Mahalingam Nagarajan[1], and Yang Bangda[2]
[1] *Intel Technology India Pvt Ltd, India*
[2] *Intel Corporation, Santa Clara* 92

[HI-7] 25-Gb/s inductorless output buffer circuit with a pre-emphasis in 65-nm CMOS
Tomoki Tanaka[1], Keiji Kishine[1], Hiromi Inaba[1], and Akira Tsuchiya[2]
[1] *University of Shiga Prefecture, Japan*
[2] *Kyoto University, Japan* 94

A2

Analog and Mixed-Signal Techniques II

13:35~15:20 Ballroom 4
Chair: Ying-Chieh Ho *(National Dong Hwa University, Taiwan)*

[A2-1] A Low Noise Class-AB Operational Amplifier with Noise Optimization Technique
Seungheun Song, Kwanseok Jung, Subin Kim, and Joongho Choi
University of Seoul, Korea 96

[A2-2] LeTourneau Motor Analyzer Senior Design Project
Ryan Tiemann, Joonwan Kim, and Benjamin Ito
LeTourneau University, USA 98

[A2-3] Low capacitance sensing circuit for fingerprint sensor integrated into display screen based on Charging and Extracting Process
Yena Yoo, Kyungmin Na, Heedon Jang, and Franklin Bien
Ulsan National Institute of Science and Technology, Korea 100

[A2-4] A design of novel transducer converting output signals of sensor modules to 4~20mA
Min-Hyeong Cho, Won-Ho Lee, Hi-Seok Kim, and Hyeong-Woo Cha
Cheongju University, Korea 102

[A2-5] Time-to-Digital Converter Architecture with Residue Arithmetic and its FPGA Implementation
Congbing Li, Kentaroh Katoh, Junshan Wang, Shu Wu, Shaiful Nizam Mohyar, and Haruo Kobayashi
Gunma University, Japan 104

[A2-6] An Open-Loop Differential Time Amplifier
Hye-Jung Kwon, Ji-Hoon Lim, Byungsub Kim, Jae-Yoon Sim, and Hong-June Park
POSTECH, Korea 106

[A2-7] The Design of 13 bits $\Sigma\Delta$ ADC for a mutual-capacitance large touch screen controller
Ihsan F. I. Albittar[1,2], Jiho Kim[1], and HyungWon Kim[1]

x x v

2014 International SoC Design Conference

[1] Chungbuk National University, Korea
[2] Mixel Incorporation, USA 108

ME
Memory Circuits and Embedded Memory / Embedded Systems

13:30~15:05 **Mara**

Chair: Ji-Hoon Kim*(Chungnam National University, Korea)*

[ME-1] Invited Paper : Low Power Memory Design for High Temperature in Ruggedized Electronics

Tony T. Kim[1], Ngoc Le Ba[1], Anh Tuan Do[1], Jayaraman K. Gopal[1], Geng Li Chua[2], and Pushpapraj Singh[2]
[1] Nanyang Technological University, Singapore
[2] A*STAR, Singapore 110

[ME-2] Comparative Analysis of Using Planar MOSFET and FinFET as Access Transistor of STT-RAM Cell in 22-nm Technology Node

Byungkyu Song[1], Taehui Na[1], Hanwool Jeong[1], Seung H. Kang[2], Jung Pill Kim[2], and Seong-Ook Jung[1]
[1] Yonsei University, Korea
[2] Qualcomm Inc., USA 112

[ME-3] Subthreshold SRAM Macro Design with Pulse-Controlled DynamicVoltage Scaling (PC-DVS)

Jun-Kai Zhao[1], Yi-Wei Chiu[1], Shyh-Jye Jou[1], and Yuan-Hua Chu[2]
[1] National Chiao Tung University, Taiwan
[2] Industrial Technology Research Institute (ITRI), Taiwan 114

[ME-4] A Gain Cell Based Embedded DRAM with Fully-Restoring Write-Back Scheme

Weijie Cheng, Hritom Das, Huarong Zheng, Baolong Zhou, and Yeonbae Chung
Kyungpook National University, Korea 116

[ME-5] An Allocation Optimization Method for Partially-Reliable Instruction Scratch-Pad Memory in Embedded Systems

Takuya Hatayama, Hideki Takase, Kazuyoshi Takagi, and Naofumi Takagi
Kyoto University, Japan 118

[ME-6] Modulo Scheduler Implementation for VLIW Processor

Jangseop Shin, Sangjun Han, Hyungyun Jung, IngooHeo, and Yunheung Paek
Seoul National University, Korea 120

SD
SoC Testing / Design Verification / Signal Integrity / Interconnect Modeling and Simulation

13:35~15:20 **Udo**

Chair: Youngmin Kim*(UNIST, Korea)*

x x vi

2014 International SoC Design Conference

[SD-1] Efficient Eye-Diagram Determination Technique of Non-Linearly-Switching Coupled-Data Links Under Power and Ground Fluctuation Noises
Junghyun Lee, Joonhyun Kim, Gihyeon Ji, Hyewon Kim, and Yungseon Eo
Hanyang University, Korea 122

[SD-2] Analysis and Reduction of Voltage Noise of Multi-layer 3D IC with PEEC-based PDN and Frequency-dependent TSV models
Seungwon Kim[1], Ki Jin Han[1], Seokhyeong Kang[2], and Youngmin Kim[1]
[1] UNIST, Korea
[2] UCSD, USA 124

[SD-3] Allocation and Optimization of Post-Silicon Tunable Buffers in TSV Based Heterogeneous 3D ICs
Sangdo Park[1], Jeongwoo Heo[2], and Taewhan Kim[2]
[1] Samsung Electronics Co., Ltd, Korea
[2] Seoul National University, Korea 126

[SD-4] Reducing the Failure Bitmap Size with a Partial Solution Search Tree for the Low Cost Automatic Test Equipment (ATE)
Keewon Cho, Woosung Lee, Jooyoung Kim, and Sungho Kang
Yonsei University, Korea 128

[SD-5] Chaotic Oscillation-based BIST for CMOS Operational Amplifier
Chatchai Wannaboon, Nattagit Jiteurtragool, and Tachibana Masayoshi
Kochi University of Technology, Japan 130

[SD-6] Analysis of Dynamic Voltage Drop with PVT Variation in FinFET Designs
Yongchan Ban, Changseok Choi, Hosoon Shin, Jaewook Lee, Yongseok Kang, and Woohyun Paik
LG Electronics, Korea 132

[SD-7] A New Redundancy Analysis Algorithm Using One Side Pivot
Jooyoung Kim, Keewon Cho, Woosung Lee, and Sungho Kang
Yonsei University, Korea 134

2014 International SoC Design Conference

Special Session

Tuesday, November 4, 2014

SS-A

Near-Threshold Voltage Circuit Design

13:35~14:50 Chuja

Chair: I-Chyn Wey *(ChangGung University, Taiwan)*

[SS-A1] Design of a Near-threshold Digital LDO with Fast Transient Response
Yunsheng Chan and Yingchieh Ho
National Dong-Hwa University, Taiwan 136

[SS-A2] Near-Threshold-Voltage Circuit Design: The Design Challenges and Chances
I-Chyn Wey, Po-Jen Lin, Bing-Chen Wu, and Chien-Chang Peng and Pin-Hsi Lin
Chang Gung University, Taiwan 138

[SS-A3] A Limited-Contention Cross-Coupled Level Shifter for Energy-Efficient
Subthreshold-to-Superthreshold Voltage Conversion
Chi-Ray Huang and Lih-Yih Chiou
National Cheng Kung University, Taiwan 142

[SS-A4] A Low Voltage Instrumentation Amplifier for Sensor Applications
Hwang-Cherng Chow, Hsin-Ti Chou, and I-Chyn Wey
Chang Gung University, Taiwan 144

[SS-A5] Comparison of Subthreshold Logic with Adiabatic Circuit Techniques
Cihun-Siyong Alex Gong, Chi-Tong Hung, Wei-Lin William Chu, Chang-Jie Lin, Yu-Fan
Luo,Chih-Yun Chien, Yu-Hung Kuo, Meng-Jung Chang, and Chin-Chih Hsu
Chang Gung University, Taiwan 146

SS-B

Image Signal Processing for Vision/Multimedia SoC

15:35~17:20 Mara

Chair: Byung-Mo Kang *(KETI, Korea)*

[SS-B1] HARDWARE FEASIBILITY ANALYSIS FOR MOTION SEGMENTATION INITIALIZATION
Subarna Tripathi[1], Youngbae Hwang[2], Sung-Joon Jang[2],
Serge Belongie[1,3], and Truong Nguyen[1]
[1] University of California San Diego, USA
[2] Korea Electronics Technology Institute, Korea
[3] Cornell NYC Tech, USA 148

x x viii

2014 International SoC Design Conference

[SS-B2] On-road Vehicle Detection with Monocular Camera for Embedded Realization: Robust Algorithms and Evaluations

Ravi Kumar Satzoda[1], Eshed Ohn-Bar[1], Jinhee Lee[2], Hohyon Song[2], and Mohan M. Trivedi[1]
[1] University of California San Diego, USA
[2] NextChip Co. Ltd., Korea ·········· 150

[SS-B3] Using Context to Improve Cascaded Pedestrian Detection

Mohammad Saberian, ZhaoweiCai, Jinhee Lee, and NunoVasconcelos
University of California San Diego, USA ·········· 152

[SS-B4] Efficient Airlight Estimation for Defogging

Yeejin Lee[1], Changyoung Han[2], Junseong Park[2], Seungkyu Park[2], and Truong Q. Nguyen[1]
[1] University of California, USA
[2] Core Logic Inc., Korea ·········· 154

[SS-B5] Depth Gradient Based Region of Interest Generation for Pedestrian Detection

Maral Mesmakhosroshahi[1], Kwang-Hoon Chung[2], Yunsik Lee[3], and Joohee Kim[1]
[1] Illinois Institute of Technology, USA
[2] Sane System, Korea
[3] Korea Elec. Tech. Inst., Korea ·········· 156

[SS-B6] Depth map estimation using modified Census transform and semi-global matching

Maziar Loghman[1], Kwang-Hoon Chung[2], Yunsik Lee[3], and Joohee Kim[1]
[1] Illinois Institute of Technology, USA
[2] Sane Systems, Korea
[3] Korea Electronics Tech. Inst., Korea ·········· 158

[SS-B7] A High Speed Pipeline Structure of Hardware Implementation for Block Classification for Distributed Video Coding

Qiang Tong and Ken Choi
Illinois Institute of Technology, USA ·········· 160

SS-C

Analog/Digital Circuits for Mobile SoC

15:35~16:50 Udo
Chair: Yunho Jung*(Korea Aerospace University, Korea)*

[SS-C1] Multi-Band Multi-Mode Wireless Connectivity SoC for 802.11 a/b/g/n, BT 4.0 and NFC

Minsu Jeong[1], Yanggun Kim[1], Jaekyung Lee[1], Deokki Ahn[1], Ensoo Lee[1], Jangsup Sohn[1], Kwanju Lee[1], Jaehun Lee[2], Jabum Gu[2], SongBum Kim[2] and Jungbo Sohn[2]
[1] RAONTECH Inc, Korea
[2] Electronics and Telecomunications Research Institute, Korea ·········· 163

[SS-C2] Design of Analog Front End for Mobile Fuel Gauge Applications

Chulkyu Park[1], Hyojae Kim[1], Jongkeun Hwang[1], Kichang Jang[1], Yeongik Yoo[2], and Joongho Choi[1]
[1] University of Seoul, Korea
[2] SiliconMitus, Korea ·········· 165

2014 International SoC Design Conference

[SS-C3] The Design of SCR-based Dual Direction ESD Protection Circuit with Low Trigger Voltage
Yong-Nam Choi, Jung-Woo Han, Hyun-Young Kim, Chung-Kwang Lee, and Yong-Seo Koo
Dankook University, Korea ·· 167

[SS-C4] A Low Power Time-Domain CMOS Temperature Sensor
Wonjong Song, Hunsik Moon, Hyeyeon Yang, and JinwookBurm
Sogang University, Korea ·· 169

[SS-C5] 2D-to-3D Conversion Using Color and Edge
Manbae Kim
Kangwon National University, Korea ······································ 171

SS-D
Design, Analysis and Tools for Integrated Circuits and Systems (DATICS)

15:35~16:50 Chuja
Chair: Ka Lok Man *(Xi'an Jiaotong-Liverpool University, China)*

[SS-D1] Simulation and Analysis of Desulfator for Smart Battery System
KaLok Man[1], Eng Gee Lim[1], Mark Leach[1], Jin Kyung Lee[2], and Kyung Ki Kim[2]
[1] Jiaotong-Liverpool University, China
[2] Daegu University, Korea ·· 173

[SS-D2] Design of An Arduino-based Smart Car
Zhao Wang, Eng Gee Lim, Weiwei Wang, Mark Leach, and KaLok Man
Xi'an Jiaotong-Liverpool University, China ······························ 175

[SS-D3] The New Wrappable Wireless Capsule Antennas
Eng Gee Lim, Zhao Wang, Tianqi Xia, Mark Leach, and KaLok Man
Xi'an Jiaotong-Liverpool University, China ······························ 177

[SS-D4] The Design of a Smart Power Conversion System as an Undergraduate Cross-Discipline Integrated Design Project
Chi-Un Lei[1], Christopher H.T. Lee[1], T.O. Kwan[1], C.K. Lee[1], K.B. Huang[1], R.Y.K. Kwok[1], and K.L. Man[2]
[1] University of Hong Kong, Hong Kong
[2] Xi'an Jiaotong-Liverpool University, China ···························· 179

[SS-D5] TAB-model for Multilevel Diagnosis and Repair of HDL SoC
Vladimir Hahanov[1], KaLok Man[2], Baghdadi AmmarAwni Abbas[3], Eugenia Litvinova[1], Svetlana Chumachenko[1], Jihyeok Ahn[4], and Kyung Ki Kim[4]
[1] National University of Radioelectronics, Ukraine
[2] Xi'an Jiaotong-Liverpool University, China
[3] Baghdad University, Iran
[4] Daegu University, Korea ·· 181

xxx

2014 International SoC Design Conference

Wednesday, November 5, 2014

SS-E
Design Challenges for Advanced Mobile Intelligence : Low Power, Object Detection, and Mobile Security

08:30~09:45 Udo
Chair: Ken Choi *(Illinois Institute of Technology, USA)*

[SS-E1] A Review on System Level Low Power Techniques
Qiaing Tong[1], Ken Choi[1], and Jun Dong Cho[2]
[1] Illinois Institute of Technology, USA
[2] SungKyunKwanUniversity, Korea 183

[SS-E2] A novel depth estimation method for uncalibrated stereo images
Maziar Loghman[1], Amin Zarshenas[1], Joohee Kim[1], Kwang-Hoon Chung[2], and Yunsik Lee[3]
[1] Illinois Institute of Technology, USA
[2] Sane Systems, Korea
[3] Korea Electronics Technology Institute, Korea 186

[SS-E3] Ring Projection Transforms by Using CUDA Implementation for Recognizing Objects in Automation System
Sang-hyeob Song, Seong-muk Kang, Hyeong-jun Cho, and Jun-dong Cho
SungKyunKwan University, Korea 188

[SS-E4] A High Performance Low Power Implementation Scheme for FSM
Shuai Li and Ken Choi
Illinois Institute of Technology, USA 190

[SS-E5] Survey on Security techniques for AMI Metering System
SungJin Kim, HyunSooChng, and Taeshik Shon
Ajou University, Korea 192

xxxi

2014 International SoC Design Conference

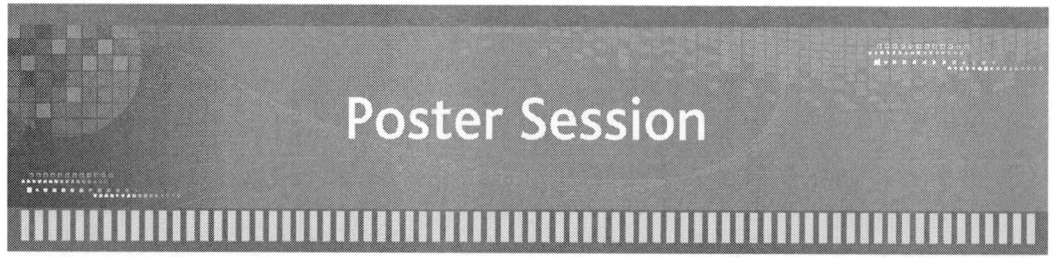

Poster Session

Wednesday, November 5, 2014

Poster Session

15:35~17:05 Lobby
Chair: Ikjun Chang*(Kyunghee University, Korea)*
Young-Chan Jang*(Kumoh National Institute of Technology)*

Analog andMixed-SignalCircuits

[PS-1] A Digital Lock Detector for a Dual Loop PLL
Chang-Hyun Bae[1,2] and Changsik Yoo[2]
[1] Samsung Electronics, Korea
[2] HanyangUniversity, Korea 194

[PS-2] Automatic Current Range Detection for Biochemical Sensor
Wei-Jun Liu, Ming-Han Yu, Bin-Da Liu, and Chia-Ling Wei
National Cheng Kung University, Taiwan 196

[PS-3] A Compact Comparator Block for Laser Radar Front-end Receiver
Jongsun An, Joo-Young Choi, Eun-Gyu Lee, and Choul-Young Kim
Chungnam National University, Korea 198

[PS-4] A 6bit 550Ms/s Small Area Low Power Successive Approximation ADCA
Zhou Peng, Chenxi Han, Dongmei Li, and Zhihua Wang
Tsinghua University, Chin 200

[PS-5] A 10-bit 20-MS/s Asynchronous SAR ADC with Controllable Analog Input Voltage
Range and Meta-stability Detection Circuit
Sang-Min Park, Yeon-Ho Jeong, Dong-Gil Jeong, Seung-Wuk Baek,
Yu-Jeong Hwang, Pil-Ho Lee, and Young-Chan Jang
Kumoh National Institute of Technology, Korea 202

[PS-6] A 24-mW 60-GHz OOK RF transceiver for 3-Gbps data communication
Hui Dong Lee[1], Tae Young Kang[1], Ki Chan Eun[2], Moon-Sik Lee[1],
and Bonghyuk Park[1]
[1] Electronics and Telecommunications Research Institute, Korea
[2] KORF Incorporated, Korea 204

[PS-7] Small size 433 MHz On-off Keying Transceiver IC for Sensor Application
Jeong-Ho Park, Han-Won Cho, Ji-Hoon Suh, and Hyung-Joun Yoo
Korea Advanced Institute of Science and Technology, Korea 206

[PS-8] A 24 GHz PMOS Body Voltage Controlled Oscillator with Transformer Coupled

2014 International SoC Design Conference

Varactor
Jae-Hoon Song[1], Byung-Sung Kim[2], and Sangwook Nam[1]
[1] Seoul National University, Korea
[2] SungkyunkwanUniversity, Korea 208

[PS-9] A Spread Spectrum Clock Generator Using Discontinuous Modulation Technique for Reduction of Time Interval Errors
TaimingPiao, Jae-Kyung Wee, Inchae Song, and Boo-Gyoun Kim
Soongsil University, Korea 210

Digital VLSI Circuitsand Embedded Systems / Application Specific SoCs& Emerging Technonogies

[PS-10] K-Critical Path Search Based Multi corner Multi mode Static Timing Analysis
Deokkeun Oh, Eunsuk Park, Myeoungwoo Jin, and Juho Kim
Sogang University, Korea 212

[PS-11] A Comparison-Free Sorting Algorithm
Saleh Abdel-hafeez[1] and Ann Gordon-Ross[2]
[1] Jordan University of Science and Technology, Jordan
[2] University of Florida, USA 214

[PS-12] A Carry Look-ahead Adder Designed by Reversible Logic
Junchao Wang and Ken Choi
Illinois Institute of Techonoly, USA 216

[PS-13] Dynamic Stability Estimation for Latch-Type Voltage Sense Amplifier
Woong Choi[1], Jongsun Park[2], and Gyuseong Kang[1]
[1] University of Minnesota, USA
[2] Korea University, Korea 218

[PS-14] CAN Data Reduction Using Three Segement Signal Decomposition
Yujing Wu[1], Jin-Gyun Chung[1], and Yinan Xu[2]
[1] Chonbuk National University, Korea
[2] Yanbian University, China 220

[PS-15] Underbody Component Monitoring System of Railway Vehicles using the Infra-red Thermal Images
Min-Soo Kim, Seh-Chan Oh, Geun-Yep Kim, and Seok-Jin Kwon
Korea Railroad Research Institute, Korea 222

[PS-16] Implementation of Efficient SHA-256 Hash Algorithm for Secure Vehicle Communication using FPGA
Chanbok Jeong and Youngmin Kim
Ulsan National Institute of Science and Technology (UNIST), Korea 224

[PS-17] An Online Test and Debug Methodology for Automotive Image Processing System
Hyunggoy Oh, Inhyuk Choi, Taewoo Han, Won Jung, Byungin Moon, and Sungho Kang
Yonsei University, Korea 226

[PS-18] Analysis of Structural Variation and Threshold Voltage Modulation in 10-nm Double Gate-All-Around (DGAA) Transistor
Myunghwan Ryu and Youngmin Kim
Ulsan National Institute of Science and Technology (UNIST), Korea 228

2014 International SoC Design Conference

[PS-19] Performance and Leakage Optimization with 22nm Bi-level FinFET
Jaemin Lee and Youngmin Kim
Ulsan National Institute of Science and Technology (UNIST), Korea 230

Nanoelectronic Devices and Circuits / Power Electronics / Energy Harvesting Circuits

[PS-20] Implicating Logic Functions with Memristors
Pravin Mane, Nishil Talati, Ameya Riswadkar, Ramesh Raghu,
and C. K. Ramesha
BITS Pilani K.K. Birla Goa Campus, India 232

[PS-21] SEU-Tolerant Active Body-Bias Inverter
Youngkyu Jang[1], Ik-Joon Chang[1], Jinsang Kim[1], and Seungjoo Lee[2]
[1] Kyung Hee University, Korea
[2] Hyejeon University, Korea 234

[PS-22] Design of Hysteretic Buck Converter with a Low Output Ripple Voltage and a Fixed Switching Frequency in CCM
Tae-Jin Jeong, Woo-Seong Kang, Ji-San Choi, and Goang-Seob Yoon
INHA University, Korea 236

[PS-23] A Buck DC-DC Converter Using Automatic PFM/PWM Mode Change for High-Efficiency Li-Ion Battery Charger
Thanh Tien Ha, Do-Young Chung, Dasom Park, Hyun-Sik Lee,
and Jong-Wook Lee
Kyung Hee University, Korea 238

[PS-24] Design of a Sub-1-V Low Dropout Regulator for Supply-Regulated Active-Loop-Filter VCOs
Young-Min Jang[1], Ji-Geun Kim[1], Jae-Hyun Cho[1], Bruce C. Kim[2],
and Sang-Bock Cho[1]
[1] University of Ulsan, Korea
[2] City University of New York, USA 240

[PS-25] Design of a Digital Controller for an LED lamp driver IC
Kilsoo Seo[1], Wonkyeong Park[2], and Jusung Park[3]
[1] KERI, Korea
[2] Silicon Mitus, Korea
[3] Pusan National University, Korea 242

[PS-26] A Low Power High Resolution Digital PWM with Process and Temperature Calibrations for Digital Controlled DC-DC Converters
Jing Lu[1], Ho Joon Lee[1], Kyung Ki Kim[2], and Yong-Bin Kim[1]
[1] Northeastern University, USA
[2] Daegu University, Korea 244

[PS-27] An Output-Capacitorless Low Dropout Regulator without Resistors
Jie MEI, Hao Zhang, and Tsutomu Yoshihara
Waseda University, Japan 246

[PS-28] Fully Digitalized Switch-mode LED Driver without ADC
Jae-Hyoun Park and Hyung-Do Yoon
Korea Electronics Technology Institute, Korea 248

2014 International SoC Design Conference

[PS-29] IC design of DMPPT controller for Cascade and MCCU topology of Photovoltaic Systems

K.H. Cho[1], S.M. Sohn[1], J.G. Jeong[1], S.I. Lim[2], I.S. Cho[2], and J.Y. Kim[1]

[1] *RTS Energy, Inc., Korea*
[2] *Seokyeong University, Korea* 250

Multimedia(A/V) Algorithm and SoCs

[PS-30] Histogram Color Correction For Multi-View Video Cording

Chan-Su Park, Hi-Seok Kim, and Hyeong-Woo Cha

Cheongju University, Korea 252

[PS-31] On FHD 300MHz@60fps, Intra/Inter CU Mode Decision Hardware Architecture for the Hypernova H.265 Encoder

Sukho Lee, Hyunmi Kim, Kyungjin Byun, and Nakwoong Eum

Electronics and Telecommunications Research Institute, Korea 254

[PS-32] Real-time HDR Multi Exposure Fusion Hardware Design

Junwon Moon and Jaeseok Kim

Yonsei University, Korea 256

[PS-33] Enhanced Tone Mapping of High Dynamic Range to Correspond to Illumination Changes

Young-Min Jang[1], Jae-Hyun Cho[1], Ji-Geun Kim[1], Bruce C. Kim[2], and Sang-Bock Cho[1]

[1] *University of Ulsan, Korea*
[2] *City University of New York, USA* 258

[PS-34] Enhanced Test Zone Search Motion Estimation Algorithm for HEVC

Nidhi Parmar and Myung Hoon Sunwoo

Ajou University, Korea 260

[PS-35] Efficient Hardware Architecture for Real-time Semi-Global Matching

Seongbo Sim[1], Kyoungwon Min[2], Seonyoung Lee[2], Haengson Son[2], and Jongtae Kim[1]

[1] *Sungkyunkwan University, Korea*
[2] *Korea Electronics Technology Institute, Korea* 262

[PS-36] Gram-Schmidt tailed High-throughput QR Decomposition Architecture for MIMO detector

Dongyeob Shin[1], Ji-Hwan Yoon[1], Jongsun Park[1], and Woong Choi[2]

[1] *Korea University, Korea*
[2] *University of Minnesota, USA* 264

[PS-37] Efficient Min-Max Nonbinary LDPC Decoding on GPU

Huyen Pham Thi, Sabooh Ajaz, and Hanho Lee

Inha University, Korea 266

[PS-38] Area-Efficient FFT Processors for OFDM Systems

Sung Kyung Shin and Myung Hoon Sunwoo

Ajou University, Korea 268

[PS-39] Implementation of Multi-Standard Digital Radio Single-Chip

Se-Ho Park and Kyung-Taek Lee

Korea Electronics Technology Institute, Korea 270

2014 International SoC Design Conference

[PS-40] Combined Channel Estimation
Kyuhoon Lee[1], Ik-Joon Chang[1], Jinsang Kim[1], and Seungjoo Lee[2]
[1] Kyung Hee University, Korea
[2] Hyejeon University, Korea 272

Memory Circuits and Embedded Memory / Processors / Multi-Core Architectures & Software

[PS-41] Ultra-low Power 12T Dual Port SRAM for Hardware Accelerators
Bo Wang[1,2], Jun Zhou[2], and Tony T. Kim[1]
[1] Nanyang Technological University, Singapore
*[2] Institute of Microelectronics, A*STAR, Singapore* 274

[PS-42] Understanding DDR4 in Pursuit of In-DRAM ECC
Sanghyuk Kwon, Young Hoon Son, and Jung Ho Ahn
Seoul National University, Korea 276

[PS-43] Ethernet, SD Card and USB Linux Driver Porting on Aldebaran SoC System
Chan Kim, KyungJin Byun, and NakWoong Eum
Electronics and Telecommunications Research Institute, Korea 278

[PS-44] Supporting Linux on Virtual Platform based on Versatile Express Board
Changmin Shin and Yongjoo Kim
Electronics and Telecommunications Research Institute, Korea 280

[PS-45] Development of a Virtual Platform for IP and Firmware Verification
Changmin Shin and Yongjoo Kim
Electronics and Telecommunications Research Institute, Korea 282

[PS-46] Design of CMOS Dual Antifuse OTP Memory Based On Gate Oxide
Seung-Youl Kim[1], Je-Hoon Lee[2], Tae-Yang Kim[2], and Younggap You[1]
[1] Chungbuk Nat'l Univ., Korea
[2] Kangwon Nat'l Univ., Korea 284

[PS-47] Design of Bio-signal Measurement System for Array Sensor and Application
Seungpyo Jung, Youngju Park, Sangman Kim, and Jusung Park
Pusan National University, Korea 286

[PS-48] Re-visit Blocking Texture Cache Design for Modern GPU
Jhe-Yu Liou and Chung-Ho Chen
National Cheng Kung University, Taiwan 288

SoC Design Methodology

[PS-49] Analysis of On Chip Decoupling Capacitor in the Double-Gate FinFETs with PEEC-based Power Delivery Network
Jaemin Lee, Yesung Kang, and Youngmin Kim
UNIST, Korea 290

[PS-50] Memory efficient data structure for graph representation of DSPF netlist
Jinwook Kim, Hae-seong Park, and Young Hwan Kim
POSTECH, Korea 292

[PS-51] Study of Sensor Fault Detection Based on Modified SVM Algorithm
Zhijia Yu[1], Yinan Xu[1], Yonghu Ma[1], Yujing Wu[2], and Jin-Gyun Chung[2]
[1] Yanbian University, China
[2] Chonbuk National University, Korea 294

[PS-52] An Optimal Power Methodology through Constrained Register Sharing
Liu Wan, Chi-Ho Lin and Su-Yeon Song
Semyung University, Korea 296

[PS-53] Procedures of the Fault Code in the Driverless Mode of the Communication based Train Control System
Min-Soo Kim
Korea Railroad Research Institute, Korea 298

[PS-54] Scan Cell Reordering Algorithm for Low Power Consumption during Scan-Based Testing
Wooheon Kang, Hyunyul Lim, and Sungho Kang
Yonsei University, Korea 300

[PS-55] Fast Allocation of Post-Silicon Tunable Buffers to Mitigate Timing Variation
Hyungjung Seo, Jeongwoo Heo, and Taewhan Kim
Seoul National University, Korea 302

12.5-Gb/s Monolithically Integrated Optical Receiver With CMOS Avalanche Photodetector

Hyun-Yong Jung,[1] Jeong-Min Lee,[1] Jin-Sung Youn,[1] Woo-Young Choi,[1] and Myung-Jae Lee[2]

[1] Department of Electrical and Electronic Engineering, Yonsei University, Seoul 120-749, South Korea
[2] Faculty of Electrical Engineering, Delft University of Technology, Mekelweg 4, 2628 CD Delft, Netherlands

hyjunghyjung@gmail.com

Abstract

We present a 12.5-Gb/s monolithically integrated optical receiver with CMOS avalanche photodetector (CMOS-APD) realized in 65-nm CMOS technology. The optical detection bandwidth limitation of CMOS-APD due to the carrier transit time is compensated by underdamped TIA. With this optical receiver, 12.5-Gb/s 850-nm optical data are successfully detected with bit-error rate less than 10^{-12} at the incident optical power of −2 dBm. The fabricated optical receiver has the core size of 0.24×0.1 mm^2 and its power consumption excluding output buffer is about 13.7 mW with 1.2-V supply voltage.

Keywords—Avalanche photodetectors (APDs); Monolithic integration; Optical interconnects; Optical receiver;

Introduction

Recently, optical interconnect technology is receiving a great amount of research attention as it can overcome the limitation of electrical interconnect bandwidth. 850-nm optical interconnects based on vertical-cavity surface-emitting lasers (VCSELs) and multimode fibers (MMFs) have found many applications for short-reach interconnects such as chip-to-chip, board-to-board and rack-to-rack interconnects [1]. With realization of high-speed photodetectors (PDs) in standard complementary metal-oxide-semiconductor (CMOS) process, monolithically integrated 850-nm Si optical receivers can be realized, which provides cost effectiveness and high-volume manufacturability as well as performance improvement without parasitic pad capacitance and bonding wire inductance.

Several monolithically integrated optical receivers on standard CMOS technology for 10-Gb/s applications have been reported [2-5]. PDs realized in standard CMOS technology do not have the optimal PD strcture and, typically, have very limited bandwidth. To overcome this, CMOS optical receivers including spatially-modulated photodetectors (SM-PDs) [3] or on-chip equalizers [5] have been reported. However, they have low responsivity and require additional power and area.

In this paper, we demonstrate another technique of overcoming the PD bandwidth. We intentionally design underdamped transimpedance amplifier (TIA) which can compensate CMOS-APD bandwidth limitation and result in enhanced overall bandwidth performance. With this design

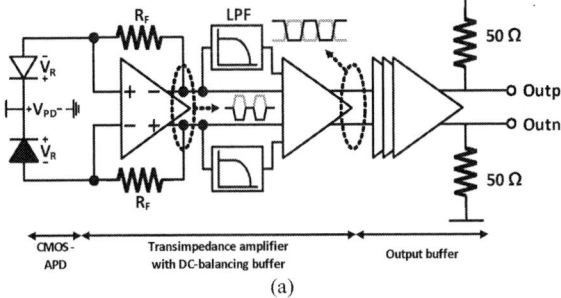

(a)

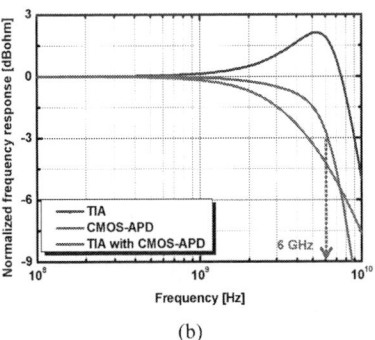

(b)

Fig. 1. (a) Block diagram of the proposed optical receiver and (b) simulated frequency responses.

approach, we successfully demonstrate 12.5-Gb/s operation.

Optical Receiver Circuit

Fig. 1(a) shows a simplified block diagram of our optical receiver. It is composed of a CMOS-APD with a dummy PD, a shunt-feedback TIA with DC-balancing buffer, and output buffer with 50-Ω load. The dummy PD provides symmetric capacitance to the differential TIA input. With CMOS-APD, the photo-detection bandwidth of our CMOS-APD is limited by the transit time of slow diffusive photocurrents. This leads to the bandwidth limit in optical receiver even with a high-speed TIA. To compensate this, we use an underdamped TIA which can be realized by decreasing the core-amplifier bandwidth of shunt-feedback TIA. Fig. 1(b) shows the simulated frequency responses for the transit time response of the PD used in our receiver, electrical response TIA with junction capacitance of used PD, and the final response with the PD and the TIA. As shown in Fig. 1(b), high-frequency peaking of the uderdamped TIA leads to bandwidth

978-1-4799-5128-4/14 $31.00 © 2014 IEEE

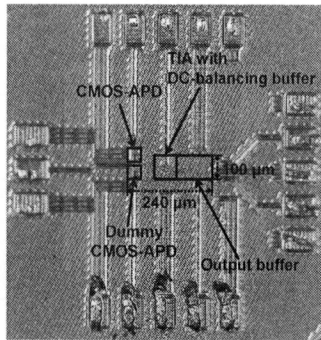

Fig. 2. Microphotograph of the fabricated optical receiver.

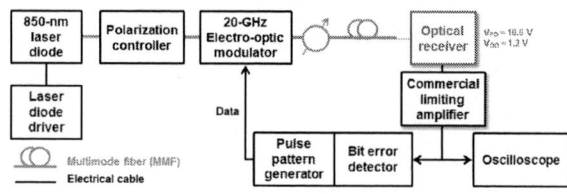

Fig. 3. Measurement setup for data transmission.

enhancement of the total receiver frequency response.

Photo-generated currents from one port of differential TIA generate TIA differential output with a DC offset which can cause decision threshold problem. To eliminate this problem, a DC-balancing buffer is added.

Experiment Result

Fig. 2 shows the micro photograph of the fabricated optical receiver in 65-nm CMOS technology. The core size is 0.24 × 0.1 mm², and the power consumption of the electronic circuit excluding output buffer is about 13.7 mW with 1.2-V supply voltage.

Fig. 3 shows the measurement setup for optical data transmission. All experiments are done on-wafer. The 850-nm modulated optical signals are generated by an 850-nm laser diode and a 20-GHz external electro-optic modulator. The modulated optical signals are transmitted through MMF and injected into the optical receiver with lensed fiber. The applied bias voltage of CMOS-APD is experimentally optimized for BER performance at 10.6 V. For bit-error rate (BER) measurement, a 12.5-Gb/s commercial limiting amplifier is used to satisfy the input sensitivity requirement of BER test equipment. Fig. 4 shows the measured BER performance with various incident optical power. The 12.5-Gb/s PRBS7 data detection is successfully achieved and the measured 10^{-12} BER is −2 dBm. The inset in Fig. 4 shows the measured eye diagram for 12.5-Gb/s data transmission with −2-dBm incident optical power.

Conclusion

A 12.5-Gb/s monolithically integrated optical receiver with CMOS-APD is realized in 65-nm CMOS technology. With careful design of TIA so that it can compensate the bandwidth limit of the CMOS-APD, the 3-dB bandwidth is enhanced and

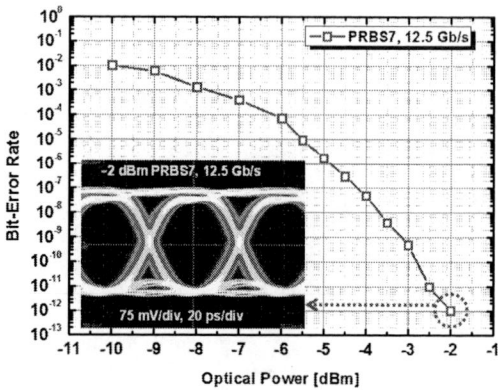

Fig. 4. Measured BER performance and eye diagram of transmitted 12.5-Gb/s data.

12.5-Gb/s optical data are successfully detected.

Acknowledgment

This work was supported by the National Research Foundation of Korea (NRF) grant funded by the Korea government (MEST) (2012R1A2A1A01009233). The authors are also thankful to IC Design Education Center (IDEC) for EDA software and MPW support.

References

[1] T.-K. Woodward and A. V. Krishnamoorthy, "1-Gb/s integrated optical detectors and receivers in commercial CMOS technologies," *IEEE J. Sel. Top. Quantum Electron.*, vol. 5, no. 2, pp. 146-456, Mar. 1999.

[2] S.-H. Huang, W.-Z. Chen, Y.-W. Chang, and Y.-T Huang, "A 10-Gb/s OEIC with meshed spatially-mudulated photo detector in 0.18-µm CMOS technology," *IEEE J. Solid-State Circuits*, vol. 46, no. 5, pp. 1158-1169, May 2011.

[3] M.-J. Lee, J.-S. Youn, K.-Y. Park, and W.-Y. Choi, "A fully-integrated 12.5-Gb/s 850-nm CMOS optical receiver based on a spatially-modulated avalanche photodetector," *Opt. Express*, vol. 22, no. 3, pp. 2511-2518, Feb. 2014.

[4] D. Lee, J. Han, G. Han, and S. M. Park, "An 8.5-Gb/s fully integrated cmos optoelectronic receiver using slope-detection adaptive equalizer," *IEEE J. Solid-State Circuits*, vol. 45, no. 12, pp. 2861-2873, Dec. 2010.

[5] J.-S. Youn, M.-J. Lee, K.-Y. Park, and W.-Y. Choi, "10-Gb/s 850-nm CMOS OEIC receiver with a silicon avalanche photodetector," *IEEE J. Quantum Electron.*, vol. 48, no. 2, pp. 229-236, Feb. 2012.

978-1-4799-5128-4/14 $31.00 © 2014 IEEE

An Area Saving Inductor Current Sensor with Load Transient Enhancement in DC-DC Converter

Ngan K. Hoang[†], Xuan-Dien Do[†], Young-Jin Woo[‡], *Member*, IEEE, and Sang-Gug Lee[†], *Member*, IEEE

[†] Electrical Engineering Dept., KAIST, Daejeon, Republic of Korea
[‡] Silicon Works Co. Ltd., Daejeon, Republic of Korea

Abstract

Among the on-chip current sensing techniques, inductor current sensing method is the most feasible one for high switching frequency converter, which aids reduction of the filter inductor size. This paper presents a reduction of area design for the combined AC and DC inductor current sensor with load transient enhancement for the systems-on-chip applications. The proposed current sensor is simulated under several switching frequencies. Finally, the transient enhancement function of the proposed circuit is simulated in a buck converter at 4.5MHz switching frequency. When the load current changes by 500mA, by increasing time-constant delay for the sensed DC current ten times, the output voltage overshoot/undershoot reduces from 126mV/229mV to 76mV/74mV.

Keywords-DC-DC converter; current sensing; inductor current sensor; transient enhancer; system-on-chip (SoC)

Introduction

Reducing area is one of the key factor for on-chip DC-DC converter design. The simplest method is to increase switching frequency to reduce required filter inductance. A feasible combined AC and DC inductor current sensing design was implemented in [4], which solved the limitations in [1], [2], and [3]. However, by analysis, the design can be further simplified to reduce chip area. In addition, load transient enhancement can be employed by a small change in the component values in [4].

Proposed Current Sensor

A. Current Sensing Principle

Shown in Fig. 1 is the overall current mode buck converter (LEFT) with the proposed inductor current sensor (RIGHT). The proposed inductor sensing circuit has the working principle similar to the one in [4], which uses a capacitor $C_{AC,buf}$ to buffer AC inductor current information using C_F voltage, and a negative loop feedback NFB to sense DC inductor current information to V_{SEN} voltage. For sensing AC, $R_{AC}C_F = L_0$ (no unit). In addition, by the principle of capacitor- charge balance, $\overline{I_{R_{AC}C_F}}$ are zero; thus, $\overline{V_{R_{AC}}} = 0$. On the other hand, by the principle of inductor volt-second balance, $\overline{V_{L_0}} = 0$. Using Thevenin theorem, $\overline{V_L} + \overline{V_{DCR}} = \overline{V_{R_{AC}}} + \overline{V_{C_F}}$. Therefore,

$\overline{V_{DCR}} = DCR.\overline{I_L} = \overline{V_{C_F}}$, i.e., C_F DC voltage does not depend on R_{AC} value. Therefore, in the proposed current sensor, DC inductor current information can be obtained using the AC sensing branch $R_{AC}C_F$.

B. DC Sensing Delay and Load Transient Enhancement

As shown in Fig. 2, when I_{LOAD} increases, if the DC component of inductor current is delayed, the slope of the sensed current reduces for some initial periods after the change. Therefore, the DUTY will be detected more slowly, which means the DUTY is larger compared to that used no DC inductor current delay. By this way, more high-side switch's current can charge more to the output. Similar explanation can be made for I_{LOAD} decreasing case, as shown in Fig. 3.

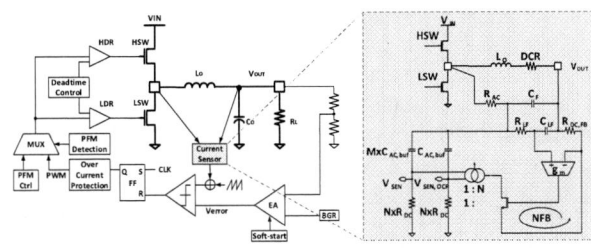

Fig. 1. Current Mode Buck Converter and the Proposed Combined AC and Delayed-DC Inductor Current Sensor.

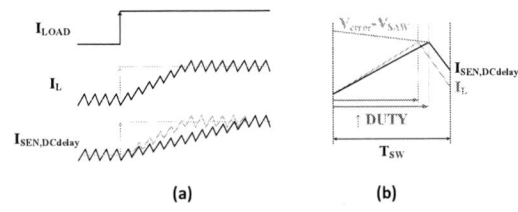

Fig. 2. (a) Sensing DC Current Delay at Rising Load Current; and (b) Its Influence to Loop Stability Explained in one period.

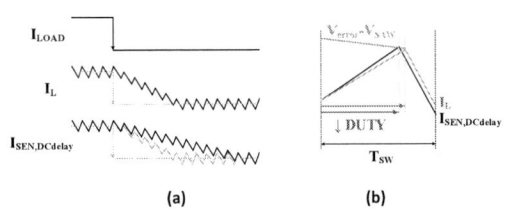

Fig. 3. (a) Sensing DC Current Delay at Falling Load Current; and (b) Its Influence to Loop Stability Explained in one period.

Therefore, by increasing delay time constant of the DC current sensing voltage by M times using $MxC_{AC,buf}$, the current compensation will be faster at the load transient response. The DC current delay should not be too much so that the different voltage of V_{error} and V_{SAW} crosses V_{SEN} within one period.

In reverse, the current sensing speed should be correct for the over current protection circuit, which limits the inductor current at an allowable level. Thus, a possible design may use different sensed DC current for different function, which can simply be accomplished by using a two-branch current mirror for each purpose as shown in Fig. 1.

Simulation Results

A. Sensed Inductor Current Response

Shown in Fig. 4 is the sensed inductor current simulated at 160MHz switching frequency: The sensed AC signal shows only maximum of 28mV error, with the mode of 7mV.

Shown in Fig. 5 (a) is the proof for the DC analysis tested at 1.2MHz switching frequency: The DC error of the voltage across V_{CF} by using the proposed structure is only 0.22% compared to the ideal $R_{AC}C_F$ time constant case ($R_{AC}C_F = L_0/DCR$). Fig. 5 (b) shows the ability of DC current tracking of the proposed current sensor tested at 5MHz switching frequency, when load current changes from 200mA to 400mA.

B. Load Transient Enhancer

Shown in Fig. 6 is the load transient response simulated with different $MxC_{AC,buf}$ capacitance to verify the transient enhancing capability explained in Fig. 2 and 3. As in Fig. 6 (a), the load transient response results a large voltage output change when load current changes from 300mA to 800mA (500mA): For the increasing load current case, the maximum change is 126mV; and for the decreasing load current case, this change is 229mV. By increasing the sensing DC current time response by ten times (i.e. increasing the $MxC_{AC,buf}$ by ten times), this maximum change reduces to 76mV for load current increasing case, and to 74mV for the decreasing case, as in Fig. 6 (b).

C. Other Recommendations

Practical implementation may consider eliminating DC OPAMP offset by using trimming bit or average sensing structure. Finally, a high switching frequency converter should also take switching loss into account.

Conclusion

An area saving current sensing inductor based on the architecture in [4] is presented and simulated under various switching frequencies. The simulation results show no difficulty in sensing current at high switching frequency. In addition, load transient can benefit from the delayed DC inductor current: By increasing the DC delay by ten times, the output voltage change reduces to maximum 3.1 times for load current change of 0.5A. Practical design may implement DC OPAMP offset cancellation. Finally, tradeoff between high switching frequency and switching loss should be considered.

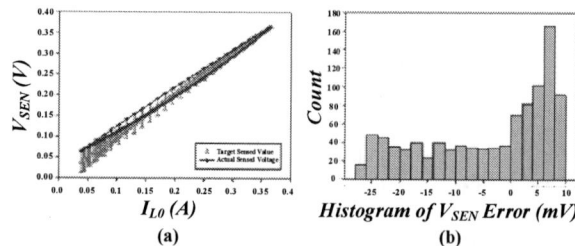

(a)   (b)

Fig. 4. Sensing AC Current: (a) Eye-Diagram of the Sensing AC Current comparing to the Ideal AC Inductor Current; (b) The Histogram of Sensing Current Error.

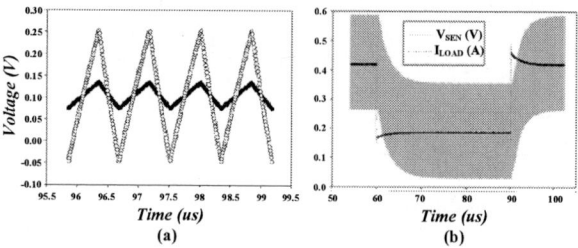

(a)   (b)

Fig. 5. Sensing DC Current: (a) V_{CF} under Different $R_{AC}C_F$ Time Constant; (b) Overall Sensing Current compared to the Load Current.

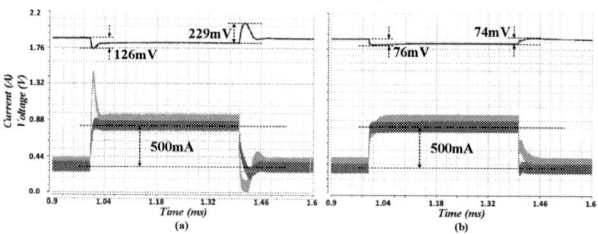

(a)   (b)

Fig. 6. Load Transient Enhancement: (a) M = 1; (b) M = 10.

Acknowledgment

This research was sponsored by the Ministry of Science, ICT & Future Planning and National Research Foundation of Korea (NRF) through the Human Resource Training Project for Regional Innovation. The simulation tool was supported by IDEC, KAIST.

References

[1] X. Zhou; P. Xu; Lee, F.C., "A novel current-sharing control technique for low-voltage high-current voltage regulator module applications," Power Electronics, IEEE Transactions on , vol.15, no.6, pp.1153,1162, Nov 2000.

[2] Chin Chang, "Combined lossless current sensing for current mode control," Applied Power Electronics Conference and Exposition, 2004. APEC '04. Nineteenth Annual IEEE , vol.1, no., pp.404,410 Vol.1, 2004.

[3] L. Hua; S. Luo, "Design considerations of time constant mismatch problem for inductor DCR current sensing method," Applied Power Electronics Conference and Exposition, 2006. APEC '06. Twenty-First Annual IEEE , vol., no., pp.7 pp., 19-23 Mar 2006.

[4] M. K. Song; Sankman, J.; D. Ma, "4.2 A 6A 40MHz four-phase ZDS hysteretic DC-DC converter with 118mV droop and 230ns response time for a 5A/5ns load transient," Solid-State Circuits Conference Digest of Technical Papers (ISSCC), 2014 IEEE International , vol., no., pp.80,81, 9-13 Feb. 2014.

A Low-IF AGC Amplifier for DSRC Receiver

Hung-Wen Lin, Wu-Wei Lin, Chun-Yen Lin

Department of Electrical Engineering, YuanZe University
Room.70629, Building No.7, 135 Yuan-Tung Road, Chung-Li, Taiwan
Tel: +886-3-463-8800 #7128 , Fax: +886-3-463-5399, hwlin@saturn.yzu.edu.tw

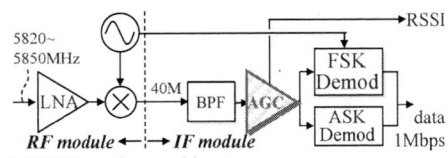

Fig. 1 DSRC receiver architecture

Abstract

This paper proposes an auto-gain-control amplifier for DSRC IF receiver. The test chip was realized in 0.18-um CMOS technology and occupied an active area of 0.064 mm². With a supply voltage of 1.8 V, the total power consumption was 10.5 mW. VGA exhibited a maximum gain of 82 dB and a dynamic range of 73 dB. For a 40-MHz of IF band, the ASK modulation index ranged from 0.23 to 1, and the data rate ranged from 10 kbps to 2.5 Mbps.

Keywords: VGA, AGC, RSSI, ASK, DSRC

Introduction

Dedicated short-range communication (DSRC) [1] is defined as a wireless link system between a toll station and a vehicle. Fig. 1 presents an example of a DSRC receiver. The 5.8-GHz *radio frequency* (RF) signal is first down-converted to a 40-MHz *intermediate frequency* (IF) signal by mixers. The in-band signal (40± 2.5 MHz) is then increased using an *auto-gain-control* (AGC) amplifier and the out-of-band noise is filtered out using a *band-pass-filter* (BPF). Finally, the enlarged IF signal is recovered to the baseband data by employing a switchable *frequency-shift-keying* (FSK)/*amplitude-shift-keying* (ASK) demodulator.

Proposed architecture

Fig. 2 shows the AGC architecture, which is composed of an eight-stage *variable-gain-amplifier* (VGA) and an AGC feedback loop. A VGA exhibiting a gain of 70 dB is easily saturated by internal or input offset noise; thus, an *offset cancellation* (OFC) feedback loop is added to cancel the noise. An ASK demodulator detects the *receive-signal-strength-indicator* (RSSI) of an AGC feedback loop to recovery binary data.

Fig. 3 shows a schematic of a single-stage VGA[2], which was designed using an n-type differential pair with a PMOS diode-connected load, and a NMOS source-degeneration resistor was incorporated to adjust the gain. Each VGA cell has its own common mode feedback circuit to enable the output common mode to equal the input common mode, so that all amplifiers demonstrate a similar performance.

For the AGC loop, the peak detector detects the

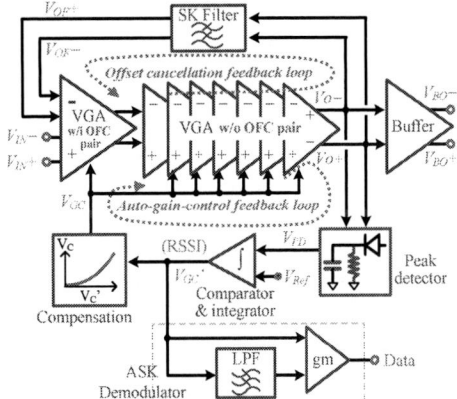

Fig. 2 Proposed VGA amplifier architecture

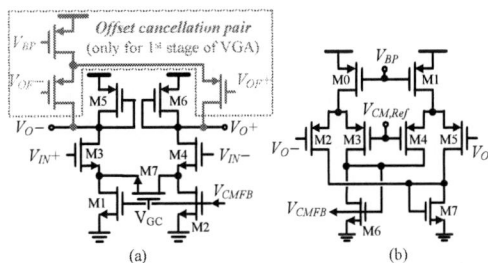

Fig. 3 (a) Schematic of VGA cell (b) Common mode feedback

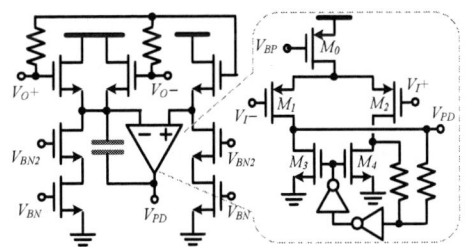

Fig. 4 Schematic of the peak detector.

swing of VGA and outputs a peak voltage(V_{PD}). Subsequently, by using an OP-based active integrator, V_{PD} was compared with a reference swing (V_{Ref}) and

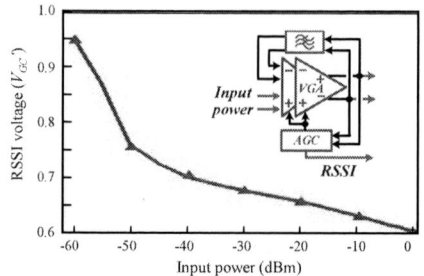

Fig. 8 Measured RSSI at different input power

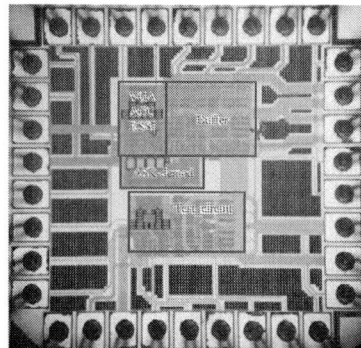

Fig. 5 Test chip photo

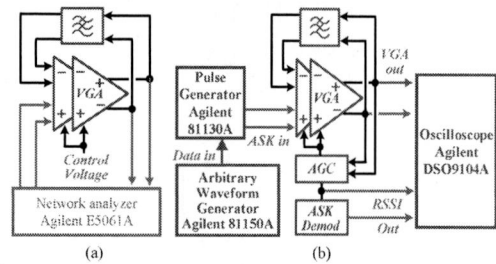

(a)   (b)

Fig. 6 (a) VGA test setup and (b) RSSI / De-ASK test setup

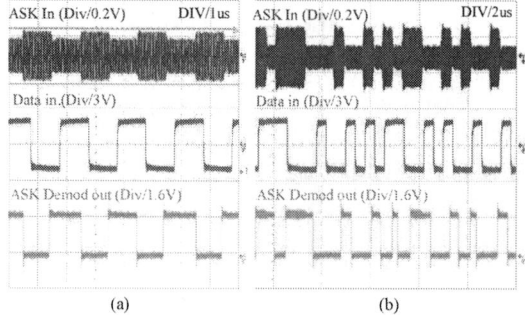

(a)   (b)

Fig. 9 De-ASK output (a)23% of modulation index(b)2.5Mbps

TABLE I PERFORMANCE COMPARISON.

	Specification	Simulation	Measurement
Technology	TSMC 0.18 μm CMOS Mixed Signal		
Chip/Active area	-	0.56 mm²/0.064 mm²	
Power consumption	-	9.5mW	10.5mW
VGA with RSSI			
Max. Gain	>70dB	83.4dB	82dB
Bandwidth	>40MHz	198MHz	101MHz
Dynamic range	>60dB	78dB	73dB
Min. Input power	<-70dBm	-78dBm	-73dBm
ASK demodulator			
Carrier frequency	40MHz	40MHz	40MHz
Data rate	1Mbps	1Mbps	1kbps~2.5Mbps
Modulation Index	0.6~0.9	0.3~1	0.23~1

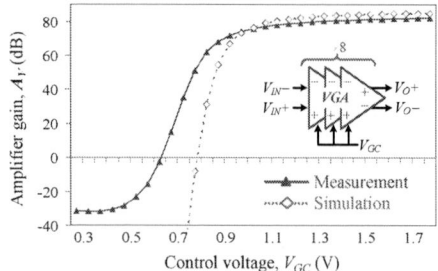

Fig. 7 Measured VGA gain at different control voltages

low-pass filtered to obtain an average value of the difference (V_{GC}') of reference swing and output swing. Thus, V_{GC}' means RSSI. Finally, V_{GC}' was enlarged using a compensator to obtain a wide gain-control range and a highly linear gain-control transfer curve.

For the OFC loop, the output offset levels were extracted, filtered and amplified to be offset controls (V_{OF}) by using a *Sallen-key* (SK) type *low-pass-filter* (LPF). V_{OF} then adjusted the input offset of a first-stage amplifier to compensate for the offset error of the overall VGA.

Test chip measurement

The AGC amplifier was realized in a 0.18-μm process, as shown in Fig. 5. The chip area and core area were 730 μm× 770 μm and 550 μm × 300 μm, respectively. Fig. 6 shows the measurement setup. The gain and bandwidth of the VGA were measured by Agilent E5061A. The ASK test input was generated by modulating 10 Kbps-2.5 Mbps PRBS from Agilent 81150A into 40 MHz of a carrier from Agilent 81130. The VGA outputs, RSSI and ASK-demodulated data were measured using Agilent DSO 9104A.

Fig. 7 shows the measured VGA gain at different controls, and the gain is varied from -30 dB to 82 dB. Fig. 8 shows the measured RSSI with varying input power. Because of the limitation of the pulse generator, the minimal test input power was -60 dB. Fig. 9(a) and 9(b) show the ASK demodulator outputs with a minimal modulation index of 23% and those operated at 2.5 Mbps PRBS, respectively. At a supply voltage of 1.8 V, the total power consumption of the overall AGC amplifier was 10.5 mW. TABLE I lists the specifications, simulation and measured results.

Acknowledgment

The authors would like to thank the National Science Council for financially supporting this research (No.102-2220-E-155-004). CIC is appreciated for supporting the chip fabrication (T18-101A-147).

References

[1] "CNS DSRC physical layer standard," http://www.ttia-tw.org/files/files/20100920103232-630782.pdf.

[2] P.-C. Huang, *et al.*, "A 2-V 10.7-MHz CMOS limiting amplifier/RSSI," *IEEE J. Solid-State Circuits*, vol.35, no.10, pp.1474-1480, Oct. 2000.

978-1-4799-5128-4/14 $31.00 © 2014 IEEE

A 10-bit Fast Lock Data Recovery Compensating Pulse-Width Distortion for Isolated Data Communications

Hironobu Akita, Takasuke Ito, Keita Hayakawa, Nobuaki Matsudaira,
Hirofumi Yamamoto, Chao Chen, Shigeki Ohtsuka, and Shinichirou Taguchi

DENSO CORPORATION, Kariya, Japan

Abstract

This paper presents a fast lock data recovery, which can compensate pulse-width distortion of the data signal waveform caused by the optocoupler. The proposed technique exploits the nature that only the rising edges of the data have a fixed amount of time delay at the output of the optocoupler. This delay is cancelled out by a recursive digital calculation after the bit number estimation using the blind oversampling. An experimental prototype fabricated in an in-house 40-V 0.8-um SOI BCD automotive process compensates more than ±50% pulse-width distortion with only 10-bit training as a preamble pattern. It can be applied with an inaccurate CR oscillator with 0.17-UI (unit interval) peak-to-peak jitter and up to ±10% frequency offset between the transmitter and the receiver.

Introduction

Isolated communications are widely used in the automotive electric components such as battery management systems [1] and motor controllers [2] especially in the recent EVs and HEVs. The optocoupler is commonly used at the insulation point, and its operation speed limits the data rate. Fig. 1 illustrates the schematic and operation waveform of the conventional optocoupler. A phototransistor switch causes pulse-width distortion, which becomes the root cause of the speed limitation. There have been several proposals for the pulse-width compensation. However, it takes a longer time to converge the compensation coefficient by statistical measurement [3] or analog threshold voltage adjustment [4]. Detection of the minimum pulse-width from the oversampling data requires the accurate reference clock, which can be generated only by the PLL type CDR of the longer lock time [5]. Therefore, these proposed techniques are not suitable for communications in the real-time control systems requiring the shorter start-up time from the no signal idle state.

This paper proposes a novel data recovery which successfully compensates pulse-width distortion by the 10-bit preamble training even with a less accurate CR oscillator reference clock.

Compensation Concept of Proposed Data Recovery

The variation time of the rising edge at the output of the optocoupler is constant for every edge as shown in Fig. 1. Therefore, time-lengths between the adjacent rising edges and the adjacent falling edges are just the same as the time-length corresponding to the bit numbers without pulse-width distortion, as illustrated in Fig. 2. The bit numbers of the measured time-lengths can be decided from the sampling numbers by the blind oversampling with the fast-lock bit number estimation technique [6]. From the calculation between the measured bit numbers, the bit numbers at the overlapped time-length can be recovered. Parameters $Cr[n]$ and $Cf[n]$ in Fig. 2 are the measured bit numbers between the adjacent rising edges and between the adjacent falling edges, respectively, in the n-th duration. Na and Nb represent the bit

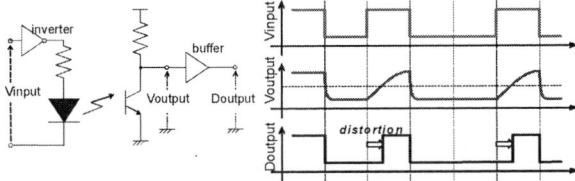

(a) Schematic of optocoupler.　　(b) Internal waveform.
Fig. 1 Cause of pulse-width distortion.

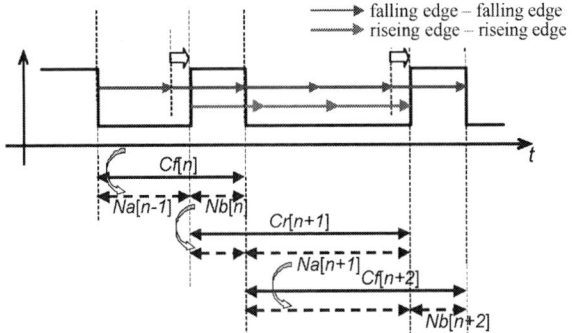

Fig. 2 Relationship between bit numbers.

numbers of the consecutive 0s and 1s, respectively. From the figure, the relationship can be derived as:

$$Nb[n] = Cf[n] - Na[n-1], \qquad (1)$$
$$Na[n+1] = Cr[n+1] - Nb[n], \qquad (2)$$
$$Nb[n+2] = Cf[n+2] - Na[n+1]. \qquad (3)$$

According to these numerical sequences, if we decide the initial bit numbers, $Na[0]$ or $Nb[0]$, all the subsequent bit numbers can be recursively derived by the subtractive calculation between the measured bit numbers Cr and Cf. By transmitting the pre-defined preamble pattern at the beginning of the communication data frame, the initial bit numbers, $Na[0]$ or $Nb[0]$, can be determined.

Circuit Design and Measurement Results

Fig. 3 depicts a circuit block diagram of the receiver with the proposed technique compensating pulse-width distortion. Two bit number decision blocks are prepared for both the adjacent rising edges and the adjacent falling edges. The sampling numbers measured by the blind oversampling at the preamble phase is stored at memory registers in the fractional resolution for both rising edges and falling edges. At the data recovery phase, sampling numbers between the adjacent rising or falling edges are compared with the corresponding values stored in the memory, deciding the number of the consecutive bits. Then, the subtraction block recovers the data for either 0s or 1s recursively from bit numbers coming from the previous blocks. The CR oscillator generates a 4-MHz clock, which is multiplied up to 20-MHz by the PLL, which realizes the blind oversampling of 10x for the 2-Mbps data rate.

978-1-4799-5128-4/14 $31.00 © 2014 IEEE

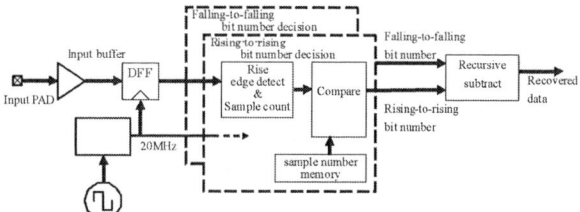

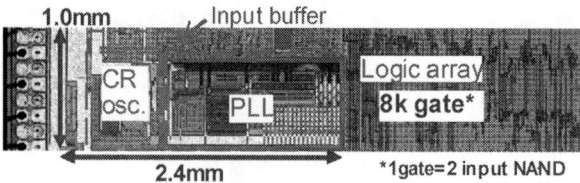

Fig. 3 Circuit block diagram of the receiver.

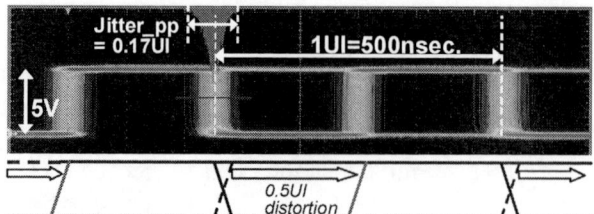

Fig. 5 Input signal with jitter and pulse-width distortion.

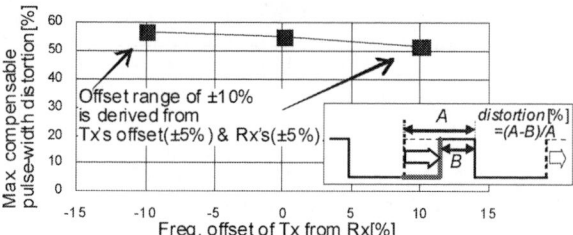

Fig. 4 Chip photomicrograph.

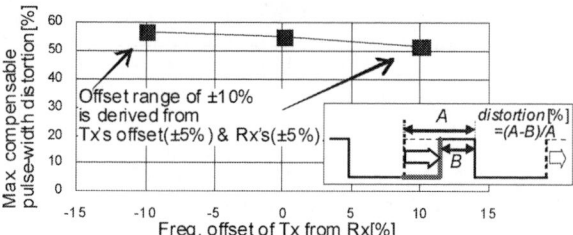

Fig. 6 Measured compensable range of pulse-width distortion.

TABLE I: Design parameter summary

Nominal data rate	2Mbps
CR oscillator accuracy	±5%
Oversampling	10 sample/bit
Data encoding	Original 4b/5b (max. run length = 3-bit)
Preamble	10-bit (Toggle pattern)

The prototype chip has been fabricated in an in-house 0.8um SOI BCD process. Fig. 4 shows the photomicrograph and the size of this receiver macro, which contains a digital data recovery core, an analog part of the input sampler, the CR oscillator, and the PLL. The overhead based on the separated bit number decision block to compensate the distortion is about 10% of the digital circuit. Table I shows the design parameter summary. The chip performance is measured with the input signal generated by the signal generator. The random data based on the original 4b5b encoding are generated with 0.17UI peak-to-peak jitter emulating the jitter of the CR oscillator at the transmitter side, following the preamble of 10-bit toggle pattern. All the rising edges are delayed as pulse-width distortion (Fig. 5). Fig. 6 is the measured compensable distortion range on the offset frequency between the transmitter and the receiver at BER<10^{-9}. Since sampling numbers are influenced by the frequency offset due to the blind oversampling, the maximum compensation range shows a slight dependency on the frequency offset. However, at least ±50% pulse-width distortion is confirmed to be compensated correctly. Table II provides a comparison of this work with published results. Our design achieves drastic improvement of the lock time over the reported techniques. Regarding the data rate, 2-Mbps is our target for optocoupler communications in the high voltage automotive system. The proposed technique can be applied easily to the higher data rate of the other applications, because of the combination of the blind oversampling and the digital calculation without using a complicated analog circuit. When applied in the systems with the accurate crystal oscillator, more than ±80% pulse-width distortion is estimated to be compensated.

In summary, a fast lock data recovery compensating pulse-width distortion has been presented for isolated communications with the conventional optocoupler. More than ±50% pulse-width distortion is compensated only with 10-bit preamble training and inaccurate CR oscillator of 0.17-UI jitter and ±10% frequency offset between the transmitter and the receiver.

References

[1] J. Chatzakis, K. Kalaitzakis, N. C. Voulgaris, and S. N. Manias, "Designing a new generalized battery management system," *IEEE Industrial Electronics*, Vol. 50, pp. 990-999, Oct. 2003.

[2] K. Ishikawa, K. Suda, M. Sasaki, and H. Miyazaki, "A 600V driver IC with new short protection in hybrid electricvehicle IGBT inverter system," *IEEE International Symposium on Power Semiconductor Devices and ICs*, pp.59-62, May 2005.

[3] R. Z. Bhatti, M. Denneau, and J. Draper. "Duty cycle measurement and correction using a random sampling technique," *IEEE Midwest Symposium on Circuits and Systems*, pp. 1043-1046, Aug. 2005.

[4] H. Tagami, et al. "A burst-mode bit-synchronization IC with large tolerance for pulse-width distortion for Gigabit Ethernet PON," *JSSC*, Vol. 41, pp. 2555-2565, Nov. 2006.

[5] J. Terada, et al. "Jitter-reduction and pulse-width-distortion compensation circuits for a 10Gb/s burst-mode CDR circuit." *ISSCC*, pp. 104-105, Feb. 2009.

[6] H. Akita, et al, "A 10-bit fast lock all-digital data recovery with CR oscillator reference for automotive network." *IEEE International Symposium on Circuits and Systems*, pp. 1171-1174, May 2013.

TABLE II: Performance comparison

	[3]	[4]	[5]	**This work**
Technology	0.13um CMOS	0.32um SiGe BiCMOS	0.25um SiGe BiCMOS	0.8um, 40V SOI BiC+DMOS for automotive
Data Rate	5Gbps	1.25Gbps	10Gbps	2Mbps
Start-up time	1024-bit ~ (Statistical measurement)	~1000-bit (CDR lock time)	~500-bit (Analog adjustment)	10-bit (Data recovery and distortion training at preamble)
compensable distortion range	−40 ~ +40%	−75 ~ +75%	−40 ~ +40%	−50 ~ +50% (with CR oscillator) −80 ~ +80% (with crystal oscillator, *estimation*)

Auto-delay offset cancellation technique for time difference repeating amplifier

In-Seok Kong, Eun-Ho Yang, Kyung-Sub Son, Young-Jin Kim, and Jin-Ku Kang

Dept. of Electronics Engineering, INHA University
Yonghyun-dong, Nam-Gu
Incheon, South korea
TEL) +82-32-860-8721 FAX) +82-32-868-3654
Email: jkang@inha.ac.kr

Abstract

This paper presents an auto-delay offset cancellation technique for time difference repeating amplifier. Pipeline time-to-digital converter (TDC) achieves fine resolution by amplifying the time residue. Therefore the linearity of the time difference amplifier (TA) is important in pipeline TDC. The pulse-train TA, time difference repeating amplifier, was proposed to improve this recently. However, it is hard to get accurate gain in TA because there are many possible mismatch issues. Our work makes the delay offset be cancelled automatically during time difference repetition. The proposed circuit is designed and simulated in 65nm CMOS process. The conversion rate is 100Msps and it has 300ps input time range. The proposed scheme shows the delay offset of about 10fs, which is much less than that of the conventional scheme (~100ps) under the equivalent device mismatch conditions.

Keywords-Time amplifier, Time-to-digital converter, All-digital PLL, Delay offset cancellation, Time-domain ADC

Introduction

Time difference amplifier(TA) is most important block in pipeline time-to-digital converter (TDC) which achieves high conversion rate and fine resolution. Especially, the accuracy and linearity in TA have a huge effect on the performance of the pipeline TDC. There were various researches on amplifying the time difference, and it could be categorized by fundamentals.

The first one is 'latch based TA' such as [1] and [2]. Latch has linear time variant (LTV) system which has time varying impulse response including delay. This makes it possible that the output time difference varies with the input time difference. While it achieves the gain, it has input time range restriction due to the cross-coupled structure. Moreover, the linearity couldn't be guaranteed over the full input time range.

Another one is 'repetition based TA' such as [3]. Contrary to voltage, amplifying time could be modeled as integration. Therefore, it is possible to make amplification by repeating and accumulating the time difference. Fig. 1 shows a pulse-train TA as an example about this. The gap between two input rising edges makes an pulse ,and the pulse repeated by buffer chain and OR logic. It obtains highly linear gain by simple mechanism. Also the input time range is only affected by conversion rates, not structure. However, the output can't be obtained exactly because delay mismatch could occur while a pulse is duplicated.

This paper suggests time difference repetition (TDR) circuit

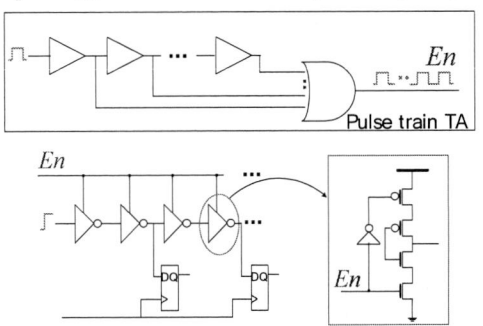

Sampling mechanism using pulse train TA

Fig. 1 An example of time difference repetition (pulse-train TA)

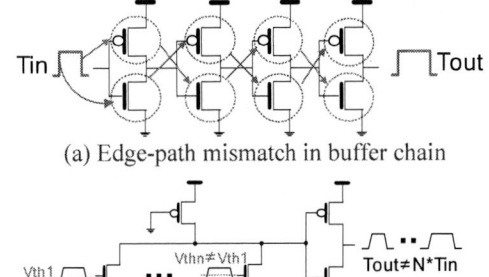

(a) Edge-path mismatch in buffer chain

(b) Threshold voltage mismatch in OR gate

Fig. 2 Possible time-offset issues by device mismatch in pulse-train TA

which has auto-delay offset cancellation. Unlike the conventional approach, the proposed repetition circuit cancels the delay offset automatically by sharing signal path. It also does not require making a pulse for repetition. Before description of the prposed work, time(delay) offset issues caused by device mismatch will be explained.

Proposed Delay-Offest Cancellation Scheme

A. Time(delay) offset issues during pulse repetition

Although the time repeating amplifier has higher accuracy, there is time-offset issue caused by various mismatches as mentioned in [3]. As shown in Fig. 2(a), the rising and falling edges of one pulse have different signal paths in buffer chain. This could make delay mismatch causing pulsewidth distortion.

978-1-4799-5128-4/14 $31.00 © 2014 IEEE

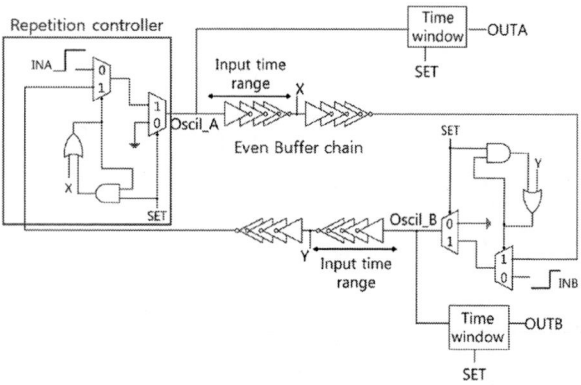

Fig. 3 The proposed time difference repetition circuit

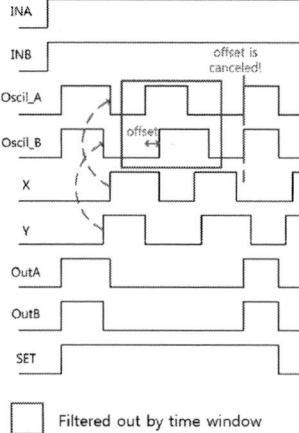

Filtered out by time window

Fig. 4 The offset cancellation process in TDR circuit

Besides, the input thresholds of OR are not guaranteed to be same each other as shown in Fig. 2(b). Those conditions will lead to time offset or delay offset.

B. Proposed time difference repetition(TDR) circuit

Fig. 3 shows the architecture of the proposed TDR. It is symmetric and includes even buffer chains and repetition controllers. Repetition controller changes the state of TDR to reset mode, input entering mode, and repetition mode. When SET is low, this circuit is in initial condition which makes the Oscil_A and Oscil_B be '0'. If SET turns to be high, input entering mode is begin and the input pulses pass through each multiplexer. After delay as the input time range, node X and Y are turned on and these switch the state of this circuit to be repetition mode. The behavior is described in Fig. 4. Even though delay mismatch occurs in buffer chain, it can not affect the gap between OutA and OutB because both signal paths are exactly overlapped. Whenever the time offset is made during a halfway, it is cancelled automatically in another half way. For filtering out redundant pulses, Time window circuit is added.

Simulation result

The proposed TDR circuit is designed in 65nm CMOS process. In order to verify delay offset cancellation effect, it was simulated under pMOS width variation. For inducing

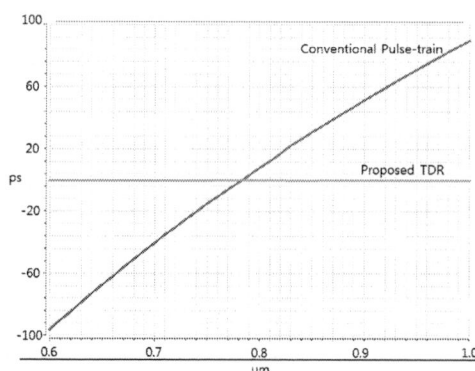

Fig. 5 The simulation result. "pMOS width variation" vs "repeated time difference offset"

delay mismatch, the variation occurs on parts of pMOS, not all of that. Fig. 5 shows the result and the output time difference offset always less than 10fs. The comparison with previous work under the equalized condition shows how well our work does. This circuit operates under 100Msps conversion rate and 300ps input time range. Due to increased the number of buffers, our work dissipates 110μ A current. This could be reduced at an expense of shortened input time range.

.

Conclusion

An time difference repeatition circuit including offset cancellation has been presented. By sharing signal paths, the delay offset could be cancelled during repeating time difference. Our work is implemented in 65nm CMOS process. The proposed scheme shows the delay offset of about 10fs, which is much less than that of the conventional scheme (~100ps) under the equivalent device mismatch conditions. The simulated conversion rate, input time range, and current consumption are 100Msps, 300ps, and 110μ A, respectively.

Acknowledgment

This research was supported by NRF (2010-0022670) and ITRC(NIPA-2013-H0301-13-1013).

References

[1] Minjae Lee, and A. Abidi, "A 9b, 1.25ps Resolution Coarse-Fine Time-to-digital converter in 90nm CMOS that Amplifies a time residue," *IEEE J.Solid-State Circuits,* vol. 43, no. 4, pp. 769-777, Apr. 2008.

[2] Seon-Kyoo Lee, Young-Hun Seo, Hong-June Park, and Jae-Yoon Sim, "A 1GHz ADPLL with a 1.25ps minimum-resolution Sub-exponent TDC in 0.18μ m CMOS," *IEEE J.Solid-State Circuits,* vol. 45, no. 12, pp. 2874-2881, Dec. 2010.

[3] Kwang-Seok Kim, Young-Hwa Kim, Wonsik Yu, and Seong-Hwan Cho, "A 7 bit, 3.75ps resolution two-step Time-to-digital converter in 65nm CMOS using Pulse-train time amplifier," *IEEE J.Solid-State Circuits,* vol. 48, no. 4, pp. 1009-1017, Apr. 2013.

A CMOS Low Power Biohybrid Lamprey Robot Controller Design

Yong-Bin Kim

Department of Electrical and Computer Engineering
Northeastern University
Boston, MA, USA
Tel: 617-373-2919 Fax: 617-363-8920 Email: ybk@ece.neu.edu

Abstract

This paper presents a CMOS low power biomimetic robot controller design capable of undulatory locomotion, initiated by light and guided by chemotaxis. This capability will be based on coupling power generation with actuation through the mitochondria of muscle engineered to respond to light and sensing by engineered cells that modulate an electronic nervous system. Engineered sensor cell responses will be connected with potentiometric interfaces to modulate the electronic central pattern generator (CPG) neurons. The CPG generates lateral undulations that propagate from front to rear and allow the two sides to be biased to mediate turning. This paper demonstates actual feasibility of the biomimetic robot's behavior using the CPG designed using subthreshold anlaog circuits design technique for low power with 180nm CMOS process and 1.8V power supply dissipating 3.28mW on 1.1mm^2 silicon area.

Keywords-component; CPG, biomimetic robot, electronic neuron

Introduction

Robots are fundamental in a broad spectrum of repetitive or dangerous work efforts ranging from industrial to field applications. There is a strong demand for consumer, industrial, and scientific applications for autonomous robots that can operate adaptively in unpredictable environments. Existing robots are commonly controlled by algorithm based systems such as finite state machines[1]. However, algorithm-based robots controlled by digital processors may adapt poorly to unstructured environments due to an inability to anticipate all contingencies. A new category of robots has been emerging with the goal of capturing the performance advantages of animal models. These robots employ the biomechanics and control strategies used in natural environments. Common features of such systems include a biomorphic body plan that captures the biomechanical advantages of the animal model, an electronic nervous system based on the command neuron, coordinating neuron CPG(Central Pattern Gennerator) model, myomorphic actuators, neuromorphic sensors and a library of behaviors reverse engineered from the animal model. A key feature of the biomimetic approach is the use of networks based on neuronal and synaptic animal models rather than algorithms. This network-based control distinguishes biological intelligence from artificial intelligence.

The innate behavior of animals is controlled by central pattern generators (CPGs) resident in central ganglia or the spinal cord[2]. It is our contention that if CPG-based controllers are imitated, many problems due to deterministic control program can be solved and it will be feasible to develop an adaptive autonomous robots that operate with flexibility in natural environment. CPGs can be constructed from electronic neurons based on non-linear dynamical models of biological neurons[3].

Biomimetic Robot Systems

Fig. 1 illustrates the frame of a bio-mimetic micro robot that adopts CMOS neuron based motion controller [4]. It is composed of biosensors, CMOS neuron based motion control unit (also called central pattern generator CPG), engineered muscle which is activated by organic light emitting diode (OLED) circuit, and powered by thin film battery, all of which are covered with Kapton chassis. In such a composition, external environment changes are received by engineered cellular sensors through exteroceptive reflexes. The sensors then release command to CPG motion control unit to generate corresponding motion patterns. OLED circuit then transfers the motion patterns and the optogenetically engineered muscles that respond to light will produce an undulatory movement of the body axis.

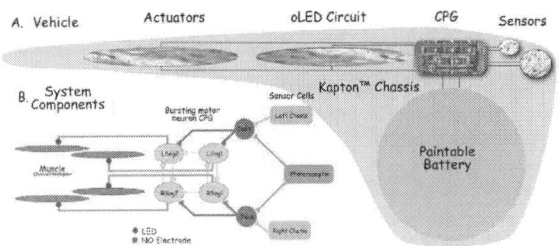

Fig. 1 Diagram of the proposed biomimetic robot system. The biosensors receive external environment signals and pass them to CMOS neuron based motion control unit (also called central pattern generator CPG).

Electronic Neurons and Synapses

We propose to build a controller based on the principles of operation of CPGs. We employ two types of controllers in our biomimetic robots. The first is based on analog computers that solve the Hindmarsh Rose equations in real time. We have recently extended this technology to subthreshold analog

Very-Large Scale Integration (VLSI). The second technology is based on a discrete time map based model that allows the operation of networks containing hundreds of neurons and synapses on a single Digital Signal Processor (DSP) chip due to being based on difference equations rather than differential equations. We have used this technology to control a lamprey-based robot with layered exteroceptive reflexes.

Electronic Neurons and Synapses

Synthetic biology seeks to build devices and systems from fungible gene parts. The fundamental scheme of synthetic biology combines parts integrated into a chassis to produce a device with properties not found in nature. The next level of organization consists of a system: an interacting set of devices. The present proposal aims at the integration of such a system. The systems in this research to synthesize focus on constitutive vs. induced expression.

Biohybrid Robots

Biohybrid robots incorporate engineered living cells with microelectronic controllers. There is a broad variety of biohybrid robots. There has been much recent work on the development of biohybrids formed from cultured muscle, although these efforts lack specific excitation/contraction coupling mechanisms to couple controllers to muscle.

Undulatory CPG VLSI Design
A. Lambrey Based Swimming Pattern

A lamprey-based swimming CPG VLSI design is presented in this paper. Fig. 2 top figure show the swimming motor pattern exhibited by lamprey swimming.

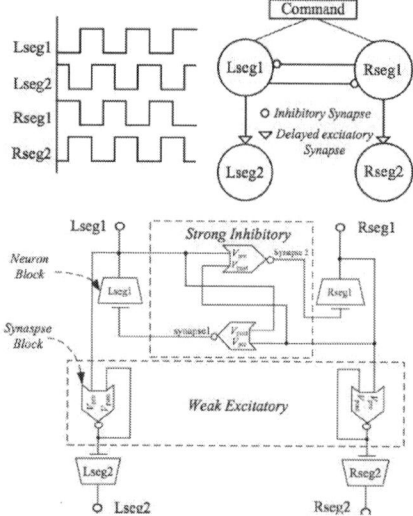

Fig. 2 Undulatpory motor program. Lamprey swimmimg pattern based CPG network(top). Electronic CPG composed of neurons and synapses(bottom).

B. Low Power Circuit Implementation

Both the electronic neuron and electronic synapse have been achieved at 1.8V supply voltage in 180nm standard CMOS process using subthreshold circuit design technique. The whole circuit has been carefully optimized under this process so as to meet the 5% parameter variation tolerance in real process. Fig. 3 shows the layout of the chip. There are 80 output pads, and some of them are internal signals, that are lead to output pads via buffer drivers for debug purposes. The electronic CPG needs to be interfaced with engineered cells to complete the input/output systems of the robot. The interface is for the communications between the electrical signals from the CPG and optical signals to the engineered cells.

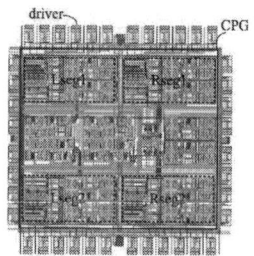

Fig. 3 Central Patern Generator Layout in 180nm CMOS process

C. Experimental Results

Fig. 4 shows the signal waveforms representing the membrane potential of a CPG neuron and the bursting pattern from the fabricated chip.

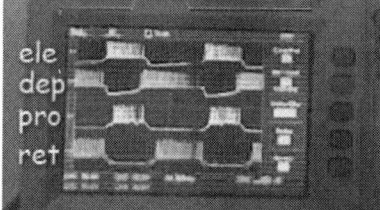

Fig. 4 Experimental results of the CMOS neuron's bursting pattern

The experimental results validate the electronic CPG performance at a 0.8V supply voltage with parameter variation tolerance of 5% dissipating 3.28mW. The die size of the chip is 1.1mm^2 including I/O pads.

References

[1] Brooks, R. "New Approaches to Robotics.", Science 253: 1227-1232. 1991.
[2] Delcomyn, F. "Neural basis of rhythmic behavior in animals.", Science 210(4469): 492-8. 1980.
[3] Pinto, R. D., P. Varona, A. R. Volkovskii, A. Szucs, H. D. Abarbanel and M. I. Rabinovich, "Synchronous behavior of two coupled electronic neurons.", Phys Rev E. 62(2 Pt B): 2644-56. 2000.
[4] J. Lu, J. Yang, Y. B. Kim, J. Ayers, "Low power high PVT variation tolerant central pattern generator design for a bio-bybrid micro robot," 2012 IEEE International Midwest Symposium on Circuits and Systems, Aug. 2012, Boise Idaho, pp. 782-785.

Asynchronous Circuit Design using New High Speed NCL Gates

Minsu Choi[1], Byung-Ho Kang[2], Yong-Bin Kim[3], Kyung Ki Kim[2]

[1] Department Electrical & Computer Engineering, Missouri University of Science & Technology
Rolla, MO, USA
[2] Department of Electronic Engineering, Daegu University
Gyeongsan, South Korea
[3] Department of Electrical and Computer Engineering, Northeastern University
Boston, MA, USA
choim@mst.edu, kissme0905@live.daegu.ac.kr, ybk@ece.neu.edu, kkkim@daegu.ac.kr (corresponding)

Abstract

The delay-insensitive Null Convention Logic (NCL) as one of innovative asynchronous logic design methodologies has many advantages of inherent robustness, power consumption, and easy design reuses. However, transistor-level topologies of conventional NCL gates have weakness of logic speed, area overhead or wire complexity. Therefore, this paper proposes a new NCL gates designed at transistor level for high-speed, low area overhead. A 4x4 multiplier using the proposed NCL gates has been compared to the multiplier using conventional NCL gates in terms of delay, area and energy consumption.

Keywords-component; asynchronous circuit; null convention logic, NCL, delay insensitive model, multiplier

Null Convention Logic

In the reliable ultra-low power design, asynchronous circuits have recently been re-considered as a solution for scaling issues [1]. Null Convention Logic (NCL) is one of the promising delay-insensitive asynchronous circuit design methodologies [2]-[5]. NCL circuits utilize threshold gates with hysteresis to maintain delay insensitivity. NCL uses delay-insensitive codes for data communication, alternating between set and reset phases. NCL uses threshold gates with hysteresis for its composable logic elements. One type of threshold gate is the *THmn* gate as shown in Fig. 1(a), where $1 \le m \le n$. A $THmn$ gate means that at least m of the n inputs has to be asserted before the output will become asserted. Threshold gate inputs and outputs can be in of two states, DATA or NULL [1].

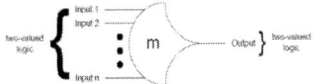

Fig. 1 THmn gate symbol

A threshold gate starting with its output in a NULL state will remain in the NULL state until the specified numbers of inputs are placed in the DATA state. Once the gate reaches the DATA state, it remains in this state until all of the inputs return to the NULL state. A dual-rail signal, D, consists of two wires, $D0$ and $D1$, which may assume any value from the set DATA0, DATA1, NULL. The $DATA0$ state ($D0 = 1$, $D1 = 0$) corresponds to logic zero, the DATA1 state ($D0 = 0$, $D1 = 1$) corresponds to logic one, and the NULL state ($D0 = 0$, $D1 = 0$)

corresponds to the empty set meaning that the value of D is not yet available. The two rails are mutually exclusive, so that both rails can never be asserted simultaneously; this state is defined as an illegal state.

New Transistor-Level NCL Gates

Several CMOS implementation schemes have been proposed for NCL gates such as dynamic, static, semi-static, and differential [3]-[5]. The dynamic implementation can be used in real-time computing applications where a minimum data rate is guaranteed so the state information can be kept on an isolated node without requiring a feedback mechanism, but this is not delay-insensitive. The static and semi-static implementations utilize feedback to maintain state information and therefore do not require a minimum input data rate [4]. The differential implementation of NCL gates has been recently introduced and discussed in [3]. The differential design is most similar to Boolean Differential Cascode Voltage-Switch Logic (DCVSL) gates, where both the output and its complement are available. NCL threshold gates are state-holding and designed to have hysteresis. As depicted in Fig. 2 (a), each static NCL gate is comprised of 4 transistor networks: *set, reset, hold1,* and *hold0*. The *set* function determines the gate's functionality as one of the 27 NCL gates. The output then remains asserted through the *hold1* function until all inputs are deasserted. The CMOS implementation of the static, semi-static, and differential TH23 is shown in Fig. 2 (a), (b), and (c), respectively. The differential implementations of NCL gates offer different advantages for design of NCL circuits. However, the differential design has weakness of wire complexity and logic delay. Static gates tend to be faster with lower voltage operation capability, while semi-static gates are more energy efficient, and differential gates are more area efficient. A comprehensive comparison of the different gate styles can be found in [4]. However, each NCL gates of the static implementation has many stack number of transistors as well as a large number of transistors, which leads large gate delay. In case of semi-static, fewer transistors are required, but the speed is slower due to the weak inverter feedback. In this paper, a new CMOS topology has been proposed for implementing gates required by NCL design at the standard-cell level as shown in Fig. 2 (d). The proposed TH23 topology is revised from the semi-static NCL gate. The semi-static NCL gate is the best transistor-level topology compared to other topology only if area overhead and wire

978-1-4799-5128-4/14 $31.00 © 2014 IEEE   13

complexity are considered. However, the main problem of the semi-static topology is the slow logic speed due to the weak inverter feedback and the number of PMOS serial stacks as shown in Fig. 2 (b). The proposed topology doesn't have any PMOS stack like the differential one, while a pseudo-NMOS logic block for a reset function is added, and a feedback memory is formed by a inverter and a NOR gate. The pseudo-NMOS logic has the advantage of speed at the cost of static power dissipation. However, the power can be saved enough to use the pseudo-type logic in a real system if the enable signal (EN) is properly controlled. In Fig. 2 (d), MP3 is turned off by EN signal generated by the NCL completion block as shown in Fig. 3, where a request-for-null (rfn) signal is asserted to the NCL completion block. That is, EN is changed from logic '0' to '1' right before NULL signals are asserted to the NCL circuits. The feedback inverter is needed to hold an output logic value, and the NAND gate is deployed to compensate for speed reduction due to the weak inverter feedback: the output switching time is reduced compared to the semi-static one because the output of the NAND gate is determined by Reset and Set block at the same time.

much smaller than other ones only if EN signal is turned off. Therefore, the average power dissipation is not much higher than expected. In case of the area overhead, the proposed NCL topology has the smallest area compared to other ones as expected because the Reset block in Fig. 2 (d) takes a very small area. Figure 4 (b) shows the layout of a 4x4 multiplier using the proposed NCL topology.

TABLE I
THE COMPARISON OF THE EXPERIMENTAL RESULTS WITH OTHER NCL GATES.

NCL topology	Power(W)	Delay(sec)	Area Overhead (normalized by Proposed NCL)
Static	1.39E-04	7.13E-10	1.50
Semi-static	1.75E-04	2.93E-09	1.25
Differential	2.32E-04	2.40E-09	1.28
Proposed (EN=ON)	4.38E-03	4.39E-10	1
Proposed (EN=OFF)	1.55E-06		

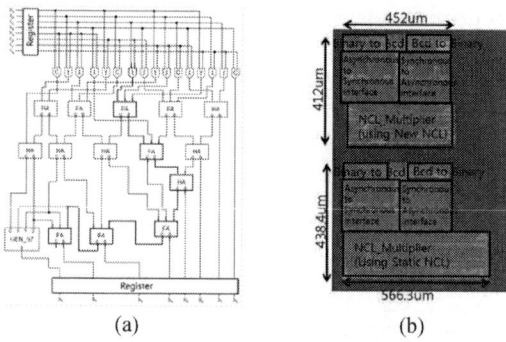

(a)   (b)

Fig. 4. (a) Block diagram of a 4x4 multiplier, (b) The layout of a 4x4 multiplier using the proposed NCL topology.

Conclusion

This paper proposes a new NCL topology for high-speed, low area overhead, and low wire complexity. The proposed topology is compared to the conventional static, semi-static, and differential topologies in terms of delay, area, and energy consumption using a 4x4 multiplier. As future work, we will improve the proposed NCL topology to reduce the static power consumption.

Acknowledgment

This work was supported by IC Design Education Center (IDEC).

References

[1] Scott C. Smith, Jia Di, "Designing Asynchronous Circuits using NULL Convention Logic (NCL)," Morgan & Claypool Publishers, 2009.

[2] F. A. Parsan, W. K. Al-Assadi, S. C. Smith, "Gate Mapping Automation for Asynchronous NULL Convention Logic Circuits," IEEE Trans. on VLSI Systems, Vol. 22, Issue 1, pp.99-112, Jan. 2014.

[3] S. Yancey and S. C. Smith, "A Differential Design for C-elements and NCL Gates," IEEE Int. Midwest Sym. on Circuits ans Systems (MWSCAS), pp.632-635, Aug. 2010.

[4] F. A. Parsan and S. C. Smith, "CMOS Implementation Comparison of NCL Gates," IEEE/IFIP Int. Conf. on VLSI and System-on_Chip (VLSI-SoC), pp.41-45, Oct. 2012.

[5] F. A. Parsan and S. C. Smith, "CMOS Implementation of Static Threshold Gates with Hysteresis: A New Approach," IEEE Int. Midwest Sym. on Circuits ans Systems, pp.394-397, Aug. 2012.

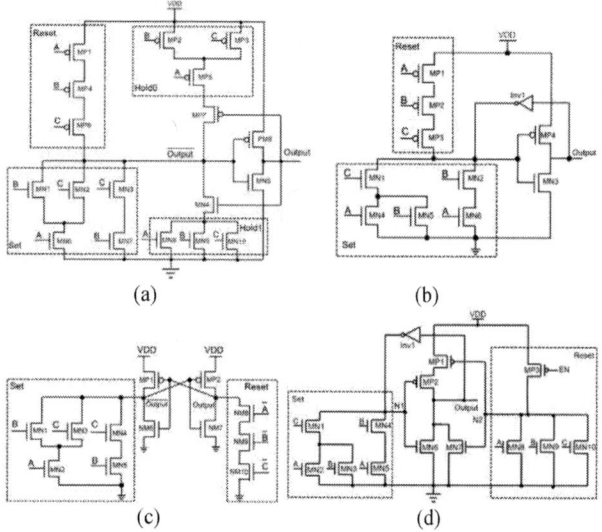

(a)   (b)

(c)   (d)

Fig. 2 CMOS implementation of (a) Static TH23, (b) Semi-static TH23, (c) Differential TH23 (d) Proposed TH23.

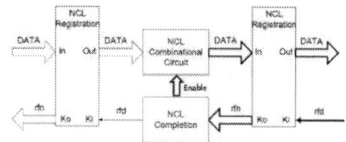

Fig. 3 Enable signal control using Completion signal.

Experimental Results

The proposed circuits have been designed and evaluated using a 0.11um MOSFET technology (VDD=1.2V). The topology has been compared to the classic static, semi-static, and differential topologies in terms of delay, area, and energy consumption using a 4x4 multiplier in Fig. 4 (a). As shown in Table 1, the simulation results show that the new topology offers faster operation and smaller area, with an increase in power. Although the power dissipation increases much more than other topologies due to the pseudo-NMOS structure, it is

AES Sbox GF(2^4) inversion functions based PUFs

Hyunmin Kim[1], Seokhie Hong[2]

Center for Information Security Technologies, Korea University
Robot Convergence building 408, 136-713, Anam-Dong, Seongbuk-Gu,
Seoul, South Korea
+82-2-3290-4756[1], +82-2-3290-4894[2] and hmkim.sec@gmail.com[1], shhong@korea.ac.kr[2]

Abstract

As the usage of PUFs such as in digital fingerprinting is growing dramatically, we've developed a new PUF based on an AES Sbox GF(2^4) inversion functions that uses differences in power consumption at output nodes due to process variation. It has several advantages such as being able to improve the properties of a PUF and it requires little additional resources.

Keywords; PUFs, AES Sbox GF(2^4) inversion, dual-rail logic

Introduction

Recently, it is becoming more and more important for users to have secure key storage in light of the identification and authentication problem. Guaranteeing the safety of integrated circuit based technology provides one core solution. These security technologies are applicable in so far as to provide anti-counterfeiting solutions against various IC chips. Also, despite using secure protocols to make the key invisible to the outside, it remains a possiblility to counterfeit by reverse engineering. The results of these vulnerabilities caused by IP leakage or the possibility to counterfeit through a reproduction of the chip design remains one of the biggest threats.

Physically unclonable functions (PUFs) are getting lots of attention as a solution to the above mentioned issues. PUF guarantees a unique way for user identification thanks to a random process variation from the IC manufacturing. On the IC chip this uniqueness is useful in mitigating the reproduction of HW IP and the technology is very innovative in so far as being impossible to reverse it after the circuit design has been leaked. In general, most popular and well known silicon PUF is delay based PUF such as arbiter PUF and ring oscillator PUF [1,2].

In this paper, we present an AES Sbox GF(2^4) inversion functions based PUFs that has a non-linear compact NMOS tree. The compact configuration increases the amount of randomness according to the process variation. Our PUF uses a characteristic difference in power consumption depending on the input transitions at the output node. To be specific, we configure it various ways by a random selection of four inversion functions. It gives much flexibility and diversity to make a PUF ID. Additionally, to increase reliability we use a sense amplifier and compare two output signals from the inversion functions.

For confirmation, our PUF has performed well in various simulation environments. The results demonst -rate that our PUF is adequated for the intended purposes.

Proposed new circuit delay based PUFs

A. Physically Unclonable Functions (PUFs)

A PUF obtains characteristics from a process variation during chip manufacturing. The most outstanding feature of this PUF is physical unclonability. This means that reproductive cloning is extremely difficult or even impossible though the chip design and technology outflows. For defining a PUF, it is necessary to have basic properties such as uniqueness, reliability and randomness. On the basis of these principles, a PUF is applied to an application and configured in a variety of constructions from many different technologies and materials with their own inherent and specific characteristics. Additionally, to reach the security aims of the constructions, it has to differentiate intrinsic PUF properties from other useful PUF-like constructions such as one-way functions which cannot be generalized to include all PUFs.

B. Arbiter PUF vs Ring Oscillator PUF

The most popular PUF are arbiter PUF and ring oscillator PUF. An arbiter PUF, using a race condition from different time arrival delay modules followed by challenge and response pairs can be used with lots of integrated digital components like RS latch. As a arbiter basic arithmetic modules are frequently used [1]. For ring oscillation PUF, entropy from differences in frequency counting forms the delay of oscillation that is exploited. Basically, the oscillator is composed of a CMOS buffer. To compensate for errors as a result of the switching count, which is caused by jitter and glitches in the oscillators, for each oscillator, helper data is made and applied to it [2].

C. AES Sbox GF(2^4) inversion functions based PUFs

Our PUF constructs AES Sbox GF(2^4) inversion functions to delay the cell. First of all, taking into consideration a dual-rail circuit, the principle at which the power consumption at the output nodes varies depends on input transitions of the circuits being used. Especially when considering that dual-rail circuits with an balanced internal node are more sensitive than unbalanced ones. In our PUF we select the AES Sbox GF(2^4) inversion functions for three reasons. First, the resource already exists in the crypto system can be made available at no extra cost. Recently, most popular application for the PUF is key storage and ID identification. These applications have to insert a crypto algorithm to be secure.

978-1-4799-5128-4/14 $31.00 © 2014 IEEE

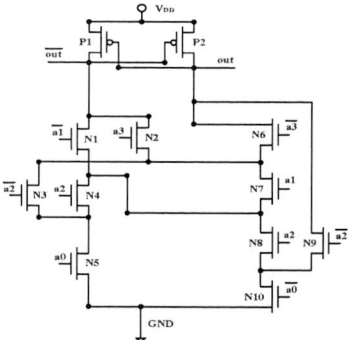

Fig. 1. q_3^{-1} inversion function

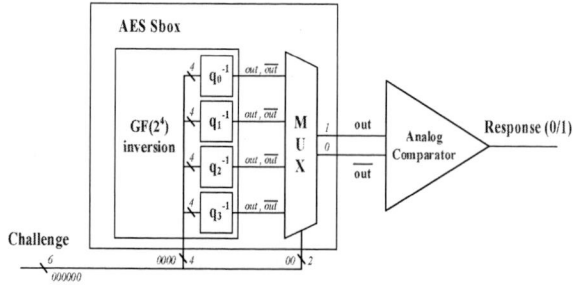

Fig. 2. proposed PUF system architecture

Therefore, our new PUF adds efficiency from sharing AES, crypto algorithms, resources to make delay paths for our PUF because it doesn't have to attach an additional module and because it uses resources already implemented in a secure module. Second, our PUF reinforces security attributes against a machine learning attack. The arbiter PUF, which is a delay based PUF with some cascaded delay cells, is vulnerable to machine learning attacks due to the delay having a feature that increases delay additively. Our PUF defeat this property just to use GF(2^4) inversion functions which is a basic module inside AES Sbox with non-linear properties. This non-linearity definitey reduces the additive delay problem. Third, our PUF compares two sources of power consumption at two output nodes in a circuit. This reduces complexity and creates more robustness leading to reliability caused by voltage/temperature variation than constructed with a separate module like a sequentially connected one with some delay cells. Also, the inversion functions with a compact NMOS tree like Fig. 1 can be influenced by process variation and input transitions because there are more transistor configurations that are more sensitive as a result of the process variation. The Fig. 2 shows a structure of our PUF system. The challenge configures a 6bit signal which is 4bit for input transition and 2bit for choosing which inversion function it uses. After receiving two output power consumptions at the analog comparator it judges a logical one to for the one with a higher rate of power consumption and a logical zero for the opposite.

Simulation Results

A. Robustness and Reliability (intra variation)

To confirm our PUF properties, we simulate for robustness and reliability in various environments and use the monte carlo simulation at several technology edges with a variation of the PMOS and NMOS width. All simulations use the M/H 0.18um technology along with HSPICE. Our PUF greatly strengthens voltage variation simulation from 1.5V to 2.1V with temperature variations from -10 to 60 ℃. On the environment variation our PUF has 1.61% error rate for reliability (98.39%) followed by (1) refered to [3]. It is available to be enough to compare with the previous PUF. (we use a notation refered to [3] for (1) and [4] for (2))

$$ ErrorRate = \frac{1}{m} \sum_{t=1}^{m} \frac{HD(R_i, R'_{i,t})}{n} \times 100\% \tag{1} $$

B. Randomness (inter variation)

For a confirmation of the uniqueness and randomness of our PUF, we calculate probability distribution function for a PUF response in all challenges. In [4] what the random probability means is described in detail. We use (2) to calculate. Our PUF has Gaussian distribution so that it has randomness as a PUF property (almost 50% distributions: 40.625% for response 1). The value approaches 35.938% to 40.625% according to the threshold level. The more balanced the inversion circuits have, the more increased the randomness has.

$$ P_n = \frac{1}{K \cdot T \cdot L} \sum_{k=1}^{K} \sum_{t=1}^{T} \sum_{l=1}^{L} b_{n,k,t,l} \times 100\% \tag{2} $$

Conclusion

Our AES GF(2^4) inversion functions based PUFs have several advantages, such as being able to use already implemented resources in crypto algorithms in security devices and being able to provide higher reliability and robustness from process variation. It is especially useful when the PUF is used with a crypto algorithm as mentioned. Although the exact comparison with other PUFs isn't an easy problem as of yet because of the novelty of our PUF configuration, we will try to compare with previous PUFs in the near future.

Acknowledgment

This research was supported by the MSIP (Ministry of Science, ICT and Future Planning), Korea, under the ITRC (Information Technology Research Center) support program (NIPA-2014-H0301-14-1004) supervised by the NIPA (National IT Industry Promotion Agency) and this work was supported by the IDEC.

References

[1] J. W. Lee et. al, "A technique to build a secret key in integrated circuits with identification authentication applications," in *Procee dings of the IEEE VLSI Ciruits Symposium*, pp. 176-179, 2004.

[2] G. E. Suh and S. Devadas, "Physical unclonable functions for device authentication and secret key generation," in *Proceedings of DAC'07*, pp. 9-14, 2007.

[3] A. Maiti, V. Gunreddy, and P. Schaumont, "A Systematic Method to Evaluate and Compare the Performance of Physical Unclonable Functions", in *Cryptology ePrint Archive*, 2011/657, 2011.

[4] Y. Hori, T. Yoshida, T. Katashita and A. Satoh, "Quantative and statistical performance of arbiter physical unclonable functions on fpgas", in *Proceedings of ReConFig*, pp. 298-303, 2010.

Modulo $2^n + 1$ Squarer Design for Efficient Hardware Implementation

Rajashekhar Modugu[1], Yong-Bin Kim[2], Kyung Ki Kim[3] and Minsu Choi[4]

[1]Qualcomm, Austin, TX, USA
[2]Dept of ECE, Northeastern University, Boston, MA, USA
[3]Dept of Electronic Eng., Daegu University, Gyeongsang, South Korea
[4]Dept of ECE, Missouri University of Science and Technology, Rolla, MO, USA

Abstract—**In this work, an efficient hardware architecture of modulo $2^n + 1$ squarer is proposed and validated. The proposed modulo $2^n + 1$ squarer use novel compressor designs and sparse tree adders as primitive building blocks for fast low-power operations in three major functional modules including partial products generation module, partial products reduction module and final stage addition module. The resulting modulo $2^n + 1$ squarer has been implemented in standard CMOS (Complementary Metal-Oxide Semiconductor) cell technology and compared both qualitatively and quantitatively with the existing hardware implementations. The unit gate model analysis and the experimental results show that the proposed implementation is faster and consume less power than existing hardware implementations.**

Keywords-modular arithmetic; modulo $2^n + 1$ squarer; residue number system (RNS); compressors; sparse Tree Adder; unit gate model

I. INTRODUCTION

RNS (Residue Number System) is widely used for various applications in digital signal processing, cryptography and residue arithmetic [1]–[6].

In residue arithmetic, the moduli set $(2^n - 1, 2^n, 2^n + 1)$ has attracted attention because of it is suitable for effective regular VLSI (Very-Large Scale Integration) implementations and easy conversions between binary and residue number system (RNS) [7]–[10]. Numerous algorithms and architectures have been proposed for this moduli set in the literature. Using this base, the input operands are $n - bit$ wide for modulo $2^n - 1$ and 2^n operations, where as for modulo $2^n + 1$ operations take inputs with $n + 1$ bits wide, which makes operations on this modular base computationally difficult and calls for special attention, especially when it comes to hardware implementations.

Even though the multiplier and squarer proposed in [11], [12] achieved considerable improvements over the previously proposed designs in terms of power consumption and delay, hardware implementation of the proposed multiplier (and the squarer extension) requires a special attention. The critical path of this multiplier depends on the partial products reduction stage which uses a Carry Save Adder (CSA) and on the final stage addition. Hence, efficient design of Carry Save adder and the final stage adder is highly needed.

In order to address this issue, a novel efficient hardware design for modulo $2^n + 1$ multiplier has been considered and a new $2^n + 1$ squarer design has been proposed and validated in this work.

II. PROPOSED FAST LOW-POWER MODULO $2^n + 1$ SQUARER IMPLEMENTATION

Hardware implementation of the modulo $2^n + 1$ multiplier and squarer are similar in all aspects except the number of partial products is less in the squarer implementation. The proposed implementation of the mod $2^n + 1$ squarer consists of three arithmetic modules. The first module is to generate partial products, the second module is to reduce the partial products to two final operands and the last module is to add the Sum and Carry operands from partial products reduction module to compute the final result. Table I summarizes the design steps for the proposed implementation.

Design steps	Proposed Implementation
A. Number system	Diminished-1 number systems, conversions to and from weighted number system are required.
B. Partial Products generation module	Basic logic gates are used to generate the partial product terms.
C. Partial products reduction module	Novel MUX-based compressors are used as the primitive blocks.
D. Final stage addition module	Sparse tree based Inverted EAC is used.

TABLE I
DESIGN STEPS OF THE PROPOSED MODULO $2^n + 1$ SQUARER
IMPLEMENTATION VS. THE EXISTING METHOD.

Fig. 1 shows the finalized 16-bit sparse tree Inverted EAC adder for the proposed modulo $2^n + 1$ squarer. Two input words A and B consist of individual bits a_i and b_i, respectively. G and P are generate and propagate terms. C is the carry and S is the sum. From Fig. 1, we can observe that all the carry-outs are computed in $log_2 n$ stages with less number of carry merge cells and reduced inter-stage wiring intensity.

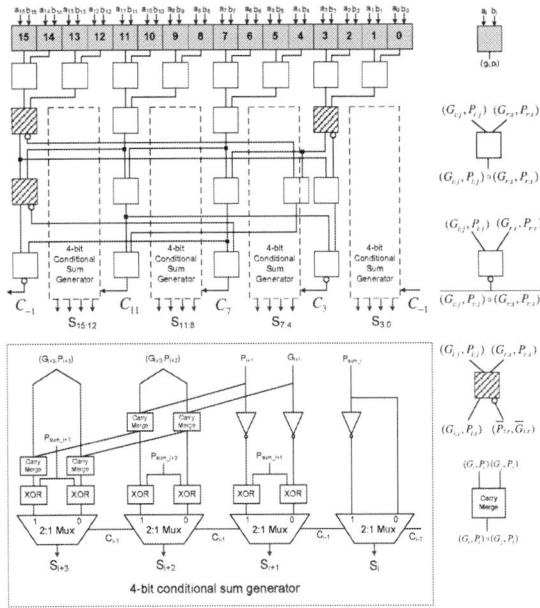

Fig. 1. 16-bit Sparse tree based Inverted EAC adder. Two input words A and B consist of individual bits a_i and b_i, respectively. G and P are generate and propagate terms. C is the carry and S is the sum.

An implementation of the modulo $2^7 + 1$ squarer has been

978-1-4799-5128-4/14 $31.00 © 2014 IEEE

designed as an example of the proposed modulo $2^n + 1$ squarer. Let $X = x_n x_{n-1} \ldots x_0$ be the input of the modulo squarer. The initial partial product matrix for this input is derived.

Then, the set of partial products are reduced into two final sum and carry vectors using a CSA network. This CSA network is composed of efficient compressors reported in [13] instead of full adders and half adders. At each level of the CSA carry and sum bits are produced and are fed to the next subsequent level. The carry outputs at the leftmost column are are fed back as carry inputs of the rightmost column of the CSA network. The sum vector and the carry vector produced by the partial products reduction module are driven to the final stage addition module (i.e., sparse tree based inverted end around carry adder) to generate the final output. The proposed implementation of modulo $2^7 + 1$ squarer is shown in Fig. 2. From the Fig. 2, we can observe that the partial reduction module consists of one $4:2$ and two $3:2$ compressors in each column for maximum possible reduction in delay, power and area.

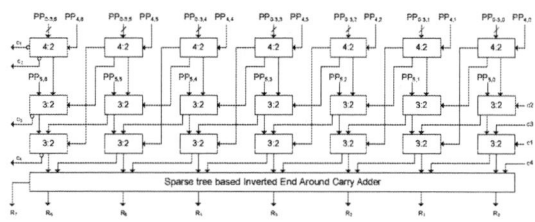

Fig. 2. Proposed implementation of the modulo $2^7 + 1$ squarer using efficient compressors

III. PERFORMANCE VERIFICATION

The standard CMOS (complementary metal-oxide semiconductor) based implementation of the proposed compressor based multiplier/squarer can yield accurate delay and power estimations. The existing compressors and the proposed compressors have same gate counts, but the laid out implementation of the new compressor performs much better because of its MUX-based implementation. The actual power and delay factors are decided based on the interstage wires and complexity of the design. The parallel prefix network based Inverted EAC has a more number of interstage connections and intensified wiring complexity. Hence, for a better perspective for the sake of comparisons, the standard cell based implementations are designed and the power, delay comparisons are carried out. The proposed multipliers/squarers for various values of n are specified using Verilog Hardware Description Language (HDL). The verilog descriptions are mapped on a 0.18 μm CMOS standard cell library using Leonardo Spectrum synthesis tool from Mentor Graphics. The design is then optimized for the highest possible speed performance. Netlists generated from synthesis tool are passed onto the route and place tool, the layouts are iteratively generated to get the circuits with minimum area. The proposed implementation and the carry save array based implementations presented in [11], [12] are implemented and compared with respect to power and delay.

The obtained experimental results are shown in Table. II. The power values are given in mW and delay values are given in ns. As clearly shown in this table, the proposed modulo $2^n + 1$ squarer outperforms the prevailing implementation in terms of delay and power for $n > 4$ when laid

	Proposed_S		Existing_S [12]	
n	Power	Delay	Power	Delay
4	0.296	1.112	0.273	0.865
8	0.935	2.038	1.169	2.352
16	1.815	3.397	2.233	3.842
24	2.563	4.895	3.315	5.102
32	5.841	6.255	6.311	7.313

TABLE II
EXPERIMENTAL RESULTS SHOWING AN AVERAGE REDUCTION OF 10%-12% IN POWER (IN mW) AND DELAY (IN ns) FACTORS.

out on a CMOS circuit. These results also agree upon the parametric simulation outcomes obtained from the unit gate model.

IV. CONCLUSIONS

The proposed design of the modulo $2^n + 1$ squarer uses compressors in the partial products reduction stage, the use of the efficient compressors in place of full adders resulted in considerable improvements in terms of delay and power. A novel sparse tree based inverted EAC adder in the final stage addition, which has less wiring complexity and sparse carry merge cells compared to parallel prefix network based implementations. The proposed squarer is compared with the most efficient modulo $2^n + 1$ squarer available in the literature to demonstrate its lower power consumption and delay.

REFERENCES

[1] A. Ghosh, S. Singha, and A. Sinha, "Floating point RNS: a new concept for designing the MAC unit of digital signal processor," *ACM SIGARCH Computer Architecture News*, vol. 40, no. 2, pp. 39–43, 2012.

[2] M. Das, A. Sinha, and N. Giri, "High speed residue number system (RNS) based FIR filter using distributed arithmetic (DA)," *ACM SIGARCH Computer Architecture News*, vol. 39, no. 5, pp. 1–4, 2012.

[3] G. Chalivendra, V. Hanumaiah, and S. Vrudhula, "A new balanced 4-moduli set {2 k, 2 n-1, 2 n+ 1, 2 n+ 1-1} and its reverse converter design for efficient fir filter implementation," in *Proceedings of the 21st edition of the great lakes symposium on Great lakes symposium on VLSI*, pp. 139–144, ACM, 2011.

[4] D. Schinianakis, A. Fournaris, H. Michail, A. Kakarountas, and T. Stouraitis, "An RNS implementation of an F p elliptic curve point multiplier," *IEEE Transactions on Circuits and Systems Part I: Regular Papers*, vol. 56, no. 6, pp. 1202–1213, 2009.

[5] W. Guo, Y. Liu, S. Bai, J. Wei, and D. Sun, "Hardware architecture for RSA cryptography based on residue number system," *Transactions of Tianjin University*, vol. 18, no. 4, pp. 237–242, 2012.

[6] S. Singh and R. Maini, "Comparison of Data Encryption Algorithms," *International Journal of Computer Science and Communication*, vol. 2, no. 1, pp. 125–127, 2011.

[7] T. Tay, C. Chang, and J. Low, "Efficient VLSI Implementation of 2n Scaling of Signed Integer in RNS {2n-1, 2n, 2n+ 1}," *IEEE Transactions on Very Large Scale Integration (VLSI) Systems*, 2012.

[8] S. Timarchi, M. Fazlali, and S. Cotofana, "A unified addition structure for moduli set {2 n- 1, 2 n, 2 n+ 1} based on a novel RNS representation," in *Computer Design (ICCD), 2010 IEEE International Conference on*, pp. 247–252, IEEE, 2010.

[9] E. Gholami, R. Farshidi, M. Hosseinzadeh, and K. Navi, "High Speed Residue Number System Comparison for the Moduli Set {2 (superscript n)-1, 2 (superscript n), 2 (superscript n)+ 1}," *Journal of communication and computer*, vol. 6, no. 3, pp. 40–46, 2009.

[10] A. Molahosseini, S. Sezavar, and K. Navi, "A new design of reverse converter for a three-moduli set," in *Intelligent Signal Processing and Communication Systems, 2009. ISPACS 2009. International Symposium on*, pp. 57–60, IEEE, 2009.

[11] H. Vergos and C. Efstathiou, "Design of efficient modulo $2^n + 1$ multipliers," *Computers & Digital Techniques, IET*, vol. 1, no. 1, pp. 49–57, 2007.

[12] H. Vergos and C. Efstathiou, "Diminished-1 modulo 2n+ 1 squarer design," *Computers and Digital Techniques, IEE Proceedings-*, vol. 152, no. 5, pp. 561–566, 2005.

[13] R. Modugu, M. Choi, and N. Park, "A fast low-power modulo 2 n+ 1 multiplier design," in *Instrumentation and Measurement Technology Conference, 2009. I2MTC'09. IEEE*, pp. 951–956, IEEE, 2009.

978-1-4799-5128-4/14 $31.00 © 2014 IEEE

Low Power Challenge and Solution for Advanced Mobile Device Design

Haiqing Nan, Dong Ke
Logical and Physical Synthesis Department
Intel Mobile Communications Technology Ltd.
Xian, China
{haiqing.nan, ke.dong}@intel.com

Ken Choi
Department of Electrical and Computer Engineering
Illinois Institute of Technology
Chicago, United States
kchoi@ece.iit.edu

Abstract— **With CMOS technology scaled down, leakage power becomes one of the most important design concerns for mobile device. In this paper, low power and leakage reduction methods in industry for 28nm CMOS technology are introduced. The methodologies which will be discussed include long poly cell ratio optimization, ultralow Vt (threshold voltage) cell ratio optimization, power switch cell insertion, daisy chain number optimization, voltage scaling as well as power island optimization. The advantages and design overhead of these approaches will be discussed as well.**

Keywords-Low power; Mobile Device; Leakage power; Nanoscale CMOS

I. INTRODUCTION

With CMOS technology keeping scaling down, leakage power issues become one of the most critical design concerns for mobile device. The threshold voltage (Vth) of the CMOS device is scaled down along with the technology scaling and this is requested by the need of the high performance. Therefore, the sub-threshold leakage current become the dominant portion for leakage power consumption.

Both in industry and academic field, there are several ways to keep the leakage power consumption within an acceptable range. The most popular and the most effective way to reduce leakage power is to use power gating structure to turn off the path to VDD or VSS. In [1], the author introduced a novel power gating footer design which operates under ultra-low voltage region and can reduce sub-threshold leakage as well as gate leakage. Different types of power gating header or footer designs have been proposed in [2] – [6]. However, these power gting designs are complicated by itself and difficult to apply them into industry standard ASIC flow. Other than power gating techniques, methods such as long poly cell usage, voltage scaling, power island usage can also reduce power and leakage power consumption.

From next chapter, leakage reduction techniques which are now using in industry will be discussed. The advantages and limitation of these methods are discussed also. Then the third chapter will give a summary of this paper and outlook of future leakage reduction methods.

II. INDUSTRY LEAKGE REDUCTION METHODS

A. Long Poly Cell Ratio optimization

One of the most effective way to reduce leakage power is to use cells which have higher threshold voltage. From physical point of view, if the cell features longer poly, then the threshold voltage of the cell will be larger. Hence, in order to achieve low leakage power consumption, it is helpful to use as many long poly cells as possible.

In the standard AISC flow, the long poly cell replacement can be done right after clock tree synthesis (CTS) is done. CTS is one of the most important step in ASIC flow. The purpose for CTS is that for a certain clock group, the clock arrives at all of its endpoint flip flops at same time ideally. However, it is almost not possible to achieve this and it will try its best to minimize the delay difference which is called skew, to reach to different end flip flop according to CTS algorithms. Hence after CTS is done, it is better do not touch CTS result for other optimization such as long poly cell replacement. Therefore, the cells not on the clock tree are selected and they are replaced to long poly cells to save leakage power. Normally, this replacement leads to over 90% of long poly cell usage. However, this happens only for standby domain but not for switch off domains. For switch off domain cells, the replacement is done with STA. By checking the timing status of the whole design, it is simple to figure out the timing paths which have positive setup time slack value. For cells on these paths, they can be replaced to long poly cell type without generating new timing violation. Hence, by taking advantages of useful slack more cells can be replaced to long poly cells as a result. Generally, after two loops replacement to long poly cells, the whole design will achieve more than 80% on average of long poly cells usage ratio. This will significantly reduce leakage power.

For timing critical blocks such as CPU, ultra-low Vt cells are used to meet the high operating frequency requirement. However, these ultra-low Vt cells feature high leakage current. Similar replacement methods as discussed above can be used to reduce ultra-low Vt cells ratio without compromising the frequency.

978-1-4799-5128-4/14 $31.00 © 2014 IEEE   19

B. Power Switch and Daisy Switch Insertion

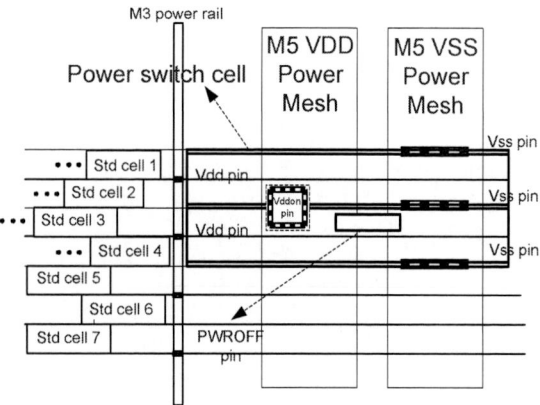

Fig. 1. Power switch cell usage for power turn on/off

Fig.1 shows a power switch cell schematic and its connection to power rail supply to turn on and off the power for normal standard cells. As it is shown in Fig.1, the power switch cell has several power related pins including vss pin, vdd pin, vddon pin and powroff pin. The vss pin is connected to global VSS through vias. In the schematic shown in Fig.1, the Vss pin which should be in metal 1 layer is connected to metal 5 layer. Then these vss pin can supply as ground for all other standard cells such as std cell 1 to std cell 4. Global VDD power mesh connected to the vddon pin of the power switch cell. Depending on the logic value of pwroff pin, the global power supply can connect to its vdd pin. Then these vdd pins can act as power supply for all other standard cells. Fig.1 also shows that a metal 3 layer vdd mesh is used to connect all other metal 1 layer supply. Hence standard cells such as std cell 5 to std cell 7, which are not directly driven by power switch cell can also be turn on and off.

Fig.1 shows a header type power switch cell, hence when the pwroff is logic 1, the connection between global VDD and vdd pin is cut off. All the standard cells driven by these vdd pin will enter sleep mode and save leakage power. When pwroff pin is logic 0, vdd pin and the power rail it connected will be charged through global VDD. There is also footer type power switch and they are used to cut off the power at VSS for leakage saving.

The number of power switch cells needed is based on the IR drop requirement. If a design has high power consumption during active mode, then it means it features high current at active mode. The IR drop at active mode can be calculated by the on-current multiply with the resistance of the power switches. In order to reduce the IR drop, the resistance of the power switches is needed to keep low which means a lot of power switch connected parallel is required.

Fig.2 shows that each of the power rail group should connect to power switch all along with the power straps. The power switch cell can save leakage power very well. However, as it is shown in Fig.1, it takes more than 10 times area compared to a normal size standard cell on average. Moreover, there are thousands of the power switch cells in a design. Hence, the area overhead to insert power switch is big. Normally, more than 10% of total standard cell area is just for power switch cells.

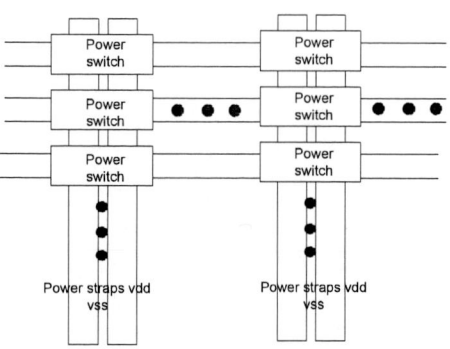

Fig. 2 Power switches for all power rails.

There is another type of cells called daisy switch cells which are used when the standard cells are turned on from sleep mode.

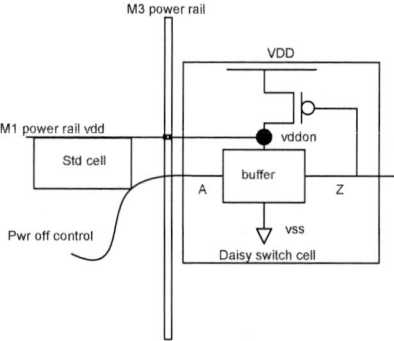

Fig. 3. Daisy switch cell and its connection

Fig. 3 shows the abstract layout and its connection. As it is shown in Fig.3, the daisy switch cell is consisted of one header and one buffer cell. The input A pin of the cell is connected to the pwr off control signal. Since the header PMOS type power switch is used, the pwr off control will issue logic 0 when the circuit is ready to be turned on. The output of the daisy switch Z will be logic 0 after a buffer delay time and turn on the PMOS header inside the daisy switch cell. Then global VDD will charge the vddon pin as well as all the metal 1 power rails to turn on all the standard cells. Fig.3 also shows that one metal 3 power mesh is used to connect all metal 1 power rail in order to charge all metal 1 rails which are not directly connect to the daisy switch cell.

The more daisy switch cell used, the faster the standard cell power rails can be charged. However, due to the rush current when power on the circuit, the number of daisy switch cell that can be used is limited.

C. Voltage Scaling and Power Island

Leakage power is proportion to the supply voltage according to the equation: $P = f(W/L, Vt, Vdd)$. However, not all the circuits in a design can be switched off ($Vdd = 0$).

For this part of circuit, leakage power can be further saved by reducing Vdd dynamically, which is so called voltage scaling. This scenario is used when the circuit is not required to be operated with its full frequency. One example is that a mobile phone user is just editing a text file instead of playing a 3D game. In this case, the supply voltage for the GPU part can be reduced. In some case, IPs/circuits need a constant supply voltage which cannot be scaled down together with other parts of the circuits. Therefore, designer has to create power islands to manage all these exceptions or scenarios.

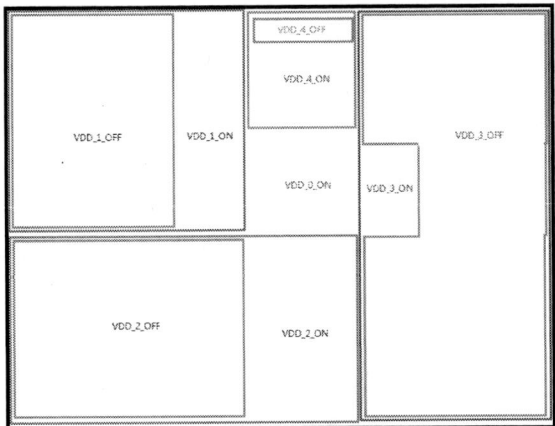

Fig. 4. Voltage scaling & Power islands

Fig.4 shows an example of voltage scaling & Power islands, VDD_0 is the top level power which can't be switched off but it could have two possible voltages: 0.99v and 0.90v; VDD_1/VDD_2/VDD_3 all contains always on domains as well as switch off domains, and all have three possible voltages for power scaling and power saving: 0.99v / 0.90v / 0.85v. VDD_4 has on and off domains, but the supply voltage cannot be scaled and has to be always 1.2v. These supply powers cannot be turn on or off simultaneously. Some of them can be scaled down to lower supply voltage and some supplies cannot be scaled. Hence, individual power islands are created for each of them so that it is feasible to achieve turning off one specific power island and others are working as desire condition.

Besides saving leakage power, another benefit of voltage scaling is saving dynamic power according to the equation: $P = aCV^2f$. Reducing the voltage of design can dramatically decrease total dynamic power. Moreover frequency scaling is also used together with voltage scaling to reduce both leakage and dynamic power.

The drawback for this solution is that: in worst case, designer could have more than 50 power islands in a design which leads to additional design effort, design time to converge design and area penalty. Level shifters are also required when the circuits have different operating voltage under a certain operating mode. Timing signoff could also become more complex; designer has to close timing for each combination of working scenarios.

III. Conclusion

In this paper, low power and leakage reduction methods in industry for 28nm CMOS technology are introduced. Long poly cell ratio optimization, ultralow Vt cell ratio optimization, power switch cell insertion, daisy chain number optimization, voltage scaling as well as power island optimization. Each of these methods has its own benefits and design overhead. The combination usage of these methods can lead to better power optimization results.

References

[1] Kyung Ki Kim, Haiqing, Nan and Ken Choi , "Power gating for ultra-low voltage nanometer ICs," IEEE Symposium on Circuits and Systems (ISCAS), pp.1472-1475, 2010

[2] S. Kim, C. Choi, D. Jeong, et. al., "Reducing ground-bounce noise and stabilizing the data-retention voltage of power-gating structures," IEEETran. on Electron Devices, Vol. 55, No. 1, pp. 197-205, Jan. 2008.

[3] S. Kim, S. Kosonocky, D. Knebel, et. al., "A multi-mode power gatingstructure for low-voltage deep-submicron CMOS ICs," IEEE Tran. onCircuits and Systems-II, Vol. 54, No. 7, pp. 327-339, July 2007.

[4] P. Royannez, H. Mair, et. al., "90nm low leakage Soc design techniques for wireless applications," IEEE ISSCC, pp. 138-139, Feb. 2006.

[5] Y. Shin, S Paik, and H. Kim, "Semicustom design of zigzag powergated circuits in standard cell elements," IEEE Tran. on CAD of Integrated Circuits and Systems, V. 28, N. 3, pp. 327-339, Mar. 2009.

[6] K. Usami, N. Kawabe, and M. Koizumi "Automated selective multithreshold design for ultra-low standby applications," in Proc. IEEE Int. Low Power Electron. Design, pp. 202-206, Aug. 2002.

Distributed Architecture of Touch Screen Controller SoC for Large Touch Screen Panels

Gyeongseop Choi[1], M.G.A. Mohamed[1,2] and HyungWon Kim[1]
[1] Dept. of Electronic Engineering, Chungbuk National University,
Cheongju, Korea
[2] Dept. of Electrical Engineering, Minia University,
Al-Minya, Egypt
gschoi@cbnu.ac.kr, mgamohamed@cbnu.ac.kr , hwkim@cbnu.ac.kr

Abstract

Currently large touch screen panels (TSP) tend to use projected capacitance technology, which allow multi touch and high sensitivity. For large TSPs with a large number of TX (driving) and RX (sensing) lines, however, it is increasingly challenging to achieve high sensitivity, high detection rate, and multi-touch. In this Paper, we propose a distributed architecture of touch screen controller where multiple controller SoCs collaborate in driving and sensing each section of a large TSP. We show that the proposed architecture and SoC design can increase the detection rate without loss of sensitivity performance. It also allows a smaller SoC implementation, while its chip expandability provides the flexibility of supporting a large range of TSP sizes. We implemented the proposed distributed SoC using TSMC CMOS 0.18um with a low power ARM core, AHB-lite bus, memories, and embedded touch algorithm software.

Keywords- Touch screen controller; Capacitive Touch screen; SPI ; SoC; Distributed architecture SoC;

Introduction

Recently the demand for large touch screen panels (TSP) are increasing. TSPs have been adopted not only for mobiles and tablets, but also for large screens such as PC monitors, large medical devices, digital white board, digital signage, etc. The larger the TSPs are, in general, the higher the number of TX and RX lines they tend to require. It is, therefore, increasing difficult to control a large TSP with one controller chip. Most of previous work, however, have been focused on a small touch screen controllers, so we claim that our distributed architecture of touch screen controller is a new approach to an efficient solve for large touch screen control.

Touch Screen Controller System

A. Touch Screen panel (TSP)

Mutual capacitance TSPs support multi-touch function and provide high touch sensitivity. They are, therefore, most common type for small TSPs and are increasingly adopted for large TSPs as well. If a point is touched on a TSP, the touch position is determined by examining the mutual capacitance change between TX and RX lines. We designed a controller SoC to support a TSP up to 23 inches with 44 TX and 78 RX lines made of ITO(Indium Tin Oxide). Our architecture can support larger TSPs by connecting multiple SoCs with serial interconnects between each SoC.

B. System Description

A Touch screen controller consists of an analog front-end (AFE) block, and a digital block (Fig. 1, Fig. 2). The AFE drives TX signals, senses RX signals, and interfaces to digital block with on-chip ADC and DAC. AFE generates square wave or sine wave signals to TX lines, and detects touch points by comparing signal's amplitudes.

It amplifies signals which come from DAC to excite TX lines, and integrates RX signal before sending to ADC to convert analog signal to digital signal. Cortex M0 is used with AHB-lite bus to control all other blocks in digital circuit. This system supports low power and high speed. It includes TSMC's SRAM IP macros to save/read sensed data. This System data to another host chip through SPI (Serial peripheral interface).

C. Proposed controller system

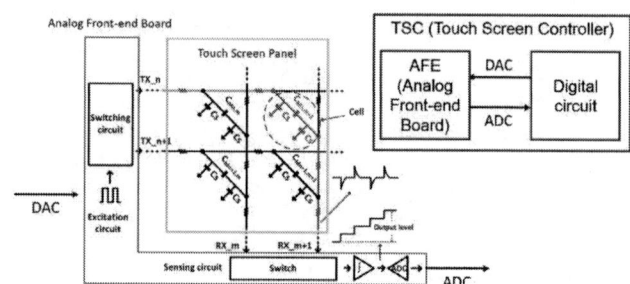

Fig. 1. Touch screen controller architecture.

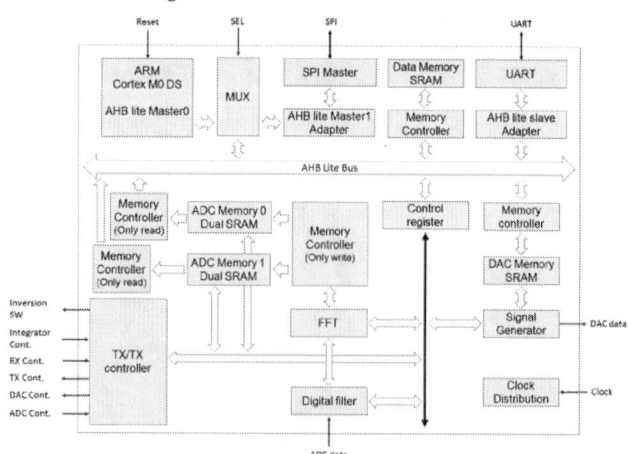

Fig. 2. Controller (Digital circuit) block diagram.

978-1-4799-5128-4/14 $31.00 © 2014 IEEE

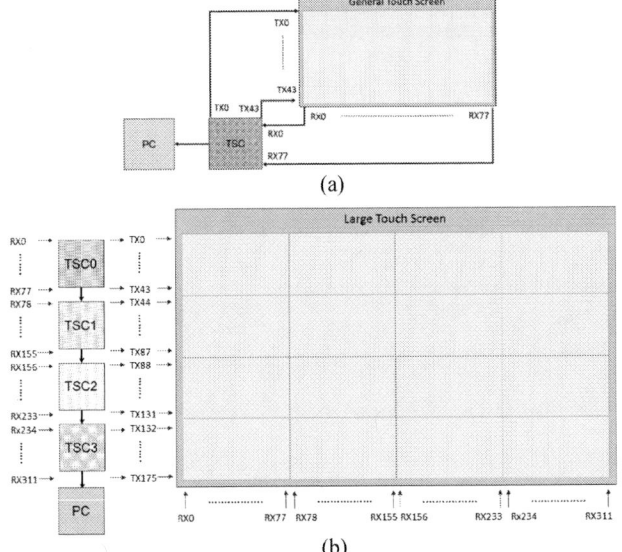

(a)

(b)

Fig. 3. (a) A unit touch screen size covered by one TSC. (b) Large Touch Screen Controller System: A large touch screen of 16 times of the unit Touch Screen size. This large TSP needs only 4 TSCs not 16 TSC.

We proposed a controller system that can support large TSPs with one SoC. The same SoC can be duplicated to control a large TSP. Our current SoC is designed for a 23" TSP with 44 TX lines and 78 RX lines. The SoC has two SPI interfaces; one is to connect SoCs together and the other is for debug purpose; See Fig. 2 and Fig. 3 (a).

In the proposed architecture, multiple Touch Screen Controller SoC (TSC) are connected in series by SPI; See Fig. 3(b) and Fig. 4 The first TSC is responsible for controlling the other chips of when to send TX and receive RX signals. Each chip sends processed touch data serially to PC. Each TSC is responsible for:

- Excitate (drive) the specified number of TX lines
- Sense the specified number of RX lines
- Run detection algorithm software on the sensed data and send touch position data to the next chip via serial link (SPI)

For example, suppose we control a very large touch screen (176x312 lines) as illustrated by Fig. 4 and Fig. 5 The operations using four TSCs (TSC0 ~ TSC3) are as follows:

1. TSC0, the Master of System controller, sends 'Start' signal to each slave (TSC1 ~ TSC3) in turn, so each TSC sends TX.
2. Only one TSC sends TX signals to its TSP section at a time.
3. All TSCs receive RX signals from their TSP section simultaneously when any TSC sends TX signals.
4. Each TSC compares its RX signals with reference. Each TSC calculates the touch positions, and sends their coordinates (x,y) to the next slave TSC.
5. Last TSC sends all the collected data to a PC (Host system).
6. When Master receives 'Finish' information from Last slave, it repeats the above process.

Even for extremely large touch screens, only a small number of TSCs are needed. The number of TSCs needed for a given touch screen size is given by the following formula.

$$(\text{Standard screen size}) \times n^2 = (\text{Large screen size})$$
$$\# \ n = \text{the number of TSC}$$

Each TSC generates its Tx signals and receives RX signals directly to/from TSP, and it also calculates the touch positions indivisually. Therefore there is very little or no latency issue caused by the proposed distributed architecture. The proposed architecture, therefore, can reduce power consumption, detection rate, and SoC costs.

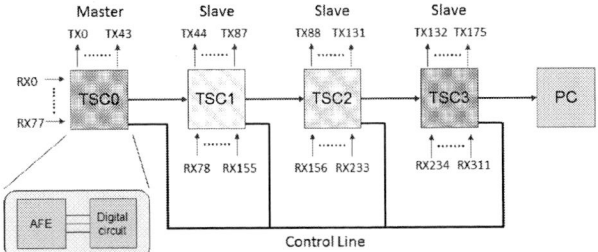

Fig. 4. Large Screen Controller System block diagram.

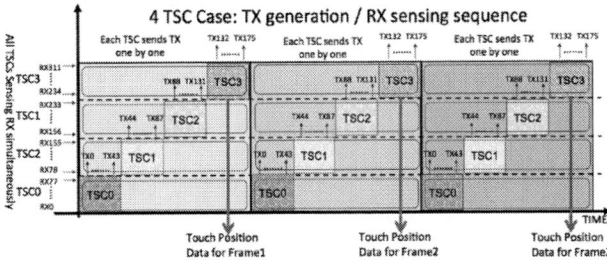

Fig. 5. TX generation/RX sensing sequence of 4 TSC case

Implementation

We implemented the proposed SoC currently in 2 chip solutions: AFE chip and digital controller chip. The AFE chp is implemented using Doungbu 0.18um BDCMOS for high voltage driver circuits. It consists of analog sense amplifiers, analog multiplexers, charge transfer and reset switches, 12bit ADC and DAC.

The digital controller chip is implemented using TSMC 0.18um, which includes ARM Cortex-M0 core and AHB-lite bus for low power and compact size. It also includes FFT for sense signal analysis, SRAM data buffers, digital filter, SPI, UART, etc. The performance of the chips has been verified using FPGA and AFE board. We developed touch detection algorithm software using Keil uVision provided by ARM. Once the chip fabrication is completed, we plan to update the test results of the chips using a large touch screen panels.

Conclusion

We proposed a distributed architecture of multi-chip touch screen controller SoC targeting large touch screen control. We have shown that the proposed design can easily support a very large touch screens with a small number of SoCs with fast detection rate, low cost, and low power without sacrificing the touch sensitivity performance, which are the crucial requirements for large capacitive touch screen panels. We implemented the SoC using TSMC 0.18um (digital block) and Dongbu 0.18um (AFE block).

Acknowledgment

This research was supported by the MSIP Korea, under the Human Resource Development Project for SoC (NIPA-2014-H0601-14-1001).

References

[1] U. Jang, T.W. Cho, H.G. Jang, S. Lee, H.W. Kim, "Architecture of Multi Purpose Touch Screen Controller with Self Calibration Scheme", *IEEK Fall Conf.*, pp. 162-166, 2013

[2] I. Seo, T.W. Cho, H.G. Jang, S. Lee, H.W. Kim, "Frequency Domain Concurrent Sensing Technique for Large Touch Screen Panels", *IEEK Fall.*, pp. 55-58, Nov. 2013.

[3] M.G.A. Mohamed, U. Jang, I. Seo, T.W. Cho, H.G. Jang, S. Lee, H.W. Kim, "Efficient Algorithm for Accurate Touch Detection of Large Touch Screen Panels," IEEE Int'l Symp. On Consumer Electronics 2014.

[4] I. Seo, U. Jang, M.G.A. Mohamed, T.W. Cho, H.G. Jang, S. Lee, H.W. Kim, "Voltage Shifting Double Integration Circuit for High Sensing Resolution of Large Capacitive Touch Screen Panels," IEEE Int'l Symp. On Consumer Electronics 2014.

Recent Advances in ASIC-compatible Circuit Techniques for a SOC in Newly Emerging Application Areas: Invited Paper

Sang H. DHONG, IEEE Fellow; Wei HWANG, IEEE Fellow

Taiwan Semiconductor Manufacturing Company; National Chiao-Tung University
Design Technology Platform, R&D; Department of Electronic Engineering
Hsinchu, Taiwan
Primary author's email address: sang_dhong@tsmc.com

Abstract

We review advances in ASIC-compatible circuits for emerging SOC areas. These applications require ubiquitously low-power consumption during standby mode while providing a required performance in active mode. Sub- or near-threshold circuits may provide a low-power solution. However, they have yet to show how they fit into overall SOC optimization including area and performance. Selectively introducing custom-circuit techniques with ASIC tool compatibility have proven very attractive in reducing both the power and the area of a SOC by extending its Dynamic Voltage-Frequency Scaling (DVFS) range down to a VDD of 0.5 V.
Keywords-SOC; ASIC; DVFS; 0.5V VDD; Pulse latch; IOT

Introduction

The importance of low-power circuits cannot be overemphasized for newly emerging SOC areas, such as the Internet of Things (IOT) and wearable computing devices. These market segments are expected to show exponential growth in the near future [1]. These applications and devices require ubiquitously low-power consumptions during a standby or when running low-intensity apps. However, they may also need a burst of high-performance or need to meet the minimum performance requirement set by a system during a wakeup or communication state. This large difference in the performance of a SOC due to its different modes of operation places very challenging requirements on the circuitry of a SOC.

Sub-threshold or near-threshold circuits [2] have been widely reported as having lowest energy consumption per bit of logical operation and have thus been proposed as the best candidate for any low-power market segments. However, the impressive low-power performance of these circuit families addresses only some of the required SOC design features. The circuit families have, in general, yet to demonstrate their capabilities with respect to the remaining aspects of the complex SOC equation, such as on-demand performance, compatibility with ASIC tool sets, and supporting ASIC design infrastructures.

Selective introduction of traditional custom-circuit design techniques to an ASIC flow, while ensuring the necessary compatibility with ASIC tools and infrastructure, has recently been proven as a very attractive alternative solution to the aforementioned problem. Advances in traditional CMOS circuitry have extended its power supply (VDD) operating range down to a Vddmin@circuit of 0.5 V. These techniques provide not only a very competitive solution in the low-power region, but also address the challenging issues of on-demand high performance through DVFS. The evolutionary nature of these improvements to traditional CMOS solutions retains the currently existing compatibility with ASIC tools and infrastructure through only a minimal investment in the newly introduced features described below.

We review four recent CMOS circuit advances, whose power is comparable to near-threshold circuits when operating close to its DVFS lower limit (≈ 0.5 V) with a smaller chip area ($\approx 20\%$), while also retaining its high performance at a higher VDD. They are: (1) a pulse latch with an improved pulse generator (*Dyskl* PG) replacing a Master-Slave Flip-Flop (MSFF); (2) SRAM operations with internal timing signals derived from the system clock in lieu of self-timed circuitry; (3) reducing usage of transmission gates for stacked CMOS gates; and (4) limiting cycle stealing during SOC integration.

Evolutionary Advances in Traditional CMOS circuitry

A. A pulse latch with Dyskl PG replacing MSFF [3]

Fig. 1 shows a pulse latch with a measured Vccmin @circuit of 0.42 V and a pulse width of ≈ 3 FO4-inverter delays. A wider operating window and reduced dependence on the input rise-time and PVT variations were obtained using a new *Dyskl* PG. A pulse in the *Dyskl* PG starts when its input crosses the switching level of its input gate, unlike in the classic text-book-style pulse-generator. An 8 to 10% improvement in power, performance, and area (PPA) of a typical digital SOC is observed when a group of pulse latches is driven by a distributed clock regenerator (DCR). The DCR has the new pulse generator at its input stage and provides pulse clocks to the pulse latches [3].

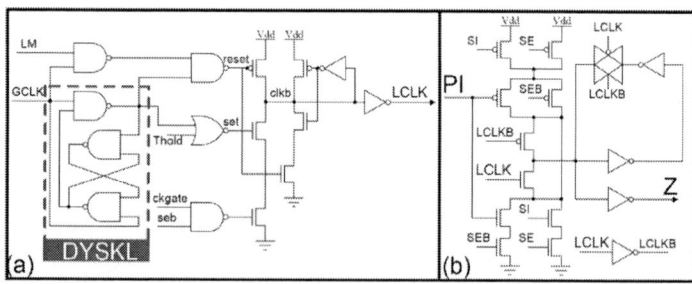

Fig.1. (a) DCR generates a pulsed clock(LCLK), which in turn drives (b) a pulse latch replacing a MSFF.

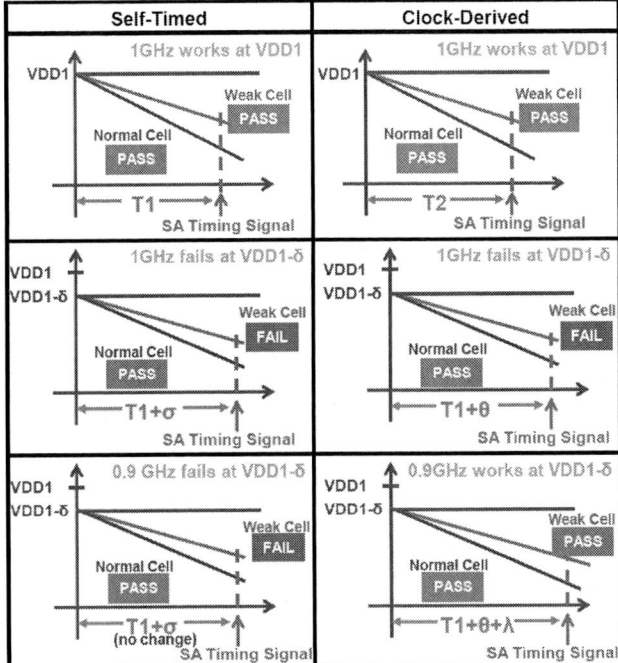

Fig.2. A clock derived signal scheme achieves a lower SRAM VCCmin at a lower frequency due to a larger sense signal.

B. Embedded SRAM operations with internal signals derived from the system clock in lieu of self-timed circuitry [4]

In a clock-derived design, activation of the sense amplifier (SA) starts with a delayed half-cycle clock as shown in Fig. 2. At a lower VDD, a slower clock delays the leading edge of the SA activation signal and increases the bit-line signal development time sufficiently to overcome the increased SA offset voltage. In a self-timed design, the SA activation signal is generated by an on-chip delay element. Any external adjustment of the signal development time is problematic at

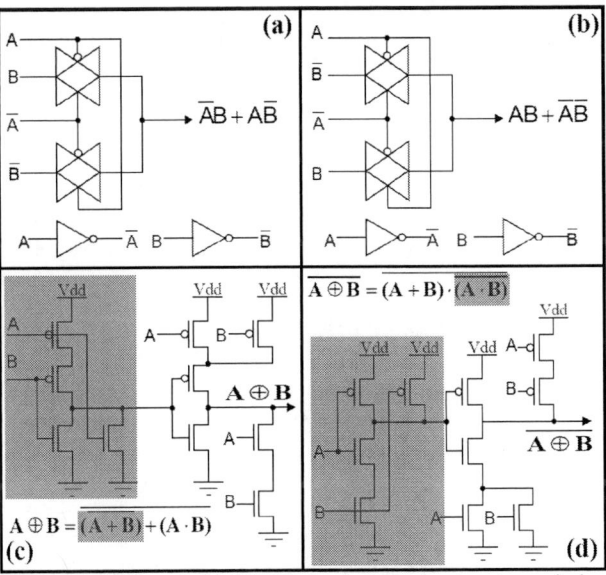

Fig.3. (a) XOR and (b) XNOR implemented using transmission gates. A stacked CMOS implementation of (c) XOR and (d) XNOR. An area reduction of 25% is achieved by replacing cell (a) and (b) with cell (c) and (d), respectively.

best, which results in a higher VCCmin.

C. Reducing the usage of transmission gates in favor of traditional stacked CMOS gates in critical standard cells [5];

As discussed in Zimmermann [5], transmission gates offer a better PPA than stacked CMOS gates. However, in 28-nm and more advanced CMOS technologies, restricted design rules have made the area of a transmission gate larger than that of a stacked CMOS gate, especially in lower power-level standard cells. This point is illustrated by Fig. 3, which is a schematic diagram of a transmission and stacked-CMOS gate of XOR and XNOR. These are used extensively in an ALU, in multipliers, and in parity generators.

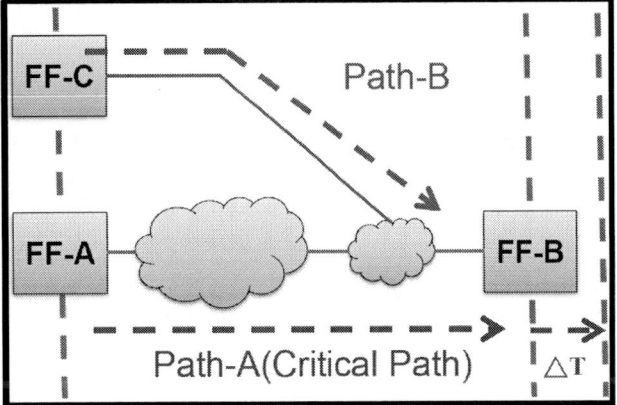

Fig.4. Improving setup time of Path-A by cycle stealing (or usable clock skew) increases the hold time of Path-B.

D. Restricting the usage of cycle stealing during a SOC build

Fig. 4 shows a Path-A timing improvement using cycle time stealing (or usable clock skew) by ΔT. However it increases the hold time by the same amount. A hold-time fix is obtained mostly by inserting an even number of inverters, which increases a chip area and power and frequently negates any benefits from a cycle time improvement. A more rigid cycle partitioning without any cycle stealing is more effective for a low-power design with a large DVFS window.

Acknowledgment

We would like to acknowledge the contributions made to this work by M.Z. Kuo, Noah Lin, Henry Hsieh, and Ryan Tseng.

References

[1] Leon Spencer, "Internet of Things market to hit $7.1 trillion by 2020: IDC", ZDNET, www.zdnet.com, June 2014.

[2] B.H. Calhoun, et al., "Sub-Threshold Design: The Challenges of Minimizing Circuit Energy," *ISLPED 2006, Proceedings of the 2006 International Symposium.* pp. 366-368.

[3] S.H Dhong. "A 0.42V Vccmin ASIC-Compatible Pulse-Latch Solution as a Replacement for a Traditional Master-Slave Flip-Flop in a Digital SOC," *CICC, 2014.*

[4] M.Z. Kuo, "A 16kB Tile-able SRAM Macro Prototype for an Operating Window of 4.8GHz at 1.12V VDD to 10 MHz at 0.5V in a 28-nm HKMG CMOS" *CICC 2014.*

[5] Reto Zimmermann, et al., "Low-Power Logic Styles: CMOS Versus Pass-Transistor Logic," *IEEE Journal of Solid-State Circuits,* pp. 1079-1090, VOL.32, No.7, July 1997

Exploring Hybrid SRAM/MRAM L2 NUCA
Stacked on 3D Chip-Multiprocessors

Seunghan Lee[1], Kyungsu Kang[2], Jongpil Jung[1], and Chong-Min Kyung[1]

[1]Smart Sensor Architecture Lab (SSAL), KAIST

[2]CAE Team, Semiconductor R&D Center, Samsung Electronics Corporation

Abstract

Non-volatile magnetic RAM (MRAM) offers high cell density and low leakage power while suffering from long write latency and high write energy, compared with conventional SRAM. The use of hybrid memories (e.g., SRAM and MRAM together) can take advantage of the best characteristics that each technology offers. In this paper, we explore the 3D-stacked SRAM/MRAM hybrid L2 cache architecture by using a design-time optimization that determined each bank capacity and a ratio between SRAM and MRAM capacities. Also, this paper proposes a runtime cache management scheme that improves the system performance. Experimental results show that the proposed method yields, on the average, 61% performance improvement in terms of instructions per second (IPS) compared to the conventional SRAM-only L2 cache or MRAM-only L2 cache.

Keywords-3D IC; hybrid cache; design-time optimization; dynamic cache management

I. Introduction

The advent of chip-multiprocessors (CMPs) has increased the pressure on achieving larger cache capacity with reasonable access latency. However, the enlargement of traditional on-chip SRAM for the use of cache memory is becoming prohibitive as the cache leakage power becomes critical and the access latency keeps increasing along with the capacity. MRAM has been explored to replace the traditional on-chip SRAM cache due to its ultra-low leakage power. However, it suffers from longer write latency and higher write energy consumption than SRAM. Therefore, we need to consider the use of hybrid SRAM/MRAM cache to improve system performance in an energy-efficient manner, since most performance-critical write accesses are accommodated by the low-latency SRAM cache while the energy efficiency is mainly determined by the MRAM cache [1]. 3D integration allows stacking disparate memory technologies (e.g., SRAM and MRAM) together in a cost-efficient manner [2].

In this paper, we provide an overall architecture exploration for 3D-stacked hybrid SRAM/MRAM L2 NUCA in order to improve performance of 3D CMPs. For that, the capacity of a bank and the numbers of stacked SRAM and MRAM layers are determined at design-time when the physical size of the whole cache architecture is fixed. In addition, a runtime cache partitioning method is proposed, which is applied to the architecture optimized at design-time for further performance improvent with knowledge of runtime behavior of executed programs. The rest of this paper is organized as follows: we introduce how a 3D CMP is optimized at design-time in section II; Section III presents the proposed runtime management policies for the performance-optimized 3D hybrid L2 NUCA; Section IV presents the experimental results followed by conclusion in Section V.

II. Design of Hybrid L2 NUCA for 3-D CMP

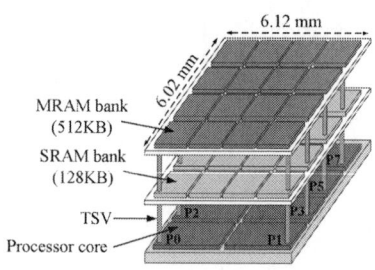

Fig. 1 3D CMP where hybrid SRAM/MRAM L2 NUCA is stacked. MRAM is placed onto the top layer and SRAM is placed on the other layer in the 3D CMP.

Fig. 1 is an illustration of the target 3D CMP. Each stacked SRAM and MRAM tier consists of multiple memory banks, forming a mesh network for planar communication. The capacities of SRAM and MRAM in 45nm technology are, respectively, 2MB and 8MB, assuming that each stacked tier has similar area to that of the Intel Atom N44 core tier (i.e., about 37mm^2) [2]. When the physical size of NUCA is fixed, the capacity of a bank determines the access latency of a cache bank itself as well as the number of cache banks in NUCA that affects the average hop distance for data traversal in the network. A trade-off between cache bank access latency and average hop distance needs to be made with respect to the cache bank capacity to minimize the average cache access time (*ACAT*), i.e.,

$$ACAT = \gamma \cdot t_{read} + (1-\gamma) \cdot t_{write} + hop \cdot t_{sw} \qquad (1)$$

where t_{read} is the the cache read latency, t_{write} is the cache write latency, *hop* is the average hop distance, and t_{sw} is the network latency, i.e., the latency owing to network router. γ is the ratio of cache read to cache access. In our target architecture, as shown in Fig. 1, the average cache access time is minimized at the bank capacity of 128KB and 512KB for SRAM cache and MRAM cache, respectively. Note that, this method is generally applicable to all 3D CMPs with 3D- stacked NUCA cache. In general, the bank sizes of L2 SRAM and L2 MRAM are determined at design-time using equation (1) without the knowledge of runtime behavior due to L2 cache partitioning. Runtime cache access behavior is dynamically taken care of by runtime algorithms running on the L2 NUCA.

III. Runtime Management of 3D-Stacked Hybrid NUCA

Given the performance-optimized 3D CMP described in Section II, our runtime solution is to find the best allocation of the SRAM and MRAM cache banks to each core, at every runtime reconfiguration time interval, such that the instruction throughput of the 3D CMP is maximized. The procedure to find the performance-maximal runtime solution consists of the following two steps; A) finding the L2 cache capacity (logically) assigned to each core to minimize the aggregated L2 cache misses and B) based on the assigned L2 cache capacity, finding the numbers and positions of SRAM cache banks and MRAM cache banks assigned to each core.

A. 1st step: Cache Capacity Allocation

This step starts with initializing the L2 cache capacity assigned to each core as the total L2 cache capacity. For each core, the increase of cache misses is estimated when the assigned L2 cache capacity is halved. After the estimation of cache misses, the core yielding the smallest increase in the L2 cache misses is assigned the cache capacity cut by half. This procedure is iterated until the sum of assigned capacity becomes equal to or less than the total L2 cache capacity.

B. 2nd step: Cache Bank Assignment

This step starts with assigning one SRAM cache bank nearest to each core, which guarantees that at least one SRAM cache bank is allocated to each core. For each unassigned bank, we assign a bank to each core which yields the largest memory access latency reduction, i.e., RL_{ij} when the bank j is allocated to core i. RL_{ij}, the estimation of reduced memory access latency when bank j is additionally assgiend to core i is given by

$$RL_{ij} = h_{ij} \cdot (L_{off} - L_{ij}) \qquad (2)$$

where h_{ij} is the number of cache hits from core i to bank j after bank j is allocated to core i. In order to estimate this number, we adopt hardware based cache utility monitors [3]. L_{off} is the latency for off-chip memory access. L_{ij} is the access latency from core i to bank j.

IV. Experimental Results

We performed experiments to compare the hybrid cache composition, i.e., *Hybrid SRAM/MRAM* L2 cache with the homogeneous cache compositions, i.e., L2 *SRAM-only* and L2 *MRAM-only* in terms of IPS. As shown in Fig. 2, *Hybrid* improves IPS over *SRAM-only* by 62.24%, because *Hybrid* yield significantly higher L2 cache hit ratios (by up to 132.58%) than *SRAM-only*. Note that the capacity of a MRAM cache bank is about 4 times larger than that of a SRAM cache bank. On the other hand, *Hybrid* improves IPS over *MRAM-only* by 61.00% due to the reduced cache access latency by accommodating the low-latency SRAM cache for most critical

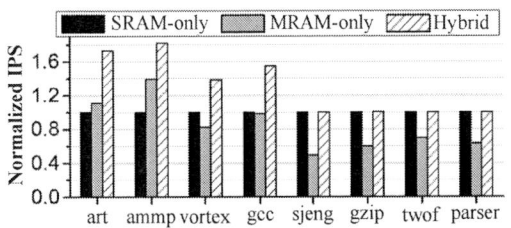

Fig. 2 IPS (Instructions per second) result of each benchmark suite, normalized with respect to *SRAM-only*.

write accesses. In *Hybrid*, thanks to runtime cache partitioning proposed in this paper, most cache accesses are made to the low-latency SRAM cache banks while larger capacity to secure low miss ratio is achieved by MRAM cache banks.

TABLE I
Effects of SRAM to MRAM ratio in the hybrid SRAM/MRAM L2 NUCA

	1S1M	1S2M	2S1M
Normalized IPS	1.000	0.933	0.826
Avg. Hop Distance	1.983	2.607	2.558
L2 Cache Miss Ratio	0.102	0.100	0.101
Avg. Temperature (°C)	60.7	62.2	65.5

The experimental results in Table I show the effects of SRAM to MRAM ratio in the hybrid SRAM/MRAM L2 NUCA on the system performance and operating temperature. In the first row of Table I, the numbers before "S" and "M" are, respectively, the numbers of SRAM and MRAM layers stacked on the processor cores. As shown in Table I, *1S2M* and *2S1M* degrades IPS by 6.7% and 17.4% compared to *1S1M*. It is because both *1S2M* and *2S1M* yield the larger average hop distances, while the L2 cache miss ratios are similar among the three cache architectures. In addition, stacking one additional cache layer incurs the higher operating temperature.

V. Conclusion

This paper has explored a design space of multi-processor with 3D-stacked hybrid SRAM/MRAM L2 NUCA by using a design-time architecture optimization and a runtime cache management. Experimental results show that the proposed scheme effectively improves IPS and operating temperature, compared with conventional homogeneous cache architecture.

Acknowledgement

This work was supported by the Center for Integrated Smart Sensors funded by the Ministry of Science, ICT, and Future Planning as Global Frontier Project (CISS-2011-0031863).

References

[1] G. Sun, et al., "A novel architecture of the 3D stacked MRAM L2 cache for CMPs," in *Proc. HPCA*, Feb. 2009, pp. 239-249.
[2] X. Dong and Y. Xie, "System-level Cost Analysis and Design Exploration for Three-Dimensional Integrated Circuits (3D ICs)," in *Proc. ASPDAC*, 2009, pp. 34-45.
[3] M. Qureshi, "Adaptive spill-receive for robust high-performance caching in CMPs," in *Proc. HPCA*, Feb. 2009, pp. 45-54.

978-1-4799-5128-4/14 $31.00 © 2014 IEEE

A New Multi-target Detection for Vechicle Radar

Zhenyi Liu, Franklin Bien, $Memeber, IEEE$

Ulsan National University of Science and Technology, Ulsan, Korea

Bien@unist.ac.kr

Abstract—A new vehicular radar model named as Double Stepped Frequency Pulse(DSFP) is presented to be able to detect both stationary and moving objects inside the observation area with short measurement time, high range and velocity resolution, and cost efficiency. DSFP transmits two stepped frequency pulses with a frequency shift and a time difference, its performance is better than conventional LFMCW radar on multi-targets with a shorter measurement time, higher resolution, and solves the Doppler shift and frequency spread over of SFP, in addition, DSFP has a simple structure based on pulse radar, so it's also very cost efficient.

Index Terms—Radar, LFMCW, Stepped Frequency Pulse, Double Stepped Frequency Pulse.

I. INTRODUCTION

Modern radar sensors offer the capability of extremely accurate measurements of target range, radial velocity, and azimuth angle for all objects inside the observation area. These target parameters must be measured simultaneously, even in multiple target situations with rapid processing speed because of short time reaction time for brake and control system. Several kinds of radar waveform are widely used as vehicular radar, the linear frequency continuous waveform (LFMCW) technique has been used extensively because it is easy to implement and has a simple detection algorithm, but the measurement time could be extended when detecting multiple targets [3]. In this paper Stepped frequency pulses(SFP) are discussed, because of its merits for high-resolution, multi- target detection and shorter measurement time. However, it causes unavoidable errors when detecting moving targets. So based on this, double stepped frequency pulse(DSFP) is presented as a new detection for vehicle radar in multi-target situation.

II. SFP& DOUBLE SFP

SFP transmits an array of short narrowband pulses divided by time intervals that are sufficient to receive echoes from the targets located at some range of interest. Once the echoes from all the pulses have been received, they are processed collectively in the receiver. Double SFP combines two stepped frequency pulses that have identical slopes and bandwidths; however, they have frequency shift and time differences. DSFP waveform is shown at fig.1.

In SFP case, if the transmitted signal at the nth pulse is $Acos2\pi(f_0 + n\Delta f)$, then the corresponding phase detector for a target at range R could be written as equation(3).

$$\Delta\phi = -2\pi(f_0 + n\Delta f)\frac{2R_n}{c} \tag{1}$$

When detecting a moving target with a velocity v, the target ranges changes with each pulse as, $R_n = R_0 + vnT$, where T

is the pulse repetition interval. With the changed target range, the equation (3) would be written as equation(4).

$$\Delta\phi = -\frac{4\pi f_0 R_0}{c} + \frac{4\pi n\Delta f R_0}{c} + \frac{4\pi f_0 vnT}{c} + \frac{4\pi n\Delta f vnT}{c} \tag{2}$$

When detecting stationary targets, only first two terms are valid. The third term represents the doppler frequency shift due to the motion of the target, and the range resolution process mistakes the frequency step as the doppler frequency shift due to the velocity on the range. The fourth term of equation (2) represents the changes in the frequency of each pulse with the targets motion, thus causing the frequency spread over the dwell, which results in negative effects, including loss of range resolution, range accuracy, and signal-to-noise ratio. The Doppler shift error and frequency spread over are important drawbacks of SFP [2].

As in DSFP case, the Double SFP gives following information of target detection as

$$\Delta f_A = \Delta f_B = \Delta f = f_D - f_B \tag{3}$$

$$\Delta\phi = \Delta\phi A - \Delta\phi B \tag{4}$$

where f_D is the Doppler frequency, which is equal to $2c/\lambda$, and λ is the wavelength of the transmitted waveform f_B, which is usually referred to as the beat frequency; it is defined as the frequency difference between the transmitted signal and the received signal, and its value is $(2Rn\Delta f)/(c \times T)$. As shown in Fig. 5, the phases of pulse A and pulse B can be expressed as

$$\Delta\phi_A = 2\pi(f_0 + n\Delta f)\frac{2R_n}{c} \tag{5}$$

$$\Delta\phi_B = 2\pi(f_0 + f_{shift} + n\Delta f)\frac{2R_n}{c} \tag{6}$$

By rewriting equations (3) and (4), following equations could be got:

$$\Delta\phi = -\frac{4\pi f_{shift}R_0}{c} - \frac{4\pi f_{shift}vnT_{CPI}}{c} \tag{7}$$

$$f_B = -\frac{2v}{c} - \frac{2n\Delta f(R_0 + vnT_{CPI})}{c \times T_{CPI}} \tag{8}$$

As the equations show, the range and velocity could be solved by the orthogonal relationship between the phase difference and the beat frequency, and as a result, the unambiguity range and velocity could be solved and expressed as:

$$R_0 = -\frac{c \times \Delta\phi}{4\pi f_{shift}} - vnT_{CPI} \tag{9}$$

978-1-4799-5128-4/14 $31.00 © 2014 IEEE

$$v = \frac{f_{sweep}c\Delta\phi}{4\pi f_{shift}} - \frac{cf_B}{2} \qquad (10)$$

The Double SFP could solve the range and velocity without the Doppler shift error and energy spread over effects, and it also provides fine resolution on multiple targets due to its frequency hopped n times, i.e., $\Delta R = c/2n\Delta f$. Besides that, the Double SFP also provides shorter measurement time than LFMCW.

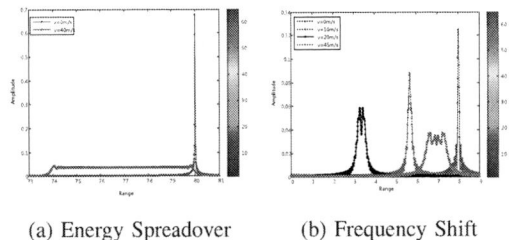

(a) Energy Spreadover   (b) Frequency Shift

Fig. 2: Stepped Frequency Pulse

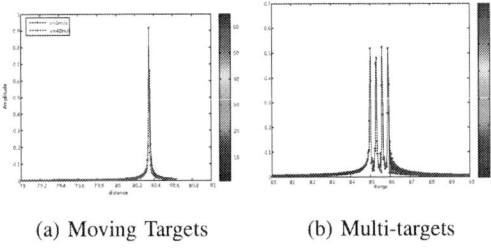

(a) Moving Targets   (b) Multi-targets

Fig. 3: Double Stepped Frequency Pulse

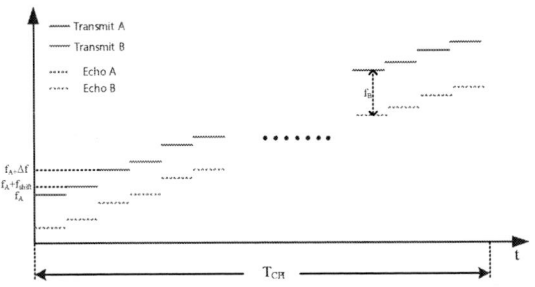

Fig. 1: Waveform of the proposed Double SFP

The structure of Double SFP is based on two pulse generators, and one direct digital synthesizer (DDS) was used to generated stable step frequency, because the total bandwidth of the frequency sweep is usually around 500 MHz, which means a 1 to 1.5-GHz reference clock could be used for DDS, thereby allowing the generation of a stable, agile, stepped frequencies. The structure of the Double SFP is simple and easy to implement, it takes shorter time to put it into production.

Simulation result show at Fig.2 and Fig.3, in SFP case, energy spread over and doppler frequency shift could be seen, Doppler frequency shift and Energy Spread over could occur. In Fig. 2, the same range information was given for four targets with different velocity information, and the results showed that only the stationary target could be detected correctly with the right range information. Due to the stepped-frequency pulse function, the energy of the signal was spread over a larger range, which gives ambiguous range information and low energy, the drawbacks are avoided by using new detection method DSFP, as show in the Fig. 3, DSFP correctly detect the two targets with different velocity, and also multi-targets situation is shown in the simulation result. The merits of the Double SFP shown in the simulation results indicate that it is a very competitive candidate among the waveforms in vehicular radar.

III. Conclusion

Stepped frequency pulse radar was discussed because of its high performance with multi-target detection situations and shorter measurement times than LFMCW. But due to the Doppler shift and energy spread over effect that could cause target detection errors when detecting moving targets, it is also suitable for detecting multiple targets. Based on stepped frequency pulse radar, double stepped frequency pulse is presented, that transmits two different stepped frequency pulses, with a time and frequency difference. Double SFP was simulated with software, indicating good resolution of a moving target and detection of multiple targets. Thus to achieve the same multi-target resolution, Double SFP only requires half of the bandwidth of LFMCW. Double SFP only needs a very easy implementation based on pulse radar, which has been a mature technique for a long time.

IV. Acknowledgement

This research was supported by Basic Science Research Program through the National Research Foundation of Korea (NRF) funded by the Ministry of Science, ICT and future Planning (NRF-2014R1A2A2A01004121) and by the MSIP (Ministry of Science, ICT and Future Planning), Korea, under the ITRC (Information Technology Research Center) support program (NIPA-2014-(H0301-14-1008)) supervised by the NIPA (National IT Industry Promotion Agency).

References

[1] Wehner, Donald R, "*High resolution radar*" Norwood, MA, Artech House, Inc., 1987, 484 p., 1987.

[2] Gill, GS, "*Step frequency waveform design and processing for detection of moving targets in clutter*" Radar Conference, 1995., Record of the IEEE 1995 International, 1995.

[3] Rohling, Hermann and Meinecke, M-M, "*Waveform design principles for automotive radar systems*" IEEE,Radar, 2001 CIE International Conference on, Proceedings, 2001.

Designing with FinFET Technology

Andrew Marshall, fellow IEEE
Dept. of Computer and Electrical Engineering
The University of Texas at Dallas, Richardson, Texas, USA
andrew.marshall@utdallas.edu

I. INTRODUCTION

The introduction of the world's first commercial silicon CMOS Tri-gate-on-bulk silicon technology in 2012 [1-3], has led to the rest of the industry evaluating this process for subsequent nodes. The new technology was introduced as a cost effective way to continue improving chip performance and maintain the Moore's law trends for the semiconductor industry. Physically tri-gate is part of the multigate FET family, which also includes FinFET and GAA (gate all around) devices. Electrically it is similar to conventional planar CMOS, but with just enough differences to make the technology interesting.

The tri-gate process uses triangular fins etched into a bulk silicon substrate (figure 1) [1]. This achieves low active power, due to lower circuit capacitances, and a higher density of device width in a given physical chip space.

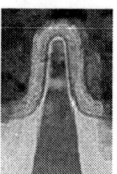

Fig. 1. TEM Image of 22nm Intel Tri-gate transistor [1]

This is illustrated in figure 2. The fin pitch defines the physical effective width of the fin, where the electrical fin width is defined by the active region of the gate under the fin. In general it appears possible to achieve electrical width approximately 15% greater than the fin-fin spacing (physical width). As a result interconnect lengths for a given width of device are reduced, which reduces electrical node capacitance. Since interconnect is the source of over 50% of the overall logic delay, any reduction in capacitance is beneficial.

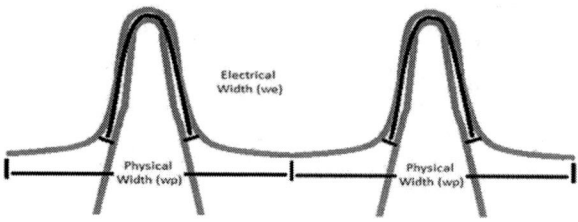

Fig. 2. Representation of Tri-gate's physical and electrical width, demonstrating trigate has more electrical width than physical width

Without a model that accurately defines the devices being modeled it is not possible to efficiently develop circuitry. Accuracy to about 5% is generally required. Anything less accurate results in designs that risk failure or overdesign. A Verilog-A based compact model for the tri-gate FET was released by the Berkeley compact modeling group in early 2012 [4]. This incorporates Multi-Gate (MG) specific transistor behavior, with volume inversion and short channel effects.

II. DESCRIPTION

CIRCUIT CONCERNS WHEN USING TRI-GATE

A. Circuit Induced Noise.

Circuit noise coupled through the substrate is likely to be somewhat lower in tri-gate, due to the additional resistance from body to substrate, which may reduce common-mode coupling (Fig. 3). Also, interconnect to gate spacing may average out to be greater in tri-gate, reducing coupling to arbitrary nodes within the circuit more than in bulk devices. It is, however, possible that gate-gate coupling may be higher in some situations in the tri-gate material.

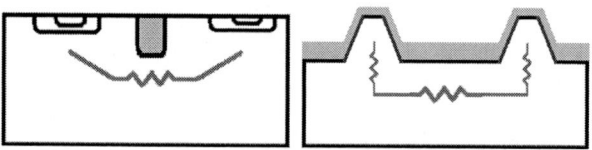

Fig. 3: Illustration of device to device coupling path in planar devices and trigate devices

B. Digital Circuit Design

The fundamental reason for switching to a 3D process from a planar CMOS is to take advantage of the extra dimension. Nowhere is this more important than with digital logic. The process is designed to give improved speed, lower active power and circuit area, without impacting leakage or off-state power consumption.

Digital logic behaves qualitatively in every way the same regardless of whether the MOS devices are planar on bulk silicon or tri-gate on bulk silicon. However, there are two differences which impact the design of digital logic. One of these is generally beneficial, the other generally detrimental.

The generally detrimental factor of tri-gate is that the devices are quantized. That is, without additional processing it is not possible to create a device width that is less than one fin or multiples of fins, so fine tuning a CMOS logic gate so the PMOS is a specific size larger than the NMOS for balanced performance is not so easy. As a practical matter most CMOS devices are larger than the minimum number of fins, and certainly critical paths through logic blocks are almost always of larger devices, and thus the quantization of CMOS is not generally an issue for logic performance.

The generally beneficial factor of tri-gate is that there is a reduced body effect as a function of voltage. As a result, in logic devices such as NAND gates, where there is a stacked gate, with conventional processes, as the operating voltage of the source of the transistor moves away from the body voltage (Vss or ground), the performance of the device degrades and the threshold voltage of the device increases. This effect is less pronounced in tri-gate devices (Fig. 4).

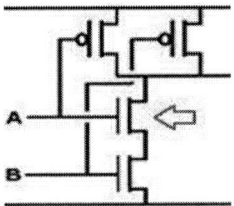

Fig. 4: Performance of the device marked, in the upper NMOS of the Nand gate shown has reduced performance due to it having a higher reverse body voltage.

A problem with the finFET devices is that they exhibit a generally higher gate-to-source and gate-to-drain capacitance than planar counterparts, due to extra coupling of the 3D capacitive structure. In order to maintain performance this must be overcome either with reduced capacitance elsewhere (such as in the interconnect), or increased drive current.

C. Analog Circuit Design

The features of tri-gate processes can have an impact on analog circuits. The areas of significant importance are fin matching, quantization of fin width and body bias effects.

Fin Quantization

It is conventional wisdom that quantization is bad for analog circuit design. Quantization implies that ratioing of device sizes cannot be achieved with the precision that could be achieved if the system were not quantized. This initial thought, however, bears some further analysis: It is extremely rare that matching of analog devices would be attempted with anything other than multiples of equally sized devices. Fins provide exactly that; equal width devices that can be used in any number combinations to provide the matching required. It is arguable that the net result may even be somewhat improved modeling, as the Spice model does not need to be accurate for a wide range of widths, for which inevitably there will be some accuracy tradeoffs, but only accurate to the fin or multiple fin level.

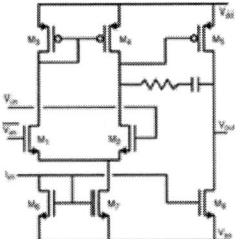

Fig. 5: Devices M1 and M2 are susceptible to body bias effects, but due to the balanced effects of the devices, it is less significant than in digital designs.

Body Bias Effects

Analog circuitry is more susceptible to the effects of body bias than digital circuits. In particular the input pair used in many operational amplifiers often relies on devices whose body contacts are connected directly to substrate or supply. In these devices the change in Vt, the threshold voltage, is dependent upon the input voltage being applied. Fortunately, the op-amp is a well-balanced design, such that the body effect in the balanced input state matches for both the input and inputb (Fig. 5). Hence, the beneficial lower body bias in tri-gate has a reduced benefit from the point of view of circuit operation. Nevertheless, reduced body bias effects are beneficial in reducing supply noise rejection, where the reduced body coupling is less likely to cause interference from either supply or ground.

Fin Matching and Variability

SEM images of the fins shows that there is mismatch between fins [1]. This affects the height of the fin, and the electrical width, and even the fin shape. The type of mismatch, may be systematic, caused by the number of nearest neighbor fins, or it may be completely random or possibly even periodic. Depending on the type of mismatch care must be taken if matched devices are needed. If random or periodic variation is excessive, it may cause design issues in small circuits such as SRAM memory cells, where individual fins are used for the highest density cells.

D. Memory Design

In principle, and demonstrated by Intel, in practice, tri-gate can be used for making conventional SRAM. Possibly the only issue with using tri-gate for SRAM is the quantization restrictions which limit the capabilities of memory design. However, by carefully controlling the bias, read and write states, a set of SRAM cells that cover the high density, high performance and low power characteristics ranges required.

III. CONCLUSIONS

Tri-gate fin-FET processes are available, and likely to be the high-performance processes for the foreseeable future at process nodes of 22nm and below. We have investigated the major differences between design using bulk silicon devices and tri-gate processes. Differences do exist for the design engineering teams using the new processes, however, these differences are readily understood and there are no significant hurdles in switching from bulk silicon designs to tri-gate and fin-fet on bulk processes.

REFERENCES

[1] D. James, "Intel Ivy Bridge Unveiled - The first commercial tri-gate, high-k, metal-gate CPU," Custom Integrated Circuits Conference (CICC), pp. 1-4, 9-12 September 2012 .

[2] A. Asenov, C. Alexander, C. Riddet and E. Towie, "Predicting future technology performance," Design Automation Conference (DAC), 2013 50th ACM / EDAC / IEEE, pp. 1-6, May 29-June 7 May/June 2013.

[3] D. Jelec, "Intel's Broadwell Manufacturing Delayed Until 1Q 2014 Due to a Defect," 16 Oct 2013. [Online]. Available: http://www.brightsideofnews.com/news/2013/10/16/intele28099s-broadwell-delayed-until-2014-due-to-a-defect.aspx.

[4] UC Berkeley Device Group: http://www-device.eecs.berkeley.edu/bsim/?page=BSIMCMG

Design and Performance Benchmarking of Steep-Slope Tunnel Transistors for Low Voltage Digital and Analog Circuits Enabling Self-Powered SOCs

Gaurav Kaushal[1], K. Subramanyam[2], Siva Nageswar Rao[2], G.Vidya[2], Radhika Ramya[2], Sadulla Shaik[2]
H. Jeong[1], S. O. Jung[1], and Ramesh Vaddi[3]

[1]School of Electrical and Electronics Enginnering, Yonsei University, South Korea
[2]Electronics and Communication Engineering Department, VFSTR University, Guntur, AP, India[1]
[3]Department of Electrical Engineering, School of Engineering, Shiv Nadar University, Dadri, UP, India

Email: kaushalg@yonsei.ac.kr, , subbuece455@gmail.com, sivaec007@gmail.com, vidya423@gmail.com, radhikaramya@gmail.com, sadulla09@gmail.com, hanwool87@yonsei.ac.kr, sjung@yonsei.ac.kr, ramesh.vaddi@snu.edu.in

Abstract

This paper presents the design insights and performance benchmarking of Tunnel FET (TFET) based low voltage digital and analog circuits to enable self-powered (energy harvesting based) wearable SOCs for vital sign monitoring etc. This work addresses some important challenges faced by nano scale CMOS digital and analog circuit designers at low voltages. This work demonstrates how TFET's device level chracterics (steep subthreshold slope, large I_{on}/I_{off} etc,) translate into favourable circuit performance metrics (power, delay and energy consumption etc, for digital and gain, g_m/I_{ds}, BW, GBW, FoM etc, for analog). TFETs are promising for designing robust, reliable and energy efficient circuits with supply voltage scaling for ultra-low power applications. The performance of TFET circuits is benchmarked with 20nm FinFET technology as base line comparison.

Keywords-Tunnel FETs, FinFETs, Steep-slope Transistors, Low voltage digital and analog designs, Self-powered SOCs, Energy harvesting, Ultra-low power.

Introduction

DESIGNING ultra-low power circuits with emerging nano scale devices for energy constrained applications such as RFIDs, implantable and wearable biomedical ICs and energy scavenging etc, is an increasing challenge. Since, any CMOS based circuits (with subthreshold swing (SS)< 60 mV/dec have energy efficiency limitations, it is necessary to look for alternative devices for future energy efficient applications. As a result steep slope transistors such as tunnel FETs, impact-ionization MOSFETs, NEMFETs, NEM relays etc, have been proposed to achieve zero standby leakage and low turn-on voltage to increase energy efficiency at low voltages. TFETs as alternative devices have attracted much attention recently for energy efficient circuit designs [1], [2], [3-6]. TFETs employ band-to-band tunneling as the carrier injection mechanism and can achieve below 60mV/decade SS at room temperature, fundamentally reducing the V_{cc} window to achieve desired on-off switch. Further, the dependence on battery as the only power source is putting an enormous burden in emerging applications like wireless micro-sensor networks, emphasizing harvesting energy from ambient sources such as light, mechanical (patient's own heart beat and body movements) and thermal (body heat) etc. The low power output from such sources [7], [8] necessitates not only the design of ultra-low power control circuits but also highly efficient power delivery interface circuits which can extract maximum power available from such sources [9]. In some recent works on emerging device technologies for energy

efficient circuit designs [10-13], the authors give some insights on FinFETs and TFETs for energy efficiency, but no specific design insights for TFET based analog and digital circuits for self-powered SOCs are highlighted to the best of our knowledge. This paper presents the design insights and performance benchmarking of TFET based low voltage digital and analog circuits to enable self-powered (energy harvesting based) wearable SOCs for vital sign monitoring etc.

Tunnel FET Device Structure, Characteristics and Models for Circuit design

Fig. 1 shows the device schematics of Si FinFET, N type P-tyep Hetero-junction TFETs used for circuit design. TFET is essentially a reverse-biased, gated p-i-n tunnel diode with asymmetrical source/drain doping. Hetero-junction TFETs enable simultaneous enhancement of the higher drive current as well as lower I_{OFF} [10]. Since compact SPICE models for III-V TFETs are not available so far, a Verilog-A model has been developed from TCAD Sentaurus device simulations [14]. Then the look-up table based Verilog-A models are generated and applied to *Cadence Spectre* for circuit evaluation. GaSb-InAs Heterojunction n-type TFET models have been calibrated with the Atomistic simulation, which is consistent with Si FinFET based circuits have been employed for baseline comparison [14]. Fig. 2. presents the steep-slope characteristics of Homo-junction and hetero-junction TFETs leading to large I_{on}/I_{off} at low V_{dd}.

Tunnel FET based Low Voltage Digital and Analog Design and Performance Benchmarking with FinFET Technology

Fig. 3 shows that with the increase in NAND gate fan-in (i.e., complexity), the energy consumption of HTFET designs increase by ~2.5x from 2-input gate to the 6-input gate, where as with FinFET designs, the increase in energy consumption is by ~3.67x, demonstrating energy efficiency of HTFET designs. Fig.3. further reveals that a 6-input HTFET gate has almost similar energy consumption as a 2-input FinFET NAND gate. This shows an important demonstration that with iso-energy consumption, with HTFETs one can design complex logic blocks by using 6-input gates than what one can achieve with 2-input FinFET gates. Fig.4. presents the LDO efficiency comparison of TFET based design with FinFET LDO designed for low input (Vin=0.3V) and light loads (few μWs). The small dropout voltages and reduced leakage power in HTFETs cause HTFET LDO to have larger peak efficiency (90%) and more than 80% efficiency over large load current range in comparison to FinFET design at such low harvested voltages. Table I and Table II summarize the performance benchmarking of HTFET based differential amlplifier and LDO designs

978-1-4799-5128-4/14 $31.00 © 2014 IEEE

(similar to basic CMOS designs, but taking into uni-directinal conduction of TFETs into account) targeted for ultra-low power applications. Due to the steep slope characteristics, HTFET designs have improved transient, AC and DC characteristics (gain, BW, GBW, ICMR, CMRR, slew Rate, FoM, etc.) in comparison to the 20nm Si FinFET designs, with slight trade-off in power consumption, and phase margin due to the enhanced on-state Miller capacitance in HTFETs. For ultra-low power applications, HTFET designs can be further optimized for a slight trade-off in FoM values with reduced power consumption.

Conclusions

This paper presents the design and performance benchmarking of Tunnel FET based low voltage digital and analog circuits targtted for energy harvesting applications. It has been demonstrated that due to the steep-slope characteristics of TFETs leading to large Ion/Ioff and g_m/I_{ds} values, TFET designs have improved transient, AC and DC characteristics (energy consumption, gain, BW, GBW, ICMR, CMRR, slew Rate, FoM, etc.) in comparison to the 20nm Si FinFET designs, with slight trade-off in power consumption, and phase margins due to increased Miller capacitances. This demonstrates Tunnel FETs promising for future energy efficient mixed signal self-powered SOC designs for vital sign monitoring etc.

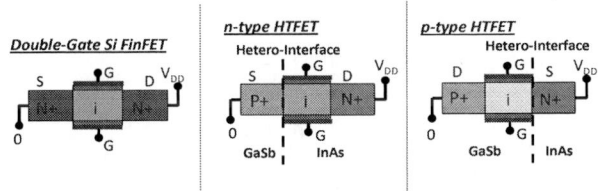

Fig. 1. Device schematics of double-gate (a) 20nm Si FinFET (b) III-V N-HTFET and (c) III-V P-HTFET.

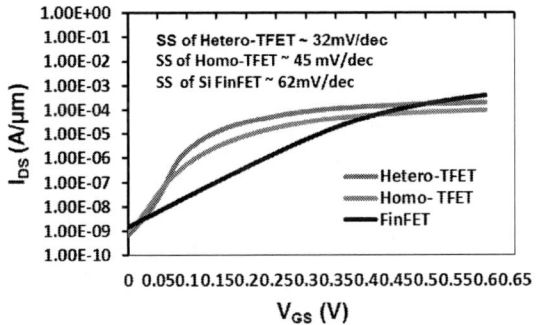

Fig. 2. I_{DS}-V_{GS} comparisons for Si FinFET, Homo-junction TFET and Hetero-junction TFETs at V_{DD}=0.2V.

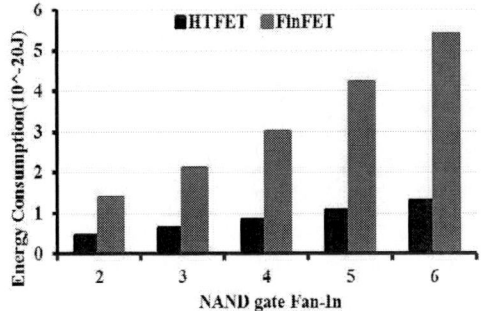

Fig. 3. Energy Consumption comparisons for Si FinFET and Hetero- TFET NAND gate Designs with increasing fan-in at V_{DD}=0.2V.

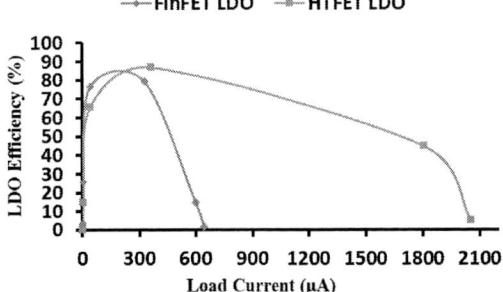

Fig. 4. LDO efficiency comparisons for Si FinFET and HTFET LDO Designs for light load and Vin=0.3V (Energy harvesting applications).

TABLE I: Performance Benchmarking of TFET low voltage Diff.amp.Design.

Performance Characteristics	HTFET	FinFET
Technology	20nm	20nm
Supply Voltage(V)	0.5	0.5
Power dissipation(µW)	0.87	0.016
Differential mode gain(dB)	33.9	13.78
CMRR(dB)	46	20
3dB Bandwidth (MHz)	13.78	1.35
Unit Gain Frequency (Hz)	712.4M	6.57M
Phase Margin(Deg)	91	103
Gain Bandwidth Product (GBW)	685.09	5.60
FOM (MHzpF/uA)	3.914	1.752
Slew Rate(V/µSec)	6.04	4.42
ICMR(V)	0.07-0.3	0.05-0.25
PSRR(dB)at 10MHz	-53.8	-16.8

TABLE II: Performance Benchmarking of TFET low Voltage LDO Design.

	FinFET	TFET
Technology (nm)	20	20
Minimum V_{dd}	0.5	0.3
Nominal V_{out}(V)	0.478	0.298
Drop out voltage (V)	0.022	0.002
Maximum load (mA)	3	3
Line regulation @ maximum load (mV/V)	0.955	0.096
Decoupling capacitor (pF) (on-chip)	1	1
Iq (µA)	20.41	18.47
PSRR (-dB) at 1MHz	67.877	73.13
Power conversion efficiency (%)	81.27	87.93

References

[1] D. K. Mohata, et. al, *IEDM Tech. Dig.*, pp. 33.5.1–33.5.4, Dec. 2011.

[2] V. Saripalli, et. al, *IEEE/ACM Int. Symp. on Nanoscale Architectures*, June 2011.

[3] V. Saripalli, et. al, *IEEE JETCAS*, vol. 1, iss. 2, pp. 109-119, June 2011.

[4] H. Liu, et. al, IEEE IEDM Technical Digest, pp. 577-580, December,2012.

[5] N. Agrawal, et. al, *IEEE DRC*, June 18–20, 2012.

[6] U. E. Avci, et. al, *IEEE Symp. on VLSI Technology*, pp. 183–184, Jun. 2012.

[7] B. Calhoun, et. al, *IEEE Trans. Comput.*, vol. 54, no. 6, pp. 727–740, Jun. 2005.

[8] M. Seeman, et al, in Proc. IEEE CICC Conf., Sep. 2007, pp. 567–570.

[9] Y. K. Ramadass, et al, *IEEE JSSC*, vol. 45, no. 1, pp. 189–204, Jan. 2010.

[10] H. Liu, et. al, *IEEE ISLPED*, September 4-6, pp.145-150, 2013.

[11] H. Liu, et. al, *IEEE ISLPED*, September 4-6, pp.157-162, 2013.

[12] K. Subramanyam, et al, 18th IEEE Int. Symp. on VLSI Design and Test (VDAT), Coimbatore, India ,16-18 July, 2014.

[13] Ramesh Vaddi, JLPEA, pp. 277-302, July 2011.

[14] H. Liu, et. al, (2014), "III-V Tunnel FET Model 1.0.0," https://nanohub.org/resources/21012.

77GHz frequency synthesizer for invention of radar systems

HyonGi Yoo and Franklin Bien, *member, IEEE*
Ulsan National Institute of Science and Technology, Ulsan, Korea
bien @ unist.ac.kr

Abstract

In this paper, the new method of 77-GHz automotive radar frequency synthesizer is introduceed. This generator is aimed of the transceiver design to implement the advanced radar module. A signal generator test verified that the developed synthesizer can generate an proper transmission signal. This development will provide the opportunity of future progress in carrying out the 77GHz automotive radar module for multi-target detection.

Index Terms — **frequency synthesizer, RF, 77-GHz automotive radar, multi-target detectuion, millimeter wave transceiver**

Introduction

For many a long year, several endeavors have been developed to promote the advanced techniques for supplying advantage[1]-[2]. Recently, the automotive radar has camed out as an advanced approach. The most generally utilized long range radar (LRR) is the frequency-modulated continuous wave (FMCW) waveform principle[3]-[4]. However, this waveform has some strong defects because of the technical reason. On the other hand, linear frequency modulation-frequency shift keying (LFM-FSK), which merge a frequency shift keying method with a linear frequency modulation waveform, is expected as a new option in this environmental due to its advanced structure[5]-[6].

This paper mainly focused on the most challenging task which generate the correct transmission waveform in the 77-GHz frequency band. The main purpose of the radar frequency synthesizer is generating the correct signal. This paper is organized as follows. Section II explains the development concept of the frequency synthesizer block, which is followed by the presentation of measurement results in section III. Conclusions are presented in section IV, the final section of this paper.

The frequency synthesizer block

In this structure, the signal was generated by the frequency synthesizer block though the phase locked loop (PLL) block with a crystal oscillator. The synthesizer part generates the waveform, and the PLL part, which generates the local oscillator (LO) frequency, along with the power divider (PDV), which sends the phase information to the voltage controlled oscillator (VCO) to generate the correct signal in the PLL. In the synthesizer part, the VCO generates a 3-GHz signal. The PLL creates a specialized shape waveform according to a

command from the PIC block, which generates the programming code from the J1 outside the synthesizer block. The lock detector controls the generated signal through J2, which verifies the operation of the PLL. Fig. 1 presents a block diagram of the synthesizer part.

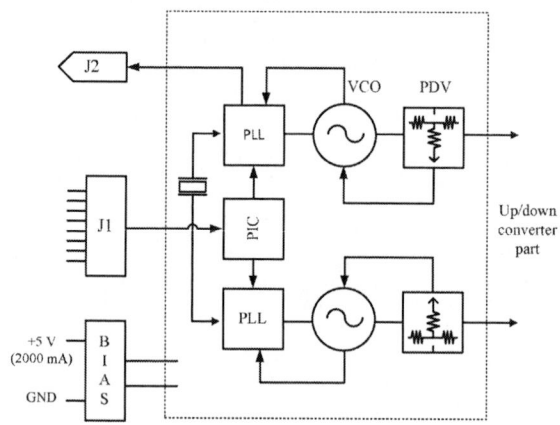

Fig. 1. Block diagram of synthesizer part

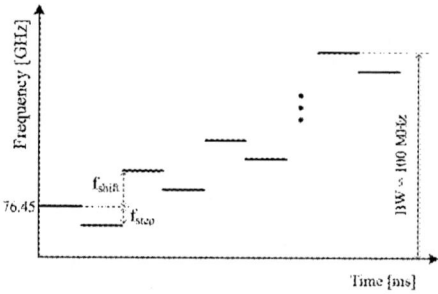

Fig. 2. The generated waveform with f_{shift} / f_{step}.

There are key concepts that enable this part to be developed effectively. First, the synthesizer frequency range is 76.5~76.6 GHz, with accurate chirp and delay time. The chirp time means the entire elapsed time period in one frequency step. The delay time is the empty time for the Digital Signal Processing (DSP) part to control the specialized algorithm, which supports the information. This frequency synthesizer also has the object of generating a waveform with enough f_{step}, which is the difference quantity among two diffreretransmitting signals. Fig. 2 shows the condition of the generated frequency waveform. In this graph, the generated signal is generated around the 76.45 GHz

978-1-4799-5128-4/14 $31.00 © 2014 IEEE

frequency band and the frequency bandwidth is around the 100 MHz.

The signal generator test

This part explains the investigation used to signal generator test the propositions. It was necessary to confirm whether the generated signal was precise by mornitoring whether the received signal supplied the correct information to the module. We used a signal generator test which measured the frequency synthesizer.

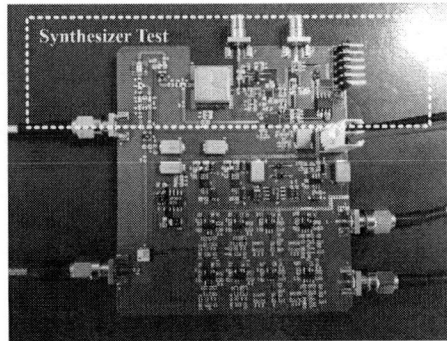

Fig. 3. Signal generator test.

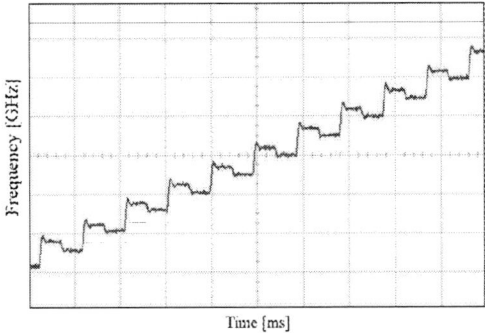

Fig. 4. V_ctrl signal waveform of the synthesizer part.

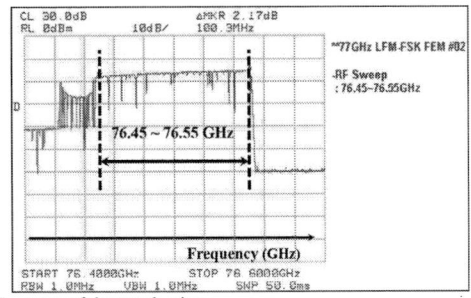

Fig. 5. RF sweep of the synthesizer.

TABLE I
MEASUREMENT RESULT OF THE SYNTHESIZER PART

parameter		unit	result
transmitter	Frequency range	GHz	76.5 ± 0.05
	Bandwidth	MHz	100
	Output power	dBm tpy.	+10
	LO frequency	GHz	73.53

The most important section in the radar module is the synthesizer part. The PCB board was invented for the signal generator test, as shown by the dotted lines in Fig. 3. The synthesizer part was measured by using this PCB board. Before frequency multiplication, the test frequency was 3 GHz in the synthesizer part.

The waveform was generated appropriately, as shown in Fig. 4, which shows the output signal of the synthesizer part. A measuring spectrum analyzer was used to obtain the RF sweep of the synthesizer, as shown in Fig. 5. This represents a transmitted signal frequency range of 76.45 to 76.55 GHz.

Conclusion

This paper delinates the development of frequency synthesizer block in the 77 GHz frequency band. The output result of the implemented synthesizer was evaluated using the signal generator test. The results of the experiment demonstrated the advantages of the proposed block.

Acknowledgment

This research was supported by Basic Science Research Program through the National Research Foundation of Korea(NRF) funded by the Ministry of Science, ICT and future Planning(NRF-2014R1A2A2A01004121) and by the MSIP(Ministry of Science, ICT and Future Planning), Korea, under the ITRC(Information Technology Research Center) support program (NIPA-2014-(H0301-14-1008)) supervised by the NIPA(National IT Industry Promotion Agency).

References

[1] Sawant, H.; Jindong Tan; Qingyan Yang; Qizhi Wang, "Using Bluetooth and sensor networks for intelligent transportation systems," Intelligent Transportation Systems, 2004. Proceedings. The 7th International IEEE Conference on, vol., no., pp.767, 772, 3-6 Oct. 2004.

[2] Kan Zheng; Fei Liu; Qiang Zheng; Wei Xiang; Wenbo Wang, "A Graph-Based Cooperative Scheduling Scheme for Vehicular Networks," Vehicular Technology, IEEE Transactions on , vol.62, no.4, pp.1450,1458, May 2013.

[3] Ramzi Abou-Jaoude, "ACC Radar Sensor Technology, Test Requirements and Test Solutions," in IEEE Transactions on Intelligent Transportation Systems, Vol. 4, no. 3, september 2003.

[4] Venkatasubramanian, V.; Leung, H., "A robust chaos radar for collision detection and vehicular ranging in intelligent transportation systems," Intelligent Transportation Systems, 2004. Proceedings. The 7th International IEEE Conference on, vol., no., pp.548, 552, 3-6 Oct. 2004.

[5] Hermann Rohling, Christof Möller, "Radar waveform for automotive radar systems and applications," in Radar Conference 2008 '08 IEEE, pp. 1-4, May 2008.

[6] Marc-Michael Meinecke, Hermann Rohling, "Combination of LFMCW and FSK Modulation Principles for Automotive Radar Systems," in German Radar Symposium GRS2000, October 2000.

Low Power Cross-Domain High-Voltage Transmitters for Battery Management Systems[†]

Chih-Lin Chen, *Student Member, IEEE,* , Yu-Hsun Su, and Chua-Chin Wang[‡], *Senior Member, IEEE*

Abstract—**This work presents a pair of low power high-voltage (HV) transmitters for battery management systems (BMS). Besides, the HV transmitter is designed using CMOS transistors without any isolator. To realize a solution on silicon, the proposed HV Transmitter shall be fabricated using an advanced HV semiconductor process, which usually is constrained by the voltage drop limitation between gate and source of HV devices. The proposed design is implemented using a typical 0.25 μm 1-poly 3-metal 60 V BCD process. The post-layout simulation results show that the HV transmitters can transmit data with HV dc level (36.4 $\sim$ 54.6 V) and the power consumption is less than 0.342 mW/Mbps.**

Index Terms—**high voltage, battery management system, high-voltage transmitter**

I. INTRODUCTION

High-voltage (HV) battery management system (BMS) is widely needed in many applications, e.g., EV and HEV, where many battery modules are assembled and integrated. One of the popular BMS architectures is the modular formation, i.e., a module monitors several batteries with a daisy-chain interface. A critical issue in the BMS is the data communication carried out by HV transceivers or digital isolators. They are usually implemented by discrete devices, e.g., optocoupler, magnetic isolator, and capacitive isolator. Several HV transceivers or digital isolators have been reported [1]-[5]. An optical coupler is used to isolate and communicate between high voltage and low voltage systems [1]. The disadvantages of optical couplers are high power consumption, poor integration, low speed, and degradation of LED (Light Emit-ting Diode). Digital isolators using magnetic coupling [2] or capacitive coupling [3] methods were proposed. However, no matter magnetic or capacitive coupling transmissions, they will generate EMI (electromagnetic interference) effect to jam other circuits or be corrupted by external RF signals to endanger the reliability. Besides, they usually need a lot of discrete components, i.e., transformer and capacitor. Therefore, it is not easily integrated in SOC (system on chip) designs. A wireless battery monitor was also proposed [4]. However, it is not cost effective to

C.-L. Chen, Y.-H. Su, and C.-C. Wang are with Department of Electrical Engineering, National Sun Yat-Sen University, Kaohsiung, Taiwan (phone: +886-7-525-2000 ext.4144; fax: +886-7-525-4199; e-mail: ccwang@ee.nsysu.edu.tw).

[†]This investigation was partially supported by National Science Council, Taiwan, under grant NSC102-2221-E-110-081-MY3 and NSC102-3113-P-110-010. The authors would like to express their deepest gratefulness to CIC (Chip Implementation Center) of NARL (Nation Applied Research Laboratories), Taiwan, for their thoughtful chip fabrication service.

[‡]C.-C. Wang is the contact author.

realize a large scale battery system, because it needs a lot of wireless modules. In other words, the cost and area efficiency will be problems. A HV transceiver without discrete devices was also proposed [5]. However, it needs to consume a large power dissipation. To resolve all the above problems, a pair of low power HV transmitters are disclosed in this paper.

II. CROSS-DOMAIN HV TRANSMITTER

A typical architecture of BMS usually consists of HVMUX (HV multiplexer), ADC, Power Management, Main Controller, High to Low Transmitter, and Low to High Transmitter. HVMUX is used to detect the voltage of battery module. ADC and other resistors serve as a current sensor. Main Controller determines the status of battery module, e.g., SOC and SOH, and Power Management generates a stable voltage, namely VDDH, to supply the High to Low Transmitter. Assume the number of batteries is 13 for E-scooters such that the voltage range of battery module is from 36.4 V to 54.6 V. Therefore, the proposed Low to High Transmitter must convert a low-voltage (LV) digital signal (0 $\sim$ 2.5 V) into a digital signal with HV dc level (36.4 $\sim$ 54.6 V). By contrast, the proposed High to Low Transmitter should convert a digital signal with HV dc level into a LV digital signal.

A. Low to High Transmitter

Low to High Transmitter is composed of M201 $\sim$ M210, as shown in Fig. 1. The upper left corner of Fig. 1 shows the symbol description of all transistors, i.e., 2.5 V NMOS, 2.5 V PMOS, HV NMOS, and HV PMOS, where the gate to source voltage of HV PMOS and HV MOS must be limited under a low voltage $\approx$ 5 V. The proposed Low to High Transmitter is basically a differential architecture. M207 $\sim$ M210 are used to limit the currents through M205 and M206. M201 $\sim$ M204 and Mc1 are in charge of keeping the voltage of Vx and Vy between VDDH and VSSH. When TxL is pulled low to VSSL, Vx is pulled high to VDDH and Vy is pulled low to VSSH. Because the Latch locks Vx and Vy, RxH is pulled low to VSSH. By contrast, when TxL is pulled high to VDDL, RxH is pulled high to VDDH. Therefore, the proposed design can convert a digital signal with LV dc level into a HV dc level.

B. High to Low Transmitter

Fig. 2 shows the schematic of High to Low Transmitter, where two bias current sources (M311 $\sim$ M315 and M322 $\sim$ M325) are used to limit the currents of M316 and M317. M318 $\sim$ M321 and Mc2 force the voltage of Va and Vb to stay

TABLE I
COMPARISON BETWEEN THE PROPOSED HV TRANSMITTERS AND PRIOR WORKS

Specifications	This work	[1]	[3]	[2]	[5]
Year	2014	2003	2005	2012	2014
Process (μm)	0.25 μm 60 V BCD	GaAs & BiCMOS	SOI	5V CMOS	0.35 μm BiCMOS
Maximum data rate	4 Mbps	25 Mbps	1 Mbps	250 Mbps	N/A
Number of isolator	0	2 opto-couplers	4 capacitors	2 transformers	0
Propagation delay	8.3 $\sim$ 62.28 ns	> 40 ns	30 $\sim$ 80 ns	5.5 ns	N/A
Power dissipation	0.342 mW/Mbps (20 pF load)	20 $\sim$ 100 mW/Mbps	40 mW/Mbps	5.4 mW/Mbps	> 24 mW
Area	0.2536 mm^2	N/A	N/A	0.12 mm^{2†}	N/A

(†) Not including the area between Tx and transformer

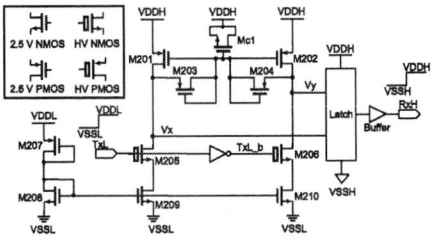

Fig. 1.   Schematic of Low to High Transmitter

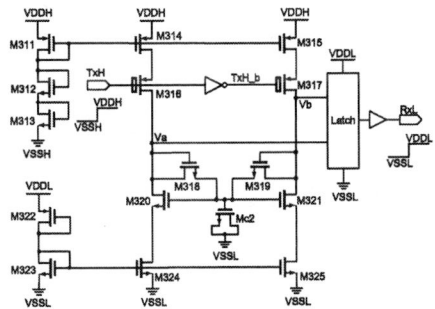

Fig. 2.   Schematic of High to Low Transmitter

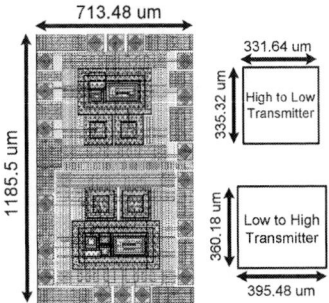

Fig. 3.   Layout of the proposed designs.

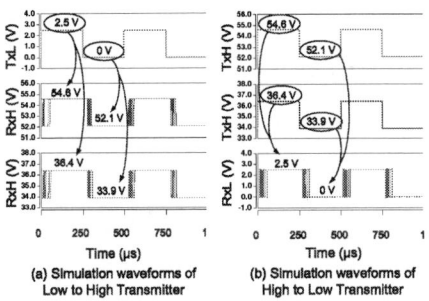

Fig. 4.   Simulation results of the proposed design.

between VDDL and VSSL. When TxH is pulled low to VSSH, Va is pulled high to VDDL and Vb is pulled low to VSSL. Because the Latch locks Va and Vb, RxL is pulled low to VSSL. By contrast, when TxH is pulled high to VDDH, RxL is pulled high to VDDL. Therefore, this design will convert a digital signal with HV dc level into a LV dc level.

III. IMPLEMENTATION AND SIMULATION

The proposed designs are implemented using the 0.25μm 1-poly 3-metal 60V BCD process to justify the performance. Fig. 3 shows the layout of the proposed designs. The chip area is 0.71348×1.1855 mm^2, where the active areas are 0.33532×0.33164 mm^2 for High to Low Transmitter and 0.36018×0.39548 mm^2 for Low to High Transmitter, respectively. Fig. 4 shows the simulation results of the proposed design at different VDDH (36.4 V and 54.6 V) and all PVT corners. The maximum propagation delay are 62.28 ns for Low to High Transmitter and 59.32 ns for High to Low Transmitter, respectively. Table I shows the comparison between the proposed HV Transmitter designs and prior works. Our design

attains the smallest power dissipation (0.342 mW/Mbps) and it is the only solution without any isolators.

REFERENCES

[1] R. Kliger, "Integrated transformer-coupled isolation," *IEEE Instrumentation & Measurement Magazine*, vol. 6, no. 1, pp. 16-19, Mar. 2003.

[2] S. Kaeriyama, S. Uchida, M. Furumiya, M. Okada, T. Maeda, and M. Mizuno, "A 2.5 kV isolation 35 kV/us CMR 250 Mbps digial isolator in standard CMOS with a small transformer driving technique," *IEEE J. of Solid-State Circuits*, vol. 47, no. 2, pp. 435-443, Feb. 2012.

[3] M. Kikuchi, T. Sase, M. Inaba, A. Watanabe, N. Akiyama, and F. Murabayashi, "On-chipı 500V capacitive isolator for 1 Mbps CAN transceiver" in *Proc. Inter. CAN Conf.*, Apr. 2002, pp. 03-09 - 03-15.

[4] M. Schneider, S. Ilgin, N. Jegenhorst, R. Kube, S. Püttjer, K.-R. Riemschneider, and J. Vollmer, "Automotive battery monitoring by wireless cell sensors," in *Proc. IEEE Inter. Instrumentation and Measurement Technology Conf.*, May 2012, pp. 816-820.

[5] K. Kadirvel, J. Carpenter, P. Huynh, J. M. Ross, R. Shoemaker, and B. Lum-Shue-Chan, "A stackable, 6-cell, Li-Ion, Battery Management IC for electric vehicles with 13, 12-bit $\Sigma\Delta$ ADCs, cell balancing, and direct-connect current-mode communications," *IEEE J. of Solid-State Circuits*, vol. 49, no. 4, pp. 928-934, Apr. 2014.

978-1-4799-5128-4/14 $31.00 © 2014 IEEE

A Single Chip Li-Ion Battery Protection IC with Low Standby Mode Auto Realease

Seunghyeong Lee*, Yongjae Jeong**, Yungwi Song**, and Jongsun Kim*

**R&D Center, SANBUD, Seoul, Korea
*Schoole of Electronic and Electrical Engineering , Hongik University, Seoul, Korea

Abstract

A fully integrated cost-effective and low-power single chip Lithium-Ion (Li-Ion) battery protection IC (BPIC) is proposed for portable devices. The control unit of the battery protection system and the MOSFET switches are integrated in a single package to prevent overcharge, overdischarge, and overcurrent of the Li-Ion battery. The BPIC supports low power standby mode and a new auto release function (ARF) is adopted for being released from the standby mode. The proposed BPIC is implemented in a 0.18-μm CMOS process and the chip size is 750×610 μm^2. It dissipates 3.0uA, while it consumes only 400nA at standby mode.

Keywords-component; Lithium-ion Battery protection; Auto release standby mode

Introduction

Lithium-ion (Li-Ion) batteries are widely used for mobile devices due to their high energy density, high cell voltage, low self-discharge rate and no memory effect. However, they have safety problems such as characteristic degradation, short life cycle, overheating or explosion when they are at the state of overcharge, overdischarge or overcurrent. To address these safety issues, battery protection circuits are included in the Li-Ion batteries [1-3]. In this paper, a cost effective and low power fully integrated single chip BPIC is presented. The proposed BPIC adopts a new auto release function (ARF) for being released from the standby mode, resulting in a ultra low stanby mode current of only 400nA. The BPIC is implemented in a 0.18-μm CMOS process and the chip size is 750×610 μm^2.

Proposed Battery Protection IC (BPIC)

Fig. 1 shows a block diagram of the proposed BPIC. The BPIC and the MOSFET switches are connected to the Li-Ion cell through RC filter and to load/charger through the +VO/-VO pins. The BPIC includes an overcharge detector (VD1), an overdischarge detector (VD2), a discharging overcurrent detector (VD3), a charging overcurrent detector (VD4), a short circuit detector (VSHORT), a reference voltage generator, an oscillator, a charger detector, and a main logic block.

The BPIC monitors the battery cell voltage and prevent over-charge, over-discharge by turning off the charging MOSFET switche if a Li-Ion cell is out of the normal operating voltage range. It also prevents overload and provides short circuit protection by monitoring the discahrge current and turning off the discharging MOSFET switch. The CMOS

BPIC and the MOSFET switches are integrated in a single chip package.

Fig. 2 shows the voltage waveforms of the Li-Ion battery (VDD) and the control signals, CO and DO, of the switching FETs at over-charge, over-discharge, discharging over-current,

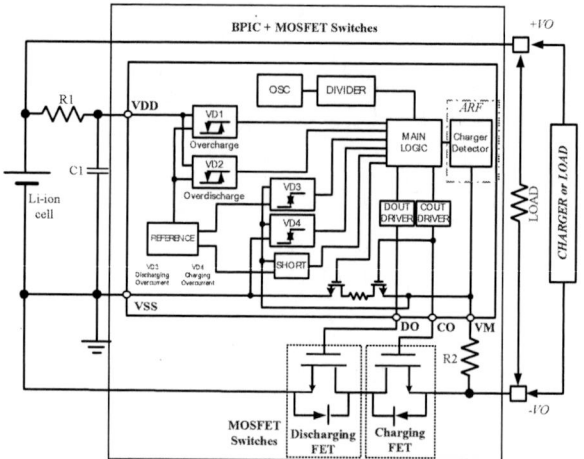

Fig. 1 Proposed Block Diagram of Li-Ion Battery Protecion Citcuit

and charging over-current. When Li-Ion battery is charged and becomes higher than the overcharge voltage (V_{DET1}), the overcharge detector compares it to a reference voltage and sends overcharge detection signal to the main logic block. Then the logic block enables the oscillator to generate proper delay amounts. Therefore, the BPIC determines proper detection delay times of tV_{DET1} and discharging over-current delay time of tV_{DET2}, In the state of over-discharge, the BPIC enters into a standby mode and turns off all the blocks except the voltage reference genetator and overdischarge detecion block, resulting in a ultra low stanby power dissipation of only 415nA

By using Rds-on resistance of two switching FETs that are connected in series, the BPIC senses and blocks over-current in VM node, When the VM nvoltage remains higher than that of discharging overcurrent detection voltage (V_{DET3}) during discharging overcurrent delay time (tV_{DET3}), discharging FET turns "Off". On the contrary, if the VM voltage remains lower than that of charging overcurrent detection voltage (V_{DET4}) during charging overcurrent delay time (tV_{DET4}), charge switch FET turns "Off" to block current path.

Fig. 3 shows the waveforms of the proposed standby mode auto release function (ARF). As can be seen in Fig. 3(a), the conventional BPIC is not released from the standby mode when the +V/-VO pins are open and the Li-Ion battery voltage remains higher than that of the overdischarge release voltage (V_{REL2}). The overdischarge release voltage of the proposed

978-1-4799-5128-4/14 $31.00 © 2014 IEEE

ARF is changed depending on the charger or the load connections by the charger detection block in Fig. 1. As shown in Fig. 3 (b), if the BPIC is connected to a load the battery voltage becomes higher than the overdischarge detection voltage (V_{DET2}), then the BPIC mode turns into a normal state. If the BPIC turns into normal state immediately when the battery voltage becomes higher than the overdischarge release voltage (V_{REL2}).

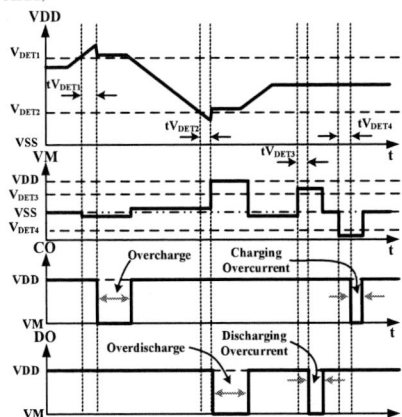

Fig. 2 Operation of the proposed BPIC

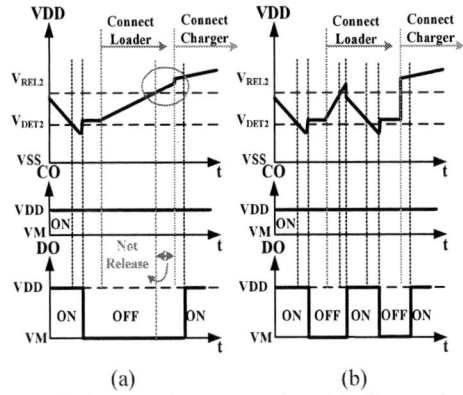

(a)   (b)

Fig.3 Overdischarge release operation (a) Conventional BPIC without auto release function (ARF) (b) Proposed BPIC with auto release function (ARF)

Experimental Results

Fig. 4 shows the chip layout and package X-ray photograph of the proposed BPIC and MOSFET switches. The chip size of the BPIC is 750×610 μm^2. The BPIC and MOSFET switches are integrated in a single package. Fig. 5 shows the measured results of the BPIC detecting overcharg/overdiscahrg voltage and charging/discharging current with overcharge detection voltage (4.25V), overdischarge detection voltage (2.3V), overcharge detection delay time (1s), overdischarge detection delay time (20ms), charging overcurrent detection voltage (0.123V), discharging overcurrent detection voltage (-0.100V), charging overcurrent delay time (12ms), and discharging delay time of overcurrent (8ms). Table I shows a performance comparison table.

Conclusion

In this paper, a fully integrated cost-effective and low-power single chip BPIC is implemented in 0.18-μm CMOS process. The BPIC and the MOSFET switches are integrated in a single package. The BPIC supports low power standby mode and a new auto release function (ARF) is adopted for being released from the standby mode. It dissipates 3.0uA at normal mode and consumes only 400nA at standby mode. Measurement results show that the proposed BPIC satisfies all the required electrical characteristics of Li-Ion battery protection systems.

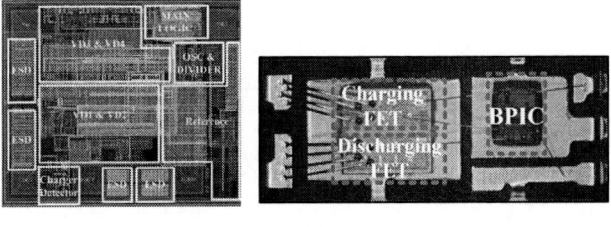

(a)   (b)

Fi. 4 (a) chip layout (b) Package X-ray Picture of the proposed BPIC and FET swithces

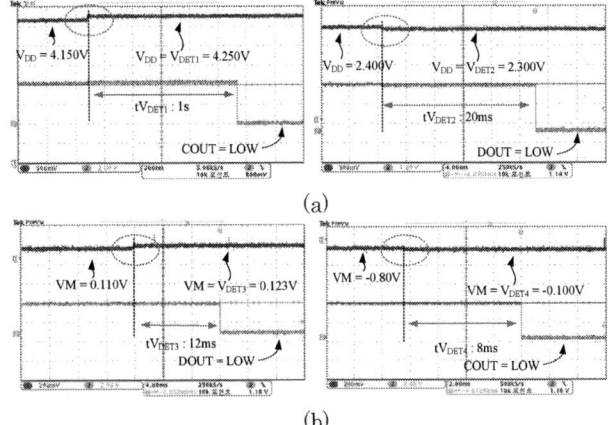

(a)

(b)

Fig. 5 Measurement Results (a) Overcharge/Overdischarge Voltage (b) Discharging/Charging Current

TABLE I
Performance comparions table

	This work	[1]	[2]
BPIC Process	0.18um CMOS	-	0.5um BCD
MOSFET switches	Packing on chip	External Component	External Conponent
Current Consumption	3 uA	6 uA	-
Standby Mode Current Consumption	400 nA	0.5 uA	-
Auto Release	O	X	X

References

[1] Lithium Ion Battery Protection, Monolithic IC MM3280 Data sheet, MITSUMI INC.

[2] Jang-Hyuck Lee and Joon-youp Sung, "Design of Battery Protection Circuit," in Proc. Intelligent Signal Processing and Communication Systems, 2004, pp. 784-786

[3] S. Matsunaga, M. Sawada, M. Sugimot, and N. Fujishima, "Low Parasitic Current Half On Operation of Battery Protection IC," in Proc. of the 1th Int. Symp. On Power Semiconductor Devices & ICs, 2007, pp. 49-52.

978-1-4799-5128-4/14 $31.00 © 2014 IEEE

A High Energy Extraction Self-controllable CMOS Resonant Rectifier Circuit for Piezoelectric Energy Scavenging System

Amad ud Din, Doyoung Chung, Dasom Park, Hyunsik Lee, and Jong Wook Lee

School of Electronics and Information
Kyung Hee University, Suwon, 446-701, Korea,
{ammad, jwlee}@khu.ac.kr

Abstract

In this study an efficient extraction energy based CMOS resonant rectifier is proposed for increasing the available output power of Piezoelectric (PE) transducer. A symmetric flipping circuit is used to extract energy from the transducer. Measured results show that the proposed rectifier significantly increases the extracted power when it is compared to the conventional full-bridge rectifier (CFBR). Moreover, results show that proposed rectifier has a flipping efficiency and power conversion efficiency of 77 % and 80 %, respectively. It is implemented in 0.18-μm CMOS process in a compact chip area of 0.072×0.136 mm^2.

Keywords-rectifier; energy harvesting; piezoelectric transducer (PE); flipping; conventional full brigde rectifier (CFBR).

Introduction

Mechanical vibrations are an attractive source of energy because it is commonly available and is ideal for piezoelectric (PE) materials, which have the ability to convert mechanical energy into electrical energy. Usually, the PE transducer's output is irregular function of time and it need to be rectified first. The commonly used AC-DC converter is conventional full-bridge rectifier (CFBR). Its drawback is low power extraction and low conversion efficiency. To improve the power extraction of the transducer, few techniques have been introduced in [1-3]. In first technique, an inductor is shunted across the PE transducer using two switches M_1 and M_2 shown in Fig. 1(a). A resonant loop including transducer's capacitor and inductor is formed to flip the voltage across the transducer [1]. This improves the power extraction, but V_{th} drop of two transistors limit the power extraction. To increase this, [2] introduces a technique using two branches with diode connected and switches M_3 and M_4 by controlling the inductor as shown in Fig. 1(b). The transducer voltage is flipped through this path. The impedance at both ends 'a' and 'b' is different; as a result flipping voltage is not symmetric. This causes the irregular flipping. This implementation also suffers the voltage drop across the diodes. The comparison between these flipping schemes is shown in Fig. 1(c). In this paper, a high extraction energy self-controllable CMOS rectifier is proposed for increasing the available output power of transducer by using symmetric flipping technique and as well as minimizes the conduction loss using active diode topology. Measurement results reveal that the proposed rectifier circuit can increase the significant amount of extracted power and minimizes the conductions loss.

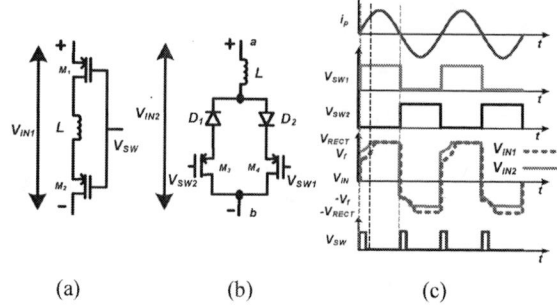

Fig. 1 Implementation of flipping techniques. (a) Using switches. (b) Using diodes and switches. (c) Comparison waveforms of techniques.

Proposed Design

Fig. 2 shows the circuit diagram of the proposed rectifier. It shows the equivalent model of Piezoelectric (PE) transducer, voltage multiplier, pulse generator circuit, flipping circuit, and comparator contained active rectifier.

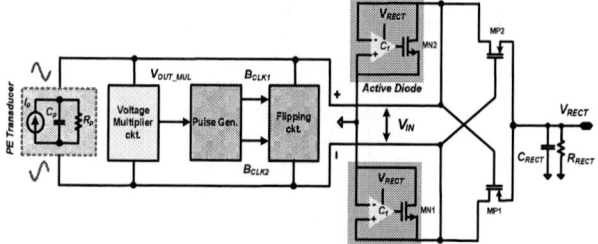

Fig. 2 Schematic of proposed design

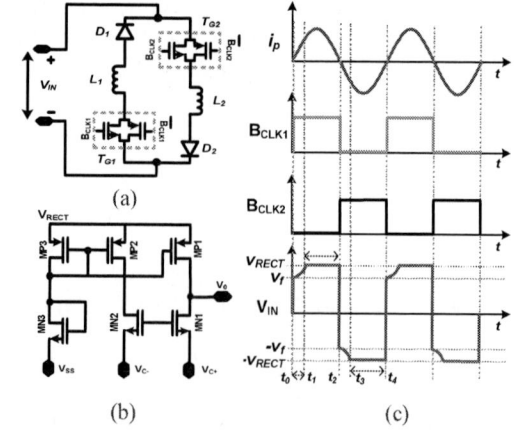

Fig. 3 (a) Schematic of flipping circuit (b) Schematic of comparator. (c) Key wave-forms of the proposed rectifier.

978-1-4799-5128-4/14 $31.00 © 2014 IEEE

The Piezoelectric (PE) transducer equivalent electrical model consists of a sinusoidal current source $i_p(t)=I_p sin(\omega_p t)$ in parallel with the capacitor C_p and a resistor R_p. The I_p is the magnitude of the current and $\omega_p=2\pi f_p$ is the excitation frequency [1]. Because of the C_p, most of the available charge from PE transducer cannot be extracted. In this proposed design, an effort has been made to reach its power extraction capability to maximum theoretical power. The core idea behind this proposed rectification scheme is to initialize self commutation of voltage across the transducer capacitor C_P. The C_p voltage is flipped across the flipping circuit which includes schottky diodes D_1, D_2, inductors L_1, L_2 and T-gates T_{G-1}, T_{G-2} shown in Fig. 3(a). The T-gates are used because of the low resistance than MOS switches. There are two symmetric flipping paths for positive and negative cycles. Positive cycle includes T_{G-2}, L_2 and D_2 while in negative cycle T_{G-1}, L_1 and D_1. The diodes used are on-chip schottky diodes, which showed superior sensitivity with a small voltage drop [4]. The pulses B_{CLK1} and B_{CLK2} are used to activate or deactivate the T-gates, which are generated by the pulse generator circuit. The pulse generator is driven by the six stage voltage multiplier circuit and it is activated from the same source and generates the voltage to bias the pulse generation circuit. To reduce the power loss, an active diode configuration is adopted instead of passive diodes. Fig. 3(b) shows the schematic of comparators C_1 and C_2 using common gate topology. These comparators are activated from the harvested output voltage V_{RECT}. The expected waveforms to drive the proposed rectifier over a complete cycle are illustrated in Fig. 3(c). It declares that C_p is pre-charged and then source current flows during t_1-t_2 and similarly for the negative cycle. The flipping efficiency of the circuit can be derived from the following equation [3].

$$\eta_f = \frac{V_f + V_{RECT}}{2V_{RECT}} \qquad (1)$$

The maximum extracted power of the proposed rectifier is be given by

$$P_{max} = C_p f_p \frac{V_p^2}{1-\eta_f} \quad \text{when } V_{RECT}=V_p/2(1-\eta_f) \qquad (2)$$

Measurement Results

The proposed rectifier shown in Fig. 2 is implemented using 180-nm CMOS process. For comparison conventional full-bridge rectifier (CFBR) is also implemented in the same die. Fig. 4 shows the chip photo of CFBR and proposed rectifier and it occupies the area of 0.37×0.412 mm^2 and 0.85×0.412 mm^2 respectively.

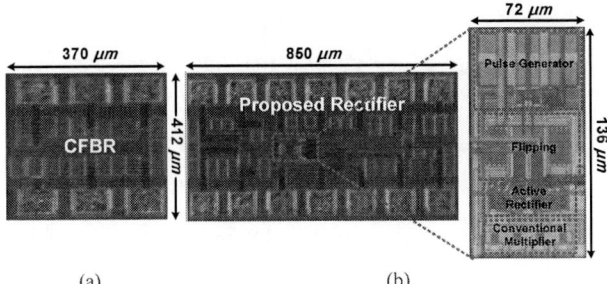

Fig. 4 Chip micrograph (a) Conventional full-bridge rectifier (CFBR). (b) Proposed rectifier.

These rectifiers are simulated and measured with the transducer parameters such that I_p=490 μA, C_p=130 nF, f_p=200 Hz, and R_p=1 MΩ, The values of Inductors L_1-L_2 and output capacitor C_{RECT} are 1000 μH and 1 μF respectively. The measurement results for the CFBR and proposed rectifier are shown in Fig. 5(a) and 5(b) respectively. Fig. 5(a) shows the input voltage V_{IN} and output voltage V_{RECT} while Fig 5(b) shows the input voltage V_{IN}, control signals B_{CLK1}, B_{CLK2} and the output rectified voltage V_{RECT}. It is obvious, when clock pulses changes, V_{IN} changes polarity and the measured flipping efficiency according to (1) is 77 %. Measured output voltage V_{RECT} of the CFBR and proposed rectifier is 2.5 V and 3.71 V respectively and measured output power is 125 μW and 809 μW. The maximum extracted power can be calculated using (2) i.e., for CFBR and proposed rectifier are 234 μW and 1017 μW, respectively which reveals that their power conversion efficiency is 53.4 % and 80 % respectively.

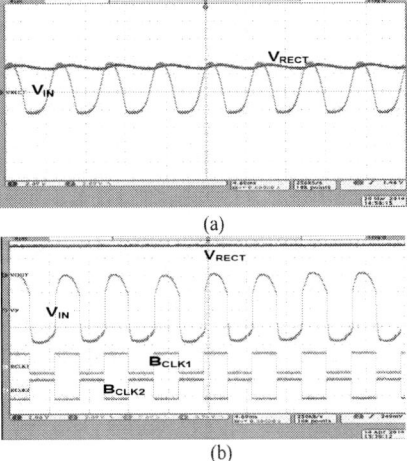

(a)

(b)

Fig. 5 Measured waveforms of (a) CFBR. (b) Proposed rectifier.

Acknowledgment

This work was supported by the Mid-career Researcher Program (No. 2012-001327) through National Research Foundation (NRF) grant funded by the Korea government. The chip fabrication and CAD tools were made available through IC Design Education Center (IDEC), Korea.

References

[1] Y. K. Ramadass and A. P. Chandrakasan, "An efficient piezoelectric energy harvesting interface circuit using a bias-flip rectifier and shared inductor", *IEEE J. Solid-State circuits,* vol. 45, no.1,pp.189-204, Jan. 2010.

[2] N. Krihely and S. Ben-Yaakov, "Self contained resonant rectifier for piezoelectric sources under variable mechanical excitation," *IEEE Trans. Power Electronics*, vol. 26, no. 2, Feb. 2011.

[3] X.-D. Do, H. H. Nguyen, S. K. Han, and S.-G. Lee, "A rectifier for piezoelectric energy harvesting system with series synchronized switch harvesting inductor," *Asian Solid Sate Circuits Conference (A-SSCC)*, Nov. 10-13, 2013.

[4] J. W. Lee, B. Lee, H. B. Kang, "A high sensitivity, CoSi$_2$-Si schottky diode voltage multiplier for UHF-band passive RFID tag chips," *IEEE Microwave and wireless components letters*, vol. 18, no. 12, Dec. 2008.

A High Efficiency Multi-Channel LED Driver Based on Converter-Free Technique and Load Adaptive Method

Si FU, Minjie CHEN, Xutao LEE, Tsutomu YOSHIHARA

Graduate School of Information, Production and Systems, Waseda University, Fukuoka, Japan. 808-0135
Email: fusi-nevergiveup@fuji.waseda.jp

Abstract—A novel high efficiency multi-channel LED using a soft self-commutating technique and load adaptive method is presented in this paper. Directly powered by a AC voltage, for a soft switching method, the proposed circuit can implement a higher efficiency as well as a lower total harmonic distortion (THD) and avoid some thorny problem such as EMI and EMC noise during hard commutations with self-commutating structure and adaptive load network. By utilizing the converter-free technique, this work can be fabricated into a BCDMOS process technology without using any inductors and capacitors. The proposed circuit is designed and simulated based on PSpice Model. The maximum output current and the efficiency for LED achieve 1.28A and 89.1% under 1KHz 220V/AC condition.

Index Terms—multi-channel, self-commutating, converter-free, adaptive load network, high efficiency, BCDMOS

I. INTRODUCTION

Recently, high-brightness LEDs have become the optimal solutions for decorating and street lighting, outdoor signage and display backlight. The brightness of an LED is related to its forward current, and LEDs should be energized by constant current. In the growing demand of backlighting application, multi-channel of the LED strings is widely used [1]. Various converter-free methods for non-isolated LED drivers with multiple LED strings connected in series have been introduced, enabling both a higher efficiency and power factor (PF) as well as lower total harmonic distortion (THD) [2-4]. The following section introduces a multi-channel LED drivers based on converter-free technique and self-adaptive load network.

II. PROPOSED STRUCTURE AND OPERATION PRINCIPLE

A. Six Channels Case and the Self-Commutating Operation

The schematic of the proposed LED driver in the six channels situation is shown in Fig.1. It is composed of a bridge rectifier, six LED strings, six Single Voltage Reference Controlled Current Source (SVR-CCS) channels and a so chosen resistors network. For the SVR-CCS structure, each consists of a high voltage n-channel MOSFET (H1), and two n-channel MOSFETs (M1, N1), an operational amplifier whose positive terminal is connected with a constant voltage reference and negative terminal is connected with the resistor network. In SVR-CCS operation, if the voltage difference between output voltage of two OPA is higher than $\sqrt{2V_{OV}}$, where

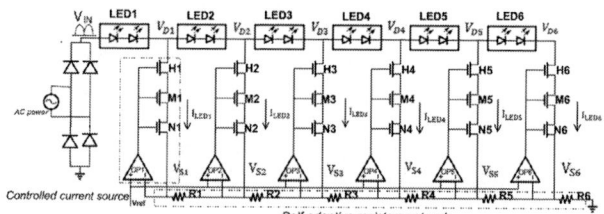

Fig. 1. Proposed six channels LED driver using SVR-CCS structure

$V_{OV} = (V_{GS} - V_T)$, the high-biased transistor carries all the current and the others have no current. This is true only when V_{D1} and V_{D2} are available at the same time [5]. There are seven work modes in the six channels case when AC wave is powered. If the $V_{IN} < V_{LED1}$, all Channels are turned off and there is no current flow. Else if $V_{IN} > V_{LED1}$ and the condition of $V_{LED(\alpha)} < V_{IN} < V_{LED(\alpha+1)}$ is satisfied, where $\alpha = 1, 2, \cdots 6$, Only Channel α is decided to be on and others is off. Under this condition, the Gate voltage of H_i, M_i, N_i are calculated as:

$$V_{gi} = A_i \times (V_{ref} - V_{si}) \tag{1}$$

where A_i is amplification coefficient of OP_i, $i = 1, 2, \cdots 6$. Because of the voltage difference between V_{gi} and $V_{g\alpha}$, the Channel i $(i < \alpha)$ carry no current. And the current of the Channel α are expressed as:

$$I_{LED\alpha} = V_{ref} / (R_\alpha + R_{\alpha+1} + \cdots + R_6) \tag{2}$$

The current through the LED strings can be expressed as:

$$I_{LED} = \sum_{i=1}^{6} I_{LEDi} \tag{3}$$

even under the commutation periods. As can be seen, the SVR-CCS multi-channel LED driver avoids current increments during commutations even under the 1 KHz AC input power while undesirable current increments are severe under this condition [5].

B. Self-Adaptive Resistors Network Design

The maximum output current is limited as 400mA because of the single resistor shared by the six channels in [5]. A little change of the value of resistor will cause mismatch in specific

978-1-4799-5128-4/14 $31.00 © 2014 IEEE

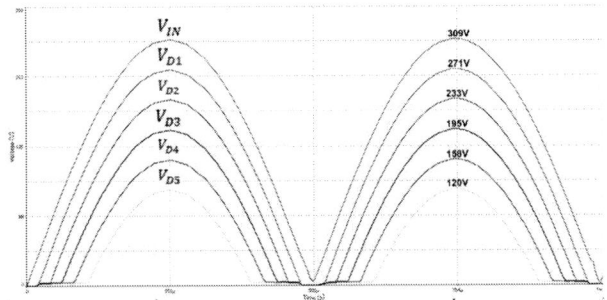

Fig. 2. Time diagram of $V_{IN}, V_{D1}, V_{D2}, V_{D3}, V_{D4}$ and V_{D5}

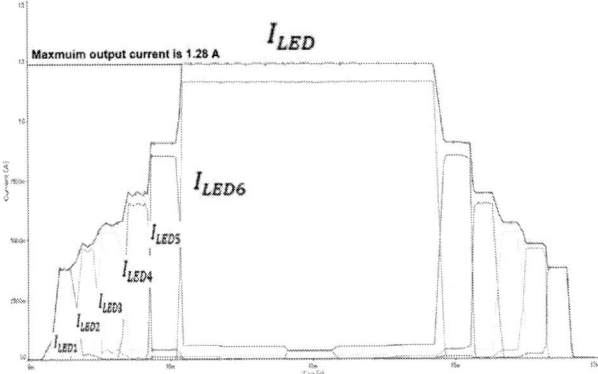

Fig. 3. Transient response of output current I_{LED} under 60 Hz 220V/AC

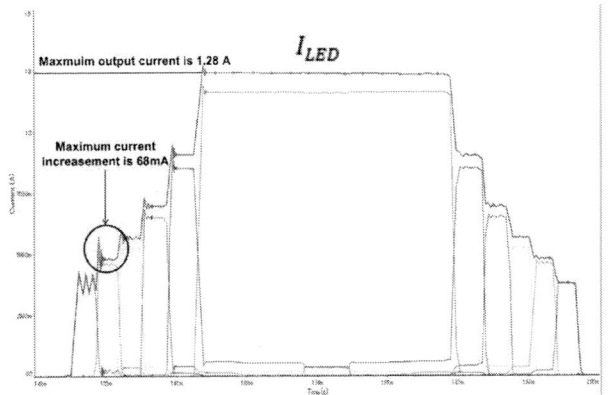

Fig. 4. Transient response of output current I_{LED} under 1KHz 220V/AC

TABLE I
COMPARISON WITH OTHER VOLTAGE REFERENCE

	This work	Ref [3]	Ref [5]
THD	7.0	13.4	8.6
PF	0.998	0.974	0.996
Efficiency(%)	89.1	85.4	92.2
Maximum Output Current(mA)	1280	115	400
Number of LED strings	6	20(interleaved)	6
Number of voltage reference	1	10	6

channel and THD will be unaccepted. Hence a self-adaptive resistors network which can increase the flexibility in the drive ability design is extremely significant. In this work the load network is designed to achieve higher efficiency, a lower THD and larger output current for the LED strings. According to Equation (2), the current of Channel 6 is only decided by the R_6, adjusting value of this resister obtains the desired current in Channel 6 first. The same procedure may be easy to obtain the satisfied output for the other Channels.

III. SIMULATION RESULTS

The proposed driver circuit is designed by BCDMOS process technology. The simulation work is based on the PSpice Model. Given 1KHz 220V/AC wave as input power, the time diagram of $V_{IN}, V_{D1}, V_{D2}, V_{D3}, V_{D4}$ and V_{D5} are shown in the Fig.2. And Fig.3 and Fig.4 illustrate the transient response of the proposed circuit under the 60 Hz 220V/AC and 1KHz 220V/AC condition.

Table I compares the performance of the proposed circuit with other multi-channel LED drivers. The circuit proposed in this brief shows the best performance in terms of maximum output current, PF and THD.

IV. CONCLUSION

A novel multi-channel LED driver in the converter-free method has been presented. With no capacitors and inductors in proposed circuit, this work can be fabricated into a 450V

1um BCDMOS process technology in the future. The proposed SVR-CCS structure extremely decreases the current increment during the commutation periods under the 1 KHz AC input. And the single voltage reference decreases the crosstalk among current sources which cause the inevitable distortion for the chip performance in actual utilization. By utilizing the load adaptive network, a lower THD and larger output drive ability have obtained. According to simulation results, the proposed circuit achieves, the maximum output current, the efficiency ,the PF and the THD are 1.28A, 89.1%,0.998 and 6.9 % under 1KHz 220V/AC conditions.

REFERENCES

[1] Luchen Yu, Yuan Zhu , Minjie Chen and T. Yoshihara, "High efficiency multi-channel LED driver based on SIMO switch-mode converter," *SoC Design Conference (ISOCC),2012 International*, vol., no., pp.483,486, 4-7 Nov. 2012.

[2] S. Cha, D. Park, Y. Lee, et al, "AC/DC Converter Free LED Driver for Lightings," *IEEE International Conference on Consumer Electronics*, vol., no., pp.706,708, 13-16 Jan. 2012.

[3] Hwu, K. I, W. C, "A High Brightness Light-Emitting Diode Driver with Power Factor and Total Harmonic Distortion Improved," in *IEEE Applied Electronics Conference and Exposition*, vol., no., pp.713,717, 6-11 March 2011.

[4] E. Kang, J. Kim, D. Oh, and D. Min, "A 6.8-W Purely-Resistive AC LightEmitting Diode Driver Circuit with 95% Power Factor," *IEEE International Conference on Power Electronics and ECCE Asia*vol., no., pp.778,781, May 30 2011-June 3 2011.

[5] Junsik Kim, Jiyong Lee and Shihong Park, "A soft self-commutating method using minimum control circuitry for multiple-string LED drivers," *Solid-State Circuits Conference Digest of Technical Papers (ISSCC), 2013 IEEE International* ,vol., no., pp.376,377, 17-21 Feb. 2013.

978-1-4799-5128-4/14 $31.00 © 2014 IEEE

A Fully Integrated AC-DC Regulator Over Wide Frequency Range for Implantable Bio-Medical Devices

Qi Cheng[1,2], Liuyan Chen[1,2], and Jianping Guo[1,2]

1, School of Physics and Engineering, Sun Yat-sen University, Guangzhou, China
2, SYSU-CMU Shunde International Joint Research Institute, Foshan, China
E-mail: guojp3@mail.sysu.edu.cn

Abstract

A fully integrated AC-DC regulator with high power conversion efficiency (PCE) is proposed in this paper. It combines the traditional rectifier and regulator together, which reduces the total output capacitance and improves the PCE. The AC-DC regulator outputs a clean DC voltage and achieves a simulated PCE of 80% at 13.56 MHz under conditions of a 1.8-vpp AC input with a 500-Ω load resistor. The circuit was implemented with 0.18-μm CMOS technology and 0.6-mm^2 chip area was estimated.

Keywords-wireless power transfer (WPT) ; AC-DC regulator; power conversion efficiency (PCE) ; bio-medical ; rectifier

Introduction

With the development of reliable wireless power supplies, wireless power transfer (WPT) system has played the essential role in special applications such as IMDs (implantable microelectronic devices), RFID (radio frequency identification), NFC (near-field communication), etc. Commonly, they can transfer AC power to stable DC power by rectifier and low-dropout regulator (LDO).

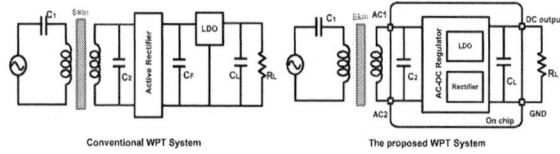

Fig. 1 Block diagrams of conventional and the proposed wireless power transfer (WPT) System.

The simplified blocks of conventional and the proposed WPT system are shown in Fig. 1. In conventional AC-DC regulator, a rectifier is adopted to convert AC power to unregulated DC power and a regulator is used to reduce output voltage ripple. A novel circuit called "rectigulator", which stands for the hybrid of rectifier and regulator, using two negative feedback channels in turn to convert AC power to DC power was verified by discrete components [1]. However, with MUXs which generate pulse controlling signal, the circuit suffers from digital noise like traditional DC-DC regulator. In this paper, by replacing MUXs with cross-couple structures, the proposed regulator not only combines the rectifier and regulator together but also fixes the noise problem. Moreover, the chip area can be saved as the output capacitance is much reduced.

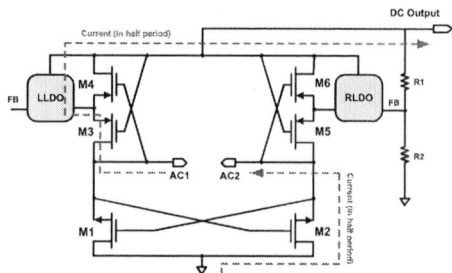

Fig. 2 Schematic of the proposed on chip AC-DC regulator

The Proposed AC-DC Regulator

In this paper, we present a power and chip-area efficient AC-DC regulator for robust wireless power transmission. As shown in Fig. 2, the regulator consists of 2 groups of symmetry part. The largest decoupling capacitor is shared by rectifier and regulator, which reduces the total chip area.

A. Rectifier

Similar to [1], the three cross-couple structures M1-M6 work as a rectifier which converts AC input to unregulated DC output. The pMOSFETs M3 and M5 conduct alternately to allow AC currents to flow from AC1 and AC2 to LDO input. The cross-couple structure consists of nMOSFETs M1 and M2 which passively switches in turn to allow the DC current to flow back from GND to AC1 or AC2. Being different from [1], the pass transistor of regulator has on-off state and the pMOSFETs M4 and M6 realize self-controlled switching.

The working flow of the regulator is described as follows. For example, in half period, when AC1 input is higher than output voltage, M3 will turn on to let current flow and LDO works to regulate output voltage. Meanwhile, M5 would turn off to avoid short current from output to ground. But when AC input is smaller than output, M3 would turn off and M4 turn on to close the main path and then the output storage capacitor discharges to maintain a stable voltage.

B. Low-dropout Regulator (LDO)

Owing to the simple form of rectifier, the voltage ripple from rectifier is large and thus we need a strong LDO for further DC regulation. The bottleneck of LDO design is to keep the system stability and improve power-supply rejection (PSR) at high frequency. Some feed-forward ripple cancellation approaches were proposed, but most of these LDOs need a big output capacitor which guarantees better PSR performance and stability [2]. Based on supply ripple subtraction, high-pass

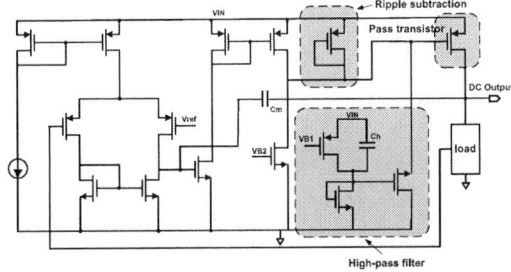

Fig. 3 Schematic of the proposed PSR enhanced LDO

filter [3] and Miller compensation technology, the LDO with high PSR and small capacitor for AC-DC regulator is proposed.

The transistor level of the proposed LDO circuit is shown in Fig. 3. It includes error amplifier and gain-boosting stages. In addition, supply ripple subtraction stage and high pass filter are used to feed-forward low and high frequency supply ripple respectively. A 1-pF Miller compensation capacitor is added to ensure system stability.

Experimental Results

The proposed AC-DC regulator was implemented with 0.18-μm CMOS technology. The estimated chip area is around 0.6 mm².

Simulation results are shown in Fig. 4 and Fig. 5. The worst-case PSR is –30dB at 13 MHz and the best-case PSR at low frequency is –102dB. Fig. 5 shows the main waveforms of the proposed AC-DC regulator.

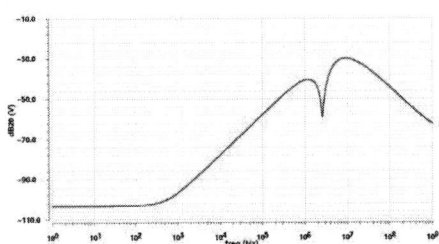

Fig. 4 Simulated PSR response of the proposed LDO

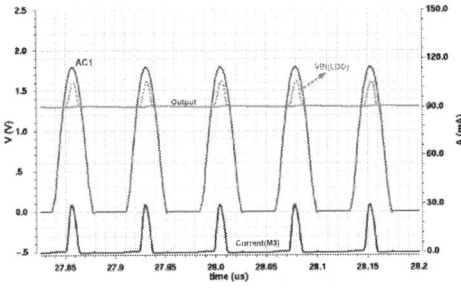

Fig. 5 Main waveforms of the proposed AC-DC regulator

Table I shows the performance comparison of previously published rectifiers and AC-DC regulators. The proposed AC-DC regulator can achieve high PCE (80%) with only 4-nF on-chip capacitance, and the frequency can be up to 2.4 GHz.

Conclusion

In this work, a fully integrated AC-DC regulator over wideband frequency is proposed. It combines rectifier and regulator together, which reduces the total capacitor and improves power efficiency. Simulation result shows that the regulator transfers 1.8-vpp AC input to 1.3V DC output with 18mV ripple voltage. The AC-DC regulator is suitable for wireless power applications.

Acknowledgment

This work was partly supported by a grant from National Natural Science Foundation of China under Project No. 61204035.

TABLE I
COMPARISON WITH PRECIOUSLY
RECTIFIER AND AC-DC REGULATOR

Parameters	2012 EL [1]	2012 T-CAS-II [4]	2012 ISSCC [5]	2013 ISSCC [6]	2014 This work
Technology	0.35μm	0.18μm	0.5μm	0.35μm	0.18μm
Structure	Rectigul-ator	Active Rectifier	Active Rectifier	Active Rectifier	AC-DC Regulator
Area(excluding pad)	N/A	0.009 mm²	0.585 mm²	1.42 mm²	0.6 mm²
Storage capacitor	N/A (off chip)	10μF (off chip)	1μF (off chip)	4nF (on chip)	4nF (on chip)
Input	5V	1.5V	2.2V	1.5V	1.8V
Output	2.37V@ R_L=500Ω	1.33V@ R_L=1KΩ	3.1V@ R_L=500Ω	1.27V@ R_L=500Ω	1.3V@ R_L=500Ω
Dropout	2.53V	~170mV	~1.3V	~230mV	500mV
Output ripple	2.4mV@ 125KHz	~100mV @13.56 MHz	~50mV @13.56 MHz	N/A	18mV@ 13.56 MHz 0.3mV@ 2.4GHz
PCE	N/A	82%	70%	81%	80%

References

[1] T. J. Sun, X. Xie, G. L. Li, and Z. H. Wang, "Rectigulator: a hybrid of rectifiers and regulators for miniature wirelessly powered bio-microsystems," Electronics Letters, vol. 48, no 19, 13th, September 2012.

[2] J. Guo, and K. N. Leung, "A 25mA CMOS LDO with -85dB PSRR at 2.5MHz," IEEE Asian Solid-State Circuits Conference, Singapore, pp. 381–384, November 2013.

[3] H. S. Jhuang, J. H. Wang, S. W. Lai and C. H. Tsai, "A high PSR over wideband frequency range low dropout voltage regulator," in Green Circuits and Systems, pp. 508–511, June 2010.

[4] H. K. Cha, W. T. Park, and M. Je, "A CMOS rectifier with a cross-coupled latched comparator for wireless power transfer in biomedical applications," IEEE Transactions on Circuit and Systems-II, vol. 59, no. 7, pp. 409–413, July 2012.

[5] H. Lee and M. Ghovanloo, "An adaptive reconfigurable active voltage doubler/rectifier for extended-range inductive power transmission," in IEEE International Solid-State Circuits Conference, pp. 286–288, February 2012.

[6] Y. Lu, X. Li, W. H. Ki, C. Y, Tsui, and C. P. Yue, "A 13.56MHz fully integrated 1X/2X active rectifier with compensated bias current for inductively powered devices," in IEEE International Solid-State Circuits Conference, pp. 66–68, February 2013.

Dynamic Performance Improvement of DC-DC Buck Converter by Slope Adjustable Triangular Wave Generator

Shu WU   Yasunori KOBORI   Haruo KOBAYASHI

Division of Electronics and Informatics, Gunma University

1-5-1 Tenjin-cho Kiryu Gunma 376-8515 Japan

Phone: 81-277-30-1789 fax: 81-277-30-1707 e-mail: t12802472@gunma-u.ac.jp

Abstract

This paper describes a simple-yet-effective control method for a DC-DC buck converter with voltage mode control (VMC). We propose to use a slope adjustable triangular wave generator (TWG) which provides a line feed-forward control and an additional duty cycle modulation effect. Then comparing with the conventional VMC, the proposed method significantly improves the line and load transient responses. This triangular wave slope regulation scheme is simple compared to digital feed-forward control scheme that requires non-linear calculation. Our SIMetrix Simulation results show the effectiveness of the proposed method.

Keywords- slope regulation; triangular wave generator; buck converter; voltage mode control; transient response

Introduction

Continuous advancement of signal processing technologies for integrated circuits has posed stringent challenges to the transient response performance of switching power supplies, especially load transient response. A lot of research indicate that wideband of the control loop can obtain fast load transient respone [1-3]. VMC is easy to design and analyze, but its bandwidth is limited by the applied op-amp. Current mode control (CMC) is easy to obtain wideband, but an accurate current sensor and a slope compensation are required. Elaborate feed-forward control schemes have been proposed. However, their main drawback is the complexity of non-linear calculation [4,5] or limited performance in large load current transient condition [6].

This paper proposes a simple control method for VMC buck converter. This approach applies an adjustable TWG, and the slope of this triangular wave is regulated based on the input and output voltages. Therefore, we obtain not only line feed-forward control, but also load transient response improvement. The proposed method is simple and does not require a current sensor and complicated calculation.

Slope Adjustabe Triangular Wave Generator

A. System Configuration and Circuit

The proposed system block diagram is shown in Fig. 1. Op-amp1 is used to generate an error signal, where Type 3 compensation is incorporated and G_c denotes the gain. Op-amp2 senses and amplifies the output deviation by G_k, and its output works as control variable for the TWG; TWG is a time-variant function of the voltage, and it is reset every period by CLK pulse signal. The slope is decided by the control variable and the input voltage.

Fig. 2 shows the principle concept of the proposed TWG, where two completely consistent NMOSFETs M_1 and M_2 in the triode region are used to configure a linear voltage

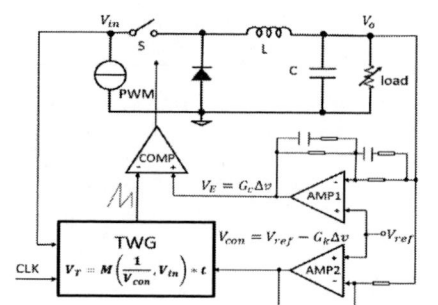

Fig. 1 DC-DC buck converter with a slope adjustable TWG.

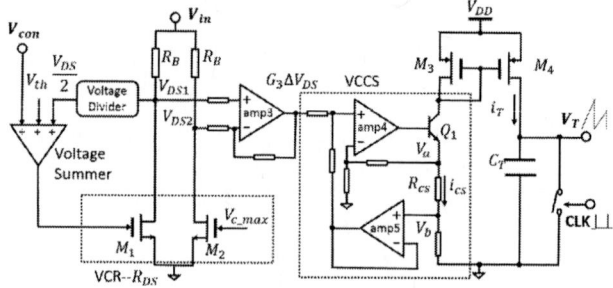

Fig. 2 Proposed triangular wave generator.

controlled resistor (VCR). The difference of two drain voltages is amplified and used to generate a proportional current signal by the voltage controlled current source (VCCS). Then this current signal is copied by the current mirror circuit. The capacitor C_T is charged by this current and reset by CLK signal. Therefore, the voltage across the capacitor is a triangular wave, which is given as

$$V_T = V_{in}\left(a\frac{1}{V_{con}} - b\right)t = M\left(\frac{1}{V_{con}}, V_{in}\right)t \qquad (1)$$

Where $a = \frac{G_3}{K_n R_B R_{CS} C_T}$, $b = \frac{a}{V_{c\_max}}$, $K_n = \mu_n C_{ox} W/L$, G_3 is the gain of amp3, V_{c_max} is the sum of the maximum control variable, V_{th2} and half of V_{DS2}. The slope is proportional to the input voltage and inversely proportional to the control variable (the output diviation).

B. Duty Cycle Modulation

Since the triangular wave slope is proportional to the input voltage, when the input voltage is changed, the peak value of the triangular wave is regulated. For the line transient response, the proposed method provides a line feed-forward control function.

For the load transient response of the VMC buck converter, the duty cycle modulation of the proposed method is shown in Fig. 3; the duty cycle variation is separated into two parts. $\Delta d_1 = \Delta t_{on1}/T_s$, which only depends on the slope variation and $\Delta d_2 = \Delta t_{on2}/T_s$ which is caused by the error signal and the slope.

Then the whole duty cycle modulation is obtained as:

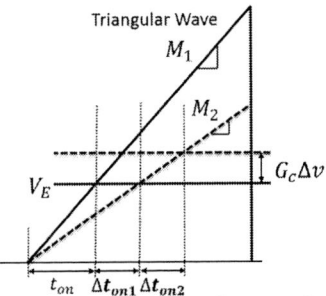

Fig. 3 Duty cycle modulation of proposed method.

$$\Delta d = \Delta d_1 + \Delta d_2 = \left(\frac{V_E}{T_s} + \frac{G_c \Delta v}{T_s}\right)\Delta \frac{1}{M} + \frac{G_c \Delta v}{T_s M_1} \quad (2)$$

Substituting the slope function of eq. (1) into eq. (2), and ignoring the second-order non-linear terms, eq. (2) can be approximated as

$$\Delta d \approx \frac{bV_E G_k \Delta v}{T_s V_{in}(a - bV_{ref})} + \frac{G_c \Delta v}{V_{p\_static}} \quad (3)$$

The first term is caused by the changed slope of the triangular wave, while the second term is just a conventional VMC. Therefore, with the same deviation in the output voltage, the proposed TWG provides faster and stronger regulation effect, which accelerates the transient response.

Simulation Results

Simulation conditions are shown in Table. 1. We have designed type 3 compensation for the conventional VMC buck converter to obtain 50kHz bandwidth and 40° phase margin.

TABLE. 1 Simulation Conditions

Buck Converter Parameter	$V_{in} = 5V$, $V_{out} = 3.5V$, $V_{p\_static} = 3V$ $L = 10\mu H$ (ESR = 10mΩ), $C = 50\mu F$(ESR = 10mΩ), $R = 35\Omega$ $f_{switch} = 1MHz$, $H = 1$
Triangular Wave Generator	$G_k = 10$, $G_3 = 100$ $K_n \approx 3$, $C_T = 50pF$ $R_B = 1k\Omega$, $R_{cs} = 100\Omega$

In our load transient response simulations with SIMetrix, the load current is changed stepwise between 100mA and 420mA. Simulation results of comparison with the conventional VMC buck converter are shown in Fig. 4. Thanks to the regulated triangular wave, the duty cycle is regulated as soon as the transient response happens. The additional effect provided by the proposed TWG prevents larger error signal development. Therefore, in step-up case, the under-shoot voltage is decreased from 14mV to 9mV, and the response time is shortened from 15µs to 6µs. In step-down case, the over-shoot voltage is decreased from 16mV to 10mV, and the response time is shortened from 20µs to 12µs.

Conclusion

This paper describes an adjustable triangular wave generator which can regulate the slope of triangular wave by the input voltage and the variation in the output voltage. The proposed method is very simple, and it does not need a current sensor and complicated calculation. Compared to the conventional VMC DC-DC buck converter, the proposed method improves both the line and load transient responses. For the line transient response, it works as a feed-forward controller to eliminate the change in input voltage. For the load transient response, the

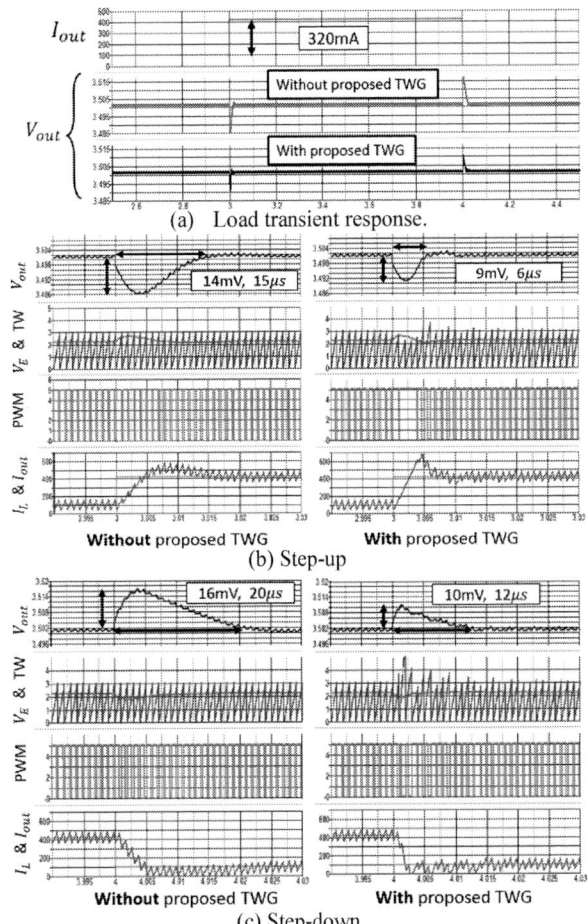

(a) Load transient response.

(b) Step-up

(c) Step-down

Fig. 4 Load transient responses (I_{out}: 100mA ↔ 420mA).

proposed method provides an additional control to the duty cycle, besides the effect of the error signal. This additional modulation effect accelerates transient response. We have verified the effectiveness of the proposed method with SIMetrix simulations.

References

[1] B. Arbetter and D. Maksimovic, "DC-DC converter with fast transient response and high efficiency for low-voltage microprocessor loads," Proc. IEEE APEC 1998, vol.1, pp.156-162, Anaheim, CA, Feb. 1998

[2] C. J. Mehas, K. D. Coonley, and C. R. Sullivan, "Converter and inductor design for fast-response microprocessor power delivery," Proc. IEEE PESC, vol.3, pp. 1621–1626, Galway, Ireland, Jun. 2000.

[3] J. Xu, X. Cao, and Q. Luo, "The effects of control techniques on the transient response of switching DC-DC converters," Proc. IEEE PEDS, vol.2, pp. 794–796, Hong Kong, China, Jul. 1999.

[4] V. Yousefzadeh, A. Babazadeh, B. Ramachandran, L. Pao, D. Maksimovic, and E. Alarcon, "Proximate time-optimal digital control for DC-DC converters," Proc. IEEE, PESC, pp. 124-130, Orlando, FL, Jun. 2007.

[5] S.Y. Chae, B.C. Hyun, W.S. Kim, and B.H. Cho, "Digital load current feed-forward control method for a DC-DC converter," Proc. IEEE APEC, pp. 498-502, Austin, TX, Feb. 2008.

[6] S. Wu, Y. Kobori, H. Kobayashi, et. al., "Design of a simple feed-forward controller for DC-DC buck converter," The 4th IEICE International Conference on Integrated Circuits Design and Verification, Ho Chi Minh City, Vietnam, Nov. 2013.

978-1-4799-5128-4/14 $31.00 © 2014 IEEE

Transfer Function Design for
Volume Rendering using K-Means Clustering

Jae Hwan Lim

Department of Creative IT Engineering
POSTECH
Pohang, Republic of Korea
thekage@postech.ac.kr

Young Hwan Kim

Division of Creative IT Engineering
POSTECH
Pohang, Republic of Korea
youngk@postech.ac.kr

Abstract

This paper presents a new transfer function design approach using boundary feature extraction in medical volume data. The proposed method applies K-Means clustering to group voxels of each similar region. Then, boundaries in volume data are extracted by cluster centroids. Opacity transfer function assigns opacity to the boundaries, and the opacity values are determined by gradient magnitude. This transfer function effectively shows boundaries which are important information in volume data. Also, this approach offers ease of use to users.

Keywords-volume rendering; transfer function design; K-Means clustering; boundary feature extraction

Introduction

Direct volume rendering is a volume visualization technique which shows 3D volume data effectively. This technique is significantly used in medical applications. A transfer function which controls optical properties has a major impact on the quality of volume rendering result. A transfer function assigns color and opacity to volume data, and it makes corresponding volume rendering result to the transfer function.

A variety of transfer function design approaches for medical image have been proposed for decades [4]. However, the approaches could not satisfy all requirements because of a characteristic of medical images. A medical image consists of 1D gray scale intensity value. Therefore, extracting features in a medical image is limited because of its insufficient information. Automatic low dimension transfer function design approaches [1, 2] could not satisfy different demands of each user, and high dimension transfer function design approaches [3, 6] gives difficulty in use because of a large number of complicated parameters. On the other hand, semi-automatic approaches [5] could not provide intuitive user interface to control optical properties.

In this paper, we propose a new transfer function design approach which emphasizes important features and hides unnecessary features efficiently.

Proposed Method

Adjacent two organs are separated by intensity values. Therefore, the boundary between two organs is detected at the edge of two different region values. For example, Fig. 1(a)

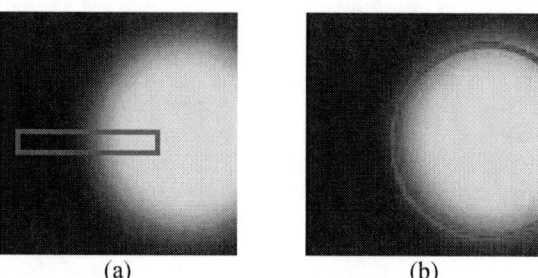

(a)   (b)

Fig. 1 (a) a circle image, (b) boundary of the circle

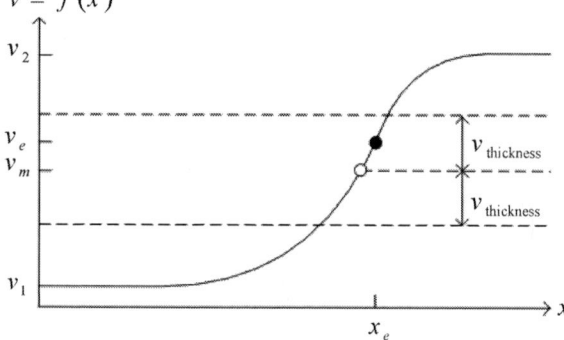

Fig. 2 Intensity value (v) vs. pixel location (x) of red rectangle in Fig. 1a (filled circle, edge point; clear circle, median point between v_1 and v_2)

shows two regions which are a white circle and black background and boundary of the circle will appear as shown in Fig. 1(b). A boundary between two regions is usually detected at the point of the largest intensity variation (x_e), as shown in Fig. 2 which shows the intensity value graph of the red rectangle in Fig. 1(a) over pixel locations. We can assume the intensity value of the edge pixel (v_e) will exist near the median value (v_m) of two intensity values (v_1 and v_2) of two regions. Therefore, a boundary between two regions can be approximately determined within the range of v_m-$v_{thickness}$ to v_m+$v_{thickness}$.

The proposed method consists of three steps. First, K-Means clustering is applied to volume data for grouping intensity values of regions. Second, boundaries of the volume data are determined using the clustered result. Finally, the proposed method generates opacity transfer function using gradient magnitude.

978-1-4799-5128-4/14 $31.00 © 2014 IEEE   48

A. Region separation using K-Means clustering

Different regions can be simply discriminated by intensity values because a medical image contains only 1D gray scale information. In case of Fig. 1(a), dominant intensity values of two regions are v_1 and v_2 in Fig. 2. For this reason, pixels which have similar intensity value are considered as the same region.

K-Means clustering is commonly used for grouping pixels in an image. In the proposed method, K-Means clustering is applied to medical volume data, and K cluster centroids are obtained. We can assume that each cluster centroid represents dominant intensity value of a region.

B. Boundary determination

A boundary was approximately determined within the range near the median value (v_m) of two region intensity values and we assumed v_m as center value of the boundary, as described above. In the previous step, we achieved K cluster centroids as intensity values of regions. K regions can form up to $K(K\text{-}1)/2$ boundaries in the volume, if all regions are adjacent each other. Therefore, $K(K\text{-}1)/2$ center values of all boundaries are defined as

$$c_{i,j}|_{i<j} = (v_i + v_j)/2, \ \text{if} \ i,j = 1,2,...,K \ , \quad (1)$$

where $c_{i,j}$ denotes center values, and v_i and v_j denote cluster centroids.

In this step, boundaries are determined with these boundary center values and a threshold value of boundary . The set of boundary voxel values is defined as

$$\mathbf{B} = \bigcup_{i<j} \{v \in \mathbf{V} \mid v_{\text{thickness}} \geq |v - c_{i,j}|\} \ , \quad (2)$$

where $\mathbf{V}$ denotes the set of all voxel intensity values, and $v_{\text{thickness}}$ denotes threshold value of boundary.

C. Gradient-based 2D transfer function design

In the previous step, we determined center values of boundaries and boundary range. To visualize volume data, opacity values must be assigned at the boundaries because important information should be emphasized. The voxels at boundaries usually have large gradient magnitude. Therefore, the opacity at the boundaries is assigned as

$$\alpha(v = f(x_i)) = \begin{cases} \beta \cdot |\nabla f(x_i)| / \max |\nabla f(x_i)|, & \text{if} \ v \in \mathbf{B} \\ 0 & , \text{otherwise} \end{cases} \quad (3)$$

where $\alpha(v)$ denotes opacity transfer function, x_i denotes voxel position in volume data, β denotes scaling factor, f denotes intensity value, and $|\nabla f(x_i)|$ denotes gradient magnitude. As shown in Eq. (3), the opacity transfer function assigns normalized gradient magnitude to the voxels in boundaries.

Results

We applied the proposed transfer function to volume rendering. We used MR head sample data of 3D Slicer for volume data set as shown in Fig. 3(a). The data set consists of 109 slices of 256×256 data. Clustering number K was set to 5, and boundary threshold $v_{\text{thickness}}$ was set to 5, and scaling factor β was set to 0.3 for the volume rendering result.

As shown in Fig. 3(b), we achieved volume rendering result

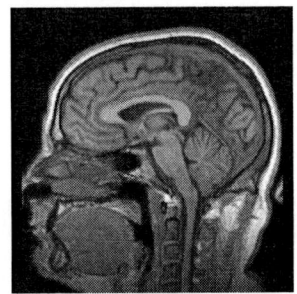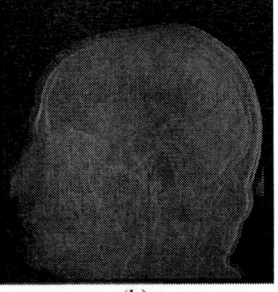

| (a) | (b) |

Fig. 3 An MR image volume rendering (a) original MR image, (b) volume rendering result using proposed transfer function

with the proposed transfer function. We could detect every boundary in the result.

Conclusion

In this paper, we proposed a new transfer function design approach for volume rendering. First, the proposed method grouped voxels in volume data using K-Means clustering to separate region. Then, it extracts boundaries using cluster centroids. Finally, it generates LUT of the opacity transfer function using gradient magnitude. As a result, the proposed method emphasized boundary features, and hided flat features effectively. Also, the proposed method can offer convenient user interface to modify the rendering result.

Acknowledgment

This research was supported by the MSIP(Ministry of Science, ICT and Future Planning), Korea, under the "IT Consilience Creative Program" (NIPA-2014-H0201-14-1001) supervised by the NIPA(National IT Industry Promotion Agency)

References

[1] C. L. Bajaj, V. Pascucci, and D. Schikore. "The contour spectrum," In *Proceedings of IEEE Visualization 1997*, pp. 167–174, 1997.

[2] M. Levoy, "Display of surfaces from volume data," *IEEE Comput. Graph. Appl.*, 8(3):29–37, 1988.

[3] J. Kniss, G. L. Kindlmann, and C. D. Hansen, "Multidimensional transfer functions for interactive volume rendering," *IEEE Trans. Vis. Comput. Graph.*, 8(3):270-285, 2002.

[4] H. Pfister, B. Lorensen, C. Bajaj, G. Kindlmann, W. Schroeder, L. S. Avila, K. Martin, R. Machiraju, and J. Lee, "The transfer function bake-off," *Computer Graphics and Applications, IEEE Comput. Graph. Appl.*, 21(3):16–22, 2001.

[5] Y. Wu and H. Qu, "Interactive transfer function design based on editing direct volume rendered images," *IEEE Trans. Vis. Comput. Graph.*, 13(5):1027–1040, 2007.

[6] H. Guo, W. Li, and X. Yuan, "Transfer function map," *IEEE Pacific Vis.*, pp.262-266, 2014.

Implementation of Pedestrian Detection using a CENTRIST-ROI in Embedded Environment

Yun-seop Hwang, Chang-min Jung, Tae-ryong Park, Kwang-yeob Lee

Department of Computer Engineering
SeoKyeong University
Seoul, Korea
vm3300@skuniv.ac.kr

Abstract

This paper proposes a pedestrian detection algorithm to which ROI was applied to implement pedestrian detection that is suitable for the embedded environment. Pedestrian detection has computations for unnecessary areas because the entire input images are computed to find pedestrians in the given images. In this paper, a pedestrian detection algorithm that is ideal for the embedded environment is proposed which reduced computations for unnecessary areas by applying ROI. The CENTIRST descriptor method was used for the pedestrian detection algorithm, which was implemented using 512x360 pixel images on an ALDEBARAN board. The proposed pedestrian detection with ROI showed a 16% improved performance of 3.6 frames per second compared to the conventional method.

Keywords-pedestrian detection; census transform histogram; feature; embedded; region of interest

Introduction

Pedestrian detection techniques for finding people in various images are applied to a wide variety of applications including driver assistance systems, smart traffic light, and security systems. However, it is very difficult to find people in various kinds of images at a high accuracy. Therefore, many algorithms for extracting the features of people have been developed to find pedestrians. Representative pedestrian detection algorithms include Haar like Feature [1], HOG (Histogram of Oriented Gradient) Feature [2], Part-based Model that improved HOG [3], and CENTRIST (Census Transform Histogram) descriptor [4]. In this paper, the CENTRIST descriptor method was used to find the features of people in images. Studies on pedestrian detection in the embedded environment as well as in the PC environment are being actively developed as its applications are increasing. However, due to many computations, it is difficult to process the pedestrian detection algorithm in real time in the embedded environment. In this paper, the ROI (Region of Interest) method was applied unlike the conventional methods that perform the pedestrian detection algorithm to the entire images. In this paper, a pedestrian detection algorithm that is ideal for the embedded environment is proposed which reduced computations for unnecessary areas by applying ROI.

Detecting Methods

A. Conventional method

The conventional pedestrian detection algorithm of the CENTRIST descriptor method uses a detect window and performs the pedestrian detection algorithm by moving in pixel units for the entire image as shown in Fig. 1 below.

Fig. 1 Conventional pedestrian detection algorithm

In this process, the algorithm is performed using the detect window even for the background where there is no pedestrian. As a result of performing the algorithm for unnecessary areas, there is the disadvantage of decreased performance.

B. Proposed method

Unlike the conventional pedestrian detection algorithm using the CENTRIST descriptor that performs pedestrian detection for the entire image, the method proposed in this paper reduced computations for the unnecessary areas by applying ROI. The proposed algorithm sets the ROI as shown in Fig. 2 and the pedestrian detection algorithm is performed by moving pixel by pixel for only the areas where there is a pedestrian.

Fig. 2 Pedestrian detection using the proposed method

978-1-4799-5128-4/14 $31.00 © 2014 IEEE

The ROI like the area 1 is set based on the pedestrian detected in the previous frame for which the pedestrian detection algorithm was performed for the entire image. In this way, because the pedestrian detection algorithm is performed only for the area 1, unnecessary computations for the background area excluding the area 1 can be reduced. However, as shown in Fig. 3, a pedestrian may not be detected if the pedestrian appears in area B outside area A to which the ROI was applied.

Fig. 3 A pedestrian appearing outside the ROI

To address this problem, the pedestrian detection algorithm was performed for the entire images for odd frames and the ROI method of the pedestrian detection algorithm was applied to the even frames only. In this way, the problem of no detection of pedestrian can be solved because even if a pedestrian appears outside the ROI, the algorithm is performed for the entire images for odd frames.

In other words, the proposed method combines the odd frames for which the pedestrian detection algorithm is performed for the entire images and the even frames for which unnecessary areas are not computed by applying ROI. An improved detection speed can be expected with the proposed method while maintaining the same accuracy as the conventional method.

Experiments

In this paper, the pedestrian detection algorithm [5] using the CENTRIST was implemented by applying the ROI method to make it more appropriate for the embedded environment. To implement the proposed algorithm, the ALDEBARAN board (300MHz) and input images of 512x360 pixels were used. Table 1 compares the performance and accuracy of the conventional method and the proposed method on the ALDEBARAN board. The conventional method that performs the pedestrian detection algorithm for the entire images had an accuracy of 63.8% and a performance of 3.1 frames per second. The proposed pedestrian detection method using ROI showed an accuracy of 63.8% and a performance of 3.6 frames per second. Therefore, the proposed method showed a performance improvement of about 16% in terms of speed while maintaining the same accuracy as the conventional method.

TABLE I

Comparison of conventional method and proposed method

	Image Size	Accuracy	Frame Per Second
Conventional method [5]	512x360	63.8%	3.1
Proposed method	512x360	63.8%	3.6

Conclusion

In this paper, a pedestrian detection algorithm using the CENTRIST descriptor was implemented by applying the ROI method to make it more appropriate for the embedded environment. Computations for unnecessary areas were reduced by applying the ROI method to the conventional pedestrian detection algorithm using the CENTRIST descriptor. It was implemented on an ALDEBARAN board which is an embedded board, and the proposed method showed a performance of 3.6 frames per second, which is about 16% improvement over the conventional method.

Recently, various pedestrian detection algorithms are being developed and better detection algorithms are researched. Like the method proposed in this paper that improved performance while maintaining accuracy by applying the ROI method to the conventional detection algorithm, studies on new algorithms that improve performance by applying various methods to conventional detection algorithms are expected. Furthermore, the enhanced pedestrian detection algorithms will be applied to more various fields by enabling real-time processing in the embedded environment as well as in the PC environment.

Acknowledgment

This research was supported by the Industrial Core Technology Development Program (10049192, Development of a smart automotive ADAS SW-SoC for a self-driving car) funded By the Ministry of Trade, industry & Energy

References

[1] Paul Viola, Michael Jones, "Rapid Object Detection using a Boosted Cascade of Simple Features," Proceedings of the IEEE Computer Society Conference on Computer Vision and Pattern Recognition, vol. 1, pp. 511-518, 2001.

[2] Navneet Dalal, Bill Triggs, "Histograms of Oriented Gradients for Human Detection," IEEE Computer Society Conference on Computer Vision and Pattern Recognition, vol. 1, pp. 886-893, 2005.

[3] Felzenszwalb, Pedro.F, Girshick, Ross.B, McAllester.D, Ramanan.D, "Object Detection with Discriminatively Trained Part Based Models",IEEE Trans, Pattern Analysis and Machine Intelligence, vol. 32, pp. 1627-1645, 2010.

[4] Jianxin Wu, James M.Rehg, "CENTRIST: A visual descriptor for scene categorization," IEEE Trans, Pattern Analysis and Machine Intelligence, TPAMI, vol. 33, pp. 1489-1501, 2011.

[5] Kwang-Yeob Lee, "Implementation of real-time pedestrian detection based on parallel processing with a dual core processor", International Journal of Control and Automation, vol.7, pp. 85-96, 2014.

Anisotropic Diffusion-based Denoising Using Residual image for Preservation of Image Details

GyuJin Bae[1], Sung In Cho[1], and Young Hwan Kim[1,2]

Department of Electrical Engineering[1], Department of Creative IT Engineering[2]
Pohang University of Science and Technology
Pohang, Korea
Contact e-mail: bgj2013@postech.ac.kr

Abstract—**This paper presents an anisotropic diffusion-based approach to noise reduction, which utilizes the residual image in the wavelet domain to improve the quality of image detail preservation. In the experimental results, the proposed method improves the quality of denoised images by increasing peak signal-to-noise ratio up to 1.17 dB and structural similarity index up to 0.026.**

Keywords— Image denoising, Noise reduction, Anisotropic diffusion, residual image

I. INTRODUCTION

Image noise is often generated during images acquisition, processing, and transmission. The image noise degrades performance of various processing applications such as image segmentation, video coding, and medical imaging. Image denoising is therefore a very important process for improving the image quality.

A variety of the image denoising methods have been proposed. They include anisotropic diffusion (AD) [1] and bilateral-filtering [2]. Among these methods, the AD model is widely used for the image denoising because of its good denoising performance with simple operations. Based on the AD model, diverse denoising approaches have been proposed [3-6]. However, the existing denoising methods do not effectively distinguish between weak image details and image noise, thereby they cannot preserve image details effectively during the noise reduction.

In this paper, we propose a novel AD-based denoising method that utilizes the residual images in wavelet-transformed domains to improve the quality of image detail preservation. This paper is organized as follows. Section II proposes the denoising framework; Section III shows experimental results, and Section IV concludes this paper.

II. PROPOSED METHOD

The proposed method consists of three steps: a noise reduction channel, a detail compensation channel, and a fusion of two channels as shown in Fig. 1. In the noise reduction channel, the typical AD is applied to eliminate the image noise. However, because the typical AD do not effectively distinguish between weak image details and the image noise, it induces the loss of image details during the noise reduction process. To compensate for the loss of image details, we extract image details from the residual image in the detail compensation channel. After extracting image details in the sub-channel, in the procedure of fusion of two

This research was supported by IDEC, and the MSIP(Ministry of Science, ICT and Future Planning), Korea, under the "IT Consilience Creative Program" (NIPA-2014-H0201-14-1001) supervised by the NIPA(National IT Industry Promotion Agency).

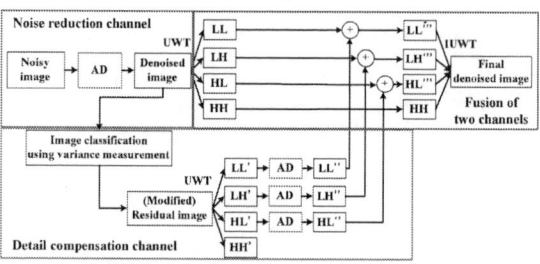

Fig. 1. Overview of the proposed method

channels, we add wavelet planes of the noise reduction channel and the detail compensation channel except the HH plane. Because the image noise is diagonal component, HH plane contains the greatest number of image noise among the wavelet planes. Hence, HH plane is excluded for compensation process. Finally, we perform the inverse wavelet transform to produce the denoised image.

The main contribution of the proposed method is the compensation of detail loss during the denoising by restoring image details from the residual images in wavelet planes. Actually, two kinds of compensation is used depending on the classification result of a given image. For the classification, we calculate the variance and classify a given image into the detail image and the smooth image. In our work, the detail image mean the image having many image details. In contrast, the smooth image means the image having a low image details. After the classification, appropriate denoising process is applied to a given noisy image depending on the classification result. For the detail image, we use the conventional residual image for compensation process. In contrast, for the smooth image, we use the modified residual images to prevent the restoration of image noise during the compensation process. The detail operations of the proposed method is described below.

A. Variance measurement

Before the variance measure, we first perform the typical AD denoising to eliminate the image noise, since the image noise increases the variance, thereby leading to the missclassification. Next, we perform the haar discrete wavelet transform (DWT) to extract image details. The DWT divides an image into four planes: LL, LH, HL, and HH. LL is a low-frequency plane. LH, HL, and HH are vertical, horizontal, and diagonal detail plane, respectively. Finally, we calculate the variance of the HH plane. Because the variance is proportional to image details, we classify a given image utializing the variance value. The threshold value for the classificaion is described in section III.

B. Detail compensation process for detail images

For the detail image, we perform the compensation process for detail images. The residual image (Res) is defined as:

$$Res = I_d - I, \qquad (1)$$

where I and I_d denote a gray level of input noisy image and the denoised image using the AD, respectively.

To extract the image details from the residual image, we transform the residual image into the haar wavelet domain. Instead of the DWT, we use the undecimated wavelet transform (UWT) to overcome the lack of translation-invariance property of the DWT [7]. The loss of the translation-invariance property leads to a large number of artifacts when an image is reconstructed after the modification of its wavelet coefficients. After transforming the wavelet domain, we perform the typical AD in each wavelet plane to remove the image noise and extract image details of each plane. Extracted image details is added with the I_d in the noise reduction channel to compensate the loss of image details which occurs in the I_d.

C. Detail compensation process for smooth images

For the smooth image, we perform the denoising framework for smooth images. For the smooth image, we modify the residual image to prevent the restoration of the image noise during the compensation process. For this, we suppress the image noise before the residual image calculation. To accomplish this, we produce I_d^{weak} by applying the AD to I utilizing weak smoothing strength compared to the typical AD which was used in the preivous step. Then, we extract the modified residual image (Res_m) from Eq. (1) by replacing the I with I_d^{weak} as follows:

$$Res_m = I_d - I_d^{weak}, \qquad (2)$$

where I_d^{weak} and I_d denote a gray level of weakly denoised image and the denoised image using the AD, respectively. By this, we can reduce the image noise components in the residual image. For the remaining frameworks, we apply the same method as denoising for the detail image except for the residual image.

III. EXPERIMENTAL RESULTS

We evaluated the image quality of the proposed approach using peak signal to noise ratio (PSNR) and the structural similarity index (SSIM). For the test image, we used the Kodak lossless true color image set, the ISO 12640-2 standard image set, and an IEC image set that was randomly captured from an IEC62087 video. The images were degraded using AWGN with 5% and 10% standard deviations (σ_ns).

For the benchmark methods, we used the typical AD (TAD) [1], Black's robust AD (BAD) [3], Chao's AD (CAD) [4], and Li's AD (LAD) method [5]. All of them are AD-based noise reduction method. In the comparisons, the parameters of each benchmark algorithms were adjusted to obtain the best PSNR and SSIM for each noise level. In the proposed method, the threshold value of the variance for classification was set 100 empirically.

As shown in Table I, the proposed method shows higher PSNR and SSIM values for each noise level, when compared

TABLE I. AVERAGE PSNRs [dB] AND SSIMs OF THE PROPOSED METHOD AND BENCHMARK METHODS

Test image (# of images)	σ_n	TAD		BAD		CAD		LAD		Proposed	
		PSNR	SSIM	PSNR	SSIM	PSNR	SSIM	PSNR	SSIM	PSNR	SSIM
Kodak (24)	5	31.18	0.902	30.92	0.901	30.78	0.905	31.35	0.906	31.45	0.907
	10	27.51	0.808	27.52	0.809	27.49	0.807	27.63	0.812	28.00	0.820
ISO (15)	5	32.63	0.911	32.33	0.910	32.31	0.915	32.72	0.912	32.93	0.915
	10	28.31	0.838	28.23	0.835	28.16	0.843	28.36	0.839	28.92	0.859
IEC (20)	5	33.42	0.926	33.28	0.926	33.10	0.933	33.25	0.928	34.13	0.936
	10	29.69	0.858	29.71	0.859	29.60	0.870	29.73	0.862	30.77	0.884

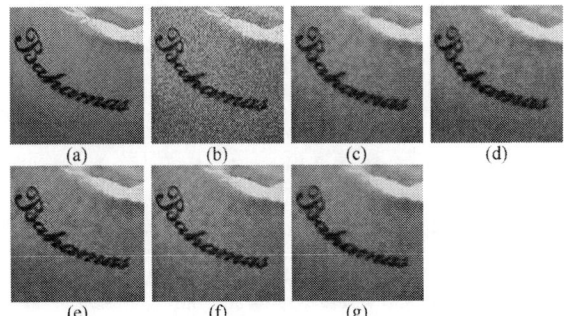

Fig. 2. Denoised results of benchmark and proposed methods for AWGN (σ_n=10%). (a) Original image (b) noisy image (c) image by TAD (d) image by BAD (e) image by CAD (f) image by LAD (g) image by the proposed method

to the benchmark algorithms. Fig. 2 showed that the proposed method suppressed the image noise effectively and preserved image details compared to the benchmark methods.

IV. CONCLUSTION

In this paper, we proposed a novel AD-based noise reduction that compensates the loss of image details using residual image. In the proposed method, AD is applied for a given noisy image to suppress the image noise. Then, residual image is decomposed into three sub-bands and AD is applied to extract image details. Finally, the loss of image details occurs during the AD-based denoising is compensated by adding the decomposed residual images in three sub-bands to its respective sub-band of the denoised image. As a result, the proposed method shows the best PSNR and SSIM compared to the benchmark methods.

REFERENCES

[1] P. Perona, J. Malik, "Scale-space and edge detection using anisotropic diffusion", *IEEE Trans. Pattern Anal. Mach. Intell.* Vol. 12, pp. 629-639, 1990.

[2] C. Tomas, R. Manduchi, "Bilateral filtering for gray and color images," *in Proc. Int. Conf. Computer Vis. Pattern Recognit.,* pp. 839-846, 1998.

[3] M.J. Black, G. Sapiro, D.H. Marimont, D. Heeger, "Robust anisotropic diffusion," *IEEE Trans. Image processing*, vol. 7, pp. 421-432, 1998.

[4] S.M. Chao, D.M. Tsai, "An improved anisotropic diffusion model for detail-and edge-preserving smoothing," *Pattern Recogn. Lett.* vol. 31, pp. 2012-2023, 2010.

[5] H.C. Li, P.Z. Fan, M.K. Khan, "Context-adaptive anisotropic diffusion for image denoising," *Electron. Lett.* vol. 48, pp. 827-829, 2012.

[6] S.I. Cho, S.J. Kang, H.S. Kim, Y. H. Kim, "Dictionary-based anisotropic diffusion for noise reduction," *Pattern Recogn. Lett.* vol. 46, pp. 36-45, 2014.

[7] J.L. Starck, J. Fadili, F. Murtagh, "The undecimated wavelet decomposition and its reconstruction," IEEE Transaction on Image Processing, vol. 16, pp. 297-309, 2007.

Propagation Delay Detection based on I-Q Modulation for Indoor Positioning System

Soo-Yong Kim, Andrey Zobenko, and Chaehag Yi
Samsung Electronics Co., Ltd.
Hwaseong, Korea
Email: odin.kim@samsung.com

Abstract—**An indoor positioning system has become a hot issue to provide a location-based service such as direction tracking, geofencing and private tracking. This paper addresses a phase-detection algorithm based on the Wi-Fi for proximity estimation between a transmitter and a receiver. In addition, the mathematical analysis of the phase-detection is shown based on the I-Q modulation. The algorithm has been designed with a combination of round-trip time measurement and quadrature modulation technique, that can have a positioning accuracy of 1m. Moreover, the quadrature modulation utilizes low frequencies, so the power consumption can be reduced. The accurate performance of the proposed estimation was verified by means of simulation results.**

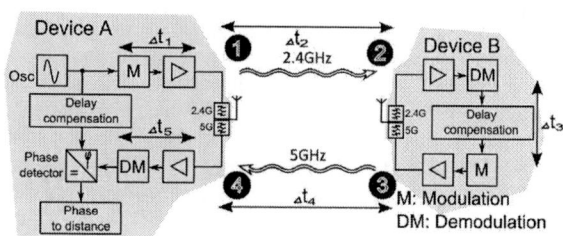

Fig. 1.   Overall system architecture.

I. INTRODUCTION

An indoor positioning system (IPS) has become a hot issue to provide location-based service (LBS) such as direction tracking, geofencing and asset tracking. The conventional IPSes consist of an inertial navigation system (INS) with MEMS sensors, fingerprinting methods based on received-signal-strength (RSS), and time-of-arrival (TOA) with access points (APs). To provide effective and efficient LBSes, such as geofencing and private tracking in crowded shopping centers, the IPSes have to supply information with localization accuracy under a few meters. However, due to severe multipath and the complexities of radio propagation in the indoor environment, it is difficult to implement these LBSes in mobile devices with low power consumption and a low cost. Practically, to improve the accuracy of TOA estimation between the receivers and the transmitters, it is necessary to measure very short timing intervals (tens of nanoseconds). Since this requires the use of high frequency for measurement of the propagation delay, it increases the power consumption by the high frequency and development complexity to secure timing margin.

This paper provides essential information about overall architecture and mathematical background of proximity estimation on the basis of the phase with the quadrature modulation technique.

II. ARCHITECTURE FOR PROXIMITY ESTIMATION

Most mobile smart-devices equipped with Wi-Fi follows the IEEE 802.11 specification in order to communicate the free APs. Especially, for implementation of features of IEEE 802.11n/ac, all up-to-date RF front-end ICs have double bands:

2.4 GHz and 5 GHz. In addition, the bandwidth of the Wi-Fi systems is from 20 MHz to 160 MHz.

Using these structures for the IEEE 802.11n/ac, a new principle for positioning estimation has been presented, as shown in Fig. 1 [1]. A sine generator in device A makes the carrier frequency of 2.4 GHz [2]. Subsequently, this carrier is fed into a quadrature modulator, together with the periodic signal having frequency 1~20 MHz. After RF modulation and amplification, the signals are transmitted to Device B through the radio channel. Device B has to demodulate the received signals and re-modulate an identical harmonic frequency, but this time with a 5 GHz carrier.

From the transmitter's point of view, the processing delay (Δt_3) of the receiver is unknown but approximately constant, as shown in Fig. 1. Therefore, Device B can be compensate for the delay via the calibration. That is, Device B modify signal to remove effects of Δt_3. In addition, the repeater has to calculate phase of received signal, then subtract from its value phase delay signal use for passing repeater. Then quadrature modulator synthesize the envelop signal with compensated phase.

III. PHASE DETECTION WITH I-Q MODULATION

A structure of the proposed algorithm is presented to measure the phase-shift by the round-trip of the signal, as shown in Fig. 2 [1]. The original signal is modulated by Device A is transmitted from the antenna and is represented as

$$S_{Tx}(t) = A_0 e^{-2\pi f_s t} \tag{1}$$

where A_0, f_s, and t are amplitude, frequency of the enveloped signal, and time, respectively.

978-1-4799-5128-4/14 $31.00 © 2014 IEEE   54

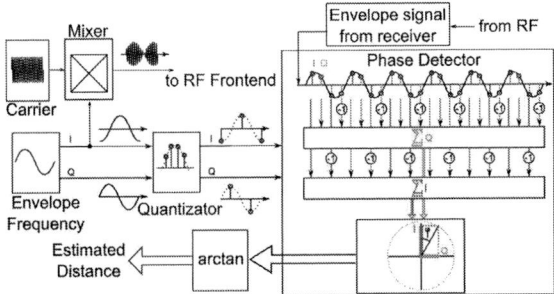

Fig. 2. Implementation to measure the phases of the signal.

Simultaneously, the envelope signal is stored through two quantizers of each I-Q branch. Subsequently, the signal sent is sent back by Device B with another carrier frequency. After demodulation of the re-transmitted signal, a phase detector estimates the distance between each device in comparison with the original and the returned signal. The received signal is expressed as

$$S_{Rx}(t) = A_0 e^{-2\pi f_s t + \varphi_d} \qquad (2)$$

where φ_d is a shifted phase by the round-trip.

The I and Q components of the received signals can be represented as

$$S_I(t) = \int_0^T S_{Rx}(t) cos(2\pi f_s t) dt = \frac{A_0}{2} T cos(\varphi_d) \qquad (3)$$

$$S_Q(t) = \int_0^T S_{Rx}(t) sin(2\pi f_s t) dt = \frac{A_0}{2} T sin(\varphi_d) \qquad (4)$$

where T is a period of the enveloped signal.

The practical receiver is a digital system which uses the A/D converted values. In the discrete domain equation (3) and (4) are expressed as

$$\hat{S}_I[n] = \frac{1}{N} \sum_{k=1}^{N} \hat{S}_{Rx}[n] \hat{S}_{cos}[n] \qquad (5)$$

$$\hat{S}_Q[n] = \frac{1}{N} \sum_{k=1}^{N} \hat{S}_{Rx}[n] \hat{S}_{sin}[n] \qquad (6)$$

where N is the number of samples during a period. Therefore, the phase by the propagation delay can be represented as

$$\varphi_d = tan^{-1}(\frac{\hat{S}_Q[n]}{\hat{S}_I[n]}). \qquad (7)$$

If the measurement distance is long, the phase will become larger in comparison with the short proximity. However, if the estimated phase exceeds 360 degrees, the frequency of the envelope signal should be decreasd to remove the ambiguity by the overlap.

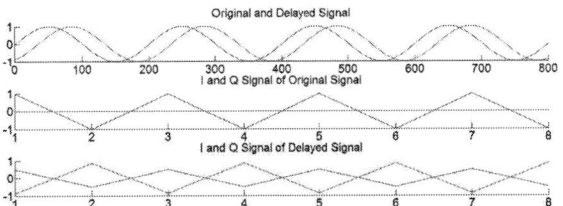

Fig. 3. I-Q Modulation. The blue lines are origanl signals and the red lines are delayed signals.

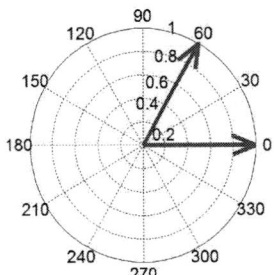

Fig. 4. I-Q Diagram.

The operation principle of the phase detector is based on a matched filter, and the distance is calculated from the angle between vectors of transmitted and received signals in I-Q surface.

IV. SIMULATION

The conditions for the simulation are an envelop signal of 5MHz and 100 times down sampling. In Fig. 3, results of the I-Q modulation are shown according to the kinds of the signal. Since the propagation delay by round-trip is 60 degrees from Fig. 4, the estimated distance becomes 4.9965 meters. However, because the true value is 5 meters, the measurement error is 0.0035 meters.

V. CONCLUSION

In this paper, we showed overall architecture and mathematical analysis of the phase detection algorithm for the indoor positioning system. Based on IEEE 802.11 standards, the proposed architecture consists of sine signal generator, an envelope detector, a phase detector, and a mixer for the frequency down conversion. The measurable distance and the accuracy of the system depend on the envelope frequency. In addition, the proposed algorithm is demonstrated by the mathematical analysis which exhibited available accuracy for measuring a few meters.

REFERENCES

[1] A. Zobenko, T. Scherrer and Soo-Yong Kim, "Proximity estimation for location-based services with round-trip time," *Electronics Letters*, Vol. 50, No. 14, pp. 1029-1031, 2014.
[2] Soo-Yong Kim *et al*, "System Identification Using Embedded Dynamic Signal Analyzer', *Jpn. J. Appl. Phys.*, Vol. 51, No. 8, pp. 08JB02 1-6, 2012.

A 10-bit 20-MS/s Dual-Channel Algorithmic ADC With Improved Clocking Scheme

Joo-Won Oh, Yong-Sik Kwak, and Gil-Cho Ahn

Dept. of Electronic Engineering
Sogang University
35 Baekbeom-ro, Mapo-Gu, Seoul 121-742, Korea
gcahn@sogang.ac.kr

Abstract

A 10-bit 20-MS/s dual-channel algorithmic analog-to-digital converter (ADC) using an improved clocking scheme is presented. The proposed ADC employs amplifier sharing technique with a conversion time scaling to reduce area and power. To achieve further improvement of conversion time scaling, dedicated MDAC sampling capacitors scaled with the accuracy requirement of each cycle are used. The ADC implemented in a 0.18μm CMOS process achieves 59.6dB SFDR and 54.3dB SNDR while consuming 8.96 mW per channel from a 1.8-V supply voltage.

Keywords - Analog-to-digital converter (ADC), algorithmic ADC, op-amp sharing, conversion time scaling, capacitor scaling

Introduction

Well known advantages of algorithmic analog-to-digital converters (ADCs) are low power and small area. However, the cyclic conversion process of algorithmic ADCs inherently reduces the sampling rate of the converters. In a conventional algorithmic ADC with two multiplying digital-to-analog converters (MDACs), the digital output of each cycle is resolved with a constant clock period. Therefore, the total conversion time is determined with overall resolution of the ADC and sub-quantizer resolution. To increase the sampling rate of the algorithmic ADC, a conversion time scaling technique has been proposed [1]. By scaling the conversion time with the accuracy requirement of each conversion cycle, it reduces the overall conversion time. In this paper, we propose an improved clocking scheme that speeds-up the conversion rate of the algorithmic ADC. The proposed ADC also employs op-amp sharing technique to achieve further reduction of power and area.

Architecture

The block diagram of the proposed 10-bit 20-MS/s algorithmic ADC is shown in Fig. 1 with a corresponding timing diagram. It consists of four stages followed by a 2-bit flash ADC, a digital correction logic (DCL), and a clock generator. These stages share one op-amp with an improved clocking scheme to reduce power consumption and area. Each stage resolves 4-3-3-2-2 bits with one bit overlap for the digital correction.

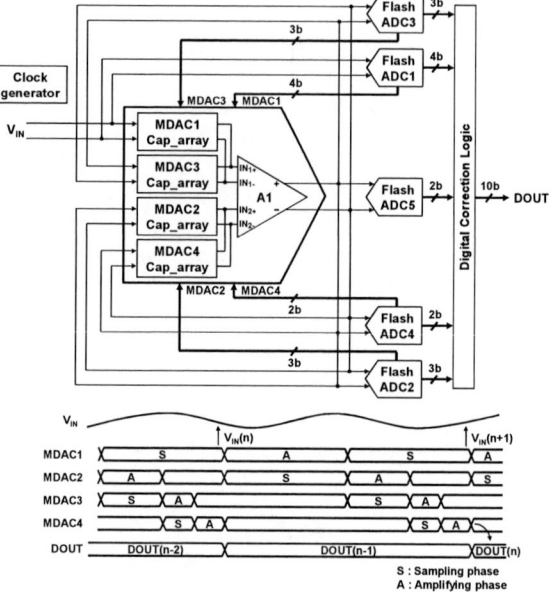

Fig. 1 Block diagram of the proposed algorithmic ADC.

Based on settling time approximation, the required transconductance of the op-amp in a pipelined stage is estimated as (1)

$$\beta g_m > (C_L / T_S) \times n \times \ln 2 \qquad (1)$$

where β is the feedback factor of MDAC, C_L is the total loading capacitance, n is the required resolution of the back-end ADC, and T_S is the available settling time [2]. In a conventional algorithmic ADC, the noise and resolution requirements of the front-end cycles are more stringent than those of the back-end cycles. Therefore, the design specification of the op-amp is determined by the front-end cycle, and thus, the settling accuracies of the back-end cycles are more than necessary if they share the same op-amp with the front-end cycle. To improve the sampling rate, conversion time scaling technique is employed as proposed in [1]. To achieve further improvement of sampling rate, dedicated MDACs sampling capacitors which are scaled with the requirements of noise and matching accuracy are used. With the proposed clocking scheme, half of the sampling clock period is assigned for MDAC1 amplification phase, and one

978-1-4799-5128-4/14 $31.00 © 2014 IEEE

fourth of the clock period is assigned for MDAC2 amplification phase. Similarly, 1/8 of the clock period is assinged for MDAC3 and MDAC4 operations. The overall conversion time of the proposed algorithminc ADC is reduced by half with the proposed clocking scheme.

Circuit Implementation

In the conventional op-amp sharing topology, the current MDAC output is affected by the previous residue since the input node of the amplifier is never reset, resulting in linearity degradation of the overall converter. To avoid these memory effect and improve the linearity, a folded-cascode amplifier using dual input ports is employed as shown in Fig. 2 [3].

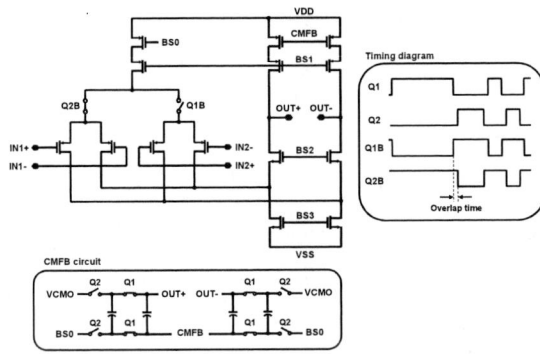

Fig. 2 Folded-cascode amplifier using dual input pairs.

The gates of inactive input pair are connected to fixed bias voltage to remove residual charge. Two pairs of input ports alternatively operate by slightly overlapped clock phase. With this clock scheme, all the transistors are always turned on so that the internal node voltage shows lower overshoot and faster settling time than the non-overlapped clock phases.

Measurement Results

A prototype 10-bit 20-MS/s dual-channel algorithmic ADC is fabricated in a 0.18μm CMOS process, and occupies 1.8mm² active die area. Fig. 3 shows the layout and die photograph of the prototype ADC. The measured differential and integral nonlinearities (DNL and INL) are 0.82 and 1.69 LSB as shown in Fig. 4. Fig. 5 shows the measured power spectrum of the prototype ADC with a 1-MHz sine wave input. The measurement results indicate a signal-to-noise and distortion ratio (SNDR) of 54.3dB, a spurious-free dynamic range (SFDR) of 59.6dB. The entire ADC dissipates 17.92 mW (8.96 mW per channel) at 1.8V supply voltage. The measured performance is compared with previous works in Table I.

Acknowledgment

This research was supported by the MSIP (Ministry of Science, ICT & Future Planning), Korea, under the University ITRC (Information Technology Research Center) support program (NIPA-2014-H0301-14-1007) and C-ITRC (Convergence Information Technology Research Center) support program (NIPA-2014-H0401-14-1002) supervised by the NIPA (National IT Industry Promotion Agency) and the IDEC of KAIST, Korea.

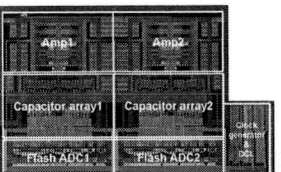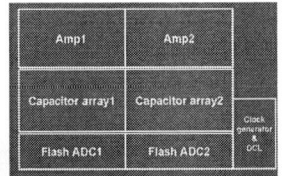

Fig. 3 Layout (left), chip die photograph (right).

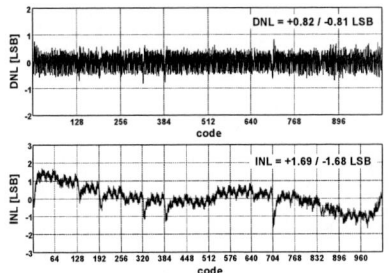

Fig. 4 Measured DNL/INL

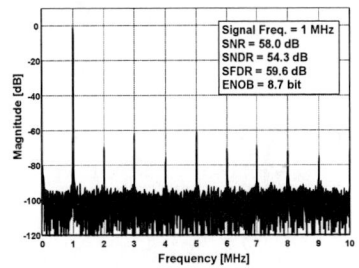

Fig. 5 Output FFT spectrum.

TABLE I
PERFORMANCE COMPARISON

	[1]	[2]	This work
Resolution (bits)	11	10	10
Conversion rate (MS/s)	10	100	20
Process (nm)	130	90	180
Supply voltage (V)	3	1	1.8
DNL, INL (LSB)	0.90 / 3.50	0.81 / 1.00	0.82 / 1.69
SNDR (dB)	56.0	55.0	54.3
Power (mW)	10.5	4.5	8.96
Size (mm²)	0.19	0.06	1.8
FoM (pJ/conversion-step)	2.08	0.10	1.08

References

[1] M. Kim, P. Hanumolu, and U. Moon, "A 10 MS/s 11-bit 0.19 mm² Algorithmic ADC With Improved Clocking Scheme", *IEEE J. Solid-State Circuits*, vol. 44, No. 9, pp. 2348-2355, Sep 2009.

[2] Y. Huang, and T. Lee, "A 10-bit 100-MS/s 4.5-mW Pipelined ADC With a Time-Sharing Technique", *IEEE Trans. Circuits Syst. I*, vol. 58, No. 6, pp. 1157-1166, Jun 2011.

[3] K. Lee, S. Lee, Y. Kim, K. Kim, and S. Lee, "Ten-bit 100MS/s 24.2mW 0.8mm² 0.18 μm CMOS pipeline ADC based on maximal circuit sharing schemes", *Electron. Lett.*, vol. 45, no. 25, pp. 1296-1297, Dec 2009.

978-1-4799-5128-4/14 $31.00 © 2014 IEEE

A 12-bit 750KS/s 69dB-SNDR 0.48mW Dual-Sampling SAR ADC with Reduced C-DAC for Wireless Charging Receiver

Hamed Abbasizadeh, Behnam Samadpoor Rikan, Ji-Hun Kang, Hyung-Gu Park and Kang-Yoon Lee

College of Information and Communication Engineering
Sungkyunkwan University
Suwon, Republic of Korea
{hamed, behnam, zpisup01, cbmass85, klee}@skku.edu

Abstract

This paper presents a 12-bit 750KS/s Successive Approximation Register Analog-to-Digital Converter (SAR ADC) for wireless portable device. The scheme of the ADC is based on a Dual-Sampling Capacitive DAC technique, low power dynamic latch comparator with Adaptive Power Control (APC) and bootstrap switch to reduce chip area and power consumption. The proposed 12-bit dual sampling CDAC topology reduces switching energy-efficient compared with 12-bit conventional SAR ADC. The prototype SAR ADC was implemented in Dongbu HiTek 0.18μm CMOS technology and occupies 0.68 mm². The post-layout simulation results show the proposed ADC achieves an ENOB of 11.196 bit at a sampling frequency 750KS/s. It consumes only 0.48mW from a 5.0V supply and achieves the INL and DNL +1.45/-0.65 LSB and +1.0/-1.0 LSB respectively, SNDR 69.16dB, SFDR 78.18dB, and figure of merit (FoM) of 273 fJ/conversion-step.

Keywords- SAR ADC; comparator; bootstrapped switch; low power; dual sampling; energy-efficient

Introduction

Nowadays, low power is one of the most relevant design concerns for energy-limited applications, such as wireless portable devices and mobile applications. Since ADCs are the key blocks of these systems, it is essential to improve the energy efficiency of the ADC to extend the system's life-span. Successive approximation register (SAR) ADC has found wide application for its moderate resolution, speed, simple structure and high energy efficiency [1,2].

This paper presents a power-efficient SAR ADC that combines several techniques to achieve low power design and speed requirements. First, dual-sampling method is used to reduces the switching energy and reduces total capacitance and also decrease the layout area. Second, bootstrapped switch are used to enhance the sampling linearity at high input frequency. Finally, the dynamic comparator is adopted to make the power consumption of the comparator scale proportional to the comparison rate.

The proposed SAR ADC architecture

Fig. 1 shows the proposed dual-sampling SAR ADC architecture which consists of a binary-weighted capacitor DAC (CDAC), a comparator, a SAR control logic, a bootstrap switch and switches, a Input buffer, a reference voltage generator and a RC-oscillator.

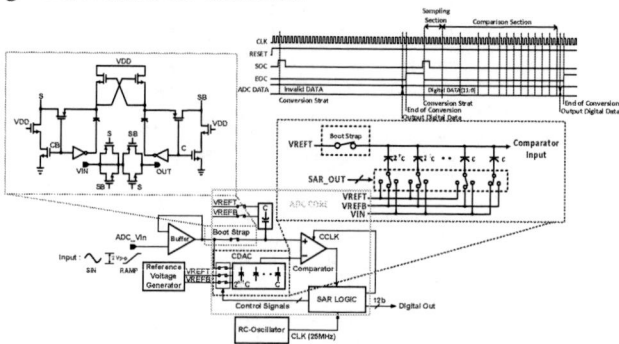

Fig. 1. Block diagram of the proposed SAR ADC

In the proposed SAR architecture, analog signals are sampled and held asymmetrically at each input side of comparator; thus it is called the dual-sampling method. This operation makes the MSB to be calculated without consuming any switching energy. In a SAR algorithm, operation is composed of sample mode, hold mode, and redistribution mode. In the sample stage, CDAC samples the analog input value. In the hold stage, one side of all capacitors is connected to ground and all capacitors hold analog input. And in redistribution stage, successive approximation process starts. One bit is determined at each cycle and n-cycle are needed for the n-bit digital code conversion [1].

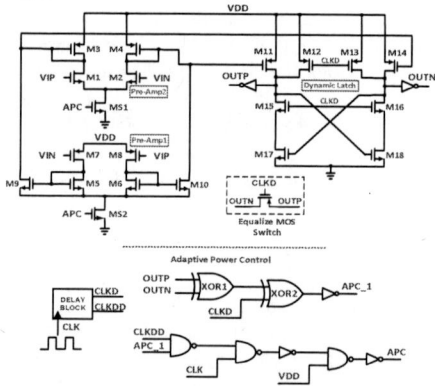

Fig. 2. Fully dynamic comparator with APC circuit

Fig 2 shows the schematic of the comparator. Low power dynamic latch comparator with APC circuit is used to increase

978-1-4799-5128-4/14 $31.00 © 2014 IEEE

the voltage gain and get the fine resolution. Fig. 3 shows the schematic of the RC-oscillator to generate a clock frequency of 25MHz for SAR logic.

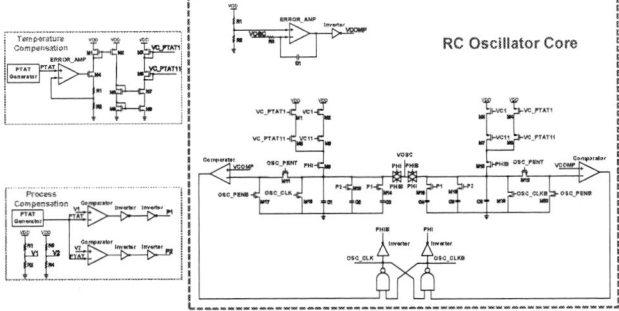

Fig. 3. RC-oscillator with PTAT generator schematic

Two buffers are used to generate the VREFT (3.5V) and VREFB (1.5V) voltages and also drive the large binary-weighted capacitor array in DAC. The schematic of the reference voltage generator is depicted in Fig. 4. The output current of the buffer should be large enough to session the fast settling time and conversion speed.

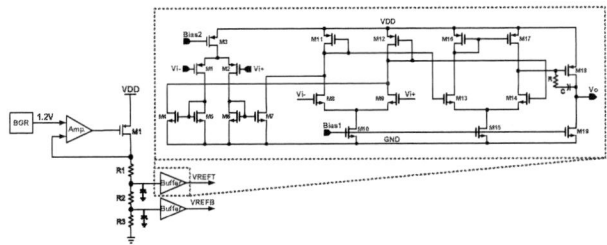

Fig. 4. Reference voltage generator with buffer schematic

Post-layout simulation results and Conclusion

The ADC is implemented in 0.18 μm 5V CMOS process. The active area of the ADC is 0.6888 mm^2, as shown in Fig. 5.

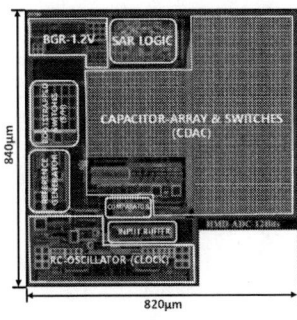

Fig. 5. Layout of the proposed ADC

The static performance is characterized through integral nonlinearity (INL) and differential nonlinearity (DNL) measurement. The measured INL and DNL are +1.45/-0.65 LSB and +1.0/-1.0 LSB (Sigma DNL: 0.50 LSB and Sigma INL: 0.44 LSB) respectively, as shown in Fig. 6.

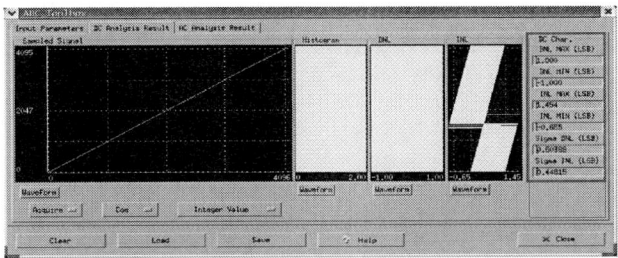

Fig. 6. Measured INL and DNL error

The 8192-point FFT spectrum of the proposed ADC is shown in Fig. 7, for a sinusoidal wave input at 90.2 Hz, sampling frequency of 750KS/s and clock frequency of 25MS/s. The SNDR is 69.163dB which equals to ENOB of 11.196 bit and the SFDR is 78.184dB. The typical figure of merit (FoM) definition of the ADCs is defined as:

$$FOM = \frac{Power}{2^{ENOB} \times f_s}$$

The FoM value calculated as 273 fJ/conversion-step by the equation.

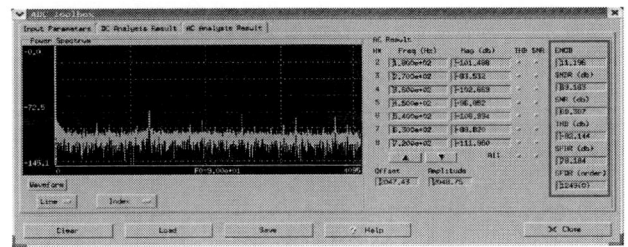

Fig. 7. FFT spectrum of the proposed ADC at 750KS/s

Table I shows summarizes the performance of the ADC.

TABLE I. Performance Summary and comparison

	This work	[2]
Technology (μm-CMOS)	0.18	0.35
Supply (V)	5	3.3
Input range (V_{P-P})	2	3.3
Sampling frequency (KS/s)	750	99
Clock frequency (MHz)	25	--
Resolution (bit)	12	12
THD (dB)/SNDR (dB)	-82.14/69.16	--/68.32
SFDR (dB)/ENOB (bit)	78.18/11.196	79/11.05
FOM(fJ/conversion-step)	273	49
Area (mm^2)	0.6888	0.7159

Acknowledgment

This research was supported by Basic Science Research Program through the National Research Foundation of Korea (NRF) funded by the Ministry of Education (NRF-2013R1A1A2010114) and also by IDEC (IPC, EDA Tool, MPW).

References

[1] Binhee Kim, Long Yan, Jerald Yoo, Namjun Cho, and Hoi-Jun Yoo, *An Energy-Efficient Dual Sampling SAR ADC with Reduced Capacitive DAC*, Circuits and Systems, Taipei, pp. 345-357, May 2009.

[2] J. Kim, S. I. Lim, K. S. Yoon, S. Lee, *Design of a 12-b Asynchronous SAR CMOS ADC*, ISOCC 2011, pp. 70-72.

Digital Calibration Algorithm for Half-Unary Current-Steering DAC for Linearity Improvement

Shaiful Nizam MOHYAR[1,2]   Haruo KOBAYASHI[1]

[1] Division of Electronics & Informatics, Faculty of Science & Technology, Gunma University,
1-5-1 Tenjin-cho, Kiryu-shi, Gunma, 376-8515 Japan
nizammohyar@unimap.edu.my, k_haruo@el.gunma-u.ac.jp
phone: 81-277-30-1789   fax: 81-277-30-1707
[2] School of Microelectronics Eng., Universiti Malaysia Perlis (UniMAP) Pauh Putra Campus, Arau, Perlis, 02600 Malaysia

Abstract

This paper introduces an algorithm called 3-stage current sorting (3S-CS) in half-unary weighted current cells to improve the linearity of a current-steering digital-to-analog converter (DAC). Based our statistical analysis and simulation results, the proposed algorithm improves the DAC static linearity as well as its dynamic performance.

Keywords- current-steering DAC, current source mismatch, linearity, current sorting, half-unary weighted

Introduction

The random error among current sources is one of fundamental problems in current-steering DACs. Its effect is deteriorating both static and dynamic DAC performance. Reduction of the current source mismatch effects is becoming important. Most of previous works have shown that the effect of these errors can be reduced by proper mapping or/and calibration techniques [1]. In this paper, we propose a half-unary current steering DAC with a 3S-CS algorithm as calibration technique to deal with this problem. Our MATLAB simulation shows that our proposed algorithm reduces both integral and differential non-linearity errors, and as a result, better spurious free dynamic range (SFDR) is achieved. The comparison has been done with a conventional unary, a unary and half-unary with 2-stage current sorting (2S-CS) in case that the static current source mismatches are considered.

DAC architecture

A unary weighted architecture of the current-steering DAC is introduced to overcome the binary-weighted disadvantages of large glitch energy and non-monotonicity. However, since the unary weighted structure has 2^N-1 unit current sources with identical weight for N-bit resolution, it suffers from low sampling speed and consumes large silicon area due to its decoding circuit. However, to increases intrinsic accuracy of this DAC architecture, we propose here another architecture called a half-unary weighted of the current steering DAC which is doubled in number of current source cells with halved weight compared to the conventional unary. This architecture offers better in term of glitch energy and redundancy than the conventional unary structure. However, since its number of current sources is double, the consumed silicon area and complexity in terms of decoder and routing is remained. Therefore, by manipulating its advantages and taking care of its weakness, we assume the combined architecture, but for

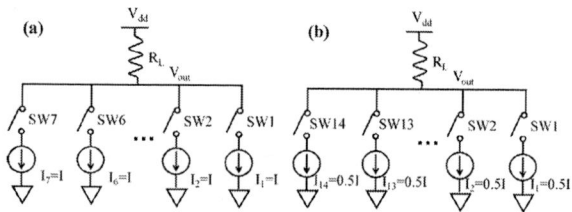

Fig. 1 A 3-bit DAC architecture. (a) Unary. (b) Proposed half-unary.

SFDR improvement, the segmented part for the higher bits which more important and hence our discussion here focuses on the half-unary structure.

DAC nonlinearity

A. Current source mismatch

In actual CMOS technologies, the current source mismatches are influenced either by the threshold matching, ΔV_{th} or by the slope mismatch, $\frac{\Delta \beta}{\beta}$. Since both are dependent on devices size ($\sqrt{WL}$), it is possible to alleviate them by changing the device size. However, without proper matching due to their limited variations, they will increase the device capacitances, which require more biasing current and thus power. So, we consider to reduce these effects after fabrication process.

Proposed algorithm

A. 3-stages current sorting

Current sorting algorithms have been used in several published works in order to reduce the effect of random static errors [2, 3]. These algorithms intend to optimize the switching selection [2]. In [3], this kind of algorithm is used to form binary weighted current cell from a group of unary current cells. In our work, we have merged these two methods, called as a 3-stage current sorting (3S-CS) algorithm. It is divided into two main stages: combine and rearrange. Fig. 2 shows an example of 3-bit of half-unary weighted current cells. Firstly, the 14 available current source cells (doubled number of conventional unary weighted structures) will be sorted by their current values. Then, by associating the smallest and the largest values (in this case, I_3 and I_7) to form a virtual unary weighted cell during conversion. Followed by the second smaller, I_{14} is combined with the second larger value, I_{10}. Next, the third weighted cells have their own pairs. The second and third stage, smaller, I_{11} and the third larger, I_4 and so on until all half-unary sequence determination. This kind of procedure is benefited for reducing the integral nonlinearity (INL) and differential

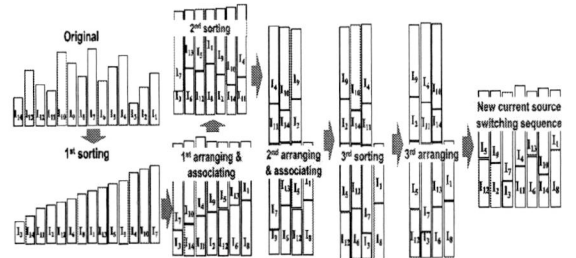

Fig. 2 Current source mismatch due to ratio of width and length of MOSFET transistor

nonlinearity (DNL) errors. Thus, a better linearity performance can be achieved.

B. Calibration

The calibration with error measurement approach is chosen to meet with our proposed techniques. Through foreground calibration mode, the overall calibration flow is shown as in Fig. 3 (left). A ring oscillator based measurement circuit located in the center of the current source is used to perform current measurement (Fig. 3 (right)). However, the accuracy of this measurement circuit becomes a trade-off with the overall area and DAC performance itself. Since, this current measurement circuit is fully digitalized implementation, it can be small size and low power consumption. After measurements, all measured values are stored in memory which later will be fed to digital calibration circuit to perform current cell sorting and switching rearrangement. The optimized switching sequence will be stored in memory and used during conversion. A look-up table (LUT) based decoder is used to decode binary code to the address of memory that contains the switch code. So, the switches will be configured as loaded codes. This process continues for all digital input. In order to reduce routing complexity and parasitic effects, a clock tree like (equal routing distance) layout has been taken place (Fig. 3 (right)). As a result, the overall dynamic performance of the proposed DAC is also improved. Fig. 3 (right) also illustrates floor plan of our proposed DAC architecture.

Results

Half-unary current-steering DACs with 8 to 12-bit has been simulated using Matlab where only static current source mismatches are considered and glitch effects are not. Fig. 4(a) and Fig. 4(b) show the static characteristics of INL and DNL which obtain better integral and differential linearity after calibration. Fig. 4(c) and Fig. 4(d) show that the SFDR performance obtains more than 15 dB while second and third harmonic distortions are significantly suppressed at condition

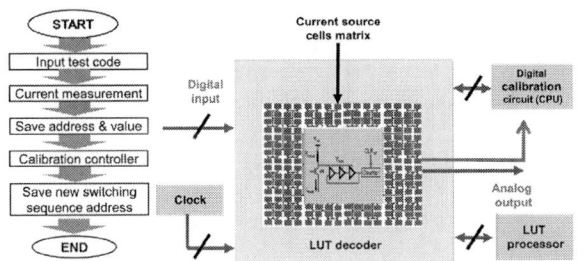

Fig. 3 Calibration procedures and floor plan of the proposed DAC.

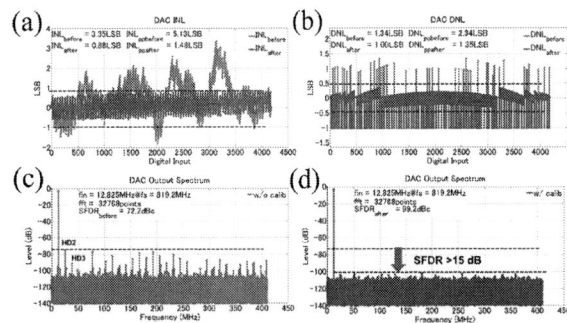

Fig. 4 A 12 bit DAC. (a) INL. (b) DNL. (c) Output spectrum before calibration. (d) Output spectrum after calibration

TABLE I
SUMMARY OF 12 BIT DAC PERFORMANCE

Architecture		Unary		This work	
Switching scheme		TC	2S-CS	2S-CS	3S-CS
Static	INL	3.99	1.06	1.01	0.93
	DNL	1.34	1.34	1.13	1.13
Dynamic	SFDR, dBc	66.6	85.8	86.3	91.6
	HD2, dB	-71.1	-88.6	-88.2	-94.6
	HD3, dB	-74.6	-96.1	-94.7	-102.0

of frequency input 12.8 MHz and sampling frequency of 819.2MHz with 0.075A of standard deviation error after calibration. For comparison, the conventional unary with thermometer-coded (TC) switching scheme, unary and half-unary with 2S-CS are also simulated. All simulation results in the same conditions are summarized in Table I. Both INL and DNL are improved compared to conventional unary and unary with 2S-CS. While, 3S-CS shows further INL improvement compared to 2-stages. The dynamic performance also improved using the proposed architecture.

Conclusion

We have proposed a 3-stage current sorting algorithm for a current steering DAC linearity improvement as well as its dynamic performance. By implementing this technique with MATLAB simulation, we have obtained SFDR of more than 15 dB compared to conventional unary with thermometer-coded switching scheme. The third of the harmonic distortion is also successfully suppressed.

Our proposed technique is simple and can be digitally implemented. The accuracy measurement circuit in this work is the key of the overall DAC performance.

Acknowledge

We would like to thank Semiconductor Technology Academic Research Center (STARC) for their valuable support on this project.

References

[1] Y. Cong and R. Geiger, "A 1.5-V 14-bit 100-MS/s Self-calibrated DAC," *IEEE J. Solid-State Circuits*, vol. 38, no. 12, pp. 2051–2060, Dec. 2003.

[2] T. Chen and G. Gielen, "A 14-bit 200-MHz current-steering DAC with switching-sequence post-adjustment calibration," *IEEE J. Solid-State Circuits*, vol. 42, no. 11, pp. 2386–2394, Nov. 2007.

[3] T. Zeng, and D. Chen, "An order-statistic based matching strategy for circuit component in data converters," *IEEE Trans. Circuit Syst. I, Reg. Papers*, vol. 60, no. 1, pp. 11–24, Jan. 2013.

Design of a low power 4th Order ΣΔ Modulator with the reused opamps

Su hun Yang, Jeong Hoon Choi, Kwang sub Yoon

Analog Circuit Design Lab, Inha Univ.
100 inharo Nam-gu Incheon 402-751, Korea
chunjasuhun@hanmail.net

Abstract

This paper describes a low power 4th order ΣΔ modulator for an implantable chip to acquire bio signals such as EEG(Electroencephalogram) or DBS(Deep Brain Stimulation) and EMG. To reduce a power consumption of the proposed modulator, only two opamps are employed for the four integrators with the KT/C noise reduction circuit. A test chip was fabricated in a 0.18um CMOS n-well 1 poly 6 metal process. The chip core area occupies 900um x 800um, and its power is 900uW with a 1.8V supply voltage. Measurement results show 90dB of SNDR and 96dB of DR. We achieve 14.8bit at the input frequency and clock frequency of 1kHz and 256kHz, respectively. FOMs are 164dB(FOM1) and 12.7pJ/step(FOM2).

Keywords- bio signal, ΣΔmodulator, implantable, low power

I. Introduction

Low power, low cost and multichannel are generally required by the bio signal processing circuits for EEG (Electroencephalogram) or DBS(Deep Brain Stimulation) and EMG(Electromyogram). These low power bio signal processing circuits employ a ΣΔ modulator with the resolution of 10-14 bits and signal bandwidth of 0.1Hz to 1kHz. The high resolution of the bio-signal processing circuits requires an order of modulator, so it may increase a number of opamps and power consumption. Reduction of the power supply voltage under 1V has been practiced in the literature [1],[2] However, these design techniques required the special switches and opamps. Another design technique practiced [3],[4] to reduce power dissipation is to reuse opamps within the integrator. This design technique suffers from lowering the resolution.

In this paper we propose a 4th order ΣΔ modulator with two clock signals to reuse two opamps with a KT/C noise reduction circuit to decrease the KT/C noise of the 1st stage of modulator.

This paper is organized as follows. In section II, the design technique of 4th order modulator and KT/C noise reduction circuit is described. The measurement results are discussed in section III. Conclusion is drawn in section.

II. The proposed modulator with KT/C noise reduction circuit

The proposed overall architecture is presented in Fig. 1. The simple feedback architecture with a single bit quantizer [5] is employed to avoid a complicated circuit structure, so it prevents the proposed circuit from degradation of the dynamic performance.

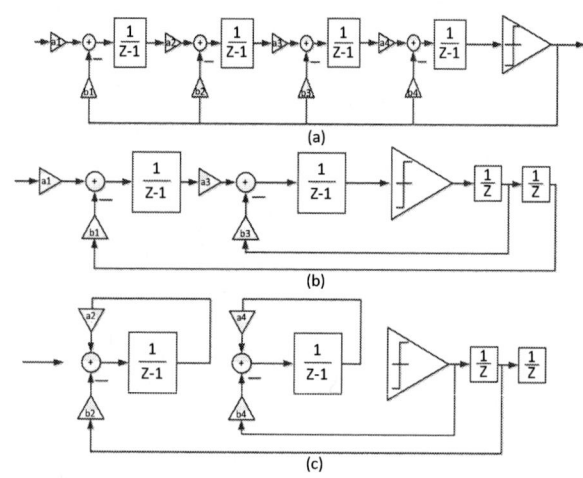

Fig. 1 (a) conventional feedback topology modulator (b) proposed modulator topology phase 1 (c) proposed modulator topology phase 2

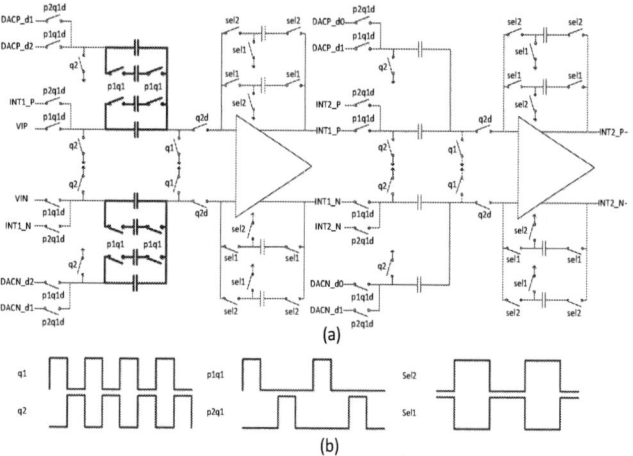

Fig. 2 (a) Circuit diagram of the proposed modulator and (b) two clock signals to control opamps to be reused.

Two clock signals to control the architecture enable the proposed modulator to reuse two opamps for 4th order modulator. The first opamp is reused within the 1st and the 2nd integrator stage. The second opamp is employed as the 3rd and the 4th integrator stage. Proposed circuit copy the signal flows of conventional 4th order modulator. Our circuit operates in the two phases. Each phase, circuit reconstructed as fig.1 (b) and fig.1 (c).

We apply a KT/C noise reduction circuit to the 1st stage

978-1-4799-5128-4/14 $31.00 © 2014 IEEE

sampling capacitor by the scheme of fig.3. In a conventional sampling sequence, total KT/C noise is sum of sampling time noise and hold time noise that is to be 2KT/C [6]. If the capacitor can be multiplied by 'n' when the sampling time, then total noise is calculated as Eq.1. If the 'n' is large enough, then the total noise will be a half of the noise of conventional circuit

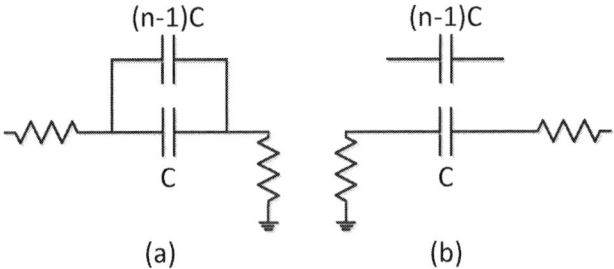

(a)   (b)

Fig. 3 scheme of KT/C noise reduction circuit in sampling time (a) and hold time (b)

III. Simulation results

Two kind of modulators are simulated by using a standard 0.18-μm CMOS process.One is the circuit with KT/C noise reduction circuit and the other one is not. Fig.4 shows a layout of implanted chip.the active acrea is 900um x 800um. The chip is tested with sine wave at 10 to 1kHz and operating clock at 512kHz. Fig.5 (a), (b) shows the measured 32k-points FFT output spectrum of each circuit. The modulator with KT//C noise reducrtion circuit achive 90dB of SNDR and 96dB of DR, the modulator without KT/C noise reducrtion circuit achive 85dB of SNDR at 256kHz sampling frequency. We can enhance the SNDR 5dB by using the KT/C noise reducrtion circuit.

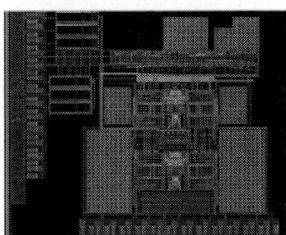

Fig.4 layout of chip

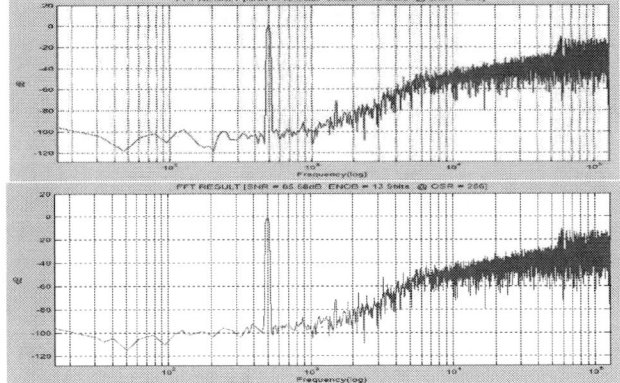

Fig.4 output FFT spectrum with KT/C noise reduction circuit (a) without KT/C noise reduction circuit (b)

IV. Conclusion

In this paper we propose the 4th order ΣΔ modulator with

KT/C nois reduction circuit for the implantable EEG and DBS applications. We achieve 90dB of SNDR and 96dB of DR at 256kHz sampling frequency whit using only two amps. SNDR is enhanced by KT/C noise reduction circuit about 5bB. The circuit consume 900 μW at the 1.8V power supply voltage and it results FOM of 164dB(FOM1) and 12.7pJ/step(FOM2). The chip is fabricated by 0.18-μm 1poly 6metal CMOS process

TABLE I
PERFOMANCE COMPARISON

Acknowledgment

	BW [kHz]	SNDR [dB]	VDD [V]	Power [uW]	FOM1 [dB]	FOM2 [fJ/step]	Process [um]
[7]	20	88.7	3	5600	161.5	5200	0.18
[1]	20	81	0.6	34	167	78.6	0.13
[2]	20	81	0.5	35.2	172	79	0.13
[3]	100	84	1.5	140	176.5	54	0.18
This	1	91.3	1.8	720	164.3	12700	0.18

"Fabrication of this chip is suppored from IDEC"
"This research was supported by the MSIP(Ministry of Science, ICT and Future Planning), Korea, under the C-ITRC(Convergence Information Technology Research Center)support program(NIPA-2014-H0401-14-1003) supervised by the NIPA(National IT Industry Promotion Agency)"

References

[1] H.D. Roh, H. J. Kim, Y. K. Choi, J. J. Roh, Y. G. Kim, J. K. Kwon, "A 0.6-V Delta?Sigma Modulator With Subthreshold-Leakage Suppression Switches" Circuits and Systems II, vol. 56, pp. 825-829, 2010

[2] Zhenglin Yang, Libin Yao, Yong Lian,"A 0.5-V 35-μW 85-dB DR Double-Sampled ΔΣ Modulator for Audio Applications" JSSC, vol.47, pp. 722-735, 2012

[3] Bonizzoni, E. ; Perez, A.P. ; Maloberti, F. ; Garcia-Andrade, M. "Third-order ΣΔ modulator with 61-dB SNR and 6-MHz bandwidth consuming 6 mW", analg Integr.Circuits Signal Process, vol. 66, no. 3, pp. 381-388, Sep. 2010

[4] Pena-Perez, A, Bonizzoni, E, Maloberti. F. "A 88-dB DR, 84-dB SNDR Very Low-Power Single Op-Amp Third-Order ΣΔ Modulator" JSSC, vol. 47, pp. 2107-2118,

[5] R. Schreier, G. C. Temes. "Understanding Delta-Sigma Data Converters" New-York, Wiley-IEEE Press 2005

[6] P. J. Quinn, Arthur H.M.Van Roermund "Switched-Capacitor Techniques for High-Accuracy Filter and ADC design" Dordrecht, Springer 2007

[7] Y.K. Choi, J.J. Roh, H. D. Roh, H. S. Nam, S. J. Lee, "A 99-dB DR Fourth-Order Delta?Sigma Modulator for 20-kHz Bandwidth Sensor Applications",Instrumentation and Measurement , vol. 58, pp. 2264-2274, 2009

Design of a Full-Swing CMOS Current Steering D/A Converter with an Adaptive High-Impedance Current Cell

Dongjoo Kim*, Woongtaek Lim*, Jongyoon Hwang*, Sangil Han**, and Minkyu Song*

*Dept. of Semiconductor Science, Dongguk University, 3-26, Phil-dong, Choong-ku, Seoul, KOREA
** Exicon Co. Ltd., 255, PanKyo-Ro, Seongnam, KOREA
mksong@dongguk.edu

Abstract

This paper describes a 6-bit full-swing CMOS current-steering digital-to-analog converter (D/A converter, DAC). Conventionally, current-steering DAC does not have a full swing analog signal, because they have an inevitable voltage drop with the output current cell. To overcome the drawbacks, a new scheme of adaptive output current cell composed of both nMOS and pMOS is proposed. Using a pMOS current cell and a nMOS current cell, the final output voltage is driven by a multiplexer which selects the optimized current cell. Both of the current cells are added at the multiplexer adaptively. A 6-bit current steering DAC has been fabricated with 0.11 μm CMOS technology to verify the performance of the full-swing DAC. From the measured results, we obtain that the DAC has a full-swing output signal.

Keywords— full-swing, current steering D/A converter, adaptive current cell

Introduction

The current trends of system-on-chip (SOC) are towards for integrating digital circuits and analog circuits in a chip. As a result, a data converter, which is part of a vital interface within those systems, is becoming an increasingly more important block. A digital-to-analog converter (DAC) is a representative circuit that a digital code is converted into an analog signal [1]-[5]. Normally, the kinds of DAC are mainly divided into two categories: one is voltage-steering type, and the other is current-steering type [1]. Since the settling time of the output voltage depends on the slew rate of the op-amp, the voltage-steering type based on the op-amp for the DAC output is not suitable for high-speed applications. In case of the current-steering type, by the way, the current generally flows directly through off-chip resistors or termination resistors inside the chip to obtain a fast operating speed. However, the output voltage at the termination resistor cannot have a full-swing, since the inevitable voltage drop is generated between the drain and the source of the output current cell. Therefore, it is almost impossible to have a full-swing output voltage at the current-steering type, and it degrades the performance of SNR, increases the linearity errors, and so on.

In this paper, a 6-b 1GSPs CMOS current-steering DAC with full-swing output voltage from the ground voltage to the power supply is proposed. The proposed DAC has an architecture that follows the thermometer code method which has excellent monotonicity and low glitch energy. In addition, the conventional logic and latch circuits have been simplified in order to reduce the power consumption and to correct mismatches. Further, the circuit uses a switching decoder and separates the power supply to prevent interference of each signal. In order to make the full-swing output voltage, the nMOS operates from the power supply to the half of the power supply, and the pMOS concurrently and independently operates from the half of the power supply to the ground voltage. After that, the final DAC output is obtained through a multiplexer which adds both signals.

Circuit Design

Fig. 1 shows the overall structure of the DAC. The 6-bit digital input codes pass through the input data sync in order to set delay time, and then the signals pass through rows & columns again. Here the binary digital code is converted into thermometer code to control switches, which are connected to the current source of the pMOS and nMOS. . Then, they pass through the proposed switching logic, giga latch, and level shifter. The output current produces an analog voltage through the termination resistors. Finally, each signal is added, and the final analog voltage is obtained. The proposed current-steering 6-bit DAC is designed with a full matrix structure for high-speed operation and accuracy. There are many advantages in such a design, including the reduction of INL, DNL error, accurate monotonicity, and low noise analog output.

Fig. 2 shows the circuit diagram of the digital switching logic [2] [3]. The switching decoder logic is designed by the Binary Decision Diagram (BDD) method. Since the BDD method uses only nMOS, the output voltage cannot have a full-swing. In order to solve this problem, it is necessary to add a level restoration circuit, Differential Cascode Voltage Switching Logic (DCVSL). The DCVSL replaces the deglitch circuit and can control the crossing point of the output pair. The proposed switching cell logic reduces the power consumption by half when compared to existing logic. Moreover, it is possible to operate at a high speed, and it is smaller than previous logic because it reduces the number of the transistors. Further, in the case of the pMOS, the crossing point is low, and in case of the nMOS, it is high. As a result, the glitch energy decreases. After the end of binary to thermometer decoder, the digital codes are holding at the giga latch. Then, the digital signals of 1.8V are shifted into 3.3V at the level shifter. In order to have the full-swing output analog signal at the current cell, a quaternary driver is proposed at the final digital block.

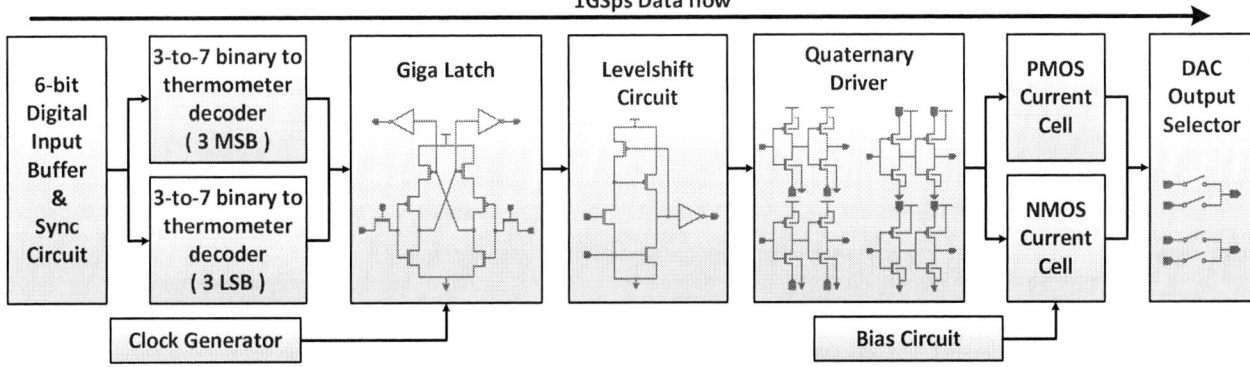

Fig. 1. Block Diagram of the Full swing DAC

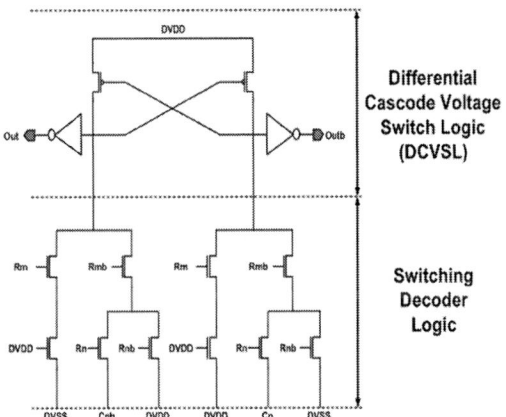

Fig. 2. Circuit Diagram of the Digital Switching Logic

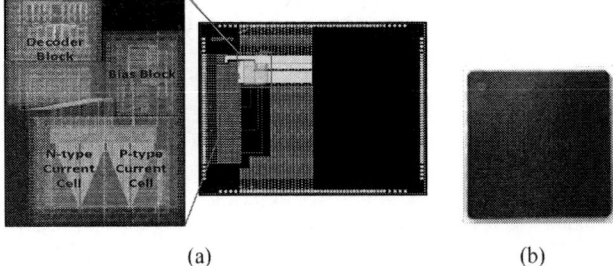

(a)　　　　　　　　(b)

Fig. 3. (a) Layout implementation (b) chip photograph

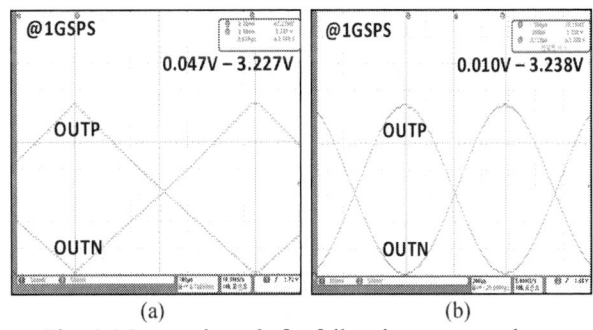

(a)　　　　　　　　(b)

Fig. 4. Measured resuls for full-swing output voltage
(a) ramp signal (b) sine waveform

Measured Results

The proposed DAC has been fabricated with Dongbu 0.11 μm 1P6M CMOS technology. Fig. 3 shows the chip layout and photograph of the fabricated DAC. For the accurate data synchronization, the clock signal uses a layout method of tree structure which is a helpful routing technique to equalize the parasitic resistance and the delay time. Excluding the input and output pads of the DAC, the area of the core is 0.46 mm^2, and the power consumption is about 19.1mW with 1.2 V for the digital and 3.3 V for the analog. Fig.4 shows the measured results of the full-swing output voltage. The measured ramp signal and sine waveforms are shown in Fig. 4(a) and (b), respectively. Thus it is verified that the proposed DAC has a full-swing output voltage.

Conclusions

With the proposed techniques, we obtained a full-swing output voltage of the CMOS current steering type DAC. The area of the designed chip was 0.46 mm^2, and the static and the dynamic performance were measured on the PCB. The measured INL/DNL were ±0.7 LSB / ±0.7 LSB, the SFDR was 37.89 dB, and the power consumption was 19.1mW, when the clock frequency was 1GS/s.

Acknowledgment

This work was supported by the MOTIE(Ministry of Trade, Industry, and Energy) Korea, under the NGPEP(New Growth Power Equipment Project) program supervised by the KEIT(Korea Evaluation Institute of Industrial Technology).

References

[1] Georgi I. Radulov et al., "An on-chip self-calibration method for current mismatch in D/A Converters," ESSCIRC., pp. 169-172, Sept. 2005.

[2] Q. Huang et al., "A 200MS/s 14b 97mW DAC in 0.18um CMOS." ISSCC Dig. Tech. Papers, pp. 364-365, Feb., 2004.

[3] J. Hyde et al., "A 300-MS/s 14-bit Digital-to-Analog Converter in Logic CMOS,"IEEE J. Solid-State Circuits, vol. 38, no. 5, pp.734-740, May, 2003.

[4] Jussi Prikkalaniemi, et al., "A14b 40 MSPS DAC with current mode deglitcher," Proc. IEEE ISCAC 2002, vol. 1, 2002. pp. 121-124.

[5] Park, K., et al., "A digital-to-analog converter based on differential-quad switching", Solid-State Circuits, IEEE Journal of, Volume 37, Issue : 10, pp 1335-1338, Oct. 2002.

978-1-4799-5128-4/14 $31.00 © 2014 IEEE

Energy-Optimal Algorithm for Dynamic Voltage Scaling with Non-Convex Power Functions

Juyeon Kim
juyeon@snucad.snu.ac.kr

Taewhan Kim
tkim@snucad.snu.ac.kr

School of Electrical and Computer Engineering, Seoul National University, Seoul, Korea

Abstract—Over the last two decades, it has been widely accepted that dynamic voltage/frequency scaling (DVS) is one of the most effective techniques of minimizing the energy consumption of the embedded systems. One common assumption of almost all of the existing DVS algorithms is that the value of power consumption monotonically increases as the level of applied voltage to the system increases, and the power is a convex function of the voltage. Theoretically, under that assumption, previous works have shown that the DVS problem for a set of tasks with continuous or discrete convex power function is energy-optimally solvable in polynomial time. However, recently it is observed that some of the DC-DC converters do not follow the convexity. Thus, the power function of the whole system including a DC-DC converter will not be convex any more. In this context, we want to answer the following question: *is there an energy-optimal polynomial-time algorithm that is able to solve the DVS problem with any non-convex power function?* The work answers 'yes', and proposes an energy-optimal polynomial-time algorithm for the new problem.

I. INTRODUCTION

As the minimization of energy consumption becomes an important issue in system design, dynamic voltage/frequency scaling (DVS) has been widely studied and utilized to reduce the energy consumption. Many previous works have developed effective DVS algorithms (e.g., [3]–[5]) to minimize the amount of energy consumption while meeting the time constraints of the input tasks. One common assumption almost all existing DVS algorithms have taken is that the power function, which describes how the power value changes as the execution speed (i.e., the level of supply voltage) varies, is convex. Further, it is assumed to be a monotonically increasing function.

However, recently it is observed that there are several types of the implementation of DC-DC converter with different power functions, and some of the power functions are not convex [1], [2]. Moreover, it is shown that to effectively reduce the energy consumption of systems which use various modes of active/sleep state combination, more than one DC-DC converter should be installed in the systems [1]. Consequently, the power function of the entire systems including DC-DC converter(s) is unlikely to be convex. For example, the curves in Fig. 1 show the power conversion efficiencies (i.e., power functions) of the three types of DC-DC converters (also called voltage regulators (VRs)). Suppose the system uses a wide range of voltage scaling from 0.1V to 1.2V. Then, one best power curve would be the one marked with dashed red line shown in Fig. 1, which means to use two DC-DC converters: On-chip VR [2] and Off-chip VR [6] such that On-chip VR is used in the output voltage range of [0.4V, 0.6V] while Off-chip VR is used in the range of [0.6V, 1.2V]. Indeed, the

corresponding power curve is not convex, which may lead to a non-convex power function of the entire system including the DC-DC converters.

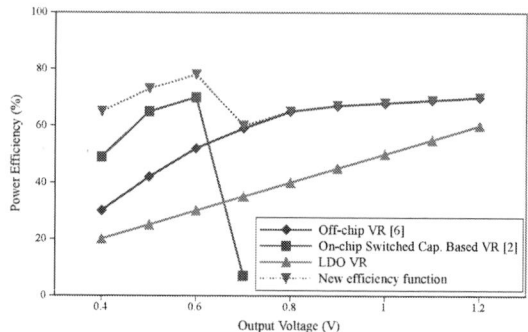

Fig. 1. An example of generating a non-convex power function (red curve) for a combination of DC-DC converters (green + blue curves), taken from the work in [1].

Unfortunately, so far no works have addressed the DVS problem with non-convex power functions while the DVS problem with convex functions has been proven to be optimally solvable for continuous functions by Yao *et al.* [3] and for discrete functions by Kwon and Kim [4]. This work answer the following question: *is there an energy-optimal polynomial-time algorithm that is able to solve the DVS problem with any non-convex power function?* This work answers 'yes', and proposes an energy-optimal polynomial-time algorithms for the new problem.

II. THE PROPOSED DVS ALGORITHM

The problem we want to solve can be described as:

Problem 1: For a non-convex power function $p(s)$ defined in voltage domain D and tasks with arrival time, deadline, and required CPU cycles, find a execution schedule and voltage levels of each task which minimize the amount of energy consumption.

Our proposed algorithm called DVS-nocvx is composed of five steps: (1) generating an initial schedule of input tasks; (2) merging time intervals; (3) converting $p(s)$ into a convex power function, say $g(s)$; (4) allocating voltage levels of tasks based on $g(s)$; (5) restoring time intervals into the original intervals obtained in (1)

Step 1. (*Generating an initial schedule*): DVS-nocvx applies Yao's algorithm [3] to the input tasks using power function $p(s)$, and produces the time intervals of the execution of tasks. For example, Fig. 2(b) shows the produced time intervals of

978-1-4799-5128-4/14 $31.00 © 2014 IEEE

the execution of the input task set in Fig. 2(a), in which the workset consists of three tasks. For example, task T_1 has arrival time of 80, deadline of 160, and workload of 400 CPU cycles.

Step 2. (*Merging time intervals*): DVS-nocvx concatenates the time intervals of the same task. For example, the two separated time intervals of task T_2 in Fig. 2(b) are merged into one as shown in Fig. 2(c).

Step 3. (*Converting $p(s)$ into a convex power function $g(s)$*): Fig. 2(d) shows an example of the conversion of power function in which the red curve is a non-convex function $p(s)$ and the blue curve is the transformed convex function, $g(s)$, of $p(s)$. As lighted in Fig. 2(d), the concave part of $p(s)$ in $2 \leq s \leq 6$ is transformed into a line to become convex in $g(s)$. We label the two end points (e.g., 2, 6) of a line in $g(s)$ with s_a and s_b.

Step 4. (*Allocating voltage levels of tasks according to $g(s)$*): The speeds of the tasks in Step 3 is checked against $g(s)$. If a voltage level $s_i \in [s_a, s_b]$ of a line of $g(s)$, the speed of the corresponding task T_i is computed in a way that T_i is run with speed s_a for $\frac{s_b - s_i}{s_b - s_a} I_i$ time units and with speed s_b for $\frac{s_i - s_a}{s_b - s_a} I_i$ time units where I_i is the range of the merged time interval of task T_i. For example, to run task T_1 whose speed s_i is set to 5 in Step 1 with the range of time interval of 80 time units, as shown in Fig. 2(c), T_1 is run with speed of 6 for 60 time units and then run with speed of 2 for 20 time units. Fig. 2(e) shows the speeds of the tasks computed from the speeds in Fig. 2(c) and $g(s)$ in Fig. 2(d).

Step 5. (*Restoring time intervals into the original intervals*): The merged time intervals are then restored into the ones in Step 1. Fig. 2(f) shows the restored time intervals from the ones in Fig. 2(e) to that in Fig. 2(b).

Theorem 1: DVS-nocvx finds an energy-optimal DVS solution. (The proof is left out due to the page limitation.)

For n tasks, k available speeds and non-convex power function $p(s)$, Step 1 of DVS-nocvx takes $O(n^3)$ time (Yao's algorithm [3]), Step 2 takes $O(n)$ and generating $g(s)$ in Step 3 takes $O(k^2)$ time. Steps 4 and 5 take $O(n)$ time. Thus, the total time of DVS-nocvx is bounded by $O(n^3 + k^2)$.

III. Conclusions

This work showed that the DVS problem of non-convex power function was solvable energy-optimally in polynomial-time. As the designs with diverse DVS capability will emerge and various combinations of DC-DC converters will be sought in the future, the power functions of the whole system will not follow the convexity property any more. In this respect, we believe this work will provide a valuable solution to the DVS problem which has never been addressed even the problem will be highly important.

Acknowledgment: This research was supported by the CISS of Global Frontier project by MSIP (CISS 2011-0031863) in Korea, the ITRC program of NIPA by MSIP (NIPA-2013-H0301-13-1011) in Korea, the Seoul R&BD Program (RI130006) in 2014, the Brain Korea 21 Plus Project in 2014,

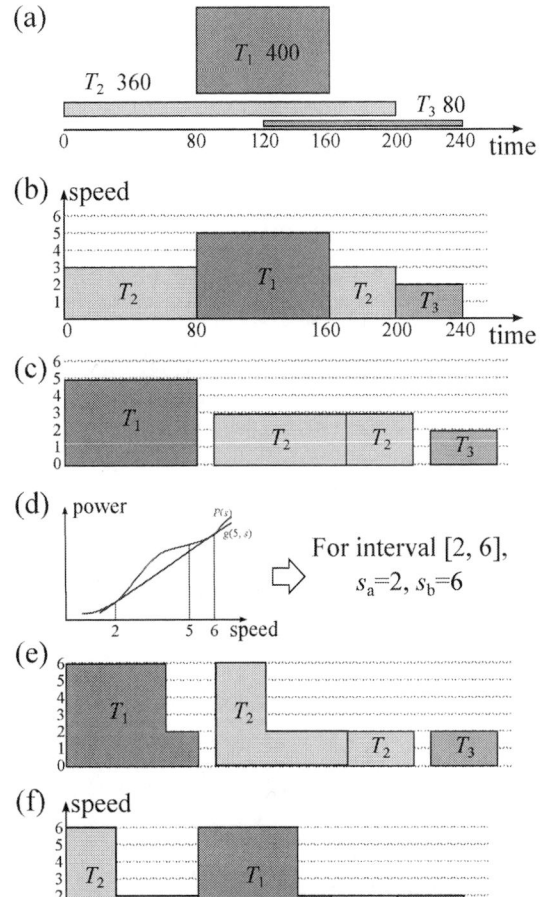

Fig. 2. (a) Input tasks T_1, T_2, and T_3. (b) Generating an initial schedule of input tasks. (c) Merging time intervals. (d) Converting $p(s)$ into a convex power function $g(s)$, (e) Allocating voltage levels of tasks based on $g(s)$. (f) Restoring time intervals into the original intervals obtained in (b)

and by the Global Ph.D Fellowship Program through the National Research Foundation of Korea (NRF) funded by the Ministry of Education (2013H1A2A1032650).

References

[1] X. He, G. Yan, Y. Han, and X. Li, "Superrange: wide operational range power delivery design for both stv and ntv computing," in *Proceedings of IEEE/ACM Design, Automation and Test in Europe Conference and Exhibition*, pp. 1–6, 2014.

[2] Y. Choi, N. Chang, and T. Kim "DC-DC converter-aware power management for low-power embedded systems *IEEE Transactions on Computer-Aided Design of Integrated Circuits and Systems*, Vol. 26, No. 8, pp. 1367–1381, October 2007.

[3] F. Yao, A. Demers, and S. Shenker, "A scheduling model for reduced cpu energy," in *Proceedings of ACM Symposium on Foundations of Computer Science*, pp. 374–382, 1995.

[4] W.-C. Kwon and T. Kim, "Optimal voltage allocation techniques for dynamically variable voltage processors," *ACM Transactions on Embedded Computing Systems*, Vol. 4, No. 1, pp. 211–230, February 2005.

[5] T. Ishihara and H. Yasuura, "Voltage scheduling problem for dynamically variable voltage processors," in *Proceedings of International Symposium on Low Power Electronics and Design*, 1998, pp. 197–202.

[6] H. P. Le, S. R. Sanders, and E. Alon, "Design techniques for fully integrated switched-capacitor dc-dc converters," *IEEE Journal of Solid-State Circuits*, Vol. 46, No. 9, pp. 2120–2131, September 2011.

On-chip Aging Prediction Circuit in Nanometer Digital Circuits

Byunghyun Jang[1], Jin Kyung Lee[2], Minsu Choi[3], Kyung Ki Kim[2]

[1]Department of Computer and Information Science, The University of Mississippi
University, MS, USA
[2]Department of Electronic Engineering, Daegu University
Gyeongsan, South Korea
[3]Department Electrical & Computer Engineering, Missouri University of Science & Technology
Rolla, MO, USA
bjang@cs.olemiss.edu, lzq881210@gmail.com, choim@mst.edu, kkkim@daegu.ac.kr(corresponding)

Abstract

In nanometer technology, accurate circuit aging prediction of MOSFET digital circuits casued by aging phenominon is one of the most critical issues for more reliable adaptive tuning system design. This paper proposes a new on-chip aging sensor circuit to predit and detect a circuit failure caused by BTI and HCI aging effects on digital circuits. The proposed circuit is based on timing warning windows to warn against a guardband violation of sequential circuits, and generates three warning bits right before circuit failures occur.

Keywords-aging sensor, guardband violation, aging effect

Introduction

As technology is scaled down, the parametric shifts or circuit failures caused by Hot Carrier Injection (HCI) and Bias Temperature Instability (BTI) have become the most critical issues for more reliables adaptive system design [1]. In CMOS circuits, aging phenomena are the process that degrade the premier characteristics of transistors during its life time. Therefore, circuit designers need to premeditate these aging effects in the early stages of circuit design to make sure there are enough margins (called guardband) for digital circuits to function correctly over their entire lifetime. However, such an approach usually bring about a waste of resources or energe. In order to reduce the waste of resoures, new circuit design techniques should be introduced for the resilient circuits. This challenge for resilient circuits will require a design paradigm shift to adaptive tuning design for overcoming the performance degradation due to aging phenomena; in addition, an accurate on-chip prediction circuit technique to monitor aging phenomena would be one of the key issues in the adaptive design techniques. The outputs of the prediction circuits can be used as control signals in the new self-adaptive system using effective methods such as adaptive body biasing, supply voltage scaling, frequency scaling, and etc. The concept of circuit failure prediction has been first proposed in Ref. [2] and [3]. In Ref. [2] and [3], distributed aging sensors was deployed and can locally measure the performance degradation due to aging phenomena. The aging sensor checks if data signals arrive within a predefined warning window in time region or not. However, the prediction circuits in Ref. [2] and [3] result in a large area overhead because the prediction circuit is added to each flip-flop. Moreover, it uses only 1-bit output to predict a circuit failure, so the circuits do not have enough information related to the circuit failure and they are not easy to apply them to a self-adaptive system. Therfore, in this paper, we propose a simple aging prediction circuit with smaller area overhead compared to the prediction circuit in Ref. [2] and [3]. The new fully digital on-chip aging prediction circuit has been implemented using a 0.11um CMOS technology and been evaluated by 4x4 mutiplier with power gating structure.

Aging Prediction Circuit

The proposed prediction circuit monitors the moment (guardband) when the critical path delays of combinational logics in a sequential design exceed a normal value which gurantees a correct circuit operation. Figure 1 shows the block diagram of the proposed aging prediction circuit. The proposed circuit has three blocks: a guardband generator (GG) to create three guardband windows, a path delay monitor (PDM) to detect data signal transistion, and a hold circuit. As showon in Fig. 1, a "Measure" signal is assered to the aging prediction circuit to turn the circuit on or off. In order to reduce a power overhead, the prediction circuit will be periodically worked for a short time and will be turned off for the most time. PDM is place on each flip-flop, while GG and "Hold Circuit" are shared for all the flip-flops, which leads to smaller area overhead. Detailed circuits of each block are shown in Fig. 2. GG block consists of a buffer chain, program- mable two skewed inverters, and an inverter chain. This block plays an important role to delay the CLK signal, and the falling transition of the delayed CLK signal is placed in the guardband region, where the delay time of CLK is controlled by the control input C0~C5. The Inv1~Inv4 play a role in the generation of three guardband windows, and G0~G2 are the final delayed CLK to make the gardband time region.

PDM block includes edge detectors consisting of a XOR gate, an AND gate, and a Pulse Generator, where the XOR gate detects the moment when the input D and the output Q of each flip-flop is different (that is, when the data signal passes through combinational logics and arrives at the input D of each flip-flop). The AND gate makes the PDM operated only at low states of each CLK signal, and the pulse generator generates a pulse with a small width to present the transistition time of each data signal. The NOR gate merges the outputs of all the edge detectors in time order.

978-1-4799-5128-4/14 $31.00 © 2014 IEEE

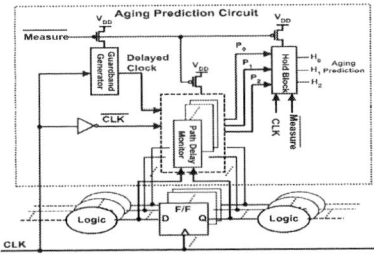

Fig. 1. The block diagram of the proposed aging sensor.

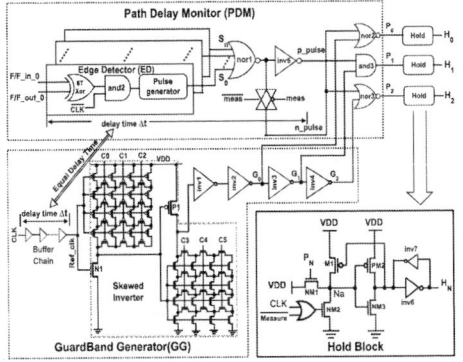

Fig. 2. The proposed aging sensor circuit.

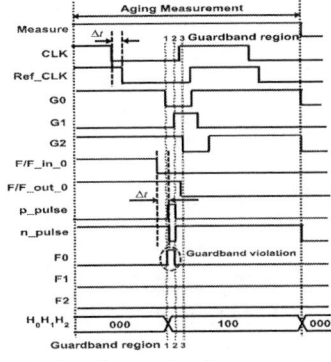

Fig. 3. Timing diagram for the proposed circuit.

The delay time Δt is the propagation time from XOR gate to Inv5 of PDM, so the generated pulse from ED is arrived at time Δt-delayed from the real transition time of the data signal. In order to compensate the delay time, the buffer chain of GG block with Δt delay time is deployed. The output P0~P2 will be pulses with a small width if each pulse reaches one of three guardband regions. Finally, the Hold block holds the 3-bit outputs depending on the outputs of PDM (P0~P2) during the measurement mode, where NM2 is used to reset the node Na. The final output H0~H2 present an aging prediction step to warn a circuit failure: H0, H1, H2 = 000 is a normal operation, 100 is the first warning step, 110 is the second warning step, and 111 is the circuit failure. Figure 3 shows a timing diagram for the proposed circuit, where a data transistion occurs at the first guardband region, so the primary output H0~H2 is "100".

Simulation Results

The proposed circuits have been designed and evaluated using a 0.11um MOSFET technology (VDD=1.2V). For a NBTI aging stress simulation, we have increased the number of cycles in the stressed input-signal with 0.5 duty cycle and 2

GHz frequency. The HCI stress time for these experiments is 400 μsec which is not ages (run time) but actual stress time (switching time). A 4x4 multiplier has been used as a benchmark circuit in our simulation. Table1 shows the comparison of the experimental results with the Ref. [2]. In this paper, because of the three aging outputs, the aging sensor have higher sensitivity to predict the degree of aging phenomena compared to the Ref. [2]. We used 4x4 multiplier as a DUT in this paper, however, Ref. [2] used OpenRISC processor as a DUT. If we use a complicated circuit with large area as a DUT, the power consumption percentage of the proposed prediction circuit will be decreased relatively. Figure 4 shows the layout of a 4x4 multiplier with the proposed prediction circuit .

TABLE I
THE COMPARISON OF THE EXPERIMENTAL RESULTS WITH REF. [2].

	Ref. [2]	Proposed circuit
# of output	1	3
Technology / DUT	65nm PTM/ OpenRISC Core	DB 0.11μm / 4 X 4 multiplier
Power penalty when the prediction circuit is off.	0.1%	1.5%
Power penalty when the prediction circuit is on.	7.5%	10%

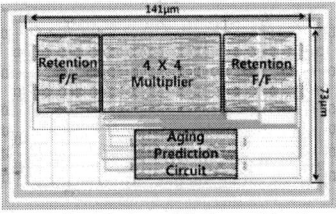

Fig. 4. The layout of a 4x4 multiplier with the proposed aging prediction circuit.

Conclusion

This paper proposes a novel on-chip aging prediction circuit in a 0.11 μm technology to monitor a guardband violation of sequential logics due to aging phenomena. The simulation results show that the proposed circuits achieve a good aging failure prediction and low overhead. For a good adaptive design technique for overcoming the performance degradation due to aging phenomena, our accurate aging prediction circuit would be a practicable solution in nanoscale CMOS circuits.

Acknowledgment

This work was supported by IC Design Education Center (IDEC).

References

[1] S. N. Wooters, A. C. Cabe, Z. Qi, J. Wang, R. W. Mann, B. H. Calhoun, M. R. Stan, And T. N. Blalock, "Tracking On-Chip Age Using Distributed, Embedded Sensors", In IEEE Transactions On Very Large Scale Integration (VLSI) Systems, Vol. No. 99, pp. 1–12, 2011.

[2] Agarwal M, Paul B, Zhang M, Mitra, S. Circuit failure prediction and its application to transistor aging. In: 25th IEEE VLSI test symposium; 2007. p. 277 –86.

[3] J. Vazquez, V. Champac, A. Ziesemer, R. Reis, I. Teixeira, M. Santos, And J. Teixeira, "Low-Sensitivity To Process Variations Aging Sensor For Automotive Safety-Critical Applications", In 28th VLSI Test Symposium (VTS), April 2010.

978-1-4799-5128-4/14 $31.00 © 2014 IEEE

A 0.5V 29pJ/Cycle Sensor Node Processor for Intelligent Sensing Applications

Jun Zhou[1], Xin Liu[1], Chao Wang[1], Kah-Hyong Chang[1], Jianwen Luo[1], Jingjing Lan[1], Lei Liao[1], Yat-Hei Lam[1], Yongkui Yang[2], Bo Wang[2], Xin Zhang[1], Wang Ling Goh[2], Tony Tae-Hyoung Kim[2], and Minkyu Je[3]

[1]Institute of Microelectronics, A*STAR (Agency for Science, Technology and Research), Singapore
[2]School of EEE, Nanyang Technological University, Singapore
[3]Department of Information and Communication Engineering, DGIST, Daegu, Korea
zhouj@ime.a-star.edu.sg, liux@ime.a-star.edu.sg

Abstract— **This paper presents a sensor node processor (SNP) with optimized energy efficiency and performance for intelligent sensing through architecture-level optimization and ultra-low voltage operation with timing-error monitoring. Two typical intelligent sensing applications are demonstrated with the proposed processor, consuming 39 and 29pJ/cycle at 0.5V respectively.**

I. INTRODUCTION

Compared to general DSP processors, higher energy efficiency is demanded for sensor node processors due to limited form factor and required long operation time. Existing sensor node processors reduces the energy mainly by applying circuit-level techniques such as power gating and ultra-low-voltage (ULV) operation [1][2][3]. However, the hardware architecture is not optimized for sensing application, resulting in limited energy efficiency and performance. In this paper, an energy-efficient SNP with optimized hardware architecture and ULV operation with timing-error monitoring is proposed. The proposed processor consumes 39 and 29pJ/cycle for two typical intelligent sensing applications (i.e. neural signal processing and tire monitoring).

II. SYSTEM ARCHITECTURE

As shown in Fig. 1, the proposed SNP incorporates a 32-bit ARM Cortex-M0 core for programmability. Versatile reconfigurable hardware accelerators (HAs) are developed to speed up the common processing tasks of sensing applications while keeping the energy consumption low. A dual-bus architecture with automatic bus occupancy detection and reconfigurable memory access is proposed to achieve energy-efficient parallel processing. ULV operation techniques with timing-error monitoring is applied to further reduce the energy consumption with variation resilience. The processor also integrates on-chip pre-amplifiers, ADC and DC-DC converters with adaptive voltage scaling.

III. PROPOSED TECHNIQUES

An energy-efficient dual-bus architecture with reconfigurable memory access is proposed (Fig. 1), which can be accessed by the ARM core and the DMA. The DMA can auto-

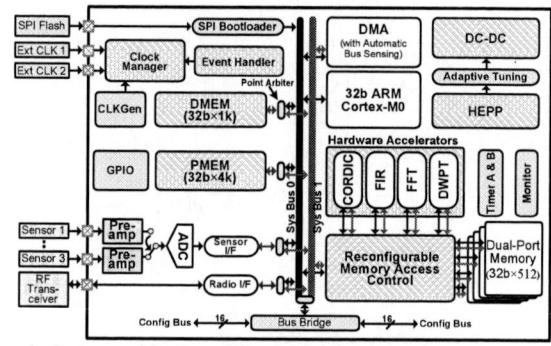

Figure 1. System architecture of the proposed sensor node processor.

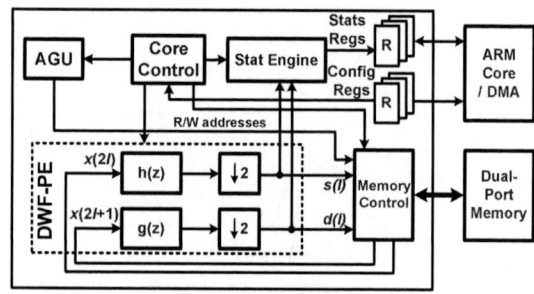

Figure 2. Proposed reconfigurable DWPT engine.

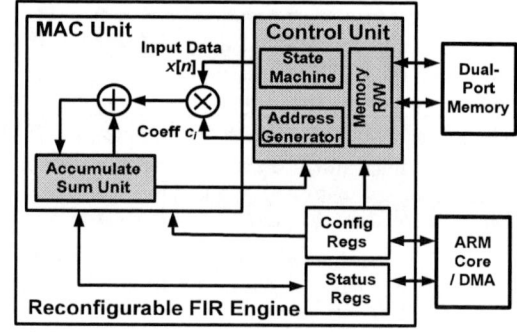

Figure 3. Proposed reconfigurable FIR filtering engine.

matically detect the bus occupancy and switch the traffic flow so that the data transfer between different components can be performed simultaneously. This allows parallel operation of

---

This work was supported by A*STAR (Agency for Science, Technology and Research) SERC (Science and Engineering Research Council), Singapore under the grant no. 122 380 6050.

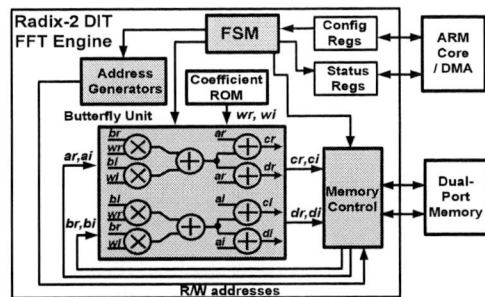

Figure 4. Proposed FFT engine.

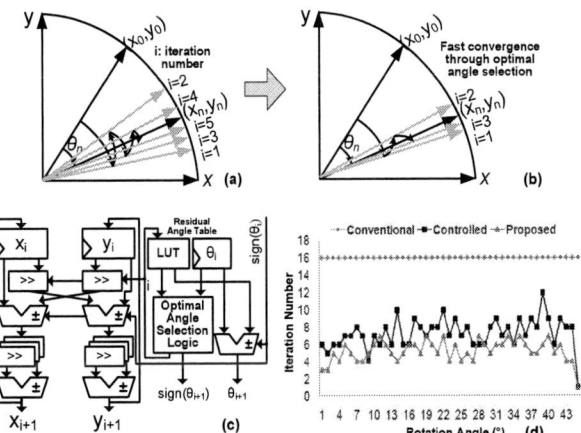

Figure 5. Proposed fast-convergence CORDIC engine: (a) operation of the conventional CORDIC, (b) operation of the proposed CORDIC, (c) architecture of the proposed CORDIC, (d) iteration comparison.

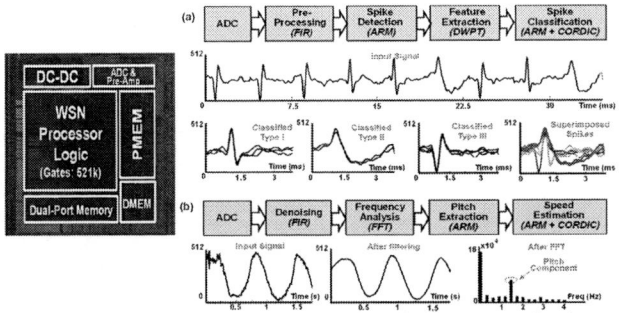

Figure 6. Chip micrograph and measurement results from two applications: (a) neural spike classification, (b) car speed estimation.

TABLE I. COMPARISION WITH THE STATE-OF-THE-ART SNPs

WSN Processor	[1]	[2]	This Work
Technology	0.13 μm	40 nm	0.18 μm
Voltage (V)	0.5 – 1	0.4 – 1.1	0.5 – 1.8
Clock Frequency (Hz)	100k – 10M	1M – 150M	500k – 48M
DSP Core	16-bit RISC	VLIW	32-bit RISC
Hardware Accelerator	Median, FFT, FIR, CORDIC	No	Reconfig FIR, Reconfig DWPT, FFT, CORDIC
On-chip Analog Components	No	DC-DC	Pre-amp, ADC, DC-DC
In-Situ Error monitoring	No	Yes, no correction	Yes, with correction
Adaptive Voltage Tuning	No	Yes	Yes
Minimum Energy/Cycle (pJ)	35 @0.5V, 100kHz (EEG)	30 @0.4V, 1MHz (ECG)	29 @0.5 V, 500kHz (Car) 39 @0.5 V, 500kHz (Neural)

deal with increasing variation at low voltages, a timing-error prediction and prevention technique [5] is applied to reduce the design margin with small overhead.

IV. EXPERIMENTAL RESULTS

A die photo of the proposed processor fabricated in 0.18μm CMOS process is shown in Fig. 6. Two typical sensing applications are implemented for demonstration and the measurement results are also shown in Fig. 6. For neural spike classification, spike signals received from neural sensors are detected and classified by the proposed processor. The algorithm is executed using the ARM core, FIR, DWPT and CORDIC engines with an average efficiency of 39pJ/cycle at 500kHz and 0.5V. For car speed estimation, frequency analysis is performed for the data received from an accelerometer sensor placed in a tire. The algorithm is mapped to the ARM core, FIR, FFT and CORDIC engines, consuming 29pJ/cycle at 500kHz and 0.5V. The benchmark table is provided in Table I. The proposed processor provides a solution for low-to-medium-data-rate intelligent sensing applications requiring scalable performance and high energy efficiency.

V. CONCLUSIONS

A SNP is proposed to employ architecture-level optimization and ultra-low voltage operation to improve energy efficiency for intelligent sensing. The proposed SNP consumes only 39 and 29 pJ/cycle at 0.5 V for neural spike classification and vehicle speed detection applications, respectively.

the ARM core and HAs as well as their data exchange with memories and sensor interfaces. Each HA has their own local memory but the memory access is reconfigurable so that the movement of intermediate data among HAs is greatly reduced for energy saving.

A reconfigurable 256-point DWPT engine is proposed to accelerate the time-frequency analysis required by many sensing applications, as shown in Fig. 2. It uses a discrete wavelet filter processing element (DWF-PE) to perform frame-based multi-scale decomposition with reconfigurable time-frequency resolution and decomposition tree structure. A reconfigurable FIR engine (Fig. 3) with memory-based processing architecture and reusable MAC unit is developed to achieve variable-length filtering for different sensing applications. A memory-based radix-2 FFT engine (Fig. 4) is developed for accelerating tasks related to frequency analysis. A fast-convergence coordinate rotation digital computer (CORDIC) engine is developed to speed up operations including sin(x), cos(x), exp(x), ln(x) and sqrt(x) (Fig. 5). An optimal-angle-selection logic is used to determine the optimal rotation angle for the next iteration by comparing the residual angle of each iteration with optional angle values stored in a look-up table. The number of iterations is significantly reduced, compared to the conventional and controlled CORDIC methods [4], as shown in Fig. 5.

ULV operation is employed to reduce the energy consumption by library re-characterization and cell pruning [2]. To

REFERENCES

[1] J. Kwong, et al., JSSC, vol. 46, no.7, Jul. 2011.

[2] M. Konijnenburg, et al., ISSCC, pp. 430–431, Feb. 2013.

[3] M. Ashouei, et al., ISSCC, Feb. 2010.

[4] S. Wang, et al., MWSCAS, pp. 236–239, Aug. 1994,.

[5] J. Zhou, et al., A-SSCC, Nov. 2013.

978-1-4799-5128-4/14 $31.00 © 2014 IEEE

Mulit-Core Architecture for Real-Time and Energy-Efficient Bearing Fault Diagnosis

Myeongsu Kang, Inkyu Jeong, and Jong-Myon Kim[*]

University of Ulsan
102 Daehak-ro Mugeo-dong Nam-gu
Ulsan, South Korea
Phone: 82-52-259-2217, Fax: 82-52-259-1687, and Email: {ilmareboy, jeonginkeyu, jongmyon.kim}@gmail.com

Abstract

This paper proposes a high-performance single-instruction, multiple-data (SIMD)-based multi-core architecture including 64 processing elements operating at 50 MHz in a Xilinx Virtex-7 field programmable gate array (FPGA) device to support online bearing fault diagnosis. The experimental results indicate that the proposed multi-core approach executes 1,293x faster than a high-performane Texas Instrument (TI) TMS320C6748 digital signal processor (DSP) by exploiting the massive parallelism inherent in the bearing fault diagnosis algorithm. In addition, the multi-core approach outperforms the equivalent sequential approach that runs on the TI DSP by substantially reducing the energy consumption.

Keywords- FPGA; multi-core architecture; online bearing fault diagnosis; SIMD

Introduction

Accurate identification of diverse faults at an incipient state can prevent severe unexpected machinery failure that can generate unanticipated interruptions on production with consequences in cost, product quality, and safety [1]. Hence, many researchers have recently investigated the development of real-time fault diagnosis systems. FPGA-based approaches are promising computational models to realize online fault diagnsosis in machinery due to the fact that they can offer high-performance digital signal processing capabilities [2],[3].

In this study, the proposed fault diagnosis approach covers a comprehensive bearing fault classification procedure that performs fault signature extraction through time-frequency analysis and reliable decision making using multi-class support vector machines. Practically, this compresentive fault diagnsosis scheme requires tremendous computatioin time and its high computational complexity may limit its use in industry. To address this issue, this paper proposes a high-performance single-instruction, multiple-data (SIMD)-based multi-core architecture including 64 processing elements (PEs) implemented in a Xilinx Virtex-7 FPGA device. It accelerates the complex fault diagnosis algorithm by exploiting the massive parallelism inherent in it.

Proposed bearing fault diagnosis methodology

A. Fault Feature Extraction with Time-Frequency Analysis

For early identification of diveser bearing failures, this paper utilizes acoustic emission (AE) signals. To record continuous AE signals, an AE sensor is generally placed on a non-rotating element of the machine, such as the bearing housing, far away from bearing failures. This results in severe attenuation in the acquired signals, and consequently, intrinsic information about bearing defects can be revealed in high-frequency bands. To address this issue, discrete wavelet packet transform (DWPT) has been widely employed for analyzing defect information inherent in the acquired AE signals due to its decomposition ability to split low- and high-frequency bands, and this paper extracts fault signatures from the 16 high-frequency bands resulting from the five-level DWPT with the Daubechies 2-tap mother wavelet function.

To classify multiple bearing defects, we utilize reliable wavelet packet energy ($RWPE$) and wavelet packet node kurtosis ($WPNK$) as fault signatures of various seeded bearing failures. First, $RWPE$ is computed as follows:

$$RWPE(m) = \left. \sum_{i=1}^{K} w_{m,i}^2 \middle/ \sum_{n=1}^{N_{node}} \sum_{i=1}^{K} w_{n,i}^2 \right. , \quad (1)$$

where N_{node} is the total nmber of high-frequency bands for feature extraction (N_{node}=16 in this study), K is the total number of wavelet coefficients in the band, and $w_{m,i}$ is the ith wavelet coefficient of the mth band. Second, $WPNK$ is defined as:

$$WPNK(m) = \left. \frac{1}{K} \sum_{i=1}^{K} \left(w_{m,i} - \overline{w_m} \right)^4 \middle/ \left(\frac{1}{K} \sum_{i=1}^{K} \left(w_{m,i} - \overline{w_m} \right)_2 \right)^2 \right. , \quad (2)$$

where $\overline{w_m}$ is an average value of wavelet coefficients in the mth band. In total, 32 fault signatures are used to diagnose beraing defects in this study.

B. Fault Classification using MCSVMs

A support vector machine (SVM) discriminates test samples into one of two classes by finding an optimal hyper-plane that correctly separates the largest fraction of data points while maximizing the distance between two classes on the hyper-plane. To design multi-class support vector machines (MCSVMs), the following three approaches can be considered: one-against-all (OAA), one-against-one (OAO), and one-acyclic-graph (OAG) [4]. Among them, this paper employs the OAA method, which is one of the most popular and simplest techniques for multi-class classifiers. In the OAA approach, each SVM structure separtes ones class from the others, and the final decision can be made by selecting an SVM structure that yields the highet value of each SVM classification function. The proposed fault diagnosis method achieves an average classification accuracy of 99.36%.

978-1-4799-5128-4/14 $31.00 © 2014 IEEE

The proposed fault diagnosis algorithm on the SIMD-based multi-core architecture

Fig. 1 illusrates the microarchitecture of the proposed SIMD-based multi-core model along with its interconnection network, which mainly consists of a two-dimensional (2D) processing elements (PEs) and local memory.

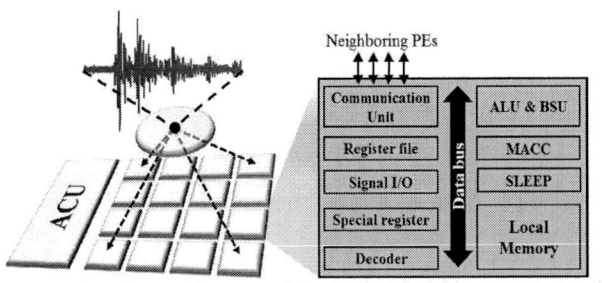

Fig. 1. The microarchitecture of the proposed multi-core array model

In this study, PEs execute a set of instructions in a lockstep fashion and are inteconected through a torus network. An array control unit (ACU) controls each PE in the array. In addition, the proposed multi-core architecture has a reduced instruction set computer (RISC) data path with the following characteristics:

- Arithmetic logic unit (ALU) performs basic arithmetic and logic operations,
- Multiply-accumulator unit (MACC) multiplies 32-bit values and accumulates them into a 64-bit accumulator,
- Barrel shifter unit (BSU) performs multi-bit logic/arithmetic shift operations,
- SLEEP unit activates/deactivates PEs based on local information,
- Communication unit allows PEs to communicate with their four nearest neighbors.

The entire multi-core design process is divided into the following three levels: application, architecture, and technology. At the application level, an instruction-level multi-core simulator is used to profile the execution statistics, such as the cycle count, dynamic instruction frequency, and PE utilization, for the multi-core architecture. At the architecture level, the Xilinx ISE Design Suite 14.2 is used to generate register-transfor level (RTL) code for each function unit of the multi-core architecture and to verify its functionality on the Xilinx Virtex 7 FPGA VC707 evaluation kit. Then, RTL-to-gates synthesis is done by the Synopsys Design Compiler with TSMC 28 *nm* technology at the technology level. Finally, a design space analysis tool collects and combines all of the paratmeters, including the cycle count and system power from the application, architecture, and technology levels, to determine the execution time and energy consumption during the execution of the proposed online fault diagnosis algorithm. This paper compares the performance of the proposed multi-core architecture with that obtained using a commercial Texas Instrument (TI) digital signal processor (DSP) in order to demonstrate its use for online bearing fault diagnosis. Table I presents a comparison of the proposed multi-core processor and a high-performance TI DSP in terms of the execution time and energy consumption. As shown in Table I, the proposed multi-core approach executes 1,293

times faster than the TI TMS320C6748 DSP by exploiting the massive parallelism inherent in the bearing fault diagnosis algorithm. In addition, the multi-core approach outperforms the equivalent sequential approach that runs on the TI DSP by substantially reducing the energy consumption. This result demonstrates that the proposed multi-core architecture can support real-time and energy-efficient bearing fault diagnosis.

TABLE I
COMPARISON BETWEEN THE PROPOSED MULTI-CORE
ARCHITECTURE AND A HIGH-PERFORMANCE TI DSP

	TI TMS320C6748	Proposed multi-core architecture
Clock frequency	456 MHz	50 MHz
CMOS technology	65 *nm*	28 *nm*
Power	509.81 *mW*	1,953 *mW*
Total # of clock cycles	$2,659.915 \times 10^6$	2.057×10^6
Execution time	5.833 seconds	0.041 seconds
Energy consumption	2,973.722 *mJ*	80.073 *mJ*

Conclusions

To support real-time baring fault diagnosis, this paper proposed an SIMD-based multi-core approach and implemented the multi-core architecture in the Xilinx Virtex-7 FPGA device. The experimental results indicated that the proposed multi-core approach outperforms the commercial high-performance TI DSP in terms of the execution time and energy reduction.

Acknowledgment

This work was supported by a National Research Foundation of Korea (NRF) grant funded by the Korean government (MEST) (No. NRF-2013R1A2A2A05004566, NRF-2012R1A1A2043644), and this work was supported by the IDEC.

References

[1] A. Soualhi, G. Clerc, and H. Razik, "Detection and Diagnosis of Faults in Induction Motor Using an Improved Artifical Ant Clustering Technique," IEEE Trans. Ind. Electron., vol. 60, no. 9, pp. 4053–4062, 2013

[2] L. M. Contreras–Medina, R. J. Romero–Troncoso, E. Cabal–Yepez, J. J. Rangel–Magdaleno, and J. R. Millan–Almaraz, "FPGA-Based Multiple-Channel Vibration Analyzer for Industrial Applications in Induction Motor Failure Detection," *IEEE Trans. Instrum. Meas.*, vol. 59, no. 1, pp. 63–72, 2010.

[3] R. J. Romero–Troncoso, R. Saucedo–Gallaga, E. Cabal–Yepez, A. Garcia–Perez, R. A. Osornio–Rios, R. Alvarez–Salas, H. Miranda–Vidales, and N. Huber, "FPGA-Based Online Detection of Multiple Combined Faults in Induction Motors Through Information Entropy and Fuzzy Inference," *IEEE Trans. Ind. Electron.*, vol. 58, no. 11, pp. 5263–5270, 2011.

[4] R. Yan, R. X. Gao, and X. Chen, "Wavelets for Fault Diagnosis of Rotary Machines: A Review with Applications," *Signal Process.*, vol. 96, pp. 1–15, 2014.

Accelerating Forex Trading System Through Transaction Log Compression

Ji Hoon Jang, Sang Muk Lee, Sang Don Kim, Oh Seong Gwon, Eunnuri Ko,
Seong Mo Lee, Jung Woo Shin, and Seung Eun Lee[*]

Dept. of Electronic Engineering
Seoul National University of Science and Technology
Seoul, Korea
+82-2-970-9021, [*]seung.lee@seoultech.ac.kr

Abstract

In this paper, we propose the hardware architecture for high-speed transaction logging of forex trading system. In forex trading market, the trading volume of currencies is growing larger every year. In order to provide real-time processing of large volume and high availability service, we focused on the two types of the workload, where the bottleneck occurs, and conducted workload analysis. The bottleneck between the application server and the internal hard disk is caused by the overhead from storing the transaction logs, due to the bandwidth limitation of a hard disk. Our key idea is to suppress an overhead of the transaction logging through the compression of the transaction logs. Implementation result demonstrates the feasibility of our proposal for increasing the bandwidth through the log compression.

Keywords: Hardware acceleration, Forex trading, High-speed logging, Compression, FPGA

Introduction

In forex trading market (a.k.a. foreign exchange trading market), high-frequency trading has been increasing. Therefore a development of real-time processing technique is needed to process the transaction in real-time in forex trading system. We conducted workload analysis of forex trading system with a financial solution company managing forex trading system.

Figure 1 shows the existing forex trading system. The forex trading system has two bottlenecks. The bottleneck between the DB server and the storage is caused by frequent updates of the specific data. The result of the workload analysis shows that 163876 SQL quries are executed during 6 minutes. From among these, 31410 SQL queries (19.17%) are repeatedly updating the specific data into the storage. This specific data indicates stock quotes, state of the stock exchange, or other real-time information of forex trading. In our previous work, we proposed novel methods to solve this bottleneck [1]. Another bottleneck between the application server and the internal hard disk is caused by the overhead of transaction logging. In financial trading system, the transaction logs are important data. Therefore using a buffered I/O in the transaction logging is inappropriate. We use a direct I/O instead of a buffered I/O for the transaction logging. In order to high-speed logging through a direct I/O, the transaction logs are compressed for lower bandwidth usage and higher

throughput. Thanks to the transaction log that is a text data, we can achieve a high compression ratio and a real-time processing. We use an additional hardware for co-processing of the hardware and the software. Required throughput of each application server is 1GB/s when the traffic volume increases exponentially. Multiprocessor systems along with hardware accelerators were proposed in different fileds such as computing, automobile, recognition, and handheld applications [2-5], accelerating the performance of systems.

Our compression accelerator uses an FPGA. An FPGA facilitates hardware parallelism and this feature is suitable for processing of big data [6]. Using an additional hardware accelerator provides an offload computation to the application server and improves the data processing performance through the co-processing of CPU and FPGA [7, 8]. Therefore, we use an FPGA in order to accelerate the compression of the transaction logs.

The rest of this paper is organized as follows. In section 2, We introduce the data compression algorithm used in our compression accelerator. Section 3 describes the architecture of the accelerator. Section 4 shows the experimental result. We conclude in section 5 by proposing the future work in this field.

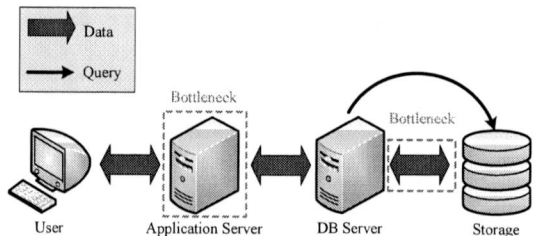

Fig. 1 The existing forex trading system

Data Compression Algorithm

In order to select a proper compression algorithm for transaction log compression, we conducted workload analysis of the transaction log data. The transaction logs are text data. Thus, the logs have a lot of redundancies due to the repetition of timestamp, variable name, frame format overhead, and so on. Therefore, the LZ(Lempel Ziv)-family compression algorithm is appropriate for our work because it is byte-oriented and dictionary-based compression. We choose a LZ4 compression scheme because LZ4 is the fastest algorithm among the LZ-family [9]. In [9], according to open source benchmark result, LZ4 is little faster than LZO, snappy, and 5-times faster than zlib (LZ77).

The System Architecture

Figure 2 shows the system architecture of our compression accelerator. Our accelerator is composed of PCIe interface unit, memory, compression engine, and controller. The PCIe interface can realize the bandwidth from 200MB/s to 6.5GB/s through a serial bus system. Therefore, PCIe interface provides enough bandwidth to the compression accelerator. In order to use an FPGA's hardware parallelism, the PCIe interface is appropriate. The memory is self-explanatory. The internal RAM of an FPGA is used for the buffer of compression engine. When the one input buffer is used by compression engine, the other is written by PCIe interface. It operates like the page-flip method (a.k.a. ping-pong buffer). These are in complementary relations. The controller controls the buffer operation for full-duplex communication of the PCIe interface and contributes to compression. We use Xilinx ZC706 board which has a hard-wired microprocessor in the FPGA. We use a microprocessor of the FPGA as a controller of our accelerator.

The transaction log compression is conducted by the co-processing of controller and compression engine. Compression engine conducts a suitable compression step for parallel processing like the sliding windows step of dictionary based compression.

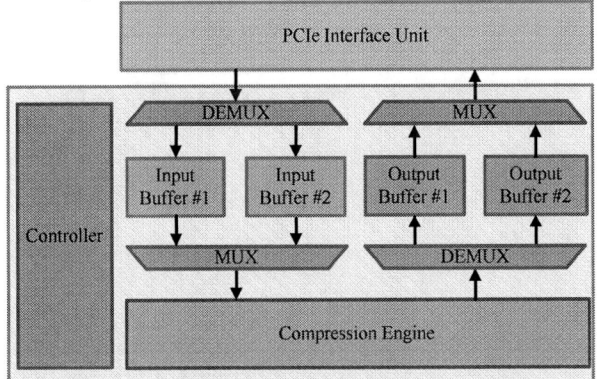

Fig. 2 The system architecture of the compression accelerator

Experimental Result

Figure 3 shows the experimental environment for our accelerator evaluation. LZ4 algorithm was used to compress the transaction logs in the current prototype. The transaction logs had been offered from financial solution company. In order to measure the throughput of our accelerator, we used SCU(Snoop Control Unit) timer built in controller of Xilinx ZC706. We set the parameters as follows. Timer load value is 666, prescaler value is 0, and operating clock frequency is 667MHz. Table I shows the experimental results. In experimental results, the compression throughput is roughly 200Mbps.

TABLE I. Throughput of each transaction log

Log size (Bytes)	Compression (Bytes)	#Ticks	Throughput (Mbps)
3424	1246	189	138.217
3413	654	113	230.434
3398	1377	203	127.707
3387	470	86	300.473
3382	588	104	248.102

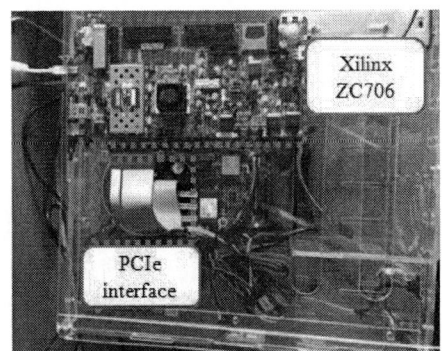

Fig. 3 The demonstration environment

Conclusion

In this paper, we proposed the architecture of compression accelerator for high-speed logging of transaction logs through direct I/O method. Our compression accelerator compresses the transaction logs for lower bandwidth usage and higher throughput. Therefore, the bottleneck between the application server and the internal hard disk is suppressed because the transaction logs are compressed. Implementation result demonstrates the feasibility of our proposal. Our compression accelerator's throughput is roughly 200 Mbps. We plan to improve the throughput by implementing the specific compression stage like finding the matched literals in dictionary into the hardware logic in order to parallel processing.

Acknowledgment

This work was supported by the IT R&D program of MSIP/KEIT. [10043896, Development of virtual memory system on multi-server and application software to provide real-time processing of exponential transaction and high availability service]

References

[1] S.J. Kim, S.M. Lee, J.H. Jang, Y.S. Jeong, S.D. Kim, S.E. Lee, "In-Time Transaction Accelerator Architecture for RDBMS", *Advanced Technologies,Embedded and Multimedia for Human-centric Computing*, pp. 329-334, 2014

[2] S.D. Kim et al., "Compression Accelerator for Hadoop Appliance", *Lecture Notes in Computer Science*, vol. 8862, pp. 416-423, 2014.

[3] S.E. Lee, K.Y. Min, T.W. Suh, "Accelerating Histograms of Oriented Gradients descriptor extraction for pedestrian recognition", *Journal of System Architecture*,vol.39, no.5, pp. 1043-1048, 2013.

[4] R. Iyer et al., "Cogniserve: Heterogeneous server architecture for large-scale recognition", *IEEE Micro*, 31(3), pp. 20-31, 2011.

[5] S.E. Lee et al., "Accelerating mobile augmented reality on a handheld platform", *Computer Design, 2009. International Conference on*, pp. 419-426, 2009.

[6] R. Guha, D. Al-dabass, "Performance Prediction of parallel Computation of Streaming Applications on FPGA platform", *12th International Conference on Computer Modeling and Simulation*, Cambridge, pp. 579-585, 2010.

[7] K. Mershad, A.R. Kaitoua, H. Artail, M.A. Saghir, "Cloud Providers Collaboration for a Higher Service level in Cloud Computing", in *3rd International Conference on Communications and Information Technology*, Beirut, pp. 109-114, 2013

[8] R. Muller, J. Teubner, G. Alonso, "Data Processing on FPGAs", in *J. Proceedings of the VLDB Endowment*, vol. 2, pp. 910-921, August 2009

[9] Y. Collet, https://code.google.com/p/lz4

FPN Correction for a Linear-logarithmic CMOS Image Sensor with a Tunable Linear Range Using Two-step Charge Transfer

Byeungseok Yoo, Inkyu Baek, and Kyounghoon Yang

Department of Electrical Engineering, KAIST
291, Daehak-ro, Yuseong-gu, Daejeon 305-701, Republic of Korea
+82-42-350-8071 and ybs1127@kaist.ac.kr

Abstract

This paper presents a fixed pattern noise (FPN) correction method for a linear-logarithmic CMOS image sensor (CIS) with a tunable linear response range. The proposed method is based on a two-step charge transfer operation to calibrate the offset FPN caused by the threshold voltage variation of the transfer gate under various linear ranges. The prototype image sensor consisting of a 320 × 240 pixel array with a 2.25 μm pixel pitch is fabricated with a 0.13 μm CIS process. It is found that the offset FPN is reduced by 90 % while the linear range is tuned by 20 % of the full output range.

Keywords- CMOS image sensor, fixed pattern noise, tunable linear-logarithmic response, wide dynamic range

Introduction

The wide dynamic range (WDR) represents an ability to obtain image scenes containing a wide range of illumination intensities which is an important factor in the CMOS image sensor (CIS) field [1]. Among various approaches to extend the dynamic range, a linear-logarithmic active pixel sensor (APS) using the subthreshold characteristics of an in-pixel transistor has the advantages of its simple operation and compatibility with the conventional 4T APS [2]. On the other hand, the conventional linear-logarithmic APS has a critical problem of the fixed pattern noise (FPN) caused by the threshold voltage variation of a transfer gate (TG). To reduce this FPN, a two-step charge transfer technique using a partial charge transfer operation was proposed in the previous work [3]. However, for the two-step charge transfer APS, a method for tuning the linear response range according to the light intensity has not been investigated yet [4].

In this paper, a FPN correction scheme for the two-step charge transfer CIS with a tunable linear range is proposed and studied. By using the two-step charge transfer operation, the effective FPN correction under tuned linear range operation is expected to be possible while keeping its system simplicity. The characteristics of the proposed method are discussed through the measured results of the prototype CIS chip with the theoretical analysis.

Operation Principle

Fig. 1 shows the pixel schematic and timing diagrams for the proposed FPN correction method with a tunable linear range. As shown in Fig. 1(a), the pixel structure is based on the conventional 4T APS. During the integration time, as shown in

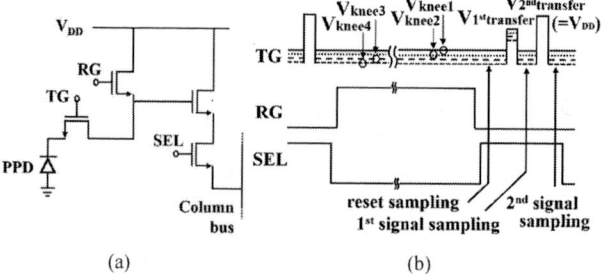

(a)   (b)

Fig. 1. The proposed (a) pixel schematic diagram and (b) timing diagram.

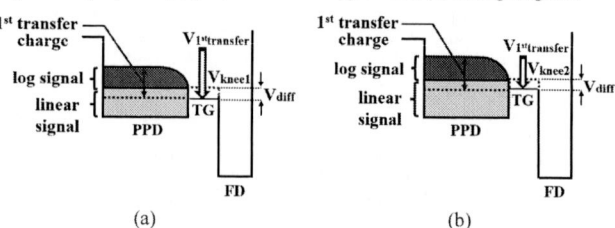

(a)   (b)

Fig. 2. Potential well diagrams of the PPD having two different linear ranges for the cases of: (a) higher knee-point voltage and (b) lower knee-point voltage.

Fig. 1(b), the TG voltage is set to the intermediate voltage (V_{knee}) which determines the transition from the linear response to the logarithmic response. This allows the knee point between the linear region and logarithmic region to be controlled by the V_{knee}. When the linear-logarithmic APS is used based on the fully transfer operation, the amount of offset FPN is significant. In order to correct this FPN arising from the knee point variation originating from the threshold voltage variation of the TG, the novel method using the two-step charge transfer operation has been proposed by the authors [3].

In this work, to reduce the offset FPN of the linear-logarithmic APS under the varied knee point condition for the tunable linear range, the two-step charge transfer operation is employed. As shown in Fig. 2, in the proposed method to reduce the offset FPN, the 1st transfer voltage also needs to be varied according to the changed knee voltage of V_{knee} in order to maintain the correlation between the knee point and 1st transfer levels. This is because the FPN performance by partial charge transfer is determined depending on the magnitude of the voltage difference (V_{diff}) between the knee point level and the 1st transfer level. Therefore, the V_{diff} is needed to be determined properly for the optimal two-step charge transfer operation with the tunable linear range. For the case of a small value of V_{diff}, the offset FPN is more reduced because the pixel to pixel variation of the well capacity is reduced due to the less significantly changed junction capacitance of PPD [4]. How-

978-1-4799-5128-4/14 $31.00 © 2014 IEEE

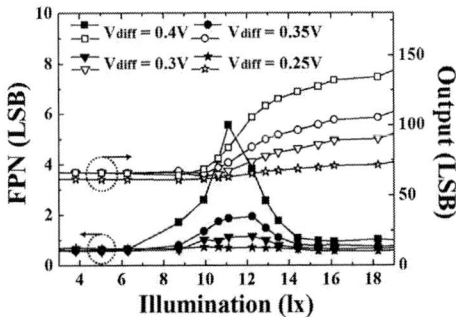

Fig. 3. Measured output and pixel FPN of the 1st transferred signal under various illumination and V_{diff}.

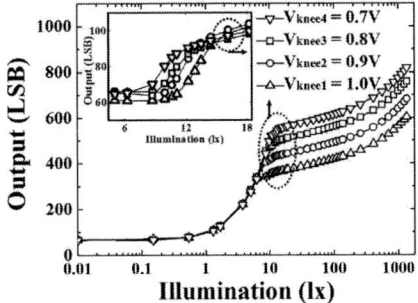

Fig. 4. Measured photo-conversion characteristics of the proposed linear-logarithmic APS with different linear ranges. The inset shows the measured output of the 1st transferred signal with the proposed V_{diff} of 0.35 V.

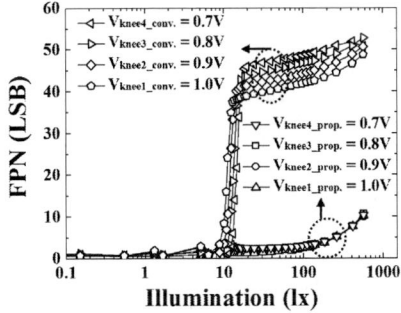

Fig. 5. Measured pixel FPN of the conventional and the proposed linear-logarithmic APSs with different linear ranges.

ever, when the V_{diff} is too low, the proper charge transfer does not occur and the mismatch between both 1st and 2nd readout signal levels is increased at the knee point [3], [5].

Experimental Results

The proposed scheme is verified by a prototype QVGA CMOS image sensor fabricated using a 0.13 μm CIS process. As shown in Fig. 3, to find out an optimal bias condition for the reduced FPN, the offset FPN and the photo-conversion characteristics of the 1st transferred signal are measured under various V_{diff} conditions. This output plot indicates the partial charge transfer properly occurs when the V_{diff} is more than about 0.3 V. This is because the amount of 1st transferred signal exceeding the threshold of TG decreases as the V_{diff} is gradually decreased from 0.4 to 0.25 V. Meanwhile, the FPN peak is measured to increase with the increase of V_{diff} as shown in Fig. 3. So, the V_{diff} needs to be slightly larger than 0.3 V for both the low FPN and the improved charge transfer capability. In addition, as shown in the inset in Fig. 4, when the V_{diff} is 0.35 V, the linear signal in the 1st readout signal is found to be transferred well at the knee point. This result shows that there

TABLE I
PERFORMANCE SUMMARY

Technology		Samsung 0.13-μm CIS
Array Format		320 × 240
Pixel Size		2.25 × 2.25 μm²
Operating Voltage		2.8 V(Analog) / 1.5 V(Digital)
ADC Resolution		10 bits
Frame Rate		36 frames/s
Sensitivity	Linear	1702 LSB/lx·s
	Logarithmic	93 LSB/dec
Dynamic Range		103 dB
FPN in logarithmic region (conv./prop.)	V_{knee4}= 0.7V	3.7 % (38 LSB) / 0.31 % (3.2 LSB)
	V_{knee3}= 0.8V	3.6 % (37 LSB) / 0.28 % (2.9 LSB)
	V_{knee2}= 0.9V	3.5 % (36 LSB) / 0.21 % (2.2 LSB)
	V_{knee1}= 1.0V	3.2 % (33 LSB) / 0.18 % (1.8 LSB)

is no noticeable mismatch between both 1st and 2nd readout signal levels at the knee point. Therefore, based on the afore-mentioned analysis, the optimal V_{diff} is verified to be 0.35 V. Figs. 4 and 5 show the photo-conversion characteristics and the pixel FPN measured by varying the V_{knee} with the optimal V_{diff} of 0.35 V, respectively. It is found that the offset FPN is less than 3.2 LSB for the change of the linear range by 20 % of the full output range with a dynamic range of 103 dB. These results indicate that the proposed method has less offset FPN over the entire dynamic range than the other FPN correction techniques [2]. The tunable linear range is expected to be further increased by optimizing the $V_{1st\,transfer}$ and V_{knee} levels [4].

Conclusion

In this paper, a FPN correction method of two-step charge transfer CIS for the tunable linear range has been proposed and demonstrated. The optimal operation of the proposed method was confirmed from its theoretical analysis of the partial charge transfer efficiency and the offset FPN. The proposed method effectively reduces the offset FPN by 90 % while the linear range is tuned by 20 % of the full output range. The proposed linear-logarithmic CIS is expected to be suitable in medical or security imaging applications requiring the tunable linear range mode with the overall system simplicity.

Acknowledgment

This work was supported by the image frontier center at Yonsei university and Samsung Electronics Co., Ltd.

References

[1] A. J. P. Theuwissen, "Better pictures through physics," *IEEE Solid-State Circuits Mag.*, vol. 2, no. 2, pp. 22-28, Jun. 2010.

[2] W. Chou et al., "A linear-logarithmic CMOS image sensor with pixel-FPN reduction and tunable response curve," *IEEE J. Sensors*, vol. 14, no. 5, May 2014.

[3] J. Lee et al., "On-chip FPN calibration for a linear-logarithmic APS using two-step charge transfer," *IEEE Transactions on Electron Devices*, vol. 60, no. 6, Jun. 2013.

[4] J. Lee et al., "Memoryless wide-dynamic range CMOS image sensor using nonfully depleted PPD-storage dual capture," *IEEE Trans. Circuits Syst. II, Exp. Brief*, vol. 60, no. 1, pp. 26-30, Jan. 2013.

[5] S. Shafie et al., "Non-linearity in wide dynamic range CMOS image sensors utilizing a partial charge transfer technique," *Sensors*, 9, pp. 9452-9467, 2009.

A RFID/NFC Based Programmable SOC for Biomedical Applications

Mayukh Bhattacharyya[1], Waldemar Gruenwald[1], Benjamin Dusch[1], Jasmin Aghassi-Hagmann[1], Dirk Jansen[1] and Leonhard Reindl[2]

[1]Institute for Applied Research, University of Applied Sciences Offenburg
[2]Department of Microsystems Engineering (IMTEK), University of Freiburg
[1]Offenburg 77652 and [2]Freiburg 79098, Germany
Tel: +49 (0) 781 205 – 365, Fax: +49 (0) 781 205 – 174 , Email: mayukh.bhattacharyya@hs-offenburg.de

Abstract

A new RFID/NFC (ISO 15693 standard) based inductively powered passive SoC (System on chip) for biomedical applications is presented here. The proposed SOC consists of an integrated 32 bit microcontroller, RFID/NFC frontend, sensor interface circuit, analog to digital converter and some peripherals such as timer, SPI interface and memory devices. An energy harvesting unit supplies the power required for the entire system for complete passive operation. The complete chip is realized on CMOS 0.18 μm technology with a chip area of 1.5 mm x 3.0 mm.

Keywords- radio frequency identification (RFID); near field communication(NFC); inductive power transfer; ultra low power; analog to digital converter (ADC).

Introduction

RFID technology for the near field communication has gained significant importance in the recent past for the development of passive sensor systems due to the possibility of harvesting energy from the RFID field [1]. Inclusion of the NFC creates new potential for the development of ultra-low power sensor systems for biotelemetry applications.

In Germany about 10 million people suffer from hypertension of which about 10% need a long term monitoring [2].The aim is to develop an implant with the proposed SoC to monitor blood pressure in arterial system (femoral), for the patients suffering from *Peripheral Artery Disease* (PAD), similar to the work presented in [3]. However the adaptable configurations of the SoC can extend the usability in a wide field of applications, e.g. for measuring ECG, blood pressure, body temperature etc.

System Overview

The basic architecture of the proposed SoC is shown in the figure 1 with all the basic blocks. The programmable feature provides a huge amount of flexibility as it increases the possible areas of application. The integrated ADC with a maximum resolution of 12 bit along with the instrumentation amplifier provides a good platform for a wide range of measurement. To measure the arterial blood pressure, a MEMS based piezo resistive type pressure sensor is to be connected externally in order to measure pressure in the range of 10 – 400 mmHg.

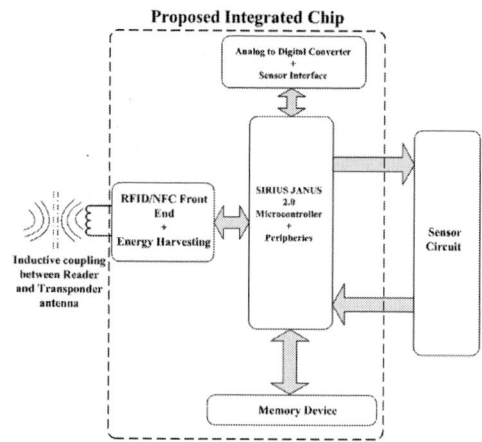

Fig. 1 Basic architecture of the proposed SoC.

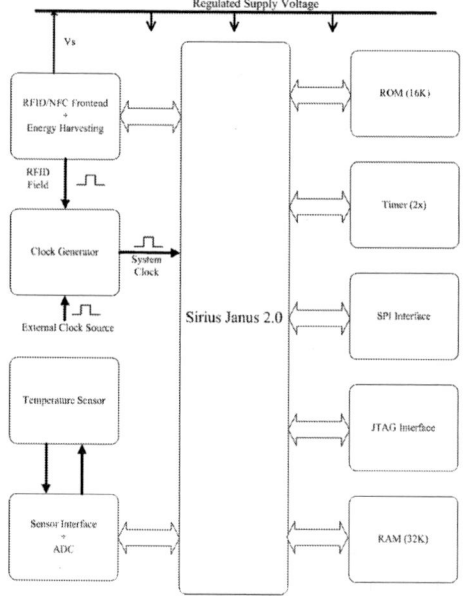

Fig. 2 Detailed block diagram of the proposed SoC.

In figure 2, the various system blocks are presented, which are discussed below in brief along with their respective functionalities.

A. RFID/NFC and Energy Harvesting

The RFID/NFC and energy harvesting block consists of the analog circuitry required for the communication and energy harvesting and the digital logic for handling the communication protocol. Figure 3 shows the measured value

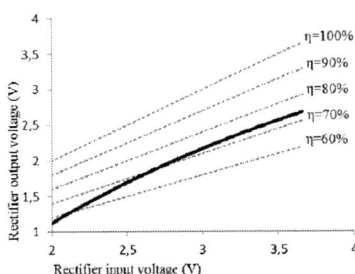

Fig. 3 Measured input-output characteristic of the energy harvesting rectifier, with the reference lines showing efficiency (η)

of the energy harvesting block rectifier's input and output voltage characteristics, where the input voltage is limited by the maximum voltage allowed for the technology. The rectifier with an average efficiency (η) of ~66% provides sufficient energy for the entire system. The low drop out regulator uses this harvested energy from the rectifier to generate a regulated power supply for the system. A temperature and supply independent reference voltage of 1.2 V ± 10 mV is provided by the bandgap.

The clock regenerator extracts the clock from the RF field which is used as the system clock. For the communication, a demodulator has been implemented with a specially designed comparator with a predefined offset voltage in order to set the working condition for demodulation. The digital logic extracts the message from the demodulated signal and prepares the required response, which is then transmitted via load modulation.

B. Microcontroller

The microcontroller *(Sirius Janus 2.0)* [4] used here is developed based on *Von Neumann* architecture. It consists of 16 bit *data* and 32 bit *address* registers along with 16 registers of which 12 are *universal* and 4 are *special* registers. The control unit is capable of handling 16 bit command format with an internal 8 bit control signal. All the arithmetic and logical operations are based on Reduced Instruction Set Computing (*RISC*).

C. Analog to Digital Converter

The analog to digital converter (ADC) proposed here is a *charge redistribution type* based on successive approximation register (SAR) logic [5], which can be operated either in single conversion or continuous mode. It can operate from 1.2 V to 1.8 V with a clock frequency of ~96 KHz to ~1 KHz. At 1.2 V it can operate with an input range of ±0.5 V with a quantization error of ± 122 µV (with12 bit resolution).

D. Sensor Interface

The sensor interface circuits consist of an instrumentation amplifier (INA) with an 8 bit programmable gain factor of 1 to 100. At 1.2 V, it has an offset voltage of ~2 mV over a temperature range of -30°C to 85°C.

E. Temperature Sensor

The internal bandgap module is modified to operate as a

TABLE I
Characteristics of the Proposed SoC

Parameter	Comment
Technology	UMC CMOS 0.18 µm
Regulated Voltage	1.2 V to 1.8 V
RFID/NFC Standard	ISO 15693
ADC	SAR 12 bit resolution (max)
ADC Channel	Two differential and four single ended
Temperature	-30°C to 85°C
Power Consumption	~120 µW (average) at 1.2 V

temperature sensor in order to measure the internal chip temperature from -30°C to 85°C which in turn also aids in temperature dependent dynamic error cancellation.

F. Memory Devices

The internal memory devices consist of random access memory (RAM) and read only memory (ROM). Internal BIOS (basic input/output) system routines, standard libraries and global variables are stored inside the ROM memory. The application specific firmware will be stored inside an external memory device e.g. SPI flash.

Conclusion

A completely passive, ultra-low power consuming SoC, for biotelemetry applications has been discussed here. The SoC consists of a 32 bit microcontroller, a reconfigurable four channel 12 bit ADC and also an instrumentation amplifier, as well as a RFID and NFC standard compliant communication interface, with the main characteristics listed in table 1. The usage of a standard RFID reader or a smart phone/ tablet with NFC capability provides a very convenient way of interacting with the system and simultaneously opens up new opportunities for next generation passive medical and healthcare systems.

References

[1] T.Volk, M.Bhattacharyya, W.Gruenwald, L.Reindl, D.Jansen, "Formal Description of Inductive Air Interfaces using Thévenin's theorem and numerical analysis", IEEE Trans on Magnetics, vol.50, no.6, June 2014

[2] K.Wolf-Maier et al, "Hypertension prevalence and blood pressure levels in 6 European countries, Canada and the United States", JAMA, vol. 289, nr.18,2003, p.2363-2369

[3] J.H Cheong, S.S Yan Ng, X.Liu, R.Xue, H. J Lim, P.B Khannur, K.L Chan, A. A lee, K.kang, L.S Lim, C.He, P.Singh, W.Park, "An Inductively Powered Implantable Blood Flow Sensor Microystem for Vascular Grafts", IEEE Trans on Biomedical Engg,vol.59,no 9, pp 2466 – 2475, September 2012

[4] D.Jansen,N.Fawaz,D.Bau, M.Durrenberger, "A small high performance Microprocessor core SIRIUS for embedded low power designs, demonstarted in a medical mass application of an electronic pill", ISBN 978-0-387-72257-3,p.363-372

[5] D.Venuto, E.Stikvoort and D.Castro, "Ultra low-power 12 bit SAR ADC for rfid Aapplications", Design, Automation & Test in Europe Conference & Exhibition, 2010 , ISBN - 978-1-4244-7054-9

Multi Sensor Voltage Signal Conditioner with Adaptive Level Shift and DC Offset Calibration on a Single Chip

Ji-Hoon Suh, Mauricio Velazquez Lopez, Jeong-Ho Park and Hyung-Joun Yoo

The Department of Electrical Engineering and the Mobile Sensor and IT Convergence Center
Korea Advanced Institute of Science and Technology
373-1 Guseong-dong, Daejeon, Republic of Korea
sj3995@kaist.ac.kr, mvl94@kaist.ac.kr, pjh0656@kaist.ac.kr, hjyoo53@kaist.ac.kr

Abstract

This paper presents an adaptive signal conditioning method for multi sensors with initial voltage setting, offset calibration and signal amplification. The initial voltage setting process to VDD/2 (600 mV) by level shifters is implemented for adapting various sensor signals, even when the signal's frequency is of several Hz or its initial voltage is of several mV. Offset calibration is implemented inside a programmable gain amplifier to compensate initial voltage setting error. The architecture can amplify various types of sensor signals below supply voltage (1.2 V) without coupling capacitors.

Keywords-Sensor signal conditioning; programmable gain amplifier; initial voltage setting; offset calibration; multi sensors

Introduction

Sensor application systems are increasingly adopting multi-sensor solutions like fire alarm systems which comprise multiple gas detectors (CO, CO_2). In this case, each sensor has different initial voltage and variation range. For instance in [1], the CO_2 sensor signal range is 380-320 mV (400 °C, 400-2000 ppm), while the CO sensor range is 3-10 mV (600 °C, 100-500 ppm). The variations of both signals are too slow to utilize coupling capacitors in the amplification stage. To amplify this kind of signals, difference amplifiers [4] are frequently used. But in this case, it is essential to set each sensor's initial voltage in order to amplify the pure signals only without offset. This can make interface circuit implementation inefficient, so the circuit needs to be reconfigurable to adapt to all these various conditions. Moreover, when the initial voltage is too low (CO sensor, 3-10 mV) to be in the linear range of an op-amp, the output of the amplifier is totally distorted. This paper proposes an architecture including an adaptive initial voltage setter and a programmable gain amplifier (PGA) with an offset calibrator, also suggesting a solution for the low initial voltage case.

System Architecture

For the conformity of various sensor signals, their initial voltage is shifted to VDD/2. This level provides maximum signal swing range and the most linear response for PGA [4]. The difference between the shifted sensor signal and the static VDD/2 is amplified by PGA as shown in Fig.1. But an inevitable offset voltage appears, so offset calibration process is implemented in the programmable gain amplifier (PGA).

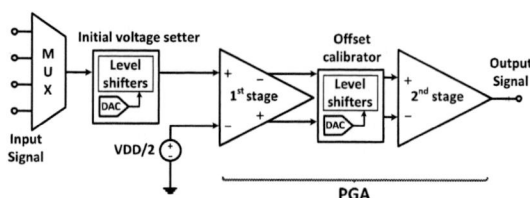

Fig. 1 Overall architecture

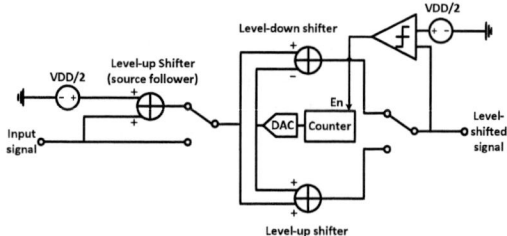

Fig. 2 Initial voltage setter

The PGA structure is based on a two-stage digitally programmable instrumentation amplifier [2] where each stage has its own gain, ranging from 19 to 60.8 dB. Level shifters consist of an op-amp feedback structure, summing and difference amplifiers [4]. Another special level shifter has a source follower structure using a pMOS for a low initial voltage case. This is adopted because the pMOS follower is adequate to transit signals which are near to 0 V. A 7 bit (binary) current steering digital to analaog converter (DAC) was implemented with unit-weighted PMOS cells [3]. Unit-weighted cells prevent mismatch errors to be heavily present on the DAC's output. DAC range goes from 0 to 640 mV with the least significant bit (LSB) of 5 mV.

A. Level Shifting Algorithm

Fig. 2 shows the initial voltage setter. The sensor signal will be shifted to VDD/2 through level up and down shifters. Additionally, in case the signal has an initial voltage of several mV, the source follower level-up shifter can be selected. After realizing where the initial voltage is, the 7 bit counter keeps counting up and the DAC output voltage goes up. Counting up process stops when the sensor signal is shifted around VDD/2.

B. Offset Calibration Algorithm

The amplified offset is also calibrated by the level shifting scheme. Largely separated differential signals after the first stage of the PGA get closer to each other by the level-up and down shifters. Fig. 3 shows the PGA with an offset calibrator.

978-1-4799-5128-4/14 $31.00 © 2014 IEEE

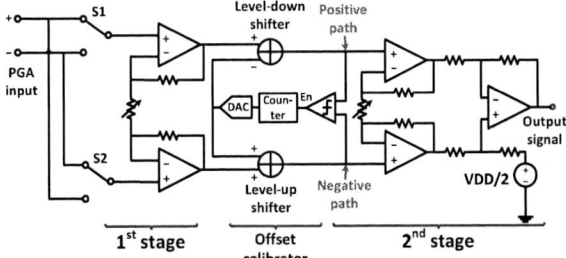

Fig. 3 PGA with offset calibrator

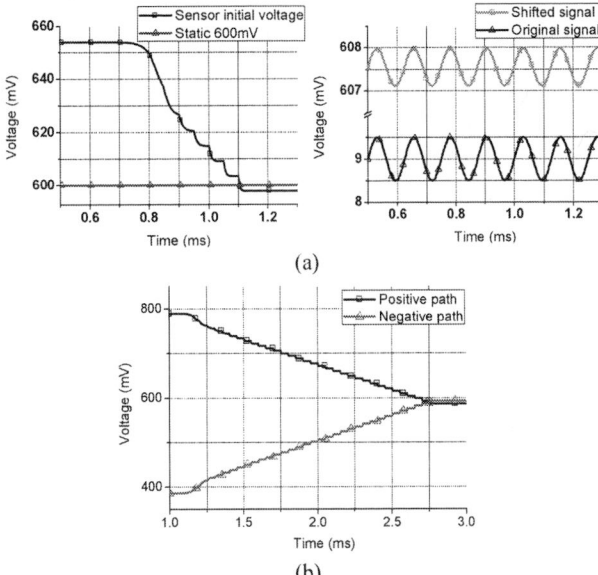

(a)

(b)

Fig. 4 Simulation results for (a) initial voltage setting and (b) offset calibration

The comparator determines when the calibration is done and stops the counting process. The second stage amplifies the calibrated signal.

Simulation Results and Discussion

The initial voltage setting and offset calibration results are shown in Fig. 4. Once an external start signal comes in, the initial voltage setting process is triggered.

The left side of Fig. 4 (a) shows an initial voltage of 655 mV which goes down toward VDD/2 (600 mV) with a step of 1 LSB of the DAC, and stops right after being below VDD/2. The right side of Fig. 4 (a) shows the shifted sinusoidal signal through the source follower level-up shifter. The original signal is out of the linear range of an op-amp inside, but source follower type shifter, not the op-amp type, shifted it from 9 to 607.5mV without distortion.

As soon as the initial voltage setting is finished, offset calibration starts. The signals at the positive and negative paths in Fig. 2 are a differential signal which is amplified by the PGA first stage. These signals are shown in Fig. 4 (b), and it has a large offset. This is because the initial voltage setting proceeds with the step of 1 LSB, and error voltage below 1 LSB amplified. It is shown in Fig. 4 (a), the difference between the shifted signal and 600 mV after 1.1 ms. The differential signals approach each other with a step of 1 LSB, and stop after they cross.

TABLE I. Comparison Table

	[5]	[6]	This work
Supply (V)	1.2	1.8	1.2
Power (mW)	6.5	0.72	2.9
Process (μm)	0.13	0.18	0.11
Size (mm²)	0.39	0.26	0.58
Max. offset (mV)	15	5	5
PGA gain (dB)	-8 to 54	0 to 18	19 to 61
Application	WLAN	Accelero.	V signal

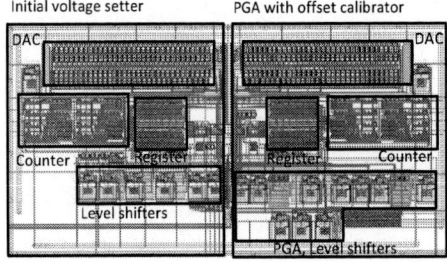

Fig. 5 Layout

Table I shows a comparison with similar works. This work carries high adaptability for input voltage signals while showing a small maximum offset and a wide gain range. Layout is shown in Fig. 5.

Conclusion

In this paper, a signal conditioner is proposed which is able to amplify various types of sensor signals below supply voltage (1.2 V) without coupling capacitors. Adapting algorithm to various sensor initial voltages is realized. The architecture above will be measured after chip fabrication.

Acknowledgment

This work was supported by IC Design Education Center (IDEC).

References

[1] C. O. Park, S. A. Akbar, and W. Weppner, "Ceramic electrolytes and electrochemical sensors," *J. Material Science*, vol. 38, pp. 4639-4660, 2003.

[2] A. T. K. Tang, "Enhanced programmable instrumentation amplifier," in *Proc. IEEE Sensors*, pp. 955-958, Nov. 2005.

[3] W.-T Lin and T.-H Kuo, "A compact dynamic-performance-improved current-steering DAC with random rotation-based binary-weighted selection," *IEEE J. Solid-State Circuits*, vol. 47, no. 2, pp. 444-453, Feb. 2012.

[4] C. K. Alexander and M. Sadiku, *Microelectronic Circuits*, 5th ed., New York: Oxford Univ. Press, 2004.

[5] X. Yao, Z. Gong and Y. Shi, "A programmable gain amplifier with digitally assisted DC offset calibration for a direct-conversion WLAN receiver," *J. Semiconductors*, vol. 33, no. 11, 115006, Nov. 2012.

[6] P. -C. Wu, B. -D. Liu, S. -H. Tseng, H. -H. Tsai, and Y.-Z. Juang, "Digital offset trimming techniques for CMOS MEMS Accelerometers", *IEEE J. Sensors*, vol. 14, no. 2, pp. 570-577, Feb. 2014.

A Transformed Radial Stub Low-pass Filter Using Defected Ground Structure for Stopband Extension

Shanshan Xu[1], Kaixue Ma[1,2], Kiat Seng Yeo[1,3]

[1]School of Electrical and Electronic Engineering, Nanyang Technological University, Singapore
[2]University of Electronic Science and Technology of China (UESTC), Chengdu, China
[3]Singapore University of Technology and Design, Singapore
sxu2@e.ntu.edu.sg

Abstract

The paper presents a low-pass filter using defected ground structure for stopband enhancement. The face-to-face E shape defected ground structure is investigated using EM simulation, which are co-designed with transformed radial stub low-pass filter to extend the stopband. The fabricated filter demonstrates a passband insertion loss of less than 1 dB, a wide stopband with more than 20 dB rejection up to $16f_c$ (f_c is the cut-off frequency of 2.4 GHz), a passband return loss of better than 20 dB, and a compact size of 5×1.3 cm^2 only.

Keywords - defected ground structrue; low-pass filter; wide stopband

Introduction

Low-pass filters (LPF) are widely used in communication systems for noise and spurious filtering. Advanced LPF designs have to accommodate for the modern mobile ICs, and feature the characteristics of low insertion loss, wide stopband, and small form-factor. Recently, LPF based on transformed radial stub (TRS) was investigated in [1]. LC structures were added at IN/OUT ports for stopband extension.

In this paper, defected ground structures (DGSs) [2-4] are investigated and implemented in the TRS LPF to enhance stopband rejection, with no additional area occupied. The fabricated filter features low insertion loss, ultra-wide stopband, and compact size.

Face-to-Face E shape DGS

The structure of the proposed DGS structure is depicted in Fig. 1. Two E shapes are defected on the ground plane in the face-to-face orientation. The middle defected slots of the two E shapes are merged together.

The DGS is investigated using 3-D full-wave EM simulator ANSYS HFSS. Fig. 2 shows the simulated frequency response. Low-pass characteristic is obtained by placing the DGS under transmission line (TL), with two transmission zeros and stopband from 17 GHz to 34 GHz.

The electric field distribution is plotted in Fig. 3. At 17.1 GHz, the resonant is created by the resonance of the collective face-to-face E shape. At 32.1 GHz, the resonant is due to resonance property of each E shape.

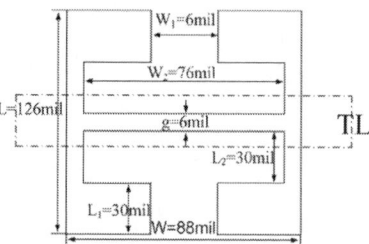

Fig. 1 Proposed face-to-face E shape DGS. (Dash line is the signal line of transmission line; the area enclosed by the solid line is the defected area on the ground plane)

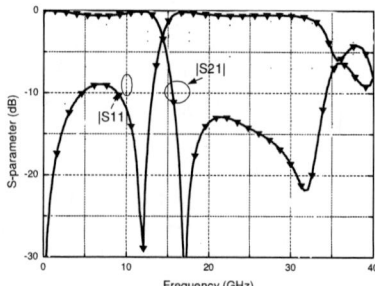

Fig. 2 Frequency response of the face-to-face E shape DGS in Fig. 1.

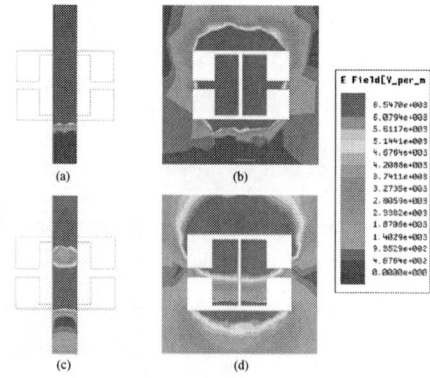

Fig. 3 Electric field distribution (a) signal line, (b) ground plane at first transmission zero (17.1 GHz), and (c) signal line, (d) ground plane at second transmission zero (32.1 GHz).

978-1-4799-5128-4/14 $31.00 © 2014 IEEE

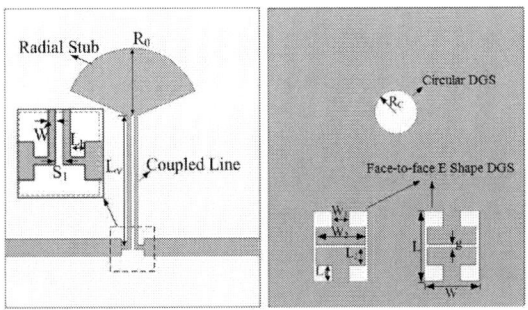

Fig. 4 Single cell of the designed LPF: top plane (left); bottom plane (right). (L_v = 231.07 mil, R_0 = 112.2 mil, W = 6mil, S_1 = 6.12mil, L_h = 11.59mil, R_C = 70 mil)

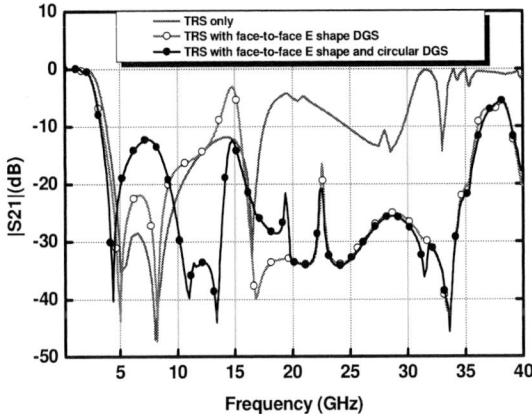

Fig. 5 EM simulated response of the TRS LPF and TRS LPF with DGS structures.

Design of the Four-Cell LPF

The proposed LPF is designed based on Rogers RT/duroid 5880 (substrate thickness = 10 mil, ε_r = 2.2, tan(δ) = 0.0009).

The single cell of the designed LPF is shown in Fig. 4. Firstly, the frequency response of the TRS LPF is simulated and shown in Fig. 5. It features low-pass response with shape roll-off and f_c at 2 GHz. But the stopband is only to 14 GHz for 10 dB rejection. Then, the proposed face-to-face E shape DGS is defected on the ground plane. The stopband rejection improves by more than 20 dB from 20 to 35 GHz. Another circular DGS shape is also defected under the coupled line to remove the undesired transmission pole at 17 GHz. Thus, the stopband is extended from 14 GHz to 35 GHz, without increasing filter size. Finally, the four-cell LPF is designed by cascading the single cell and inserting via walls to avoiding leakage modes caused by cross coupling between different cells.

The fabricated filter is only 5×1.3 cm^2 including the testing structures, that is shown in Fig. 6. The scattering parameter of the filter is measured using Rohde & Schwarz VNA ZVA-67, after standard SOLT calibration. As shown in Fig. 7, the four-cell LPF achieves low-pass response with f_c at 2.4 GHz, and stopband rejection better than 20 dB up to 38.5 GHz, which is 16 times the f_c. The passband insertion loss is less than 1 dB and return loss is better than 20 dB.

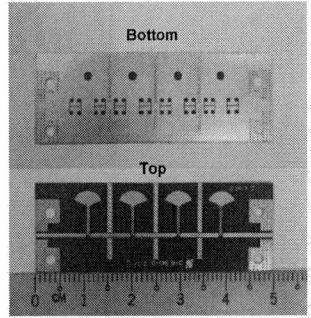

Fig. 6 Photograph of the fabricated four-cell LPF.

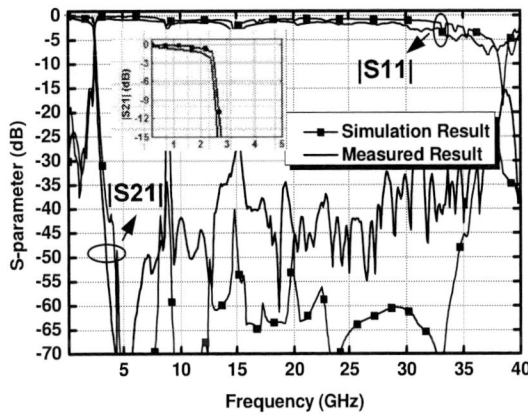

Fig. 7 Measured and simulated results of the fabricated four-cell LPF.

Conclusion

This paper proposed the face-to-face E shape DGS structure. It is studied and investigated using EM simulation with the assist of electric field distribution. The proposed structure was defected under the TRS to improve the stopband rejection, with no additional area occupation. The four-cell filter was designed and verified experimentally. Low insertion loss, shape roll-off, and ultra-wide stopband was demonstrated in the fabricated filter.

References

[1] K. Ma and K. S. Yeo, "New ultra-wide stopband low-Pass filter using transformed radial stubs," *IEEE Trans. Microw. Theory Tech.*, vol. 59, pp. 604-611, 2011.

[2] Ahn, D., Park, J.-S., Kim, C.-S., Kim, J., Qian, Y., and Itoh, T, "A design of the low-pass filter using the novel microstrip defected ground structure," *IEEE Trans. Microw. Theory Tech.*, vol. 1, pp. 86-93, 2001.

[3] A. B. Abdel-Rahman, A. K. Verma, A. Boutejdar, and A. S. Omar, "Control of bandstop response of Hi-Lo microstrip low-pass filter using slot in ground plane", *IEEE Transactions on Microwave Thoery and Techniques*, pp. 1008–1013, March 2004.

[4] J. Xiao, Y. Zhu, and J. Fu, "Non-uniform DGS Low Pass Filter with Ultra-wide Stopband", *International Symposium on Antennas Propagation and EM Theory (ISAPE)*, pp.1216-1219, November 2010.

Design of 80 – 1150 MHz CMOS LNA-less Receiver for Long-Range Wireless Sensor Network Applications

Seunghyeon Kim and Hyunchol Shin

High-Speed Integrated Circuits and Systems Lab.
Kwangwoon University, Seoul 139-701, Korea
seung_hyeon@kw.ac.kr

Abstract

A UHF/VHF LNA-less receiver is designed in 65nm CMOS for TV white-space long-range wireless sensor network applications. Noise cancelling and blocker-tolerant architecture is adopted for wide-band low-noise performance without a narrow-band LNA. The receiver consists of main- and auxiliary-path passive mixers, TIA, harmonic recombination block, and baseband filters, all of which are configured for noise cancellation. Operating between 80MHz and 1.15GHz, noise figure with 1GHz LO frequency is 1.9 dB. Blocker noise figure with 80MHz-offset 0dBm-blocker is 3.2 dB.

Keywords-component; TV White-space, Long-range wireless sensor network, CMOS Receiver

Introduction

Along with the analog-to-digital transition of television broadcasting, various possible uses of TV white space (TVWS) have been investigated for disaster communications, environmental monitoring network, public services, etc [1]. TVWS is also seen as a useful frequency band for the upcoming internet-of-things and long-range wireless sensor networks [2].

In order to cover the ultra-wideband UHF and VHF bands with a single receiver, a blocker-tolerant noise-cancelling receiver introduced in [3] is suitable. The receiver does not need a narrowly-tuned blocker-susceptible LNA, but only needs a main and auxiliary mixer-TIA path. Compared to the 40nm CMOS implementation and no baseband filter inclusion in [3], this work presents a full RF receiver chain design including a tunable baseband filter in a lower-cost 65nm CMOS technology, which makes this results more suited for complete radio implementation.

Design

Fig. 1 shows the noise cancelling receiver architecture. The main (MAIN) and auxiliary (AUX) paths amplifies the desired signal in opposite polarity but the input-referred noise in same polarity, by which the noise is cancelled out at the output. A large blocker beside the desired signal does not compress the passive mixer's gain due to its high linearity and is filtered out by the subsequent trans-impedance amplifier (TIA).

Fig. 2 shows the block diagram of the designed receiver. RF signal is down-converted by the passive mixers at MAIN and

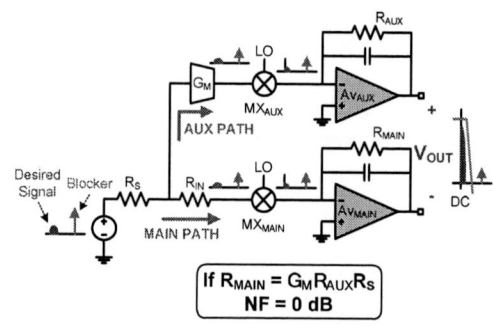

Fig. 1 Noise-cancelling receiver architecture.

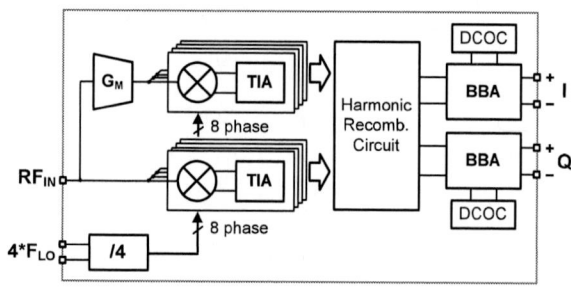

Fig. 2 Receiver block diagram.

AUX paths with 8-phase LO signal. After TIA, the 8-phase baseband signals are recombined for harmonic rejection, which helps further reduction of the output noise caused by the higher harmonics aliasing. Baseband amplifier (BBA) with dc offset cancellation circuit (DCOC) filters out-of-channel components. BBA is based on the active-RC 6th–order Chebyshev-II filter [4].

Fig. 3 shows the detailed circuit schematic of the RF front-end. It consists of MAIN and AUX paths. In the AUX path, class-AB RF G_M-cell amplifier is highly linear not to deteriorate the overall linearity, thus able to handle up to 0dBm blocker. Four-phase passive mixers with 0°/45°/90°/135° LO signals are followed by TIA. The four-phase down-conversion is needed for the 3rd and 5th harmonic rejection and I/Q generation at the subsequent harmonic recombination block. TIA is a pseudo-differential class-AB inverter type amplifier with common-mode feedback. The simple structure of TIA also helps lower the input-referred noise at the AUX path. The input impedances of the mixer and TIA chain at MAIN and AUX paths are set different. The MAIN-path impedance is

978-1-4799-5128-4/14 $31.00 © 2014 IEEE

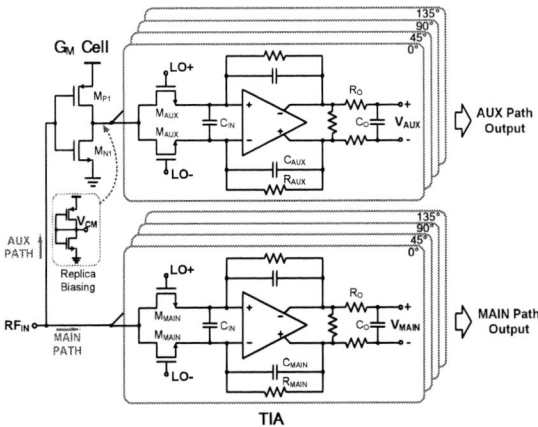

Fig. 3 RF front-end circuit.

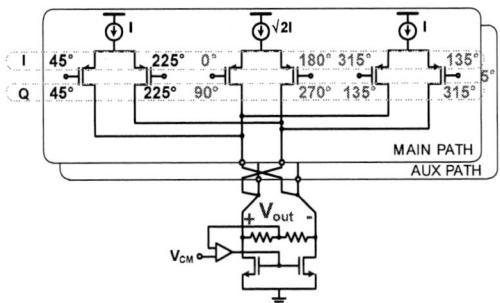

Fig. 4 Harmonic recombination circuit.

50Ω for input matching, and the AUX-path impedance is set at much lower 15Ω for achieving high linearity in front of RF G_M cell. As shown in Fig. 1, the noise cancelling condition is achieved by properly setting RF G_M value to $R_{MAIN}/(R_{AUX} \times R_S)$. In this design, $G_M = 142mS$, $R_{MAIN} = 14k\Omega$, $R_{AUX} = 2k\Omega$ are used.

Harmonic recombination circuits are shown in Fig. 4. It is based on three PMOS differential g_m stages, whose g_m's are set to be $1:\sqrt{2}:1$. The g_m stages accept the 8-phase TIA output signals. This scheme provides 3rd and 5th harmonic rejection. The I/Q baseband signals are generated by combining $0°/45°/315°$ and $90°/45°/135°$ signals, respectively.

LO signal is generated by divide-by-4 shift registers with 12.5% duty cycle. The register cell is based on dynamic logic for minimizing uncorrelated noise. Input frequency is 4 times higher than LO frequency. Simulated phase noise of LO generation block is -173 dBc/Hz at 80 MHz offset of 1 GHz LO.

Results

Circuit design and simulations are carried out by using 65nm CMOS with Cadence RF spectre. LO generation block is operated from 80 MHz to 1150 MHz, over which the receiver operation is confirmed. Current dissipation is 19.7 mA from 1.1V for the front-end up to TIA, and 24.4 mA from 1.8V for the harmonic recombination and baseband filter, and 12 mA from 1.1V for the LO generation block. Maximum overall gain of receiver is 66 dB. Fig. 5 compares the total noise figure when the noise-cancelling path is turned on and off, in which

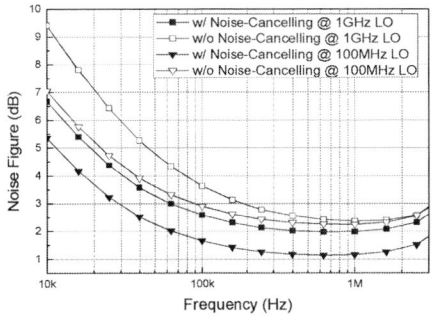

Fig. 5 Noise figure simulation result.

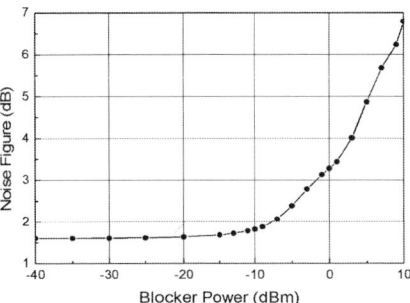

Fig. 6 Blocker noise figure at 1MHz baseband with 80MHz offset blocker.

LO frequency is 100 MHz and 1 GHz. As can be seen, noise figure is greatly improved, particularly at the low frequency region. The improvement is 3 dB at 10 kHz and 0.5 dB at 1 MHz with 1 GHz LO frequency. It is also improved by 1.7 dB at 1 MHz 10kHz and 1.1 dB at 1MHz with 100 MHz of LO frequency. Fig. 6 shows the noise figure at 800 MHz when a blocker at 80MHz offset come in together with the desired signal. This blocker noise figure is 3.2 dB at 0 dBm blocker power.

Acknowledgment

This work has been supported in part by ICT R&D program of MSIP/IITP (Development of quasi-millimeter-wave channel adaptive antennas and transceivers) and IT R&D program of MOTIE/KEIT (10044092, 60GHz RF transceiver IP development).

References

[1] *A study on plan of use and service activation in TV white space*, Korea Communications Commission (KCC), Dec. 2011.
[2] *www.weightless.org*, Weightless SIG, UK.
[3] D. Murphy *et al.*, "A Blocker-Tolerant, Noise-Cancelling Receiver Suitable for Wideband Wireless Applications," *IEEE JSSC*, pp. 2943-2963, Dec. 2012.
[4] H. Shin *et al.*,"A CMOS Active-RC Low Pass Filter with Simultaneously Tunable High- and Low-Cutoff Frequencies for IEEE 802.22 Applications," *IEEE T-CAS II*, pp. 85-89, Feb. 2010.

978-1-4799-5128-4/14 $31.00 © 2014 IEEE

A U-band VCO in 65nm CMOS with 0.44dBm output power

Jongsuk Lee[a], Sangho Shin[b] and Yong Moon[a]

[a]School of Electronic Engineering, Soongsil University, Seoul, Korea
[b]University of California Santa Cruz, USA
ljs1385@ssu.ac.kr

Abstract

A high output power VCO (voltage controlled oscillator) in U-band is implemented using 65nm CMOS process. The proposed VCO uses MTM(meta-material) technique with transmission line to increase output voltage swing and overcome the limitation of the CMOS technologies. Two varactor banks widen frequency tuning range upto 5%. The VCO operates at 51.55~54.18GHz and the measured phase noise is -100.67dBc/Hz at 10MHz offset. The chip area is 0.16x0.16mm^2 and the output power is 0.44dBm. The power consumption is 33.6mW with 1.2V supply voltage. The measured FOM$_P$ is -181dBc/Hz.

Keywords-CMOS, voltage controlled oscillator (VCO), U-band, transformer, transmisson line

I. Introduction

VCO is an inevitable component to implement a single-chip radio in communication system [1]. But CMOS VCO has the limitation of low Q-factor in on-chip passive components [2-3]. So transformer feedback technique is used to increase the output voltage swing, and transmission line theory is used to get the inductance of LC resonator enhance the high frequency performance [1-4]. The high output power could correctly transfer data but deteriorates phase noise performance. The frequency tuning range of VCO is also important factor and the capacitance value of LC resonator decides the tuning range of VCO in U-band [3]. So the proposed VCO uses two varactor banks to widen tuning range.

The remainder of this paper is organized as follows. In Section II, the proposed transformer VCO topology and the circuit design are presented. The circuit implementation and experimental results of VCO are provided in Section III. Finally, the conclusion is given in Section IV.

II. VCO Architecture and Circuit Design

The proposed VCO architecture is NMOS cross-coupled pair differential LC type and the transformer inductors using MTM concept are spiraled by L_{P1}, L_{P2} and L_{S1}, L_{S2} as shown in Fig. 1. The transformer inductors can provide drain-to-source feedback and voltage swing below ground. The two varactor banks, C_{VAR1} and C_{VAR2}, are used to widen tuning range of VCO. The value of C_{VAR1} is twice that of C_{VAR2}. The output buffers, MN3 and MN4, are added to isolate the output of VCO from other block but output swing is reduced somewhat. And

Sangho Shin is the corresponding author for this paper.

the output buffer includes 50Ω- matching network.

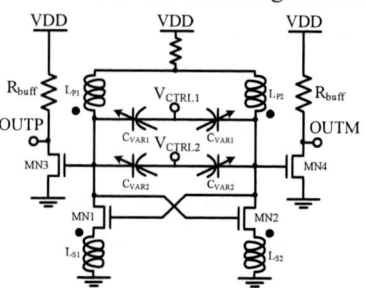

Fig. 1 Proposed VCO schematic

The cross-coupled pair of MN1 and MN2 ensures the differential mode operation and compensates loss from the passive components. The size of MN1 and MN2 is decided to sufficient gain for oscillating condition and better phase noise performance. The negative resistance of cross-coupled pair cancels the positive resistance of LC tank for oscillation. Additionally, as the mobility of PMOS transistors is lower than that of NMOS, so NMOS core is suitable for operation at high frequency.

III. Circuit Implementation and Experimental Results

The designed VCO was implemented by 65nm CMOS process and verified from cadence spectre RF simulator. The VCO chip microphotograph is shown in Fig. 2.

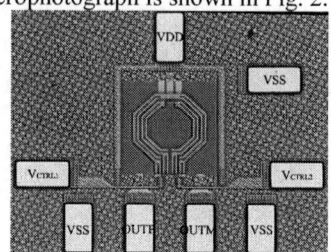

Fig. 2. VCO Chip micrograph

The four inductors are implemented by transmission line and L_{P1} (or L_{P2}) is placed close to L_{S1} (or L_{S2}) to get the high mutual inductance and high-Q. By doing this, we get the advantage in area and the area of VCO including buffer is only 0.16×0.16mm^2. The inductors in inner circle have wider width to get small inductance. L_{P1} and L_{P2} are placed at outer circle to have more inductance. We optimized the width and length of inductance to get output power as large as possible. If coupling coefficient K is 1, the output swing is shown in Eq. (1).

$$\text{Output swing} = \left(1 + \sqrt{\frac{L_P}{L_S}}\right) \bullet \text{VDD} \qquad (1)$$

978-1-4799-5128-4/14 $31.00 © 2014 IEEE

So the source inductor can swing below ground, on the other hand, if the value is too large, the oscillation is suppressed.

In measurement, on-wafer probing was carried out using a probe station, and N9010A spectrum analyzer, 11970V external mixer, N9029AE13 diplexer and dual power supply are used. Fig. 3 shows the measured frequency tuning range versus V_{CTRL1} and V_{CTRL2}.

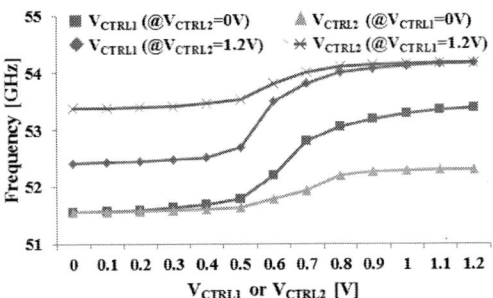

Fig. 3. The measured oscillating frequency of VCO

The FTR (Frequency Tuning Range) is from 51.55 to 54.18GHz, and the variation of V_{CTRL1} is larger than that of V_{CTRL2} as C_{VAR1} has more capacitance compared to C_{VAR2}. For the performance verification of VCO, the output power and phase noise are measured and shown in Fig. 4.

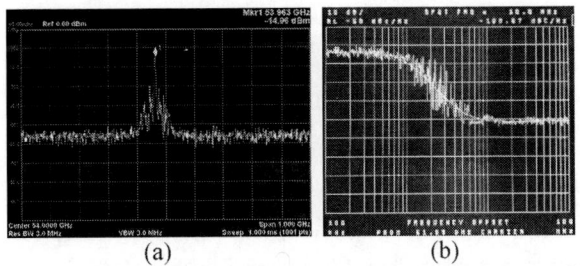

Fig. 4. (a) Measured output power and (b) phase noise versus offset frequency

The losses of measurement environment exist and we compensate those and the values are as follows. 1.2dB at Probe, 2.2dB of two Cables, 0.5dB of Adapter and 1.5dB of Diplexer are given from datasheet. So the compensated output power is 0.44dBm (ATTEN:10dB) and which is very high output voltage swing in millimeter wave data transfer system. And the measured phase noise is -100.67dBc/Hz at 10MHz offset. By using these measured values, the proposed VCO are compared to other works to verify the performance. FOM_P is used for evaluation and defined as [4],

$$FOM_P = L(\Delta f) - 20\log(\frac{f_0}{\Delta f}) - \log(\frac{P_{RF}}{1mW}) - 10\log(\eta) \quad (2)$$

Where $L(\Delta f)$ is the phase noise, Δf is offset frequency, f_0 is the oscillation frequency, P_{RF} is RF output power and η is the dc-to-RF efficiency as

$$\eta = \frac{P_{RF}}{P_{DC}} \bullet 100\% \quad (3)$$

In equation (3), P_{DC} is DC power consumption. The proposed VCO shows superior performance compared with existing VCO's which uses the same technologies. TABLE I summarized the performance comparison between the proposed work and recently reported works.

TABLE I. Performance and summary and comparisons

	[1]RFIC 2012	[2]ESSCIRC 2012	[3] TCSI 2014	This work
Process	65nm CMOS	65nm CMOS	65nm CMOS	65nm CMOS
Supply Voltage[V]	0.5	1.2	1	1.2
Operating Frequency[GHz]	37~44.1	48.8~62.3	57~65.5	51.55~54.18
Phase Noise [dBc/Hz]	-96@1MHz	-94@1MHz	-108.3@10MHz	-100.67@10MHz
Output Power[dBm]	-10	-43	-20.97	0.44
FOM_P[dBc/Hz]	-178	-163	-157	--181
P_{DC}[mW]	10	30	6	33.6

IV. CONCLUSION

A U-band 65nm CMOS MTM VCO having high output power and excellent FOM_P is designed. The VCO topology is NMOS cross-coupled pair differential LC type using transmission line and drain-to-source feedback structure for high output swing. The inductors are located close to each other to derive high Q-factor and large inductance. This structure has the advantage in area and the area of proposed VCO is $0.16\times0.16mm^2$. To widen tuning range, two varactor banks is used and the measured operating range is from 51.55 to 54.18GHz. The measured output power is 0.44dBm and phase noise is -100.67dBc/Hz at 10MHz offset. The power consumption is 33.6mW with 1.2V power supply and FOM_P is -181dBc/Hz which shows better performance than previous works. In this study, the proposed VCO is useful for millimeter wave data transfer systems. CAD tool and MPW was supported by IDEC.

Acknowledgment

This work was supported Basic Research Laboratories (BRL) through NRF grant funded by the MEST (No.20110020262) and Development Program funded by the Ministry of Trade, Industry, and Energy(MOTIE, Korea)" (NO.10048747)

References

[1] Vishal P. Trivedi and Kun-Hin To, "A Novel mmWave CMOS VCO with an AC-Coupled LC Tank," IEEE, Radio Frequency Integrated Circuits Symposium (RFIC), pp. 515-518, Jun. 2012.

[2] Liang Wu, Howard C. Luong, "A 49-to-62GHz CMOS Quadrature VCO with Bimodal Enhanced Magnetic Tuning," in Proc. ESSCIRC, pp.297–300, Sept. 2012.

[3] Wei Fei, HaoYu, Haipeng Fu, Junyan Ren, and Kiat Seng Yeo, "Design and Analysis of Wide Frequency-Tuning-Range CMOS 60 GHz VCO by Switching Inductor Loaded Transformer," Circuit and Systems I: regular papers, IEEE Transactions on (TCSI), vol. 61, no. 3, pp. 699-711, 2014.

[4] Che-Chen Lee, Shu-Yan Huang, and Hong-Y eh Chang, "A 44-49 GHz Low Phase Noise CMOS Voltage-Controlled Oscillator with 10-dBm Output Power and 16.1 % Efficiency," Microwave Symposium (IMS), 2014 IEEE MTT-S International, pp. 1-4, Jun. 2014.

Multiobjective Optimization of Input Low Noise Amplifier for Common GPS/Galileo/GLONASS/Compass Satellite Navigation System Receiver

Josef Dobeš, Jan Míchal, František Vejražka, Jakub Popp, and Václav Paňko

Czech Technical University in Prague, Department of Radio Engineering, Technická 2, 16627 Praha 6, Czech Republic
Phone: +420-2-24352207, Fax: +420-2-33339801, E-mails: {dobes, michal, vejrazka, poppjaku, pankovac}@fel.cvut.cz

Abstract

As all the four main navigation systems (GPS, Galileo, GLONASS, and Compass) work in similar frequency bands, it is reasonable to create a common input low noise amplifier for all of them. Although the whole chip including a lot of correlators and other digital circuits is quite complicated, a common low noise antenna preamplifier operating at the frequencies from 1.1 to 1.7 GHz could be quite simple and efficient. We have proposed finding a compromise between the amplifier's amplification and noise figure under several natural constrains by multiobjective optimization. Moreover, we have utilized our enhancement of known optimization algorithm (the goal attainment method) to improve its efficiency, which led to finding a very good design tradeoff for the amplifier.

Keywords-navigation systems, low noise amplifier; noise figure; SOC; CAD; multiobjective optimization; goal attainment method

Introduction

The process of the design automation of electronic circuits strongly relies on the use of computer algorithms. One class of methods not only uses them as a circuit simulation tool, but also uses numerical optimization algorithms as a means of determining parameter values in order to bring the designed circuit as close as possible to some prescribed behavior. In this way, an unusually good design tradeoff can be achieved.

Description of Algorithm of Multiobjective Optimization

A. Multiobjective Optimization Problem

There are often multiple mutually contradicting requirements on the circuit. In such cases, our aim is to solve the multiobjective optimization problem which can be written as

$$\underset{\boldsymbol{x} \in S}{\text{minimize}} \; \{f_1(\boldsymbol{x}), f_2(\boldsymbol{x}), \ldots, f_k(\boldsymbol{x})\}. \quad (1)$$

The decision vectors $\boldsymbol{x}$ belong to the feasible region S, $S \subseteq \Re^n$, which can be defined by a number of equality constraints, inequality constraints, or bounds on the decision variables x_i.

The vector of the objective functions will be denoted by $\boldsymbol{f}(\boldsymbol{x}) = (f_1(\boldsymbol{x}), f_2(\boldsymbol{x}), \ldots, f_k(\boldsymbol{x}))^{\text{t}}$, and the image of the feasible region, also called the feasible objective region, will be denoted by $Z = \boldsymbol{f}(S)$, $Z \subseteq \Re^k$. The elements of Z are called objective vectors and denoted by $\boldsymbol{z} = (z_1, z_2, \ldots, z_k)^{\text{t}}$, where $z_i = f_i(\boldsymbol{x}) \; \forall i = 1, 2, \ldots, k$ are objective values. The geometrical representation can easily be shown on a two-dimensional case, as it is presented in Fig. 1 for $n = k = 2$.

B. Pareto Optimality

We want to minimize all the objective functions simultaneously. However, because of the contradiction between them, it is not possible to find a single solution that would be optimal for all of them simultaneously. The concept of noninferiority must be used to characterize the objective vectors. A noninferior solution is the one in which any improvement in one objective requires a deterioration of another. The set of all the noninferior solutions is also called the Pareto front – in Fig. 1, it is marked by the thick curve segment between $\boldsymbol{z}_{\text{A}}$ and $\boldsymbol{z}_{\text{B}}$.

C. Goal Attainment Method (GAM)

The goal attainment method (primal in [1], improved in [2]) is defined as a scalar constrained optimization problem

$$\underset{\gamma \in \Re, \; \boldsymbol{x} \in S}{\text{minimize}} \quad \gamma$$
$$\text{subject to} \quad f_i(\boldsymbol{x}) - w_i \gamma \leqq z_i^*, \; i = 1, \ldots, k, \quad \text{(Fig. 1) (2)}$$

where f_i are the k objective functions to be minimized (design goals), S is the set of acceptable solutions (feasible region), z_i^* are predefined design goals associated with the functions f_i, $w_i \in \Re$ are predefined weighting coefficients, and γ is an auxiliary variable making the new single objective function.

The method (2) requires $2k$ input parameters (k goals z_i^* and k weights w_i) but only uses $2k - 1$ degrees of freedom, which becomes obvious from the geometrical representation in Fig. 1. The goal vector $\boldsymbol{z}^* = (z_1^*, z_2^*)^{\text{t}}$ represents a goal point in the objective space, either feasible ($\boldsymbol{z}^* \in Z$) or infeasible ($\boldsymbol{z}^* \in Z$).

TABLE I

DESIGN GOALS FOR LOW NOISE ANTENNA PREAMPLIFIER

No.	Symbol	Type	Direction	Optimum/ Bound	Unit
1	A_{pt}	Objective	Maximum	21.56	dB
2	F_n^{dB}	Objective	Minimum	0.7176	dB
3	i_{d}	Constraint	$\leqq$	120	mA
4	P_{diss}	Constraint	$\leqq$	725	mW
5	K_{Rs}	Constraint	$\geqq$	0.5	–
6	$|\Delta|$	Constraint	$\leqq$	0.9	–

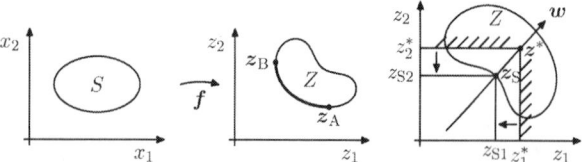

Fig. 1. Feasible region (S), feasible objective region (Z), Pareto front, and a geometrical representation of the goal attainment method (GAM).

978-1-4799-5128-4/14 $31.00 © 2014 IEEE

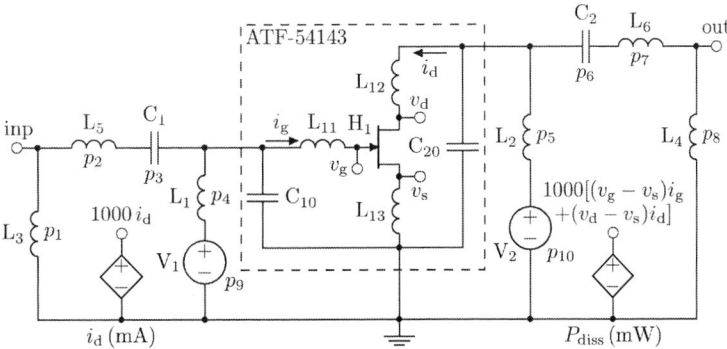

Fig. 2. Low noise antenna preamplifier for the general GPS/Galileo/GLONASS/Compass satellite navigation system receiver. The controlled sources monitor maximum allowed drain current and dissipation power of the transistor. Totally, ten parameters $p1, \ldots, p10$ can be selected by the multiobjective optimization.

The weight vector $\boldsymbol{w} = (w_1, w_2)^{\mathrm{t}}$ defines the direction of movement from the goal point (when $\gamma = 0$) to the unique solution point $\boldsymbol{z}_{\mathrm{S}} = (z_{\mathrm{S1}}, z_{\mathrm{S2}})^{\mathrm{t}}$ achieved by minimizing γ.

Finding Good Tradeoff by Multiobjective Optimization

As a demonstration of this procedure, consider a low noise preamplifier in Fig. 2. As it is shown, the circuit has ten parameters $p1, \ldots, p10$, i.e., even the biasing sources are selectable. The nominal antenna and load impedances were 50 Ω. Table I summarizes all the design goals used: transducer power gain A_{pt}, noise figure F_n^{dB}, transistor maximum ratings represented by the DC drain current i_{d} and total dissipated power P_{diss} (the two controlled sources evaluate the checked current and power), and stability ensured by the Rollett factor K_{Rs} and determinant of the S-parameter matrix $|\Delta|$.

For finding the optimal compromise between A_{pt} and F_n^{dB}, our original improvement of the goal attainment method [2] was used with the result in Fig. 3 – the selected red point in the center of the Pareto front corresponds to the two values at 1.7 GHz in Fig. 4. Please note that the maximum ratings of the transistor were not exceeded: $i_{\mathrm{d}} = 95.8$ mA and $P_{\mathrm{diss}} = 479$ mW, which is well below the manufacturer's limit [3].

We created and measured a few SMD versions of the proposed LNA, and the circuits had adequate properties including IP2 and IP3 points. For the IP3 point, the difference between the measurement and simulation was only 0.5 dB. For the optimal (red) operating point, however, a small cooler for the device is appropriate because we noticed its apparent heating.

Conclusion

Obviously, the multiobjective optimization can help to find a suitable balance among various contradictory requirements.

Acknowledgments

This work has been supported by the Technology Agency of the Czech Republic, grant No. TE01020186, and by the internal grants of the Czech Technical University in Prague Nos. SGS14/082/OHK3/1T/13 and SGS13/206/OHK3/3T/13.

References

[1] B. Kolo, *Single & Multiple Objective Optimization.* Weatherford, 2010.

[2] J. Dobeš, J. Míchal, V. Paňko, and L. Pospíšil, "Reliable procedure for electrical characterization of MOS-based devices," *Solid-State Electronics*, vol. 54, no. 10, pp. 1173–1184, Oct. 2010.

[3] "ATF-54143 low noise enhancement mode pseudomorphic HEMT data sheet," Avago Technologies, Tech. Rep. AV02-0488EN, June 2012.

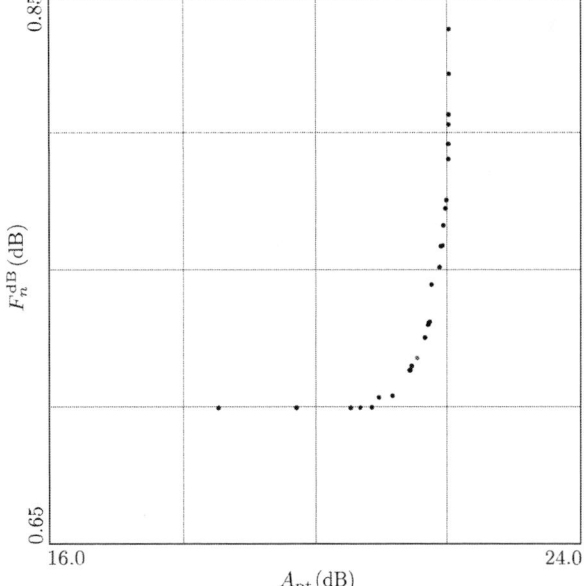

Fig. 3. Fragment of the Pareto front in the two-dimensional objective space, a detailed frequency analysis for the selected red point is shown in Fig. 4.

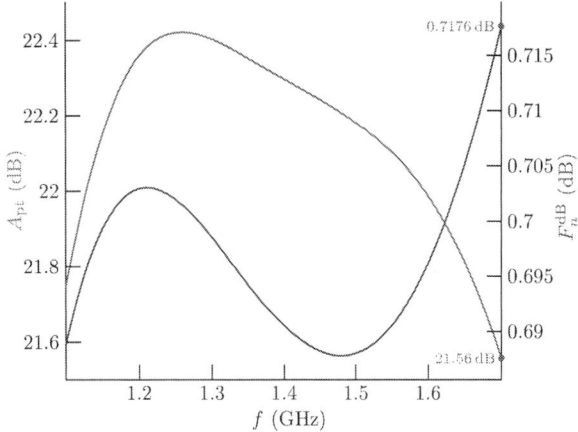

Fig. 4. Transducer power gain and noise figure for the selected tradeoff – as it is shown on the vertical axes, A_{pt} and F_n^{dB} do not change too much.

An 10-Gb/s Pulse-Mode I/O for On-Chip 5-mm interconnect

Hung-Wen Lin, Guan-Ru Wu, Zhi-Xiang Shao, Yong-Hsin Huang

Department of Electrical Engineering, YuanZe University
Room.70629, Building No.7, 135 Yuan-Tung Road, Chung-Li, Taiwan
Tel: +886-3-463-8800 #7128 , Fax: +886-3-463-5399, hwlin@saturn.yzu.edu.tw

Abstract

This paper proposes a low-power pulse-mode I/O for an on-chip data link. To use low-area MOS-type AC-coupling capacitors and the high-bandwidth source-follower-type receiver front-end, the common-mode levels of the channels are set higher than the supply voltages of I/Os. The signal amplification and high-pass filtering function at the receiver were designed using standard logic gates and MOS-type resistors only. The proposed I/Os were realized in a 90-nm CMOS process at 10Gbps and with 5-mm of on-chip microstrip channels. A test chip revealed that the I/O occupies a total area of 0.0025 mm^2, consumes 3.4 mW at 1.2 V of supply voltage, equal to a power efficiency of 0.068 pJ/bit/mm

Keywords: input/output (I/O), pulse signaling

Proposed I/O architecture

Fig. 1 shows the schematic of the pulse-mode driver, which includes four-stage tapered buffers with a multiple ratio of 2, NMOS-type AC-coupling capacitors, voltage doublers and biasing resistors. The area of MOS-type AC-coupling capacitors is one-fifth smaller than that of MIM-type capacitors, but the capacitance of an MOS is subject to high variations according to its operation regions (on or off). To achieve a stable capacitance, the MOS capacitors are set to an "ON" state during all output levels. A voltage doubler is thus added to generate the bias voltage (V_{DT}) which is a one-time threshold voltage higher than the supply voltage ($V_{DT}>V_{DD}+V_{TH}$). The voltage doubler determines the common mode level of the channel according to the biasing resistors.

Fig. 2(a) and (b) show the channel architecture and the parasitic parameters, respectively. Differential microstrip lines were placed in Metal 7, and Metal 5 and 9 were grounded planes for shielding. The line width was 0.42 μm, and the line-to-line space was 0.7 μm. The parasitic parameters of 500-μm channel were extracted for the lump model cell by using one-order curve fitting. By cascading the segments of the lump model cells, the characteristics of a 5-mm channel were thus approximated. Fig. 3 simulates the driver with the channels at 10Gbps and 1.2 V of supply voltage. The driver output levels ranged from 1.3 to 2 V, equal to 1.4 V of a differential swing. After the channels, the eye diagram at

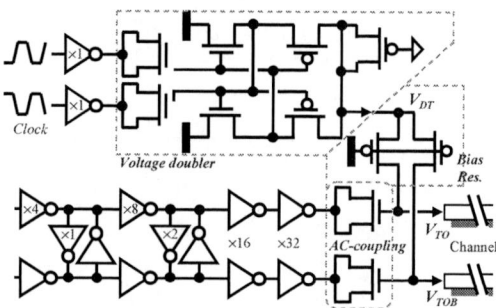

Fig. 1 Schematic of the pulse-mode driver

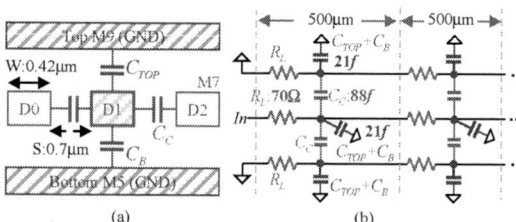

Fig. 2 On-chip channel (a)Architecture (b) Parasitic parameters

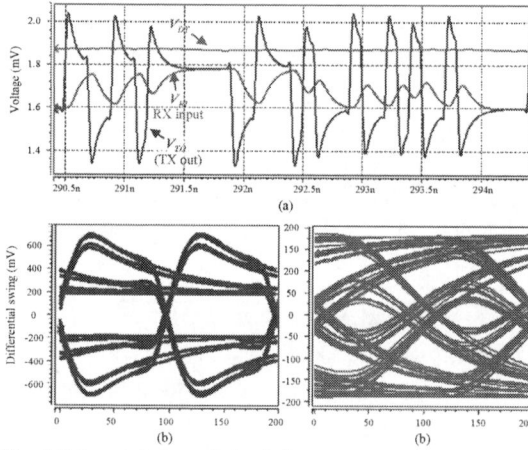

Fig. 3 Driver and channel simulations.

the receiver input was 60 mV and 0.5 UI of the vertical and horizontal openings, respectively. The resistance of PMOS resistors connected to V_{DT} varied substantially at different output levels, causing a non-zero net current from V_{DT} to V_{TO} and V_{TOB}. Thus, the common mode level of the channel was lower than instead of equal to V_{DT}.

Fig. 4(a) shows the receiver architecture which includes an N-type source follower and four stages of a gain amplifier. As shown in Fig. 4(b), the source follower

978-1-4799-5128-4/14 $31.00 © 2014 IEEE

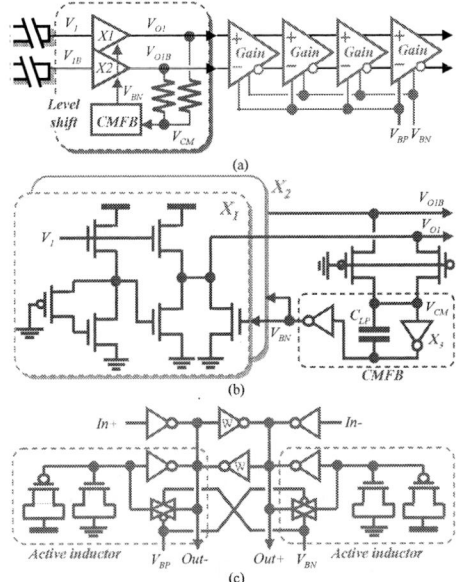

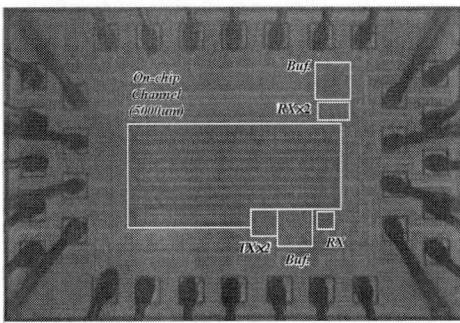

(a)

(b)

(c)

Fig. 4 Schematic of the pulse-mode receiver.

Fig. 5 Test chip photo.

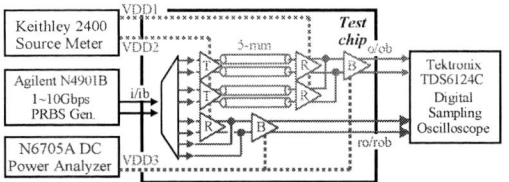

Fig. 6 Test chip measurement setup.

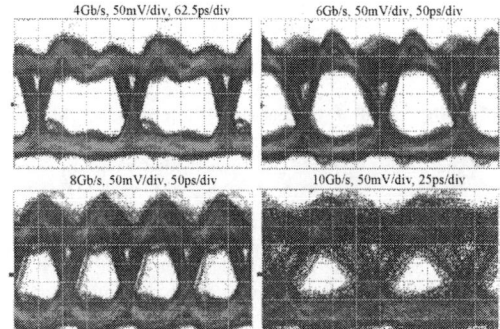

Fig. 7 Measured eye diagrams at different data rates.

TABLE I PERFORMANCE COMPARISON.

Reference	This work	JSCC'10[1]	JSCC'10[2]	ISSCC'13[3]
Process (um)	0.09	0.09	0.09	0.65
Data rate (b/s)	10G	6G	2G	4G
Channel (mm)	5	10	10	10
Width (um)	0.42	0.6	0.54	0.5
Space (um)	0.70	0.4	0.32	1
Energy per bit (pJ/bit)	Tot:0.34	Tot:0.36	Tot:0.28	Tot:0.11
	TX:0.19	TX:0.28	TX:0.15	TX:NA
	RX:0.15	RX:0.08	RX:0.13	RX:NA
Jitter (UI)	0.65	0.40	0.47	0.65
VDD (V)	1.2	1.2	1.2	1
Core area (mm²)	Tot:0.0024	Tot:0.00176	Tot:0.00035	Tot:0.0033
	TX:0.001	TX:0.00112	TX:0.00023	TX:0.013
	RX:0.0014	RX:0.00064	RX:0.00012	RX:0.002
Energy per bit per channel length (pJ/bit/mm)	0.068	0.0635	0.028	0.011

provided a high driving current and low capacitive loads, and shifted the common mode levels form ($V_{DD}+V_{TH}$) to the threshold voltage of the inverter through the common mode feedback loop. The inverter $X3$ and capacitors C_P formed the Miller capacitors to reduce the area of the loop filter. As shown in Fig. 4(c), the tunable active inductors were designed using transmission gates, complementary MOS capacitors and input-output-shorted inverters. The high-pass band of the gain amplifier can be adjusted using V_{BN} and V_{BP} to compensate for the attenuation of the channels.

Test chip measurement

The pulse mode I/O was realized in 90-nm CMOS technology, as shown in Fig. 5. The chip area was 1050 μm × 700 μm. The driver and receiver cells occupy an area of 50 μm × 21 μm and 40 μm × 36 μm, respectively. Fig. 6 shows the test setup. Agilent N4901B was used to generate random data with a swing of 600mV. Tektronix TDS6124C were used to measure the eye diagrams. Keithley 2400 were used to generate three supply voltages of 1.2 V to the driver, receiver and output buffer.

Fig. 7 shows the measured eye diagrams from 4 to

10 Gbps. At 10 Gbps, the peak-to-peak jitter of the overall system was 65 ps and the current consumption was 2.8 mA. TABLE I presents a comparison of the proposed I/O and another long-channel on-chip link[1]-[3]. This work operated at the highest data rate, but the power efficiency was poor. The circuit area was low because of digitalized circuitry.

Acknowledgment

The authors would like to thank the National Science Council for financially supporting this research (No.102-2220-E-155-004). CIC is appreciated for supporting the chip fabrication (TN90RF-102B-A0026).

References

[1] B. Kim *et al.*," An Energy-Efficient Equalized Transceiver for RC-Dominant Channels," *IEEE JSSC*, vol.45, no.6, June 2010.

[2] E. Mensink *et al.*, "Power Efficient Gigabit Communication Over Capacitively Driven RC-Limited On-Chip Interconnects," *IEEE JSSC*, Vol.45, No.2, Feb. 2010.

[3] S.-K. Lee *et al.*," A 95fJ/b Current-Mode Transceiver for 10mm On-Chip Interconnect," *ISSCC Digest Tech. Papers*, pp. 262-263, Feb. 2013.

978-1-4799-5128-4/14 $31.00 © 2014 IEEE

A Low Power High Linearity Phase Interpolator Design for High Speed IO Interfaces

Siddharth Katare[*], Sitaraman V. Iyer[†], Guluke Tong[†], Lasya R Munagala[*], Mahalingam Nagarajan[*] and Yang Bangda[†]

[*]Intel Technology India Pvt Ltd, India, [†]Intel Corporation, Santa Clara

Abstract—In clock and data recovery system of high speed IO, the phase of the clock for data sampler needs fine resolution control so that the incoming data can be sampled at a time point with the best signal-to-noise ratio. A phase interpolator (PI) is normally used as a phase shifter (or phase rotator) to generate an output clock whose phase is precisely controlled. In this paper we present a novel control code generation algorithm which can improve the linearity of PI without increasing the complexity of analog circuit used in linearly coded PI.

Index Terms—Clock data recovery, analog phase interpolator, digital encoding

I. INTRODUCTION

In clock and data recovery system of high speed IO, the phase of the clock for data sampler needs fine resolution control so that the incoming data can be sampled at a time point with the best signal-to-noise ratio. The clock phase is either controlled by a local PLL or a phase interpolator. In high speed serial IO such as PCIe, it is a common practice to share one PLL across multiple transmit/receiver pairs so as to save area and power. PLL generates in-phase ($\mathbf{I}$) and quadrature($\mathbf{Q}$) clocks to be delivered to each receiver.

PI controls the weight of input $\mathbf{I}$ and $\mathbf{Q}$ clocks and mix them to generate an output clock with the expected phase shift. If the input $\mathbf{I}$ and $\mathbf{Q}$ clocks are approximated by sinusoidal waveform, PI function can be described by the following trigonometric function:

$$
\begin{aligned}
CLK_O &= A.CLK_I + B.CLK_Q \\
&= A.cos(2\pi ft) + B.sin(2\pi ft) \\
&= \sqrt{A^2 + B^2}cos(2\pi ft - \theta)
\end{aligned} \tag{1}
$$

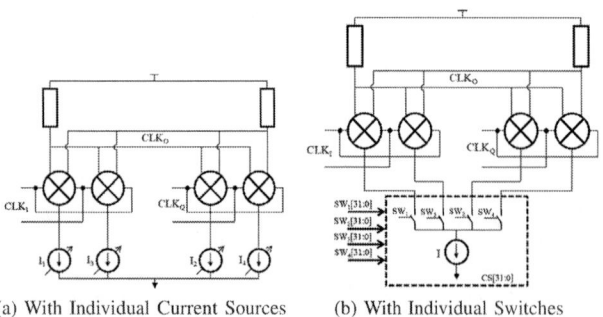

(a) With Individual Current Sources (b) With Individual Switches

Fig. 1. Linear Phase Interpolator

where $\theta = tan^{-1}(B/A)$ is phase of output clock, A and B are coefficient for $\mathbf{I}$ and $\mathbf{Q}$. The Fig. 1 shows the circuit implementation of the PI. In Fig. 1a, $A = I_3 - I_1$ and $B = I_2 - I_4$. In Fig. 1b, the $SW_1[31:0], SW_2[31:0], SW_3[31:0]$ and $SW_4[31:0]$ are thermally encoded bits and one of $SW_1[i], SW_2[i], SW_3[i]$ and $SW_4[i]$ is high at a time. The Fig. 1b shows the scheme for generating 128 different phases in one full clock cycle.

In the linear coded PI [1] [2] [3] [4], the value of A and B for a given PI code n can be determined using following equation:

$$
A(n) = \begin{cases} n & \text{if } 0 < n \le 32 \\ 64 - n & \text{if } 32 < n \le 96 \\ n & \text{if } 96 < n \le 127 \end{cases} \tag{2}
$$

$$
B(n) = \begin{cases} 32 - n & \text{if } 0 < n \le 64 \\ n - 96 & \text{if } 64 < n \le 127 \end{cases} \tag{3}
$$

The relation between A and B with $SW*$ can be given by following equation:

$$
\begin{aligned}
&\text{if } A \ge 0, SW_1[31:0] = A, SW_3[31:0] = 0 \\
&\text{if } A \le 0, SW_1[31:0] = 0, SW_3[31:0] = -A \\
&\text{if } B \ge 0, SW_2[31:0] = B, SW_4[31:0] = 0 \\
&\text{if } B \le 0, SW_2[31:0] = 0, SW_4[31:0] = -B
\end{aligned} \tag{4}
$$

For example, for $n = 5$, $A = 5$ and $B = 27$. This results in $SW_1[31:0]$=0x0000001F, $SW_2[31:0]$=0xFFFFFFE0, $SW_3[31:0]$=0x00000000 and $SW_4[31:0]$=0x00000000.

The main limitation with such linear coded PI is the phase no longer the linear function of code. [1] [2] and [4] suggest the way to over come this problem using analog solution which results in increased analog design complexity.

II. PROPOSED SCHEME FOR PI CONTROL CODE GENERATION

In this section we present the novel algorithm to generate the PI control codes without increasing the analog complexity. Fig. 2 shows the modified scheme of linear PI to implement the new algorithm subsequently referred as sine coding. Two additional half current cells *CS_half[1:0]* are added to linear PI to implement the new sine coding scheme

In the sine coded PI, the value of A and B for a given PI code n can be determined using following equation:

978-1-4799-5128-4/14 $31.00 © 2014 IEEE

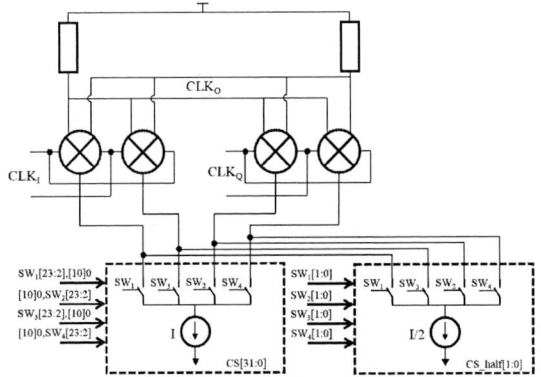

Fig. 2. Proposed Phase Interpolator

$$A(n) = \left\lfloor \frac{32}{\sqrt{2}} \cos(2\pi \frac{n}{128}) \right\rfloor, 0 \leq n \leq 127 \quad (5)$$

$$B(n) = \left\lfloor \frac{32}{\sqrt{2}} \sin(2\pi \frac{n}{128}) \right\rfloor, 0 \leq n \leq 127 \quad (6)$$

where $\lfloor x \rfloor$ is floor of x. The relation between A and B with $SW*$ can be given by following equation:

$$\begin{aligned}
&\text{if } A \geq 0, S_1 = A, S_3 = 0 \\
&\text{if } A \leq 0, S_1 = 0, S_3 = -A \\
&\text{if } B \geq 0, S_2 = B, S_4 = 0 \\
&\text{if } B \leq 0, S_2 = 0, S_4 = -B \\
&\text{if } S_1 + S_3 \leq 16, R_1 = \frac{32 - |A| - |B|}{2}, R_2 = 0 \\
&\text{if } S_2 + S_4 \leq 16, R_1 = 0, R_2 = \frac{32 - |A| - |B|}{2}
\end{aligned} \quad (7)$$

$$\begin{aligned}
SW_1[23:2] = S_1 + \lceil R_1 \rceil, SW_3[23:2] = S_3 + \lfloor R_1 \rfloor \\
SW_2[23:2] = S_2 + \lceil R_2 \rceil, SW_4[23:2] = S_4 + \lfloor R_2 \rfloor
\end{aligned} \quad (8)$$

$$\begin{aligned}
&\text{if } R_1 > 0, SW_1[1:0] = 1 - (\lceil R_1 \rceil - \lfloor R_1 \rfloor) \\
&\quad, SW_3[1:0] = 1 + (\lceil R_1 \rceil - \lfloor R_1 \rfloor) \\
&\quad, SW_2[1:0] = 0, SW_4[1:0] = 0 \\
&\text{if } R_1 \leq 0, SW_2[1:0] = 1 - (\lceil R_2 \rceil - \lfloor R_2 \rfloor) \\
&\quad, SW_4[1:0] = 1 + (\lceil R_2 \rceil - \lfloor R_2 \rfloor) \\
&\quad, SW_1[1:0] = 0, SW_3[1:0] = 0
\end{aligned} \quad (9)$$

where $\lceil x \rceil$ is ceil of x and $|x|$ is absolute value of x. For example, for $n = 5$, $A = 5$ and $B = 21$. This results in $SW_1[23:0]$=0xFE2002, $SW_2[23:0]$=0x7FFFFC, $SW_3[23:0]$=0x01C001 and $SW_4[23:0]$=0x000000.

III. SIMULATION RESULTS

This section describes the simulation setup and results from the simulation. The digital encoding for sine coding was implemented using a lookup table to simplify the digital logic.

The linearity is calculated for PCIe-Gen3 link (8Gbps). The

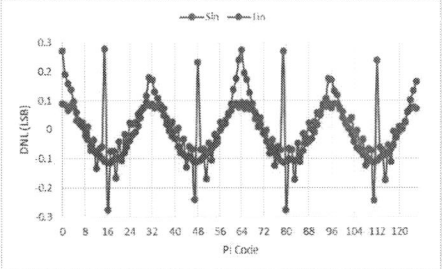

Fig. 3. Differential Non-Linearity of PI

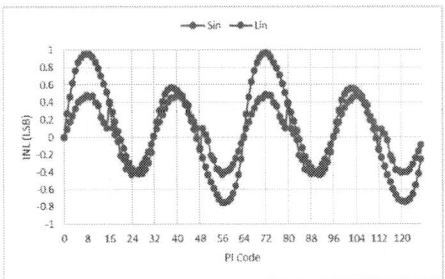

Fig. 4. Integrated Non-Linearity of PI

differential non-linearity (DNL) of the PI is shown in Fig. 3 while the integral non-linearity (INL) is shown in Fig. 4.

IV. CONCLUSION

In this paper we have presented a novel scheme for generating the control codes for PI to improve the linearity. With this new coding scheme, the DNL is similar to linear coding while the INL improves significantly. The analog circuit complexity is same as linear coding with only 3% increased power due to additional half current sources. The additional complexity arises in digital domain due to new coding scheme but it can be easily implemented using the lookup tables. This tradeoff in digital complexity help is maintaining simple analog circuitry and thus can be scaled easily to next generation CMOS process.

REFERENCES

[1] R. Kreienkamp, U. Langmann, C. Zimmermann, T. Aoyama, and H. Siedhoff, "A 10-gb/s cmos clock and data recovery circuit with an analog phase interpolator," Solid-State Circuits, IEEE Journal of, vol. 40, no. 3, pp. 736–743, March 2005.

[2] M. Meghelli, S. Rylov, J. Bulzacchelli, W. Rhee, A. Rylyakov, H. Ainspan, B. Parker, M. Beakes, A. Chung, T. Beukema, P. Pepeljugoski, L. Shan, Y. Kwark, S. Gowda, and D. Friedman, "A 10gb/s 5-tap-dfe/4-tap-ffe transceiver in 90nm cmos," in Solid-State Circuits Conference, 2006. ISSCC 2006. Digest of Technical Papers. IEEE International, Feb 2006, pp. 213–222.

[3] S. Hu, C. Jia, K. Huang, C. Zhang, X. Zheng, and Z. Wang, "A 10gbps cdr based on phase interpolator for source synchronous receiver in 65nm cmos," in Circuits and Systems (ISCAS), 2012 IEEE International Symposium on, May 2012, pp. 309–312.

[4] Z. Zhihui, W. Yuan, Z. Junlei, Y. Hailing, and J. Song, "A clock and data recovery circuit for 3.125gb/s rapidio serdes," in Electron Devices and Solid-State Circuits (EDSSC), 2010 IEEE International Conference of, Dec 2010, pp. 1–4.

25-Gb/s inductorless output buffer circuit with a pre-emphasis in 65-nm CMOS

Tomoki Tanaka,
Keiji Kishine
and Hiromi Inaba
University of Shiga Prefecture
2500, Hikone Hassaka,
Shiga, 522-8533, Japan
Email: oz23ttanaka@ec.usp.ac.jp

Akira Tsuchiya
Kyoto University
Yoshida Honmachi, Sakyoku
Kyoto, 606-8501, Japan

Abstract—A 25-Gb/s inductorless output buffer circuit with a pre-emphasis is proposed. We designed the circuit parameters according to the frequency characteristics of the emphasized signal. To confirm the advantages of the emphasis circuit, we fabricated an output buffer IC in a 65-nm CMOS process. The proposed circuit has a control voltage to adjust the emphasis amplitude according to the load outside the chip. Measurement results showed that the jitters were 40% lower with the emphasis circuit than without, indicating that our proposed configuration can be applied to the design of output buffer circuits for higher operation speed.

I. INTRODUCTION

Optical communication systems transmit and receive a great deal of information using higher speed signals. One of the key issues with high speed communication circuits is signal degradation at the I/O circuit caused by large loads such as a long transmission line outside the chip. An I/O circuit that can compensate for this degradation is strongly required [1]-[3].

In this paper, we propose a 25-Gb/s inductorless output buffer circuit with a pre-emphasis circuit to compensate for the degradation. The proposed circuit consists of an output buffer circuit with an emphasis circuit based on a high-pass filter (HPF). The degradation in the main path signal is compensated for by the signal generated in the emphasis circuit. To clarify how well the emphasis circuit worked, we fabricated an output buffer IC with a pre-emphasis circuit using a 65-nm CMOS process. The IC achieved a 40% lower jitter operation with the emphasis circuit than without at 25 Gb/s. The supply voltage is 1.2 V and the power consumption for the emphasizing is 15% of the output buffer.

II. ANALYSIS OF EMPHASIS CIRCUIT FOR CML BUFFER CIRCUIT

Generally, current mode logic (CML) is used for the output buffer to provide higher speed operation. However, the long transmission line outside the chip degrades the high-frequency components in the output signal and leads to unstable operation in the following chips. To avoid this, we use a technique to emphasize the high-frequency components in the I/O circuit. An additional path (emphasis path) generates the high-frequency components in the signal (Fig. 1(a)) [3]. The

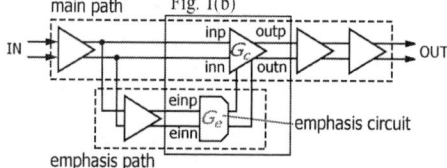

(a) Block diagram of output buffer circuit with a pre-emphasis.

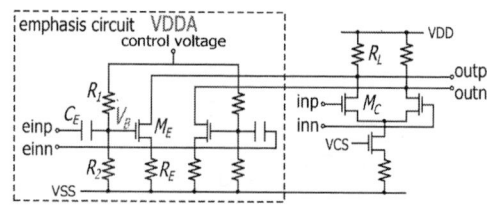

(b) Schematic of emphasis circuit and CML circuit.

Fig. 1. 25Gb/s output buffer circuit with a pre-emphasis.

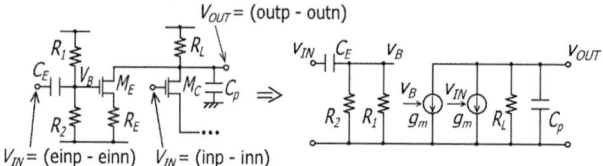

Fig. 2. Small-signal equivalent-circuit model.

signal generated in the emphasis path is combined with the signal in the main path to compensate for the degradation. We adopt a high-pass filter (HPF) circuit for the emphasis path, in order to amplify the higher frequency components in the transmitted signal(Fig. 1(b)). These circuits were designed without an inductor to ensure a small area. As shown in Fig. 1(a), we set the gains of the main and emphasis paths to G_c and G_e, respectively, and the total gain can be expressed as

$$G = G_c + G_e. \tag{1}$$

In this way, the gains in the main and emphasis paths can be controlled individually. In this work, to find out the effect of the circuit parameters on the frequency characteristics of the emphasis signal, we used a small-signal equivalent-circuit model (Fig. 2). We assume $V_{IN} = (\text{einp}-\text{einn}) = (\text{inp}-\text{inn})$,

978-1-4799-5128-4/14 $31.00 © 2014 IEEE

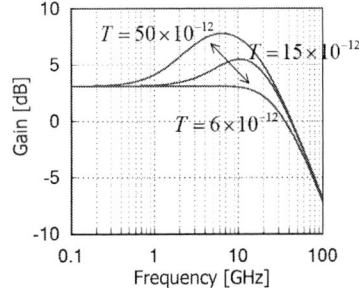

Fig. 3. Frequency characteristics of the gain estimated using (2).

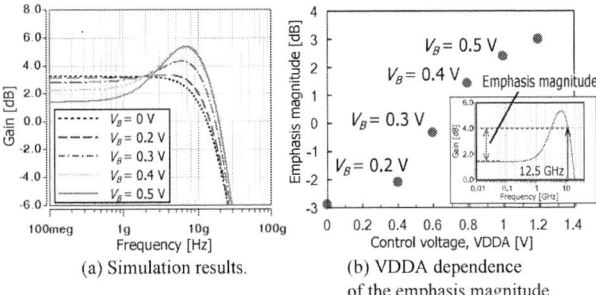

(a) Simulation results.

(b) VDDA dependence of the emphasis magnitude.

Fig. 4. Emphasis magnitude.

to obtain the expression for the gain easily in this equivalent-circuit. Here, the total gain is given by

$$G_e + G_c = -g_m R_L \frac{1 + j\omega 2T}{(1 + j\omega C_p R_L)(1 + j\omega T)}, \quad (2)$$

where $T = C_E \left(\frac{R_1 R_2}{R_1 + R_2} \right)$.

Fig. 3 shows the frequency characteristics of the gain estimated by using (2). It shows the characteristics are changed by the time constant T. When the T value becomes smaller, the gain becomes smaller and the frequency at the peak gain becomes higher. Furthermore in the detailed design, it is crucial to take the MOSFET characteristics into account, particularly in operation around the frequency of 12.5 GHz to meet the target input data speed of 25 Gb/s. Therefore, in this work, we determined T as 15.8×10^{-12} using HSPICE simulation to make the gain at 12.5 GHz the highest.

In a practical implementation of this emphasis circuit, we determined the gain by adjusting V_B with the determined T (Fig. 4(a)), which can be controlled by the voltage VDDA (Fig. 1(b)). Fig. 4(b) shows the VDDA dependence of the gain at the frequency of 12.5 GHz which is defined as the emphasis magnitude. On the basis of these characteristics, we adjust the appropriate emphasis magnitude by the voltage VDDA according to the load outside the chip.

III. EXPERIMENTAL RESULTS

To confirm the advantages of the emphasis circuit, we fabricated an output buffer IC with a pre-emphasis circuit using the 65-nm CMOS process. The area for the output buffer circuit with a pre-emphasis was $230 \times 112 \ \mu m^2$ and the emphasis circuit occupied an area of $40 \times 105 \ \mu m^2$. The supply voltage was 1.2 V. The power consumption for

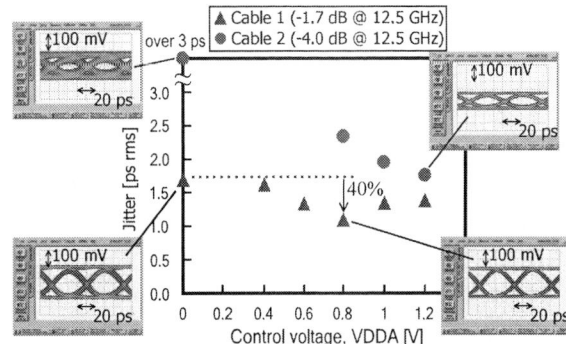

Fig. 5. VDDA dependence of measured jitter at 25 Gb/s.

emphasizing was 15% of a CML buffer circuit consuming 74 mW. To evaluate the IC, we measured the jitter from an eye diagram with a 25 Gb/s input. Fig. 5 shows the VDDA dependence of the measured jitter, in which the measured jitter is 40% lower with the emphasis circuit than without in cable 1, which shows a loss of 1.7 dB. The measured jitter differed depending on the VDDA voltage. For example, in cable 1, the minimum jitter is 1.100 ps rms (@$2^9 - 1$ PRBS data input) with VDDA of 0.8 V, and the emphasis magnitude at VDDA of 0.8 V is almost identical to the degradation of cable 1 at 12.5 GHz (Fig. 4(b)). In cable 2, which shows a loss of 4.0 dB, the minimum jitter is 1.752 ps rms (@$2^9 - 1$ PRBS data input) with VDDA of 1.2 V. These results demonstrate that we can provide the appropriate emphasis against various loads by adjusting the VDDA voltage.

IV. CONCLUSION

In this paper, a 25-Gb/s inductorless output buffer circuit with a pre-emphasis circuit was presented. We determined the circuit parameters using a small-signal equivalent-circuit to obtain the characteristics of the emphasis signal and fabricated an IC in the 65-nm CMOS process. Experimental results demonstrate that we can provide the appropriate emphasis against various load by adjusting the VDDA. The IC achieves a 40% lower jitter with the emphasis circuit than without at 25 Gb/s. The area of the emphasis circuit is $40 \times 105 \ \mu m^2$ and its power consumption is 15% of the output buffer.

ACKNOWLEDGMENT

The authors would like to thank H. Katsurai and M. Nogawa of NTT Corp. for helpful discussions and suggestions. A part of this research was supported by Grants-in-Aid for Scientific Research, The Japan Society for the Promotion of Science.

REFERENCES

[1] T. Takemoto, et. al., "A 4× 25-to-28Gb/s 4.9mW/Gb/s -9.7dBm High-Sensitivity Optical Receiver Based on 65nm CMOS for Board-to-Board Interconnects," International Solid-State Circuit Conference, pp. 118-119, Feb. 2003.

[2] P. Westergaard, et al.,"A 1.5V 20/30 Gb/s CMOS backplane driver with digital pre-emphasis," Custom Integrated Circuit Conference, pp. 23-26, Oct. 2004.

[3] H. Wakita, et. al.,"A study of pre-emphasis circuit for 28-Gbit/s-class driver IC," The Institute of Electronics, Information and Communication Engineers, p. 70, Mar. 2013.

A Low Noise Class-AB Operational Amplifier with Noise Optimization Technique

Seungheun Song, Kwanseok Jung, Subin Kim, and Joongho Choi

Department of Electrical and Computer Engineering
University of Seoul, Seoul, Korea
E-mail: jchoi@uos.ac.kr

Abstract

This paper presents a low noise operational amplifier with noise optimization and reduction techniques. It consists of a folded-cascode amplifier and a class-AB output stage with a simple minimum selector. The input-referred noise of an input transistor can be optimized through utilization of the relationship between input parasitic capacitance and flicker noise. The other main noise sources in a folded-cascode amplifier are reduced by increasing the size, adding resistors or by stacking FETs at the cascade biasing circuitry. The proposed amplifier is implemented in a 0.18-μm CMOS process.

Keywords- operational amplifier, class-ab amplifier, input referred noise, low noise technique

I. Introduction

In many applications, two-stage opamps are widely used owing to their large DC gain and wide output swing at low-supply voltages [1]. However, the noise sources from transistors might decrease their performance. The input-referred noise has to be minimized in order to improve signal-to-noise ratio. The flicker noise source from the input differential pair and the current sources are mainly dominant in a low-frequency band where thermal noise is not significant. Although low-frequency noise can be shifted to a higher frequency band using chopper stabilization [2], additional circuitry, which consumes power and area, such as a clock generator and switches are required for discrete signal processing.

In this paper, a noise optimization technique is presented which optimizes the input transistor size in terms of input parasitic capacitance and flicker noise. A noise reduction technique which reduces the noise from the current source of the amplifier is also described. A simple class-AB output stage is included for achieving higher driving capability.

II. Low Noise Amplifier Design Technique

A. Input Gate Size Optimization

In a signal processing circuit with capacitors, the input parasitic capacitance of the amplifier is critical unless the sampling or sensing capacitor is large enough so that the relevant parasitic capacitance becomes negligible. To reduce the input parasitic capacitance, the transistor size should be kept small. However, reducing the size might cause significant flicker noise, as shown in Fig. 1, where the input sampling capacitor

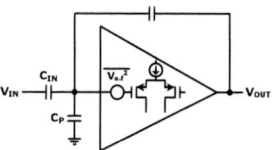

Fig. 1 Input-referred noise in sampling or sensing application.

and parasitic capacitances are denoted as C_{IN} and C_P, respectively. PMOS transistors are preferred over NMOS transistors due to their lower noise characteristics. The input-referred noise resulting from the flicker noise property can be approximated by

$$\overline{V_{N.IN}^2} = \int \frac{K}{C_{OX}WL} \cdot \frac{1}{f} df \cdot \left(1 + \frac{C_P}{C_{IN}}\right)^2 \qquad (1)$$

where K is a process-dependent constant. From (1), the trade-off between the flicker noise and parasitic capacitance can be optimized. The optimized transistor size is shown in Fig. 2. As input capacitance decreases, the input-referred noise increases and the optimized size decreases. The minimum input-referred noise occurs at $WL = 16 \ \mu m^2$ when C_{IN} for the application is 1.25pF.

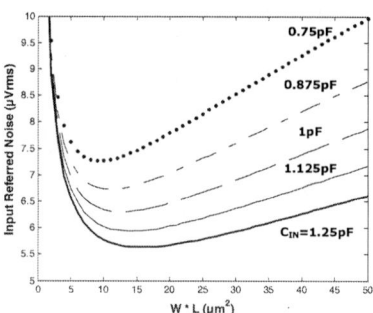

Fig. 2 Calculated and simulated optimized input gate size.

B. Noise Reduction Technique

To satisfy noise performance and input voltage range requirements, the folded-cascode amplifier with a PMOS input differential pair is chosen. The amplifier circuit is shown in Fig. 3(a), where the FETs in circles are the dominant sources of noise. The input gate size is optimized through the given size. The flicker noise of current sources is minimized by increasing

978-1-4799-5128-4/14 $31.00 © 2014 IEEE

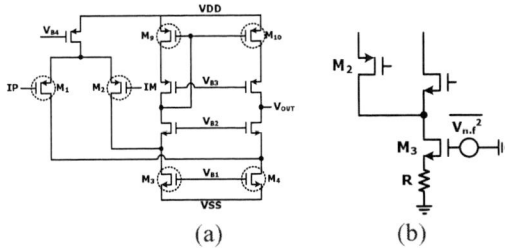

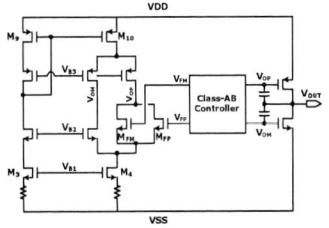

(a)   (b)

Fig. 3 (a) Folded-cascode opamp. (b) Noise reduction technique.

the size. In addition to increasing the FETs, flicker noise can be further reduced with the technique shown in Fig. 3(b).

The equivalent input-referred noise from M_3 for the conventional current mirror and proposed circuit are expressed below in (2) and (3), respectively.

$$\overline{V_{N,IN,M3.Conv}^2} = \frac{g_{m3}^2}{g_{m2}^2} \cdot \overline{V_{n.f}^2} \qquad (2)$$

$$\overline{V_{N.IN.M3.Prop}^2} = \frac{1}{g_{m2}^2} \cdot \left(\frac{g_{m3}}{1+g_{m3}R}\right)^2 \cdot \overline{V_{n.f}^2} \qquad (3)$$

As seen in the equation, the resistor R behaves as a source degeneration resistor. The input-referred noise can be reduced as this resistance increases or equivalently as transconductance of M_3 decreases. A reduction of operating voltage range due to the voltage drop across the resistor is negligible.

III. Hardware Implementation

In addition to a simple folded-cascode amplifier configuration, a class-AB output stage with a feedback minimum selector in [3] is incorporated to improve output driving capability. The control circuit for determining an output quiescent current consists of current mirrors. The class-AB output stage with a minimum selector monitors the output current and prevents the output FETs from being in the cut-off region. The output stage is shown in Fig. 4. The monitored current is converted into voltage. This voltage steers the minimum current of the output transistors.

The operational amplifier is designed and fabricated in a 0.18-μm CMOS process. The supply voltage range can be reduced to 1.5 V for low-voltage applications. Simulation results at the supply voltage 1.5 V are summarized in Table I. The value of the input-referred noise with the input sensing capacitor is 4.87 μVrms. The frequency response of the amplifier and the input noise spectrum are shown in Fig. 5. The layout of the amplifier is shown in Fig. 6, where the active area is 0.048 mm².

IV. Conclusion

In this paper, we have presented circuit design techniques for a low noise class-AB operational amplifier implemented in a 0.18-μm CMOS process. The proposed noise optimization technique is suitable for capacitor sampling or sensing applications. More than a 10% decrease in noise can be achieved by using the noise reduction technique.

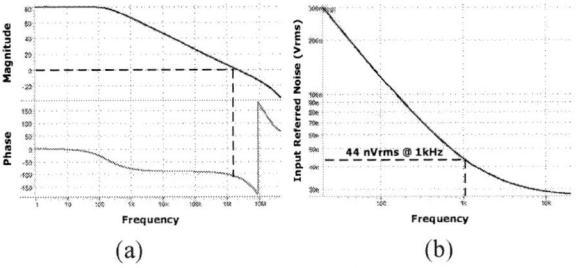

Fig. 4 Class-AB output stage with the controller.

(a)   (b)

Fig. 5 (a) Frequency response. (b) Input noise spectrum.

Fig. 6 Layout of the proposed amplifier.

TABLE I
SUMMARIZED PERFORMANCES

Parameter	Simulated	Unit
Die area	0.048	mm²
Supply voltage	1.5 – 3.3	V
Quiescent current	< 100	μA
Input noise voltage (20 Hz – 20kHz)	4.87	μVrms
Output voltage range	0.9	V_{PP}
DC-gain	> 80	dB
Phase margin	> 60	°
Unity-gain frequency	> 2	MHz

Acknowledgment

This research was supported by the MSIP (Ministry of Science, ICT and Future Planning), Korea, under the ITRC (Information Technology Research Center) support program (NIPA-2014-(H0301-14-1008) supervised by the NIPA (National IT Industry Promotion Agency).

References

[1] Jirayuth Mahattanakul and Jamorn Chutichatuporn, "Design Procedure for Two-Stage CMOS Opamp With Flexible Noise-Power Balancing Scheme," *IEEE Trans. Circuits Syst. I,* Vol. 52, pp. 1508-1514, Aug. 2005.

[2] Kuo-Chiang Hsieh, Paul R. Gray, Daniel Senderowicz, and David G. Messerschmitt, "A Low-Noise Chopper-Stabilized Differential Switched-Capacitor Filtering Technique," *IEEE J. Solid-State Circuits,* Vol. SC-16, pp. 708-715, Dec. 1981.

[3] Klaas-Jan de Lange and Johan H. Huijsing, "Compact Low-Voltage Power," *IEEE J. Solid-State Circuits,* Vol. 33, pp. 1482-1496, Oct. 1998.

978-1-4799-5128-4/14 $31.00 © 2014 IEEE

LeTourneau Motor Analyzer Senior Design Project

Ryan Tiemann, Dr. Joonwan Kim, Benjamin Ito

LeTourneau University
2100 S. Mobberly Ave.
Longview, Texas, U. S. A.
ryantiemann, joonwankim, benjaminito@letu.edu

Abstract

Faulty electric motors can cause many problems for a company, including increased maintenance cost, shortened motor lifespan, and serious accidents. Some companies have produced motor analysis devices to predict and prevent these harmful events from occurring. However, these devices are expensive and require an in depth understanding of electric circuits and motors. When a motor is suspect, the company must call in experts to analyze the motor. This is very expensive and time consuming. With the advancement of processing power in microcontrollers and the development of integral systems on chips, it is possible to produce a small unit that will be able to perform the functions of the more expensive analysis devices. The device would be a handheld tester can diagnose the health of the motor by using resistance, inductance and high voltage "megger" testing, giving a user-friendly pass/fail result, allowing non-skilled personal to conduct tests on motors. It utilizes modern circuitry and micro-controllers to perform the tests, resulting in an affordable accurate tester. The increase in motor testing availability, affordability, and frequency will decrease the chances of catastrophic accidents that can cost time, money, and lives.

Keywords: motors, sensors, signal processing, MEMS, SoC.

Introduction

There are numerous motor analyzers being produced by companies like PdMA, SKF, and others. However, these devices are very expensive, and require that the user have an in-depth knowledge of circuits and motors to be used effectively. This project will address both the problems of cost and user-friendliness that exist in current analyzers as through implementing modern technology. The LeTourneau Motor Analyzer (LEMA) was instigated by Flanders Electric who approached LeTourneau University with the goal of researching and prototyping a micro-controller device that will be able to accomplish the same accurate testing capabilities of current electro-mechanical testers.

Project Goal and Objectives

The goal of this project is to create a three phase electric motor analyzer that is user-friendly and capable of accurate motor health diagnoses. In achieving the project goal this product must fulfill the following objectives:

A. *Design and build a circuit that can measure, store and display complex impedance.*
B. *Design and build a circuit to power the unit.*
C. *Design and build a case to contain the unit.*
D. *Design and build a circuits that accomplish high voltage "megger" testing for insulation breakdown.*

Component Research

In order to fulfill the design requirements, research must be done into each component's capabilities and limitations. Components such as: batteries, SoC design and construction and microcontroller selection, were needed to implement this project.

The research conducted resulted in more options with which to proceed with the project. The use of an impedance measuring chip was discovered and one of two methods was implemented. Research also provided updated information on the resources available for the microcontroller. This revealed a newer, faster alternative to the often used Arduino. Also investigation into battery construction and types revealed where the strengths and short-comings were.

Design

Prototype design would entail building an oscillator circuit for measuring impedances, microcontroller selection and interfacing, hardware component selection in connectability to the microcontroller, and programming.

In order to measure the impedance of a circuit, an oscillator must be implemented. A pulse-oscillation inductance/capacitance (LC) Test proved to be inadequate as a stable way of measuring impedance, therefore a simple LC oscillator was implemented.

The inductance test was performed by running an LC oscillator circuit with a known capacitance, using the motor windings as the inductance element, and measuring the frequency response. This circuit shown in Fig. 1, produces a steady stream of data points, increasing the reliability of the inductance measurement.

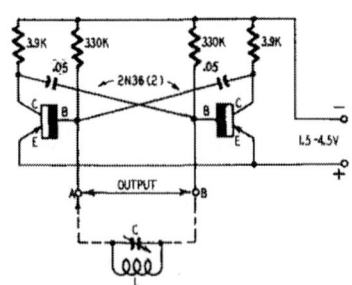

Fig. 1: Colpitts LC Oscillator Circuit

978-1-4799-5128-4/14 $31.00 © 2014 IEEE   98

A basic Arduino micro controller was initially implemented for rapid prototyping and circuitry validation. This was then upgraded to an Arduino ATMega328, since much of the circuitry that needs to interface wth the microcontroller required 3.3V logic. Although a more conventional microcontroller was initially planned for, difficulties occurred in the implementation of the program, so an Arduino based microcontroller was decided upon for the final prototype. In order to bypass having to set the individual fuse bits in the microcontroller, the data from the componants was flashed onto the chip via the serial peripheral interface (SPI) bus with Arduino Pro bootloader. This decreased the number of pins required for the microcontroller.

A single Arduino microcontroller proved to be inadequate for the amount of code needed to perform the tests, therefore, a second Arduino was implemented. One microcontroller was used for all input/output functions, and the other was assigned to test implementation.

A Secure Digital (SD) card was also implemented for memory storage capabilities so the user has the ability to log the data for each individual motor. The controller formats the data into an excel spreadsheet for ease of tracking motor performance. The spreadsheet includes all the specific values for the resistance, inductance and high-voltage megger tests.

The final tester includes: Two Arduino micro-controllers, high voltage board, inductance and resistance testing board, I/O board, case, universal banana adapters, user keypad and output screen.

Test Operation

The resistance test pulses a DC voltage into the motor phase being tested to both degauss the motor case and to determine the resistance of the motor winding via voltage division. The microcontroller unit (MCU) takes a reading at both the high and low segments of the pulse using the analog to digital converter (ADC). After a predetermined number of pulses, the program averages a certain percentage of the samples at the beginning and ending of the data run. If the difference between these two averages is above a certain percentage then a mechanical issue with the machine or motor is suspected, such as bearings, misalignment or external rubbing. The final average is the actual resistance of the winding, and is used to calculate the percent difference between the phases.

The inductance test uses the frequency outputted by the LC oscillator. When a direct current (DC) voltage is applied to the LC oscillator, it produces a wave frequency that is proportional to the inductance of the motor according to the equation:

$$f = 1/\left(2\pi\sqrt{LC}\right)$$

The MCU has an interrupt that triggers on every falling edge of the wave causing the timestamp of the interrupt to be stored. Once the function has collected a number of samples, the difference between each sample and the next is found and averaged to find the average period of the wave. The period can then be used to find the frequency of the wave and inductance of the coil.

The high voltage "megger" test utilizes a pulse-width modulated (PWM) signal from the MCU to create a variable voltage for the DC/DC converter. After it sets the voltage, the MCU waits for a voltage value from the ADC. The value read

from the ADC is proportional to the resistance of the coil insulation, and is used to determine the integrity of the insulation on the windings.

Conclusion

A prototype was produced that is able to measure resistance and inductance in three phases, measure leakage current at high voltages, and store the results of each test to a SD card. A file must be created for each specific motor before it is tested. When the test is run, the file for the specific motor is located on the SD card, then the test is run. The values returned by the test are stored onto the SD card under that motor. The controller displays a pass or fail based on the values.

The final prototype shown in Fig. 2, was able to meet all the design criteria by performing resistance, inductance and megger tests returning values that were within 4% of a comparable electro-mechanical tester. It also displays a pass/fail to the user.

Flanders Electric is funding another year of research to streamline the prototype by implementing these recommendations listed:
A. Implement a single, more powerful micro-controller.
B. Fully automate testing. Allowing for ease of user operation giving single analysis feedback to the operator.
C. Test for induced voltage before testing to ensure no residual voltage is present in the windings.

Acknowledgment

This research was done by the LEMA Senior Design Team 2013-2014. Members: Benjamins Ito, Aaron McConnell, Mario Perretta, Alex Rickards, Zac Slade, Nathan Lee (Junior member).

Funded by Flanders Electric with Mr. Robert Stokes as acting field supervisor.

References

[1] Ito, Benjamin. *Final Report.* Rep. no. LEMA. Longview: LeTourneau University School of Engineering and Engineering Technology, 2014. Print.

[2] Sedra, Adel S., and Kenneth Carless. Smith. *Microelectronic Circuits.* New York: Oxford UP, 2010. Print.

[3] Edvard. "Megger Tests." *EEP Electrical Engineering Portal RSS.* Electrical Engineering Portal, 26 Nov. 2010. Web. 10 July 2014.

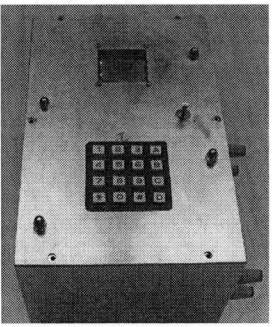

Fig. 2: LEMA Prototype

Low capacitance sensing circuit for fingerprint sensor integrated into display screen based on Charging and Extracting Process

Yena Yoo, Kyungmin Na, Heedon Jang and Franklin Bien

School of Electrical and Computer Engineering
Ulsan National Institute of Science and Technology
Ulsan, Republic of Korea
bien@unist.ac.kr

Abstract

In this paper, a sensitive and simple structure to measure low mutual capacitance of fingerprint sensor in 130-nm CMOS technology is presented. The proposed fingerprint sensor is integrated with display screen. Present sensing structures use projective capacitive touch technique. However, the new structure applies charging and extracting process (CEP). As using the CEP, the circuit challenges another different fingerprint sensor combined with display screen. Also, the current difference between ridge and valley is about 25 nA.

Keywords: fingerprint sensor, charging and extracting process, low capacitance

1. Introduction

A. Importance of Fingerprint Sensor intergrated into display screen

For the security of electronic devices, fingerprint sensor is an upcoming important topic nowadays. The most popular type, capacitive type, of fingerprint sensor is equipped in smart phone and arranged separately with touch screen. The fingerprint sensor segregated with screen may be out of simplicity trend and cause inefficient usage of screen.

Better situation is that people start using their device as soon as they touch any part of its screen, especially not being checked by another apparatus outside of screen. Therefore, to suggest combining the sensor with display screen is a natural trend.

B. Facing problems to be solved in this research

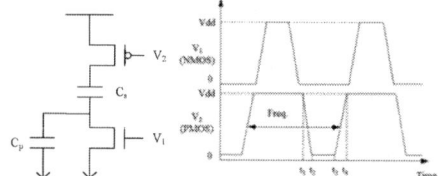

Fig. 1 Fundamental scheme for CEP

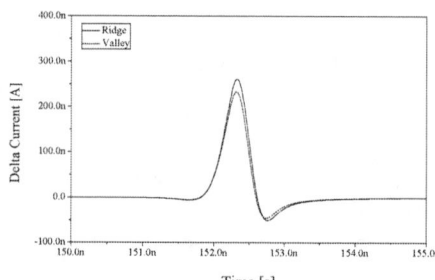

Fig. 2 Current difference between valley and untouched, ridge and untouched metal lines

The purpose of this research is to identify positions of valley and ridge despite their low capacitance difference.

To scan fingerprint by sensor, which is similar to present TSP, the lines should be denser to get accurate fingerprint measurement since the length of valley and ridge is so short. Naturally existing ITO panel should be narrower. However, if the lines become denser, there would be 2 problems.

At first, if the panel become thinner, its resistance would be larger. Then, high resistance makes the voltage drop increase. Nowadays, metal mesh has become trend for its solution.

Another problem is that the sensor couldn't guarantee proper transparency in display screen if the lines become denser. Lines should be thinner enough to be invisible by the naked eye. Metal mesh becomes solution again.

The remaining problem is to sense low mutual capacitance caused by thin metal structure. By applying [1] whose capacitance sensitivity is 0.01 fF, the CEP is used for a proposed circuit. Cs simulates the capacitance between 2 lines, and Cp symbolizes the capacitance between sensing line and grounded finger.

When V_2 is zero, PMOS become turned on and charge the C_P and C_S. When V_1 is high, NMOS connects the capacitors, C_P and C_S, with ground. At that time, the current flows differently with respect to valley or ridge in figure 2. The proposed sensing circuit is demonstrated to estimate the extent of real current size when nmos is on.

2. Simulation to estimate the capacitance

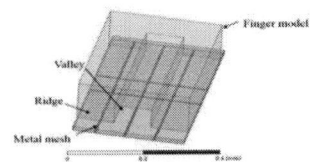

Fig. 3 Metal mesh and fingerprint model

Capacitance between 2 metal lines is calculated to 2.54fF. It is given by the following equation.

$$C = \varepsilon \times (s / d) \qquad (1)$$

C_S is set to 2.54 fF. Finally, capacitance between metal line and ridge is 7.3 fF. Capacitance between metal line and valley is 6.6 fF.

3. Proposed sensing circuit using CEP

Meanwhile, projective capacitive touch technique is used, which stimulates the capacitance continuously. However, the CEP stimulates the capacitance temporarily. It is the biggest difference between 2 processes. The current difference is too small to detect, so the switched-capacitor integrator is used to integrate the current difference.

In figure 3, the big capacitor, which symbolizes capacitor between lines, and the small capacitor, which represents capacitor between sensor and finger, connects with the voltage source during charging process. At that time, the finger capacitor voltage is different with respect to ridge and valley. This is the most important concept in the circuit, and the feedback capacitor's charge in switched-capacitor integrator is redistributed, and the voltage input in the op-amp is set to V_{REF}.

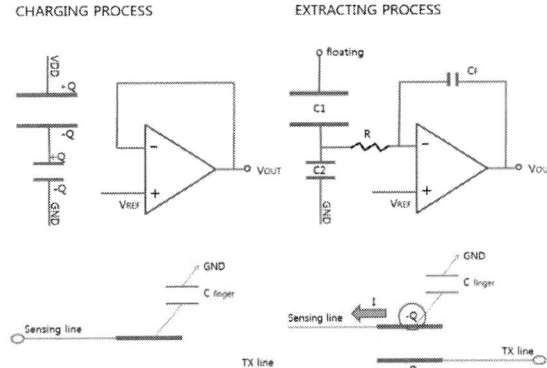

Fig. 4 The simple schematic of the circuit based on CEP

During the extracting process, in an instant the charge from the sensing line is extracted by connecting to V_{REF}. The important thing is that current size is dependent on the voltage difference between V_{REF} and voltage through the finger capacitor, so there is no reason to connect with ground.

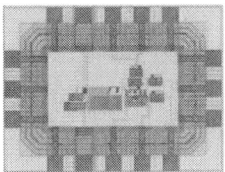

Fig. 5 Final chip for the sensing circuit

4. Conclusion

The resistor and feedback capacitor make the voltage increase by decreasing the value of resistor and feedback capacitor. However, there is limit to decrease the values, so the simulation in the any circuit design program is needed to set the smallest value of the resistor and feedback capacitor. This calculating process is regarded as the most important thing.

Measured current difference is almost 25 nA at 10 MHz. The value is suitable for relative capacitance measurement. Furthermore, the magnitude of 25 nA is enough to distinguish either valley or ridge. The proposed circuit using CDT is a new technique which could use the advantage of the relationship, that C_S is not as big as C_P.

References

[1] J. chen, D. Sylvester, C. Hu, H. Aoki, S. Nakagawa, and S. Oh, "An On-Chip, interconnect Capacitance Characterization Method with Sub-Femto-Farad Resolution," IEEE A-SSCC Papers, pp. 77-80, Mar. 1997.

A design of novel transducer converting output signals of sensor modules to 4~20mA

Min-Hyeong Cho, Won-Ho Lee, Hi-Seok Kim, and Hyeong-Woo Cha

Department of Electronics Eng., Cheongju University
298 Deaseong-ro Cheongwon-gu, Cheongju-city, Chungbuk, 360-764, Republic of Korea
Tel : 043-229-8441, E-mail : jmh35795@naver.com, aultra@nate.com, khs8391@cju.ac.kr, hwcha@cju.ac.kr

Abstract

A novel transducer converting output signals of sensor modules to 4~20mA was designed. The interface was composed with digital controllable universal voltage-to-current converter(VIC) with a current gain amplifier. The VIC used seven analog switches and six weighted resistors for selecting universal input signals. The simulation result shows that the circuit has good output current characteristics of 4~20mA for the signal 1~5V, 0~5V, -10~+10V, -5~+5V, and 0~20mA. The nonlinearity error is less than 0.1% for the input signals.

Keywords: standard signal, 4-20mA converter, analog signal process, transducer

Introduction

The transducers for sensor signals is essential device for industry measurement system such as weight, temperature, humidity, distance, and etc.[1]. Many kinds of interface transducers for accurate instrumentation and measurement system are produced by several companies [2]-[3].

The industry standard specification for the transducers have been had input and output signal of 4~20mA, 0~20mA, 1~5V, 0~5V, and -5~+5V, -10~10V from a long time ago. Therefore, the company has developed many kinds of interface circuits suitable to the industry standard specifications, respectively[4]-[5]. Because the transducers have fixed for each sensor signal output, the circuit configuration of the transducer was very complex and new circuit design is needed for the each transducer.

The transducer with current output signal of 4-20mA is used various field of long distance measurement. because it has characteristics of noise attenuation. The conventional interface for current output of 4-20mA consists of sensor module, instrumentation amplifier (IA), and voltage-to-current converter(VIC). In this case, the circuit for the IA has to design newly because the sensor module has different signal output. Therefore, the conventional transducer has demerit of circuit complex and high cost.

In this paper, new transducer converting industrial standard signal inputs such as 1~5V, 0~5V, -10~ +10V, -5~+5V, and 0~20mA to 4~20mA is presented.

Circuit Descriptions

The block and circuit diagram of the proposed transducers converting output signals of industry standard sensor module to

4~20mA is shown in Fig. 1. The block diagram of transducer consists of universal signal converter with switch control block and current gain amplifier.

The VIC composed with $R_1 \sim R_7$, $S_1 \sim S_6$, A_1, Q_1, and current source I. The current gain amplifier consists of $Q_1 \sim Q_2$, $R_7 \sim R_8$, and A_2.

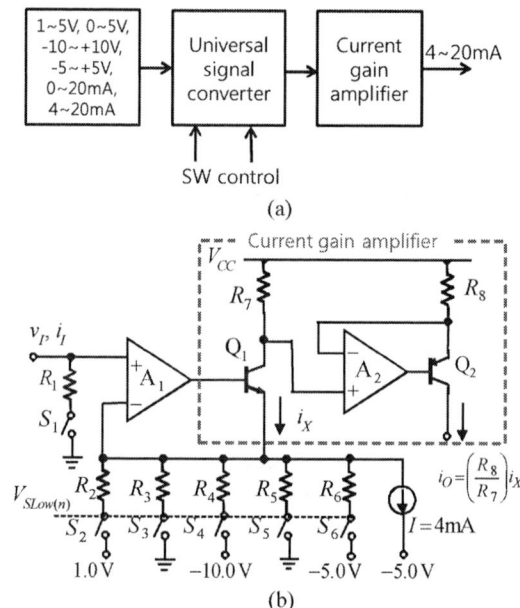

Fig. 1. (a) Block and (b) circuit diagram of principle of the proposed transducer converting output signal of sensor to 4~20mA

If the op-amp and transistor are identify with ideal characteristics, (-) terminal voltage at op amp A_1 is same to (+) terminal voltage due to negative feedback. Therefore, the current i_X for voltage signal inputs can be written by

$$i_X = \frac{v_I - V_{SLow(n)}}{R_n} + I \tag{1}$$

where, R_n is $R_1 \sim R_7$ selected by switch $S_1 \sim S_6$ and $V_{SLow(n)}$ is lower terminal voltage of switch $S_1 \sim S_6$ when the switch is ON. Because of the equation i_O for i_X $i_O = (R_8/R_7)i_X$, the final output current i_O for voltage signal v_I and current signal i_I can be written by

978-1-4799-5128-4/14 $31.00 © 2014 IEEE   102

$$i_O = \frac{R_8}{R_7}\left\{\frac{v_I - V_{SLow(n)}}{R_n} + I\right\} \tag{2}$$

$$i_O = \frac{R_8}{R_7}\left\{\frac{i_I R_1 - V_{SLow(n)}}{R_n} + I\right\}. \tag{3}$$

Table 1 shows ON switches and selected resistors for the input signals in the proposed interface.

Table I: Parameters values for input signals at the proposed transducer
($R_7 = 100\Omega$, $R_8 = 100\Omega$)

Input signal	ON Switch	Gain	Resister
1V ~ 5V	S_2	1	$R_2 = 250\Omega$
0V ~ 5V	S_3	0.8	$R_3 = 312.5\Omega$
-10V ~ +10V	S_4	0.2	$R_4 = 1.25k\Omega$
0V ~ 10V	S_5	0.4	$R_5 = 625\Omega$
-5V ~ +5V	S_5	0.4	$R_6 = 625\Omega$
0mA ~ 20mA	S_1, S_3	0.8	$R_1 = R_2 = 250\Omega$

The voltage and current reference circuit for 1.0V, -10.0V, -5.0V, and 4mA are realized by inverting or non-inverting amplifier and voltage-to-current converter based on the bandgap reference circuit shown in Fig. 2[6].

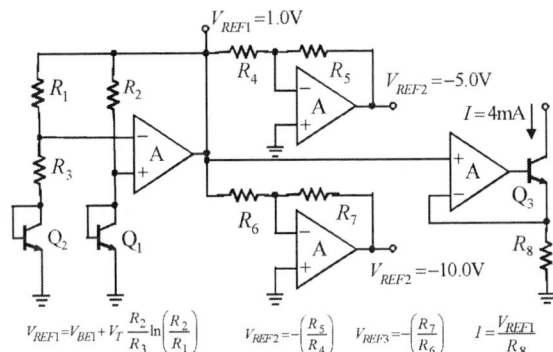

$$V_{REF1} = V_{BE1} + V_T\frac{R_2}{R_3}\ln\left(\frac{R_2}{R_1}\right) \quad V_{REF2} = -\left(\frac{R_5}{R_4}\right) \quad V_{REF3} = -\left(\frac{R_7}{R_6}\right) \quad I = \frac{V_{REF1}}{R_8}$$

Fig. 2. Voltage and current reference circuit for 1.0V, -10.0V, -5.0V, and 4mA.

Simulation results and discussion

The operation of the proposed interface circuit shown in Fig. 1 was simulated by using OrCAD PSPICE with actual devices such as OP07(op amp; A_1, A_2), ADG844(SW; $S_1 \sim S_6$), and BJT transistor Q2N2222(Q_1) and Q2N2907 (Q_2), respectively. The supply voltages $V_{CC} = +15.0\,\text{V}$ and $V_{EE} = -15.0$ V. The load resistor is $R_L = 250\Omega$. Other resistors have the value shown in table1 for switch operating.

To realize voltage and current reference for 1.0V, -10.0V, -5.0V, and 4mA, we used the resistors values of $R_1 = 7k\Omega$, $R_2 = 880\Omega$, $R_3 = R_4 = R_6 = 1k\Omega$, $R_7 = 2R_5 = 10k\Omega$, and $R_8 = 250\Omega$ at the Fig. 2

Fig. 2 shows output current waveforms for DC voltage and current input signal of 1~5V, 0~5V, -10~+10V, -5~+5V, and 0~20mA based on Fig. 1. The characteristics of the output were a good agreement with a theoretical equation (2) and (3) for input signals. The nonlinearity error was less than 0.1% for the six types of input signals.

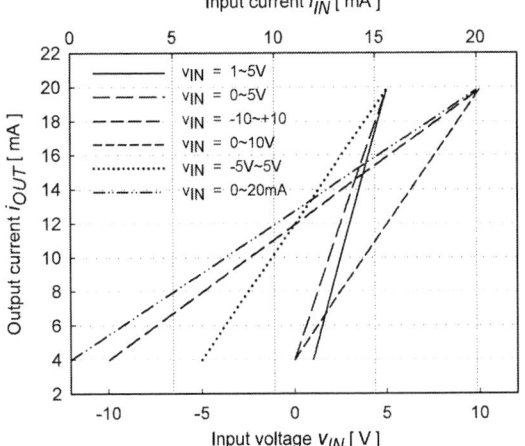

Fig. 2. Current output waveforms at DC input signals.

Conclusions and future subject

A novel digital controllable transducer circuit with current output of 4~20mA for industry standard signal processing was presented. Simulation has demonstrated that the circuit has good characteristics between input and output. Its input voltage can be controlled by digital signal for switch operation. Therefore, the interface will be a building block of instrumentation system between industry sensor modules.

In the future, we will optimize the transducer circuit using the fabrication model parameters of 0.35 μm TSMC CMOS process and fabricate chip.

Acknowledgment

This work was supported by the Industrial Core Technology Development Program(10049192, Development of a smart automotive ADAS SW-Soc for a self-driving car) funded By the Ministry of Trade, industry & Energy.

References

[1] J. J. Carr, Sensors and circuits ; sensor tranducers, and supporting circuits for electronic instrumentation measurement and control, PTA Prentice Hall, 1993.
[2] http://www.analog.com/
[3] http://www.linear.com/
[4] http://www.maxim-ic.com/
[5] http://www.sensorland.com/
[6] Alan B. Grebene, Bipolar and MOS analog integrated circuit design, Wiley Interscience, chap. 4, 1983.

Time-to-Digital Converter Architecture
with Residue Arithmetic and its FPGA Implementation

Congbing Li   Kentaroh Katoh   Junshan Wang   Shu Wu   Shaiful Nizam Mohyar   Haruo Kobayashi

Division of Electronics and Informatics, Gunma University 1-5-1 Tenjin-cho Kiryu Gunma 376-8515 Japan

Phone: 81-277-30-1789   fax: 81-277-30-1707   e-mail: t13802483@gunma-u.ac.jp

Tsuruoka National College of Technology, Japan   email: k-katoh@tsuruoka-nct.ac.jp

Abstract

This paper describes a time-to-digital converter (TDC) architecuture with residue arithmetic or Chinese Remainder theorem. It can reduce the hardware and power significantly compared to a flash type TDC while keeping comparable performance. Its FPGA implementation and measurement resuts show the effectiveness of our proposed architecture.

Keywords-   Timing Measurement, Time to Digital Converter, Residue, Chinese Remainder Theorem, FPGA

Introduction

A Time-to-Digital-Converter (TDC) measures the time interval between two edges, and time resolution of several picoseconds can be achieved when the TDC is implemented with an advanced CMOS process. TDC applications include phase comparators of all-digital PLLs, sensor interface circuits, modulation circuits, demodulation circuits, as well as TDC-based ADCs. The TDC will play an increasingly important role in the nano-CMOS era, because it is well suited to implementation with fine digital CMOS processes. [1,2,3].

There are various kinds of TDC circuits, and here we focus on a flash-typeTDC (Fig.1) [1]. It uses a delay line which consists of CMOS inverter buffer delays. Baesd on this flash-type TDC, we will introduce a new type TDC---Residue Arithmetic TDC to reduce the hardware and power significantly compared to a flash-type TDC while keeping comparable performance.Then we have implemtened it on an FPGA to verify the operation and performance.

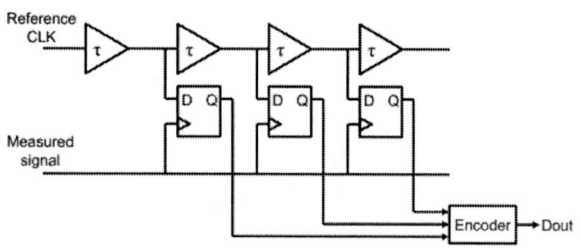

Fig 1. Flash-type TDC

Residue Arithmetic

Suppose that m1,...,mr are positive integers and coprime each other. Then there is unique positive interger x for given integers (a1,...,ar) which satisfy the following:

$$x \equiv ak \pmod{mk}, \quad k = 1, 2, \dots, r$$

where $0 \leq ak < mk$, $0 \leq x < N$ ($N = m1 \cdot m2 \cdots mr$). Table I shows the case of m1= 2, m2= 3, m3 = 5 and N=2 x 3 x 5=30, and we see that each k is mapped to residues of (m1, m2, m3) one to one [3,4].

Table I. An integer k and residues of (m1, m2, m3)

m1	m2	m3	k	m1	m2	m3	k
0	0	0	0	1	0	0	15
1	1	1	1	0	1	1	16
0	2	2	2	1	2	2	17
1	0	3	3	0	0	3	18
0	1	4	4	1	1	4	19
1	2	0	5	0	2	0	20
0	0	1	6	1	0	1	21
1	1	2	7	0	1	2	22
0	2	3	8	1	2	3	23
1	0	4	9	0	0	4	24
0	1	0	10	1	1	0	25
1	2	1	11	0	2	1	26
0	0	2	12	1	0	2	27
1	1	3	13	0	1	3	28
0	2	4	14	1	2	4	29

Residue Arithmetic TDC Architecture

We consider to use this residue arithmetic for TDC implementation, because obtaining the residue is relatively easy for time signal (used in TDC design) while it is difficult for voltage signal. (used in ADC design). Fig.2 shows the proposed residue arithmetic TDC in the case of m1= 2, m2= 3, m3 = 5 and N=2 x 3 x 5=30, where the residues a (mod 2), b

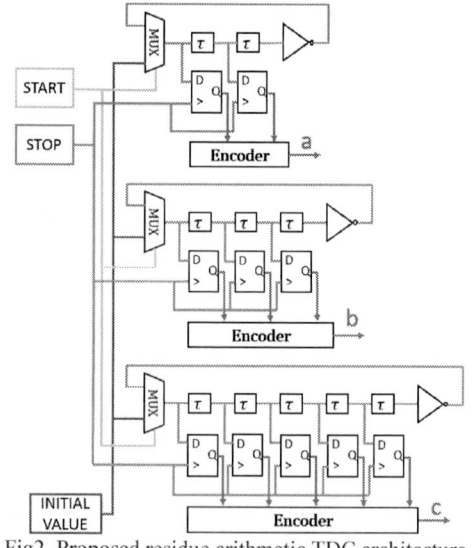

Fig2. Proposed residue arithmetic TDC architecture.

978-1-4799-5128-4/14 $31.00 © 2014 IEEE

(mod 3), c (mod 5) are obtained with ring oscillators.

Note that the proposed TDC uses only 10 delay cells and 10 flip-flops (because 2+3+5=10) while the corresponding flash TDC requires 30 delay cells and 30 flip-flops; in general, the proposed TDC uses M delay cells and M flip-flops (where M=m1+m2+···+mr) while the corresponding flash-type TDC uses N delay cells and N flip-flops (where N = m1·m2···mr), and hence the circuit and power reduction of the proposed TDC can be significant for a large N with proper choice for M<<N compared to the flash TDC.

FPGA Implementation

We have implemented our proposed TDC with an FPGA (Fig.3) [5, 6, 7], and Table II and Fig.4 show its measurement results. We see that the proposed TDC works with good linearity as expected.

Conclusion

This paper describes residue arithmetic TDC and its FPGA implementation, and the measurement results verify its operation principle.

Acknowledgements

We would like to thank STARC which supports this project.

References

[1] S. Ito, S. Nishimura, H. Kobayashi, S. Uemori, Y. Tan, N. Takai, T. Yamaguchi, K. Niitsu, "Stochastic TDC Architecture with Self-Calibration," IEEE Asia Pacific Conference on Circuits and Systems (Dec. 2010).

[2] K. Katoh, Y. Doi, S. Ito, H. Kobayashi, E. Li, N. Takai, O. Kobayashi, "An Analysis of Stochastic Self-Calibration of TDC Using Two Ring Oscillators", IEEE Asian Test Symposium (Nov. 2013).

[3] William A. Chren Jr., "Low-Area Edge Sampler Using the Chinese Remainder Theorem", IEEE T. Instrumentation and Measurement 48(4): 793-797 (1999).

[4] http://www.ndl.go.jp/math/s1/c2.html

[5] J. Xilinx, San Jose : "Virtex-5 LX FPGA ML501 Evaluation Platform", http://www.xilinx.com/products/boards-and-kits/H W-V5-ML501-UNI-G.htm.

[6] Xilinx, San Jose : "Virtex-5 user guide", 2010. [Online]. Available: www.xilinx.com.

[7] Xilinx, "Using Xilinx ChipScope Pro ILA Core with Project Navigator to Debug FPGA Applications". [Online]. Available: www.xilinx.com.

Table II Measurement results of the proposed TDC.

Sample in Window	Elapsed Time (ns)	a	b[0]	b[1]	c[0]	c[1]	c[2]	k
0	0.00	0	0	0	0	0	0	0
3	30.30	1	1	0	1	0	0	1
6	60.60	0	0	1	0	1	0	2
9	90.90	1	0	0	1	1	0	3
12	121.20	0	1	0	0	0	1	4
15	151.50	1	0	1	0	0	0	5
18	181.80	0	0	0	1	0	0	6
21	212.10	1	1	0	0	1	0	7
24	242.40	0	0	1	1	1	0	8
27	272.70	1	0	0	0	0	1	9
30	303.00	0	1	0	0	0	0	10
33	333.30	1	0	1	1	0	0	11
36	363.60	0	0	0	0	1	0	12
39	393.90	1	1	0	1	1	0	13
42	424.20	0	0	1	0	0	1	14
45	454.50	1	0	0	0	0	0	15
48	484.80	0	1	0	1	0	0	16
51	515.10	1	0	1	0	1	0	17
54	545.40	0	0	0	1	1	0	18
57	575.70	1	1	0	0	0	1	19
60	606.00	0	0	1	0	0	0	20
63	636.30	1	0	0	1	0	0	21
66	666.60	0	1	0	0	1	0	22
69	696.90	1	0	1	1	1	0	23
72	727.20	0	0	0	0	0	1	24
75	757.50	1	1	0	0	0	0	25
78	787.80	0	0	1	1	0	0	26
81	818.10	1	0	0	0	1	0	27
84	848.40	0	1	0	1	1	0	28
87	878.70	1	0	1	0	0	1	29

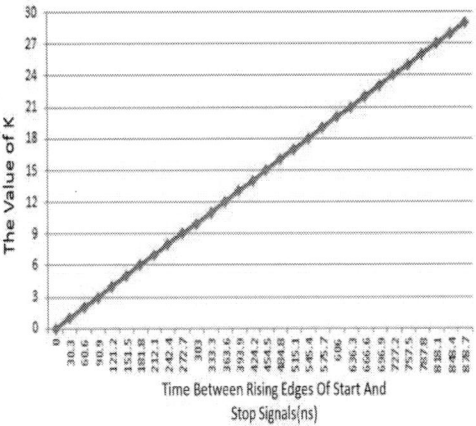

Fig.4 Measurement results of the proposed TDC.

Fig.3 Proposed TDC implementation on FPGA.

An Open-Loop Differential Time Amplifier

Hye-Jung Kwon[1], Ji-Hoon Lim[1], Byungsub Kim[1], Jae-Yoon Sim[1], and Hong-June Park[1,2]

Department of Electronic and Electrical Engineering[1],
Division of IT Conversion Engineering, WCU program[2],
Pohang University of Science and Technology (POSTECH), Pohang, Korea
Email: hjpark@postech.ac.kr

Abstract

An open-loop differential time amplifier is proposed to achieve a large time gain up to 114 and a wide range of input time difference up to 3.6ns. The time gain is determined by the ratio of two bias current values with the slew-rate control method. The differential architecture reduces the rms noise of the output time difference by 45% and the mismatch calibration reduces the time offset by 85%. The time amplifier consumes 1.5mW at 100MS/s and 1.2V supply in a 0.13μm CMOS process.

Keywords- time amplifier, open-loop, time gain, slew-rate control.

Introduction

In recent CMOS circuits, the time resolution is better than the voltage resolution due to the improved operation speed of transistors and the reduced supply voltage. Recently, the on-chip time measurement circuits such as the time-to-digital converter(TDC) starts to replace the analog-to-digital converter (ADC). To enhance the timing resolution of TDC, time amplifiers(TA) are used.

The conventional time amplifiers published in the literature[1,2] have a small time gain and a narrow range of input time difference because they employ a positive-feedback closed-loop architecture. To solve this problems, the open-loop time amplifier was proposed[3]. It gives both a large time gain and a wide input time range. Also, the time gain is easily controlled by the ratio of bias current. But the time amplifier in [3] employs a single-ended architecture and the output time difference is rather noisy.

In this work, a differencial open-loop time amplifier is proposed. The proposed time amplifier employs the advantage of open-loop architecture and reduces the noise of output time difference by utilizing the differential architecture. Also, the mismatch calibration reduces the time offset of output time difference.

Circuit implementation

A. Open-loop differential TA

Fig. 1(a) shows the proposed open-loop differential TA. The left-hand and the right-hand circuits are symmetric to each other. The circuit consists of two pre-charged capacitors(C_{AN}, C_{BN}), two pre-discharged capacitors(C_{AP}, C_{BP}) and current sources. When the $IN1$ signal goes high earlier than the $IN2$ signal, as shown in Fig. 1(b), the left-hand side capacitors

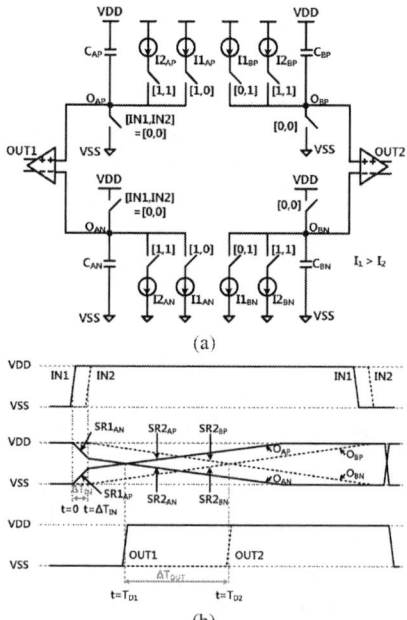

(a)

(b)

Fig. 1. Concept of differential TA. (a) circuit (b) timing diagram

(C_{AN}, C_{AP}) are discharged or charged at the fast slew-rate of SR1$_{AN}$ or SR1$_{AP}$, respectively. After the $IN2$ signal goes high($t > \Delta T_{IN}$), the fast slew-rate paths are cut-off and all capacitors are dischareged(C_{AN}, C_{BN}) or charged(C_{AP}, C_{BP}) at the slow slew-rate. Two comparators compare the node voltage of O_{AP} with O_{AN} and O_{BP} with O_{BN} and generate $OUT1$ and $OUT2$, respectively. By comparing the differential voltage instead of comparing a single-ended voltage with the reference voltage(V_{REF}) as in [3], the proposed differential TA removes the reference voltage of comparator which dominates the output noise. Fig. 2 shows the simulation results of the rms noise of output time difference. The noise of the differential TA is less than that of the single-ended TA with the reference voltage noise of 0.4mV.

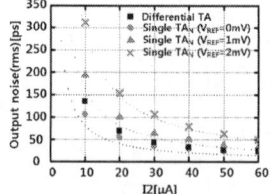

Fig. 2. Output noise of differential and single TA(ΔT_{IN}=100ps, σT_{IN}=1ps)

978-1-4799-5128-4/14 $31.00 © 2014 IEEE

Since the sum of the voltage drop of O_{AP} and the voltage increase of O_{AN} from $t = 0$ to T_{D1} and the sum of the voltage drop of O_{BP} and the voltage increase of O_{BN} from $t = \Delta T_{IN}$ to T_{D2} are the same as VDD, the time gain G can be derived as

$$G = \frac{T_{D2} - T_{D1}}{\Delta T_{IN}} = \frac{SR1_{AN} + SR1_{AP}}{SR2_{AN} + SR2_{AP}} = \frac{I1}{I2} . \quad (1)$$

The time gain G is determined by the ratio of bias current only, it can be easily controlled in a wide range of the input time difference.

B. Mismatch calibration

Each branch of the time amplifier shown in Fig. 1(a) generates a time offset T_{OS} in ΔT_{OUT} due to the mismatch in various parameters, such as the input offset voltage mismatches between two comparators and the slew-rate mismatches. Because T_{OS} is a DC phenomenon, it can be eliminated by the mismatch calibtation.

The comparator offset calibration is performed first before the normal operation. For this, the input nodes of the comparator are shorted and the source current is controlled until the output of comparator is revesed. In this way, the effect of input offset voltage is eliminated. From the monte-carlo simulation result of comparator offset voltage of with 1000 iterations, this scheme can reduce the input offset voltage from 4.3mV to 1.8mV.

Fig. 3 shows the calibration procedure of slew-rate mismatch. After the comparator mismatch calibration, the $SR1$ calibration is performed first. Two comparators compare the O_{AN} and O_{AP} node voltage with V_{REF}. $I1_{AP}$ is contolled untile the $OUT2$ voltage is closest to the $OUT1$ voltage, as shown in Fig. 3(a). This eliminates the mismatch($\Delta SR1_{AP}$) between $SR1_{AP}$ and $SR1_{AN}$. The $SR2$ calibration is similar to the $SR1$ calibration, as shown in Fig. 3(b).

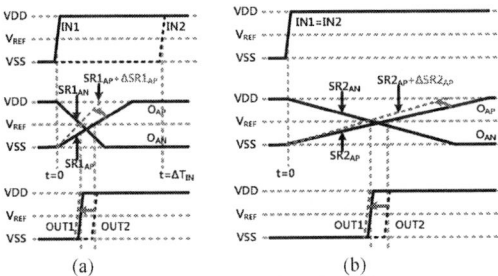

Fig. 3. Slew-rate mismatch calibration (a) SR1 (b) SR2

Measurement results

The chip was fabricated in a 0.13µm CMOS process. Fig. 4 show the layout and the microphotograph of the test chip. Fig. 5 shows the comparison between the measurements(mark) and calculations(dashed (1)) of ΔT_{OUT} versus ΔT_{IN} for different values of $I1$ and $I2$. The vertical axis represents the digital output code from 6-bit TDC. The resolution of TDC is 50ps. Fig. 6(a) shows the time offset with and without mismatch

Fig. 4. Layout and chip photograph.

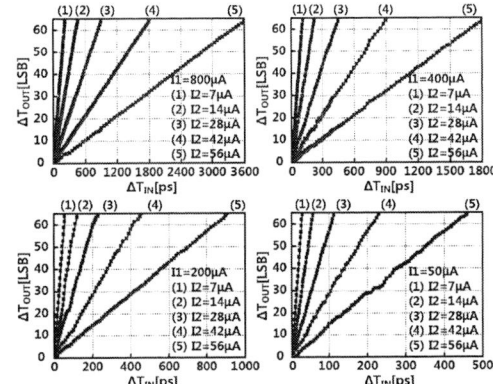

Fig. 5. Measured TA output versus input time difference.

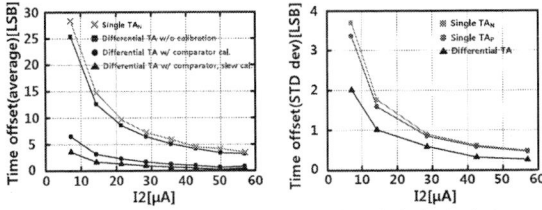

Fig. 6. Time offset (a) average(DC), (b) standard deviation(noise)

calibration. T_{OS} is reduced by 70% after the comparator calibration, and reduced by 85% after the comparator and slew calibration. Fig. 6(b) compares the output noise of the proposed differential TA with with the single-ended TA. Because The differential architecture elmininate the V_{REF} noise, the rms noise of output time difference is reduced by 45% compared to the single-ended TA. The DC time offset and the output noise are inversely proportional to $I2$ and independent of $I1$. The time amplifier consumes 1.5mW at 100MS/s.

Conclusion

The open-loop differential time amplifier is proposed to achieve a large time gain and a wide range of input time difference. The time gain is determined by the ratio of two bias currents. The differential architecture eliminates the dominant noise source of the single-ended TA. The rms noise is reduced by 45%. The DC time offset is also reduced by 45% by using the mismatch calibration.

Acknowledgement

This research was supported by the NRF funded by the MEST (2012005294),ITRC(NIPA-2013-H0301-13-1007),and IDEC.

References

[1] M. Lee and A. Abidi, "A 9b, 1.25 ps resolution coarse-fine time-to-digital converter in 90 nm CMOS that amplifies a time residue," *IEEE J. Solid-State Circuits*, vol. 43, no. 4, pp. 769–777, Apr. 2008.

[2] S.-K. Lee, Y.-H. Seo, Y. Suh, H.-J. Park, and J.-Y. Sim, "A 1GHz ADPLL with a 1.25ps Minimum-Resolution TDC in 0.18µm CMOS," *IEEE Int. Solid-State Circuits conf.(ISSCC)*, pp. 482-483, Feb. 2010.

[3] H.-J. Kwon, J.-S. Lee, Byungsub Kim, J.-Y. Sim, and H.-J. Park, "Aanalysis of and Open-Loop Time Amplifier with a Time Gain Determined by the Ratio of Bias Current," *IEEE Tran. Circuit and Systems II*, vol. 61, no. 7, pp. 481-485, July. 2014.

978-1-4799-5128-4/14 $31.00 © 2014 IEEE

The Design of 13 bits $\sum\Delta$ ADC for a mutual-capacitance large touch screen controller

Ihsan F. I. Albittar[1,2], Jiho Kim[1], HyungWon Kim[1]

Dept. of Electronics Engineering, Chungbuk National University[1]

Cheongju, South Korea

Mixel Incorporation, California, USA[2]

Abstract

Touch sensing circuits for large touch screens require high resolution ADCs with high data rate for accurate and fast touch detection. This paper presents a third-order $\sum\Delta$ ADC design with an input bandwidth of 480 KHz at 61.44 MHz sampling frequency, the SNR and ENOB are measured to be 85 dB and 13 bits respectively. We present an efficient mixed signal design methodology consisting of circuit design, integration, and verification with analog and digital simulations. Using the proposed design methodology, the target $\sum\Delta$ ADC design is implemented with Dongbu 0.18-μm BDCMOS technology. The design results show its suitability for large mutual-capacitance touch screen controller applications.

Keywords- Sigma-delta ADC, touch screen controller, desgin methodology.

Introduction

Nowadays, capacitive touch screens are the most popular touch screen panels (TSP). Among different types of capacitive TSPs, projected mutual capacitance TSPs are most widely used for the middle-sized and small-sized mobile products. They have superior visibility and durability and also exhibit a multi-touch function. In mutual capacitance TSPs, touch detection is determined by measuring the change in the mutual capacitance [1] [5] [6] [7] which is susceptible to noise. Such TSPs, therefore, often have a very low signal-to-noise ratio (SNR), and so require high resolution ADCs to distinguish the mutual capacitance changes from their high ambient noise. Fig. 1 shows a TSP and its excitation and sensing circuits with an ADC performing detection operations. Among variety of ADC architectures, Sigma-Delta ($\sum\Delta$) ADC is often considered as best for high resolution results.

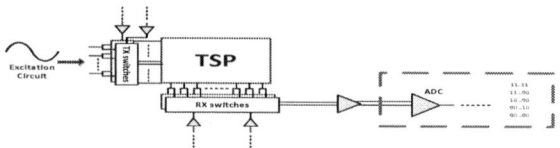

Fig. 1 TSP and its excitation and sensing circuits with ADC

Archecticture

Fig. 2 shows a general structure of a $\sum\Delta$ ADC. We designed the analog modulator with a third-order fully differential switched-capacitor implementation, which provides good common-mode noise rejection and cancelation for the even order harmonics.

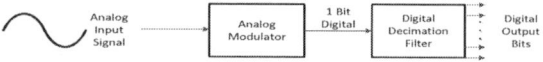

Fig. 2. Block diagram of sigma-delta ADC

Also, switched-capacitor implementation is used for its high resolution superiority. Since the multi-order modulators shape the quantization noise, and push it to even higher frequencies than do the lower-order modulators. Equation (1) is the SNR of a modulator loop architecture, which is a function of over-sampling ratio (OSR) ,modulator order (L), and the B-bit quantizer[2]. For the target ADC, we chose L = 3 to meet the required SNR specifications.

$$SNR = (20L+10)\log OSR + 10\log(2L+1)+6.02B+1.76 - 9.94L \quad (1)$$

Fig. 3 shows our analog modulator architecture. The modulator design consists of three parasitic-insensitive switched–capacitor integrators, a comparator that serves as a 1-bit ADC, and a distributed two-level (1-bit) DAC. The modulator operates on a two-phase nonoverlapping clock. During phase 1, switches S1 and S3 are on, while during phase 2, switches S2 and S4 are on [3].

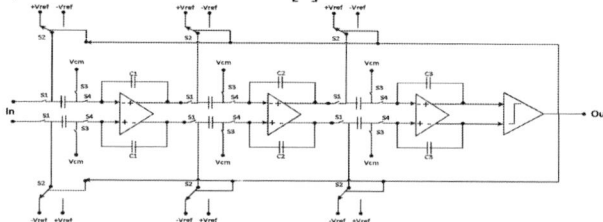

Fig. 3 The proposed third-order analog modulator

The modulator output is filtered by a digital decimation filter. In our design, we chose a 4th order decimation filter. It has multi-functions, which includes to achieve a multi-bit digital output by down-sampling the signal, to attenuate the quantization noise and remove aliases from the side band, and as a result, to complete the process of analog to digital conversion [3]. $Sinc^K$ filter has area advantage since it eliminates the needs for digital multipliers [4] lead to a relatively small size. We implemented a $Sinc^K$ filter by cascading K stages of accumulators operating at the high sample rate, followed by K stages of cascaded differentiators operating at the lower sample rate[4]. Fig. 4 shows the digital decimation filter that we implemented for the presented $\sum\Delta$ ADC design.

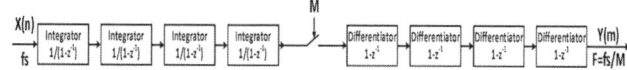

Fig. 4 The proposed digital filter block diagram

Proposed Design Methodology

In Fig. 5, we present our design methodology of $\sum\Delta$ ADC, which use two flows of simulations: Analog Simulation and digital simulation. Analog simulation uses Cadence Spectre for $\sum\Delta$ modulator and VerilogA for decimation filter, whereas digital simulation uses Verilog for decimation filter implementation and MATLAB for verification of the output correctness. Using this automated analog-digital mixed signal design methodology, we can converge the real design and target spec quickly and shorten the overall design cycles.

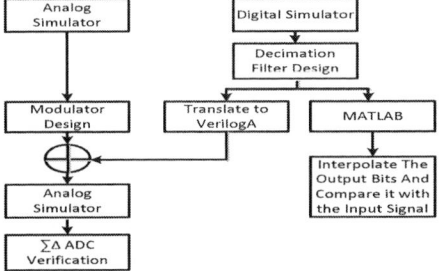

Fig. 5 The proposed design methodology

Implementation and Simulation Results

We implemented the target $\sum\Delta$ ADC design using Dongbu 0.18-μm BDCMOS technology, and measured performance using Cadence Spectre simulations. Simulation results in Fig. 6 show that the third order $\sum\Delta$ modulator can operate at a sampling frequency 61.44 MHz for a sin wave input of 800 mVpp and 480 KHz bandwidth. Fig. 7 shows a frequency domain spectral power analysis of the modulator's output, which confirms that the quantization noise is shifted towards higher frequency, which is to be removed by the decimation filter.

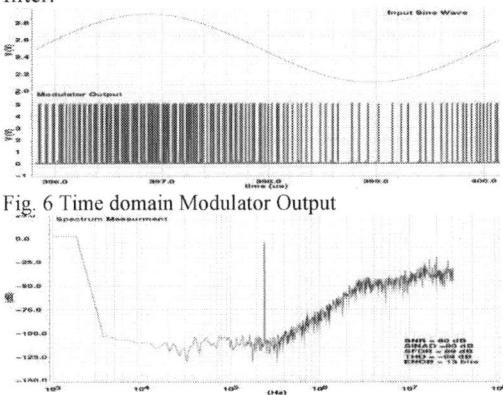

Fig. 6 Time domain Modulator Output

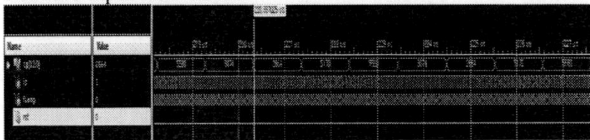

Fig. 7 Spectral power analysis of the modulator's output

Fig. 8 shows the digital outputs in decimal numbers. Our proposed design methodology (Digital simulation flow) proved that these digital outputs (red wave) exactly match the input (blue wave) by generating interpolated wave form (black wave) and compare it with the input. Fig. 9 shows the verification process of the digital flow of our design methodology, which proved that the digital output of the whole ADC matches up with the input sine wave.

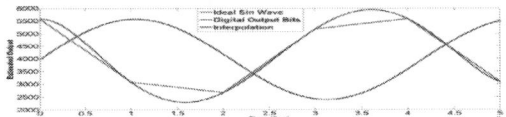

Fig. 8 Decimal numbers represent the digital output bits from the decimation filter

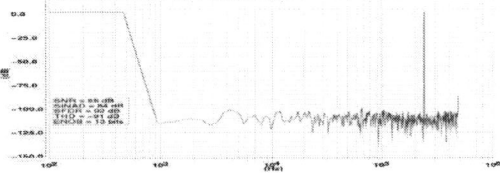

Fig. 9 MATALB result verification for the output digital bits

Fig. 10 shows the spectral power of the decimation filter. This is the result of the analog simulation flow of our design methodology, which integrated VerilogA model of decimation filter with the analog modulator design. The complete ADC performance is summarized in Table (1).

Fig. 10 The output spectrum for the whole system

Table (1) : Summary of $\sum\Delta$ ADC

Parameters	Value
Signal bandwidth	480 KHz
Sampling Frequency	61.44 MHz
Over sampling ratio	128
Modulator order	3
Filter order	4
ENOB	13 bits
SNR	85 dB
SINAD	84 dB
SFDR	92 dB
THD	-91 dB

Conclusion

We designed a high speed high resolution $\sum\Delta$ ADC design using an efficient design methodology. We have demonstrated that the implemented ADC meets the stringent specifications of target large touch screen (23") controller SoC (Sample rate of 61.44MHz, SNR of 85dB, input B/W of 480 KHz). We also showed that our design methodology verifies the performance of mixed signal designs in analog and digital simulations without converting the whole digital circuits to analog circuits, and thus allowing fast design cycles. We actually used this $\sum\Delta$ ADC IP in the sensing block of a touch screen controller SoC implemented in Dongbu 0.18um BDCMOS technology.

Acknowledgment

This research was supported by the MSIP Korea, under the Human Resource Development Project for SoC (NIPA-2014-H0601-14-1001).

References

[1] U. Jang, T.W. Cho, H.G. Jang, S. Lee, H.W. Kim, "Architecture of Multi Purpose Touch Screen Controller with Self Calibration Scheme", *IEEK Fall Conf.*, pp. 162-166, 2013

[2] Chris Hutchens, Chia-Ming Liu "Pow/bit optimization of Δ-Σ ADCs using multi-bit quantizers", IEEE Conf., 2002.

[3] Steven R. Norsworthy ,Richard Schreier ,Gabor C. Temes, "Delta-Sigma Data Converters," IEEE Press, 1997.

[4] Texas Instruments ,"Combining the ADS1202 with an FPGA Digital Filter for Current Measurement in Motor Control Applications" ,2003.

[5] I. Seo, T.W. Cho, H.G. Jang, S. Lee, H.W. Kim, "Frequency Domain Concurrent Sensing Technique for Large Touch Screen Panels", *IEEK Fall.*, pp. 55-58, Nov. 2013.

[6] M.G.A. Mohamed, U. Jang, I. Seo, T.W. Cho, H.G. Jang, S. Lee, H.W. Kim, "Efficient Algorithm for Accurate Touch Detection of Large Touch Screen Panels," IEEE Int'l Symp. On Consumer Electronics 2014.

[7] I. Seo, U. Jang, M.G.A. Mohamed, T.W. Cho, H.G. Jang, S. Lee, H.W. Kim, "Voltage Shifting Double Integration Circuit for High Sensing Resolution of Large Capacitive Touch Screen Panels," IEEE Int'l Symp. On Consumer Electronics 2014.

Low Power Memory Design for High Temperature in Ruggedized Electronics

Tony T. Kim[1], Ngoc Le Ba[1], Anh Tuan Do[1], Jayaraman K. Gopal[1], Geng Li Chua[2], Pushpapraj Singh[2]

[1]VIRTUS, School of Electrical and Electronic Engineering, Nanyang Technological University, Singapore
[2]Institute of Microelectronics, A*STAR, Singapore
Email: thkim@ntu.edu.sg

Abstract

Applications such as Logging-While-Drilling (LWD), automotive, and aerospace systems require electronics whose operating temperature is much higher than that of conventional consumer electronics. One of the most critical functional blocks for high temperature operation is memory due to the significantly increased leakage. This paper explains two different memories (SRAM and NEMS NVM) for high temperature operation. In the SRAM, circuit techniques for improving sensing margin under the leaky bitline condition will be discussed. In the NEM NVM, two adhesion-force-based NEM NVM devices will be introduced.

Keywords-high temperature; leakage; memory; static random access memory; nano-electro-mechanical switch

Introduction

Integrated circuits for high temperature operation has gained the interest of researchers due to the recent development of various ruggedized electronics including automotive, aerospace, and Logging-While-Drilling (LWD) systems [1],[2]. One of the most challenging issues in the high temperature operation is leakage. Various solutions in the technology level have been employed to improve the leakage. Technologies such as silicon-on-insulator (SOI), silicon carbide (SiC) CMOS and tungsten interconnection have been adopted for lowering the amount of leakage from devices [3]. However, integrated circuit design techniques are equally important for reliable high temperature operation.

Various integrated circuits such as $\Sigma\Delta$ modulator and EEPROM have been developed for high operating temperature. But, very few memory solutions supporting high temperature operation have been introduced. In the market, only EEPROMs are available as a high temperature memory solution. However, random access memories are indispensable for computational power. Nonvolatile memories are also highly required to achieve low power with long idle and power-down modes. However, commercially available nonvolatile memories cannot retain data at high temperature due to the leakge at the storage layer. Nano-electro-mechanical (NEM) devices have been explored for high temperature operation since they have ideally zero leakage current [4],[5]. However, conventional NEM devices are difficult to be employed as a part of integrated circuits due to the large physical dimension compared with CMOS devices. To address the above issues, we present two different memory solutions for high temperature operation.

Low Power SRAM Design for High Temperature Operation

A. Selection of Operating Point

To achieve a wide range of operating temperature, the impact of temperature change on the operation of integrated circuits should be carefully investigated. Fig. 1 explains the impact of temperature change on the delay of digital circuits at various supply levels. Lowering supply increases the delay regardless of the operating temperature. At high V_{DD}, higher temperature shows a larger delay while lower temperature produces a larger delay at low V_{DD}. As a result, an operating point with minimum variation point is formed in the near-threshold region. Fig. 1 shows that the minimum variation (μ/σ) point is at $V_{DD} = 2.5$ V. This minimum variation point is used for the SRAM design since it minimizes the required timing margin. The nominal V_{DD} of the adopted SOI technology is 5 V, and V_{DD} of 2.5V represents near-threshold operation.

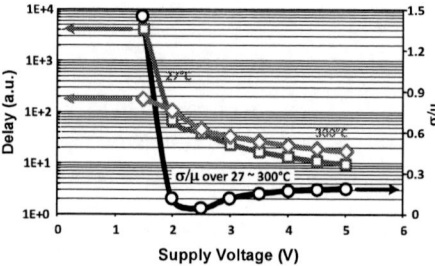

Figure 1. Optimal operating point with minimum performance variation under wide temperature range [6].

B. Sensing Margin Improvement Technique

At high temperature, bitline sensing is the most critical issue due to the substantial amount of leakage. Conventional sensing scheme based upon dynamic operation (precharge and conditional discharge) is highly likely to fail to provide acceptable sensing margin. To address this, *Static Read Bitline (S-RBL)* was proposed in [6] whose principle is illustrated in Fig. 3. During read operation, a read wordline (RWL<0>) is activated like the conventional decoupled SRAMs. However, the PMOS devices (M1, M2, M3, and M4) provide pull-up reference current (I_{ref}) whose strength is adjusted by Bias_Out. The pull-up reference current (I_{ref}), the pull-down RBL leakage current (I_{lkg}), and the pull-down cell current (I_{read0} and I_{read1}) generate the RBL levels for data '1' and '0'. Fig. 2 presents a sample simulation result comparing the proposed *S-RBL* with the conventional scheme. It can be seen that the pull-up current prevents RBL from being fully discharged to GND in *S-RBL*

(RBL1 and RBL0 in Fig. 3(Left)). *S-RBL* also provides a larger RBL swing and a wider sensing timing window (Fig. 3(Right)). To maximize the *S-RBL* sensing margin, Bias_Out should be carefully controlled. Fig. 4 illustrates the block diagram of the Bias_Out controller. The optimal Bias_Out is generated by averaging the Bias_Out for the minimum I_{ref} for sensing data '1' (Lower Loop) and that for the maximum I_{ref} for sensing data '0' (Upper Loop).

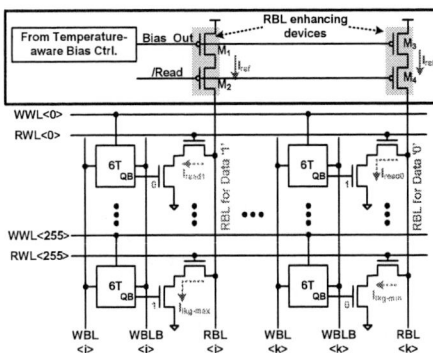

Figure 2. Simplified circuit of the proposed static read bitline (*S-RBL*) [6].

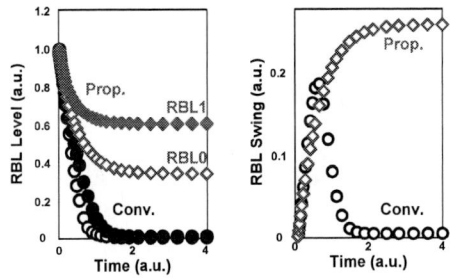

Figure 3. Comparison of the proposed S-RBL and conventional RBL. RBL swing is defined as V(RBL of data '1') – V(RBL of data '0') [6].

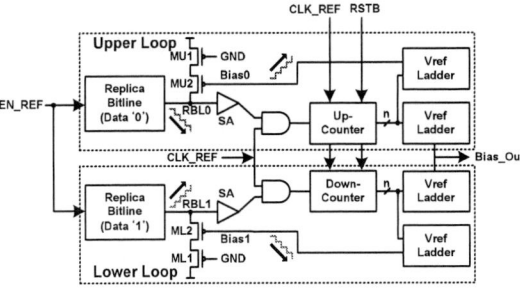

Figure 4. Block diagram of the proposed Bias_Out controller [6].

Nano-Electro-Mechanical Non-volatile Memory

NEM-based NVM devices can be good memory solutions for high temperature operation. However, the large NEM device sizes are obstacles to overcome for adoption in actual systems. Fig. 5 shows an anchored NEM NVM structure for high temperature operation [7]. It has three terminals that are named *Contact, Actuator* and *Cantilever*. A potential difference between *Cantilever* and *Actuator* forms electrostatic force, which actuates the *Cantilever* terminal to touch *Contact*. Deactuation is done by applying AC signal to *Actuator*. The nonvolatile memory operation is achieved by the *Van der Waal* adhesive force. However, the anchored structure is difficult to scale. For scalability, anchor-free structure are demanded. Fig.

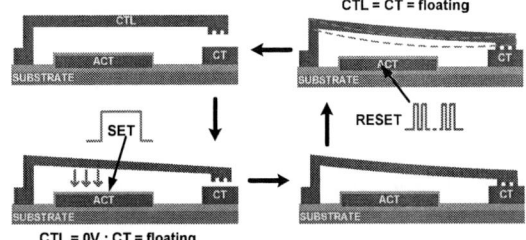

ACT: Actuator, CT: Contact, CTL: Cantilever

Figure 5. Operation of the proposed anchored NEM memory device [7].

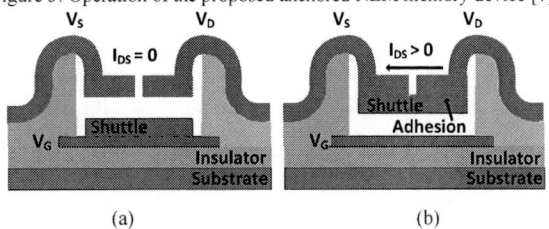

Figure 6. Cross section of the anchor-free NEM memory device: (a) Down state and (b) Up state [8].

6 shows the anchor-free NEM NVM device with high scalability [8]. It also has three terminals (V_S: *Source*, V_D: *Drain* and V_G: *Gate*). For writing data, a potential difference is applied between *Gate*, and *Source* and *Drain*. This generates electrostatic force and moves the shuttle to the other side. Similar to the anchored NEM NVM device, the *Van der Waal* adhesive force also forms nonvolatile memory operation in the anchor-free NEM NVM device.

Conclusions

Memory design for high temperature is increasingly demanded due to the development of various ruggedized electronics. This paper explained an SRAM and NEM NVM devices as memory solutions for high temperature operation.

References

[1] D. Gogl *et al.*, "A 1-Kbit EEPROM in SIMOX Technology for High-Temperature Applications up to 250 °C," *IEEE Journal of Solid-State Circuits*, vol. 35, no. 10, pp. 1387-1395, Oct. 2000.

[2] C. Davis *et al.*, "A 14-Bit High-Temperature ΣΔ Modulator in Standard CMOS," *IEEE Journal of Solid-State Circuits*, vol. 38, no. 6, pp. 976-986, June 2003.

[3] J.-S. Chen *et al.*, "A Silicon Carbide CMOS Intelligent Gate Driver Circuit with Stable Operation over a Wide Temperature Range," *IEEE Journal of Solid-State Circuits*, vol. 34, no. 2, pp. 192-204, Feb. 1999.

[4] V. Pott *et al.*, "Mechanical computing redux: relays for integrated circuit applications," *Proc. of the IEEE*, vol. 98 (12), pp. 2076-2094, Dec. 2010.

[5] R. Gaddi *et al.*, "Reliability and performance characterization of a MEMS-based non-volatile switch," *Proc. of IEEE International Reliability Physics symposium*, pp. 171-176, Jun. 2011.

[6] T. Kim *et al.*, "A Low Voltage 8-T SRAM with PVT-Tracking Bitline Sensing Margin Enhancement for High Operating Temperature (up to 300°C)." *IEEE Asian Solid-State Circuits Conference* (A-SSCC), pp. 233-236, Nov. 2013

[7] J. Gopal *et al.*, "A Cantilever-based NEM Non-volatile Memory Utilizing Electrostatic Actuation and Vibrational Deactuation for High Temperature Operation," *IEEE Transactions on Electron Devices* (TED), Vol. 61, pp. 2177-2185, June 2014

[8] V. Pott *et al.*, "The shuttle nano-electro-mechanical non-volatile memory," *IEEE Transactions on Electron Devices* (TED), Vol. 59, pp. 1137-1143, Apr. 2012

Comparative Analysis of Using Planar MOSFET and FinFET as Access Transistor of STT-RAM Cell in 22-nm Technology Node

Byungkyu Song[1], Taehui Na[1], Hanwool Jeong[1], Seung H. Kang[2], Jung Pill Kim[2] and Seong-Ook Jung[1]

[1]Department of Electrical and Electronic Engineering
Yonsei University, Seoul, Korea
[2]Qualcomm Inc., San Diego, USA
sjung@yonsei.ac.kr

Abstract

As technology node scaling, FinFET becomes the substitution for planar MOSFET due to several merits of FinFET such as superior gate controllability, large on-current, and low variability. For the reasons, using FinFET as access transistor of spin-transfer-torque random access memory (STT-RAM) cell should be analyzed. This paper compares using planar MOSFET and FinFET as access transistor of STT-RAM cell and concludes which device is more proper solution for high write and read yields.

Keywords-Access transistor; FinFET; MRAM; Read yield; STT-RAM; Technology node scaling

Introduction

Currently, Spin-transfer-torque random access memory (STT-RAM) uses a planar MOSFET. However, as technology node scales down, the high doping in channel region is necessary in order to suppress a short channel effects (SCE), resulting in the increase in random dopant fluctuation (RDF) which becomes a serious problem due to the increase in process variation [1]. Planar MOSFET which is vulnerable to RDF is replaced by FinFET below 22-nm technology node because FinFET has a superior gate controllability, which makes SCE suppressed without high doping to the channel, leading to the tolerance of process variation. Thus, when compares to planar MOSFET, FinFET having larger on-current and lower variability has the higher write and read yield. However, FinFET can be more affected by the read disturbance as well. In this paper, we compare using planar MOSFET and FinFET as access transistor of STT-RAM cell in aspect of write and read yields.

1T1MTJ Cell and Size

Fig. 1(a) shows that 1T1MTJ cell structure of STT-RAM which is composed of one transistor and one MTJ. The 1T1MTJ cell structure is widely used in STT-RAM because of the compact area. Fig. 1(b) and (c) show the layouts of 1T1MTJ cells which use planar MOSFET and FinFET as access transistors, respectively. In case of planar MOSFET, the cell size is determined by the width of access transistor (W_{Access}) and the active spacing, whereas the cell size of FinFET is determined by N_{fin} and fin pitch. Both cases of planar MOSFET and FinFET, the minimum cell size is

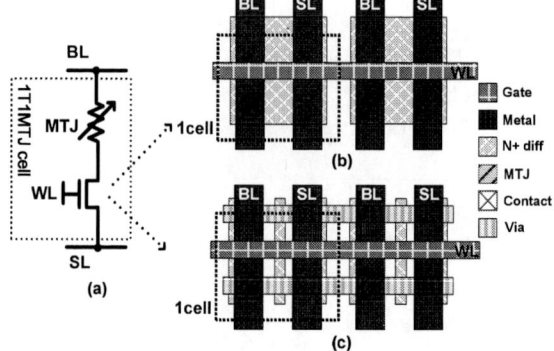

Fig. 1 (a) 1T1MTJ cell structure and layout of 1T1MTJ cell when using (b) planar MOSFET and (c) FinFET as access transistor. this.

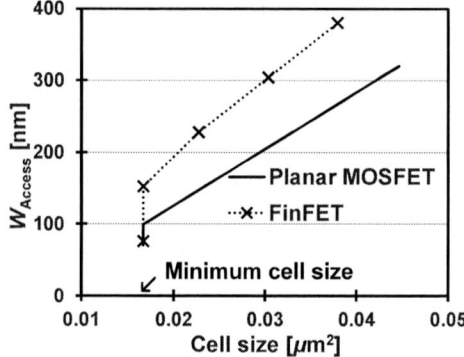

Fig. 2 W_{Access} of each structure according to cell size.

determined by the design rule of metal pitch which is related to the minimum distance from bit line (BL) metal and source line (SL) metal. Fig. 2 shows the W_{Access} of planar MOSFET and FinFET according to the cell size which is estimated by the lambda rule [2] in 22-nm technology node. The cell size and W_{access} of FinFET are quantized because these are determined by the N_{fin}. In same cell size, FinFET has larger W_{Access} than planar MOSFET because the channel of FinFET is surrounded by the gate, leading to larger on-current and smaller turn-on resistance.

Comparison

HSPICE Monte Carlo simulations are performed using the LPPTM V2.1 model and BSIM-CMG 107.0 model for planar MOSFET and FinFET in 22-nm technology node, respectively. The nominal voltages of planar MOSFET and FinFET are 0.9

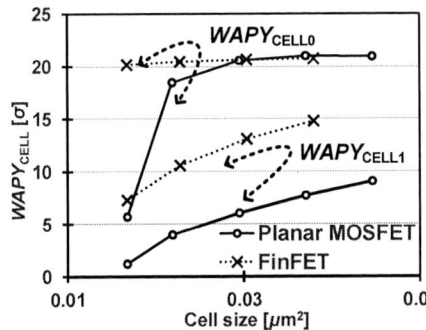

Fig. 3 $WAPY_{CELL}$ of each structure according to cell size.

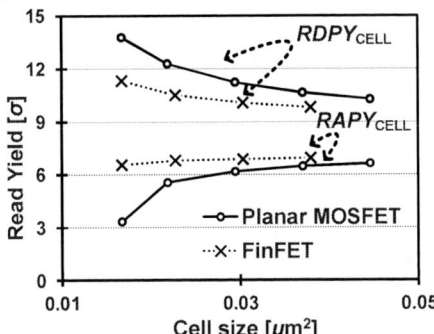

Fig. 4 Read yield of each structure according to cell size.

V and 0.8 V, respectively. The simulation performed with MTJ model in [3], which has a tunnel magnetoresistance ratio (TMR) of 150%, R_L of 3 kΩ, and R_H of 7.5 kΩ. Additionally, the critical switching current (I_C) based on [4] is used and a standard deviation of MTJ is assume to be 4%. To estimate the write and read yield estimation, a read access pass yield ($RAPY_{CELL}$), a read disturbance pass yield ($RDPY_{CELL}$), and a write access pass yield ($WAPY_{CELL}$) based on [5] are used in this work. For read and write operations, the source degeneration sensing circuit in [5] and conventional write driver in [6] are used in this work.

Fig. 3 shows that the $WAPY_{CELL0}$ and $WAPY_{CELL1}$ of each structure according to the cell size. As the cell size increases, the $WAPY_{CELL}$ increases because the larger W_{Access} can induce the larger write current. In all cases of cell size, FinFET has higher $WAPY_{CELL}$ than planar MOSFET because of larger on-current and lower variability. In general, the $WAPY_{CELL1}$ is much smaller than the $WAPY_{CELL0}$ because of the source degeneration effect, besides the I_C of state 0→1 (I_{AP}) is larger than that of state 1→0 (I_P). For the reasons, the $WAPY_{CELL1}$ determines the overall $WAPY_{CELL}$, regardless of cell size.

The smaller turn-on resistance of access transistor, which corresponds to larger on-current, acquires the larger effective TMR leading to the higher read yield. For the reasons, in all cases of cell size, the $RAPY_{CELL}$ of FinFET having small turn-on resistance is much higher than that of planar MOSFET. Besides, FinFET has lower threshold voltage variation than planar MOSFET, leading to the higher $RAPY_{CELL}$. As shown in the Fig. 4, the $RAPY_{CELL}$ of planar MOSFET increases according to the increase in cell size because the large W_{Access} improves the effective TMR. On the other hand, the $RAPY_{CELL}$ of FinFET is almost saturated according the cell size because the turn-on resistance of access transistor becomes negligible in large W_{Access}. Fig. 4 shows that the $RDPY_{CELL}$ of each structure according to the cell size. Even though FinFET has

TABLE I
COMPARISON BETWEEN PLANAR MOSFET AND FINFET

Cell size (μm^2)	Planar MOSFET		FinFET	
	Read yield (σ)	Write Yield (σ)	Read yield (σ)	Write Yield (σ)
0.016	5.6	4.0	6.6	7.3
0.023	6.2	6.1	6.8	10.6
0.031	6.5	7.7	6.9	13.1
0.038	6.6	9.1	6.9	14.8

higher $RAPY_{CELL}$ than planar MOSFET, the $RDPY_{CELL}$ of FinFET is lower than planar MOSFET due to larger on-current of FinFET. Additionally, increasing the cell size reduces the $RDPY_{CELL}$ due to the increases in sensing current. However, in both cases of planar MOSFET and FinFET, the $RAPY_{CELL}$ determines the overall read yield (= min($RAPY_{CELL}$, $RDPY_{CELL}$)) because the I_C is large enough to immune to the read disturbance, thus read disturbance is not less critical than read access to determine the read yield. For the reasons, using FinFET as access transistor can acquire the maximum read yield.

As shown in Table. I, FinFET has the higher write and read yields than planar MOSFET in all cases of cell size. FinFET satisfies the target read yield of 6σ with the minimum cell size. Thus, using FinFET as access transistor in 22-nm technology node has advantages to improve the write and read yields.

Conclusion

FinFET has a number of desirable features including higher on-current with same cell size and lower variability. These two facts make the write and read yields of STT-RAM improved when a FinFET is used as an access transistor instead of planar MOSFET. For the reasons, using FinFET as access transistor of 1T1MTJ cell is the proper solution for high write and read yields in 22-nm technology node.

References

[1] H. Mahmoodi, S. Mukhopadhyay, and K. Roy, "Estimation of delay variations due to random-dopant fluctuations in nanoscale CMOS circuits," *IEEE J. Solid-State Circuits*, vol. 40, no. 9, pp. 1787–1796, September 2005.

[2] S. K. Gupta, SP. Park, N. N. Mojumder, and K. Roy, "Layout-aware optimization of STT MRAMs," in *Proc. Conference on Design, Automation and Test in Europe (DATE)*, 2012, pp. 1455-1458.

[3] C. J. Lin *et al.*, "45 nm low power CMOS logic compatible embedded STT MRAM utilizing a reverse-connection 1T/1MTJ cell," in *IEEE Int. Electron Devices Meeting (IEDM) Tech. Dig.*, 2009, pp. 279-282.

[4] S. H. Kang, "Embedded STT-MRAM for advanced mobile system-on-chips," in *2nd CSIS Int. Symp, Spintronics-based VLSIs* (Japan), 2012.

[5] J. Kim, K. Ryu, S. H. Kang, and S.-O. Jung, "A novel sensing circuit for deep submicron spin transfer torque MRAM (STT-MRAM)," *IEEE Trans. Very Large Scale Integr. (VLSI) Syst.*, vol. 20, no. 1, pp. 181-186, January 2012.

[6] T. Na, K. Ryu, J. Kim, S. H. Kang, and S.-O. Jung, "A comparative study of STT-MTJ based non-volatile filp-flops," *IEEE Int. Symp. Circuits and Systems*, 2013, pp. 181-186.

978-1-4799-5128-4/14 $31.00 © 2014 IEEE

Subthreshold SRAM Macro Design with Pulse-Controlled Dynamic Voltage Scaling (PC-DVS)

Jun-Kai Zhao, Yi-Wei Chiu, and Shyh-Jye Jou
Department of Electronics Engineering
National Chiao Tung University
Hsinchu, Taiwan, R.O.C.

Yuan-Hua Chu
Industrial Technology Research Institute (ITRI)
Hsinchu, Taiwan, R.O.C.

Abstract

In this paper, we propose a pulse-controlled dynamic voltage scaling (PC-DVS) scheme in an SRAM macro to suppress leakage power consumption to reduce total power. The proposed SRAM macro is capable of operating in low-voltage regime with high variation immunity. The proposed PC-DVS scheme reduces the array power up to 62% at 500 kHz while the selected sub-bank operating at 0.5 V and the unselected sub-banks Hold data at 0.35 V.

Introduction

Internet of Things (IoT) and bio-medical systems become popular recently. The devices in the systems are required to be portable and wearable. Therefore, energy efficiency is an important issue for those devices to last the life time of batteries. Reducing voltage level of supply voltages is an effective way to reduce both dynamic and static power. Moreover, dynamic voltage scaling is a common technique to save power consumption [1]–[5]. However, the traditional DVS technique is implemented in operation-mode-based systems. Those systems change the voltage levels only between high throughput (active) mode and low throughput (standby) mode. For wearable and bio-medical applications, the operation frequency is about tens of Hz to several kHz. The clock period is relative long. To further reduce the power consumption in one clock cycle, we propose a pulse-controlled DVS (PC-DVS) scheme in our SRAM macro.

Subthreshold Voltage Operation SRAM Macro with Pulse-Controlled Dynamic Voltage Scaling

A single-ended disturb-free 9T SRAM bitcell which can operate at a subthreshold supply voltage is used in the SRAM macro [6]. Fig. 1 shows the circuit schematic of the bitcell. Fig. 2(a) shows the block diagram of the SRAM macro. The peripheral circuits, control circuits, finite state machine (FSM), decoder, data-in and data-out (DIDO) and dummy cell, are connected to the supply voltage for peripheral circuits (VDDP). The SRAM cell array is connected to two supply voltages, one is VDDH and the other is VDDL, through power PMOSs (PP1 and PP2). The control signals of PP1 and PP2 are determined by the control circuit and FSM. The supply voltages of VDDP, VDDH and VDDL are set to the voltage levels V_P, V_H and V_L, respectively. We set V_P is equal to V_H, and V_P and V_H are larger than V_L.

The operation period of SRAM macro can be partitioned into two periods, one period is to Read/Write data and the other period is to Hold data. The minimum supply voltage of an SRAM bitcell is limited by Read/Write operation, which is always higher than the minimum supply voltage of Hold operation (V_{DDMIN_HOLD}). During the access data period, we use higher voltage to Read/Write data to make sure the Read and Write ability and operation speed. However, the Read/Write time only occupies a small portion of clock cycle, especially in a slow operation frequency. The proposed PC-DVS scheme adjusts the supply voltages within one clock cycle. To make sure the Read/Write operation is finished, we use a tracking circuit implemented by a dummy cell array to generate the required pulse-width. We scale down the supply voltage after the Read operation is done. During the Hold data period, V_{DDMIN_HOLD} is used to maintain the correctness of the stored data.

The timing diagram of the proposed SRAM macro is shown in Fig. 2(c). When the clock signal (CK) raises to "1", the corresponding VDDC is pulled up to V_H before the word line enable (WLE) raises to "1" for Read/Write operation. Then, the WLE is disabled by the signal (DBL) from the tracking circuit, which is discharged to ground by dummy cell. After the WLE pulse is disabled, the VDDC of selected sub-banks is switched to V_L for retaining data. The VDDCs of other unselected sub-banks are kept at V_L for the whole clock cycle to reduce the leakage power.

Performance Evaluation and Simulation Results

In [5], the work indicated that the DVS would introduce energy overhead while charging/discharging the power node between two different voltage levels. In our work, the energy overhead is relative to the capacitance of the VDDC node of a sub-bank. The power consumption of the selected sub-bank is divided into active and non-active periods. The selected bitcells consume dynamic power when the word-line is enabled (active period). After the access is done, the bitcells consume only static power for the rest of clock period (non-active period). Other unselected sub-banks consume static power all the time. Fig. 2(b) shows the array leakage power at 0.35 V is 49% less than that at 0.5 V. The array power is dominated by leakage power when the clock period is longer. Fig. 3(a) shows the array power under different clock periods. With PC-DVS, there is a 36% reduction in array power as compared to without DVS when the clock period is 2 μs.

Another way to improve the power reduction is decreasing the the capacitance of power nodes by dividing the SRAM array into smaller sub-banks. Although increasing the segments results in some area and energy overhead induced by additional peripheral circuits. The energy overhead is very small as compared to the saved power. To illustrate the trend of power saving by increasing the number (N) of sub-banks of an SRAM array. We use the simulated results of "N = 4" to derive the power consumption at different configurations. For simplicity, only array power is taken into consideration. The array power consumption of the SRAM macro with PC-DVS (P_{PC-DVS}) and without DVS (P_{non_DVS}) are shown in Fig. 3(b). It shows the ratio of power with to without the PC-DVS versus clock periods with different numbers of sub-banks in SRAM array. The ratio is decreasing as N is increasing. When the ratio is less than 1, it represents that the SRAM macro with PC-DVS consumes less power than without PC-DVS. When we divide the cell array into more sub-banks (larger N), PC-DVS scheme is able to save power even at a higher operating frequency.

Implementation Results

We implement an 8 kbits SRAM macro which is accompanied with PC-DVS scheme to verify the performance. The SRAM macro is fabricated in 65nm LP CMOS technology. The die photo of the test chip combined with layout is shown in Fig. 4. The V_{DDMIN} of previous work which limited by Write operation is 0.35 V, and the Hold V_{DDMIN} is even lower. To demonstrate the advantage of our PC-DVS scheme and ensure the correctness of the test chip, we leave some margin to voltage levels for all operations. In this work, when using PC-DVS, V_H is chosen to be 0.5 V which is dependent on the application supplied by energy harvesting and the voltage is about 0.5 V. V_L is chosen to be 0.35 V for the Hold stability concern of bitcells. This technique would induce energy overhead which comes from charging/discharging power nodes of the selected sub-bank,

978-1-4799-5128-4/14 $31.00 © 2014 IEEE

and the area overhead is less than 1%. However, the energy overhead can be diluted by a long clock period or dividing the SRAM array into more sub-banks. Measurement results show that 62% power reduction at 500 kHz while utilizing PC-DVS as shown in Fig. 5. In contrast, only 36% power reduction at 500 kHz while utilizing traditional DVS. TABLE I shows the characteristics of the SRAM macro.

Acknowledgment

The authors would like to thank CIC for CAD tools supporting, ITRI for research funding and technical support, and TSMC for the University Shuttle Program.

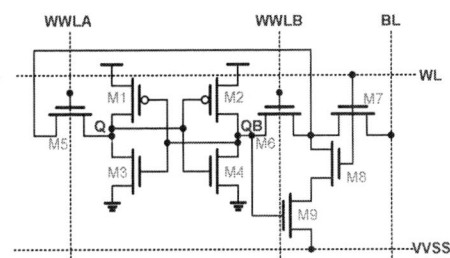

Fig. 1. Schematic of disturb-free 9T SRAM bitcell.

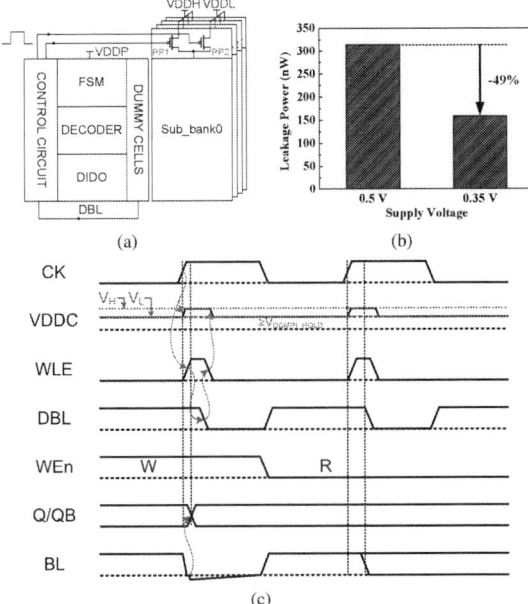

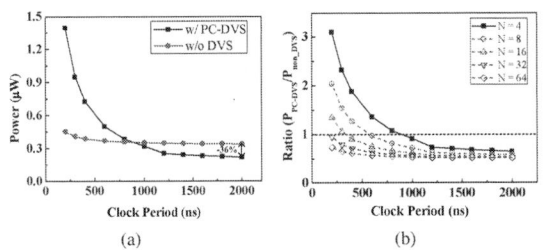

Fig. 2. (a) Block diagram of SRAM macro. (b) Simulated comparison of leakage power consumptions at different supply voltages. (c) Timing diagram of PC-DVS with timing tracking scheme.

Fig. 3. (a) Simulated comparison of array power consumptions with two schemes at different clock periods. (b) Ratio of P_{PC-DVS} to P_{non_DVS} versus clock periods with different numbers of sub-banks.

References

[1] T. Burd, T. Pering, A. Stratakos, and R. Brodersen, "A dynamic voltage scaled microprocessor system," *IEEE J. Solid-State Circuits*, vol. 35, no. 11, pp. 1571–1580, Nov. 2000.

[2] S. Lutkemeier *et al.*, "A 65 nm 32 b subthreshold processor with 9T multi-Vt SRAM and adaptive supply voltage control," *IEEE J. Solid-State Circuits*, vol. 48, no. 1, pp. 8–19, Jan. 2013.

[3] Y. Shakhsheer *et al.*, "A 90nm data flow processor demonstrating fine grained DVS for energy efficient operation from 0.25V to 1.2V," in *Proc. IEEE Custom Integrated Circuits Conf. (CICC)*, Sep. 2011, pp. 1–4.

[4] S.-Y. Peng *et al.*, "Instruction-cycle-based dynamic voltage scaling power management for low-power digital signal processor with 53% power savings," *IEEE J. Solid-State Circuits*, vol. 48, no. 11, pp. 2649–2661, Nov. 2013.

[5] M. E. Sinangil, N. Verma, and A. P. Chandrakasan, "A reconfigurable 8T ultra-dynamic voltage scalable (U-DVS) SRAM in 65 nm CMOS," *IEEE J. Solid-State Circuits*, vol. 44, no. 11, pp. 3163–3173, Nov. 2009.

[6] M.-H. Tu *et al.*, "A single-ended disturb-free 9T subthreshold SRAM with cross-point data-aware write word-line structure, negative bit-line, and adaptive read operation timing tracing," *IEEE J. Solid-State Circuits*, vol. 47, no. 6, pp. 1469–1482, Jun. 2012.

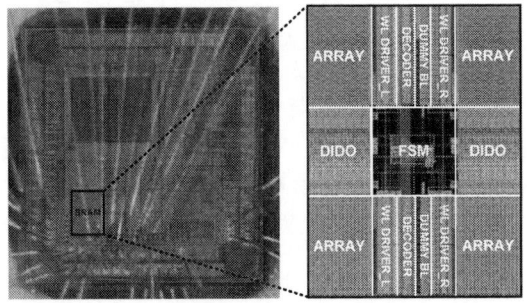

Fig. 4. Die photo combined with layout for 8 kbits SRAM with 9T bitcell and PC-DVS scheme.

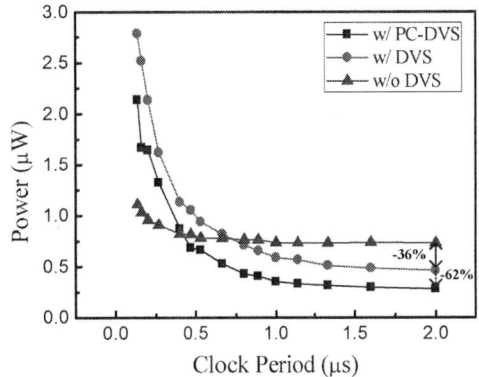

Fig. 5. Measurement results of array power.

TABLE I. MEASUREMENT RESULTS OF THE SRAM MACRO

Technology	65nm Low Power Process
Supply Voltages	V_H: 0.5 V; V_L: 0.35 V
Bitcell Size	$1.565 \times 1.19\ \mu m^2$
Macro Size	$195 \times 281\ \mu m^2$
Configuration	8 kbits (4 sub-banks) 256 rows × 32 bits
Dynamic Power	14.4 µW @ 5 MHz 2.67 µW @ 500 kHz
Static Power	1.39 µW
Array Power Reduction	-62% @ 500 kHz

978-1-4799-5128-4/14 $31.00 © 2014 IEEE

A Gain Cell Based Embedded DRAM with Fully-Restoring Write-Back Scheme

Weijie Cheng, Hritom Das, Huarong Zheng, Baolong Zhou, and Yeonbae Chung

School of Electronics Engineering, Kyungpook National University, Daegu, Republic of Korea
E-mail: ybchung@ee.knu.ac.kr

Abstract

In this paper, we present a hybrid 2T gain cell based embedded DRAM with body-voltage controlled technique. The memory bit-cell is composed of a high-V_{TH} write transistor and a standard-V_{TH} read transistor. The negative cell-body toggle signal couples up the data '1' storage level after data write. It results in an enhanced data retention time. Moreover, the proposed technique exhibits much strong immunity on write disturbance since the subthreshold leakage through the write device is drastically reduced. Simulation results from a 64-kbit eDRAM implemented in a 130 nm triple-well logic CMOS technology demonstrate the effectiveness of the proposed embedded memory technique.

Keywords-gain cell; embedded DRAM; data retention; SoC

Introduction

Gain cell eDRAMs are receiving a growing attention due to their preferable embedded attributes over currently dominant 6T SRAMs and 1T/1C DRAMs. Featuring logic CMOS compatibility, compact bit-cell area, decoupled read-write path and non-destructive read-out, the gain cell memories are becoming a strong contender for future embedded appications [1-4]. Improving bit-density over 6T SRAM and attaining practical retention time remain as key challenges. In this work, we explore a body-controlled write-back technique for 2T gain cell eDRAM which provides much improved data retention and strong immunity to the write disturbance.

Boby-Controlled Write-Back Technique

Fig. 1 shows the proposed 2T gain cell schematic, layout and its device cross-sectional view implemented in triple-well CMOS technology. The memory cell consists of a write access high-V_{TH} NMOS and a read access standard-V_{TH} NMOS. Both transistors are made in pocket p-well. A unique feature of the proposed technique is that the memory operation utilizes the reverse-biased junction capacitance of a parasitic diode formed between the pocket p-well and the cell data-node (DN). The parasitic junction capacitance is controlled by the cell-body (BD_C) signal.

Fig. 2 illustrates the cell bias conditions during the standby mode, read access and write access. In the standby, both the read-bitline (RBL) and the write-bitline (WBL) are sustained to V_{DD}, and the write-wordline (WWL) and the read-wordline (RWL) are set to the ground and V_{DD} respectively. BD_C is tied to the ground. For the read operation, RWL is pulled down to the ground while WWL is kept low. The read path either turns on or stays off depending on the DN status. If DN is low, the read-NMOS does not turn on so that RBL will be maintained to V_{DD}. If DN is high, the read-NMOS turns on the

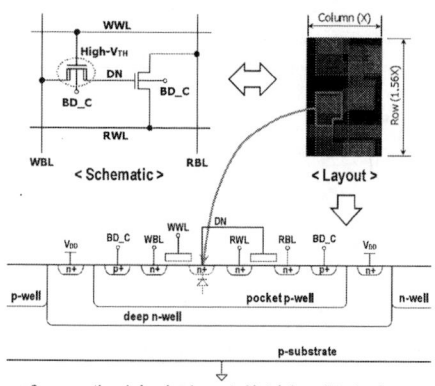

Fig. 1 Proposed 2T gain cell schematic, layout and its device cross-sectional view. (RWL: read-wordline, RBL: read-bitline, WWL: write-wordline, WBL: write-bitline, BD_C: cell-body, DN: data-node)

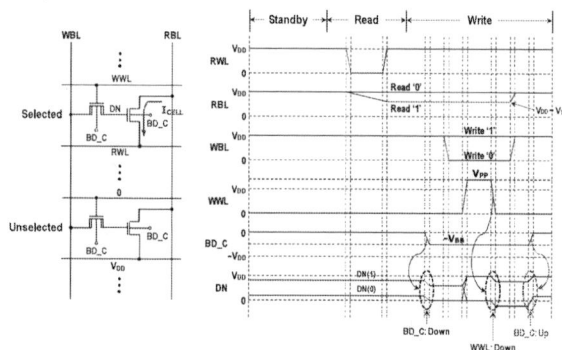

Fig. 2 Memory cell bias conditions for standby, read and write access. (I_{CELL}: read cell current)

read path and thus a discharging current will flow from RBL to RWL. This lowers the voltage level of RBL. Results of voltage difference in RBL can be sensed and amplified by a voltage sense amplifier. For the write operation, BD_C is set to a negative voltage -V_{BB} at the beginning. This couples down both initially stored data '1' and '0' voltages. Then, WWL is switched to a boosted voltage V_{PP} while RWL is kept to V_{DD} to allow new data from WBL to be written into the cell. However, when WWL is switched back to the ground, there is a coupling loss on the data storage voltages by the gate-overlap capacitance of the write transistor. To resolve this issue, BD_C is finally switched to the ground after the write operation is accomplished. This couples up both data '1' and '0' storage voltages, fully restoring the cell data '1' storage level.

Fig. 3 shows a simplified configuration of memory array.

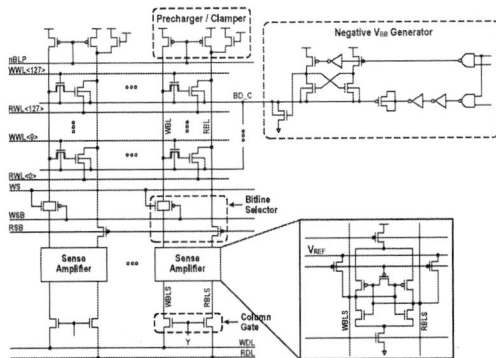

Fig. 3 Simplified memory array circuitry.

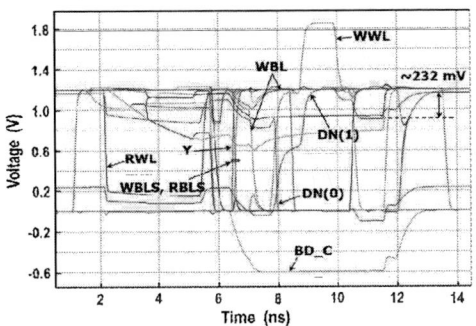

Fig. 4 Signal waveforms for write access at V_{DD} = 1.2 V.

The memory operation activates a row and enables the corresponding bitline sense amplifier. A feature of our sensing scheme is generation of a reference voltage V_{REF} to determine whether the state of the accessed cell is a '0' or a '1'. V_{REF} is an intermediate voltage between data '1' and '0' signals for the sense amplifier to correctly sense both data states. Fig. 4 shows waveforms for write access. The actual write access begins with an initial read operation from the cells, which activates the sense amplifiers and drives WBLSs for unselected columns to reflect the cell contents. Then the external data for the selected columns are developed on the sense amplifier bitlines RBLS and WBLS when the column decoding signal Y is enabled. The sense amplifiers flip if the external data are different from the initial data of the accessed cells. Then, BD_C is set to a negative voltage $-V_{BB}$. Here, $-V_{BB}$ is -0.6 V for V_{DD} = 1.2 V. Next, WS and WSB are triggered to connect WBL and WBLS together. WWL is then set to a boosted voltage V_{PP} allowing a new data to be written into the cell. The simulation indicates a coupling loss on the data '1' voltage when WWL is switched back to the ground. BD_C is switched to the ground immediately after data write to restore the coupling loss. The restoring voltage level is about 232 mV.

Primary benefit of the proposed cell-body control technique is an increase of data retention time. Fig. 5 shows the simulated data retention at 1.2 V and 85 °C. Data '0' and '1' are written into the cells respectively, then both data after 200 µs are read continuously with 50 µs period. The retention time of eDRAM with -0.6 V body toggle is 300 µs, which is improved by 66 % compared to that (180 µs) of eDRAM without cell body control.

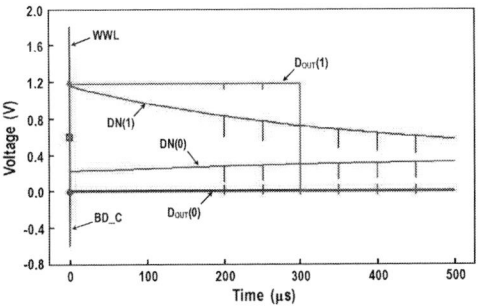

Fig. 5 Data retention of 2T gain cells with -0.6 V body toggle at 1.2 V and 85 °C.

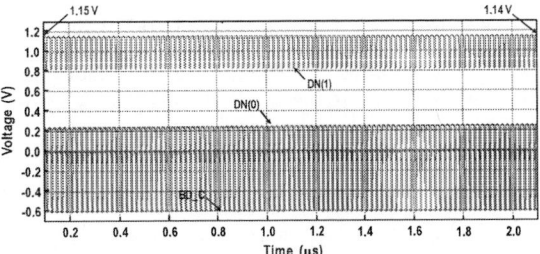

Fig. 6 Data disturbance of unselected cells under 1.2 V, 85 °C and 100 body-controlled write '0' cycles.

The data retention under write disturbance is also shown in Fig. 6. Under 1.2 V, 85 °C and 100-cycle write '0' disturbance, the data '1' loss of the body-controlled eDRAM measures 10 mV, which is 70 % smaller compared to that (33 mV) of eDRAM without cell body control. Owing to the relatively low subthreshold-leakage in the write transistor, which is caused by a body-to-source bias of -0.6 V, the proposed eDRAM exhibits much strong write disturbance immunity.

Conclusion

We have addressed a hybrid 2T gain cell eDRAM with body-controlled write-back scheme. Simulation results from a 64-kbit prototype implemented on 130 nm generic logic CMOS technology exhibits much improved data retention and stronger immunity to the write disturbance.

Acknowledgments

This research was supported by Samsung Electronics Corporation and IDEC in Korea. This research was also supported by Basic Science Research Program through the National Research Foundation of Korea (NRF) funded by the Ministry of Education (2014R1A1A4A01008225).

References

[1] D. Somasekhar et al., *IEEE J. Solid-State Circuits*, vol. 44, pp. 174-185, January 2009.

[2] K. C. Chun et al., *IEEE J. Solid-State Circuits*, vol. 46, pp. 1495-1505, June 2011.

[3] K. C. Chun et al., *IEEE J. Solid-State Circuits*, vol. 47, pp. 2517-2526, October 2012.

[4] W. Cheng and Y. Chung, *IET Circuits, Devices and Systems*, vol. 8, pp. 107-117, March 2014.

An Allocation Optimization Method for Partially-Reliable Instruction Scratch-Pad Memory in Embedded Systems

Takuya Hatayama*, Hideki Takase*, Kazuyoshi Takagi* and Naofumi Takagi*

*Kyoto University

Abstract—In this paper, we propose the use of a memory system which has partially reliable SPM in order to optimize energy consumption while ensuring required reliability. An allocation method for such memory system is formulated as integer linear programming whose solution leads to energy-optimal consumption while ensuring required reliability. Evaluation shows that proposed method is effective when overhead for error correction is large.

Keywords—SPM; ECC; low energy consumption

I. INTRODUCTION

Scratch Pad Memory (SPM) is often employed as an on-chip memory in modern embedded systems due to its efficiency in terms of area, energy and predictability compared with cache [1]. However, designers or software must determine the allocation of instruction and data to SPM.

In recent years, the rise in memory soft error rate (SER) has been a major concern. A soft error means transient fault in the circuits caused by alpha ray or neutron. Especially, instruction memory needs to enhance soft error tolerance because instruction memory is accessed in almost every cycle. One way to enhance soft error tolerance of memory is adoption of error correcting code (ECC). In modern semiconductor manufacturing technology, the SERs in FIT/bit for DRAM with ECC, SPM with ECC and SPM without ECC are 10^{-12}, 10^{-7} and 10^{-4}, respectively [2], [3]. However, adopting ECC in a simplistic form may cause excessively increase of energy consumption.

The purpose of this paper is energy optimization of embedded systems while ensuring required reliability. We propose the use of a partially reliable SPM. Partially reliable SPM is SPM that ECC is adopted for a part of region. The allocation optimization method for partially reliable SPM is formulated as an integer linear programming (ILP). An optimal solution of the proposed ILP problem corresponds to the instruction allocation which minimizes energy consumption of the system while ensuring required reliability.

II. TARGET SYSTEM

Fig. 1 shows the proposed memory system which has partially reliable SPM. Instructions which require high reliability are allocated to the region of reliable SPM. In contrast, instructions not requiring high reliability are allocated to the region of normal SPM. The proposed method can contribute reduction of energy consumption while ensuring required soft error tolerance.

III. INSTRUCTION ALLOCATION OPTIMIZATION METHOD

A. Problem definition

This section describes an ILP problem of proposed allocation method. The purpose of this paper is minimization of energy consumption while ensuring required reliability. We

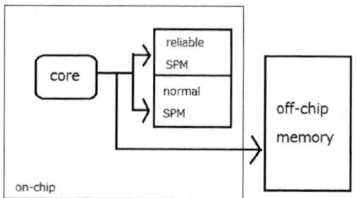

Fig. 1. Target system which has partially reliable SPM

propose an instruction allocation method for partially reliable SPM to achieve the purpose.

Energy consumption of instruction memory is calculated by energy consumption per memory access and the number of fetch. In this paper, instructions which require high reliability are defined as frequently fetched instructions. Vulnerability of the whole system is determined by vulnerability of instructions and SER of memory. Vulnerability of instructions is defined by proportion of the number of fetches, because errors in instructions frequently fetched widely propagates. Ensuring required reliability means that vulnerability of the whole system should not be more than given value by instruction allocation.

B. Preparations

Constants of our ILP problem in Table I. In this paper, F_i means i-th function in the program. S_i and N_i are obtained by task profiling, and the other constants are determined by the target system configuration. The decision variables of our ILP problem, which take binary values shown in Table II, indicate that the allocation in which memories F_i is allocated to. RSPM means the region of SPM where ECC is adopted.

C. Vulnerability

The vulnerability of the whole system in this paper is defined as follows.

$$V_{system} = \sum_i V_i \cdot SER(F_i) \qquad (1)$$

$$V_i = (N_i)/(\sum_k N_k) \qquad (2)$$

Here, V_i means vulnerability of F_i, that is the rate of frequency for F_i. $SER(F_i)$ means SER of F_i.

TABLE I. CONSTANTS

Constant	Definition
S_i	Memory size of F_i
N_i	Number of fetches of F_i
$E_{MM}, E_{RSPM}, E_{SPM}$	energy consumption per access of main memory, RSPM and SPM
C_{RSPM}, C_{SPM}	Capacity of RSPM and SPM
$R_{MM}, R_{RSPM}, R_{SPM}$	SER of main memory, RSPM and SPM

TABLE II. BINARY VARIABLES

x_i	$x_i = 1$	$\iff$	F_i is allocated in main memory
y_i	$y_i = 1$	$\iff$	F_i is allocated in RSPM
z_i	$z_i = 1$	$\iff$	F_i is allocated in SPM

978-1-4799-5128-4/14 $31.00 © 2014 IEEE

$$SER(F_i) = x_i \cdot R_{MM} \cdot S_i + y_i \cdot R_{RSPM} \cdot S_i + z_i \cdot R_{SPM} \cdot S_i \quad (3)$$

D. Objective function

Objective function is the energy consumption.

$$\textbf{minimize} : E_{system} = \sum_i E(F_i) \quad (4)$$

$E(F_i)$ means energy consumption of F_i.

$$E(F_i) = x_i \cdot E_{MM} \cdot N_i + y_i \cdot E_{RSPM} \cdot N_i + z_i \cdot E_{SPM} \cdot N_i \quad (5)$$

E. Constraints

There are three constrains in our ILP problem. First is about allocation. Each F_i must be allocated in only one memory.

$$\forall i, \ x_i + y_i + z_i = 1 \quad (6)$$

Second is about SPM capacities. Sum of S_i allocated on RSPM or SPM cannot exceed the capacity of them.

$$\sum_i y_i \cdot S_i \le C_{RSPM} \quad (7)$$

$$\sum_i z_i \cdot S_i \le C_{SPM} \quad (8)$$

Third is about vulnerability. Vulnerability of the system V_{system} should not be more than the given vulnerability V_{max}.

$$V_{system} \le V_{max} \quad (9)$$

IV. EVALUATION

To evaluate the effectiveness of proposed method, we employed SkyEye [4] as the simulator of ARM. We executed benchmark programs and profiled execution logs in order to obtain constants in Table I. Benchmark programs were basicmath, bitcount, susan and dijkstra from MiBench [5]. We used CPLEX as a ILP solver. All ILP problems were solved in 0.1 seconds by CPLEX.

We firstly used CACTI6.5 [6] for the model as the amount of energy consumption per access to each memory (Table III). However, CACTI ignores ECC coding and decoding to calculate energy consumption. Reference [7] reported that these process for a cache line of 70 nm process consumes 40.783 pJ of energy. We gave parameters which corresponded to configuration of [7] to CACTI. 40.783 pJ is approximately 10 times as large as energy consumption in Table III. Therefore, we also evaluated the case for $E_{RSPM} = E_{SPM} * 10$.

Tables IV and V show the evaluation results with the values of CACTI and with the assumption of $E_{RSPM} = E_{SPM} * 10$. We varied the given vulnerability that means what times vulnerability is allowed from original value. The baselines for energy consumption are the case of 4KB RSPM and 0KB SPM.

From Table V and IV, proposed method is effective in case overhead for coding and decoding are large in all programs.

Therefore, proposed method is effective in case overhead for error detection and correction are large. Additionally, proposed method is effective in programs which have high locality of memory access. In fact, the proportion of fetch from SPM and RSPM of bitcount and susan is more than 90%.

From evaluation results, we can observed that optimal size of adopting ECC is varied by programs and given value. It is important to determine the appropriate configuration according to programs and given value.

V. CONCLUSION

In this paper, we proposed memory allocation method for partially reliable SPM. Proposed method optimize energy consumption while ensuring required reliability. From evaluation, proposed method is effective in case ECC overhead is large and programs have high spatial locality. In future, we will extend proposed method for data SPM.

ACKNOWLEDGMENT

The part of this work was supported by JSPS KAKENHI Grant Number 26870303.

REFERENCES

[1] R. Banakar, S. Steinke, Bo-Sik Lee, M. Balakrishnan and P. Marwedel, "Scratchpad memory: a design alternative for cache on-chip memory in embedded systems", In *Proc. of CODES*, pp. 73–78, 2002.

[2] C.W. Slayman., "Cache and memory error detection, correction, and reduction techniques for terrestrial servers and workstations", *IEEE Trans. on DMR*, Vol. 5, No. 3, pp. 397–404, 2005.

[3] M. Sugihara, T. Ishihara and K. Murakami, "Task scheduling for reliable cache architectures of multiprocessor systems", In *Proc. of DATE*, pp. 1–6, 2007.

[4] Skyeye_wiki, http://skyeye.sourceforge.net/.

[5] MiBench Version 1.0, http://www.eecs.umich.edu/mibench/.

[6] N. Muralimanohar, R. Balasubramonian and Norman P. Jouppi, "Cacti 6.0: A tool to understand large caches", *technical report*, university of utah and hewlett packard laboratories, 2009.

[7] V. Degalahal, Lin Li, V. Narayanan, M. Kandemir and M.J. Irwin, "Soft errors issues in low-power caches", *IEEE Trans. on VLSI Systems*, Vol. 13, No. 10, pp. 1157–1166, 2005.

TABLE IV. RESULTS IN CASE OF VALUES OBTAINED BY CACTI

Given value	Configuration RSPM/SPM	Normalized energy consumption			
		basicmath	bitcount	susan	dijkstra
1e0	4KB/0KB	1.00	1.00	1.00	1.00
1e1	3KB/1KB	1.01	1.12	0.99	1.08
	2KB/2KB	1.33	4.03	2.18	1.46
	1KB/3KB	2.95	11.28	10.34	2.61
	0KB/4KB	4.26	47.66	11.66	6.97
1e2	3KB/1KB	0.99	0.91	0.98	0.99
	2KB/2KB	0.99	0.96	1.03	1.01
	1KB/3KB	2.54	6.87	9.96	1.85
	0KB/4KB	3.77	37.73	9.97	5.81
1e3	3KB/1KB	0.99	0.82	0.98	0.98
	2KB/2KB	0.99	0.86	1.02	0.98
	1KB/3KB	0.99	0.83	0.98	0.98
	0KB/4KB	1.00	0.97	0.99	1.00

TABLE V. RESULTS IN CASE OF $E_{RSPM} = E_{SPM} * 10$

Given value	Configuration RSPM/SPM	Normalized energy consumption			
		basicmath	bitcount	susan	dijkstra
1e0	4KB/0KB	1.00	1.00	1.00	1.00
1e1	3KB/1KB	0.97	0.93	0.93	1.00
	2KB/2KB	1.16	1.31	1.38	1.18
	1KB/3KB	2.27	2.44	5.04	1.81
	0KB/4KB	3.20	8.71	5.63	4.41
1e2	3KB/1KB	0.94	0.73	0.88	0.89
	2KB/2KB	0.90	0.74	0.82	0.88
	1KB/3KB	1.96	1.61	4.82	1.32
	0KB/4KB	2.83	6.89	4.81	3.68
1e3	3KB/1KB	0.88	0.32	0.88	0.72
	2KB/2KB	0.77	0.21	0.56	0.65
	1KB/3KB	0.75	0.18	0.48	0.64
	0KB/4KB	0.75	0.18	0.48	0.63

TABLE III. ENERGY CONSUMPTION OBTAINED BY CACTI

Memory	Size	[pJ]
Main Memory	4MB	456.266
RSPM	1KB	3.019
	2KB	3.838
	3KB	4.499
	4KB	4.952
SPM	1KB	2.861
	2KB	3.604
	3KB	4.227
	4KB	4.714

978-1-4799-5128-4/14 $31.00 © 2014 IEEE

Modulo Scheduler Implementation for VLIW Processor

Jangseop Shin, Sangjun Han, Hyungyun Jung, Ingoo Heo and Yunheung Paek

Dept. of Electrical and Computer Engineering, Seoul National University
1 Gwanak-ro, Gwanak-gu
Seoul, South Korea
Phone: +82-2-880-1742, E-mail: jsshin, sjhan, hgjung, igheo@sor.snu.ac.kr, ypaek@snu.ac.kr

Abstract

For VLIW processors, compiler must statically schedule instructions since there are no hardware for detecting hazards and reordering instructions at runtime. Thus, instruction scheduling techniques for VLIW processors have critical influence on correct execution and effective utilization of hardware resources. Software pipelining is a popular instruction scheduling technique which enables overlapped execution of successive loop iterations. We implemented modulo scheduler, which is a widely used technique of obtaining software pipelined schedule. Experiments on a set of multimedia applications show that performance is increased up to 2.6x compared to simple list scheduling implementation.

Keywords-VLIW; software pipelining; modulo scheduling; instruction scheduling; instruction level parallelism

Introduction

VLIW processors employ multiple function units(FUs) to exploit instruction level parallelism. Unlike superscalar processors, they have no hardware to check instruction dependency violations and resource conflicts and schedule instructions dynamically. Thus, compiler's role as an instruction scheduler is critical for VLIW processors. Without elaborately implemented scheduler, compiler-generated code either won't utilize multiple FUs effectively or produce wrong results due to scheduling constraint violation that cannot be detected by the hardware.

The simplest instruction scheduling strategy is to schedule instructions within the basic block boundary. Basic block is the stream of instructions which are executed sequentially regardless of control flow. It doesn't require complicated algorithm to implement this kind of scheduler but as a tradeoff, there is a fundamental limit to the performance of generated code. To overcome this limit, many scheduling techniques consider instructions across the basic block boundary. They are much complicated because scheduling across basic block boundary means executing some instructions ahead of time even though it is not sure that corresponding program path will be taken. The scheduler must make sure that it is safe to execute those instructions ahead of time.

Software pipelining is one of the popular instruction scheduling techniques that schedules instructions across loop boundary. Instead of executing loop iterations one after another, software pipelining allows successive loop iterations to overlap in time, just like hardware pipelining where executions of successive instructions are overlapped in time.

Since most programs spend their time in loops, this technique greatly improves performance of the generated code, and thus is popular choice for optimizing compilers.

We have implemented modulo scheduler, which is the most popular kind of software pipelining technique, based on swing modulo scheduler[1]. This paper is organized as follows. First, background knowledge on modulo scheduling technique is explained. Second, key characteristics of swing modulo scheduling are described. Lastly, experimental results and analysis will be followed by the conclusion.

Background

In software pipelining terminology, "initiation interval(II)" means the interval in which new iteration will be initiated. Since loop iterations start every II cycles, total execution time of the loop will be approximately proportional to II. Thus, minimizing value of II is the number one goal in software pipelining.

Modulo scheduling tries to achieve this by trying to find valid schedule given some starting value of II, and then incrementing II until it succeeds. To calculate starting value of II, we consider resource constraint and data dependency constraint separately. If we have 4 instructions that use a function unit, and we have 2 function units, then II should be at least 2. This bound is called "resource MII(minimum II)." To see another kind of MII, suppose we have a loop consisting of 3 instructions, as in Fig. 1. In this graph, instructions are represented as nodes and dependencies between them are represented as edges. It is easy to see that the second iteration can only initiate after at least 2 cycles (assuming each instruction takes 1 cycle,) since otherwise the value of R1 will be overwritten by the second iteration before it can be used by the third instruction. This bound is called "recurrence MII." To simplify the representation, edges representing cross-iteration dependencies are drawn together with intra-iteration

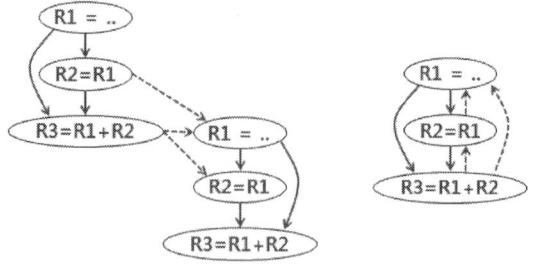

(a) Data dependence between two iterations   (b) Data dependence graph.

Fig. 1. Example showing how recurrence can bound II

dependencies as shown on the right. This kind of graph is called "data dependence graph." If you see carefully, you can see that edges contributing to the bound form a cycle. This is called "recurrence." Recurrences are not only used to determine the bound on II but also play the key role in increasing the chance to succeed scheduling at a given II, as explained later. The starting value of II will be the maximum of these two MIIs.

There cannot always be a valid schedule for this starting value of II, since MIIs are found by only considering one of the constraints. Moreover, computing resource MII like above can be incorrect, since there can be another kind of resource conflicts such as register read/write port conflicts. Finding optimal II considering all of the constraints is a NP-complete problem, so heuristic modulo scheduling is used in production compilers.

Swing Modulo Scheduling

In order to find a valid schedule for a given II, swing modulo scheduler adopts several heuristics. To reduce the scheduling time, it determines the order of nodes to schedule just once during the whole process, and when it fails to find a valid schedule, the same order is used to schedule at an increased II.

To maximize the chance of finding a valid schedule, nodes are grouped into ordered sets, and then nodes in each set are ordered to get a final node order. In computing partial order of the nodes, recurrence information is used. Recurrence which gives higher bound on II is given higher priority. The reason for this is that, if some node in the bigger recurrence could not be scheduled at some cycle due to already scheduled instruction, all nodes that are chained to this node will be delayed, making it harder to find a valid schedule for a given II. In most of the times, smaller recurrences can be scheduled in left over slots after bigger recurrences are scheduled.

Another important heuristic in determining partial order is when there is some path from current recurrence to the recurrence having higher priority. In this case, all the nodes on the path are scheduled before the current recurrence, since we may fail to find valid position to put those nodes if we try to schedule after scheduling recurrences on both sides.

Experimental Results

We implemented our modulo scheduler on VLIW compiler based on LLVM compiler infrastructure[2]. A set of multimedia applications are compiled and tested with processor simulator. Fig. 2 shows the comparison of II between implemented modulo scheduler and default list scheduler provided in LLVM for major loops of the applications. MIIs are also shown. Our modulo scheduler could obtain II that is a little larger than the MII whereas schedule obtained by the list scheduler has II that is much larger than the MII. The reason is that in many cases, there is a long chain of instruction dependencies which don't form a cycle, which limits the list scheduler's performance since iteration can only begin after the previous iteration is finished.

Fig. 3 shows the gain in performance with modulo scheduler as compared to the list scheduler in terms of processor cycles. The gain is a little less than the gain in terms of II, since programs don't consist only of loops. The more the program spends time on loops, the bigger is the gain obtained by modulo scheduling.

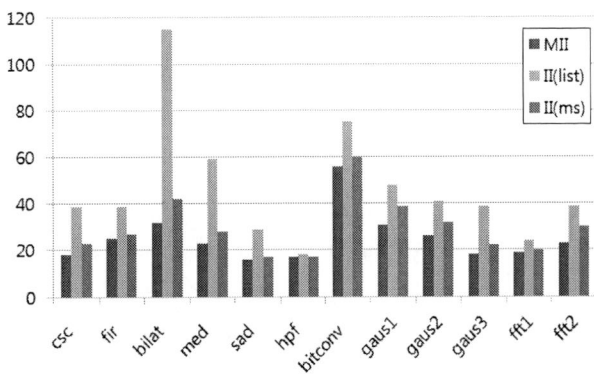

Fig. 2. Comparison of initiation interval between simple list scheduler and implemented modulo scheduler

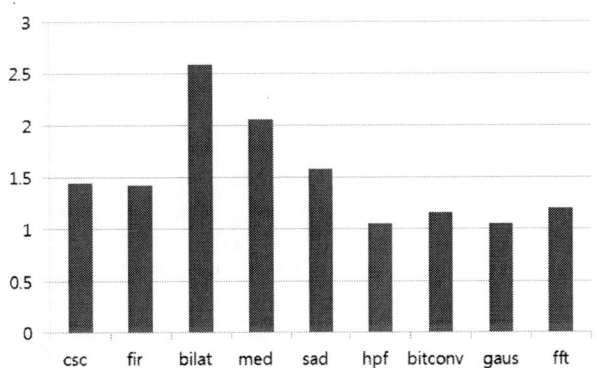

Fig. 3. Gain in cycle count with modulo scheduler as compared to simple list scheduler

Conclusion

Instruction scheduling techniques play a critical role in maximizing utilization of multi-issue processor resources. Software pipelining is one of the famous instruction scheduling technique that can accelerate execution of loops. In this research, we have implemented modulo scheduler and validated its effectiveness through the experiment on several applications.

Acknowledgment

This research was partly supported by the Engineering Research Center of Excellence Program of Korea Ministry of Science, ICT & Future Planning (MSIP) / National Research Foundation of Korea (NRF) (Grant NRF-2008-0062609), the IT R&D program of MSIP/KEIT[K10047212, Development of homomorphic encryption supporting arithmetics on ciphertexts of size less than 1kB and its applications], the Brain Korea 21 Plus Project in 2014 and IDEC.

References

[1] J. Llosa, A. Gonzalez, E. Ayguade, and M. Valero, "Swing modulo scheduling: a lifetime-sensitive approach," *IFIP WG10.3 Working Conference on Parallel Architectures and Compilation Techniques,* pp. 80-86, Boston (USA), October 1996.

[2] C. Lattner, V. Adve, "LLVM: a compilation framework for lifelong program analysis & transformation," *Proceedings of the 2004 International Symposium on Code Generation and Optimization,* pp. 75-86, March 2004.

Efficient Eye-Diagram Determination Technique of Non-Linearly-Switching Coupled-Data Links Under Power and Ground Fluctuation Noises

Junghyun Lee, Joonhyun Kim, Gihyeon Ji, Hyewon Kim, and Yungseon Eo

Dept. Electrical and Computer Engineering
Hanyang University
Ansan, Korea
E-mail : {jhlee, jhkim, ghji, hwkim, eo}@giga.hanyang.ac.kr

Abstract

A novel eye-diagram determination technique for coupled data links driven with the non-linear drivers is proposed. The non-linear characteristics of the driver are pre-characterized for a given circuit topology using SPICE step responses. Input test vectors that constitute eye-boundaries are found. The eye-diagram is determined with the test vectors. It is shown that the proposed technique is much more efficient than that of the PRBS-based generic technique in the order of two, whereas its accuracy is similar to that of the PRBS-based SPICE simulation (less than 5% error in both eye-height and jitter).

Keywords-component; eye-diagram; inter-symbol-interference; simultaneous switching noise.

Introduction

As the switching frequency and the level of integration of integrated circuits drastically increase, data links between the circuit blocks become one of the crucial system design issues [1]. The signal integrity degradation of the data links is mainly due to i) signal loss in interconnect lines, ii) reflections due to discontinuity, iii) electromagnetic coupling, and iv) power/ground supply noise during the circuit switching. Conventional piecemeal timing analysis and electromagnetic coupling analysis of data links are not sufficient enough to fully reflect the aforementioned physical phenomena.

In order to evaluate the holistic circuit performance, the inter-symbol-interference (ISI) is usually investigated. The ISI can be readily evaluated with the eye-height, eye-width, and jitter using the eye-opening at the destination node of a channel [2], [3]. Generically, the eye-diagram can be determined with SPICE simulation, applying numerous pseudo random bit sequence (PRBS) to the input of the circuit. Even if the technique is accurate, however, it requires huge amount of computation time and hardware resources. For example, in order to determine the accurate eye-diagram, PRBS input data more than "2^{17}~2^{18}" are at least required. In practice, SPICE simulation using such large number of input data may not be acceptable in multi-coupled data links because of the large computation time and hardware resources. In order to overcome such an impractical problem, numerous techniques that are computationally efficient have been developed [2]-[6].

The peak distortion analysis (PDA) algorithm (or similar techniques) [4] that approximates the input driver circuit as a voltage source with a linear resistor is a representative technique. However, it has a fundamental limitation to be employed for non-linear circuits. As a TX (input driver) is inherently non-linear and asymmetrically switching, the PDA analysis is too inaccurate to be acceptable in practical circuits.

Although the double-edge response (DER) approach [5], [6] improves the PDA problems a bit, the accuracy is still not sufficient enough to be employed for the evaluation of the practical circuits that are electromagnetically coupled and contaminated with various noises during circuit switching.

In this work, an accurate and efficient eye-diagram determination technique for the signal integrity verification of electromagnetically coupled data links with non-linear drivers that include simultaneous-switching-noise (SSN) is proposed. The inherent non-linear properties of drivers are pre-characterized with step responses using SPICE simulation. Next, possible input test vectors (i.e., bit streams) that constitute eye boundaries are found. Then, the eye-diagram for the channel is determined using SPICE simulation with the input test vectors. Finally, for a practical test circuit that is comprised of non-linear drivers with SSN, the accuracy and efficiency of the proposed technique are verified.

Test Vectors and ISI Determination

Signal transients in coupled lines are strongly dependent upon input switching patterns. Since a signal for each line may have one of the three switching states, (i.e., rising state, quiet state, and falling state), the switching pattern for the coupled lines is one of the 9 possible patterns. Given a circuit topology, these switching patterns should be pre-characterized with step responses using SPICE simulation that reflect the almost all of the non-linear characteristics for the given system. With the pre-characterized step responses, the input test vectors that constitute the eye-diagram boundaries are determined as follows. First, the number of the significant input bit stream is determined. A step response is settled down after a time elapses. Thus, since a present signal is affected by the tail signals of the previous signals until all the previous signals are settled down, the length of the input bit patterns (m) is defined as

$$ m \equiv \lceil t_{settle} / T_b \rceil, \qquad (1) $$

where t_{settle} is the settling time and T_b is an one bit duration time. Thus, the significant input bit stream vector of the i-th line can be represented by

$$ \left[B_i \right]_z = \left[b_{im} \quad \cdots \quad b_{ij} \quad \cdots \quad b_{i1} \quad b_{i0} \right], \qquad (2) $$

where the b_{i0} indicates the present bit of the i-th line and the b_{ij} indicates the j-th previous bit of the i-th line. Note that an eye-diagram is composed of eight eye-boundaries, which is indicated with the subscript $z \in \{1, 2, \cdots, 8\}$. Thus, in two coupled lines, the significant input bit stream can be represented with a matrix form,

$$[B]_z = \begin{bmatrix} B_1 \\ B_2 \end{bmatrix}_z = \begin{bmatrix} b_{1m} & \cdots & b_{1j} & \cdots & b_{11} & b_{10} \\ b_{2m} & \cdots & b_{2j} & \cdots & b_{21} & b_{20} \end{bmatrix}. \quad (3)$$

Secondly, in the interested line, since the present bit state may be either of 1 (logic high) or 0 (logic low) state, the state of the present bits of the two coupled lines are one of the 4 possible states. The response signals of the first two previous bit pairs, $\{b_{10}, b_{11}\}$ and $\{b_{20}, b_{21}\}$, can be readily evaluated using the pre-characterized step responses. Since the worst response or the best response may be eye-boundaries, they are set as input test vectors. Similarly, using the greedy algorithm for all the previous bit stream, their switching patterns are iteratively evaluated with the T_b-delayed step responses using the same decision criteria. Then the final input test vectors are used for SPICE input test vectors for the eye-diagram determination.

Verification

In order to verify the accuracy and efficiency of the proposed technique, a test circuit as described in Fig. 1 is evaluated. The eye-diagrams for the test circuit using both PRBS-based SPICE simulation and the proposed technique are determined (see Fig. 2) and summarized in Table 1. The proposed technique shows an excellent agreement with the PRBS-based SPICE simulation using 2^{17}–PRBS input bits as shown in Fig. 2. The eye-opening data (height and jitter) using the proposed technique show excellent agreement with those of PRBS-based SPICE simulation (less than 5% error), whereas the computation time is much faster in the order of two. Note that other conventional techniques show large discrepancies although the computation time is reasonable.

Conclusion

In this paper, a new efficient eye-diagram determination technique for the signal integrity verification of non-linearly switching coupled data links is proposed. The non-linear characteristics of drivers are pre-characterized with the switching-dependent step responses using SPICE simulation. Using the step responses, the input test vectors for the worst eye-diagram are determined with a newly developed algorithm. Finally, the worst eye-diagram can be accurately and quickly determined using SPICE simulation with the input test vectors.

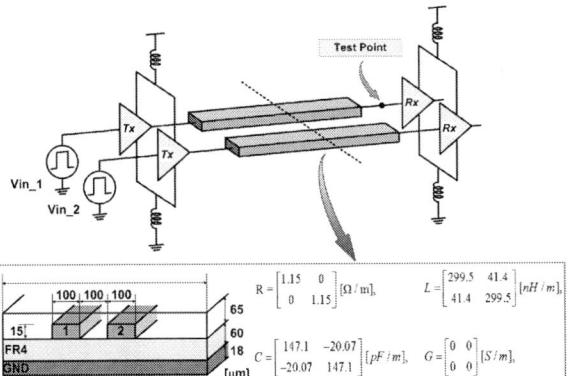

Fig. 1. A practical 5 Gb/s data link comprised of non-linear drivers under SSN noise. The line length is 5cm.

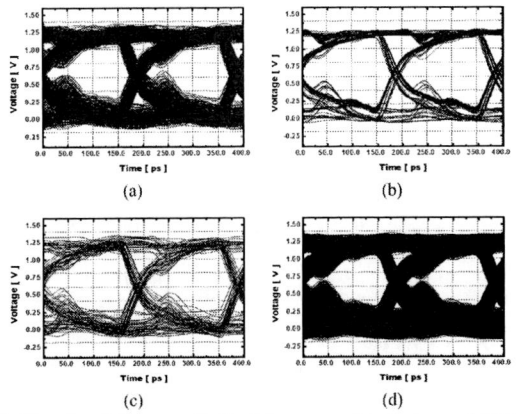

Fig. 2. The simulation results for (a) proposed technique, (b) DER[5], (c) DER[6], and (d) 2^{17}-PRBS SPICE simulation.

TABLE I
SIMULATION RESULT COMPARIOSN

Tech. Eye Opening	Eye-Determination Tech.			
	proposed	DER[5]	DER[6]	SPICE (2^{17}-bits)
Height[mv]	752.0 (+4.7%)	1006.9 (+40.2%)	852.9 (+18.8)	718.1
Jitter[ps]	35.7 (-4.5%)	19.8 (-47.1%)	27.0 (-27.8)	37.4
Time [s]	64.2	53.9	54.9	1543.2

The efficiency and accuracy of the proposed technique was verified with a practical test circuit with non-linear drivers under SSN noises. The simulation results show excellent agreement with those of PRBS-based SPICE simulation, whereas the computation time is much faster in the order of two. Therefore, the proposed technique can be efficiently exploited for the high-speed data link design.

Acknowledgment

This work was supported by the project, Development of Technologies for Next-Generation Electromagnetic Wave Measurement Standards, of the Korea Research Institute of Standards and Science under Grant 12011016.

References

[1] V. Stojanovic and M. Horowitz, "Modeling and analysis of high-speed links," in *Proc. IEEE Custom Integrated Circuits Conference*, pp. 589-594, Sept. 2003.

[2] N. Ou, T. Farahmand, A. Kuo, S. Tabatabaei, and A. Ivanov, "Jitter models for the design and test of Gbps-speed serial interconnects," *IEEE Design and test of computers*, vol. 21, no. 4, pp. 302-313, Jul.Aug. 2004.

[3] G. Breed, "Analyzing signals using the eye-diagram," *High Frequency Electronics*, vol. 4, no. 11, pp. 50-53, Nov. 2005.

[4] B. K. Casper, M. Haycock, and R. Mooney, "An accurate and efficient analysis method for multi-Gb/s chip-to-chip signaling schemes," in *IEEE Symp. Very Large Scale Circuits*, pp. 54-57, Jun. 2002.

[5] R. Shi, W. Yu, Y. Zhu, E. S. Kuh, and C.-K. Cheng, "Efficient and accurate eye diagram prediction for high speed signaling," in *Proc. IEEE Int. Conf. Computer Aided Design*, pp. 655-661, Nov. 2008.

[6] F. Lambrecht, C.-C. Huang, and M. Fox, "Technique of determining performance characteristics of electronic systems," U.S. Patent 6775809, Mar. 2002.

Analysis and Reduction of Voltage Noise of Multi-layer 3D IC with PEEC-based PDN and Frequency-dependent TSV models

Seungwon Kim[1], Ki Jin Han[1], Seokhyeong Kang[2], and Youngmin Kim[1]

[1]School of Electrical and Computer Engineering, UNIST, Unist-gil 50, Ulsan 689-798, Republic of Korea
[2]Department of Electrical and Computer Engineering, UCSD, La Jolla, CA 92093-0404 USA
{kskyh002, kjhan, youngmin}@unist.ac.kr, shkang@vlsicad.ucsd.edu

Abstract

Three dimensional (3D) integrated circuit (IC) technology has been proposed and used to reduce the delay among layers by shortening interconnection with TSVs. However, large power and ground TSV structures generate voltage noise and cause additional IR-drop in power delivery network (PDN). In this work, we investigate and analyze the voltage noise in multiple layers 3D IC stacking with PEEC-based on-chip PDN and frequency-dependent TSV models. Then we propose multi-paired on-chip PDN structure for reducing voltage noise in a 3D IC. Our proposed PDN architecture can achieve approximately maximum 19% IR-drop reduction. In addition, layer dependency of 3D IC between the conventional and the proposed PDN models is analyzed.

Keywords-component; voltage noise, IR drop, 3D IC, TSV, power delivery network (PDN), PEEC, S-parameter

Introduction

Recently, 3D IC stacking technology has been proposed and considered as a promising solution for more than Moore era. 3D IC effectively reduces propagation delay by shortening wirelength with vertical interconnections (e.g., TSVs). However, TSV related power integrity issues still require accurate understanding in the multi-layers stacking [1, 2]. In this study, we investigate the power delivery network of the 3D IC stacking structures. The on chip metal PDN is extracted with PEEC-based equivalent models [3] and frequency-dependent TSV parameters including C4 bumps are generated by EM modeling method [4, 5]. We combine the on-chip PDN and TSV in multi-layer 3D IC for power integrity analysis of static and dynamic voltage drop. In addition, various PDN structures are investigated in the proposed 3D IC analysis methodology and an optimal PDN architecture is generated for reducing the IR drop. A general TSV-based 3D IC stacking methodology is shown in Fig. 1. The multiple dies are stacked in a way of face to back bonding. We vary the number of stacking layers to identify the impact of the multi-layer on the power integrity.

Power Delivery Network of the TSV-based 3D IC

As the power delivery network (PDN) has become complicated and the wire resistance has increased owing to the interconnect scaling, the supply voltage fluctuations due to the IR drop have become a significant problem in the PDN design.

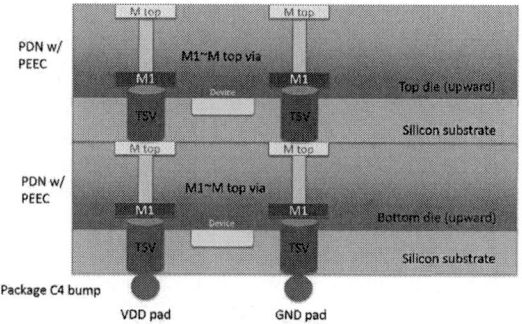

Fig. 1. Face to back bonding of multi-layer 3D IC

TABLE I
Structural parameters of the metal interconnect and TSV.

Parameter	Value (μm)
Metal 5 width	5
Metal 5 pitch	200
Metal 6 width	10
Metal 6 pitch	200
TSV diameter	10
TSV height	50
TSV pitch	200
TSV liner thickness	0.1
TSV pad size	15

A partial electrical equivalent circuit (PEEC) method [3] has been introduced for an accurate extraction of RCL components of wires. We use the PEEC method to extract RLC components from a PDN having nine VDD and nine GND pads. To obtain a frequency-dependent model of TSV structures, we use a 3D electromagnetic (EM) method [4]. Frequency-dependent S-parameters of the TSV structures are extracted from a 3D EM modeling method based on the mixed-potential integral equation [4]. We then perform a dynamic transient analysis with clock buffers to simulate the IR-drop-related voltage fluctuations and analyze the PDN architecture in 3D IC stacking.

The metal interconnect parameters and TSV dimensions used in this study are summarized in Table I [1]. The on-chip PDN consists of top two metal lines (i.e., M6 and M5) and the power supply of the logic block is connected through vias to supply required current. Fig. 2(a) shows the 3D visualiza-

978-1-4799-5128-4/14 $31.00 © 2014 IEEE

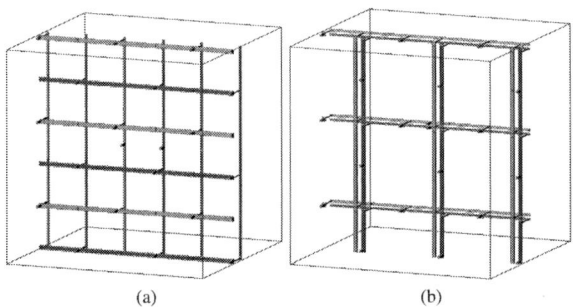

Fig. 2. (a) 3D view of the regular structure. PDN is generated using M6 and M5 with 3x3 VDD and GND. (b) 3D view of the multi-paired structure.

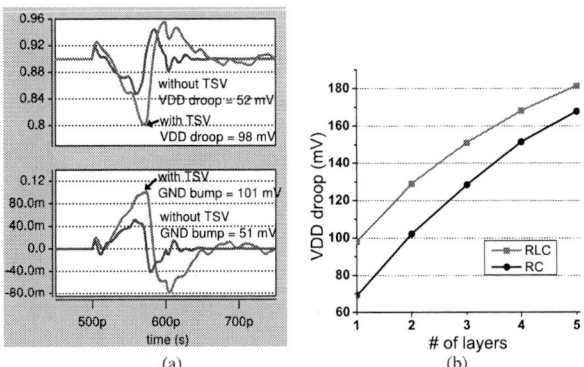

Fig. 3. (a) VDD droop (top) and GND bump (bottom) comparison with TSV and without TSV in a single layer of die. (b) Impact of the number of layers and the inductance (L) on the PDN voltage noise.

tion of the PEEC elements in the regular on-chip PDN structure. All the wire segments and vias are plotted and ports for the VDD and GND are shown in red color. We assume that $0.5\,mm \times 0.5\,mm$ die contains nine VDD and nine GND. And all these ports are connected to the corresponding S-parameter TSV ports to construct entire PDN of 3D IC. We apply ideal DC voltage source simultaneously to the all power/ground TSVs. The center VDD and GND ports are connected to the power supply pins of the logic gates. The logic consists of 40 clock buffers to mimic the current load in the simulations and is implemented in 22 nm [8]. The nominal VDD value is set to 0.9 V and the size of clock buffer is $W_n = 4\,\mu m$ and $W_p = 8\,\mu m$ We then measure power supply fluctuation for various configurations of 3D IC when the internal logic makes transitions in SPICE transient analysis [6]. At first, we stack 1 layer of die with TSV and C4 bump. Then we vary the number of stacking layers to identify the impact of the multi-layer on the IR-drop.

Analysis and Reduction of Voltage Noise

Fig. 3(a) shows the voltage fluctuation results of the 3D IC with a single layer TSV. As shown, the TSV attributes 89% additional voltage droop and 98% ground bump in a single layer.

The impact of the number of 3D IC layers on the voltage

TABLE II
The worst VDD droop in conventional regular PDN and proposed multi-paired PDN.

# of layer	regular structure VDD droop (mV)	proposed structure VDD droop (mV)	reduction (%)
1	98.07	79.65	18.8
2	129.00	114.83	11.0
3	151.05	139.77	7.5
4	168.14	159.02	5.4
5	181.35	173.05	4.6

drop is shown in Fig. 3(b) both by RL and RLC PEEC models. As shown, lager voltage fluctuation is expected as the number of layer increases and RLC model shows more impact on the variation than only RL model. Therefore, the inductance component should be considered in high-frequency 3D IC for accurate voltage fluctuation analysis such as Ldi/dt noise.

Shorter wire space between VDD and GND can effectively reduce mutual inductance of wire [7]. We propose and investigate a multi-paired PDN structure as shown in Fig. 2(b) and compare the voltage fluctuation with that of the conventional regular structure (e.g., Fig. 2(a)). We measure the benefit of the proposed PDN architecture with increasing the number of layers in 3D IC stacking. The comparison of the VDD drop between the regular PDN and multi-paired PDN is summarized in Table II. As shown, there is up to 19% reduction in VDD droop for a single layer by the proposed PDN architecture and the benefit is attenuated as the number of layers increases.

Conclusion

In this paper, first we investigate the voltage noise of multi-layer 3D IC with PEEC-based PDN and frequency-dependent TSV models. We find out that TSV affects voltage fluctuation in PDN significantly. Moreover, the inductance component and frequent-dependent TSV models are vital for accurate understanding of the power integrity of the 3D IC. Second, we propose and investigate the multi-paired on-chip PDN structure for reducing voltage noise of the PDN in 3D IC. Multi-paired structure effectively reduce IR-drop approximately maximum 19% and the advantage decreases as the number of layers increases.

References

[1] ITRS 2012, [online], http://public.itrs.net.

[2] G. Huang, *et al.*, "Power Delivery for 3D Chip Stacks: Physical Modeling and Design Implication," *IEEE TCMPT*, vol. 2, no. 5, pp. 852–859, 2012.

[3] A. E. Ruehli, "Equivalent circuit models for three dimensional multiconductor systems," *IEEE TMTT*, vol. MTT-22, pp. 216–221, 1974.

[4] K. J. Han, *et al.*, "Electromagnetic Modeling of Through-Silicon Via (TSV) Interconnections Using Cylindrical Modal Basis Functions," *IEEE TAP*, vol. 33, no. 4, pp. 804–817, 2010.

[5] K. J. Han, *et al.*, "Inductance and resistance calculations in three-dimensional packaging using cylindrical conduction-mode basis functions," *IEEE TCAD*, vol. 28, no. 6, pp. 846–859, 2009.

[6] Hspice reference manual, ver. H-2013.03-SP1, [Online]. Available: http://www.synopsys.com.

[7] W. H. Lee, *et al.*, "Analysis and reduction of on-chip inductance effects in power supply grids," *IEEE ISQED*, pp. 131–136, 2004.

[8] 22nm PTM HP model, [Online]. Available: http://ptm.asu.edu.

Allocation and Optimization of Post-Silicon Tunable Buffers in TSV Based Heterogeneous 3D ICs

Sangdo Park[1]
sangdo0.park@samsung.com

Jeongwoo Heo[2]
jwheo@snucad.snu.ac.kr

Taewhan Kim[2]
tkim@snucad.snu.ac.kr

[1]System LSI, Samsung Electronics Co., Ltd
[2]School of Electrical and Computer Engineering, Seoul National University, Seoul, Korea

Abstract—Through-silicon via (TSV) based 3D IC design is a promising solution to reducing the length of interconnects and improving the power and speed. However, when heterogeneous dies are stacked together to form a 3D IC, a considerable timing discrepancy among the layers could happen since the devices in different layers might have been affected quite differently by process variations. With this respect, this work makes two contributions: (1) proposing a PST buffer allocation scheme in 3D ICs to resolve the timing discrepancy between dies; (2) with the proposed allocation scheme, proposing a technique that is able to minimize the total cost of PST buffer implementation.

Keywords: Post-silicon tunable buffer; 3D IC; process variation; yield; allocation.

I. INTRODUCTION

Through-silicon via (TSV) based 3D IC design has been regarded as one of the doable solutions to the next generation design paradigms [1]. Generally, 3D ICs are made by splitting a large design into several sub-designs and assigning them to multiple dies i.e., *homogeneous* integration or by integrating more than one different designs into a large system i.e., *heterogeneous* integration [2]. Contrary to the homogeneous integration, the dies assembled in a 3D package by heterogeneous integration could arouse a considerable timing mismatch among the devices in the dies since the devices are affected differently by process variations, exhibiting different process corners [3].

Conventionally, it is known that post-silicon tunable (PST) buffer, which can adjust its delay statically or dynamically, can be used to control the timing in a circuit to mitigate the timing variation in the device [4]. A PST buffer consists of a set of capacitors or chained inverters and a control circuitry that sets ON/OFF to each capacitor or inverter according to the signal from an external port [5]. However, even though there exist several works on solving the problem of PST buffer allocation in a single device of 2D ICs, no work has addressed the problem of resolving the timing discrepancy between the devices in different dies in 3D ICs. In this work, we propose a solution of PST buffer allocation to cope with the timing discrepancy due to die-to-die variation in 3D ICs. More precisely, (1) we propose a PST buffer allocation scheme in 3D ICs to resolve the timing discrepancy between dies and (2) with the proposed allocation scheme, we propose a technique that is able to minimize the total cost of PST buffer implementation. It should be noted that, to the best of our knowledge, this is the first work that addresses the PST buffer allocation in 3D ICs for mitigating on-package variation.

II. THE PROPOSED PST BUFFER ALLOCATION AND OPTIMIZATION IN 3D ICs

A. PST Buffer Allocation

The left part of Fig. 1 shows an example of clock network in a heterogeneous 3-layered IC, in which it is seen that each layer has its own style of clock tree that synchronizes the timing of the circuit on the layer. During the pre-bonding stage, individual dies will be tested using their local clock networks. Once all known-good-dies (KGDs) are identified, those are bonded together to form a 3D IC. Let us assume that the clock source of the 3D clock network after bonding in the example comes from layer 2. Then, since the devices in layers exhibit on-package variation as illustrated on the right part in Fig. 1, the timing of the 3D clock tree is required to be adjusted, so that all the clock sinks (i.e., FFs or latches) in 3D ICs should be synchronized at the same time. To achieve this, we allocate a clock TSV that spans all layers, together with a PST buffer to each location intersected by the TSV and layers, and adjust the delay of PST buffers. For example, in Fig. 1 the PST buffers in red color indicate lengthening the delay while the PST buffer in black color indicates shortening the delay.

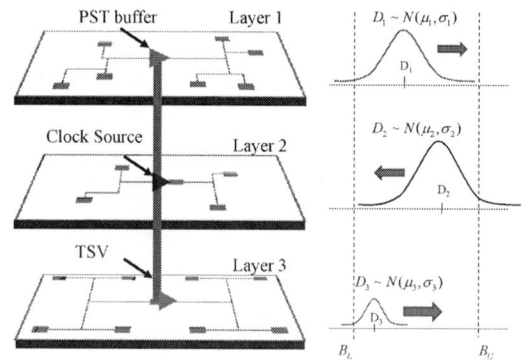

Fig. 1. An example of proposing PST buffer allocation in 3D clock network in a heterogeneous 3-layered IC. For K-layer ICs, K PST buffers are allocated, one for each layer.

Since the implementation cost of PST buffers closely relies on the number of capacitors or chained inverters deployed in the PSTs, which in turn is linearly proportional to the adjustable delay range of PST buffers, in the following subsection, we propose a technique of determining the delay range of each PST buffer that leads to a minimal total delay range.

978-1-4799-5128-4/14 $31.00 © 2014 IEEE

B. PST Buffer Optimization

Clock skew $\kappa_{i,j}$ between two layers i and j can be expressed as:

$$\kappa_{i,j} = (x_i + D_i) - (x_j + D_j) \qquad (1)$$

where D_i and D_j are random variables representing the clock delays from the source in layers i and j to sinks in the same layers, respectively. x_i and x_j are variables representing PST buffer delays to be adjusted in layers i and j, respectively. For a skew bound Γ, the yield $Y_{i,j}$ of clock skew $\kappa_{i,j}$ is

$$Y_{i,j} = Prob(\kappa_{i,j} \leq \Gamma) = Prob(D_i - D_j \leq \Gamma + x_j - x_i). \quad (2)$$

Let Y be the clock skew yield to be met for every pair of clock trees in the layers of 3D IC. Then, by $Eq.(2)$, we want to find values of x_i and x_j that satisfy $Prob(D_i - D_j \leq \Gamma + x_j - x_i) \geq Y$: we compute a value, $U_{i,j}(Y)$, that satisfies $Prob(D_i - D_j \leq U_{i,j}(Y)) = Y$, from which we derive the following inequality that should be satisfied:

$$\Gamma + x_j - x_i \geq U_{i,j}(Y). \qquad (3)$$

Thus, by $Eq.(3)$, the PST buffer optimization problem can be formulated into a linear programming (LP) formula as follows:

$$\begin{aligned} \text{minimize} \quad & \sum_{i=1}^{K} x_i \\ \text{subject to} \quad & x_j - x_i \geq U_{i,j}(Y) - \Gamma \quad \forall i,j \in \{1, \cdots, K\}, \\ & x \geq 0. \end{aligned}$$

III. EXPERIMENTAL RESULTS

The proposed algorithm has been implemented using C++ and run on Linux machine with 2.66 GHz Intel Core 2 Duo Processor and 2GB memory. We evaluate the efficiency of our PST buffer optimization technique, called PST-3D, by using the clock networks synthesized by [6] for the circuits in ISPD2009 clock network synthesis contest [7]. The delays in the benchmark circuits are modeled as a normal distribution and their variations are set to 5% of the nominal value as in [8]. Heterogeneous 3D ICs used in our experiments are created by integrating the clock networks in [7] with a single clock TSV, and Mosek [9] program is used for an LP solver. The skew bound is set to 15% of the latest clock period. Table I shows a comparison of the total size of the PST buffers optimized by our PST-3D with the PST buffer unoptimized result under the yield constraints of 99% and 95%. Note that we only consider the boundary points of each distribution to satisfy the given skew bound in the unoptimized method. Delay in Table I represents the average root-to-sink latency in each layer, and 4 right most columns mean the total amount of inserted delays by PST buffers. In summary, our PST-3D is able to reduce the size of PST implementation by 32% on average.

IV. CONCLUSION

This work addressed a new problem of PST buffer allocation and optimization in 3D IC designs to mitigate the on-package variation. The contributions were (1) *proposing a PST buffer allocation scheme* in 3D ICs to resolve the timing discrepancy between dies, and with the proposed allocation scheme, (2)

TABLE I
COMPARISON OF PST BUFFER SIZES OPTIMIZED BY OUR PST-3D AND THE UNOPTIMIZED ONES FOR 2~4-LAYERED 3D ICS.

Benchmark		Delay (ps)	99% yield		95% yield	
			UnOpt (ps)	PST-3D (ps)	UnOpt (ps)	PST-3D (ps)
tb1	ispd09f11	660	140	93	101	63
	ispd09f12	581				
tb2	ispd09f12	581	164	117	125	89
	ispd09f21	685				
tb3	ispd09f22	531	496	445	449	410
	ispd09f31	979				
tb4	ispd09f35	853	650	626	617	599
	ispd09fnb1	212				
tb5	ispd09f11	660	346	198	271	120
	ispd09f12	581				
	ispd09f21	685				
tb6	ispd09f21	685	746	434	647	330
	ispd09f31	979				
	ispd09f32	918				
tb7	ispd09f22	531	1425	1131	1350	1068
	ispd09f31	979				
	ispd09fnb2	292				
tb8	ispd09f11	660	2362	1413	2260	1317
	ispd09f21	685				
	ispd09f31	979				
	ispd09fnb1	212				
tb9	ispd09f31	979	750	493	594	275
	ispd09f32	918				
	ispd09f33	909				
	ispd09f34	837				
Ratio			1	**0.68**	1	**0.67**

proposing a technique called PST-3D *that is able to minimize the total cost of PST buffer implementation.* Through experiments, it was shown that our PST-3D was able to reduce the implementation cost of PST buffers by 32% on average.

Acknowledgment: This work was supported by the CISS of Global Frontier project by MSIP (CISS 2011-0031863) in Korea, the ITRC program of NIPA by MSIP (NIPA-2013-H0301-13-1011) in Korea, the Seoul R&BD Program (RI130006) in 2014, and the Brain Korea 21 Plus Project in 2014.

REFERENCES

[1] D. H. Kim, K. Athikulwongse, and S. K. Lim "A study of through-silicon-via impact on the 3D stacked IC layout," *ICCAD*, pp.674–680, Nov. 2009.

[2] F.-W. Chen and T.-T. Hwang, "Clock tree synthesis with methodology of reuse in 3D IC," *DAC*, pp.1094–1099, June 2012.

[3] S. Ozedemir, Y. Pan, A. Das, G. Memik, G. Loh, and A. Choudhary, "Quantifying and coping with parametric variations in 3D-stacked microarchitectures," *DAC*, pp.144-149, June 2010.

[4] B. Li, N. Chen, and U. Schlichtmann, "Fast statistical timing analysis for circuits with post-silicon tunable clock buffers," *ICCAD*, pp.111-117, Nov. 2011.

[5] J. Tsai, D. Baik, C. C. Chen and K. K. Saluja, "A yield improvement methodology using pre- and post-silicon statistical clock scheduling," *ICCAD*, pp.611-618, Nov. 2004.

[6] T. -Y. Kim, and T. Kim, "Clock tree synthesis for TSV-based 3D IC designs," *ACM Trans. on Design Automation of Electronic Systems*, vol. 16, no. 4, pp. 48:1-48:21, Oct. 2011.

[7] ISPD.(2009) ISPD 2009 clock network synthesis contest. Available:http://ispd.cc/contests/09/ispd09cts.html.

[8] X. Zhao, S. Muhopadhyay, and S. K. Lim, "Variation-tolerant and low-power clock network design for 3D ICs," *ECTC*, pp. 2007-2014, May. 2011.

[9] APS Mosek, "The MOSEK optimization software," online at http://www.mosek.com

Reducing the Failure Bitmap Size with a Partial Solution Search Tree for the Low Cost Automatic Test Equipment (ATE)

Keewon Cho, Woosung Lee, Jooyoung Kim, and Sungho Kang

Department of Electrical and Electronic Engineering
Yonsei University
Seoul, Korea
ckw1505@soc.yonsei.ac.kr, uoos@soc.yonsei.ac.kr, kimjy9850@soc.yonsei.ac.kr, and shkang@yonsei.ac.kr

Abstract

In general, the external automatic test equipment (ATE) is used to progress a redundancy analysis (RA) process in industrial semiconductor memories. However, many researches have pointed out that a high price of the external ATE can be a huge burden on testing costs. This paper presents a new concept which reduces the failure bitmap size in the external ATE without an any extra hardware overhead. Despite the reduction of the failure bitmap size, converting some parts of the faulty information to a partial solution search tree prevents unwanted yield drops.

Keywords: yield, testing costs, redundancy analysis (RA), bitmap and automatic test equipment (ATE)

Introduction

To gain a reasonable manufacturing yield, one thing that cannot be missed in the memory repair area is a redundancy analysis (RA) technique. At the beginning, all the faulty information is transferred to the external automatic test equipment (ATE) and stored in the failure bitmap. Then based on the properly selected RA algorithm, solution addresses are collected from the external ATE. A faulty memory is then repaired by allocating fault-free spare lines to solution addresses. Various RA algorithms have been studied steadily. Repair-most (RM) is a simple algorithm that allocates spare cells from the most faulty line [1]. Intelligent-solve (IS) and intelligent-solve-first (ISF) use a compressed binary search tree with backtracking processes [2].

The most crucial problem of using the external ATE is that the cost of the device is way too expensive. Various research topics have been studied to solve this problem. In [3], the author says that the external ATE can be substituted with built-in self repair (BISR) which consists of built-in self test (BIST) and built-in redundancy analysis (BIRA) by performing the functions of the ATE in the chip itself. Although this previous work economically relieves the burden of the external ATE, preparing a space in the chip for the extra built-in logics can be a trouble to chip designers.

This paper presents a simple concept which can reduce the failure bit map size using a partial solution search tree based on a exhaustive search RA algorithm. Because a RA algorithm is inserted into the ATE in a form of software, any extra logic is not required. The rest of this paper shows details of the proposed idea.

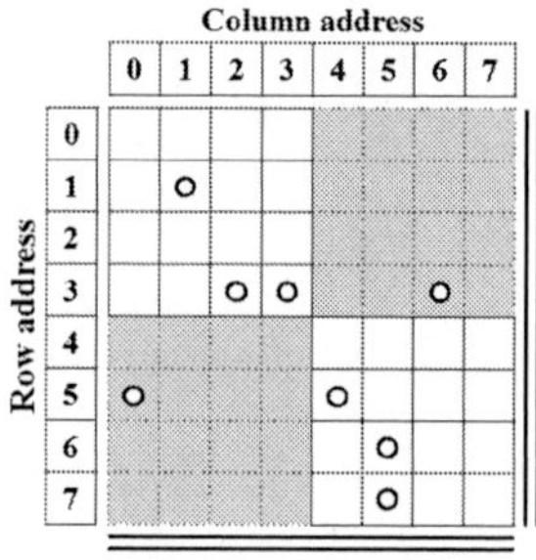

Fig. 1 Example of a faulty memory block.

Proposed Idea

The key point of the proposed idea is converting some faulty addresses to a binary search tree so that some part of the failure bitmap can be omitted. Fig. 1 is an example of an 8×8 faulty memory block with 2 row spares and 2 column spares. This memory block is divided into four quadrants. First and third quadrants of the failure bitmap, shaded areas on Fig. 1, are the target areas to be omitted.

Faults are sorted into three groups. Faults which are placed in first or third quadrants belong to group 1. Because first and third quadrants of the failure bitmap will be omitted, group 1's faulty addresses should be stored in a different way. Whenever faults in group 1 are detected, a partial solution search tree is updated. As a result of the fault collection phase, a partial solution search tree which contains all kinds of solutions that can cover group 1's faults is completed. The faulty information of group 1 is no longer needed because each branch of a partial solution search tree can be a solution of group 1. After fault collection, faults in second or fourth quadrants are split into two types. Faults which cannot be covered by the information of a partial solution search tree belong to group 2. Remaining faults and then belong to group 3.

The proposed idea uses a simple RA process. Since a parital solution search tree cannot cover group 2's faults, another binary solution search tree is needed for group 2. Then, the ATE finds a solution between the combination of two solution search trees using the exhaustive search. There are some conditions that a solution must be satisfied. The number of used spare lines should not exceed that of prepared spare lines. And if the combination of two solution search trees cannot cover all group 3's faults, remaining spare lines should cover group 3. Fig. 2 describes the entire process of the proposed idea.

978-1-4799-5128-4/14 $31.00 © 2014 IEEE

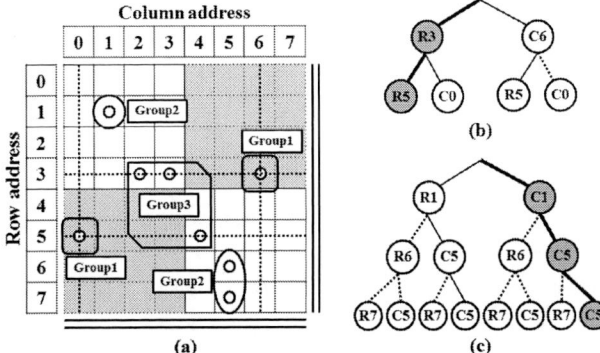

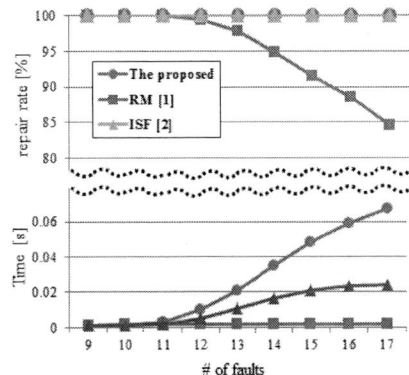

Fig. 2 Solution search process of Fig. 1. (a) Fault grouping. (b) Partial solution search tree of group 1. (c) Solution search tree of group 2.

Fig. 4 Comparison of test time and repair rate (Row / Column spares = 5 / 5).

Fig. 2(a) is a consequence of the fault grouping process. Faulty cells on (3, 6) and (5, 0) are included in group 1 since they are placed in first or third quadrants of the failure bitmap. Based on group 1, the combination of (row 3, column 6) and (row 5, column 0) becomes a partial solution search tree which is shown in Fig. 2(b). The dotted line on Fig. 2(a) denotes the paths that might be covered by a partial solution search tree. Faulty cells on (1, 1), (6, 5) and (7, 5) are then gathered in group 2. Fig. 2(c) shows a solution search tree for group 2. The rest of faulty cells belong to group 3.

After all solution search trees are constructed, the ATE finds a final solution that meets the conditions. At first, the ATE checks all the candidates which use fewer spare lines than prepared spare lines. There are three candidates which are depicted with solid lines in Fig. 2(b) and (c). In this case, all the candidates, (R3, R5, C1, C5), (R3, C0, R1, C5) and (C6, R5, R1, C5), use 2 row spares and 2 column spares. Because there is no extra spare lines, only (R3, R5, C1, C5) can cover group 3's faults among the candidates. So a final solution for this example is row three, row five, column one and column 5. It is depicted with bold lines in Fig. 2(b) and (c).

Another strength of the proposed idea is that the rate of reduction can be adjustable. In the above example, only 50% of memory cells are omitted. But the failure bitmap size can be reduced further while core cells are maintained. Any cells can be selected to core cells but the whole set of core cells should contain every address in a memory block at least one time. However, the more the bitmap area is saved, the more the analysis time is spent. So the rate of reduction should be carefully chosen to make the best outcome. Fig. 3 is one kind of the reduction process. For better visual image, diagonal cells are selected to core cells. Every row(or column) address can be covered by diagonal cells. While cells on (0, 0) to (7, 7) are maintained, shaded areas can be freely selected for saving the bitmap size.

Experimental Results

In the experiments, a 1024×1024 memory array is assumed. Fig. 4 shows test time and repair rate when a memory has 5 row (and column) spares. Each RA algorithm was repeated 10,000 times with the faults using polya-eggenberger distribution. Though RM has the smallest test time, its repair rate is quite low. Unlike with RM, ISF can achieve 100% repair rate. However, both preivous algorithms should use the full bitmap. Otherwise, a graph of the proposed algorithm in Fig. 4 uses the completely compressed failure bitmap. This result means that the proposed idea greatly lessens the burden of the memory size and maintains 100% repair rate at the same time by spending some additional analysis time for RA.

Conclusion

A novel method which reduces the failure bitmap size in the external ATE by changing some faulty information to a partial solution search tree was proposed in this paper. At first, target cells are selected and faults are classified into three groups. And then, using a generated partial solution search tree, an exhaustive search algorithm is applied to the RA phase. Eventhough it took a little more RA time, effective results of reduction were gained without additional built-in logics. This idea leads semiconductor industries to the conclusion that the low cost ATE is very helpful to save the overall cost of test.

Acknowledgment

This work was supported by the National Research Foundation of Korea (NRF) grant funded by the Korea government (MEST) (No. 2012R1A2A1A03006255).

References

[1] M. Tarr, D. Boudreau, and R. Murphy, "Defect analysis system speeds test and repair of redundant memories," *Electronics,* vol. 57, pp. 175–179, Jan. 1984.

[2] P. Öhler, S. Hellebrand, and H.-J. Wunderlich, "An integrated built-in test and repair approach for memories with 2D redundancy," in *Proc. Eur. Test Symp. (ETS)*, May 2007, pp. 91–96.

[3] W. Jeong, I. Kang, K. Jin, and S. Kang, "A fast built-in redundancy analysis for memories with optimal repair rate using a line-based search tree," *IEEE Trans. Very Large Scale Integr. Syst.*, vol. 17, no. 12, pp. 1665-1678, Dec. 2009.

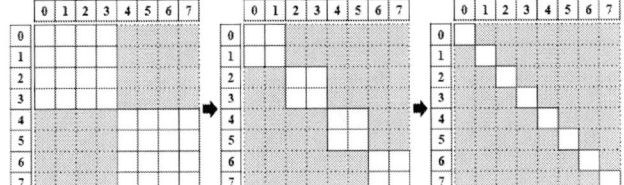

Fig. 3 Example of the additional reduction process of the failure bitmap

Chaotic Oscillation-based BIST for CMOS Operational Amplifier

Chatchai Wannaboon, Nattagit Jiteurtragool and Tachibana Masayoshi

Kochi University of Technology
Tosayamada, Kami-City, Kochi, 782-8502, Japan
177002d@kochi.ac.jp, 177001i@kochi-tech.ac.jp, tachibana.masayoshi@kochi-tech.ac.jp

Abstract

This paper presents chaotic oscillation-based built-in self-test (BIST) for CMOS operational amplifier. The proposed BIST technique is based on the use of designed operational amplifier (op-amp) in the unity gain buffer of discrete time chaotic oscillator as circuit under test (CUT) of BIST. The presented BIST detected faults by using a differentiation of chaotic output signal among fault free and faulty CUT. The circuit is simulated in 0.18μm CMOS technology. The simulation results on circuit-level are presented to examine the feasibility and efficiency to detecting faults in op-amp.

Keywords-component; analog BIST, oscillation-based test, chaotic oscillator.

Introduction

The widely design and development of System-on-a-Chip (SoC) design technology have made a feasibility of integrated millions of component to a single chip which consists of analog and digital component. Due to the rising of mixed signal systems developing, the test circuit has become an unavoidable to examine and eliminate defective circuits [1].

The test circuits that may be designed and placed on the same chip with circuit under test are well known as built-in self-test (BIST). Generally, BIST circuits are designed to use as measurement tool for the circuit under test (CUT) through the comparison among measured value and reference value.

Chaotic oscillator has been widly use for many decade of years. Wherewhich the propertys of chaotic which can provide true random number and severe sensitivity to initial condition and parameter [2].

Consequently, this paper proposed the chaotic oscillation-based BIST for op-amp testing. Fault decting using chaotic sensivity to the changed in circuit by injecting catastrophic and parametric fault [3]. The BIST is used the designed op-amp in unity gain buffer of BIST as the CUT. The circuit is designed and simulated in 0.18μm CMOS technology which shows efficiency of faults detection.

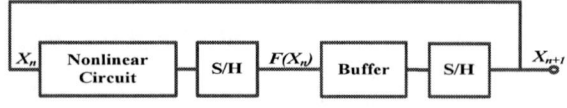

Fig. 1 Simple block diagram of discrete-time chaocuit circuit

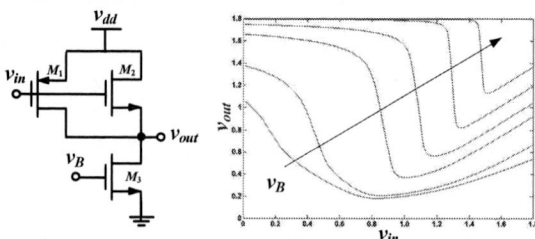

Fig. 2 (a) Chaos map schematic and (b) DC characteristic of circuit

Discrete-time Chaotic Circuits

Discrete-time Chaotic Circuits provide the stochastic signals though an iteration and can be described by a simple one-dimensional equation as follows:

$$X_{n+1} = f(X_n) \tag{1}$$

The iteration equation in (1) can be implemented in a simple structure circuit as shows in Fig. 1. The nonlinear circuit represents the chaos map. The two sample-and-hole circuits (S/H) are used to track and store the signals as a memory. The signals can be carried to the next stage by a buffer. After that, the output signals are fed back into the input of chaos map and all of process continue iteratively.

Typically, chaotic signals can be generated, on a condition that a characteristic of map circuit should be V or N-shape [4]. Therefore, this paper has constructed chaos map circuit by approximating the V-shape characteristic as illustrates in Fig. 2 (a). Where M_1 is operated in low voltage region, while M_2 is operated when input of chaos map circuit is greater than threshold voltage. In addition, the shape of chaos map circuit can be modified by adjusted the voltage value V_B at M_3. The DC characteristic of chaos map circuit with various biasing V_B is depicted in Fig. 2 (b).

Overall, the chaotic signal can be generated by replacing nonlinear function with chaos map circuit as described in section A. The buffer and S/H circuit are implemented by a two-stage operational amplifier (op-amp) and complementary switches, respectively. The output signal in time-domain of chaotic oscillator is shown in Fig. 3.

Proposed Chaotic Oscillation-based BIST

Fig. 5 illustrates overall BIST system, in normal mode (TEST = 0), the op-amp is operated as a general designed.

978-1-4799-5128-4/14 $31.00 © 2014 IEEE   130

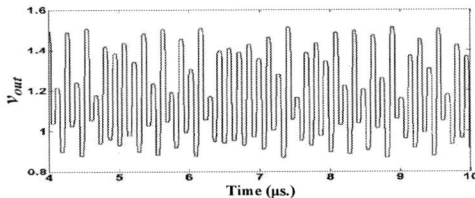

Fig. 3 The output of chaotic oscillator with $V_B = 0.68$ V.

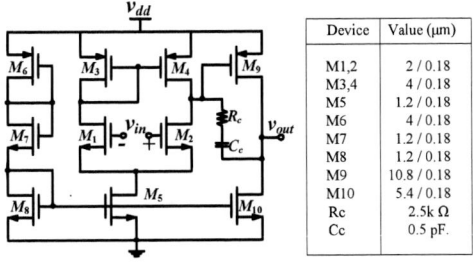

Device	Value (μm)
M1,2	2 / 0.18
M3,4	4 / 0.18
M5	1.2 / 0.18
M6	4 / 0.18
M7	1.2 / 0.18
M8	1.2 / 0.18
M9	10.8 / 0.18
M10	5.4 / 0.18
Rc	2.5k Ω
Cc	0.5 pF.

Fig. 4 Circuit-under-test with value of all MOS transistors.

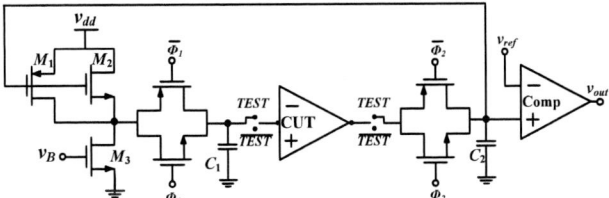

Fig. 5 Block diagram of CUT tramformed into the chaotic oscillator.

In the test mode (TEST = 1), the BIST is performed by transforming the general op-amp into the chaotic oscillator as buffer. The two-stage comparator is connected through the output stage of BIST and converts the signals into bits stream. The system is achieved by a complementary clock Φ_1 and Φ_2. The size of MOS transistors M1, M2 and M3 are 4/0.18 μm, 2/0.18 μm and 1.2/0.18 μm, respectively.

Simulation Results

The proposed BIST system in Fig. 5 was designed using Cadence and simulated using Hspice simulations in 0.18 μm CMOS technology at 25 MHz for 1.8 V of voltage supply. Fig. 6 depicts output signals for fault free circuit (a) compared with faulty signal where the capacitor value is reduced by 10% (b), which clearly showed the distinction signal.

Consequently, for the efficiency examination of the purposed BIST, faults models were injected to the op-amp. Table 1 shows the simulations results of CUT where the used of catastrophic faults (Short, Open) and parametric fault. Short faults were performed and simulated by a serial connection of resistors (R_S) and open faults were performed by a parallel connection of resistors (R_O) and capacitor (C_O). Value of resistors R_S, R_O and capacitors C_O were 1Ω, 10MΩ and 1fF respectively. Parametric faults were performed by deviated 10% values of resistor and capacitor.

Conclusion

Chaotic oscillation-based BIST for op-amp testing was proposed. The proposed method can be use as buit-in self-test for op-amp in many mixed-signal systems. The proposed BIST

TABLE 1
Simulated results of injected and defected fault

Fault Types	Injected Faults	Detected Faults
Gate-Drain Shorts	8	8
Gate-Source Shorts	6	6
Drain-Source Shorts	10	10
Gate Opens	10	10
Drain Opens	10	10
Source Opens	10	10
Resistor Short/Open	2	2
Capacitor Short/Open	2	2
Reduced Value Resistor	2	2
Reduced Value Capacitor	2	2
Total Faults	62	62

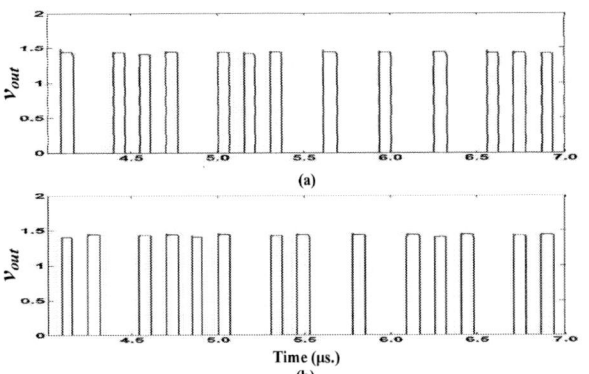

Fig. 6 Output signals from BIST (a) fault free circuit (b) faulty circuit

was designed and simulated in 0.18 μm CMOS technology. According to the chaotic oscillator sensitivity, simulation results is demonstated the efficientcy to detected all injected catastrophic faults and parametric faults.

Acknowledgment

The authors are grateful to Kochi-University of Technology (KUT) for research fund supports. This work was supported by VLSI Design and Education Center (VDEC), the University of Tokyo in collaboration with Synopsys, Cadence Design System, Mentor Graphics, Rohm Corporation and Toppan Printing Corporation. And it was also supported by KAKENHI (23500067) Grant-in-Aid for Scientific Research.

References

[1] M. Burns and G. W. Roberts, "An Introduction to Mixed-Signal IC Testand Measurement," The Oxford Series in Electrical and Computer Engineering. Oxford Univ. Press, 2000.

[2] Salih Ergüna, Serdar Özo`guz, "Truly random number generators based on a non-autonomous chaotic oscillator," AEU - International Journal of Electronics and Communications, vol. 61, pp. 225-242, April 2007.

[3] M. Bushnell and V. Agrawal, "Essentials of Electronic Testing for Digital, Memory, and Mixed-Signal VLSI Circuits," Berlin, Germany: Springer-Verlag, 2000.

[4] Angel Rodriguez-Vazquez, Jose L. Huertas, Adoracion Rueda, Belen Perez-Verdu, and Leon O. Chua, "Chaos from Switched-Capacitor Circuits: Discrete Map", Proceedings of the IEEE, vol. 75, pp. 1090-1161, 1987.

Analysis of Dynamic Voltage Drop with PVT Variation in FinFET Designs

Yongchan Ban, Changseok Choi, Hosoon Shin, Jaewook Lee, Yongseok Kang and Woohyun Paik

System IC R&D Lab., LG Electronics
19, Yangjae-daero 11gil, Seocho-gu
Seoul 137-130, South Korea
Phone: +82-10-5046-8692, E-mail: yc.ban@lge.com

Abstract

In this paper we have analyzed vectorless dynamic voltage (IR) drops and characterized the impact of transistor-level PVT (process-voltage-temperature) variation and metal-level RC (resistance-capacitance) corner variation in FinFET SoC (system-on-a-chip) designs. The impact of systematic process variations on signal interconnects is also considered for analyzing the worst case voltage drop in our design. Then, we discuss factors that can affect the design closure metrics for power and voltage integrity.

Keywords; Dynamic voltage drop, IR drop, power, vectorless, PVT variation, systematic variation, FinFET, SoC

I. Introduction

Designing a power grid has a direct impact on chip performance and reliability [1-2]. It should be guaranteed that the power grid provides enough power supply voltage to all transistors in cells in order to work the chip well with a certain noise margin for all possible operations [2-3]. However, since the threshold voltage of transistors does not scale down as fast as the supply voltage with technology shrinking, circuit tolerance to voltage drop becomes highly decreasing, which results in smaller logic noise margins, chip performance degradation, and even IC functional failures [4].

In addition, the impact of process, voltage, and temperature (PVT) variation and inter die/within die process variation in nanometer node design makes designers difficult to analyze the design margin and estimate the voltage drop [5-6]. Moreover, 3D-type FinFET transistors facilitate voltage drops due to their large driving strength. In this reason, a robust analysis of voltage drop is crucial in high performance and ultra-low power SoC (system-on-a-chip) designs.

Static IR drop computes an average voltage drop in an early design stage to roughly estimate the IR drops and to provide some feedback in an initial power grid design. However, the verification with static voltage drop is not enough to ensure power integrity because it does not consider power density, switching activity, and the impact of inductance and decoupling capacitances [4].

In this paper, we have analyzed vectorless dynamic voltage (IR) drop (DVD) in an early FinFET design stage. The impact of process, voltage, temperature variation and metal RC corner variation has taken into consideration. In addition, we also compute the impact of lithographic, etch and CMP (chemical-mechanical polishing) systematic variations for analyzing the worst case voltage drop.

II. Calculation of Dynamic Voltage (IR) Drop

A. Power Estimation

Dynamic Power is comprised of internal power and switching power as shown in Eq. (1).

$$P_{dynamic} = p_{internal} + P_{switching} = C_{eff} \times V^2 \times f_{clk} \quad (1)$$

where the internal power is dissipated due to charging and discharging of the internal load and the switching power is dissipated by the charging and discharging of the load capacitance at the output of a cell.

B. Dynamic Voltage Drop

Dynamic voltage drop evaluates the IR drop for peak current demand causing by the impact of decaps, inductance, and spatial and temporal circuitry switching [3-4]. Vectorless DVD uses current constraints to capture the circuit uncertainty at an early design stage, and evaluates the worst-case voltage drops at each node on the power grid [1].

C. Impacts of Process Variations

Interconnect distortions caused by process variations might impact on voltage drop by changing the internal and load capacitance. Since the patterning of signal interconnects still use ArF (193nm wavelength) lithography to print sub-20nm node, there are many unwanted effect resulting in large distortions for the shapes on the wafer. The etch loading effect induces etch bias variation which is not a constant value for all kinds of features. While CMP varies the interconnect thickness which leads to electrical shorts or increased wire resistance.

III. Experimental Condition and Results

For analyzing DVD in 16nm node FinFET design, we have designed a CPU with an ARM Cortex-A57 CPU as shown in Fig. 1. Standard cell library and memory macro library were given from TSMC 16nm foundry. The maximum frequency target in our CPU design is 1.2GHz at setup timing condition with 9 track standard V_{th} and low V_{th} cell libraries.

 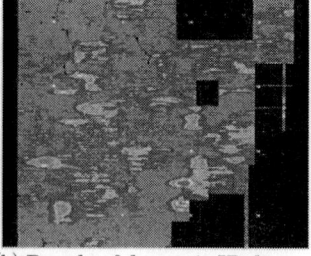

(a) CA57 CPU floorplan   (b) Result of dynamic IR drop
Fig. 1. Cortex-A57 CPU floorplan and a map of IR drop

TABLE 1. DVD WITH INPUT VOLTAGES

Voltage (Vdd)	# DVD Max.	# DVD Min.	Worst Instance	Worst Wire
0.72 V	4256	1837	52.1%	50.2%
0.8 V	4122	1586	45.4%	43.5%
0.88 V	4543	1086	40.3%	38.5%

Simulation of dynamic IR drop was done with Apache Redhawk. We extracted the routed design netlist in DEF format and the library technology in LEF format. Timing information with SDC format, the instance-specific toggle information and the technology file given from foundry were used. In this experiment, since in an early design stage we did not place low power architectures and power switches. We are going to focus on the trends of variation to DVD in this paper.

We swept PVT (process, voltage, and temperature) corners for the front-end FinFET transistor devices. The temperature across the chip was analyzed from -40℃ to 125℃. The corner files of metal line parasitic were extracted from Synopsys Star-RCXT. In addition to the conventional RC corners (*Typical*, *RCbest* and *RCworst*), we further analyzed DPT (double pattern technology) lithography induced RC corners: *RCbest_DPT* and *RCworst_DPT*.

Table 1 shows DVD violations in terms of input voltage. The PVT for the front-end devices was set to a typical process corner and 125℃. RCworst parasitic corner is used for metal interconnects. The limitation of voltage fluctuation is 15% of Vdd. **#DVD Max** denotes the amount of instances where DVD is in beteween Vdd and the maximum DVD point. **#DVD Min** is vice versa. **Worst Instance** and **Worst Wire** are the portion of maximal DVD at the level of VDD in instance and wire, respectively. As the input voltage decreases, the DVD violation and the portion of instance and wire violations are increased. This is due to that theat at the lower input voltage, a small voltage fluctuation could induce huge impact.

TABLE 2. DVD WITH CHIP TEMPERATURES

Temperature	# DVD Max.	# DVD Min.	Worst Instance	Worst Wire
-40 ℃	1413	84	32.6%	29.8%
0 ℃	1760	275	33.3%	30.5%
85 ℃	2028	388	34.8%	31.7%
125 ℃	4543	1086	40.3%	38.5%

DVD on signal interconnects in terms of ambient chip temperature is shown in Table 2. RCbest parasitic corner is used for metal lines. When the higher temperature is applied, the DVD violation increases. This is because the instances and wires have more resistive as temperature increases. Moreover, instances induce more current due to temperature inversion at 16nm node and cause more spontaneous voltage drops.

Table 3 shows the DVD violation with the back-end parasitic RC corners. The PVT for the front-end devices was set to a 0.88V Vdd. Compared to Typical corner, the number of DVD violation at the RCWorst corner has higher (worse). This is because the interconnects have more resistive just like as temperature. At RCBest corner, the number of DVD violation is smaller than Typical corner.

TABLE 3. DVD WITH METAL RC CORNERS

RC corners	# DVD Max.	# DVD Min.	Worst Instance	Worst Wire
RCBest_DPT	531	426	27.8%	27.0%
RCBest_dPV	712	433	29.1%	27.6%
RCBest	561	433	28.4%	27.6%
Typical	1537	513	32.9%	31.6%
RCWorst	4722	1199	40.5%	38.7%
RCWorst_dPV	2365	400	33.5%	30.5%
RCWorst_DPT	4543	1086	40.3%	38.5%

RCBest_DPT corners, where "_DPT" is a new corner induced by double patterning lithography, shows better for DVD compared to RCBest, meanwhile RCWorst shows worse DVD than RCWorst_DPT.

The impact of design retargeting due to lithography/etch (width) and CMP (thickness) process variation was also considered at DVD simulation. By directly applying bias table in extraction, R&C values for metal/via layers are updated. We implemented the design retargeting due to process at RCBest and RCWorst corners which are renamed with RCBest_dPV and RCWorst_dPV, respectively. As shown in Table 3, at RCBest_dPV corner DVD increases because the design target is thinner (higher R&C) than one at RCBest corner. Meanwhile, DVD decreases at RCWorst_dPV corner due to thicker design target from process variations. Due to the metal retargeting, DVD on wire is highly reduced.

IV. Conclusion and Future Works

Dynamic voltage drops with PVT and RC corner variations are reported. As the input voltage decreases, the temperature increases, and at RCWorst parasitic corner, the DVD on instances and wires increases in 16nm FinFET design. Vector DVD with VCD (value changed dump) will be analyzed for better accuracy for power sign-off. The optimal power switch will be also proposed in FinFET designs.

References

[1] W. Zhao , Y. Cai and J. Yang, "Fast Vectorless Power Grid Verification using Maximum Voltage Drop Location Estimation," *Asia and South Pacific Design Automation Conference*, 2014.

[2] S. Lin and N. Chang, "Challenges in power-ground integrity", *International Conference on Computer Aided Design*, 2001.

[3] N SK, G. Shanmugam, S. Chandrasekar, "Dynamic voltage (IR) drop analysis and design closure: Issues and challenges," *International Symposium on Quality Electronic Design*, 2010.

[4] K. Arabi, R. Saleh and X. Meng, "Power Supply Noise in SoCs: Metrics, Management,and Measurement," *IEEE Publication Design & Test of Computers*, vol. 24, pp. 236-244, 2007.

[5] Y. Ban, S. Sundareswaran, and D. Pan, "Total Sensitivity Based DFM Optimization of Standard Library Cells," *International Symposium on Physical Design*, 2010.

[6] Y. Ban et al., "Layout induced variability and manufacturability checks in FinFETs process", *International Symposium on SPIE Advanced Lithography*, 2014.

[7] Y. Ban et al., "Analysis and Optimization of Process-Induced Electromigration on Signal Interconnects in 16nm FinFET SoC," *International Symposium on SPIE Advanced Lithography*, 2014.

978-1-4799-5128-4/14 $31.00 © 2014 IEEE

A New Redundancy Analysis Algorithm Using One Side Pivot

Jooyoung Kim, Keewon Cho, Woosung Lee, and Sungho Kang

Department of Electrical and Electronics Engineering
Yonsei University
Seoul, Korea
{kimjy9850, ckw1505, uoos}@soc.yonsei.ac.kr and shkang@yonsei.ac.kr

Abstract

It is important to test the memory and repair faults for improving the memory yield. Many redundancy analysis (RA) algorithms have been developed to repair the memory faults. However, it is difficult to achieve high repair rate and fast analysis speed. The previous RA algorithms do not achieve both high repair rate and fast analysis speed. To overcome this problem, a new RA algorithm called one side pivot (OSP) is proposed. Using the property of pivot fault and its repair priority, the analysis time to find a solution can be reduced. The experimental results show that the proposed algorithm is efficient in terms of repair rate and analysis speed.

Keywords: redundancy analysis (RA) and one side pivot

Introduction

With developing of technology these days, as the density of memory increases, the probability of occurring faults in memory also increases. Thus, it is important to repair the faulty cells using redundancy memories which are used to replace the faulty cells to increase the yield of memory. Many RA algorithms have been proposed to achieve a high repair rate which affects directly to the yield of memories. Since the redundancy analysis (RA) is known to be NP-complete, it is very difficult to achieve both a high repair rate and a fast analysis speed [1].

Repair-most (RM) algorithm is a greedy algorithm [2]. It checks all the counters of row or column lines and finds a location of redundancy memories by the counters. However, RM algorithm has a low repair rate as the number of the faults increases. Intelligent solve first (ISF) provides an optimal repair rate based on the binary search tree, but it is very slow due to the binary search tree structure [3]. Essential spare pivoting (ESP) algorithm is a fast RA algorithm using threshold numbers to find a solution but shows a low repair rate as the number of faults increases [4]. Similarly, FAST algorithm has a fast analysis speed using fault grouping to find a solution [5]. It shows an optimal repair rate when the number of faults is small but the repair rate decreases as the number of faults increases.

Thus, the previous algorithms show a solution for small number of faults, but are not appropriate for memories with many faults. Therefore, a new RA algorithm is required to derive a high repair rate and fast analysis speed when there are lots of faults occurred in the target memory.

Proposed algorithm

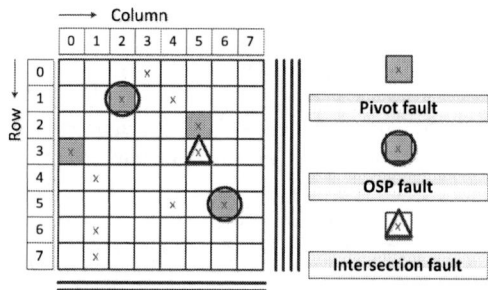

Fig 1. Example of proposed algorithm.

In order to explain the proposed algorithm, several terms are defined. A pivot fault is defined as a fault in a faulty line which is not included in any other faulty lines. An intersection fault is defined as a fault which is included in both a column and a row faulty line. A one side pivot (OSP) fault is defined as a pivot fault which is not included in a faulty line which does not have an intersection fault. For example shown in Fig. 1, fault (3,0) is a pivot fault, fault (2,2) is a OSP fault, and fault (3,5) is an intersection fault, respectively.

The flow chart of the proposed algorithm is shown in Fig. 2. The proposed algorithm consists of three phases. In the first phase, the must repair analysis is performed [2]. The must-repair analysis is based on the number of row spares and the number of column spares in the memory. If the number of faults in a row (column) line in the target memory is greater than the number of the available column (row) spares, that row (column) line must be repaired by the row (column) spare.

In the OSP analysis phase, the OSP faults are searched and the faulty lines which include the OSP faults. If the pivot fault which is included in a faulty line with an intersection fault, it is not a OSP fault. Among the pivot faults, the OSP faults are selected. Only OSP faults are repaired in the OSP analysis phase. The row faulty line that includes the row OSP fault is repaired by the row spare and the column faulty line that includes the column OSP fault is repaired by column spare.

To explain the phase, consider the example shown in Fig. 1. The fault (1,2), (3,0) and (5,6) is included in a row faulty line but not in a column faulty line. These faulty cells are row pivot faults. Likewise, the fault (2,5) is a column pivot fault. In Fig. 1, the row pivot fault (3,0) and the column pivot fault (2,5) have intersection fault (3,5). Thus, the row pivot faults (1,2) and (5,6) are selected as the OSP faults and repaired in the OSP analysis phase.

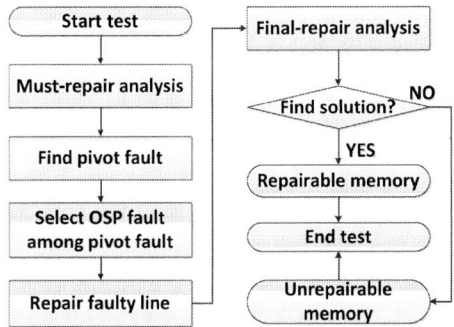

Fig 2. Flow chart of proposed algorithm.

In the final analysis phase, the faulty line which has the highest number of faults among remaining faulty lines is repaired first. This is repeated according to the descending order of the fault counters until all faults are repaired or all spares are used. If the target memory cannot be repaired using the spares, the memory is regarded as an unrepairable memory.

Experimental results

For the experiments, a 1024 X 1024 memory is used and faults are injected using the Polya–Eggenberger distribution [6]. For each number of faults and number of spares, 10,000 experiments are executed and average values are derived. The repair rate comparison of the proposed algorithm with the previous algorithms is shown in Fig. 3. The repair rate of the proposed algorithm is 100% when the number of faults is small and more than 99% when the number of faults is large. However, the repair rates of RM, ESP, and FAST algorithms decrease drastically compared to the proposed algorithm.

Fig. 4 shows analysis speed comparison of RM, ESP, ISF and the proposed algorithm. Analysis speed shown in Fig. 4 is the summation of 10,000 RA execution times of these algorithms. In Fig. 4 (a) and (b), analysis speeds of other algorithms are almost the same as the number of faults increases. However, analysis speed of ISF algorithm drastically increases as many faults occur in the target memory. As expected, analysis speed of ISF algorithm is the slowest and analysis speed of the proposed algorithm is slightly slower than that of RM and ESP algorithms.

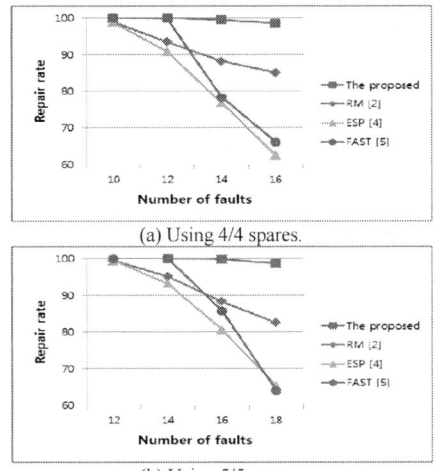

Fig 3. Repair rate comparison.

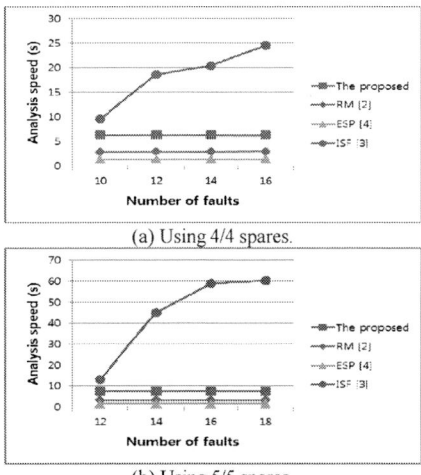

(a) Using 4/4 spares.

(b) Using 5/5 spares.

Fig 4. Analysis speed comparison.

Conclusion

As the memory density increases, it becomes more important to use an efficient RA for memory repair in terms of the optimal repair rate and fast analysis speed. The previous approaches have a difficult to achieve both when there are many faults in a memory. In order to overcome this problem, a new RA algorithm is proposed which is based on the property of pivot faults. The experimental results show that the performance of the proposed algorithm in terms of the repair rate and analysis speed is better than that of the previous works. Therefore, the proposed algorithm can be a practical solution for memory repair specially in the case of repairing many faults in the target memory.

Acknowledgment

This work was supported by the National Research Foundation of Korea (NRF) grant funded by the Korea government (MEST) (No. 2012R1A2A1A03006255).

References

[1] S.-Y. Kuo and W.K. Fuchs, "Efficient spare allocation for reconfigurable arrays," *IEEE Design and Test of Computers*, vol. 4, no. 1, pp. 24-31, Feb. 1987.

[2] M. Tarr, D. Boudreau, and R. Murphy, "Defect analysis system speeds test and repair of redundant memories," *Electronics*, vol. 57, pp. 175-179, Jan. 1984.

[3] P. Öhler, S. Hellebrand, and H.-J. Wunderlich, "An integrated built-in test and repair approach for memories with 2D redundancy," *Proc. IEEE European Test Symposium (ETS)*, May 2007, pp. 91-96.

[4] C.-T. Huang, C.-F. Wu, J.-F. Li, and C.-W. Wu, "Built-in redundancy analysis for memory yield improvement," *IEEE Transactions on Reliability*, vol. 52, pp. 386-399, Dec. 2003.

[5] H. Cho, W. Kang, and S. Kang, "A fast redundancy analysis algorithm in ATE for repairing faulty memories," *ETRI J.*, vol. 34, no. 3, pp. 478-481, Jun. 2012.

[6] C. Lee, W. Kang, D. Cho, and S. Kang, "A new fuse architecture and a new post-share redundancy scheme for yield enhancement in 3-D-stacked memories," *IEEE Transactions on Computer-Aided Design of Integrated Circuits and Systems (TCAD)*, vol. 33, no. 5, pp.786-797, May 2014.

Design of a Near-threshold Digital LDO with Fast Transient Response

Yunsheng Chan and Yingchieh Ho
Department of Electrical Engineering
National Dong-Hwa University, Taiwan, R.O.C
ycho@mail.ndhu.edu.tw

Abstract—**A near-threshold digital LDO (DLDO) with fast transient response is presented in this paper. In order to improve settling time, a voltage-controlled delay line (VCDL) with a new proposed delay cell is developed to enhance conversion gain. In addition, Vernier-delay-line-based time-to-digital convertor (TDC) is used to quantize phase from phase frequency detector (PFD). Furthermore, digital coarse and fine tunings are developed to control PMOS array. The test chip is designed in 90nm SPRVT CMOS and operates at 0.3-0.5V with output voltage of 0.25-0.45V. Simulation results show that current efficiency at 2.8mA is 99.6%. The settling time is only 8us and 12us when load current I_{LOAD} is 200uA and 2.8mA, respectively.**

Keywords—*Digital LDO, voltage control delay line, time-to-digital converter*

INTRODUCTION

In the recent years, near-threshold-supply circuits have become popular in low power CMOS VLSIs. Many important designs, such as processor, memory, PLL, and other basic macros, have been successful developed with a near-threshold supply. Unfortunately, circuits which operate at near-threshold region suffer much more power noise than operating in nominal supply. As a result, a near-threshold circuit with a proper power manager becomes an important issue.

Low-dropout (LDO) regulators play an important role in power management circuits. As compared to an analog LDO regulator, digital LDO (DLDO) provides another scalable solution that has more noise margin with ultra-low-voltage operation. A DLDO at 0.5V has been reported and it has good current efficiency [1]. However, this design costs large area by using shift register and its maximum I_{LOAD} is 200uA as well. Accordingly, PMOS array is controlled using shift register that limits its settling time. In [2], a 0.38V LDO has been presented to improve noise rejection ability of power supply.

PROPOSED DLDO

In this paper, we propose a near-threshold DLDO which can operate at 0.3-0.5V. Fig. 1 schematically depicts the proposed DLDO which consists of a pair of voltage-controlled delay lines (VCDLs), a phase frequency detector (PFD) to detect the phase error, TDC to convert the phase error into digital code and control circuit to switch PMOS array.

The authors are grateful to United Microelectronics Corporation (UMC), Taiwan, for technology supports. The work is supported by Ministry of Science and Technology (MST), R.O.C.

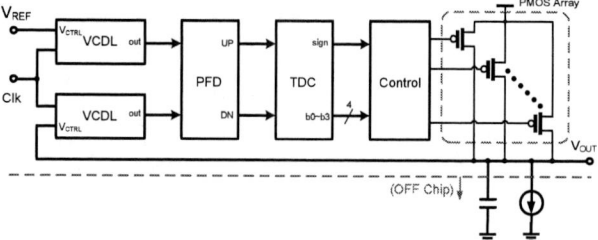

Fig. 1. Block diagram of the proposed DLDO.

In order to provide both heavy I_{LOAD} and good current efficiency at the near-threshold supply, it has several features. First of all, output voltage is compared with V_{REF} using two VCDLs and convert voltage difference between V_{REF} and output voltage to timing. In our VCDL, a new voltage-controlled delay cell is proposed, which enlarges gain of voltage to timing. To improve settling time of DLDO, TDC is used to sense the quantized phase error and then controls the PMOS array immediately. There are two VCDLs, one VCDL's V_{CTRL} is connected to V_{REF}, and the other is connected to V_{OUT}. If there is a voltage between V_{REF} and V_{OUT}, it will cause different delay time about VCDL and generate different phase between two VCDL's outputs. Then PFD detects the difference accordingly and generates UP and DN signals to TDC. TDC can quantize the signals from outputs of PFD and generates digital code to control circuit. Accordingly, the control circuit distinguishes that output voltage is higher or lower than V_{REF}, and then adjust output voltage to equal V_{REF}. Each block in detail will be discussed as follows.

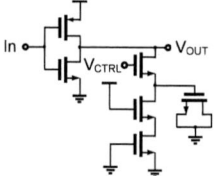

Fig. 2. Schematic of the proposed voltage-controlled delay cell.

In our DLDO, a new VCDL is developed by a proposed voltage-controlled delay cell. Fig. 2 shows the proposed delay cell which is designed with two NMOS devices to increase its voltage controlled resistance. In the conventional delay cell, the NMOS devices serve as a RC loading, as shown in Fig. 3. One is a voltage-controlled resistor controlled by V_{CTRL}, and the other one is a NMOS capacitor. As compared to conventional one, our design has two more NMOS devices to enhance the RC value. As a result, the proposed VCDL has high sensitivity to delay time while V_{CTRL} is changed. Fig. 3 also shows the results of conversion gains. As compared to conventional VCDL, our design performs obviously high conversion gain at the range of 0.3-0.5V.

978-1-4799-5128-4/14 $31.00 © 2014 IEEE

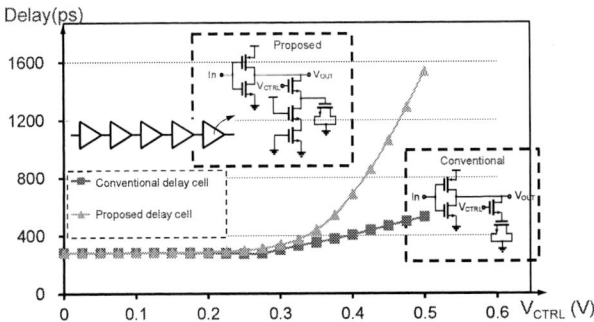

Fig. 3. Comparisons transfer curve of two delay cells.

Besides, a 4-bit TDC is used to quantize phase error from outputs of PFD. In order to achieve high resolution of TDC when it operates at near-threshold voltage, Vernier delay line is used. According to the simulation results at 0.5V and TT corner, the average resolution of the TDC in the proposed DLDO is about 15ps. Based on fine tune and coarse tune circuits, our design can reduce large ripple on output node to achieve target current and voltage. Coarse tune circuit consists of 5-bit shift register and is controlled by the sign bit only. Fine tune is controlled by 4-bits and a sign bit, which are generated by TDC.

SIMULATION RESULTS AND DISCUSSION

The proposed DLDO is verified by HSPICE using a 90nm process. The DLDO is superior to previous designs in terms of maximum current and settling time. According to the simulation results, when the load current I_{LOAD} is 200uA, its settling time is 8us and the ripple is 5.05mV, as shown in Fig. 4. When the maximum load current I_{LOAD} is 2.8mA, its settling time is 12us and the ripple is 0.75mV, as shown in Fig. 5. According to our simulation results, when I_{LOAD} is changed from 0uA to 2.8mA, it has overshoots of 5.85mV and undershoots of 13.5mV. Fig. 6 illustrates the current efficiency curve of the proposed DLDO, which is highly relative to I_{LOAD}. When I_{LOAD} is 2.8mA, the current efficiency can achieve 99.6%.

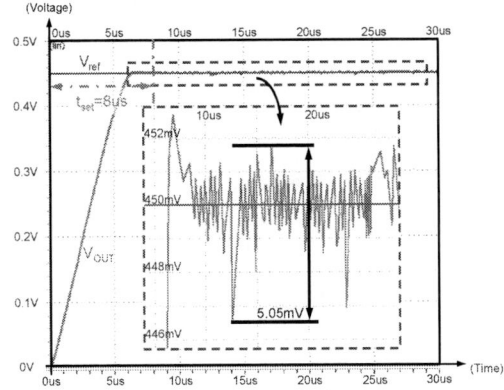

Fig. 4. Simulation results of settling time and ripple at 200uA.

Table I lists performance comparisons as compared to similar works. With a 100nF load, our design has very short settling time. Also it can work at only 0.3V to 0.5V and has very small active area of 0.012mm². At 0.5V, it provides 2.8mA output current. Even at 0.3V, it can output regulated 0.25V and 300uA output current.

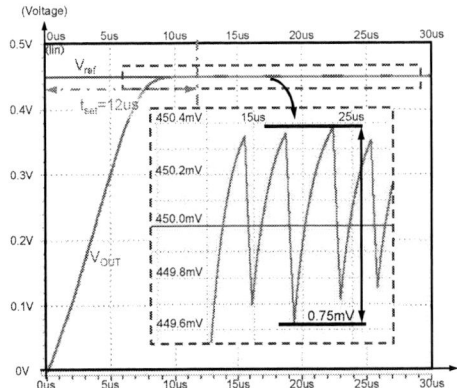

Fig. 5. Simulation results of settling time and ripple at 2.8mA.

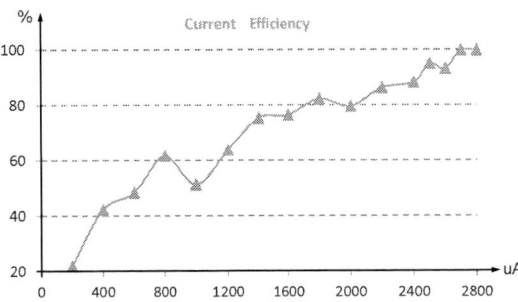

Fig. 6. Current efficiency as a function of output current.

Table I. Comparisons

	Unit	[2]	[3]	[1]	[4]	This work
CMOS Technology	-	180nm	90nm	65nm	90nm	90nm
Control	-	Analog	Analog	Digital	Digital	Digital
Area	mm²	0.122	0.019	0.042	0.026	0.012
V_{IN}	V	0.65-0.95	0.75-1.2	0.5	0.38	0.5
V_{OUT}	V	0.5	0.5-1.0	0.45	0.12-0.32	0.45
$I_{LOAD,MAX}$	mA	50	100	0.2	1	2.8
Line regulation	mV/V	4.6	3.78	3.1	2.9	1.14
Load regulation	mV/mA	0.38	0.1	0.65	0.11	0.07
Settling time	us	10	N/A	240	104	8
Quiescent current	uA	12.7	8	2.7	6.85	11.2
Current efficiency	%	99.9	99.9	98.7	99.3	99.6

CONCLUSIONS

In this paper, we have described a near-threshold DLDO which can operate at 0.3-0.5V. The proposed delay cell and the high-resolution TDC result in improved settling time. The simulated settling times are 12us and 8us, when the load current I_{LOAD} is 2.8mA and 200uA, respectively. According to simulation results, our design at 0.5V has 2.8mA I_{LOAD} with the ripple of 0.75mV and 200uA with the ripple of 5.05mV, respectively. This DLDO can even work at 0.3V with regulated output 0.25V.

References

[1] Y. Okuma, *et. al*, "0.5V input digital LDO with 98.7% current efficiency and 2.7uA quiescent current in 65nm CMOS," *IEEE CICC*, 2010.

[2] W. J. Huang, *et. al*, "Sub-1V capacitor-free low-dropout regulator," *IET Electron. Lett.*, pp. 1395-1396, 2006.

[3] J. Guo, *et. al*, "A 6-uW chip-area-efficient output-capacitorless LDO in 90nm CMOS technology," *IEEE JSSC*, pp. 793-803, 2010.

[4] Y. Kim, *et. al*, "A 0.38V near/sub-V_T digitally controlled low-dropout regulator with enhanced power supply noise rejection in 90nm CMOS process," *IET*, pp. 31-41, 2013.

[5] M. G. Johnson, *et. al*, "A variable delay line PLL for CPU-coprocessor synchronization," *IEEE JSSC*, pp. 1218-1223, 1988.

[6] P. Dudek, *et. al*, "A high-resolution CMOS time-to-digital converter utilizing a Vernier delay line," *IEEE JSSC*, pp. 240-247, 2000.

978-1-4799-5128-4/14 $31.00 © 2014 IEEE

Near-Threshold-Voltage Circuit Design: The Design Challenges and Chances

I-Chyn Wey, Po-Jen Lin, Bing-Chen Wu, and Chien-Chang Peng, and Pin-Hsi Lin

Department of Electrical Engineering, School of Electrical and Computer Engineering,
College of Engineering, Chang Gung University
Taoyuan, Taiwan, R.O.C.
E-mail: ichynwey@gmail.com

Abstract

NTV is a new low power design concept for the pursuit of the highest power usage efficiency. The characteristics for each logic family are quite different under NTV while comparing to its operation under normal supply voltage. The circuit/architecture design policy under NTV is also different from its normal supply voltage operation. Process variation, performance degradation, and noise-interference are the three major design challenges in NTV design. In this paper, some effective candidate design solutions are presented to overcome these crucial NTV issues.

Keywords-near-threshold-voltage; low power; process-variation; noise-tolerant

I. Introduction

With VLSI technology advances, various digital electronic products flourish, such as 4G wireless systems, wireless LAN, wireless sensor networks, digital TV broadcasts, Internet of Things (IOT) iPhone, etc. Among a variety of applications, because the volume of electronic products continues to shrink, battery capacity sizes are inevitably limited.

In order to extend battery time of wireless devices, how to effectively reduce power consumption of the core CPU, ALU, and DSP is the serious challenge of SoC design. In particular, such demands will be even more intense in handheld products. Intel pointed more directly that power consumption will be actually the only design bottleneck based on the Moore's Law in the next ten years [1]. In order to solve the crisis of power consumption, the operation of the system at a low voltage circuit is the general direction to save power in circuit and system design. In recent years, Near Threshold Voltage (NTV) circuit design technique is the most representative which is vigorously promoted by Intel since 2012 [2].

NTV is a new low power design concept. Unlike the previous low power design for the purpose of the minimum operating voltage or the minimum power consumption in individual operation, the purpose of NTV is for the pursuit of the highest power utilization efficiency. In other words, NTV is in the pursuit of the optimization of system power reduction without excessively sacrificing system performance at the same time. NTV operating voltage is set near the threshold voltage of opening transistors, instead of reducing operating voltage without restriction. As shown in Figure 1 (a), the NTV range is

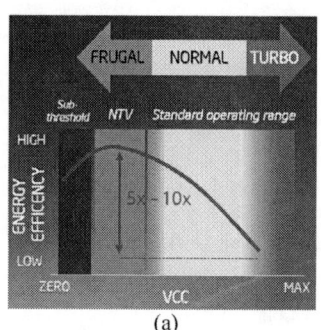 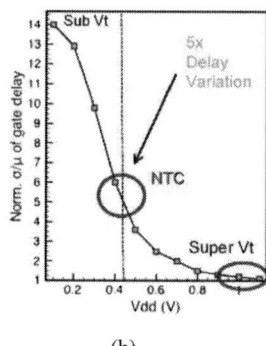

(a)   (b)

Fig. 1 (a) The most energy-efficient state of CMOS circuits is in the operation in the range of NTV. (b) The relationship diagram of operating voltage and delay variation [2].

the optimal energy efficiency range, which is because when standard operating voltage is reduced to the NTV range, operating frequency decreases linearly with the reduction of voltage supply, but power consumption exponentially declines so that efficiency gradually increases as voltage reduces, achieving 5-10 times improvement (as shown in Figure 1 (b)). However, in the sub-threshold range, the power consumption of the leakage current decreases moderately. Compared to the leakage current, the operating frequency decreases faster, leaving efficiency to become low. Therefore, NTV is the best operation range for low power consideration.

II. Design Challenge and Chances of NTV Circuits

Although operating under NTV environment improves energy use efficiency, it also brings some design challenges. The three most important challenges are performance variability increases, performance degradation, and serious noise interference.

● Performance Variation

In performance variation, in the NTV environment, circuits are affected by process variation, threshold voltage, operating voltage, and temperature more seriously than their operations under normal voltage, so as to cause severe performance uncertainty on NTV design. If the circuit works in the NTV range, process drifting will cause five-time global variance of delay [3].Compared to the normal operating voltage, if the parameters of PVT variation are evaluated together, the 20-time variation will be resulted in [3], which is more serious than the variation of general operating voltage. Moreover, differences of threshold voltage cause very different impacts under normal voltage and sub-threshold voltage. In the condition under normal operating voltage, processing and threshold voltage variation only results in 3-time variation difference of the extreme value of the driving current (I_{on}). However, process changes bring in the 700-time difference in the extreme value of the sub-critical current (I_{sub}) and 900-time difference in the extreme value of the leakage current (I_{OFF}) [3]. Therefore, process and critical voltage changes only cause approximately 10% of the difference in driving current variation of transistors (σ/μ) under normal operating voltage. However, under NTV with 400mV supply voltage, differences will substantially increase to 54% [4]. Under NTV, a simple string of the 30-level inverter serial circuit can conduct 1000 cycles of the Monte Carlo simulation analysis as the operating voltage is fixed at 265mV, showing that the differences of 7 times between the longest delay and the shortest delay [4].

In current literatures, the solutions of process variation under NTV can be divided into static, dynamic, and alarm solutions. Static solutions include cascade connection of transistors [5], increasing the transistor size [5], transistor size optimization under NTV [6], logic family selection, applications of reverse short channel effects [7], vector flip-flop, stacked min-delay buffer, ganged logic and vector control signal [8], super-pipeline[9], Schmitt-Trigger [10].

Most of these methods are simple and effective, without restrictions of application occasions. Among them, logic family selection is currently the most viable and safest choice under the NTV environment; however, there are still many inadequacies in its effectiveness and power consumption. For example, the general methods in literatures are in the use of the Schmitt-Trigger architecture to improve the ratio of Ion/Iof. If the Schmitt-Trigger architecture can be further applied to the leak current suppression of the non-critical path, the circuit performance improvement under NTV will get more significant upgrade. However, there are still some difficulties mainly because the Schmitt-Trigger structure may produce an additional leakage path.

In the past, circuits such as dynamic CMOS logic and pass-transistor logic had good performance under normal voltage. However, floating node leakage, signal contention, no driving capability, and other issues let these designs less likely to survive in the NTV environment. Even though the problem of signal contention cannot be tolerated, the remaining two issues can still be overcome as long as there is proper design. As a result, NTV circuit performance can be further enhanced through adopting dynamic CMOS logic families.

Besides, the path delay difference existing in a chip inherently distributes much wider than process variation induced delay difference, as shown in Fig. 2. Once we can pay more attention on controlling the delay distribution profile, the process variation induced design difficulties can be lowered.

- **Performance Degradation**

Performance degradation is one big issue in NTV applications and will determine the promotion range of NTV products. Recently, the adaptive remedial style design adopting voltage-over-scaling with error correction under low
v o l t a g e

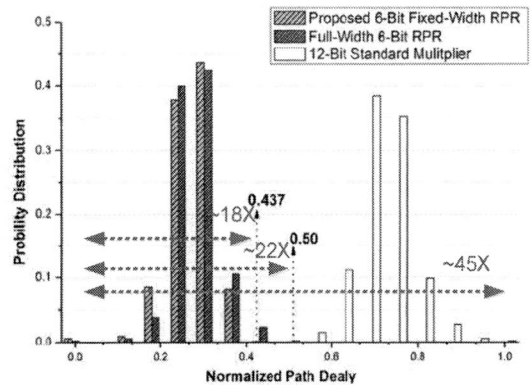

Fig. 2 Delay distribution difference inherently exists in a chip. By taking a 12-bit standard multiplier and 6-bit fixed-width multiplier as an example, the path delay difference in an arithemic circuit is much worse than its process variation induced difference even under NTV environment [11].

environment, such as the algorithmic noise tolerant (ANT) design [12], the fixed-width reduced precision replica redundancy block based ANT design [11], can provide advantages in the speed and process variation in NTV environment. Among them, because in the fixed-width reduced precision replica redundancy block based ANT design [11], the computing speed is fast, error compensation is precise, and the error correction circuit is simple (as shown in Figure 3 (a)), it can perform superior power saving efficiency in low-voltage environment. More importantly, ANT based designs can achieve superior power saving efficiency under NTV environment while maintaining it operating speed without significant performance degradation. Such superior characteristics make ANT architecture family to be a good candidate for NTV applications.

Dynamic solutions include Adaptive Body Bias (ABB) [13]-[16], Adaptive Beta Ratio Modulation (ABRM)[13],[16], the threshold voltage balancer with ABB, Ultra-low voltage split-output level shifter [2], ReVIVaL (voltage interpolation and variable latency)[17], and Soft-edge flip-flop (SFF) [18][19]. The performance of such designs can be more optimized and resistance to process variation can also be improved. However, more control variables exist, such as the situations that multi-group or adjustable operating voltage, multiple or adjustable clock, or adjustable body voltage is relatively complex in design. Moreover, how to achieve global optimization, not just local optimization, how to avoid additional performance overhead, and how to accurately and timely conduct dynamic performance tuning are the places to be improved in such designs.

In alarm solutions, such as Razor pipeline [20], Canary signaling [21], and asynchronous with critical-path-replica [22], which achieve process variation resistance upgrade mainly from the system/architecture level and are usually used simultaneously with static and dynamic techniques. Both Razor and Canary circuits can adjust system operation timing based on actual system computing delay requirements to make the system successfully resist operation error caused by process variation. However, they are required to match system-level control signals to design the alarm system.

978-1-4799-5128-4/14 $31.00 © 2014 IEEE

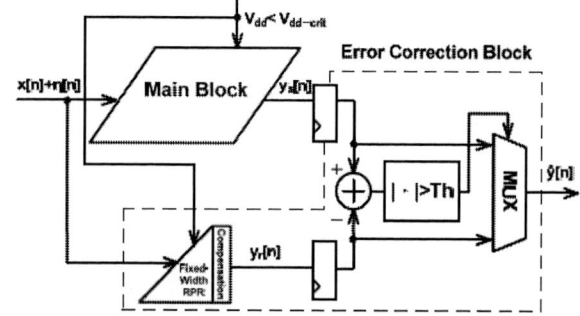

Fig. 3 The fixed-width RPR based ANT architecture [12]

- Noise Interferences

The noise margin CMOS circuits rely on is mainly built in voltage differences between VDD and Vth. Vth is a strong noise interference protection barrier in the CMOS circuit. Under low-voltage operating condition, noise interference is originally an important factor affecting digital circuits operating and an important consideration resulting that the normal operating voltage and the threshold voltage cannot continue to decrease. (Another consideration is the drain current.) Nowadays, in the NTV environment, the voltage difference between VDD and Vth is already compressed to the limit. Difficulty challenges in design due to noise interference are urgent and will directly affect the feasibility in NTV technological development.

In the past, CMOS circuits had many technologies of noise reduction or resistance to noises, published in the literatures [23]-[27]. These techniques were mostly designed to reduce or eliminate crosstalk noises, substrate noises, power supply noises, signal integrity, or temporary soft errors. However, in such designs, the possibility of operation under low voltage conditions was not taken into account.

In the past, in literatures [28]-[30], noise-tolerant Markov Random Field based design has been confirmed to have a superior noise-resistance capability in the NTV environment. However, its design is very complex, its area cost is too high, and its speed is relatively too low, which are the biggest bottlenecks to promote such design for commercial production. Therefore, over the past few years, many studies have been committed to circuit simplification and performance improvement of the anti-noise Markov Random Field (MRF) [31]-[33].

In order to further upgrade noise resistance capabilities of circuits under limited hardware costs and performance, new design directions , as shown in Fig. 4, are to adopt temporal redundancy and spatial redundancy in the mixed use and to

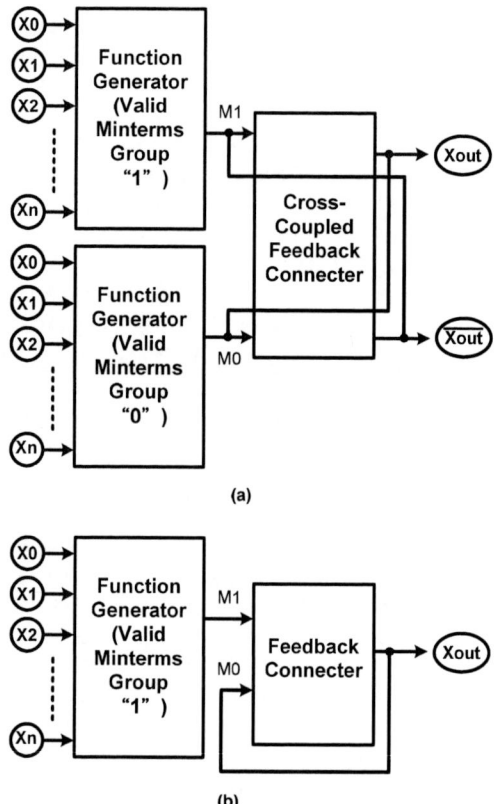

Fig. 4 Construction of spatial redundancy and temporal redundancy in the MRF designs (a) Spatial redundant MRF design (b) Temporal redundant MRF design.

replace the previous dual-rail design by the single-rail design, making the arithmetic circuit maintain excellent noise-tolerance capabilities in NTV environment. As shown in Table 1, the temporal redundant based MRF noise-tolerant circuit can perform superior noise-tolerance with less hardware overhead and less performance sacrifice in terms of power-delay-product.

Conclusion

In the NTV circuit deisgn, how to lower its process variation is the most important issue. Dynamic logic family circuits may still have good chance because of high speed and delay variation insensitive superiority; however, its signal contention and leakage problems should be solved in advance. The ANT-based architecture may be a good candidate for NTV deisgns to avoid its perforamcne degradation. To overcome noise-interference in a hardware efficient way, mixed spatial redundant and temporal redundant is needed.

Table 1 Performance comparisons for various MRF noise-tolerant circuits designs

		Output's SNR (dB)				Circuit Performance			
		SNR2	SNR3	SNR4	SNR5	Power (μW)	Delay (ps)	PDP (fW*s)	Transistor
AND	Direct-Mapping MRF [30]	25	25	25	25	5.65	430.0	2.43	60
	ML-MRF [33]	23.5	24.7	25	25	27.1	340.0	9.21	28
	M&S MRF [29],[31]	16.2	19.9	25	25	0.92	213.0	0.196	28
	Temporal-Redundant MRF	23.5	24.1	25	25	0.67	184.5	0.124	16
	CMOS	6.39	7.46	9.01	11.10	0.06	30.0	0.002	6

References

[1]   http://www.itrs.net

[2] Hsu, S.K. et al., "A 280 mV-to-1.1 V 256b Reconfigurable SIMD Vector Permutation Engine With 2-Dimensional Shuffle in 22 nm Tri-Gate CMOS," IEEE Solid-State Circuits Society, vol. 48, no. 1, pp. 118-127, 2013.

[3] F. Moradi, D. T. Wisland, H. Mahmoodi, A. Peiravi, S. Aunet, and T. V. Cao, "New subthreshold concepts in 65nm CMOS technology," in Proceedings of the 10th International Symposium on Quality Electronic Design (ISQED '09), pp. 162–166, San Diego, Calif, USA, 2009.

[4] B. Zhai, S. Pant, L. Nazhandali et al., "Energy-efficient subthreshold processor design," IEEE Transactions on Very Large Scale Integration (VLSI) Systems, vol. 17, no. 8, pp. 1127–1137, 2009.

[5] J. Kwong and A. P. Chandrakasan, "Variation-driven device sizing for minimum energy sub-threshold circuits," in Proceedings of the 2006 International Symposium on Low Power Electronics and Design (ISLPED '06), pp. 8–13, 2006.

[6] S. Y. Yang, "Performance study of applying RSCE and ABB techniques in nano-CMOS circuits," National Chung-Cheng University, Graduate Institute of Electrical Engineering, Master Thesis, 2010.

[7] T. H. Kim, J. Keane, H. Eom and C. H. Kim, "Utilizing Reverse Short-Channel Effect for Optimal Subthreshold Circuit Design" IEEE Transactions on Very Large Scale Integration (VLSI) Systems, vol. 15, no. 7, 2007.

[8] Hsu, S.K. et al., "A 280 mV-to-1.1 V 256b Reconfigurable SIMD Vector Permutation Engine With 2-Dimensional Shuffle in 22 nm Tri-Gate CMOS," IEEE Solid-State Circuits Society, vol. 48, no. 1, pp. 118-127, 2013.

[9] D. Jeon, M. Seok, C. Chakrabarti, D. Blaauw, and D. Sylvester, "A Super-Pipelined Energy Efficient Subthreshold 240 MS/s FFT Core in 65 nm CMOS," IEEE JSSC, vol. 47, no. 1, pp. 23-34, 2012.

[10] N. Lotze, and Y. Manoli, "A 62 mV 0.13 □m CMOS Standard-Cell-Based Design Technique Using Schmitt-Trigger Logic," IEEE JSSC, , vol. 47, no. 1, pp. 47-60, 2012.

[11] I. C. Wey, C. C. Peng, and F. Y. Liao, "Reliable Low-Power Multiplier Design by Using Fixed-Width Replica Redundancy Block", in IEEE Transactions on Very Large Scale Integration Systems, in publish, Feb. 13, 2014.

[12] B. Shim, S. Sridhara and N. R. Shanbhag, "Reliable Low-Power Digital Signal Processing via Reduced Precision Redundancy," IEEE Trans. on VLSI Systems, vol. 12, no. 5, pp. 497-510, May. 2004.

[13] K. Roy, J. P. Kulkarni, and M. E. Hwang, "Process-tolerant ultralow voltage digital subthreshold design," in Proceedings of IEEE Topical Meeting on Silicon Monolithic Integrated Circuits in RF Systems (SiRF '08), pp. 42–45, 2008.

[14] Y. Pu, J. P. De Gyvez, H. Corporaal, and Y. Ha, "An ultra-low-energy multi-standard JPEG Co-processor in 65 nm CMOS with sub/near threshold supply voltage," IEEE Journal of Solid- State Circuits, vol. 45, no. 3, pp. 668–680, 2010.

[15] S. Hanson, B. Zhai, M. Seok et al., "Exploring variability and performance in a sub-200-mV processor," IEEE Journal of Solid-State Circuits, vol. 43, no. 4, pp. 881–890, 2008.

[16] M. E. Hwang ,and K. Roy, "ABRM: adaptive β-ratio modulation for process-tolerant ultradynamic voltage scaling," IEEE Transactions on Very Large Scale Integration (VLSI) Systems, vol. 18, no. 2, pp. 281–290, 2010.

[17] X. Liang, G. Y. Wei, and D. Brooks, "ReVIVaL: a variation tolerant architecture using voltage interpolation and variable latency," in Proceedings of the 35th International Symposium on Computer Architecture (ISCA '08), pp. 191–202, June 2008.

[18] M. Ghasemazar, B. Amelifard, and M. Pedram, "A mathematical solution to power optimal pipeline design by utilizing soft edge flip-flops," in Proceedings of the 13th ACM/IEEE International Symposium on Low Power Electronics and Design (ISLPED '08), pp. 33–38, Bangalore, India, August 2008.

[19] V. Joshi, D. Blaauw, and D. Sylvester, "Soft-edge flip-flops for improved timing yield: design and optimization," Proc. Of International Conference on Computer-Aided Design, 2007, pp. 667- 673.

[20] S. Das, C. Tokunaga, S. Pant, Wei-Hsiang Ma, S. Kalaiselvan, K. Lai, D. M. Bull, and D. T. Blaauw, "RazorII: In Situ Error Detection and Correction for PVT and SER Tolerance," IEEE Journal of Solid-State Circuits, vol.44, no.1, pp.32–48, 2009.

[21] A. Drake, R. Senger, H. Deogun, G. Carpenter, S. Ghiasi, T. Ngyugen, N. James, and M. Floyd, "A distributed critical-path timing monitor for a 65 nm high-performance microprocessor," in IEEE ISSCC, pp. 398–399, 2007.

[22] I. J. Chang, S. P. Park, and K. Roy, "Exploring asynchronous design techniques for process-tolerant and energy-efficient subthreshold operation," IEEE JSSC, vol. 45, no. 2, pp. 401-410, 2010.

[23] M. Omana, D. Rossi and C. Metra, "High-Performance Robust Latches," IEEE Transactions on Computers, vol. 59, no. 11, pp.1455-1465, Nov. 2010.

[24] E. Ibe, H. Taniguchi, Y. Yahagi, K. Shimbo, and T. Toba, "Impact of Scaling on Neutron-Induced Soft Error in SRAMs From a 250 nm to a 22 nm Design Rule," IEEE Transactions on Electron Devices, vol. 57, pp. 1527-1538, 2010.

[25] F. Frustaci, P. Corsonello, S. Perri, G. Cocorullo, "High-performance noise-tolerant circuit techniques for CMOS dynamic logic," IET Circuits, Devices & Systems, vol. 2, pp. 537-548, 2008.

[26] I. C. Wey, Y. S. Yang, B. C. Wu, and C. C. Peng, "A Low Power-Delay-Product and Robust Isolated-DICE Based SEU-tolerant Latch Circuit Design", in Microelectronics Journal, 45(1): pp. 1-13, 2014.

[27] I. C.Wey, C. C. Peng, H. J. Chou, and P. C. Chen, "Reliable and Low Error Dual Modular Redundancy FIR Filter with Wide Protection Window", in IEICE ELEX, vol. 11, no. 9, pp. 1-6, 2014.

[28] I. C. Wey, Y. G. Chen, C. Yu, J. Chen, and A. Y. Wu, "Design and Implementation of Cost-Effective Probabilistically-Based Noise-Tolerant Circuits," IEEE Transactions on Circuits and Systems Part-I, vol. 56, no. 11, pp. 2411-2424, 2009.

[29] I. C. Wey, Y. G. Chen, C.H. Yu, J. Chen, A. Y. Wu, "A 0.18 lm probabilistic-based noise tolerate circuit design and implementation with 28.7 dB noise-immunity improvement," Proceeding of IEEE Asian solid-state circuits conferences pp. 291-294, 2006.

[30] I. C. Wey, "Design and Implementation of Noise-Tolerant Digital CMOS Circuits," National Taiwan University, Graduate Institute of Electronics Engineering, PhD Disserion, 2008.

[31] K. Nepal, R. I. Bahar, J. Mundy, W. R. Patterson, and A. Zaslavsky, "Optimizing Noise Immune Nanoscale Circuits using Principles of Markov Random Fields," In Proc. of Great Lakes Symposium on VLSI, April 2006, pp. 149-152.

[32] K. Nepal, R. I. Bahar, J. Mundy, W. R. Patterson, A. Zaslavsky, "Techniques for Designing Noise-Tolerant Multi-Level Combinational Circuits," In Proc. of Design, Automation and Test in Europe, March 2007, pp.576-581.

[33] I. C. Wey, and Y. J. Shen, "Hardware-Efficient Common-Feedback Markov-Random-Field Probabilistic-based Noise-Tolerant VLSI Circuits", in Integration, the VLSI Journal, in publish, Jan. 8, 2014.

A Limited-Contention Cross-Coupled Level Shifter for Energy-Efficient Subthreshold-to-Superthreshold Voltage Conversion

Chi-Ray Huang and Lih-Yih Chiou

Dept. Electrical Engineering, National Cheng Kung University, Tainan, Taiwan

Email: {n28981264, lihyih}@mail.ncku.edu.tw

Abstract—The subthreshold level shifter is an indispensable circuit for ultra-low voltage systems to communicate with different power domains. In this paper, we present a limited-contention level shifter using non-ratioed cross-coupled structure instead of ratioed one. Experimental results show that the proposed circuit can robustly convert subthreshold input up to superthreshold output. About 55% static power reduction is achieved without sacrifing performance when compared with the cross-coupled structure that needs diode-connected transistors and MTCMOS technique to converter ultra-low voltage signals.

I. INTRODUCTION

Subthreshold circuits are indispensable for ultra-low power applications, such as wireless sensor networks and biomedical systems [1]. In ultra-low voltage digital systems, numerous level shifters are needed to communicate with high supply voltage blocks or IO pads. Conventional level shifters exploit the cross-coupled structure to achieve the high speed and low power consumption requirements for converting from low superthreshold signal to high superthreshold signal. Unfortunately, when a subthreshold signal is applied to the pull-down transistors, a level shifter fails to convert up to superthreshold output due to the extremely strong contention between pull-down transistors (weak inversion) and pull-up transistors (strong inversion). Therefore, a novel level shifter that has wide voltage conversion range is necessary for power-aware multiple-supply voltage LSI systems.

Several subthreshold level shifters have been proposed to address the wide voltage conversion range challenge. Limiting the pull-up current driving ability and increasing the pull-down current driving ability are two effective ways to combat severe contention current. Multi-threshold voltage (MTCMOS) technique is a common approach to fulfill the requirement. Generally, pull-up transistors use high-V_{th} devices (HVT) to suppress the charging current, but are still insufficient for converting subthreshold signals. Some structures thus use diode-connected transistors or even stack in a series to suppress the current sourced from VDDH [2-6]. On the other hand, pull-down transistors usually require low-V_{th} devices (LVT) or zero-V_{th} devices to increase the driving strength [2, 4-6]. However, the multi-Vth technique is less attractive when considering severe process variations

[7]. Moreover, the use of diode-connected transistors [4-5] cause voltage drop of the internal output node which is connected to output buffers and subsequently induces large static current from VDDH. Works in [2] and [6] utilize diode-chains to weaken the pull-up current. However, it costs large area overhead and is hard to scalable with different VDDH. In [8], PMOS current limiters are proposed to bias the pull-up networks, but the VDDL-dependent bias current increases the static power dissipation of this circuit.

The current-mirror (CM) based structure is another type of level shifter. The work in [7] indicates that CM structures have superior robustness for converting subthreshold signals. In [9], a feedback transistor is used to cut off the static current when the level conversion is completed. However, it induces voltage drop of the output node because the feedback transistor reduces the current that charges the output high. The voltage drop issue causes large static current in the output buffer.

In this paper, we propose a novel cross-coupled level shifter which is capable of converting subthreshold input to superthreshold output robustly without the need of diode-connected transistors or MTCMOS technique. The rest of the paper is organized as follows. Section II shows the proposed subthreshold level shifter. Section III describes the experimental results of the proposed design. Finally, the conclusions are provided in Section IV.

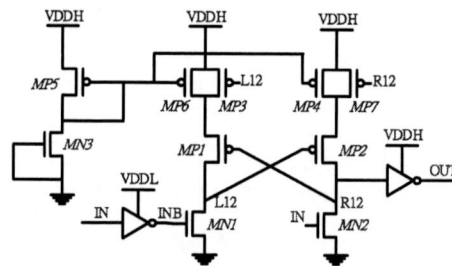

Fig. 1. Proposed limited-contention cross-coupled level shifter

II. PROPOSED DESIGN

The schematic of the proposed level shifter is shown in Fig. 1. It consists of an input inverter powered by VDDL, a cross-coupled level conversion stage, current mirror circuitry and an output buffer. Conventionally, the

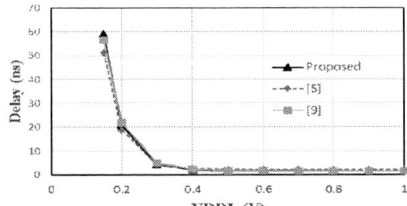

Fig. 2. Propagation delay versus VDDL

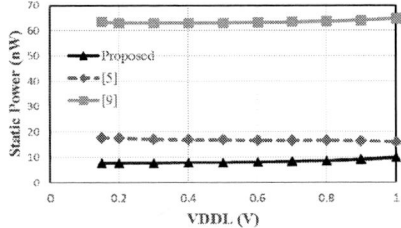

Fig. 3. Static power dissipation versus VDDL

internal node L12 (/R12) suffers from severe contention between strong cross-coupled pair and weak pull-down network. To eliminate the large contention current and convert input signals successfully through the cross-coupled pair, the best situation is that the pull-up transistors are off and the discharging current is relatively large enough to fulfill the level conversion procedure. However, turning off pull-up transistors makes the internal node L12 (/R12) vulnerable to the pull-down leakage current, and eventually would lose the previous data without transition. Therefore, in the proposed design, the current mirror circuitry (MP4 , MP5 & MP6) which mirrors the pull-down leakage current (MN3) to cross-coupled pair is exploited to limit the contention current during transition procedure and also maintain the output state when transition procedure is done.

The reduced contention current, which is in leakage current level, enables the cross-coupled structure convert subthreshold signal to superthreshold signal robustly. Moreover, PMOS MP3 and MP7 are used to boost the pull-up procedure and turned off again when the internal node is charged.

III. EXPERIMENTAL RESULTS

The post-layout experimental results are obtained from HSPICE simulations using TSMC 90nm technology under room temperature. The VDDH is set to 1V. For the sake of fair comparison, level shifters in [5] and [9] are implemented by following the sizing strategy as suggested in the original paper for the purpose of minimum energy-delay-product. Additionally, the output buffer of the level shifter, which has large effect on power consumption due to the voltage drop issues, is also designed as described in the original paper. The input signal frequency is set to 1MHz while the output load is set to 100fF. Fig. 2 shows the propagation delay of three level shifter designs as a function of VDDL. As can be seen from Fig. 2, the propagation delay increases exponentially as the VDDL is in deep subthreshold region. Fig. 3 shows the static power consumption comparison as a function of VDDL. The proposed level shifter has the lowest static power as compared to the other two level

shifters. That is because no voltage drop of the high output node is occurred in the proposed level shifter. And the current mirror limits the biasing current of the cross-coupled pair in leakage current level. No propagation delay has been compromised to achieve lower leakage power.

Fig. 4 shows the energy-delay-product comparison between proposed level shifter and the level shifter in [5]. When the input signal is 0.4V for conversion up to 1V, the EDP of the proposed level shifter is 24.6% lower than the level shifter in [5]. As the VDDL goes down to the subthreshold region, the EDP of [5] close to the EDP of proposed level shifter due to faster delay. That is because the level shifter in [5] use the MTCMOS technique and the pull-down transistor is equipped with LVT devices to increase the discharging speed.

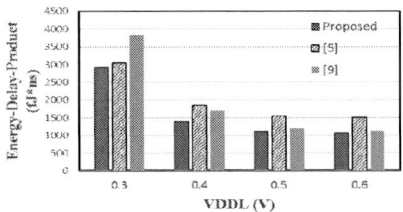

Fig. 4. Energy-Delay-Product comparison

IV. CONCLUSIONS

A limited-contention cross-coupled level shifter is proposed and validated using 90nm CMOS technology. The circuit exploits the leakage current mirror technique to address the severe pull-up and pull-down current imbalance issue. The wide voltage range conversion ability thus is achieved without the need of MTCMOS technique and dioded-connected transistors. In addition, no voltage drop due to dioded-connected transistors is occurred and the static power consumption is improved. According to the post-layout simulation results, about 55% static power reduction is obtained without sacrifing the propagation delay when compared to the level shifter in [5]. The proposed level shifter is suitble for power-aware LSI systems that has both low and high supply voltage domains.

ACKNOWLEDGEMENT

The authors would like to thank the National Science Council of Taiwan (NSC) and the National Chip Implementation Center (CIC) for their support. This work was supported in part of the projects NSC 102-2221-E-006-280.

REFERENCES

[1] M. Ashouei et.al., ISSCC, pp. 332-334, 2011.
[2] Y.-S. Lin et.al., ISLPED, pp. 197-200, 2008.
[3] S. N. Wooters et.al., TCASII, pp. 290-294, Apr. 2010.
[4] M. Lanuzza et.al., TCASII, pp. 922-926, 2012.
[5] M. Lanuzza et.al., TVLSI, 2014.
[6] Y. Kim et.al., VLSIC, pp. 188-189, 2011.
[7] S.-C. Luo et.al., ISCAS, pp. 2553-2556, 2012.
[8] T.-H. Chen et.al., J. Low Power Electron, pp. 251-258, 2006.
[9] S. Lutkemeier et.al., TCASII, pp. 721-724, Sep. 2010.

A Low Voltage Instrumentation Amplifier for Sensor Applications

Hwang-Cherng Chow, Hsin-Ti Chou and I-Chyn Wey*

Department and Graduate Institute of Electronics/Electrical* Engineering
Chang Gung University
259 Wen-Hwa 1st Road, Kwei-Shan, Tao-Yuan 333
Taiwan, Republic of China
Email: hcchow@mail.cgu.edu.tw

Abstract

A low voltage instrumentation amplifier for sensing applications such as the detection of biomedical signals is proposed. Since biomedical signals are of low frequency and small amplitude, noise has to be taken into consideration. In addition, the bio-signal amplifier should meet low power and low noise demand. Based on the 1.8V post-layout simulations of the overall circuit, the power consumption, gain and equivalent input referred noise are 25.92μW, 39.76dB and 3.44μVrms, respectively.

Keywords-low voltage; amplifier; bio-signal;low noise

Introduction

Recording biomedical signals is one of the challenges in a biomedical electronics detection system, because biomedical signals have very small amplitude and low frequency, usually of few milli-volts or less and the frequency below 1 kHz [1]. A biomedical electronics detecting system is shown in Fig. 1, which consists of electrodes, amplifier, LPF, sample and hold (S/H) and ADC.

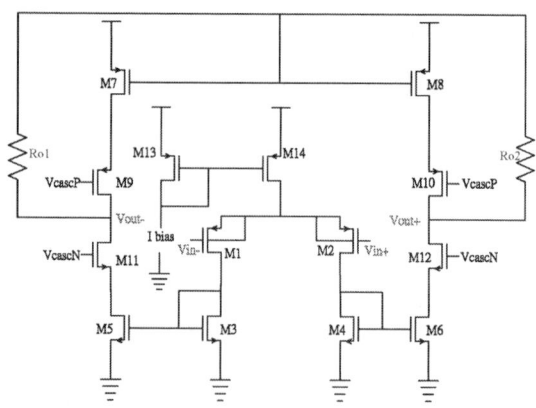

Fig. 1 The biomedical system block for signal detection

Biomedical signals, such as EEG/ECG signals, are characterized by their low voltage-levels and very low frequencies. Thus, an instrumentation amplifier (IA) must exhibit very low input-referred noise [2-4]. These biomedical signals are acquired and transferred to the voltage type signals

with amplitude of several milli-volts. The used IA must have high input impedance, low output impedance, limited bandwidth, and low power consumption [1-6]. Additionally, it must have adequate gain, high power-supply rejection ratio (PSRR) and high common-mode rejection ratio (CMRR) to suppress noise. In this paper, a new low noise, low-power and high CMRR IA for the biomedical signal processing is proposed. The presented IA is able to work under a single 1.8V power supply with low power consumption.

Proposed Design

The proposed instrumentation amplifier is shown in Fig. 2. To reduce the noise of the amplifier, the input PMOS transistors are biased under sub-threshold which effectively reduces power consumption and noise. To stabilize the output signals, a resistive feedback circuit is used instead of the traditional common-mode feedback circuit to further reduce the power consumption. Besides, extra capacitances have been added, as shown in Fig. 3, to generate a band-pass filter to filter out low frequency noise. This closed loop circuit can have precise gain which is mainly determined by the ratio of C1/C2.

Fig. 2 The proposed instrumentation amplifier for open loop sensing applications

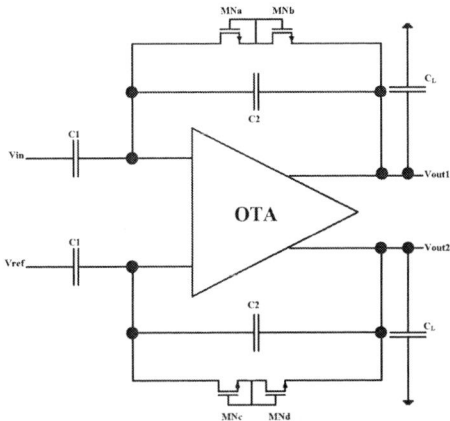

Fig. 3 The proposed instrumentation amplifier for closed loop sensing applications

Simulation Results

The achieved test chip layout with on-chip capacitors is shown in Fig. 4. The post-layout simulations for closed loop sensing function is demonstrated in Fig. 5. As for the illustration of open loop sensing function, it is shown in Fig. 6. Finally the summary results are given in table 1.

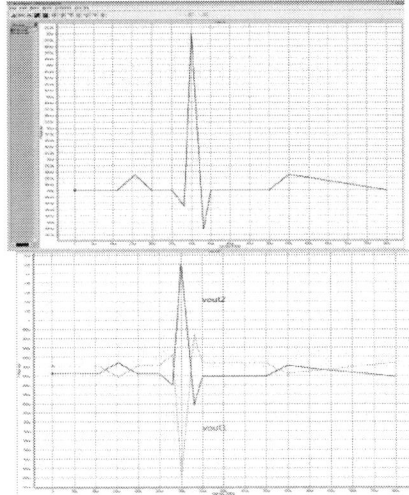

Fig. 4 Test chip layout

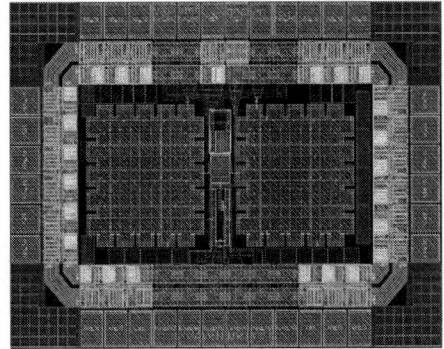

Fig. 5 Closed loop function (gain=100, upper: input @5mV, lower: differential outputs)

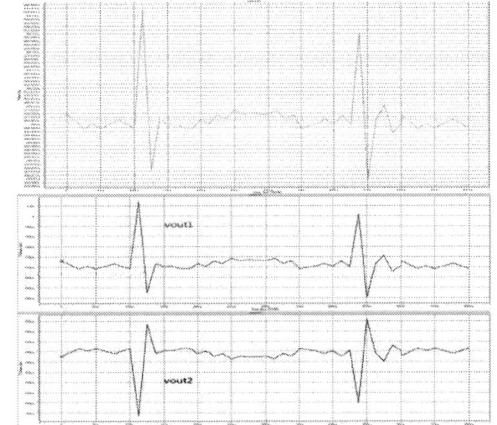

Fig. 6 Open loop function (gain> 70dB, upper: input @62.5uV, lower: differential outputs)

Table 1 Performance summary (tsmc 0.18um CMOS)

corners	TT	TT
VDD (V)	1.8	1.0
Power (uW)	25.92	2.09
Gain (dB)	39.76	39.4
BW (kHz)	~11.6	~5
CMRR (dB)	>80	>110
NEF	3.18	2.74
Input noise (600u~250Hz,rms)	3.44uV	1.83uV 0.5-200Hz
Input noise (600u~7kHz,rms)	5.03uV	4.88uV 0.5-5kHz

Acknowledgment: This research was partly supported by the National Science Council of Taiwan, ROC under contract no. NSC 101-2221-E-182-076 and NSC 102-2218-E-182-001. The software and chip fabrication support from the Chip Implementation Center is also acknowledged.

References

[1] K. A. Ng and P. K. Chan,, "A CMOS analog front-end IC for portable EEG/ECG monitoring applications", IEEE Trans. on Circuits and Systems I, Vol. 52, Issue 11, pp. 2335 – 2347, Nov. 2005.

[2] Edward K.F.Lee ,E. Matei and R. Ananth "A 0.9V rail-to-rail Constant gm Amplifier for Implantable Biomedical Applications",IEEE ISCAS, pp. 653-656, 2006.

[3] Honglei Wu and Yong-Ping Xu, "A 1V 2.3μW Biomedical Signal Acquisition IC", IEEE International Solid-State Circuits Conference, 2006.

[4] R.R. Harrison and C. Charles, "A low-power low-noise CMOS amplifier for neural recording applications" IEEE Journals of Solid-State Circuits, pp.958-965, June 2003.

[5] V. Majidzadeh, A. Schmid and Y. Leblebici, "Energy Efficient Low-Noise Neural Recording Amplifier with Enhanced Noise Efficiency Factor" IEEE Journals of Biomedical Circuits and System, pp.262-271. June 2011.

[6] F. Shahrokhi, K. Abdelhalim, D. Serletis, P.L. Carlen and R. Genov,"The 128-channel fully differential digital integrated neural recording and stimulation interface" IEEE Journals of Biomedical Circuits and Systems, pp.149-161, June 2010.

Comparison of Subthreshold Logic with Adiabatic Circuit Techniques

Cihun-Siyong Alex Gong, Chi-Tong Hung, Wei-Lin William Chu, Chang-Jie Lin, Yu-Fan Luo,
Chih-Yun Chien, Yu-Hung Kuo, Meng-Jung Chang, and Chin-Chih Hsu

Department of Electrical Engineering, School of Electrical and Computer Engineering,
College of Engineering, Chang Gung University
Taoyuan, Taiwan, R.O.C.
E-mail: alex.mlead@gmail.com

Abstract

Among the reviews and discussions demonstrated in the literature, very little attention was paid to which logic style is the most for adiabatic and subthreshold techniques. This study tries to make up such a deficiency, with particular emphasis on two representative circuit structures. All the comparisons were based on 0.18-μm standard CMOS process.

Keywords-CMOS; ultra-low voltage; ultra-low power; subthreshold; adiabatic; energy recovery

Introduction

Driven by the ever increasing demand in eco-friendly electronics and products with extended life span, the semiconductor foundries have been making the process shrinked to an extent where the Moore's law will come to the end of observation within the next decade. Ultra-low voltage (ULV) stemming from the requirement of ultra-low power (ULP) (consumption) becomes dramatically important for the ongoing and oncoming ultra large-scale integration ear. We have witnessed an explosive increase in the development of ULP techniques towards extremely energy-efficient electronics during the past years. Two renowned circuit families were developed and have shown that they are promising- ULV logic and energy recovery (adiabatic) logic. This paper tries to choose some technically sound circuit structures of them to make comparison in selected aspects.

Adiabatic vs. ULV

Adiabatic techniques are based on a truth that switching power dissipation of logic circuit can be dramatically decreased by keeping the voltage source faced small and constant when a path of circuit wire is turned conductive. Among the energy recovery logic (ERL) techniques, a quasi-static energy recovery logic (QSERL) was proposed in order to overcome the drawbacks of the irreversible ERL families [1]-[3], as shown in Fig. 1(a). QSERL features simplicity and staticlogic-resembled characteristics, which substantially decreases the complexity and switching activity. The employment of only complementary sinusoidal power clocks makes it be provided with higher energy efficiency compared to those of the prior arts utilizing trapezoid or triangular clocking scheme. Despite the advantages, QSERL suffers from inrobustness caused by output floating associated with the alternate hold phases in operation. Although the floating can be eliminated by adding clocked feedback keeper to each logic (keeper is turned ON only when QSERL is in the hold phase), there is still unwanted power loss. Also, the additional area cost as well as control signals would restrict its application. Motivated by this, a complementary energy path adiabatic logic (CEPAL), shown in Fig. 1(b), was proposed [4]. CEPAL not only inherits all the advantages of QSERL, but also eliminates the hold phase for the same operation conditions, thereby not only improving the robustness but also drastically increasing the throughput. For ULV, operating a circuit in the MOS subthreshold region has been a feasible approach so far. The work in [5] continued those prior works to analyze the performances of both the regular static CMOS and its Pseudo-NMOS counterpart operating in the subthreshold region. The performance derived from the regular sub-Vt CMOS was compared to its counterpart operated in strong inversion region and both the circuits (sub-CMOS and sub-PseudoNMOS) were compared to QSERL. In [6], two efficient body biasing approaches were proposed to improve robustness and tolerance to temperature and process variations of the regular subthreshold logic families. In addition to the minimization in the ill-effects, the subthreshold logic families adopting the two biasing approaches in [6] have energy (power-delay product) efficiency better than the regular sub-Vt counterpart as the decrease in delay outweighs the increase in power. In addition, similar to [5], the work presented in [6] also analyzed Pseudo-NMOS structures operating with one of the biasing approaches. A recently published paper complemented those existing works [7]. In [7], it was noted that previous studies emphasized device sizing, ultra-low power gain, and most importantly -- how to bias bulk terminals to obtain best results. Very little attention was paid to which logic style is the most efficient one when operating in the subthreshold region. The simulation results can be used in consort with [5] and [6] for concluding that which logic and biasing approach are the ones best suited for the comparison as they were simulated in the same process (TSMC 0.18-μm).

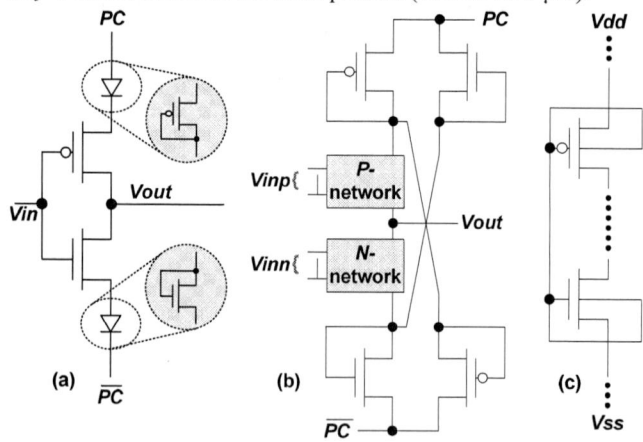

Fig. 1 Representative adiabatic and subthreshold circuit structures.

978-1-4799-5128-4/14 $31.00 © 2014 IEEE   146

Considering whether a biasing approach is absolutely demanded by the subthreshold logic, it can be seen clearly from Fig. 6 in [7] that using body biasing approach can considerably decrease the delay for complex gates operating in the subthreshold region, especially for the PMOS stacked logic with large number of inputs. As can be seen from Fig. 2 in [7], the swapped body bias (SBB) and the tunable body biasing (TBB) yield the smallest delays in terms of the test subjects (gates). However, owing to the invariably lowest threshold voltages in SBB, the approach suffers from reliability concern during the off state. The TBB in [7] on the other hand might not be well explained about how its body bias came from. By contrast, the dynamic threshold CMOS (DTMOS), which was first introduced in [6], has inherently no the reliability concern, and needs no exclusive biasing circuits, greatly reducing the system complexity in effect. Moreover, DTMOS has performance comparable to the subthreshold logic with SBB or TBB in spite of its larger input capacitance. With regards to the choice of logic style, it has been mentioned in [7] that no matter what choice of biasing approaches, the Pass Transistor Logic (PTL) has fast speed. Nevertheless, its delay is heavily dependent on the input pattern. Referring to Fig. 4 in [7], we can see the Domino logic has the longest delay while the regular CMOS structure has the best performance in terms of the speed. Notably, [7] shows that for complex gates, the Pseudo-NMOS logic style does not achieve speed better than its regular CMOS counterpart on the biasing approaches when operating in the subthreshold region, which is somewhat different from the conclusion deduced from [6]. As a result, we chose DTMOS with regular CMOS logic style as the subthreshold logic style used in the comparison with the CEPAL. All the referred ground Vss was connected with 0 V.

Comparison Results and Conclusion

Several aspects have been explored for DTMOS and CEPAL. To assure that DTMOS operates indeed in the subthreshold region even if there is about 10% (absolute) variation in the power supply, a 0.35-V supply voltage was chosen. As to CEPAL, we introduced 1.8-V power clocks (PCs) to support its operation. The delay as a function of operation frequency for the DTMOS and CEPAL inverters has been analyzed. Both the inverters had the same output load 200 fF for the analysis of driving ability, corresponding to the equivalent fanouts of approximately 57 and 22 for CEPAL and DTMOS (with identical MOS dimensions) individually. It can be discovered that DTMOS has a uniform delay over the frequencies, but CEPAL does not. The delay on CEPAL heavily depends on the PCs speed, making it have delays much longer than DTMOS at lower frequencies (less than 500 KHz). At higher frequencies, the PCs-driven CEPAL achieves better performance in delay. Following the analysis of delay versus frequency, we considered the impact of variation on the two logic- the inverter delay (with the output loaded with the same inverter) as a function of supply voltage variation. Simulations were carried out under conditions of 500-KHz frequency, ~10% supply voltage variations, process corner TT (Typical (NMOS) - Typical (PMOS)), and 25 °C. The results show that delay variation of CEPAL is from -21.05% -to- 24.21% where the variation of DTMOS is from -50% -to- 150%. The inverter delays as a function of temperature and process variations have shown that CEPAL has better tolerance to the PVT variations than DTMOS as a result of its relatively large switching current. It has also discovered from our experiments that for different DTMOS gates, the energy is subject to variances and the performance is dominated by either delay or power. The CEPAL counterparts on the other hand are with better stability. Comparing the energy consumption of CEPAL with that of DTMOS is a bit unfair as the longer delay and larger power of CEPAL are, respectively, governed by its PCs speed and higher signal swings. The simulations of inverter delay versus fan-out number were conducted under a 400-KHz operation frequency in order to obtain an intersection point. For the fan-out numbers below the intersection point (=40), DTMOS outperforms CEPAL. However, an almost linear increase in delay was observed with the fan-out number of DTMOS. By contrast, CEPAL has only a slight change in terms of the delay, which again demonstrates its driving ability. Moreover, the single-source noise margin (SSNM) for the subthreshold logic is better than for the adiabatic counterpart due to the exponentially high gain resulting from DC power supply. Nevertheless, it suffers from skewed noise margins causing it to identify HIGH more easily for efficient body biasing [7]. Lastly the major difficulties with respect to implementations of the two logic styles are compared. To assure acceptable performance of the subthreshold logic, variation in its supply voltage must be kept within an extremely small range, resulting in challenge in designing the on-chip regulator with small die size and accurate output level for fully integrated mixed-signal systems. As to the adiabatic logic, the use of inductive circuits are generally necessary to generate the PCs, making it relatively poor in aspect of manufacturability. Nevertheless, in some applications where inductive circuits are inherently involved, such as miniature radio frequency transponder system, the disadvantageous factor could not exist and adiabatic logic becomes preferable.

Acknowledgments

This study was supported by the Ministry of Science and Technology, Taiwan, and Chang Gung University under Grant numbers MOST 103-2221-E-182-070 - and UERPD2D0051.

References

[1] V. De and J. D. Meindl, "Complementary adiabatic and fully adiabatic MOS logic families for gigascale integration," Proc. ISSCC Dig. Tech. Papers, pp. 298 - 299, 1996.

[2] Y. Ye and K. Roy, "QSERL: quasi-static energy recovery logic," IEEE J. Solid-State Circuits, vol. 36, pp. 239 - 248, Feb. 2001.

[3] M.-E. Hwang, A. Raychowdhury, K. Roy, "Energy-Recovery Techniques to Reduce On-Chip Power Density in Molecular Nanotechnologies," IEEE Trans. Circuits Syst. I, vol. 52, pp. 1580 - 1589, Aug. 2005.

[4] Cihun-Siyong Alex Gong et al., "Analysis and design of an efficient complementary energy path adiabatic logic for low-power system applications," Proc. IEEE SoCC, pp. 247-250, 2007.

[5] H. Soeleman and K. Roy, "Ultra-low power digital subthreshold logic circuits," Proc. Int. Symp. Low Power Electron. Design, pp. 94 - 96, 1999.

[6] H. Soeleman, K. Roy, B. C. Paul, "Robust subthreshold logic for ultralow power operation," IEEE Trans. VLSI Syst., vol. 9, pp. 90 - 99, Feb. 2001.

[7] J. Nyathi and B. Bero, "Logic Circuits Operating in Subthreshold Voltages," Proc. Int. Symp. Low Power Electron. Design, pp. 131 - 134, Oct. 2006.

HARDWARE FEASIBILITY ANALYSIS FOR MOTION SEGMENTATION INITIALIZATION

Subarna Tripathi*, Youngbae Hwang†, Sung-Joon Jang†, Serge Belongie*,‡, Truong Nguyen*

* University of California San Diego
† Korea Electronics Technology Institute
‡ Cornell NYC Tech

ABSTRACT

We analyze and evaluate different initialization methods for motion layers segmentation, a powerful mid-level vision tool. We estimate 6-D affine motion models from the optical flow corresponding to different over-segmentations. Over-segmentations can be either uniform rectangular blocks; or adaptive sized rectangular blocks; or super-pixels; or based on any other clustering methods. We present performance analysis of motion segmentation initialization algorithms on video sequences and discuss the relative pros and cons of these methods in terms of hardware implementation issues.

Index Terms— video segmentation, affine motion, layered representation, superpixel, optical flow

1. INTRODUCTION

A mid-level vision tool such as motion layer is useful in video analysis applications like segmenting moving scenes, 2D-to-3D video conversion, and efficient video compression. However, the success of these segmentation or representation methods depends on the tractable initialization methods. We analyze different types of over-segmentations as the initialization step and evaluate their performances. We have explored different cases of initial over-segmentations in the form of (1) non-overlapping uniform blocks [1], (2) color-biased or super-pixel based [2], (3) motion-influenced adaptive block-based such as quadtree [3] and (4) motion influenced stability-based segmentations [4] by using optical flow [5] similarity. However, the 2D translation model cannot describe complex motions present in the scene. For natural video, most often, the parametric motion model corresponding to different entities or objects present in the scene are well approximated by 6-D affine model. We evaluate performances of the above methods qualitatively and quantitatively. To the best of our knowledge, there has been no comparative evaluation performed on initialization methods. We also analyze the feasibility of the their hardware implementations.

This work is supported by the Technology Development Program for Commercializing System Semiconductor funded by the Ministry Of Trade, Industry and Energy (MOTIE, Korea). (No 10041126, Title: International Collaborative R&BD Project for System Semiconductor). Authors thank Zucheul Lee for creating the synthetic dataset with ground truth.

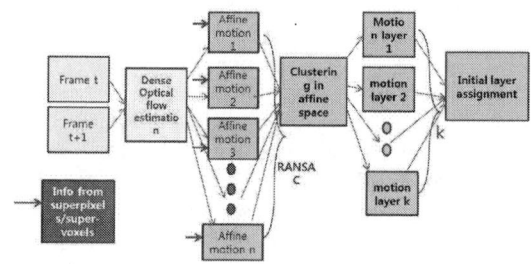

Fig. 1. Block-based and super-pixel based over-segmentation generation using clustering in parametric space without and with the information bias from superpixel respectively.

2. INITIALIZATION METHODS

The block diagram of uniform block-based and superpixel based initialization method has been shown in Figure 1. The inherent assumption is all the pixels in a block or superpixel belong to a single object and exhibit similar motion. However, the estimation of superpixel or other measure of color-bias is not trivial. Another method for motion based over-segmentation is achieved by partitioning the 2-D space into quadrants using quadtree decomposition similar to [3], but instead of partitioning on image intensity we apply the same in optical flow space. Lastly, we apply stability-based segmentation [4] method on optical flow instead of colors to select stable motion segments as an initializer towards motion layers estimation.

3. RESULTS

Unlike, low-level or high-level vision analysis, ground truth for mid-level vision tools are often unavailable or not well defined. To compare performances of the above algorithms quantitatively (Table 1), we use a synthetic video created using Autodesk 3DS MAX software for which we have the ground truth for the four motion layers corresponding to the background, person, bird and dog. Figure 2 shows the qualitative performance results on three real sequences.

978-1-4799-5128-4/14 $31.00 © 2014 IEEE

Table 1. F-measure values on the synthetic video

Layer	Block-based	Superpixel	Quad-tree	Stability-based
background	0.9942	0.9916	**0.9947**	0.9945
person	0.8873	0.8368	**0.888**	0.8589
bird	0.5821	**0.7674**	0.6613	0.6721
dog	0.8417	0.8171	0.8726	**0.8747**

Fig. 2. Layer initialization for Stefan, Lovebird, and Coast-guard (top row). Top to bottom: block-based, quadtree-based, superpixel-based, and one of the stability-based segments respectively

4. FEASIBILITY OF H/W IMPLEMENTATION

Initialization of motion segmentation involve three primary operations - optical flow estimation; over-segmentation generation; and affine parameter estimation from optical flow corresponding to different over-segmentations. In spite of difficulties of the implementation, attempts towards dedicated VLSI seem to gather attention with the demands by various applications. A dedicated VLSI with all functions is still in a research stage. Different real-time VLSI/FPGA system for hierarchical dense optical flow estimation have already been proposed such as [6]. Affine parameter estimation from optical flow values corresponding to a specific number of pixels using least square has a closed-form solution; thus easy to implement in hardware and fully parallelizable. As block-based over-segmentations perform their operations by a block of a fixed size, its VLSI design is relatively clear. They can be fully pipelined. With a design to enable a block size to change and hardware friendly quadtree-decomposition method, the block-based method can be extended to the quadtree-based one. Unlike these two methods, superpixel-based method relies on graph-cut algorithm. Even efficient polynomial-time graph-cut algorithms need the entire image in memory which makes its hardware implementation very challenging. For stability-based approach, even the graph-cut is performed multiple times. Thus, only block-based and quadtree-based approaches are suitable to real time implementation; where quad-tree based method achieves the best quality.

5. CONCLUSIONS

We analyze different over-segmentations as initialization methods for motion based segmentation and their hardware implementation issues. Based on the quantitative performance analysis on synthetic data and qualitative analysis on real sequences, we can conclude that generally uniform block-based or adaptive block-based (quadtree decomposition) perform well, the later being always better than the former as it suffers more from blocking artifacts. Color-bias or superpixel based initialization method works best provided the color image segmentation is accurate, which is a difficult task. Similar to the superpixel based method, performance of stability-based approach also depends on the accuracy of the clustering as the pre-processing step. From hardware feasibility analysis, we see only block-based and quadtree based methods are amenable to efficient hardware implementations. Overall, thus, quad-tree based approach performs best in quality-complexity trade-off.

6. REFERENCES

[1] J Y A Wang and E H Adelson, "Representing moving images with layers," *IEEE Trans. on Image Processing*, vol. 3, no. 5, pp. 625–638, 1994.

[2] Pedro F. Felzenszwalb and Daniel P. Huttenlocher, "Efficient graph-based image segmentation," *Int. J. Comput. Vision*, vol. 59, no. 2, pp. 167–181, Sept. 2004.

[3] M. Khan and Y. Ohno, "A hybrid image compression technique using quadtree decomposition and parametric line fitting for synthetic images," *Advances in Computer Science and Engineering*, , no. 3, pp. 263–283, 2007.

[4] Andrew Rabinovich, Serge Belongie, Tilman Lange, and Joachim M. Buhmann, "Model order selection and cue combination for image segmentation.," in *CVPR*. 2006, pp. 1130–1137, IEEE Computer Society.

[5] Ce Liu, *Beyond Pixels: Exploring New Representations and Applications for Motion Analysis*, Ph.D. thesis, Cambridge, MA, USA, 2009.

[6] F. Barranco, M. Tomasi, J. Diaz, M. Vanegas, and E. Ros, "Parallel architecture for hierarchical optical flow estimation based on fpga," *Very Large Scale Integration (VLSI) Systems, IEEE Transactions on*, vol. 20, no. 6, pp. 1058–1067, June 2012.

On-road Vehicle Detection with Monocular Camera for Embedded Realization: Robust Algorithms and Evaluations

Ravi Kumar Satzoda[*], Eshed Ohn-Bar[*], Jinhee Lee[+], Hohyon Song[+], and Mohan M. Trivedi[*]

[*]Laboratory for Intelligent and Safe Automobiles,
University of California San Diego, La Jolla, CA-92093
rsatzoda@eng.ucsd.edu, eohnbar@ucsd.edu, mtrivedi@ucsd.edu
[+]NextChip Co. Ltd.
jini92.lee@nextchip.com, hodal@nextchip.com

Abstract

Vehicle detection is critical operation in automotive active safety systems. Although there are a number of vehicle detection techniques available in literature, computationally efficiency for realization on embedded platforms is not explored and addressed in most existing works. In this paper, we present a computationally efficient vehicle detection algorithm that is particularly designed for architectural translation into efficient embedded hardware. The proposed method uses camera calibration to derive the appropriate window scales that must be used for vehicle detection, resulting in a computational cost reduction of over 10 times. In addition to reduction in sampling windows, the proposed vehicle detection technique uses a novel multi-part based vehicle detection method which detects the vehicles that pose the highest risk to the ego-vehicle. The proposed method is evaluated using different datasets and computational savings are seen in orders of magnitude as compared to conventional sliding window approaches, without compromising on accuracy.

Keywords- vehicle detection, computational efficiency

I. Introduction

Although there is a rich amount of literature on the different tasks involved in drive analytics [1]-[3], and advanced driver assistance systems (ADAS) such as lane detection [4]-[6], and vehicle detection [7]-[9], few works address computational efficiency of constituent algorithms [5][6]. Among the different ADAS operations, detection of on-road vehicles is a critical operation in advanced driver assistance systems (ADAS) such as collision avoidance, lane change assistance etc. [7]. Use of monocular cameras in order to detect vehicles has been explored in studies such as [7]-[9]. While [7] and [8] describe techniques to detect the rear-views of the vehicles using forward looking cameras, [9] describes techniques to detect overtaking vehicles from the rear-view of the ego-vehicle.

In this paper, we present a novel vehicle detection technique that is particularly aimed at improving the computational efficiency of the vehicle detection process. Additionally, considering that vehicle detection is an operation for active safety, it is critical that robustness of the vehicle detection process is not compromised. The proposed method involves part-based detection of vehicle, while leveraging on the camera calibration to reduce the computational complexity.

II. Proposed Method

A. Contextual Information for Detecting On-road Vehicles

The proposed method uses the camera calibration and context for processing in order to detect the vehicles using a forward looking camera from the ego-vehicle. Fig. 1 shows the contextual information for vehicles as seen from the forward looking camera. It can be seen that the vehicle in front of the ego-vehicle is of highest risk to the ego-vehicle followed by the vehicle that is farther away. Also the vehicles appear in a

Fig. 1. Contextual information about vehicles seen from forward looking images.

particular manner, i.e. the closest vehicles appear first, followed by the next vehicle, and so on, if the input image frame is scanned from the closest point to the ego-vehicle towards the vanishing point.

B. Vehicle Detection Method

This contextual information is used to devise the proposed method in the following way. Instead of scanning the entire image using a sliding window of different sizes, which is conventionally done in existing methods, the proposed method first uses the camera calibration to generate an LUT of window sizes. Given the homography matrix H that maps the image space to world plane (inverse perspective mapping - IPM) , a given axle width (w_A) of the vehicle, two points

$P_1^W(x_1^W, y_1^W)$ and $P_2^W(x_2^W, y_1^W)$ in IPM such that $x_2^W - x_1^W = w_A$, we get the window width in the image domain at a given y position (vertical axis along the image domain) using the following equation:

$$\begin{bmatrix} x_1 & y_1 & 1 \end{bmatrix}^T = \chi H^{-1} \begin{bmatrix} x_1^W & y_1^W & 1 \end{bmatrix}^T$$

$$\begin{bmatrix} x_2 & y_1 & 1 \end{bmatrix}^T = \chi H^{-1} \begin{bmatrix} x_2^W & y_1^W & 1 \end{bmatrix}^T$$

$$w(y_1) = x_2 - x_1; h(y_1) = \alpha w(y_1)$$

where $w(y_1)$ and $h(y_1)$ are the width and height of the sliding window to detect the vehicle at $y=y_1$ (vertical axis) in the image domain). An LUT for all y in the image domain for an $\mathcal{M} \times \mathcal{N}$ (columns x rows) sized image is generated, which will be used in the next step.

Given the LUT of window sizes, a sliding window I_V with size defined in the LUT is scanned over the image. Each window is divided into two parts P_1 and P_2 such at the height of P_1 is 1/3-rd of the window height. P_1 corresponds to the lower part of the vehicle from the bottom of the vehicle till the bumper of the vehicle. P_2 captures the upper part of the vehicle.

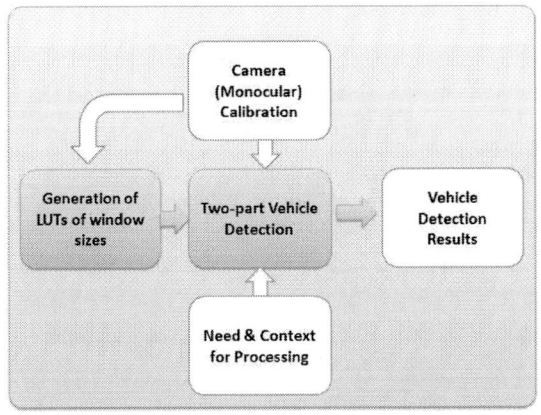

Fig. 2. Proposed vehicle detection method.

HOG features are computed for P_1 first, and a support vector machine (SVM) is used to classify whether it is the lower part of the vehicle. If the classifier detects the lower part, then HOG features are computed for the upper part, and SVM classifier is used to detect the upper part. A vehicle is considered detected if the following condition is met:

$$I_V = vehicle \ if \ p_1 \times p_2 > T$$

where p_1 and p_2 are the classifier scores for P_1 and P_2 parts of the sliding window I_V, and T is the classification threshold.

III. Performance Evaluation

The proposed method was evaluated for robustness using the LISA datasets in [8], and a comparison was done with the passive learning recogntion in [8]. Table I lists the true positive rates (TPR) and false detection rates (FDR) for the two methods. It can be seen that the proposed method gives similar or better TPR but with nearly 30-40% less false alarms.

In terms of computational complexity, the proposed method gives orders of magnitude lesser computational cost. The computational cost is defined relatively as the total number of windows that are processed for classification. Table I lists the total number of windows that are processed in each method. In

TABLE I
PERFORMANCE EVALUATION
Accuracy Evaluation

	[8]		This Work	
	LISA-2	LISA-3	LISA-2	LISA-3
TPR	83.5%	98.1%	100%	97.5%
FDR	79.7%	45.8%	53.1%	26.7%
Computational Complexity Analysis				
	Conventional		This Work	
Num. windows	12,943,00		124,692	

conventional sliding window approaches, 10 scales of windows are assumed for an image region given by 1280x301 pixels, with a horizontal skip of 3 pixels. It can be seen that the proposed method involves 1/10-th number of windows, thereby resulting in 10x reduction in computation cost.

IV. Conclusions

In this paper, a novel technique to detect vehicles is proposed that uses the need for detecting vehicles, and the camera calibration to efficiently and robustly detect vehicles. The proposed method is shown to detect vehicles with less than 10 times the computations, while reducing false alarms by at least 30%.

Acknowledgment

We would like to thank our sponsors, particularly Korea Electronics Technology Institute (KETI) and NextChip Co. Ltd.

References

[1] R. K. Satzoda, S. Martin, L. Mihn Van, P. Gunaratne and M. M. Trivedi, "Towards automated drive analysis: A multimodal synergistic approach", *16th Intl. IEEE Conf. on Intell. Trans. Sys. - (ITSC)*, pp.1912-1916, Oct. 2013.

[2] R. K. Satzoda, P. Gunaratne, and M. Trivedi, "Drive analysis using lane semantics for data reduction in naturalistic driving studies", *IEEE Intell. Veh. Symp.*, pp. 293-298, 2014.

[3] R. K. Satzoda and M. M. Trivedi, "Drive Analysis using Vehicle Dynamics and Visual Lane Semantics", *IEEE Transactions on Intelligent Transportation Systems*, 2014 (To Appear).

[4] J. C. McCall, and M. M. Trivedi, "Video-based lane estimation and tracking for driver assistance: survey, system, and evaluation," *IEEE Trans. on Intell. Trans. Sys.*, 7(2):20-37, 2006.

[5] R. K. Satzoda, and M. M. Trivedi, "Selective salient feature based lane analysis," *IEEE Intl. Conf. on Intell. Trans. Sys.*, pp.1906-1911, 2013.

[6] R. K. Satzoda, and M. M. Trivedi, "Vision-Based Lane Analysis: Exploration of Issues and Approaches for Embedded Realization," *2013 IEEE Conference on Computer Vision and Pattern Recognition Workshops (CVPRW)*, pp.604-609, 2013.

[7] R. K. Satzoda, and M. M. Trivedi, "Efficient Lane and Vehicle Detection with Integrated Synergies (ELVIS)", *2014 IEEE CVPRW on Embedded Vision*, pp.694-699, 2014.

[8] S. Sivaraman and M. M. Trivedi, "A General Active-Learning Framework for On-Road Vehicle Recognition and Tracking", *IEEE Trans. on Intell. Transp. Sys.*, 11(2):267–276, 2010.

[9] A. Ramirez, E. Ohn-Bar, and M. Trivedi, "Integrating motion and appearance for overtaking vehicle detection," *IEEE Intelligent Vehicles Symposium*, pp. 96-101, 2014.

Using Context to Improve Cascaded Pedestrian Detection

Mohammad Saberian, Zhaowei Cai, Jinhee Lee, Nuno Vasconcelos

University of California San Diego
9500 Gilman Drive
La Jolla, California, USA
saberian@ucsd.edu, zwcai@ucsd.edu, jini92.lee@nextchip.com, nvasconcelos@ucsd.edu

Abstract

The design of a fast and accurate pedestrian detector is considered. A system combining a fast cascaded pedestrian detector and a pedestrian validator is proposed. The detector first scans the image of interest and proposes a set of candidate bounding boxes. The pedestrian validator then decides if each proposed bounding box is consistent with a true pedestrian, based on scene context. Experiments show that the resulting system is faster and more accurate than current approaches to pedestrian detection.

Keywords-component; pedestrin detector; cascade detector; pedestrain validator and boosting

Introduction

In recent years, significant attention has been devoted to driver assistance systems and self-driving vehicles. An important problem for these systems is to detect and localize pedestrians that could be harmed by a vehicle. The design of such pedestrian detectors is challenging because 1) real-time operation requires very fast detectors and 2) the detection must be very accurate. Most notably, these systems must detect all pedestrians in the field of view while guaranteeing a very low false positive rate, so as to avoid false alarms that can ultimately lead drivers to ignore them.

While many methods have been proposed for pedestrian detection, most pedestrian detectors with acceptable accuracy and real-time performance are based on the cascade structure of [1]. In this architecture, a fast pedestrian classifier is trained on a large dataset of carefully cropped pedestrian examples. The detector is then applied in a sliding-window fashion, i.e. evaluated at all possible candidate regions of the image where the detection is to be performed. The main difficulty of this approach is that the decision about each candidate region is independent of the information outside of that region. However, this information can be a very useful aid to the detection process, especially when the image is of a structured scene. For example, Fig 1. shows a street scene along with detection outputs of a typical cascaded pedestrian detector. The detector was able to successfully detect two pedestrians, but also detected two false-positives. However, in the context of the larger scene, these false-positive cannot be valid pedestrians: the left one would corresponds to a 3-meter tall person, the one in the center to a pedestrian that walks 2 meters above ground.

In this paper, we propose a system to include context information in the detector so as to identify invalid

Fig. 1, detection result of a typical pedestrian detector

false-positives. In the proposed system, the pedestrian detector first scans the image and proposes a set of candidate bounding boxes. A pedestrian validator then decides if the proposed bounding boxes are consistent with real pedestrians, in the context of the whole image. Experiments show that the use of an effective pedestrian validator helps to eliminate false positives. In addition it is shown that, by integrating the pedestrian validator within the pedestrian detector, it is possible to shrink the scanning region and speed up the pedestrian detector itself.

Proposed System

The proposed system is illustrated in Fig. 2. The pedestrian detector first scans an image and proposes a set of candidate bounding boxes for the locations most likely to contain a pedestrian. The pedestrian validator then uses information from image context to decide if the proposed bounding boxes are consistent with true pedestrians. For applications like driver assisted systems, where pedestrian locations have high regularity across scenes, the validator can be trained a priori. This results in a set of scale-dependent locations for the bounding box, as shown in Fig. 3. The detector can, in turn, use this information to avoid scanning regions that do not conform to valid pedestrians.

A. System implementation

The cascaded detector is implemented with the algorithm of [6]. To design an accurate yet simple pedestrian validator we used a data driven approach. Rather than hard coding parameters of the scene, such as camera location inside the vehicle, we trained a classifier to discover the properties of valid bounding boxes from a training set. To train this classifier, we collected a set of random bounding boxes as negative examples. For positive examples, we collected the bounding box information for all non-occluded pedestrians in the training set of Caltech Pedestrian dataset [2]. From each bounding box we extract the following feature vector

$$\left[\frac{i}{H}, \frac{j}{W}, \frac{h}{H}, \frac{w}{W}\right], \tag{1}$$

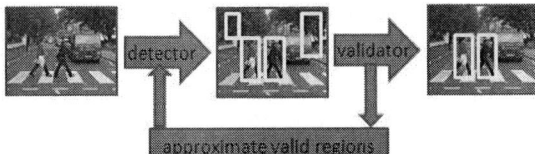

Fig. 2, proposed system for pedestrian detection

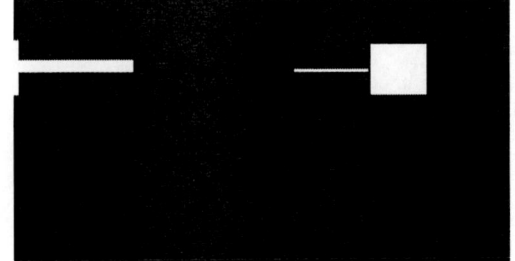

Fig. 3, white pixels show the possible locations for upper-left corner of pedestrians of height 260 pixels in a 640×480 images

where (i,j) are coordinates of the upper left corner of the bounding box, (h, w) its height and width and (H, W) the height and width of the image. This feature vector encodes relative size and location of pedestrians observed by a camera mounted on a vehicle. While the training process should be repeated whenever the camera position changes, this is not expected to be a problem for driver-assistance, where pedestrian detection is based on cameras installed by the manufacturer during vehicle assembly.

The main difficulty in training the classifier to discriminate between the sets of random and true bounding boxes is that the two sets have a significant overlap. In fact, the second set is almost completely inside the first, as most pedestrian locations are also possible locations for pedestrian absence. Hence, training a validator with 100% accuracy is impossible. To overcome this problem, we propose to design a classifier that, instead of minimizing the error rate, accepts all positive examples and minimizes the false-positive rate. In this way, the combined pedestrian+validator system accepts all pedestrians and rejects as many false positives as possible. Since this is a special case of cost-sensitive classification, we have used the cost-sensitive boosting approach of [7] to train the validator. The white pixels of Fig. 3 show the learned locations for the upper-left corner of pedestrians with height of 260 pixels in a 640×480 image. This is roughly the size of the left false-positive in Fig 1. As the map of Fig.3 suggests, to correspond to a true pedestrian, the bounding box should have a much higher upper-left corner. Hence, the validator rejects the bounding box. In addition, the regions of acceptance by the validator can be approximated by rectangles and the detector itself applied only inside these rectangles. This shrinks the scanning region and speeds up the overall detection.

Experiments

We compared the proposed system to a set of state-of-the-art pedestrian detectors on the Caltech Pedestrian dataset [2]. Similarly to Dollar et al. [3] we adopted an image representation based on a 10 channel decomposition. This included 3 color channels (LUV color space), 6 gradient orientation channels, and a gradient magnitude channel. The performance of the proposed system was evaluated with the toolbox of [2]. The comparison is based on detecting

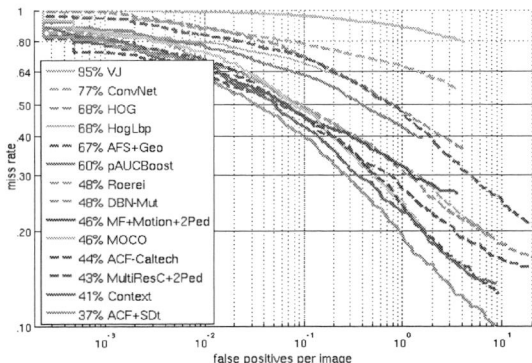

Fig. 4 miss rate vs. FPPI rate for of various pedestrian detectors. The number on the left of each legend is the log-average miss-rate.

pedestrians of height at least 50 pixels in 640×480 images. This is equivalent to detecting pedestrians about 40m away from the vehicle. Fig. 4 presents the results of this comparison. The numbers shown on the left of the legend summarize the detection performance by the log-average miss-detection rate. The set of benchmark detectors includes popular architectures, such as HOG [4] or the deformable part model of [5]. The proposed method outperforms all detectors other than ACF+SDt [6], which (unlike ours) uses motion information. The most direct comparison is ACF-Caltech [6], which uses a similar detector but ignores context information (no validator). The proposed system has better accuracy, i.e. 44% vs. 41% log-average miss-detection. In addition, by using the context information inside the detector and shrinking the scanning area, we were able to speed up the detector by 25%.

Acknowledgment

This work was supported by NSF grant (NSF IIS-1208522) and the Technology Development Program for Commercializing System Semiconductor funded By the Ministry of Trade, industry \& Energy(MOTIE, Korea), [No. 10041126, Title: International Collaborative R\&BD Project for System Semiconductor].

References

[1] P. Viola and M. Jones. "Robust real-time object detection". Workshop on Statistical and Computational Theories of Vision, 2001

[2] P. Doll′ar, C. Wojek, B. Schiele, and P. Perona. "Pedestrian detection: An evaluation of the state of the art". IEEE Transactions on Pattern Analysis and Machine Intelligence, 34(4):743–761, 2012.

[3] P. Doll′ar, Z. Tu, P. Perona, and S. Belongie. "Integral channel features". In Proceedings of British Machine Vision Conference, 2009.

[4] N. Dalal and B. Triggs. "Histograms of oriented gradients for human detection". In Proceedings of IEEE Conference on Computer Vision and Pattern Recognition, pages 886–893, 2005.

[5] P. Felzenszwalb, R. Girshick, D. McAllester, and D. Ramanan. "Object detection with discriminatively trained part-based models". IEEE Transactions on Pattern Analysis and Machine Intelligence, 2010.

[6] P. Doll′ar, R. Appel, S. Belongie, P. Perona. "Fast Feature pyramids for Object Detection". IEEE Transactions on Pattern Analysis and Machine Intelligence, 2014.

[7] P. Viola and M. Jones. "Fast and robust classification using asymmetric adaboost and a detector cascade". In Proceedings of the Neural Information Processing Systems Conference, pages 1311–1318, 2002.

978-1-4799-5128-4/14 $31.00 © 2014 IEEE

Efficient Airlight Estimation for Defogging

Yeejin Lee*, Changyoung Han[†], Junseong Park[†], Seungkyu Park[†], and Truong Q. Nguyen*

*University of California, San Diego
*ECE Dept., UCSD, La Jolla, CA 92093-0407, USA
[†]Core Logic Inc.
[†]670 Sampyeog-dong, Bundang-gu, Seongnam-si, Gyeonggi-do, Korea
yel031@eng.ucsd.edu, changyoung.han@gmail.com, junseong.park@bokwang.com, seungkyu.park@bokwang.com,
tqn001@eng.ucsd.edu

Abstract

This paper presents a new technique to efficiently estimate atmospheric light of visibility restoration algorithms from a hardware implementation perspective. The quality of defogged images depends on the accuracy of estimated atmospheric light, where it is typically estimated from the brightest portion of foggy images. In most existing algorithms, long execution time and large memory resources are required to find the largest intensity values on hardware platform. We explore an efficient method to measure atmospheric light by detecting sky region.

Keywords - defogging/dehazing; airligh estimation; FPGA implementation; image enhancement/restoration; advanced driver assistance systems

Introduction

Advanced Driver Assistance Systems (ADAS) is one of the fastest growing areas in automotive electronics due to increasing road safety concerns and customer demands. ADAS technology is based on vision systems, sensor technology, and car data network. Vision systems of ADAS applications are utilized to enhance safety on the road such as road sign recognition, pedestrian detection, lane departure warnings, and collision avoidance. High quality images or videos are required to successfully perform such functionalities in any environmental conditions.

Images and videos captured in bad weather suffer from contrast degradation and low visibility. It is necessary to employ visibility restoration algorithm in order to improve the performance of outdoor applications in real time. However, most existing defogging algorithms are computationally expensive, particularly in estimating atmospheric light. Moreover, most algorithms focus on improving the accuracy of transmission estimation despite the fact that inaccurate atmospheric light causes color artifacts in defogged images. In this paper, we investigate an efficient and robust technique for measuring airlight in foggy images.

The remaining of this paper is organized as follows. The proposed methods is discussed in the next section. In section III, experimental results verify that the proposed method outperforms the other methods and can reduce computational complexity. Finally, we conclude this paper in the last section.

Proposed Method

A. Atmospheric Scattering Model

In the presence of small particles such water droplets or dust, scene radiance I coming into a camera is expressed as the interaction of light scattering phenomenon:

$$ I(x) = I_o(x)e^{-\beta d(x)} + A\left(1 - e^{-\beta d(x)}\right). \quad (1) $$

By direct scattering, the surface radiance I_o is exponentially decayed with respect to the distance d from a camera to an object and the scattering coefficient β. The fraction of the original surface radiance transmitted to the camera is called as transmission. Another effect is that scattered light along an atmospheric path acts like light source under the assumption the atmospheric light A is invariant.

Defogging problem is composed of estimating unknown transmission and airlight for a given foggy image I. Various approaches have been proposed for the purpose of estimating transmission map more accurately, whereas there have been few studies on atmospheric light estimation. Therefore, in this section, we explore a new method to solve airlight estimation problem.

B. Atmospheric Light Estimation

Atmospheric light is the perceived intensity at a point infinitely far away toward the horizon. Atmospheric light is typically estimated from the most haze-opaque regions using bright channels or dark channels [1]. This method includes a step of sorting all pixels in an image to select proper regions, which introduces long execution time and requires large memory resources on hardware implementation. To be more precise, suppose we have dark or bright channels of an image containing n pixels in a color plane. Measuring atmospheric light in an image requires $(n - 1)$ comparison operations to find p% of the most haze-opaque pixels, and $(3\lfloor pn \rfloor - 3)$ addition operations and a division operation to compute their mean values.

In addition, inaccurate atmospheric light estimation causes color artifacts and unrealistic appearance of scenes in defogged images [2]. The accuracy of measuring atmospheric light depends on the predefined bright portion (p) as well as the regions where selected pixels are located. One limitation of this estimation method is that selected regions may not represent sufficiently the intensity of entire regions caused by

978-1-4799-5128-4/14 $31.00 © 2014 IEEE

atmospheric light [3] . Another limitation is that bright pixels can come from bright objects because they are indistinguishable from fog. Due to such considerations, we measure atmospheric light by detecting sky area and taking the intensity average as airlight estimate. In order to detect sky area, we employ simple region growing scheme taking into account computational complexity.

The proposed sky area detection method consists of two simple steps. The first step is to collect pixels of the top row considered part of the same region. The collection begins with a top-middle pixel based on the observation that the sky region is normally placed in the top-middle part of a scene from driver's view. Neighbor pixels are examined to determine whether a pixel should be included to the region on the basis of intensity similarity. Then, we search the first highest gradient pixel in the bottom direction for the collected pixel locations. Sky area is the region in which objects are indistinguishable. Thus, the horizon can be automatically detected through a search for high gradient pixels assumed to be the border line between sky and other objects [3] . The magnitude of gradient is computed as the sum of absolute gradients in x and y directions to avoid square operations.

Since this method measures the atmospheric light without comparing intensities of all pixels, the number of computational operations can be reduced.

C. Robust Atmospheric Light Estimation

We found that sky attribute is more distinct in bright channel maps measured as the maximum intensity out of RGB values. To verify this observation, we collect outdoor images from Flickr.com and measure bright channels and dark channels. The values of the complement of bright channels in sky regions are closer to zero than dark channels:

$$1- \max_{c\in\{r,g,b\}} (x^c) \leq 1- \min_{c\in\{r,g,b\}} (x^c) \leq 1-\min_{x\in\Omega}\left(\min_{c\in\{r,g,b\}} (x^c) \right) \quad (2)$$

where x^c is the cth color component of a pixel and is a local patch centered at x . This means that sky area in bright channels is more recognizable from the land than in dark channels as shown in Fig. 1 (b) and (c).

Experimental Results

We utilize real-world image of outdoor scenes that contains a portion of sky to evaluate the performance of the proposed method. The performance of the proposed method is compared against conventional method using Tarel's defogging algorithm [4] . The results show that sky area can be detected more robust using the characteristics of bright channels in Fig. 1 (d) and (e). Fig. 1 (f) and (g) demonstrate that the quality of defogged images varies depending on estimated atmospheric light.

Conclusion

We proposed an efficient approach to improve the quality of defogged images by measuring accurate atmospheric light from hardware implementation perspective. We have shown

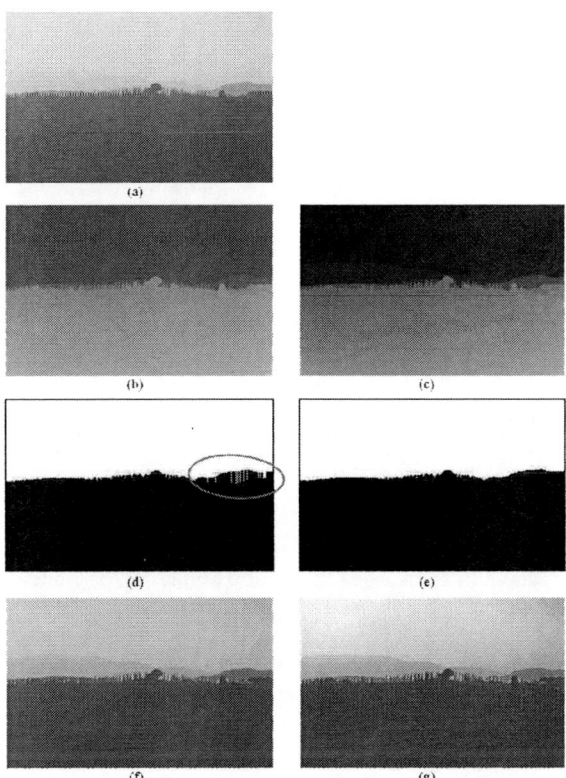

Fig. 1 Examples of defogged images using different atmospheric colors. (a) Foggy image. (b) Complement of dark channels. (c) Complement of bright channels. (d) Detected sky area using dark channels. (e) Detected sky area using bright channels. (f) Defogged image using the atmospheric light used in Tarel's method. (g) Defogged image with the atmospheric light measured using the proposed method and bright channels.

that the proposed method is effective in measuring atmospheric light with less computational complexity.

Acknowledgment

This work is supported by the Technology Development Program for Commercializing System Semiconductor funded by the Ministry of Knowledge Economy (MKE, Korea). (No. 10041126, Title: International Collaborative R&BD Project for System Semiconductor).

References

[1] K. He, J. Sun, and X. Tang, "Single image haze removal using dark channel prior," *IEEE Trans. PAMI*, vol. 33, no. 12, pp. 2341-233, Dec. 2011.

[2] M. Pedone and J. Heikkilä, "Robust airlight estimation for haze removal from a single image," *CVPRW*, 2011, pp. 90-96.

[3] F. Cozman and E. Krotkov, "Depth from scattering," *CVPR*, 1997, vol. 31, pp. 801-806.

[4] J. -P. Tarel and N. Hautière, "Fast visibility restoration from a single color or gray level image," *ICCV*, 2009, pp. 2201-2208.

978-1-4799-5128-4/14 $31.00 © 2014 IEEE

Depth Gradient Based Region of Interest Generation for Pedestrian Detection

Maral Mesmakhosroshahi

Department of Elec. and Comp.
Engineering
Illinois Institute of Technology
Chicago, Illinois, USA
mmesmakh@hawk.iit.edu

Kwang-Hoon Chung

Sane System
439 Dongahn-Gu,
Anyang-su
Gyungki-do, South Korea
jkh@sanesys.com

Yunsik Lee

Korea Elec. Tech. Inst.
Bundang-gu,
Sungnam-si
Gyungki-do, South Korea
leeys@keti.re.kr

Joohee Kim

Department of Elec. and Comp.
Engineering
Illinois Institute of Technology
Chicago, Illinois, USA
joohee@ece.iit.edu

Abstract

In this paper, we present a novel region of interest (ROI) generation method for stereo-based pedestrian detection systems. In the proposed algorithm, the vertical gradient of the clustered depth map is used to find the flat regions and variable-sized bounding boxes are used to extract the ROIs on the boundary of these regions. The ROIs are then classified into the pedestrian and non-pedestrian classes. Simulation results show that our proposed algorithm outperforms the existing monocular and stereo-based methods.

Keywords- pedestrian detection; ROI generation; advanced driver assistance; stereo vision

Introduction

Due to the development of the advanced driver assistance systems, on-board pedestrian detection has become an important research area in recent years which aims at detecting and tracking static and moving pedestrians on the road and warning the driver about their location. Early pedestrian detection methods used monocular cameras for detecting pedestrians [1]. Recently, several attempts have been made to employ stereo vision in order to improve the performance of pedestrian detection. Some of the stereo-based approaches use a disparity map to extract ROIs for pedestrian detection [2]. In several stereo-based pedestrian detection systems, dense or sparse depth map is used to extract information about the geometric features of the objects and generate ROIs [3]. In our previous works [4], [5], we used depth layering and skeleton extraction for ROI generation. In [6], stixel world is used for ground plane estimation and Integral Channel Feature [7] is used for pedestrian classification.

In this paper, we propose a stereo-based ROI generation algorithm for pedestrian detection by fusing the depth and color information obtained from a stereo vision camera to locate pedestrians in challenging urban scenarios by extracting the flat regions and generating variable-sized ROIs.

Proposed Algorithm

In this section, we explain the proposed ROI generation method in detail. In our stereo-based ROI generation framework, we first cluster the depth map using image quantization and extract the flat regions using the vertical gradient of the clustered depth map. The boundary of the flat regions is then considered as an area where pedestrians can stand and variable-sized bounding boxes are used to search for pedestrians and generate the candidate regions for pedestrian detection.

A. Flat Region Extraction

Several methods have been proposed for flat region extraction using monocular cameras. These methods are not robust against illumination variations and incur huge computational complexity. In this section, we propose a fast flat region extraction method which uses a depth map to detect the flat area and is robust against illumination variations.

The data obtained from a camera installed in a vehicle provides the perspective view of the scene in which, the distance from the camera changes in vertical direction. Therefore, in a depth image, the depth values of the flat regions decrease in vertical direction from the nearest to the farthest part of the regions. In the proposed method, we first quantize a depth map and generate several depth layers in order to cluster the objects based on their distance from the camera and take the vertical gradient of the clustered depth map in order to classify the scene into the flat and non-flat areas. The vertical gradient of the depth image can be computed using the Sobel gradient operator given by (1),

$$\nabla_y d = \begin{bmatrix} -1 & 0 & 1 \\ -2 & 0 & 2 \\ -1 & 0 & 1 \end{bmatrix} * d, \quad (1)$$

where d and $\nabla_y d$ are the depth image and its vertical gradient, respectively and * denotes the 2-D convolution. In order to extract the flat area in the image, we threshold the gradient values and the distance between the two consecutive nonzero gradient values using T_{val} and T_{dist}, respectively.

B. ROI Extraction

In most of the conventional pedestrian detection methods, fixed-sized windows and exhaustive scanning are used to search for pedestrians in the entire image which incur huge computational complexity. Fixed-sized ROIs are unable to detect pedestrians with different sizes as well. In order to overcome these limitations, we propose a stereo-based ROI generation method for pedestrian detection. In our proposed

978-1-4799-5128-4/14 $31.00 © 2014 IEEE

method, we first find the boundary of the flat regions extracted as an area where a pedestrian can stand. To extract the regions of interest, we use the depth values of the boundary pixels to estimate the size of the pedestrians. Since the size of the object in pixel is proportional to the disparity value, we can estimate the height and width of the bounding box using (2),

$$\begin{bmatrix} h \\ w \end{bmatrix} = \frac{d_b}{255} \times \begin{bmatrix} h_1 \\ w_1 \end{bmatrix}, \quad (2)$$

where the initial bounding box size is $w_1 \times h_1$ and d_b is the disparity value of the pixel on the boundary. To ensure that pedestrians with different poses are detected, we extract three bounding boxes for each boundary pixel (i,j) where the location of the top-left corners are $(i-h,j-w/2)$, $(i-h,j)$ and $(i-h,j+w/2)$. Then, we threshold the area of the foreground object inside the window in order to extract the ROIs

Experimental Results

To evaluate the performance of our method, we use the Daimler pedestrian benchmark [3]. Among several stereo matching algorithms [8], we use block matching method for depth estimation by setting the block size and the maximum disparity to 10×10 and 32, respectively. In experiments, images are downsampled to 320×240 and the quantization level, T_{val} and T_{dist} are set to 15, 60 and 30, respectively. Also, the initial window size is set to 125×250.

First, we present the performance of our ROI generation algorithm by computing the average number of ROIs per frame, the detection rate and the processing time of each step for the 21790 frames in the test set which are shown in Table I and II. Results show that the probability of missing pedestrians in our method is very low, while reducing the computational complexity significantly compared to the exhaustive search.

Then we classify the ROIs using the HOG/Linear SVM [1] and ICF/Adaboost [7]. Fig. 1 shows the ROC curves that compare the performance of our pedestrian detection system and the pedestrian detection methods introduced in [3]. Simulation results show that our proposed method outperforms the Daimler's monocular pedestrian detection method and provides competitive results with their stereo-based method.

Conclusion

In this paper, a stereo-based ROI generation method has been proposed for stereo-based pedestrian detection. To extract the ROIs, the vertical gradient of the clustered depth map is used to find the flat regions and variable-size bounding boxes are used to search for pedestrians on the boundary of the flat regions. Extracted ROIs are then classified into the pedestrian and non-pedestrian classes. Simulation results show that our method improves the performance of pedestrian detection in challenging urban environments compared to the existing systems.

TABLE I
PERFORMANCE OF THE PROPOSED ROI GENERATION.

Method	Detection rate	#ROIs
Proposed ROI generation	98.8%	630

TABLE II
PROCESSING TIME FOR OF THE ROI GENERATION STEPS.

Step	Proc. Time (ms)
Flat region extraction	2.71
ROI generation	20.8

Acknowledgment

This work was supported by the Technology Development Program for Commercializing System Semiconductor funded By the Ministry of Trade, industry & Energy (MOTIE, Korea). (No. 10041126, Title: International Collaborative R&BD Project for System Semiconductor).

References

[1] N. Dalal, "Histograms of Oriented Gradients for Human Detection," In *Proc. of the Computer Vision and Pattern Recognition (CVPR)*, vol. 1, pp. 886-893, Jun. 2005.

[2] R. Labayrade, D. Aubert and J. P. Tarel, "Real time Obstacle Detection on Non flat Road Geometry Through disparity Representation," In *Proc. of the IEEE Intelligent Vehicles symposium (IVS)*, 2002.

[3] C. G. Keller, M. Enzweiler, and D. M. Gavrila, "A New Benchmark for Stereo-based Pedestrian Detection," In *Proc. of the Intelligent Vehicles Symposium (IVS)*, pp. 691-696, Jun. 2011.

[4] M. Mesmakhosroshahi, J. Kim, Y. Lee and J-B. Kim, "Stereo based Region of Interest Generation for Pedestrian Detection in Driver Assistance Systems," In *Proc. on the Intl. Conf. on Image Processing (ICIP)*, pp. 3386-3389, 2013.

[5] J. Kim and M. Mesmakhosroshahi, "Stereo-based Region of Interest Generation for Real-time Pedestrian Detection," In *Peer-to-Peer Networking and Applications*, pp. 1-8, Sep. 2013.

[6] R. Benenson, M. Mathias, R. Timofte, L. Van Gool, "Pedestrian Detection at 100 Frames per Second," In *Proc. of the IEEE Conf. on Computer Vision and Pattern Recognition (CVPR)*, pp. 2903-2910, Jun. 2012.

[7] P. Dollar, Z. Tu, P. Perona, and S. Belongie, "Integral channel features," In *Proc. of the British Machine Vision Conf. (BMVC)*, 2009.

[8] M. Loghman and J. Kim, "SGM-based Dense Disparity Estimation Using Adaptive Census Transform," In *Proc.of the IEEE Intl. Conf. on Connected Vehicles and Expo (ICCVE)*, pp. 592-597, Dec. 2013.

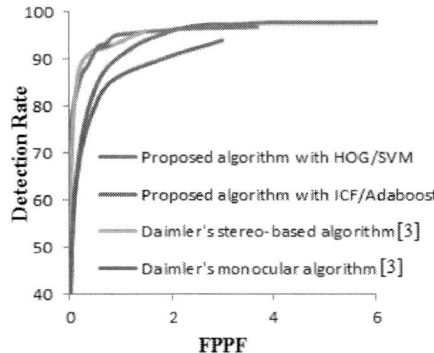

Fig. 1. Performance comparison between the proposed method and the methods introduced in [3].

Depth map estimation using modified Census transform and semi-global matching

Maziar Loghman

Illinois Institute of Technology
3301 S. Dearborn St.
Chicago, U.S.A.
mloghma1@hawk.iit.edu

Kwang-Hoon Chung

Sane Systems
439 Dongahn-Gu, Anyang-So
Gyunki-do, South Korea
jkh@sanesys.com

Yunsik Lee

Korea Electronics Tech. Inst.
Bundanga-gu, Sungnam-si
Gyungki-do, South Korea
leeys@keti.re.kr

Joohee Kim

Illinois Institute of Technology
3301 S. Dearborn St.
Chicago, U.S.A.
joohee@ece.iit.edu

Abstract

Generating a dense disparity image is one of the essential prerequisite for many applications such as rendering virtual views, 3D scene reconstruction, and advanced driver assistance systems (ADAS). In this paper, a depth estimation technique is proposed which is based on non-parametric Census transform and semi-global optimization. Simulation results indicate that the proposed method fulfills the aims of the algorithm by enhancing the quality of the estimated depth maps and reducing the computational complexity.

Keywords-Stereo vision; Census transform; semi-global optimization; curvature

Introduction

Depth maps are essential in a variety of applications such as 3D scene reconstruction, view synthesis, and advanced driver assistance systems [1]. Over the past several years, stereo-based methods have attracted a lot of attention in the research community. A survey study of different stereo matching algorithms is available in [2].

The stereo matching techniques can be classified into two groups, namely local and global techniques. The local methods [3] consider a finite neighboring window to estimate the disparity. The local methods are fast and computationally simple but they are highly error-prone. On the other hand, in global techniques an energy function is globally optimized to find the disparity [4]. Global depth estimation techniques can generate high-quality depth maps. However, due to the computational complexity of such algorithms, it is not feasible to exploit them in real-time applications. Combining the concepts of local and global stereo matching methods was first introduced as semi-global matching (SGM) [5]. SGM performs pixel-wise matching based on mutual information and the approximation of a global smoothness constraint and a good trade-off between accuracy and runtime is obtained. However, it achieves limited performance under illumination changes.

Calculating accurate depth values using point correspondence is a difficult task in many situations, especially for textureless regions, occluded regions and under illumination changes. There exist a trade-off between the accuracy of the obtained depth map and the computational complexity of the depth estimation algorithm.

In this paper, we propose a depth estimation algorithm based on modified Census transform and semi-global optimization. The proposed algorithm focuses on reducing the computational complexity for real-time processing while improving the quality of the depth maps under challenging conditions.

Proposed algorithm

The proposed method consists of three steps: down-sampling and mask generation, cost calculation and aggregation, and semi-global optimization. The Census transform is used as the matching criterion and adaptive window patterns that exploit the local characteristics of a depth image are proposed in calculating the matching cost.

A. Down-sampling and mask generation

In an ideal case, pixels belonging to the same object should have the same depth value. However, it is not occurring all the time due to the erroneous mismatches, change of illumination, etc. We use the curvature of the pixels in the given color images to distinguish between the smooth area and sharp edges of the objects in the scene. In order to reduce the computational complexity of the algorithm, the first step is to down-sample the stereo color images by a factor of 4. Then, we obtain a mask which indicates the smoothness of different regions in the reference image. The curvature is calculated using the first and second order gradients of each pixel given by (1).

$$k = \frac{u_{xx} u_x^2 - 2 u_x u_{xy} u_y + u_{yy} u_y^2}{(u_x^2 + u_y^2)^{3/2}}, \quad (1)$$

where u_x and u_{xx} are the first and second order gradients, respectively. Subscripts indicate the direction of gradient. The Prewitt kernel is used to find the gradient. After computing the curvature, we aggregate the values over a 5×5 windows and store it in a curvature map. A binary mask is generated using the curvature map. When the aggregated curvature of a pixel is less than a threshold, a zero value is assigned to the mask.

B. Cost calculation and aggregation

The non-parametric Census transform [6] is used to calculate the cost function which maps the surrounding block of a pixel into a bit stream. We use a simple Census window pattern for the smooth regions to reduce the computational

978-1-4799-5128-4/14 $31.00 © 2014 IEEE

TABLE II
COMPUTATIONAL TIME COMPLEXITY

Stereo image	Size (pixels)	Time(ms)
Tsukuba	384 × 288	61.5
Cones	450 × 375	71.1
Teddy	450 × 375	72.3
Venus	434 × 383	64.4

complexity and use a more complex pattern for the non-uniform regions which usually contain edges and object boundaries. For the pixel $I_c(x, y)$, the Census transform is calculated using (2).

$$R_T(x, y) = \otimes_{x \in N} \xi(I_c(x, y), I_c(x+i, y+j)), \quad (2)$$

where N is the neighborhood of the current pixel within the Census transform, ξ is the step function and $\otimes$ is bit-wise concatenation.

The binary mask generated in the previous step is used to decide which pattern to use. If the number of mask pixels in the neighborhood of the reference pixel is less than a predefined threshold, the simple pattern is used. The cost function is calculated by finding the Hamming distance between the obtained bit streams of the left and right reference images using (3).

$$C((x, y), d) = \sum_{x_i, y_i \in W} d_H(BS_l(x_i, y_i), BS_r(x_i, y_i)), \quad (3)$$

In (3), BS is the calculated bit stream. d is the disparity and d_H is the Hamming distance function and the subscripts l and r refer to the left and right reference images, respectively.

In order to get the final matching cost, we aggregate the cost over a 5 × 5 window. The calculated cost function is used for optimization.

C. Semi-global optimization

At this stage of the algorithm, we obtain the best match for each pixel and calculate the disparity. The optimum disparity is calculated by minimizing the energy function. The energy function consists of three terms. First term is the matching cost from the previous step which is based on Census transform. The other two terms are the smoothness energy terms. We add two penalty terms to the matching cost function to take into account slight and abrupt changes in the disparity of neighboring pixels. The 8-direction optimization path is used

TABLE I
PERCENTAGE OF BAD PIXELS FOR THE MIDDLEBURRY DATASET[7]

Stereo image	Proposed method	Adaptive Census [8]
Tsukuba	2.43 %	2.76 %
Cones	7.86 %	8.57 %
Teddy	11.9 %	12.6 %
Venus	0.80 %	0.81 %

to reach the optimum value.

Experimental results

In order to evaluate the performance of our depth estimation algorithm, we used the Middlebury dataset [7]. The workstation runs the Windows 7 operating system with Intel Xeon Quad-Core processor and 8 GB RAM. Four different image pairs from the Middlebury dataset [7] are chosen for the evaluation. Table I indicates the error statistic of percentage of bad pixels with respect to the provided ground truth depth map. The processing time by running the algorithms on Middlebury dataset [7] using C programming on CPU is shown in Table II.

Conclusion

In this paper, a novel depth estimation method have been proposed. By down-sampling the reference images, the computational complexity of the whole algorithm is reduced. The proposed method uses the curvature of pixels to create a mask and by applying census transform, the cost function is calculated. Simulation results indicate that the proposed methods fulfills the aims by improving the quality of the generated depth maps and reducing the computational complexity.

Acknowledgment

This work was supported by the Technology Development Program for Commercializing System Semiconductor funded By the Ministry of Trade, industry & Energy (MOTIE, Korea). (No. 10041126, Title: International Collaborative R&BD Project for System Semiconductor).

References

[1] M. Mesmakhosroshahi, J. Kim, Y. Lee, and J-B. Kim, "Stereo based Region of Interest Generation for Pedestrian Detection in Driver Assistance Systems," in *Proc. of the International Conference on Image Processing (ICIP)*, pp. 3386-3389, 2013.

[2] D. Schartsein, and R. Szeleiski, "A taxonomy and evaluation of dense two frame stereo correspondence algorithm," in *International Journal of Computer Vision*, vol. 47, no. 1, pp. 7-42, May 2002.

[3] Y. Boykov, O. Veksler, and R. Zabih, "A variable window approach to early vision," in *IEEE Trans. Pattern Analysis and Machine Intelligence*, vol. 20, no. 12, pp. 1283–1294, 1998.

[4] J. Sun, N. N. Zheng, and H. Y. Shum, "Stereo matching using belief propagation," in *IEEE Trans. on Pattern Analysis and Machine Intelligence*, vol. 25, no. 7, pp. 399–406, 2003.

[5] H. Hirschmuller, "Accurate and efficient stereo processing by semi-global matching and mutual information," *in Proc. of IEEE Conference on Computer Vision and Pattern Recognition (CVPR)*, pp. 807–814, 2005.

[6] R. Zabih, and J. Woodfill, "Non-parametric local transform for computing visual correspondence," in *Proc. of European Conference on Computer Vision (ECCV)*, pp. 151-158, May 1994.

[7] D. Scharstein, and R. Szeliski, "High-accuracy stereo depth maps using structured light," in *Proc. of IEEE Intl. Conf. on Computer Vision and Pattern Recognition (CVPR)*, vol. 1, pp. 195-202, Jun. 2003.

[8] M. Loghman, and J. Kim, "SGM-based dense disparity estimation using adaptive census transform,", in *Proc. of the IEEE Intl. Conf. on Connected Vehicles and Expo (ICCVE)*, pp. 592-597, Dec. 2013.

A High Speed Pipeline Structure of Hardware Implementation for Block Classification for Distributed Video Coding

Qiang Tong and Ken Choi

Department of Electrical and Computer Engineering,
Illinois Institute of Technology.
3301 S. Dearborn Street,
Chicago, Illinois, United States,
qtong@hawk.iit.edu, kchoi@ece.iit.edu

Abstract

Distributed video coding (DVC) is a video coding paradigm that aims at shifting the complexity from the encoder to the decoder. Traditionally, the existing DVC codecs encode the video frames as either key frame or Wyner-Ziv (WZ) frame. In WZ frames, all blocks are encoded as WZ blocks. We observed that, in WZ frame, there exists few blocks having intense motions, while many blocks have scarce motions. To improve the higher Rate-Distortion performance, we need to classify the blocks in a WZ frame into three categories, namely key block, WZ block and skip block. And encode each category with different scheme. So, a high speed classifier is very important to improve the throughput of the codec. In this paper, we propose a high speed pipelined block classifier for distributed video codec, which is suitable to work in real time video processing.

Keywords- Distributed Video Coding (DVC); Pipelining; Very Large Scale Integrated (VLSI) Circuits; High Speed Circuits; Block Classification)

Introduction

H.264/AVC is the most popular and most widely used video compression standard. Within same quality, H.264 can achieve much higher compression rate compared with previous video compression standard (e.g. H.263 and MPEG4)[1]. However, H.264 is computation intensive at the encoder end. This is because the motion estimation and block prediction, which are complex, are all done at the encoder. To implement such an encoder, powerful processors, such as DSP, are needed. Some System-on-chips (SOCs) even integrate hardcore H.264 encoder to accelerate the video compression process. With the diverse application of video coding, we cannot just use H.264 standard to satisfy all the needs, even if there several profiles in H.264 to meet the needs of different situations. Distributed video coding (DVC) is an alternative of H.264 for the use of light devices. DVC is based on Slepian and Wolf's and Wyner and Ziv's theories[2-3] which were proposed in 1970s. It was only in the last decade that emerging applications have motivated serious attempts at practical techniques[4].

The basic idea of the DVC is shifting the computation complexity from the encoder to the decoder[7]. To be more specific, motion estimation and block prediction are not done at the encoder, instead motion estimation is done at the decoder side. There are several DVC solutions proposed recently. They have been proved to be superior to H.264 with intra mode only. PRISM[6] and DISCOVER[5] are two pioneers of DVC codec. The two codecs classify the frames of video stream into key frame and WZ frame. Key frames are encoded by H.264 intra prediction encoder. While the WZ frames are encoded by DVC encoder. One major drawback of these two DVC codec is that they both need feedback channel from the decoder to encoder when encoding the video stream. This feedback scheme make the coding process suffering large delay problem. In [7], the authors proposed a DVC architecture without feedback channel. And in [8], the author proposed a hardware implementation of the DVC architecture in [7], which is based on finite state machine (FSM). The FSM implementation is not fast enough for high speed application scenarios. In this paper, we extract the block classification function module, and implement it in a pipelining structure.

Block Classifier

Unlike the PRISM and DISCOVER codec, the architecture proposed in [7] classify the macroblocks in WZ frame. Three types of block are used, namely key block, W-Z block and skip block. The key blocks are encoded by H.264 (intra mode) encoder, WZ blocks are encoded by DVC encoder, the skip blocks are skipped without any processing. In W-Z frame, coding some blocks as key blocks can improve the video quality, while coding some blocks as skip blocks can improve the compression rate. The following figure 1 shows the first two frames of foreman video sequence (qcif size).

Fig. 1 First two frames of foreman video

In video sequences with few motions, the skip blocks can account for over 70 percent of the total blocks. We can see

978-1-4799-5128-4/14 $31.00 © 2014 IEEE   160

from figure 1 that the difference between two consecutive frame is very trivial. Figure 2 shows the reconstructed frame 2 with only skip block.

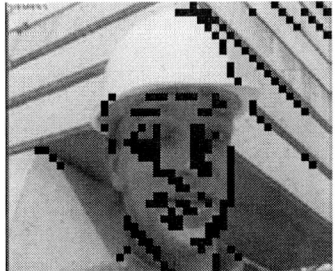

Fig. 2 Skip blocks of frame 2

The block classifier is very simple. We just need to set two thresholds of the SAD: T1 and T2.

$$block \quad type = \begin{cases} skip, & SAD < T1 \\ WZ, & T1 \le SAD < T2 \\ key, & T2 \le SAD \end{cases}$$

Hardware implementation

A. Video stream reordering

Most video capture sensor transmit the video stream in raster scanning order. That is line by line. In our application, we are processing the macroblock which is fixed 4x4 size. So, to implement the pipelining structure, we need to reorder the video streaming sequence. Figure 3 shows an example of what the reordering process do. Figure 3(a) shows the original video sequence (each square denotes one pixel data), Figure 3(b) shows results of reordering process.

1	2	3	4	5	6	7	8
9	10	11	12	13	14	15	16
17	18	19	20	21	22	23	24
25	26	27	28	29	30	31	32

(a) Original block sequence

1	2	3	4	17	18	19	20
5	6	7	8	21	22	23	24
9	10	11	12	25	26	27	28
13	14	15	16	29	30	31	32

(b) Reordered block sequence

Fig. 3 Illustration of reordering process

Our solution for reordering the video sequence is using 2 buffers to cache the video data alternately. For example, buffer 1 firstly store pixel data from line 1 to line 4, and then buffer 2 store pixel data from line 5 to line 8, while simultaneously, data in buffer 1 are read out. Usually, we need two separate systems to generate the address for the two buffers. To keep the two buffer synchronized, we make them share a single address generator. The idea is using 2

combined counters as address generator, and the two different addresses are generated by rearrange the order of the address lines.

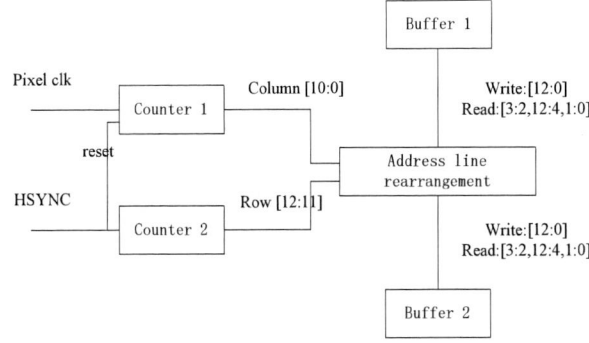

Fig. 4 Address generation for reordering

Figure 4 shows how the address is generated. Two counters which are driven by HSYNC and pixel clock signal from the camera sensor. HSYNC is usually the synchronous signal of each line in the frame, and the pixel clock is the synchronous signal with each pixel. We assume the number of pixels in one line is smaller than 2^{11} (if we need to process video with larger size, we just need to increase the line number of the counter output). The counter1 is reset by the HSYNC signal, that means each line is counting from 0. Counter 2 is used to index the rows. When writing the buffer, we just need to arrange the order of the address line as [12:0], while reading the buffer, the address line is rearranged as [3:2,12:4,1:0], which is the block wise order.

A. Video stream reordering

The architecture of block classification module is shown in figure 5. The key point is to make the structure pipelining. First, we need to use a FIFO buffer to store previous frame which will be used as reference to calculate the SAD. And the comparator will compare the SAD with predefined thresholds, and give the results as selection signals. To make the comparison results synchronized with the data stream, we need to make the data stream delaying 16 cycles, which is the pixel numbers of one macroblock.

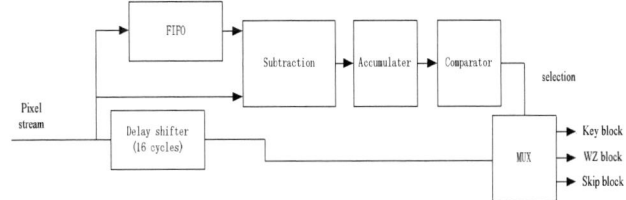

Fig. 5 Architecture of the block classification

Experiment

We implement the block classifier in Verilog HDL code, and tested in on Xilinx Virtex5 FPGA. This block classifier is also integrated in a low power video coding ASIC, which is taped out as a chip. We are testing the chip now. Before we have the testing result, we summarize the synthesis and simulation reports here.

A. Logic Synthesis

The implementation is synthesized by Design Compiler

978-1-4799-5128-4/14 $31.00 © 2014 IEEE   161

from Synopsys. The approximate gate count is 295376. Note, to speed up the video sequence reordering process, we are using two on chip memory as buffers, these two buffers account for about half of the logic cells.

The synthesis result shows that the design can work at frequency as high as 200Mhz.

B. FPGA Simulation

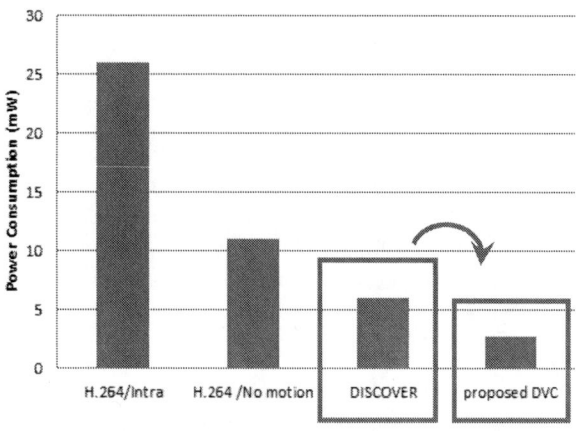

Fig. 6 Simulation results of block classifying

The implementation is also simulated with Xilinx ISE. Figure 6 shows the simulation results of block classifying. The bottom wave in the figure show the classifying results. '00' combination denotes key block, '01' combination denotes WZ block, and '10' combination denotes skip block.

C. Comparison Results

Acknowledgment

This work was supported by the Technology Development Program for Commercializing System Semiconductor funded By the Ministry of Trade, industry & Energy (MOTIE, Korea). (No. 10041126, Title: International Collaborative R&BD Project for System Semiconductor).

Conclusion

We implemented a pipelining block classifier for distributed video coding. Compared to the existing finite state machine based implementation, our proposed implementation has a higher throughput and fewer hardware resources. We demonstrate that the proposed pipelining block classifier is suitable for real time distributed video coding.

References

[1] Thomas Wiegand, Gary J. Sullivan, Gisle Bjontegaard, et al. "Overview of the H.264/AVC Video Coding Standard," IEEE Trans. on Circuits and Systems for Video Technology. Vol. 13, No. 7, July 2003.
[2] David Slepian, Jack K. Wolf. "Noiseless Coding of Correlated Information Sources," IEEE Trans. on Information Theory, Vol. 19, Iss. 4, July 1973.
[3] Aaron D. Wyner, Jacob Ziv. "The Rate-Distortion Function for Source Coding with Side Information at the Decoder," IEEE Trans. on Information Theory, Vol. 22, Iss. 1, Jan 1976.
[4] Bernd Girod, Anne Margot Aaron, Shantanu Rane, et al. "Distributed Video Coding," Proceedings of the IEEE, Vol. 93, Iss. 1, Jan 2005
[5] X. Artigas, J. Ascenso, M. Dalai et al."The DISCOVER Codec: Architecture, Techniques and Evaluation," Picture Coding Symposium, 2007.
[6] Rohit Puri, Kannan Ramchandran. "PRISM: An Uplink-Friendly Multimedia Coding Paradigm," Proceedings, IEEE International Conference on Acoustics, Speech, and Signal Processing. Vol. 4, April 2003.
[7] Krishna Rao Vijayanagar, Joohee Kim. "Dynamic GOP Size Control for Low-delay Distributed Video Coding," IEEE International Conference on Image Processing. 2011.
[8] Li Li, Krishna Rao Vijayanagar, K. Choi, Joohee Kim. "Energy efficient Encoder Design of Distributed Video Coding for Wireless Video Sensor Network," WTA 2011.

Multi-Band Multi-Mode Wireless Connectivity SoC for 802.11 a/b/g/n, BT 4.0 and NFC

Minsu Jeong[1], Yanggun Kim[1], Jaekyung Lee[1], Deokki Ahn[1], Ensoo Lee[1], Jangsup Sohn[1], Kwanju Lee[1], Jaehun Lee[2], Jabum Gu[2], SongBum Kim[2], Jungbo Sohn[2]

[1]RAONTECH Inc,
Seongnam, Korea
[2]Electronics and Telecomunications Research Institute
Daejeon, Korea

Abstract

The 65nm CMOS wireless connectivity SoC which support 802.11 a/b/g/n, BT and NFC is implemented. The SoC cover 13.6MHz band for NFC, 2.4GHz and 5GHz band for BT and 802. 11a/b/g/n. The SoC includes RF Transceiver, Baseband modem for 802.11 a/b/g/n, BT and NFC and ARM9 core with various peripherals. The Soc is fabricated using 65nm CMOS process with 5.5 x 5.5 mm² die area. The power consumption of 802.11n is about 160mW and 180mW for 2.4GHz receiver mode and transceiver mode except PA respectively and 170mW and 190mW for 5GHz case. The 802.11n maximum throughput is measured to 150Mbps. The 802.11 a/b/g/n receiver noise figure is about 4~5dB at 2.4~2.5GHz band and 4.8~5.7GHz band. The sensitivity is about -88.7dBm at MCS0 mode. The sensitivity of BT is about -94dBm. NFC operation range is up to 3.5cm.

Keywords-802.11a/b/g/n; bluetooth; NFC; wireless connectivity

Introduction

The 65nm CMOS multi-band and multi-mode SoC for 802. 11a/b/g/n, BT 4.0 and NFC composed of RF/Analog front-end (RF/AFE), power management system, digital MODEM and MCU with peripherals is introduced. It covers three-band of 2.4GHz ISM Band(2.4~2.5GHz), 5GHz 802.11n band(4.9~5.8GHz) and 13.6MHz ISM band for NFC. and supports 3 communication standards of 802.11 a/b/g/n, BT 2.0 to 4.0 and NFC which supports ISO-14443, ISO-15693, ISO-18092 and is compatible with existing RFID[1][2][3]. To minimize RF/AFE area and power consumption as digital in 65nm, the receiver system budgeting removed a lot of baseband analog processing blocks while high performance ADC, DAC and digital filters take over that functions and VCO and mixer covers whole band using only one core.

the Multi-Band Multi-Mode Wireless Connectivity SoC

A block diagram of SoC architecture is shown in Fig. 1. The SoC is composed of WLAN RF/MODEM, BT RF/MODEM, NFC RF/MODEM and MCU, memory and peripherals.

The WLAN has three parts – RF/Analog , PHY and MAC. The RF conversion architecture is zero-IF. It consists LNA, Mixer, Baseband Filter and delta sigma modulator ADC for receiver and DAC, BBA, up-mixer and power amplifier for transmitter and LO synthesizer covers from 2.4~2.5GHZ and

4.8~5.9GHz. The WLAN MODEM supports HT20, HT40 with 1 ~ 150Mbps various data rate. The MODEM is compatible to 802.11 a/b/g/n/h/i/e and supports WEP, WPA/WPA2(AES and TKIP) security options.

The BT RF is more simple compare to WLAN due narrow signal bandwidth and higher SNR ADC and DAC. The conversion architecture is low-IF to avoid 1/f noise. The BT MODEM of the SoC can support BT 2.0 EDR and BT 4.0 LE.

The NFC part of the SoC consists of a analog transceiver and digital modem. Transceiver has been designed as a transmitter is possible 0 ~ 100% ASK modulation with adjustable output current. And receiver having a dynamic range of input of 50dB, dc offset cancel loop. Digital modem works through state machine so as not to require another MCU.

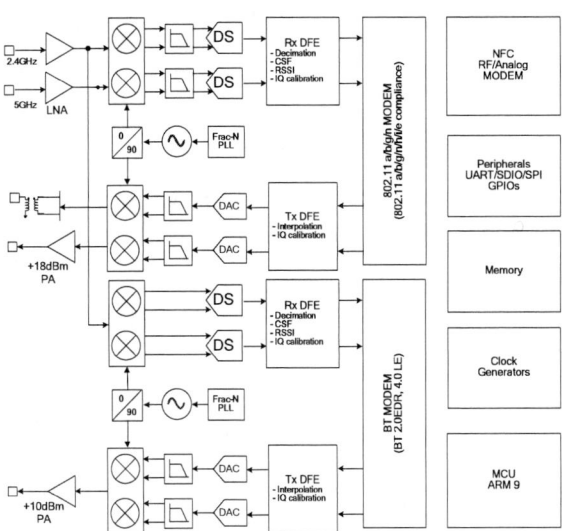

Fig. 1 Block Diagram of the Multi-mode Multi-band Connectivity SoC.

Digital switching noise can degrade RF/AFE performances significantly in SoC. The digital and RF/AFE isolation is very important and should be considered carefully [4][5]. To prevent that switching noise affects to RF/AFE, deep n-well isolates between digital baseband and RF/AFE and the switching circuits and clock oscillator is located far from RF input stage. The induced noise level from switching regulator and digital is measured very low sufficiently which not affect to RF performance.

In the RF/Analog of the SoC, the many potion of analog

baseband signal processing is replaced to digital signal processing such as wide gain dynamic range with fine gain step, high stop-band rejection ratio, high image rejection ratio and DC offset calibration loop with high SNR ADC. The ADC of WLAN receiver is continuous-time delta sigma modulator which maximum sampling rate is 640Msps and 65dB SNR with 20MHz signal bandwidth. Using this high SNR ADC, the WLAN baseband analog filter is reduced to 3rd order low pass filter and removed baseband gain stage. BT ADC is also CT-DSM ADC and it's SNR is over 70dB.The RF/AFE maximum gain step is reduced 3dB gain step but the receiver gain step is 0.25dB with digital gain control. The image rejection and DC offset calibration moves to digital. In low-IF system for BT, the mixer and baseband analog circuits is suffered from devices imperfections which make IQ imbalance. This imbalance can be removed by digital real-time IQ calibration.

Process	65nm LP CMOS	
Die Area	5.5mmx5.5mm	
Supply Voltage	1.2V(core), 1.8~3.6V(I/O)	
Communication Standard	802.11a/b/g/n/h/i/e, BT 4.0, ISO-14443, ISO-15693, ISO-18092	
WLAN Performance		
Maximum Throuput	150	Msps
Sensitivity (802.11n MCS0)	-88.5	dBm
PA max output(802.11n HT40)	18	dBm
BT Performance		
Sensitivity(DQPSK)	-94	dBm
NFC Performance		
Maximum throughput	424	Kbps
Operating distance	3.5	*cm*

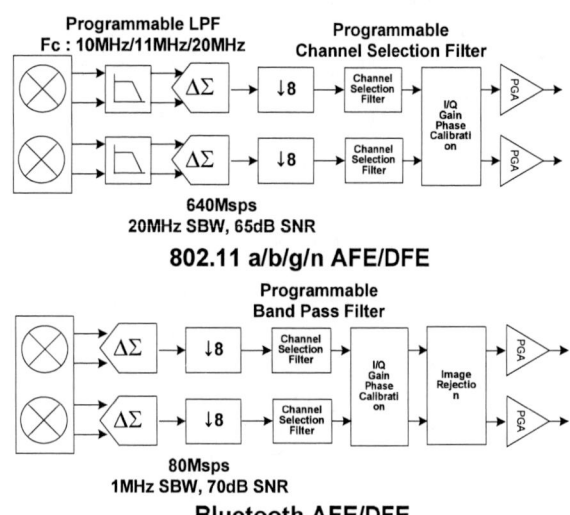

Fig. 2 WLAN and BT baseband analog and digital front-end

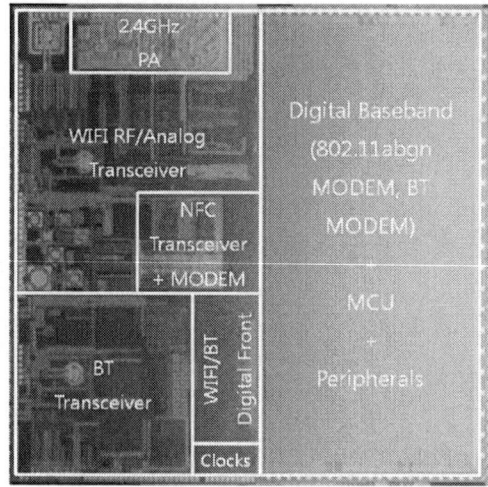

Fig. 3 Microphotograph of the SoC.

Implementation and Conclusions

The multi-mode multi-band wireless connectivity SoC for 802.11 a/b/g/n, BT and NFC is fabricated in a 65nm CMOS process. The die size is 5.5x5.5 mm2. The SoC power consumption is 160mW and 180mW for WLAN Rx/Tx, while RF/AFE and digital front-end consume 100mW and 130 mW. The sensitivity is -88.5dBm and -94dBm a for 802.11n MCS0 and BT EDR QPSK mode respectively. The EVM with 18dBm PA output and Rx middle power input are 3.6% and 2.2% respectively. High performance ADC and DAC reduces baseband analog porting and most of baseband signal processing moves to digital. To prevent digital switching noise injection to RF/AFE block, the digital portion and clock oscillator are located far from RF input and LO generator. Various methods such as deep n-well isolation and careful layout floor planning is choosed to reduce digital induced noise. Measured spurious or digital switching noise power in RF inputs is less than -100dBm. RF and system performance is shown in Table I. The chip micrograph is shown in Fig 7.

TABLE I
Performance Summary

Acknowledgment

This research was supported by Korean Government Fund - System IC 2015 project.

References

[1] International Standard ISO/IEC 18092:2004(E), "Information technology - Telecommunications and information exchange between systems - Near Field Communications- Interface and Protocol (NFCIP-1)", Geneva, 2004.

[2] International Standard ISO/IEC 14443:2000(E), "Identification cards - Contactless integrated circuit(s) cards - Proximity cards -", Geneva, 2000.

[3] International Standard ISO/IEC 15693:2000(E), "Identification cards - Contactless integrated circuit(s) cards - Vicinity cards -", Geneva, 2000.

[4] B. Kim, et. al., "A 9dBm IIP3 Direct-Conversion Satellite Broadband Tuner-Demodulator SOC", IEEE International Solid-State Circuits Conference, pp. 446-507, Feb. 2003.

[5] Minsu Jeong, et al., "A 65nm CMOS Low-Power Small-Size Multistandard, Multiband Mobile Broadcasting Receiver SoC", ISSCC Digest of Technical Papers, Feb. 2010, San Francisco, USA.

Design of Analog Front End for Mobile Fuel Gauge Applications

Chulkyu Park, Hyojae Kim, Jongkeun Hwang, Kichang Jang, Yeongik Yoo*, and Joongho Choi

University of Seoul, Seoul, Korea,
*SiliconMitus, Kyounggi-do, Korea
jchoi@uos.ac.kr

Abstract

A high-resolution second-order integrating sigma-delta analog-to-digital converter (ADC) using double-sampled integrators is presented whitch performs two times faster sampling than conventional modulator. The modulator has been designed in a 0.18-um CMOS technology. It acheives a signal-to-noise and distortion ratio (SNDR) of 93.03 dB at a conversion rate of 64 sample/s. Power dissipation is 36μW for a single supply voltage of 2.0V. Active area of prototype ADC is 0.396μm^2.

Keywords; Battery Management IC(BMIC), Integrating Sigma-Delta ADC, Single-Sampled Integrator, Double-Sampled Integrator

I. Introduction

As requirements of versatility for mobile devices have been increased in order to incorporate more functions, their computational capacity has been continuously increased as well. Low-power dissipation and efficient power control should be the most stringent issues in hardware implementation to prolong the time of battery usage while satisfying high-performance specifications [1]. Elaborate monitoring of battery status must be one of the most significant operations for efficient power control scheme [2].

ADC is used to monitor battery status such as voltage, temperature and current in the digital signal domain. A high-resolution sigma-delta modulator that is suitable for battery management integrated circuit (BMIC) application is used. A micro-controller unit (MCU) that uses the gauge algorithm can express battery usage and remaining capacity with digital data converted by ADC. For accurate sensing with digital calibration, an integrating sigma-delta ADC is used to monitor the current flowing through the battery.

II. Double-Sampling Integrating Sigma-Delta ADC

Fig. 1 shows the second-order integrating sigma-delta ADC block biagram suitable for high-resolution sensing applications [3]. In this paper, the second-order modulator with OSR of 512 is presented to achieve resolution higher than 16 bits. The modulator mainly consists of two integrators with reset scheme for every sample, one quantizer, feedback digital-to-analog coverter (DAC), and simple digital filter made of two accumulators.

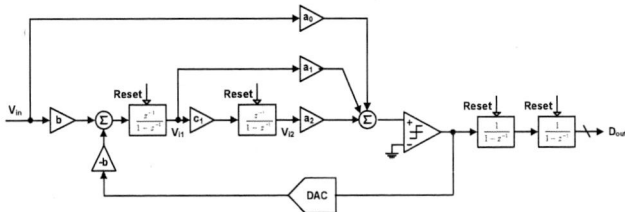

Fig. 1. Block diagram of the second-order integrating sigma-delta ADC.

Fig. 2 shows the circuit schematic of integrators which are included in the integrating sigma-delta ADC. Conventional integrator in Fig. 2(a) samples the input signal at the end of clock ϕ_1, and amplifies during the phase of ϕ_2. In this single-sampling scheme, the operational amplifier doesn't need to be working during the sampling phase of ϕ_1, so it is in idle state while holding the previous integrating output value. One the other hand, the integrator shown in Fig.2 (b) uses double-sampling scheme [3]. The integrator is equipped with double input-sampling branches so that input sampling operation and integrating operation can be processed at every clock of ϕ_1 and ϕ_2. Correspondingly the operational amplifier can be operating for both clock phases and integration can be performed in twice faster clock rate. It makes to possible to reduce the operating time of ADC by half so that average operating current of the entire analog front end can be signicantly reduced.

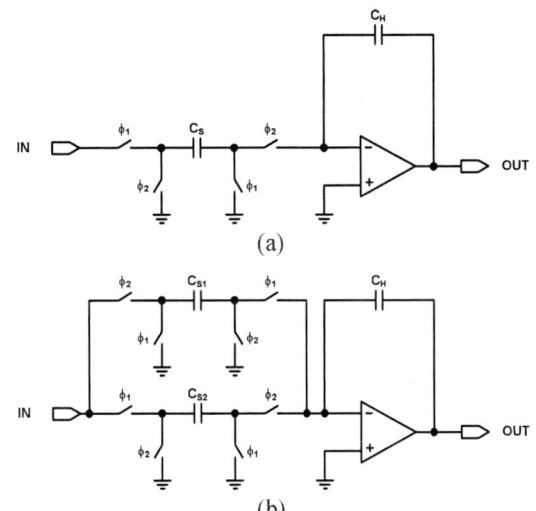

Fig. 2. (a) Single-sampled integrator (b) Double-sampled integrator.

978-1-4799-5128-4/14 $31.00 © 2014 IEEE 165

III. Hardware Implementaion

Block diagram of the proposed integrating sigma-delta ADC is shown in Fig. 3. The analog core is composed of two double-sampled integrators, comparators and relevant bias block. During ϕ_1, the comparator 1 operates and produces quantizer output. Likewise, the comparator 2 does during ϕ_2. The digital block consists of one up/down counter (operating as the 1st-order filter) that receives the modulator output, accumulator (operating as the 2nd-order filter), and a clock generator.

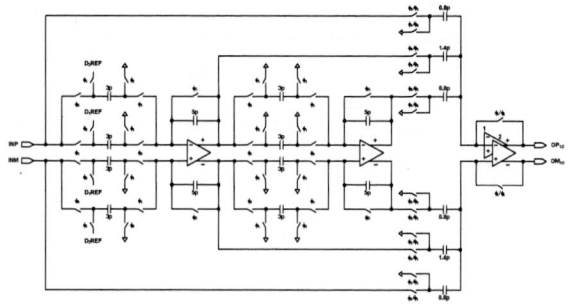

Fig. 3. Block diagram of the proposed ADC.

Simulation results are shown in Fig. 4. The modulator runs at a clock frequency of 32kHz. The single-sampling modulator is also designed for performance comparison. Fig.4 (a) shows the simulation result with single-sampled integrators, and achieved SNDR is about 93dB at 32 sample/s. Fig.4 (b) shows the simulation results of the modulator with double-sampled integrators. It is shown that nearly identical perofmrance can be achieved with doubled sampling rate of 64 sample/s. The layout of the proposed ADC is shown in Fig. 5, where active area is about 0.4mm^2. Table 1 summarizes the performance of the proposed integrating sigma-delta ADC.

IV. Conclusion

The second-order integrating sigma-delta ADC with double-sampled integrators is presented in this paper. By doubling the sampling rate, the operating time of the ADC and corresponding power consumption can be reduced by half. The low-power modulator can be incorporated into BMIC system for mobile applications.

Acknowledgment

This work was supported by the Technology Innovation Program of MOTIE/KEIT. [10041135, Development of Multi Functional Power Management IC for Smart Mobile Devices]

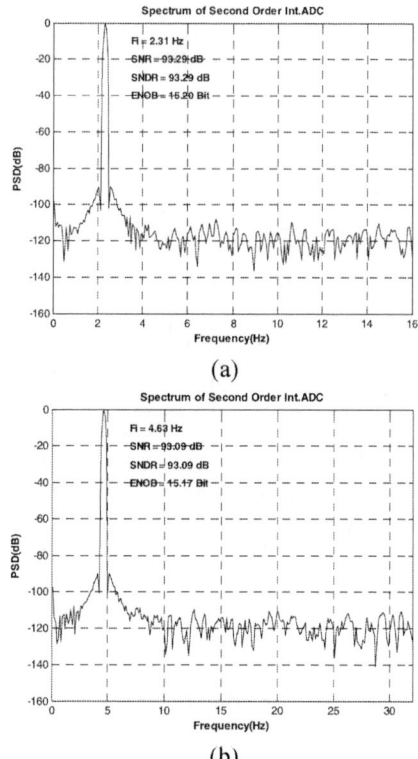

(a)

(b)

Fig. 4. (a) Simulation result with single-sampled integrators (b) Simulation result with double-sampled integrators.

Fig. 5. Layout of the proposed ADC.

Table 1. Summarized of the proposed ADC

Technology	0.18-μm CMOS
Supply Voltage	2.0 V
Clock Frequency	16,384 Hz
Conversion Rate	64 Sample/s
ENOB	15.17 bit
Power Consumption	36 μW
Active Area	550μm X 720μm

References

[1] J. Robert, G. C. Temes, V. Valencic, R. Dessoulavy, and P. Deval, "A 16-bit Low-Voltage CMOS A/D Converter," *IEEE J. Solid-State Circuits*, vol. 22, no. 2, pp. 157-163, Apr. 1987.

[2] J. Márkus, J. Silva, and G. C. Temes, "Theory and Applications of Incremental ΔΣ Converters," *IEEE Trans. Circuit Syst. 1, Reg. Papers*, vol. 51, no. 4, pp. 678-690, Apr. 2004.

[3] M. Kim, G. Ahn, P. K. Hanumolu, S. Lee, S. Kim, S. You, J. Kim, G. C. Temes, and U. Moon, "A 0.9 V 92 dB Double-Sampled Switched-RC Delta-Sigma Audio ADC," *IEEE J. Solid-State Circuits*, vol. 43, no. 5, pp. 1195-1206, May 2008.

The Design of SCR-based Dual Direction ESD Protection Circuit with Low Trigger Voltage

Yong-Nam Choi, Jung-Woo Han, Hyun-Young Kim, Chung-Kwang Lee,
Yong-Seo Koo* (Corresponding Author)
Department of Electronics and Electrical Engineering, Dankook University
152 Jukjeon-ro, Suji-gu
Yongin-si, Gyeonggi-do, Korea
+82-31-8005-3643, supremeyn@gmail.com

Abstract

In this paper, a SCR(Silicon Controlled Rectifier)-based dual directional ESD protection circuit is proposed. The proposed protection circuit can provide effective protection for ICs against ESD (Electrostatic Discharge) in the both positive and negative directions. To analysis its electrical characteristics, it is verified through the TLP(Transmission Line Pulse) system. Also thermal reliability under high temperature conditions (300K-500K) is verified. In the measurement results, it has trigger voltage of 10.16V and holding voltage of 4.64V. It is shown that the proposed protection circuit has superior characteristics compared to the conventional SCR. In addition, these characteristics make the proposed protection circuit has area efficiency. As a result, the proposed dual directional ESD protection circuit can discharge ESD current in four stress mode (PD, ND, PS, NS) with superior characteristics and reliability.

Keywords-ESD; SCR; ggNMOS; Trigger Voltage; Holding Voltage;

Introduction

As the development of semiconductor processing technology has been advanced, damage caused by Electrostatic Discharge (ESD) has become serious problem in reliability of integrated circuits (IC) [1]. Thin gate oxide and shallow junction depth by scaling down process integration makes susceptible to ESD damage. The grounded-gate NMOS (GGNMOS) is the most generally used ESD protection structure because it discharges ESD current effectively through parasitic bipolar junction transistor (BJT) and has compatibility to CMOS process technologies [2]. However, GGNMOS consumes relatively large area due to its low current driving capability and high current density at the surface [3]. For this reason, GGNMOS is drawn with multi-finger layout style to reduce the total occupied silicon area [4]. However, it has been reported that multi-finger GGNMOS can't be uniformly turned on under ESD stress. Therefore expected ESD level couldn't be realized [5].

Silicon Controlled Rectifier (SCR) has been considered as an on-chip ESD protection device for its high robustness and low on-resistance. However, conventional SCR has snapback characteristics in one direction. So, in order to provide dual-directional path, it is needed more area. Also conventional dual-directional SCR has high trigger voltage [6, 7].

In this paper, therefore, to reduce the trigger voltage of the conventional dual-directional SCR, a dual-directional SCR with NMOS structure is proposed.

Proposed ESD Protection Circuit

Fig. 1 shows the cross-sectional view of the proposed ESD protection circuit. The dual-directional ESD protection circuit consists of one lateral NPN transistor (Q1), two vertical PNP transistor (Q2, Q3) and two parasitic resistors (R1, R2). The external resistors RG1 and RG2 are connected between each terminals and gate to protect gate oxide by limiting gate current. Its operation mechanism is as follows. When positive ESD surge is applied to terminal 1 with respect to terminal 2 as a ground in PS mode, the junction between N+ diffusion region and P-well which is base-collector junction of Q1 is reverse-biased. When it goes into avalanche breakdown, large amount of electron-hole pairs (EHP) are generated. Generated excess hole current flows into terminal 2, building up a voltage drop across the parasitic resistor R2, and makes forward bias across the base-emitter junction of Q1. The parasitic NPN Q3-Q1 starts to operate and forms an active low impedance discharge path. It enables ESD current to discharge safely and to clamp the pad voltage to sufficiently low level. When the ESD surge is over, the SCR turns itself off as the current decreased to below its holding current level. Even if NS mode event is occurred, the mechanism of the operation is the same by forming discharge path through the SCR of Q2-Q1 in the opposite direction.

Therefore, the proposed ESD protection circuit provides

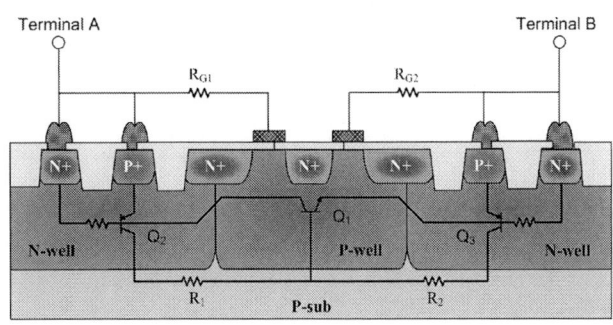

Fig. 1 Proposed dual-directional SCR with NMOS structure.

978-1-4799-5128-4/14 $31.00 © 2014 IEEE

effective protection for ICs against ESD in the two opposite directions.

Measurement Results

TLP measurement result is shown in Fig. 2. Measurement is performed in the two opposite directions. As shown in Fig. 2, the proposed ESD protection circuit presents same snapback I-V characteristics symmetrically. It has 10.16V of trigger voltage and 4.64V of holding voltage.

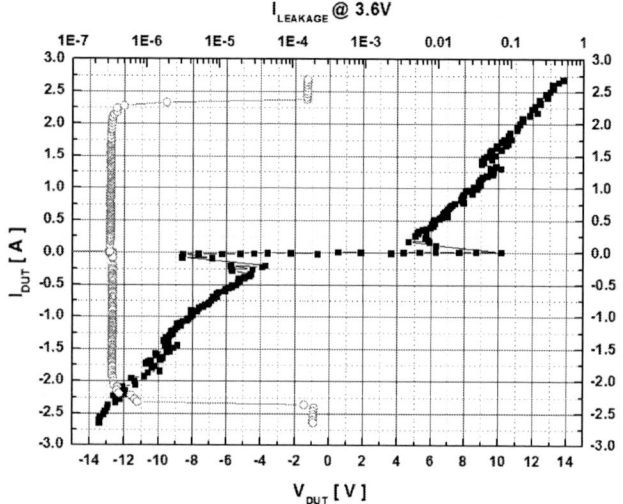

Fig. 2 TLP I-V Characteristics of the proposed dual-directional ESD Protection Circuit

In addition, a high temperature characteristic is measured to verify thermal reliability under high temperature (300-500K). As temperature increases, the holding voltage is reduced. It is because diode saturation current of bipolar transistor increases at high temperature. As a result, voltage across the base-emitter junction is decreased and also holding voltage is decreased

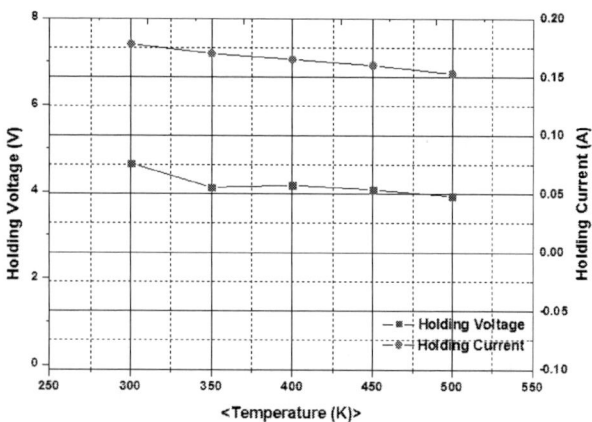

Fig. 3 Measurement of holding voltage and current dependence on temperature (300-500K)

Conclusions

This paper presents a dual-directional SCR-based ESD protection circuit with low trigger voltage. The proposed protection circuit can provide effective protection for ICs against ESD (Electrostatic Discharge) in the both positive and negative directions. As a measurement results, it has 10.16V of trigger voltage and 4.64V of holding voltage. Also, thermal reliability at high temperature is measured. Its holding voltage and current is decreased at high temperature but its difference is slight. Therefore, the proposed ESD protection circuit can be used effectively in a wide range of analog system.

Acknowledgment

This research was supported by the Technology Innovation Program, 10041135,Development of Multi Functional Power Management IC for Smart Mobile Devices funded by the Ministry of Knowledge Economy(MKE, KOREA), the IT R&D program of MKE/KEIT [10035171, Development of High Voltage/Current Power Module and ESD for BLDC Motor] and the Ministry of Science, ICT & Future Planning, Korea, under the University ITRC support program supervised by the National IT Industry Promotion Agency (NIPA-2014-H0301-14-1007)

References

[1] A. Amerasekera and C. Duvvury, "The impact of technology scaling ESD robustness and protection circuit design," in Proc. EOS/ESD Symp., 1994.

[2] R. G. Wagner, J. Soden and C. F. Hawkins "Extend and Cost of EOS/ESD Damage in an IC Manufacturing Process," in Proc. EOS/ESD Symp., pp.49-55, 1993.

[3] K. Bock, et al., "Influence of Gate Length on ESD Performance for Deep Submicron CMOS Technology," in Proc. EOS/ESD Symp., pp.95-104, 1999.

[4] C. Jiang, E. Nowak and M. Manley, "Process and design for ESD robustness in deep submicron CMOS technology", in Proc. IEEE Int. Reliability Physics Symp. (IRPS), 1996, pp. 233-236.

[5] C. Russ, K. Bock, M. Rasras, I. D. Wolf, G. Groeseneken and H. E. Maes, "Non-uniform triggering of gg-nMOSt investigated by combined emission microscopy and transmission line pulsing", in Proc. EOS/ESD Symp., 1998, pp. 177-186.

[6] AZ, H. Wang., "An On-Chip ESD Protection Circuit with Low Trigger Voltage in BiCMOS Technology", IEEE Journal of Solid-State Circuits, Vol.36, No.1, pp.40-45, January 2001.

[7] Zhiwei et al., "An Improved Bi-directional SCR Structure for Low-Triggering ESD Protection Applications", IEEE electron device letters, Digital Library:pp.1-3, January 2008.

A Low Power Time-Domain CMOS Temperature Sensor

Wonjong Song, Hunsik Moon, Hyeyeon Yang and Jinwook Burm

Dept. of Electronics Engineering, Sogang University
121-742, Sinsu-dong, Mapo-gu
Seoul, Korea
burm@sogang.ac.kr

Abstract

The proposed temperature sensor is composed of serial inverter chain and time-to-digital converter(TDC) instead of proportional to absolute temperature(PTAT) and analog-to-digital converter(ADC) for reducing current consumption. The generated temperature dependent delay is compared with temperature insensitive clock. The proposed temperature sensor's temperature resolution is under 0.2°C and draw current is under 800 nA from a 1.8V supply voltage. The simulated temperatre range is 0~100°C and temperature error is –within ±2°C. The simulated results are trimmed by 2-point calibration. The chip is fabricated using a 0.18 µm CMOS process.

Keywords-Temperature sensor, thermal sensor, time-to-digital converter (TDC), 2-point calibration

Introduction

Temperature sensors are used widely in various electronic applications including, especially, smart mobile devices. because battery of mobile device is limited, Temperature sensor is required high resolution low power consumption and small area [1].

This paper proposes a method of measuring temperature using a time-to-digital converter with time delay inverter chain. In comparison to another type of temperature sensors, this method has high resolution and lower power consumption than voltage domain temperature sensors.

Implementation of proposed temperature sensor

Instead of using voltage-domain for measuring temperature by using PTAT and ADC, proposed work is composed of inverter chains for temperature sensing and a TDC, because this sheme has lower power consumption than voltage-domain temperature sensors [2][3].

Fig. 1 descirbes the block diagram of this work. Unlike other time-domain structure, the proposed scheme uses time difference between external reference clock(CLK$_{ref}$) and delayed clock (CLK$_{delay}$) which was influenced temperature.

A. Temperature sensing inverter chain

This inverter chain generates transmission delay. In Fig.1 the temperature sensing block consists of inverter chains. These are placed by pairs which make CLK$_{delay}$ signal, but not inverted after passing the block [4].

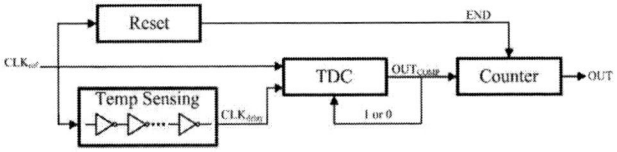

Fig. 1. Block diagram of proposed temperature sensor

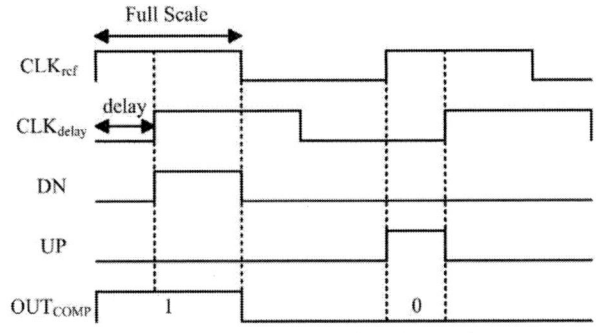

Fig. 2. Timing diagram of proposed internal TDC

In the proposed temeperature sensor CLK$_{delay}$ will be lagged from CLK$_{ref}$ because temperature delay is proportional to number of inverters and this delay will be reflected by the temperature.

Two clocks generated by the inverter chains and the direct input will be inserted to TDC to calculate digital code output.

B. Time-to-Digital Converters

Fig. 2 shows timing diagram of TDC, which describes how TDC works in this scheme. With CLK$_{ref}$ and CLK$_{delay}$ which generated by each block modify TDC, TDC determines DN signals and UP signals which reflected time difference because of temperature. Difference of UP and DN signal's delay is generated by output signal(OUT$_{COMP}$). If output value is low, UP signal is faster than DN signal. Otherwise output value is high, UP signal is slower than DN signal.

Counter determines output digital code based on temperature degree and it stops when END signal is generated by Reset block.

Since a digital code is needed, counter output signal will pass quantizer and decimation filter [5]. The counter works as low-pass filter, hence it integerates the comparator output at the conversion time [6]. Finally, the converted digital code will be displayed by two elements. When the conversion time is reached, the reset block will generate flag signal to start converting the next output to digital code.

978-1-4799-5128-4/14 $31.00 © 2014 IEEE   169

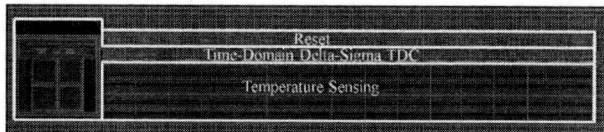

Fig. 3. Layout of proposed temperature sensor

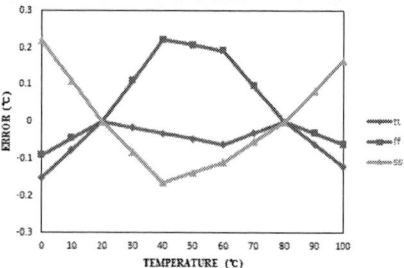

Fig. 4. Simulation results of error code versus temperature

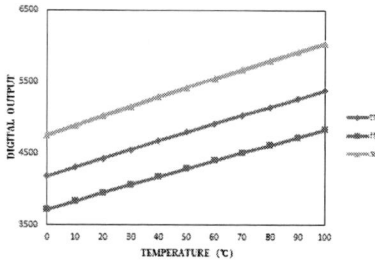

Fig. 5. Simulation results of digital output code versus temperature

Simulation Results

The temperature sensor is fabricated in a 0.18-μm CMOS technology and its area size is 0.089mm². Fig. 3 shows a layout of the proposed temperature sensor.

Fig. 4 shows that the output results of test samples are proportional to temperature. 10 test samples are simulated and performance is measured from 0°C to 100°C. Fig. 5 is the simulation error code after the two-point calibration of 10 sample time. The results showed the deviation from the linearity, which should be compensated to have more accurace code. 20 °C and 80 °C are used for two-point calibration temperature to minimize the measurement errors. The reference curve that was composed of the average slope and the curvature of measured curve results [7].

Conclusion

It is very important for smart mobile devices to sustain their performances in wide-range temperature condition. Because of that reason, low-power, small-size, and high-resolution with low error temperature sensor is needed for mobile device.

This work proposes a low power temperature sensor with time-domain. This temperature sensor configuration has inverter chain and time-to-digital converter instead of PTAT and CTAT because of power consumption. The simulated temperature sensor shows its resolution under 0.2°C, with power consumtion under 800 nA, and an inaccuracy of ±2°C for the temperature variation of 0°C to 100°C after 2-point calibration. The chip area is 0.089 *mm²*. The simulation results of the performance of the sensor is shown at Table I. In

TABLE I
SIMULATION RESULTS OF TEMPERATURE SENSOR

Parameter(Unit)	Value
Resolution (°C)	0.1
Power Consumption (nA)	267
Area (mm²)	0.089
Error (°C)	-0.16~0.22
Temperature Range (°C)	0~100
CMOS Technology (μm)	0.18

conclusion, the proposed temperature sensor shows that this work has low power consumption and small chip area with temperature-invariant temperature measuring performance.

Acknowledgment

This research was supported by R&D program of the Ministry of Knowledge Economy [K10041135, Development of Multi-Functional Power Management IC for Smart Mobile Devices]. This research was supported by the MSIP(Ministry of Science, ICT & Future Planning), Korea, under the C-ITRC(Convergence Information Technology Research Center) support program (NIPA-2013-H0301-13-1007) supervised by the NIPA(National IT Industry Promotion Agency). This research was supported by Basic Science Research Program through the National Research Foundation of Korea(NRF) funded by the Ministry of Education (NRF-2013R1A1A2011943). Also this work was supported by IDEC.

References

[1] Minsoo Kang, and Jinwook Burm, "Time-domain Temperature Sensor using Two Stage Vernier type Time to Digital Converter for Mobile Application," *Int. SoC Design Conference(ISOCC),* pp. 431-434, Nov. 2012.

[2] Young-Jae An, Kyungho Ryu, Dong-Hoon Jung, Seung-Han Woo, and Seong-Ook Jung, "An Energy Efficient Time-Domain Temperature Sensor for Low-Power On-Chip Thermal Management," . *IEEE Sensors Jornal,* vol. 9, no. 1, pp. 104-110, Jan. 2014

[3] Michiel A. P. Pertijs, Kofi, A. A. Makinwa, and Johan H. Huijsing, "A CMOS Smart Temperature Sensor With a 3σ Inaccuracy of 0.1°C From -55°C to 125°C," *IEEE Jornal of Solid-State Circuits,* vol. 40, no.12, pp. 2805-2815, Dec. 2005.

[4] Poki Chen, Chun-Chi chen, Yu-Han Peng, Kai-Ming Wang, and Yu-Shin Wang, "A Time-Domain SAR Smart Temperature Sensor With Curvature Compensation and a 3σ Inaccuracy of -0.4°C~+0.6°C Over a 0°C to 90°C Range," *IEEE Jornal of Solid-State Circuits,* vol. 45, no.3, pp. 600-609, Mar. 2010.

[5] Omar Fathy, Ahmed Abdallah, Amr Wassal, and Yehea Ismail, "Counter Based CMOS Temperature Sensor for Low Frequency Applications," *IEEE Int. Thermal Issues in Emerging Technologies Theory and Applications(ThETA),* pp. 103-109, Dec. 2010.

[6] Anton Bakker, and Johan H. Huijsing, "Micropower CMOS Temperature Sensor with Digital Output," *IEEE Jornal of Solid-State Circuits, ,* vol. 31, no.7, pp.933-937, Jul. 1996.

[7] Poki Chen, Shou-Chih Chen, You-Sheng Shen, and You-Jyun Peng, "All-digital time-domain smart temperature sensor with an inter-batch inaccuracy of −0.7°C − +0.6°C after one-point calibration," *IEEE Trans. Circuits Syst.* vol. 58, no. 5, pp. 913-920, May. 2011.

978-1-4799-5128-4/14 $31.00 © 2014 IEEE

2D-to-3D Conversion Using Color and Edge

Manbae Kim

Dept. of Computer and Communications Eng., IT College, Kangwon National University
192-1, Hyoja2dong, Chunchon, Gangwondo, Republic of Korea, 200-701
+82-33-250-6395, manbae@kangwon.ac.kr

Abstract

The 2D-to-3D conversion technique plays an important role in the development and promotion of three-dimensional television (3DTV), because it can easily provide 3D program contents. This paper focuses on the real-time autmatic 3D conversion method. Depth map is made by the efficient combination of luminance, color, visual saliency, caption localization and final surface modeling. Then the 2D image is converted to a stereoscopic one by DIBR (Depth Image Based Rendering). The proposed method has been performed on diverse 2D video clips and it has been verified that the 3D perception is stable and acceptable.

Keywords-component; 2D-to-3D Conversion; saliency; edge; depth map;

Introduction

Following the commercial success of 3DTV, 2D-to-3D conversion has been incorporated into 3DTV, 3D displays, 3D cameras and so forth. Even though most conversion algorithms do not always deliver perfect 3D such as 3D movies, the 3D quality has gained much attraction of 3D viewers. Fundamentally, the conversion methods extract a depth map from a single 2D image.

Basically, generation of 3D video from monoscopic 2D video input source have been investigated for many years [1-6]. Most of them are based on an estimated depth map of each frame and then use DIBR (Depth Image Based Rendering) to synthesize steroscopic views.

To derive a depth from a given image, diverse methods have been proposed according to the principle of human visual system including depth from motion, e.g. optical flow, depth from defocus, e.g. inverse filtering, depth from geometrical perspective, e.g. vanishing line detection and gradient plane assignment and depth from shading, etc. Recently, some researchers have used saliency to support the creation of a depth map. The usage of the saliency requires a careful manipulation due to the inherent property of the saliency. Each algorithm has its own strengths and weakness. Most depth extraction algorithms make use of multiple depth cues from combining two or more depth cues to generate depth map. The aforementioned methods are expected to work well for images that are suitable to each associated algorithm. In other words, if the depth cues are not sufficient, the algorithms might fail, producing uncomfortable 3D perception. This observation results in the design of the global real-time conversion methods.

The proposed method is composed of four main components; (1) visual saliency estimation, (2) affinity model and binomic filter, (3) edge detection, and (4) depth generation. The overall block diagram is pictorized in Fig. 1.

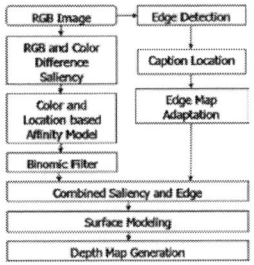

Fig. 1 The overall block diagram of the proposed method

This paper is organized as follows. In the next section, saliency will be introduced followed by affinity model, binomic filter, edge detection, caption localization, and depth map generation. The experimental results are provided and finally, we summarize our work.

Saliency Map Generation

The saliency map is a core component of the proposed method. The important fact is that only saliency containing distance information is useful. Otherwise it is difficult to utilize the saliency for depth map. One of the saliency estimation methods is as follows:

$$
\begin{aligned}
S_r &= |R - (G+B)/2| \\
S_g &= |(R+G)/2 - |R-G| - B| \\
S_b &= |B - (R+G)/2|
\end{aligned}
\tag{1}
$$

In other method, the mean values of R_i, G_i and B_i are calculated for an RGB image to obtain R_{ave}, G_{ave} and B_{ave}. The saliency in each pixel is then calculated by (1).

$$
S_{RGB} = \sqrt{(R_{ave} - R_i)^2 + (G_{ave} - G_i)^2 + (B_{ave} - B_i)^2}
\tag{2}
$$

The weighted combination of the two saliencies produces the saliency maps as in Fig. 2.

Fig. 2 Examples of saliency map

978-1-4799-5128-4/14 $31.00 © 2014 IEEE

Affinity Model

The pixels with same RGB have identical saliency values. Therefore, two neighboring pixels with different colors, but belonging to the same object might have different saliency values that hampers the consistent depth. To solve this problem, we employ an affinity model in order to alleviate the boundary discontinuities at the identical object region.

In the usage of saliency data, defining the affinity model gained by local saliency values at the boundary is important. As mentioned, a single object can be represented by multiple different saliency data, resulting in different depth data. The affinity model used in the segmentation can solve the aforementioned problem. Close-by pixels with similar saliency values likely belong to the same segment. The color-based affinity model is defined by

$$\psi_p(i,j) = \exp(-\frac{\| x_i - x_j \|}{\sigma_p}), \quad \psi_s(i,j) = \exp(-\frac{\| s_i - s_j \|}{\sigma_s}) \quad (3)$$

where x_i, s_i denote the position and saliency values of pixel i, respectively. σ_p, σ_s control the weights of the two factors.

The combined model is used to design a better affinity model. Two models can simply be combined with a parameter α for the combined affinity model ψ_m as follows:

$$\psi_m = \alpha\psi_p + (1-\alpha)\psi_s \quad (4)$$

Binomic Filter

Since the saliency uses a color, different saliency values are spread over an identical object (e.g., human). To alleviate such problem, one of effective methods is a binomic filter, where its elements are binomic numbers which are created as a sum of the corresponding two numbers in Pascal's triangle. We expand this filter to an image as follows; Given N x N pixel block B(i,j), then we convolve B(i,j) with its scaled B(i,j).

$$B(i,j) = B(i,j) \otimes B_s(i,j) \quad (5)$$

where B_s is the intensity-scaled block. The scale factor s varies at [0, 1]. The larger it is, the more the output is saturated.

Edge Map

Edge plays an important role in the procedure in terms of boundary correction as well as the refinement of saliency data. Let I(i,j) be the image intensity at (i,j). Then the gradient vectors are computed by convolving the image block with edge convolution mask. The edge magnitude is strong on the high illumination and weak on the low illumination due to the property of the gradient. To solve this, Robinson's edge detection [7] utilizes a temporal local threshold using masking method when a pixel is decided as edge or non-edge.

Depth Map Generation

The final depth generation being composed of saliency, edge, detected caption and surface modeling is shown in Fig.3.

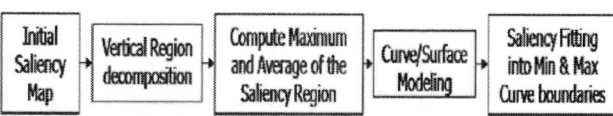

Fig. 3 Depth map generation procedure

Experimental Results

The proposed method has been tested on diverse video clips. The image caption shows stable 3D, which is one of merits. 3D stereoscopic images made by this method show satisfactory 3D perception as well as stable stereo. Fig. 4 shows test images as well as their associated depth maps.

(a)   (b)

Fig. 4 RGB test images and depth maps.

Acknowledgment

This work was supported by the MKE (The Ministry of Knowledge Economy)/KEIT [10041082, System and Semiconductor Application Promotion Project].

References

[1] W. Tam and L. Zhang, "3D-TV Content Generation: 2D-To-3D Conversion," Proc. of IEEE ICME, 2006.

[2] M. Kim and et al., "Stereoscopic conversion of monoscopic video by the transformation of vertical-to-horizontal disparity," SPIE, vol. 3295, pp. 65-75, Jan. 1990.

[3] M. Kim, S. Park, and Y. Cho, "Object-Based Stereoscopic Conversion of MPEG-4 Encoded Data," LNCS, vol. 3, pp. 491-498, Dec. 2004.

[4] A. Kim, A. Baik, Y. Jung and D. Park, "2D-to-3D image/video conversion by using visual attention analysis," ICIP, 2009.

[5] I. Ideses, L. Yaroslavsky, B. Fishbain, "Real-time 2D to 3D video conversion," Journal of Real-Time Image Processing, vol. 2(1), pp. 2-9, 2007.

[6] F. Xu, G. Fr, X. Xie, and Q. Dai, "2D-to-3D Conversion Based on Motion and Color Mergence," 3DTV Conference: The True Vision-Capture, Transmission and Display of 3D Video, pp. 205-208, May 2008.

[7] G. Robinson, "Edge detection by compass gradient masks," Computer Graphics and Image Processing, Vol. 6, pp. 492-501, 1977.

Simulation and Analysis of Desulfator for Smart Battery System

Ka Lok Man[1], Eng Gee Lim[2], Mark Leach[3], Jin Kyung Lee[4], Kyung Ki Kim[5]

[1]Department of Computer Science and Software Engineering, [2,3]Department of Electrical and Electronic Engineering,
[1,2,3]Xi'an Jiaotong-Liverpool University, Suzhou, China

[4]Department of Electronic Engineering, [5]College of Information and Communication Engineering, School of Electronic and Electrical
Engineering, [4,5]Daegu University, Gyeongsan, South Korea

[1]ka.man@xjtlu.edu.cn, [2]enggee.lim@xjtlu.edu.cn, [3]Mark.Leach@xjtlu.edu.cn, [4]jklee@live.daegu.ac.kr, [5]kkkim@daegu.ac.kr

Abstract

Lead-acid batteries play a crucial role in our daily life. However, their efficiency deteriorates due to the accumulation of chemical substance, known as sulfate (SO^2) on the surface of batteries' plates. Hereby, the methods to overcome this problem is investigated in this work. A workable desulfator found online is adopted in this study. Further, a new and more efficient circuitry is proposed. The advantages and disadvantages of the new circuit is also outlined.

Keywords-component; batteries, battery management system, electrochemical processes, pulse circuits

Introduction

The advent of the electricity is one of the greatest achievements in human history and has tremendously improve the life of mankind. Accompanying with the discovery of electricity, Lead-Acid (LA) batteries were invented to store this new form of energy. The LA batteries opened the door for the rapid development of industries, the birth of the electric vehicles and the modern life of human beings. Although the LA batteries have been developed by the accumulation of new technologies and the discovery of new materials for over 150 years, the short lifetime, severe pollutions and small capacity of the LA batteries still limit its applications. Nowadays, the revolutionary developed valve regulated lead-acid (VRLA) batteries is widespread and used in transport vehicles, information technologies and energy storage systems[1] because of its stable output and light pollution. However, the life span of the VRLA batteries is also short due to the sulfate accumulation problem.

Although the compositions of the chemical electrolyte and the structure has been improved, the working principle and the major reasons causing the battery failures remains unchanged [3]. A previous report written by Jiramoree et al.[4] mentioned that the sulfation which was caused by the sulfate crystal on the electrodes of the batteries contributed over 80% of all LA battery failures. An early literature [3] reported that the sulfation could give rise to corrosion of the battery electrodes, increasing the internal resistances of batteries, and also reducing the effective area of electrodes. These effects would result in a large amount of heat during the charging process and the attenuation of the chemical reaction between electrolyte and active materials, thereby influenced the capacity and lifetime of the LA batteries [5]. Therefore, it is essential to investigate the root cause of sulfation and finding an effective method to overcome this problem.

Working principle of lead acid battery

The working principle is essentially the chemical reaction shown:
$$PbO_2 + 2H_2SO_4 + Pb \rightleftharpoons 2PbSO_4 + 2H_2O \qquad (1)$$

When the battery is in discharging state, the active materials of positive electrode, lead dioxide, will be converted into lead sulfate. Similarly, the negative electrode which is made up of lead will be converted into lead sulfate as well. The electrolyte, sulfuric acid, will be consumed by reacting with the active materials [2]. The electrode reaction equations of positive electrode and negative electrode in discharging state are demonstrated below.

$$
\begin{aligned}
Positive: &\quad PO_2 + 4H^+ + 2e^- \rightleftharpoons Pb^{2+} + 2H_2O \\
Negative: &\quad Pb \rightleftharpoons Pb^{2+} + 2e^-
\end{aligned}
\qquad (2)
$$

When the battery is in the charge state, the lead sulfate will be converted back to the active materials.

$$
\begin{aligned}
Positive: &\quad PbSO_4 + 2H_2O \rightleftharpoons PbO_2 + HSO_4^- + 3H^+ + 2e^- \\
Negative: &\quad PbSO_4 + H^+ + 2e^- \rightleftharpoons Pb + HSO_4^-
\end{aligned}
\qquad (3)
$$

Theoretically, a LA battery can be used for a very long time, if the chemical reaction equation as shown in the Eq. (1) can completely proceed. It is shown in Eq. (1) that the chemical reaction proceeding in the LA batteries is kind of reversible, as a result, the lead sulfate will gradually crystallize and deposit on the electrodes. This phenomenon is known as the sulfation. This usually leads to a short lifetime of the LA batteries [3].

A. Charging Strategies

Several charging strategies have been investigated. The most common charging method is found to be the constant trickle current (CTC) method [6]. The cost of CTC charger is very low, however, the CTC charger has a significant drawback as it requires a long time to charge the batteries due to the small current imposed in the charging process. A new strategy called constant current (CC) method is used to charge the batteries with a higher current compared to CTC [6]. Although the charging time could be reduced by high charging current, it may cause overcharge and damage the battery. In addition to CC and CTC methods, the constant voltage (CV) is also widely used [7]. It uses a constant voltage which is higher than the electromotive force of the battery to charge the batteries.

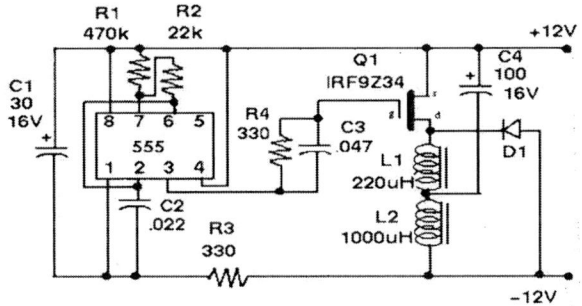

Fig. 1. Desulfator for a 12 V LA battery [9].

978-1-4799-5128-4/14 $31.00 © 2014 IEEE

Methodology

A typical desulfator is simulated by using both software and real hardware implementation.

A. Circuit Analysis and Simulation

Fig. 1 shows the configuration of the typical desulfator used to rejuvenate a 12 V LA batteries. The first part consists of a 555 timer circuitry that generates a pulse with relevant frequency and duty cycle in order to control the transistor Q1. The second part is the speed-up capacitor which is used to enhance the switching speed of the transistor. The third part is the current pulse generator. When the transistor is turned down by the input pulse from the timer, the current will be drawn from the battery and stored in the capacitor. The diode D1 is used to prevent the current flowing through the inductor L1 when the transistor is turned off. Then, the energy stored in the capacitor will be transferred into the inductor L1 at the time when Q1 is turned on. Once this transistor is turned off again, the energy stored in the inductor will be transferred back to the battery as a high current pulse. This high current pulse will be able to dissolved the sulfated the batteries. For diode D1, a fast recovery diode with model G1826CT is used in the circuit. The values of other components are available in the circuit diagram of Fig. 1.

Before testing is carried out, the value of the peak current frequency pulsing back to the battery is calculated at 1240 Hz. Using an oscilloscope, the peak current measured was 6.5 A. The peak value is further verified through circuit simulation of Fig. 1 in MultiSim, depicting a peak value of 6.54. However, in the case of severely sulfated battery, this current magnitude may not be efficient to remove the sulfate quickly. As such, hereby we propose some improvements to the typical desulfator.

B. Amplification of the voltage

The suitable voltage range for desulfation purposes could be 60 V-300 V. However, in this work, we set this voltage peak to 40 V based on the ability to produce an appropriate duty cycle from calculation. Also, a lower level of voltage is safer and this prevents damage caused by overvoltage during desulfation process. Voltage by the differential amplifier and a negative signal will be feed into PWM, reducing its duty cycle. Thus the gain of the boost circuit will be reduced. As such, the output voltage will be re-adjusted to 40 V.

C. Output capacitor

The output of the boost circuit is the voltage of the capacitor, which is connected in parallel. The ripple voltage of the output capacitor is caused by the ESR (equivalent series resistance) of the capacitor. The magnitude of the ripple voltage is a function of the output voltage, output current and the capacitance of the capacitor [8]. As mentioned, the capacitance of the output capacitor is proportional to the turn on period of transistor and this turn on duration is determined by the duty cycle and frequency of the controlling input pulse. Therefore, when the ripple voltage was fixed, the capacitance of the output capacitor was a function of frequency, duty cycle, output current, output voltage and the ripple voltage.

D. MOSFET and Fast Switching Diode

The chosen MOSFET has a maximum drain current which is higher than the peak current flowing through it. In addition, the turn on voltage V_{DS} of the transistor is small enough to cut off the diode when the transistor is turned on. Finally, the MOSFET with model number IRF540 is adopted. Similarly, the fast switching plastic rectifier diode chosen is GI826 after considering the peak current.

The maximum peak current of the GI826 is 300 A, with a reverse breakdown voltage of 600 V.

Results and calculations

For simulation, input voltage of 18 V is set with a load of 15 kΩ. The peak of the voltage pulse was recorded at 40.338 V, which is very close to the expected result. From the results above, it is evident that the output of the designed circuit is close to the pre-specified output. Also, this improved desulfator is able to perform charging and desulfation process simultaneous with higher peak current. This means that the improved circuit has greater efficiency than its predecessor. The error due to the deviation in the frequency is 5.5%. This deviation may be caused by the low accuracy of the 555 timer, impedance, and turn-on voltage of the added diode.

Conclusions

In this work, the background information and working principle of the lead-acid batteries are discussed, accompanied by a careful survey of the previous works concerning battery desulfator. A typical desulfator is adopted and the drawback analysis is outlined. Further, we propose an improved version of battery desulfator circuitry. The circuit design and the relevant parameter calculations are validated through real hardware simulation. In reality, the improved version is able to achieve better efficiency in removing sulfate chemicals on the surface of lead-acid plates.

For further research on hardware implementation, an electric double layer capacitor could be added into the improved desulfator to stabilize its output. A state of charge detector of the batteries will be expected to avoid the overcharge which can shrink the lifetime of the batteries.

Acknowledgment

This work is supported by Xi'an Jiaotong-Liverpool University under RDF-13-01-13.

References

[1] Detchko. Pavlov, *LeadeAcid Batteries: Science and Technology*, First edition. Amsterdam: Elsevier, 2011.

[2] Z. M. Zhou and A. H. Ji, *The Repair and recycling of LA Battery*. Beijing: Posts and Telecom Press, 2010 (in Chinese).

[3] Schilling, S. *Ensuring lead-acid battery performance with pulse technology*. in Battery Conference on Applications and Advances, 1999. The Fourteenth Annual. 1999.

[4] Jiramoree, T., P. Paisuwanna, and S. Khomfoi. *A multilevel converter charger utilizing superimposed pulse frequency method for prolonging lead-acid battery lifetime*. in Electrical Engineering/Electronics, Computer, Telecommunications and Information Technology (ECTI-CON), 2011 8th International Conference on. 2011.

[5] Yan, Z., et al. *A high current pulse activator for the prolongation of Lead-acid batteries*. in Vehicle Power and Propulsion Conference, 2008. VPPC '08. IEEE. 2008.

[6] Liang-Rui, C., *A Design of an Optimal Battery Pulse Charge System by Frequency-Varied Technique*. Industrial Electronics, IEEE Transactions on, 2007. 54(1): p. 398-405.

[7] Li, S., C. Zhang, and S. Xie. *Research on Fast Charge Method for Lead-Acid Electric Vehicle Batteries*. in Intelligent Systems and Applications, 2009. ISA 2009. International Workshop on. 2009.

[8] Abraham I. Pressman et al., *Switching Power Supply Design*, Third Edition. New York, McGraw-Hill Companies, 2011.

[9] Lead Acid Battery Desulfator. Available from: http://www.alton-moore.net/graphics/desulfator.pdf

[10] Liang, T.J., et al. *Implementation of a regenerative pulse charger using hybrid buck-boost converter*. in Power Electronics and Drive Systems, 2001. Proceedings., 2001 4th IEEE International Conference on. 2001.

Design of An Arduino-based Smart Car

Zhao Wang, Eng Gee Lim, Weiwei Wang, Mark Leach, Ka Lok Man.

Xi'an Jiaotong-Liverpool University, Suzhou, China

Enggee.lim@xjtlu.edu.cn

Abstract

Remote-controlled cars are one of the most popular toy products currently on the mass market. Each series of car has a specific remote-control unit. This presents the consumer with a critical problem; obtaining a substitution controller where the original control unit has broken down. In this work a robot based on an external Arduino microcontroller, controllable by an Android application via Bluetooth, which can be recommended as a prototype for the combination of embedded systems with Android mobile devices is investigated.

Keywords- Remote-controlled; smart car; Android.

Introduction

Android smartphones command 81% of the global market share despite their relatively short history. In comparison to the iPhone operating system (iOS) and Windows phones which occupy 12.9% and 3.6% of the overall market respectively [1]. With the rapidly growing number of smartphone owners, millions of Android applications (Android apps) have been developed aimed at improving every aspect of life, including, home automation systems, daily arrangements, mobile databases and entertainment systems. As a consequence of this functional integration in portable devices, there is a crucial need for human presence in digital and vertical environments. One possibility is the fusion of embedded systems and manual operation to control external devices.

Arduino is designed as an open-source electronics prototyping platform providing schematics and flexible development kits for enthusiastic users who intend to produce interactive objects or environments. Arduino can be used to sense surroundings by utilising various transducers to read and interpret inputs in order to make responses for example through the controlling of motors or transferring of data [2].

Since 2010, Android has been regarded as the most widely-used operating system across the whole spectrum of PC's, handsets and other terminal devices ahead of iPhone, Black Berry and other Operating Systems. It is an open-source mobile platform developed by Google mainly for touchscreen mobile devices with direct operations that are like actions in the real world such as tapping, sliding and shaking. These direct manipulations are attributed to internal components including proximity sensors, accelerometers, gravitational sensors and gyroscopes which may be suitable for user input. For example, the suspension of the alarm clock can be manipulated by shifting the phones orientation [3]. The source

code, released under the Apache License, permits any enthusiast developers' free modification of the software on the Android platform. Therefore, a large number of applications have been written to extend the functionality of touchscreen mobile devices. Software alterations are based on the Java programming language. In addition, Android operating systems are commonly described as a low-cost, pre-existing, light and customizable choice [4]. As seen in Figure 1, the modified Linux kernel is the essential basis for the Android operating system. On top of the Linux kernel, application software runs on a virtual machine, application programming interface (API), middleware and libraries written in C, C++, Java and XML are contained in this software stack [5].

The Bluetooth core system contains an RF transceiver, baseband and protocol stack and provides services allowing the connection of various devices and the exchange of different data classes [6]. Bluetooth products can be divided into slave and master, which are able to actively initiate and negotiate with other Bluetooth modules. They have various effective ranges in different situations due to the influence of the environment, material coverage, battery's power and antenna configuration. The range is lower than the theoretical distance if propagation is indoors due to attenuation caused by signal reflections. For a variety of classes of device and maximum power, their specified ranges are distinct. The general distances for Class 3, 2 and 1 radios are up to 1, 10 and 100 meters. Mobile devices based on class 2 radios while class 3 radios are mostly found in industrial environments [7].

Smart Car Design

As A smart mobile device and an embedded system side are involved in this project. The overall framework should contain a user, an Android smartphone field, an Arduino-based car namely the embedded system field and a personal computer as shown in Fig. 1.

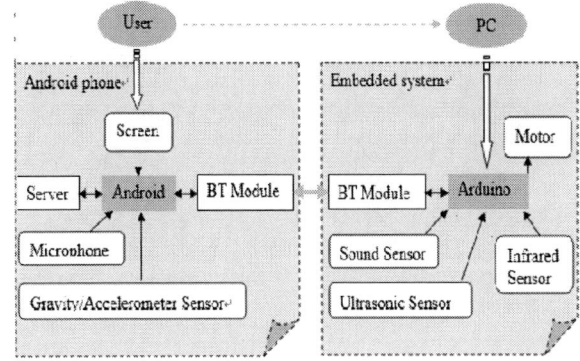

Fig. 1. Overall design of this project

978-1-4799-5128-4/14 $31.00 © 2014 IEEE

With assistance of the Arduino integrated development environment (IDE) in the PC, sketches are compiled and uploaded into the Arduino board via a USB transmission line. The car and mobile phone are linked via wireless communication. By touching or pressing on the screen or user interface (UI) of an Android phone, a manipulator can send commands to the Arduino microcontroller on the car through Bluetooth and observe the corresponding executions accomplished by actuators, for example motors.

Two gear motors, two wheels, a universal wheel, a battery holder, two 18650 lithium batteries, a switch and two baseboards compose the chassis of the car as shown in Fig. 2.

Fig. 2. Hardware realization process

Results

Two modes have been designed in this paper. The first mode is wireless control and the second mode is for obstacle avoidance. When the car is operated in mode 1, the only method for controlling the car is by operation from the smartphone via Bluetooth communication. Four functions are achieved here. The fundamental ones are forward, left, right and reverse movements as well as a stop action based on the touching of arrows or sliding a white ball as shown in Fig. 3.

Another form of control is assisted by the gravitational sensor which refers to the accelerometer sensor built-in to the Android smartphone. It is possible to display, store, show and clear measurements sent by Arduino microcontroller. A voice control interface has also been included based on Google Voice Search, it can handle receiving data and sending some instruction messages.

In mode 2, the car keeps going forward until an obstacle appears within a defined threshold distance. After exploring the barrier, it will stop and detect distances in the front-left and front-right directions, to determine which offers greater clearance and then proceeds in that direction.

Conclusion

The proposed design is simple and low-profile. Control and monitoring of many embedded systems can be achieved by combining mobile devices, which provide general manipulators with an easy and user-friendly interaction with embedded systems in industry and domestic areas. One of the crucial features of products is accessibility. Uncomplicated operations and compact user interfaces are preferred. Connection within a system composed of embedded devices and mobile applications lends itself to wireless communication. The outcome of this paper is a small car that is powered by an external Arduino microcontroller with a motor shield and controllable by an Android application via Bluetooth and can be recommended as a prototype remote-control system following a successful trial of this concept.

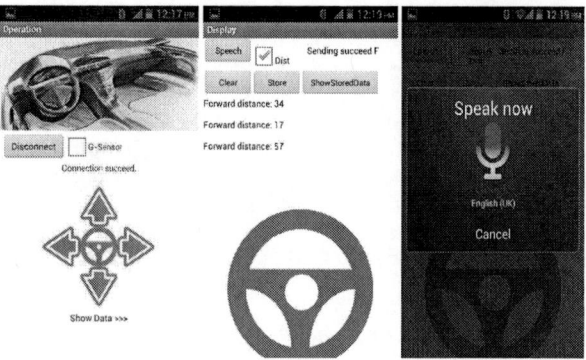

Fig. 3. Android application

Acknowledgment

This work is partially supported by the Natural Science Foundation of Jiangsu province (BK20131183) and Suzhou Science and Technology Bureau (SYG201211).

References

[1] Data sheet *"Top Four Operating Systems, Shipments, and Market Share, Q3 2013 (Units in Millions)"*. International Data Corporation (IDC), Available from: http://www.idc.com/getdoc.jsp?containerId =prUS24442013, 2013.

[2] Arduino. *What Arduino can do*. Available from: http://arduino.cc/ , 2014

[3] Arduino, *"Arduino Motor Shield"*. Available from: http://arduino.cc/en/Main/ArduinoMotorShieldR3, 2014.

[4] J.Brodkin. *"On its 5th birthday, 5 things we love about Android"*. Available from: http://arstechnica.com/gadgets/2012/11/on-androids-5th-birthday-5-things-we-love-about-android/, 2013.

[5] M. Butler. (2011) Android: Changing the Mobile Landscape. IEEE Xplore. Available from: http://ieeexplore.ieee.org/stamp/stamp.jsp? tp=&arnumber=5676144&tag=1 , 2014.

[6] Anon. Bluetooth SIG. *"How Bluetooth Technology Works"*. Available from: http://web.archive.org/web/20080117000828/ http://bluetooth.com/Bluetooth/Technology/Works/, 2013.

[7] Anon. Bluetooth SIG. *"A Look at the Basics of Bluetooth Wireless Technology"*. Available from: http://www.bluetooth.com/Pages/ Basics.aspx , 2013

The New Wrappable Wireless Capsule Antennas

Eng Gee Lim, Zhao Wang, Tianqi Xia, Mark Leach, Ka Lok Man.

Xi'an Jiaotong-Liverpool University, Suzhou, China

zhao.wang@xjtlu.edu.cn

Abstract

A general introduction to wireless capsule endoscopy (WCE) and WCE antennas are firstly presented. The most important part of this paper lies in the investigation of the inductively loaded wrappable patch antenna. The original design is 18.5 mm in width and 31 mm in the length. The substrate material is LCP and the antenna radiates at 433 MHZ. In total 39 pairs of notches are averagely subtracted from the patch. Software-based simulation and optimization is performed in accordance with Rogers Duroid 5870 as a substrate material. The 5870-based study model is focused on theoretically exploring the properties of the structure. Additionally, some innovative structures are designed, including a log-periodic structure and miniaturised antenna for WCE applications for children.

Keywords- Wireless Capsule Endoscopy; patch antenna; miniaturised.

Introduction

The concept of wireless capsule endoscopy was first introduced at the beginning of the 21st century, offering a new way to diagnose gastrointestinal (GI) track diseases, including cancer and ulcers [1]. This new technology can help surgeons to visualize the entire GI track in a non-invasive way and thus overcome the drawbacks of traditional examination instruments, for example, gastroscopy and colonoscopy, which are normally torturous and cannot detect small intestine problems [2]. In 2000, Given Image, an Israel company, released their first WCE product: M2A [3]. The device was 30 mm in length and had a radius of 5.5 mm; it could be easily ingested by the patient and was able to continuously deliver more than 50000 pictures over the following 7-hour period. In the next decade an increasing number of institutions started to investigate and promote their own capsule endoscopy systems: Olympus Inc., Japan and Jinshan Science and Technology, China. Their products received positive feedback from clinical tests and played an important role in capturing the attention of the international market [4].

Owing to their low profile and huge flexibility, microstrip antennas are widely adopted in the design of wireless capsule endoscopy systems. Many innovative structures have been proposed by senior researchers, like conformal planar meandered dipole antenna or even "waffle-type" antenna based on genetic algorithms [5-7]. As one of the major parts of this paper, a software-based experiment is implemented on one novel WCE antenna design: the inductive loaded wrappable patch antenna [8]. The antenna model is investigated and simulated using a different substrate material:

the Rogers Duroid 5870 laminate. The resonance point and return loss are recorded and analysed. Various characteristics of the structure are studied.

Antenna Design

As the core part of this paper, an omnidirectional wrappable compact patch antenna proposed by [8] is investigated as shown in Fig.2.

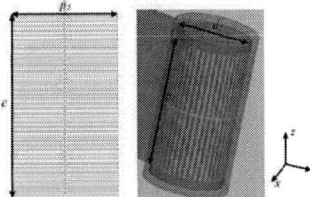

Fig. 2. Wrappable compact patch antenna [8]

The design comprises a compact microstrip antenna with equally divided inductive notch structure. The patch is fabricated on a flexible liquid crystalline polymer (LCP) substrate and then wrapped into a cylindrical shape to fit the size requirement of WCE applications. The operating frequency is targeted around the 433 MHZ Industrial, Scientific and medical (ISM) band while an omnidirectional radiation pattern is exhibited. [8]

This paper extensively investigates the characteristics of this structure then proposes feasible miniaturisation strategies, while maintaining optimal antenna properties within the appropriate frequency band.

Simulation

A. Log periodic structure

The initial design pertains to the "fish-bone" antenna or Yagi-Uda antenna to a certain extent. This expedites a reasonable assumption that the log-periodic framework is probably a promising orientation for this design (as shown in Fig. 4.).

The antenna has a -21.3 dB return loss at 321 MHz. Compared with the original structure, the log periodic design has a significant frequency-shift effect. Miniaturisation is also achieved as the modified new design occupies merely 1/3 of the space. Further investigation by decreasing the number of notches as well as reducing the width of the patch, as shown in Fig. 5 and 6, is also carried out.

978-1-4799-5128-4/14 $31.00 © 2014 IEEE

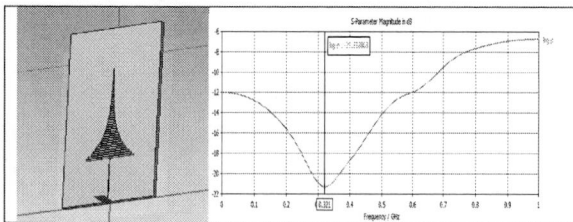

Fig. 4. log-periodic model and simulation result

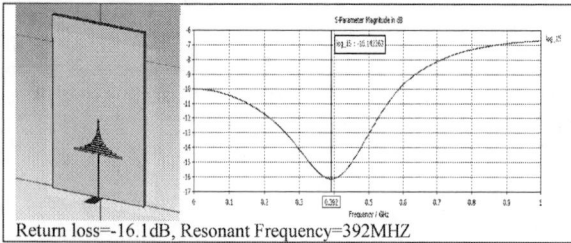

Return loss=-16.1dB, Resonant Frequency=392MHZ

Fig. 5. New design 1

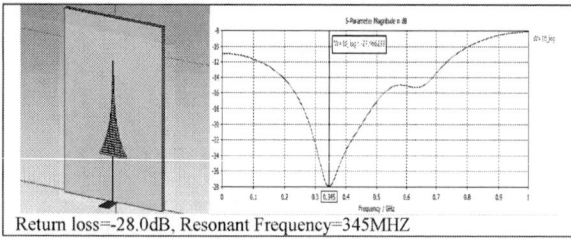

Return loss=-28.0dB, Resonant Frequency=345MHZ

Fig. 6. New design 2

B. Miniaturized design

The optimal structure of the antenna is found to be 8.3 mm in diameter, which fits the size of the smallest WCE on the market. Further miniaturisation can be achieved by decreasing the notch number. The new design, with 30 pairs of notches (as shown in fig. 7.) could possibly be adopted for children's WCE applications, which is another new research hot spot in this area.

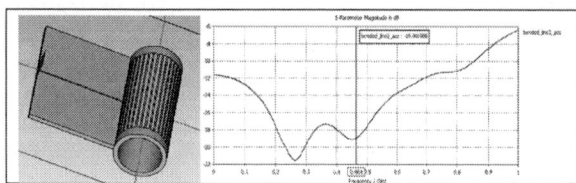

Fig. 7. Miniaturized design and the simulation result

With 30 pairs of notches remaining, the miniaturized antenna can radiate at 464 MHz with an acceptable return loss (-19dB).

Conclusion

A general understanding on the frontiers of WCE and WCE antennas are developed thorough the literature reviews. One inductively loaded WCE patch antenna is simulated based on referenced paper [8] but using different substrate material - Rogers Duroid 5870. In the case of Duroid 5870, the frequency shift effect on the folded antenna is demonstrated and miniaturization is achieved. The recommended number of notches is 30, depending on the specific requirement of the design. The shortened patch length also implies the diameter of the folded antenna is reduced. For the 30-pair structure, the diameter can be controlled within 8.5 mm, which fits the smallest WCE capsule size on the market. The optimised antennas can be even considered for children WCE applications.

Acknowledgment

This work is partially supported by the Natural Science Foundation of Jiangsu province (BK20131183) and Suzhou Science and Technology Bureau (SYG201211).

References

[1] D. Adler, C. Gostout, "Wireless Capsule Endoscopy", *Hospital Physician*, pp. 14-22, May 2003.

[2] Z. Wang, E. Lim, T. Tillo and F. Yu, "Review of the Wireless Capsule Transmitting and Receiving Antennas, *Wireless Communications and Networks - Recent Advances*", Dr. Ali Eksim (Ed.), ISBN: 978-953-51-0189-5, InTech, Available from: http://www.intechopen.com/books/wirelesscommunications-

[3] G. Pan and L. Wang, "Swallowable Wireless Capsule Endoscopy Progress and Technical Challenges", *Gastroenterology Research and Practice*, vol. 2012, pp. 1-9, 2011.

[4] G. Ciuti, A. Menciassi and P. Dario, "Capsule Endoscopy: From Current Achievements to Open Challenges", *IEEE Reviews in Biomedical Engineering*, vol. 4, pp. 59-72, 2011.

[5] I. Bahl, P. Bhartia and S. Stuchly, "Design of Microstrip Antenna Covered with a Dielectric Layer", *IEEE transactions on antennas and propagation*, vol. ap-30, no. 2, march, 1982

[6] P. Soontornpipit, C.M. Furse and Y.C. Chung, "Design of Implantable Microstrip Antenna for Communication With Medical Implants," *IEEE Transactions on Microwave theory and techniques*, vol. 52, no. 8, pp. 1944-1951, August 2004.

[7] V. Shirvante, F. Todeschini, X. Cheng and Y. Yoon, "Compact Spiral Antennas for MICS Band Wireless Endoscope Toward Pediatric Applications", *Antennas and Propagation Society International Symposium*, July, 2010.

[8] X. Cheng, J. Wu, E. David and Y. Yoon, "An Omnidirectional Wrappable Compact Patch Antenna for Wireless Endoscope Applications", *IEEE Antennas and Wireless Propagation Letters*, vol.11, pp. 1667-1670, 2012.

The Design of a Smart Power Conversion System as an Undergraduate Cross-Discipline Integrated Design Project

Chi-Un Lei[1], Christopher H.T. Lee[1], T.O. Kwan[1], C.K. Lee[1], K.B. Huang[1], R.Y.K. Kwok[1], K.L. Man[2]

[1] Department of Electrical and Electronic Engineering, University of Hong Kong, Hong Kong
[2] Department of Computer Science and Software Engineering, Xi'an Jiaotong-Liverpool University, China P.R.C
Email: culei@eee.hku.hk

Abstract

The paper describes the design of a smart power conversion system as a project vehicle in a cross-discipline integrated design project (IDP) course. This project allows senior undergraduate students to consolidate their technical knowledge and design skills of feedback control systems, energy storage systems, big data analysis systems and high-power electrical systems. Since students learn experientially, the project is also useful to motivate students towards adapting electrical system design as well as provide solid learning evidence to their future employers.

Keywords-design project; cross-discipline learning; project vehicle; project-based learning; energy conversion; data sciences

I. Introduction

Recently, emerging student learning technologies and process have been introduced to reshape the scope of engineering education [1–5]. In particular, in order to assess students' competence comprehensively, various project-based learning curricula have been revamped. In our department, we have also reformulated the cross-discipline integrated design project (IDP) course. We hope that through the revamped IDP, senior undergraduate students in teams can have an opportunity to apply and integrate their knowledge and skills in practices, to implement a practical electrical/electronic system. In order to assist students learn effectively and solidly, project vehicle (i.e. the product to be designed) has to be designed carefully.

In this paper, we discuss how a smart energy converter can be used as a project vehicle in our IDP course. Contributions of our paper are as follows:

- Pedagogical requirements and technical requirements of the project vehicle have been described.
- Design of the smart power conversion system has been outlined, based on an intelligence hierarchy.

II. Requirements of the Project Vehicle

According to Kolb's learning cycle [6], for learning to take place in experiential learning courses (e.g. design courses), the project vehicle should be designed carefully. In particular, the vehicle should be designed, such that the following events can be accomplished: i) concrete experiencing (i.e. working actively, instead of simply watching or reading), ii) reflective observation (i.e. observe and reflect what has been done and experienced), iii) abstract conceptualization (i.e. frame the observation), and iv) active experimentation (i.e. put what they have learnt into practice).

Meanwhile, in order to help students equip product design skills, the project vehicle should help students master the following contents: i) design principles of an integrated system, ii) design techniques of electronic systems, iii) use of computer-aided design tools and equipment for building electronic systems, iv) integrated knowledge and skills from different electrical and electronic engineering disciplines (i.e., computer engineering, electronic engineering and electrical engineering), and v) techniques of problem solving and project management in a mixed design team. Thus, the project vehicle should have a clear design purpose, functionality and hierarchy, such that students are assisted to conceive, design and implement an electronic system. In the course, we define the project vehicle should contain the following modules:

- circuits that interacts with physical quantities, including high-power quantities
- feedback control systems that integrate actuators, sensors and microcontrollers for processing
- decision support systems that turn external and internal data into actionable knowledge for the system

III. Design of the Smart Power Conversion System

Based on requirements described in Section II, a smart energy conversion system for temperature regulations in smart laboratories [3] has been proposed as the project vehicle. Besides circuits for sensing and actuating, feedback control and decision support have been added to enhance the functionality of the system. The block diagram of the system is shown in Fig. 1. Through designing the system, students can explore the following topics:

- Control of thermoelectrical cooling systems
- Power storage and regulations
- Feedback control via open-source hardware
- Decision support via big data analysis

A. Circuit Level

In order to regulate the environment, a solid-state thermoelectric heat pump (i.e. a peltier device) is used here to transfer heat from air in metal ventilation ducts to the ambient. Two blower fans are placed at the opening of ventilation ducts, in order to bring air out from the ducts. The current/power supplied to electrical devices (i.e. the system cooling strength) is controlled by a pulse-width modulation power driver.

For safety reasons, the "high-voltage" input power source is

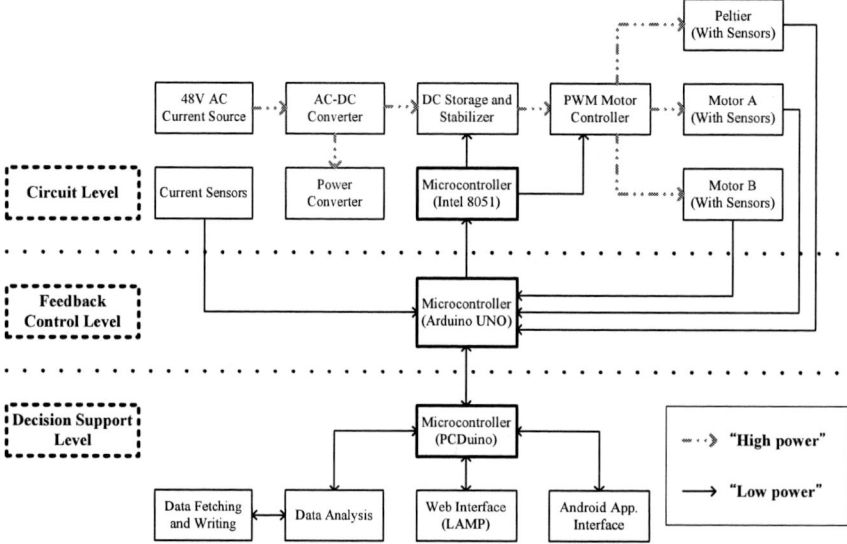

Fig. 1 Block diagram of the proposed energy conversion system.

replaced by a 48V AC current source. Furthermore, an array of large capacitors is used as a temporarily power boosting device. In normal situations, supplied power is not enough for the peltier and blower fans to perform a full-load operation, but can charge up capacitors one by one. However, when the temperature has to be lowered in a short period of time, charged capacitors can be discharged to provide an extra power for fans and the peltier, in order to perform a full-load operation temporarily (e.g. 15 seconds). Operation control of capacitors, fans and peltier are handled by an Intel 8051 microcontroller, through a battery management system configuration.

B. Feedback Control Level

An open-source microcontroller board Arduino is used for providing feedback to circuits and establishing a closed-loop feedback control. In the system, sensors are connected to the board for measuring temperatures, currents and other physical quantities. Based on the measurement, feedback control signals (e.g. desired fan speed and delivery power of the peltier) are sent from the Arduino board to inform the Intel MCS-51 microcontroller board. Multi-thread control and external interrupt control are used with teacher/community guidance.

C. Decision Support Level

An advanced microcontroller board PCduino is used to analyze data and make high-level control judgments. Data can be collected from a variety of online sources (e.g. weather information, user schedules). Furthermore, the system should have an Android application and a web server interface for notifications and manual control. Through these interfaces, analytic charts and construct exploratory summaries of data can be shown to users in support of manual control decisions.

In order to control over the network, a LAMP (Linux + Apache + MySQL + PHP) web server has to be constructed through the PCduino and its in-built WIFI module.

Furthermore, in the project, PCduino is required to communicate with other Arduino broads through the USB interface or with other hardware (e.g. infra-red remote controller) via other input/output interfaces.

IV. Conclusions

In this paper, we have shown how a smart energy conversion system (a smart cooler for smart laboratories) can be used as a project vehicle for an integrated design project. We hope that through designing the conversion system, students are better equipped to learn other advanced EEE materials and practically design electronic systems for their future career development.

Acknowledgement

This research is partially supported by the Research Development Fund (RDF-13-01-13) from Xi'an Jiaotong-Liverpool University, China.

References

[1] N. Paulino, J.P. Oliveira, R. Santos-Tavares, "The design of an audio power amplifier as a class project for undergraduate students," in Proc. IEEE ISCAS, pp.2565,2568, May 2013

[2] C.-U. Lei, H. K.-H. So, E. Y. Lam, K. K.-Y. Wong, R. Y.-K. Kwok, and C. K. Chan, "Teaching introductory electrical engineering: a project-based learning experience," in Proc. IEEE TALE, Aug. 2012

[3] C.U. Lei, K.L. Man, H.-N. Liang, E.G. Lim, K. Wan, "Building an Intelligent Laboratory Environment via a Cyber-Physical System," International Journal of Distributed Sensor Networks, vol. 2013, Article ID 109014, 9 pages, 2013

[4] C.-U. Lei, "Teaching Introductory Circuits and Systems: Enhancing Learning Experience via Iterative Design Process and Pre-/Post-Project Learning Activities," in Proc. IEEE ISCAS, pp.2413-2416, Jun 2014.

[5] C.-U. Lei, N. Wong, K.L. Man, "Integration of a Wireless Sensor Network Project for Introductory Circuits and Systems Learning," in Proc. IEEE ISCAS, pp. 2569-2572, May 2013.

[6] David A. Kolb, "Experiential learning: Experience as the source of learning and development." Englewood Cliffs, NJ: Prentice-Hall, 1984.

TAB-model for Multilevel Diagnosis and Repair of HDL SoC

[1]Vladimir Hahanov, [2]Ka Lok Man, [3]Baghdadi Ammar Awni Abbas,
[4]Eugenia Litvinova, [4]Svetlana Chumachenko, [5]Jihyeok Ahn, [6]Kyung Ki Kim

[1,4]National University of Radioelectronics, Kharkov, Ukraine
[2]Xi'an Jiaotong-Liverpool University, Suzhou, China
[3]Baghdad University, Iran
[5,6]Daegu University, Gyeongsan, South Korea

[1]hahanov@kture.kharkov.ua, [2]ka.man@xjtlu.edu.cn, [5]ajh3166@live.daegu.ac.kr, [6]kkkim@daegu.ac.kr

Abstract

This paper describes technology for diagnosis SoC HDL-models, based on transaction graph. Diagnosis method is focused on decreasing the time of fault detection and memory for storage of diagnosis matrix by means of forming ternary relations between test, monitor, and functional component. A method for analyzing the activation matrix to detect the faulty blocks with given depth and synthesis logic functions for subsequent embedded hardware fault diagnosis is given.

Keywords-component; diagnosis; faulty blocks detection; HDL SoC model; transaction graph.

Faulty SoC components TAB-Model diagnosis

The main objective is the realization of TAB-matrix model (Tests – Assertions – Blocks functional model) and diagnosis methods to reduce the time of testing and memory for storage by means of forming ternary relations (test – monitor – functional component) within one table.

The challenges involve:

1) Development of digital system HDL-model in the form of a transaction graph for diagnosing functional blocks by using assertion set [1-6,15];

2) Development method for analyzing TAB-matrix to detect minimal set of fault blocks [4-7,13];

3) Synthesis of logic functions for embedded fault diagnosis procedure [8-11,14].

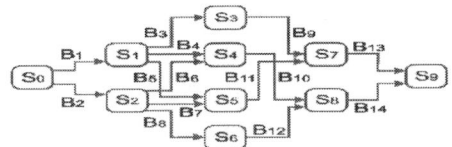

Fig. 1   Example of ABC-graph of HDL-code

The ABC-graph model of HDL-code describes both the software structure, and test segments of the functional coverage, generated using software blocks, incoming to the given node. Collectively, all nodes have to be full state coverage space of software variables, which determines the test quality, equal to 100%: $Q = \text{card} \bigcup_{i=1}^{m} S_i^r / \text{card} \bigcup_{i=1}^{m} S_i^p = 1$. Furthermore, the assertion set $<A, S>$ that exists in the graph, allows monitoring arcs (code-coverage) $B = (B_1, B_2, ..., B_i, ..., B_n)$ and nodes (functional coverage) $S = \{S_1, S_2, ..., S_i, ..., S_m\}$.

The ABC-graph makes the following possible: 1) to minimize the costs for generating tests, diagnosing and correcting the functional failures by using assertions; 2) to estimate the software quality via diagnosability design; 3) to optimize test synthesis via coverage all arcs and nodes by a minimum set of activated test paths.

Detecting faulty functional blocks is based on xor-operation between the real assertion response vector and TAB-matrix columns

$$A^* \oplus [M_1(B_1) \vee M_2(B_2) \vee ... \vee M_j(B_j) \vee ... \vee M_n(B_n)].$$

Where monitors $M_j = B_j(T_j, A_j)$. The features mean: Every row of the matrix is a subset of the Cartesian product between test and monitor, all matrix rows are unique – eliminating the test redundancy, and the number of matrix rows must be greater than the binary logarithm of the number of columns that determines the potential diagnosability of every block. The diagnosis function of every block depends on the complete test and monitors, which must be minimized without diagnosability reduction.

Design for diagnosability

Diagnosability is the relationship $D = N_d / N$ between the recognized faulty blocks amount N_d, (when there are not equivalent components, or the diagnosis depth is equal to 1), and the total number N of HDL-blocks.

For the expense E evaluation of the TAB-matrix model for detecting functional failures, it can use the pair test-assertions efficiency for a given diagnosis depth. Criterion E functionally depends on the relation between the ideal $]\log_2 N[\times N$ and real $|T| \times |A| \times N$ memory sizes or resources (where $|T|$ – the test length, $|A|$ – a number of assertions).

The diagnosis quality criterion of the ABC-graph takes the form:

$$Q = E \times D = \frac{]\log_2 N[}{|T| \times |A|} \times \frac{N - N_n}{N}.$$

The expression above produces practical rules for synthesis of diagnosable HDL-code: 1) Test must create a minimal number of single activation paths, and cover all the nodes and arcs in the ABC-graph. 2) Base number of monitors equals the end node number of the graph with no outgoing arcs. 3) Additional monitors can be placed on each non end node. 4) Parallel independent code blocks must have n monitors and a single concurrent test, or one integrated monitor and n serial tests. 5) Serially connected blocks have one activation test for serial path and n-1 monitor, or n tests and n monitors. 6) The

978-1-4799-5128-4/14 $31.00 © 2014 IEEE

graph nodes, which have more than 1 number of input and output arcs, create good conditions for the diagnosability of the current section by single path activation tests without installation additional monitors. 7) The test pattern or testbench has to be 100% functional coverage for the nodes of the ABC-graph. 8) Diagnosis quality criterion as a function depending on the graph structure, test and assertion monitors can always be increased close to the 1-value. For this purpose there are two alternative ways. The first one is increasing test segments by activating new paths for recognition equivalent faulty blocks without increasing assertions, if the software graph structure allows the potential links. The second way is adding assertion monitors on transit nodes of the graph.

Multilevel diagnosis method of digital system

Method for faulty blocks diagnosis Hardware-Software HS-system, based on multi-tree model, allows creating the universal engine in form of algorithm (Fig. 2, block 6) for traversal of tree branches on the depth, specified a priory:

$$B_j^{rs} \oplus A^{rs} = \begin{cases} 0 \to \{B_j^{r+1,s}, R\}; \\ 1 \to \{B_{j+1}^{rs}, T\}. \end{cases}$$

Here $A_i^{rs} = m_i^{rs} \oplus g_i^{rs}, i = \overline{1, k_{rs}}$. If all coordinates of vector xor-sum $B_j^{rs} \oplus A^{rs} = 0$, then one of the following action is performed: the transition to the activation matrix of the lower level $B_j^{r+1,s}$ or repair of the functional block $B = B_j^{rs}$.

One of two analysis ways is executed, what is the most important: 1) the time (t>m, block 10) – then repair of faulty block is performed; 2) the money (t<m) – then a transition down to specify a more exact fault location, because replacement of smaller block decreases the repair cost. So, the TAB-engine has four end-nodes, where one of them is B-good which indicates successful finishing of the testing, and the other three means the intermediate results in the test process, which is necessary to take into account for the increasing a test quality and diagnosis depth by using extra assertions and/or additional test segments generation.

High performance and technological combination of

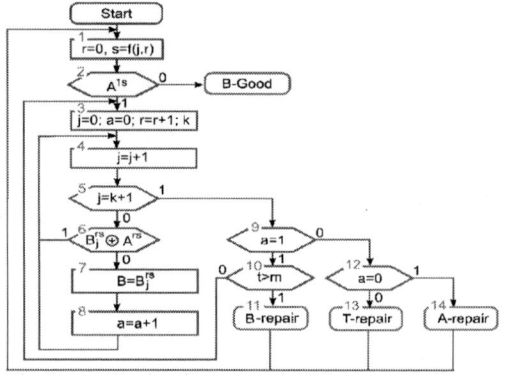

Fig. 2   Engine for traversal of diagnosis multitree

assertion analysis system and HDL-simulator of Aldec Company is largely achieved through integration with the internal simulator components, including HDL-language compilers. Processing the results of the assertion analysis system is provided by a set of visual tools of the Aldec Riviera

environment to facilitate the diagnosis and removal of functional failures.

The assertion analysis model can also be implemented in hardware with certain constraints on a subset of the supported language structures. Products of Riviera including the components of assertion temporal verification, which allow improved design quality for 3-5%, currently, occupies a leading position in the world IT market with the number of system installations of 5,000 a year in 200 companies and universities in more than 20 countries.

Conclusion

The Proposed transactional graph model and method for diagnosis of digital systems-on-chips are focused to considerably reducing the time of faulty blocks detection and memory for storing the diagnosis compact matrix describing ternary relations in format: the monitor-oriented test-segments which detect faulty functional components of the Hardware-Software system. A model for diagnosing the functionality of system-on-chip in the form of multi-tree and tree traversal, implemented in the engine for detecting faulty blocks with given depth, are developed. These considerably increases the performance of software and hardware Infrastructure.

References

[1] P.P. Parhomenko, "Technical diagnosis basics" Moscow: Energy, 1976.

[2] P.P.Parhomenko, and E.S. Sogomonyan, "Technical diagnosis basics (Optimization of diagnosis algorithms, hardware tools)", Moscow: Energy, 1981.

[3] M.F. Bondarenko, O.A. Guz, V.I. Hahanov, and Yu.P. Shabanov-Kushnarenko, "Infrastructure for brain-like computing", Kharkov: Novoye Slovo, 2010.

[4] V.I. Hahanov, I.V. Hahanova, E.I. Litvinova, and O.A. Guz, "Design and Verification of digital systems on chips", Kharkov: Novoye Slovo, 2010.

[5] V.V. Semenets, I.V. Hahanova, and V.I. Hahanov, "Design of digital systems by using VHDL language", Kharkov: KHNURE, 2003.

[6] V.I. Hahanov, and I.V. Hahanova, "VHDL+Verilog = synthesis for minutes", Kharkov: KHNURE, 2006.

[7] IEEE Standard for Reduced-Pin and Enhanced-Functionality Test Access Port and Boundary-Scan Architecture IEEE Std 1149.7-2009.

[8] F. Da Silva, T. McLaurin, and T. Waayers, "The Core Test Wrapper Handbook. Rationale and Application of IEEE Std. 1500™", Springer, 2006, XXIX.

[9] E.J. Marinissen, and Yervant Zorian, "Guest Editors' Introduction: The Status of IEEE Std 1500', IEEE Design & Test of Computers, No26(1), pp.6-7, 2009.

[10] A. Benso, S. Di Carlo, P. Prinetto, and Y. Zorian, "IEEE Standard 1500 Compliance Verification for Embedded Cores", IEEE Trans. VLSI Syst., No 16(4), pp. 397-407, 2008.

[11] V.I. Hahanov, E.I. Litvinova, S.V. Chumachenko, and O.A. Guz, "Logic associative computer", Electronic simulation, No 1, pp.73-83, 2011.

[12] Ngene Christopher Umerah, Hahanov V. A diagnostic model for detecting functional violation in HDL-code of SoC // Proc. of IEEE East-West Design and Test Symposium.– Sevastopol, Ukraine.– 19-20 September, pp. 299-302, 2011.

[13] Ubar R., Kostin S., Raik J. Block-Level Fault Model-Free Debug and Diagnosis in Digital Systems. DSD '09. 12th Euromicro Conference 2009, pp. 229 – 232.

[14] Benabboud Y., Bosio A., Girard P., Pravossoudovitch S., Virazel A., Bouzaida L., Izaute I. "A case study on logic diagnosis for System-on-Chip,"Quality of Electronic Design, 2009, pp.253,259.

[15] Datta K., Das P.P. Assertion based verification using HDVL. Proceedings 17th International Conference VLSI Design. 2004, pp. 319 – 325.

A Review on System Level Low Power Techniques

qtong@hawk.iit.edu

jdcho@skku.edu

Abstract

With the shrinking integrated circuits (IC) technology and the popularity of portable electronic devices, power dissipation has become a critical issue in very large scale integrated (VLSI) circuits design. To reduce the power dissipation, much attention had been drawn to the techniques in process technology, circuits level, gates level and register-transfer (RT) level. However, from the experience with the traditional approach, it is clear that system level design optimization have the greatest influence on power consumption. And hence, power optimization techniques in system level are able to gain considerable power reduction, compared with techniques in other level. In this review, some general state-of-art system-level low power techniques are tested and reviewed, and some guidelines for system level low power design are generalized.

Keywords-Low power; Very Large Scale Integrated (VLSI) Circuits; system level; electronic devices

Introduction

It is tempting to suppose that only hardware dissipates power, not software. However, that would be analogous to postulating that only automobiles burn gasoline, not people. In microprocessor, micro-controller, and digital signal processor based systems, it is software that directs much of the activity of the hardware. Consequently, the software can have a substantial impact on the power dissipation of a system[1].

For complex project, like h.264 encoding, we usually prefer implementing it in system level using high level programming languages like C/C++, Java or Python. This is because high level programming languages are more abstract and describe an event in a more human friendly way. However, when we are writing high level programming language, we are prone to ignoring the power efficiency of the our code in the sense of hardware. In this review, some general state-of-art system-level low power techniques are tested and reviewed, and some guidelines for system level low power design are generalized.

In a computer system, the CPU is the core and most active part. Each instruction will be processed by CPU. So, to optimize the power consumption in system/software level, we have to take the CPU architecture and processing mechanism into consideration.

One of the major contributors to power dissipation of computer system that can be influenced significantly by software is the memory system. Two reasons account for this phenomenon, one is that the memory system is usually an external device which need additional effort to access, unlike the cache and registers in the CPU; The other one is that the memory system has hug load capacitance, one access may occur large charging or discharging current. The memory system takes of a substantial fraction of the power budget (on the order of one tenth to one fourth) for portable computers[2] and it can be the dominant source of power dissipation in some memory intensive DSP applications such as video processing[3]. So, we should be aware when we are implementing system level design.

Branching is also a troublesome problem in computer system. The modern CPUs are all implemented in a pipeline architecture. Usually, the execution of instruction is divided into 5 stages: Instruction fetch (IF), Instruction decoding(ID), Execution (EX), Memory access (MEM) and Write-back (WB). (Note, many newer CPU break some of these stages apart into even smaller stages. However the 5-stage model is the basic and general form, which is enough for explaining the CPU working mechanism.) The 5-stage model makes the pipeline structure feasible. In a pipelining CPU, one instruction can start executing even before the previous instruction ends. As a result, the code sequence can be executed faster. The pipeline structure is efficient to execute instruction sequence without branching. When the code sequence includes branching instructions, the pipeline structure CPU needs to stall several cycles and flush the buffer and load new data. To improve the capability of dealing branching instructions, the modern CPU also integrate hardware for branch prediction. It is good to use branch predictor since it usually has a high probability to make the right prediction and keep the pipeline processing rolling with less stalls. But there are also odds that the branch predictor may make wrong predictions, in which case a mis-prediction penalty will be applied[4].

Although researchers have noticed that different instructions have different power dissipation in same CPU[5]. And the difference might vary dramatically. However, in [6], the authors found that shortest code sequence was invariably the lowest energy code sequence for a variety of microprocessor and DSP devices. In no case was the lower average power dissipation of a slightly longer code sequence enough to overcome the overhead energy costs associated with the extra execution cycles[1].

Testing Platform

In this report, all the experiments are operated on a software simulator named WATTCH[7].

WATTCH is an architectural simulator that estimates CPU

power consumption. The power estimation is based on a suite of parametrizable power models for different hardware structures and on per-cycle resource usage counts generated through cycle level simulation. The power models have been integrated into the SimpleScalar architectural simulator tool kit.

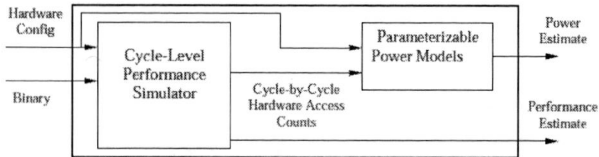

Fig. 1 Overall Structure of the WATTCH

The SimpleScalar[8] tool set performs fast, flexible, and accurate simulation of modern processors that implement the SimpleScalar architecture (a close derivative of the MIPS architecture). It is used to collect results by WATTCH. SimpleScalar provides a simulation environment for modern out-of-order processors with 5-stage pipelines: fetch, decode, issue, writeback and commit. The power oriented modifications provided by WATTCH (whose modules are integrated within SimpleScalar) track which units are accessed on each cycle and compute the power values associated with those units accordingly. These power modules have been verified against industrial circuits and have been found to be within 10% for low level power estimation.

System level power reduction techniques

A. Memory access reduction

There are many ways to reduce the memory access in high level programming language.

(1). Pointer chain substitution

Pointer is widely used in C/C++ to construct complex data structure or perform efficient argument passing. Pointer is good to use, but it also very expensive in terms of power. Since each time referring a pointer, memory access is committed.

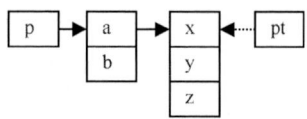

Fig. 1 Pointer chain illustration

Figure 1 shows how the pointers are used to access the member variable 'x' in a compound data structure. We can see that when we are using the pointer chain, we need to access the memory once for each stage of the pointer chain. One more pointer reference need one more memory access operation, which is time and power consuming. In this case, we can create a temporary pointer 'pt' which points to the variable 'x'. if 'x' is frequently accessed, the latter way would have a lot power saving. In our experiment, using a temporary pointer substituting a pointer chain has 28.3965% power saving.

	before	after

Using a temporary pointer to point the target data member instead of using pointer chain	struct user_str1
	{ int x;
	      int y;
	      int z;
	};
	
	struct user_str2
	{
	      struct user_str1* p_obj1;
	      struct user_str1* p_obj2;
	};
	
	int main()
	{
	      int i =0;
	      struct user_str1 objx1, objx2;
	      struct user_str2 objx;
	      struct user_str2* p_objx = &objx;
	
	      objx.p_obj1 = &objx1;
	      objx.p_obj2 = &objx2;
	      for(i=0; i<100000; i++){
	      p_objx->p_obj1->x = i;
	      p_objx->p_obj1->y = i;
	      p_objx->p_obj1->z = i;
	      }
	      return 0;
	}
		struct user_str1
		{ int x;
		      int y;
		      int z;
		};
		
		struct user_str2
		{
		      struct user_str1* p_obj1;
		      struct user_str1* p_obj2;
		};
		
		int main()
		{
		      int i =0;
		      struct user_str1 objx1, objx2;
		      struct user_str2 objx;
		      struct user_str1* temp;
		
		      struct user_str2* p_objx = &objx;
		
		      objx.p_obj1 = &objx1;
		      objx.p_obj2 = &objx2;
		
		      temp = p_objx->p_obj1;
		
		      for(i=0; i<100000; i++){
		      temp->x = i;
		      temp->y = i;
		      temp->z = i;
		      }
		      return 0;
		}

Avg. power per instruction	46.0893	38.1948
Total instruction executed	2206039	1906042
Total power	101690912.2505	72814245.3569
Power saving	N/A	28.3965%

(2). Function cost

Function is a very useful in programming. However, it is also power-consuming. There are two reasons that make the function call expensive. First, the before entering the function, many historical data in register file need to be pushed into stack, and after the function returns, these data need to be pop out; second, the argument passing is also need to access the memory. A nicely designed function with power aware consideration can save a lot power. Here are a few techniques to optimize the function.

a) Declare a function as MACRO or 'inline' in C/C++

As ways to reduce power consumption, the MACRO and 'inline' work in the same way. They are used in function definition. They tell the compiler not to compile the code section as 'function', instead the target code section will be just deployed at compilation time. At the execution stage, the code sections are not executed as function, so there is no function call cost.

In our testing, declaring a function as inline, the total instruction executed decreases a lot. And the average power reduce by 3.6887 percent.

b) Argument passing

Function argument passing has two ways in C/C++ language: pass-by-value and pass-by-reference.

Pass-by- reference is much more efficient than pass-by-value if the arguments are very large. Especially for

978-1-4799-5128-4/14 $31.00 © 2014 IEEE

some complex structure. Pass-by-value makes a copy of each arguments, which needs to access memory. This copying process makes pass-by-value both time consuming and power consuming. Pass-by-value is more efficient, since it only pass the address to the parameters.

In our testing, change the function with 2 24-byte structure arguments from pass-by-value to pass-by-reference. We observe 21.2793% power saving.

	before	after

Argument passing	struct node
	{
	    int x;
	    int y;
	    double value;
	    double key;
	};
	
	int test1(struct node node1, struct node node2){
	    if(node1.x == node2.x && node1.y == node2.y) {
	    node1.value = node2.value;
	    node1.key = node2.key;
	    }
	}
		struct node
		{
		    int x;
		    int y;
		    double value;
		    double key;
		};
		
		int test1(struct node* node1, struct node* node2){
		    if(node1->x == node2->x && node1->y == node2->y)
		{
		    node1->value = node2->value;
		    node1->key = node2->key;
		    }
		}

Avg. power per instruction	45.2772	46.4114
Total instruction executed	60051	46051
Total power	2735604.4463	2153487.2440
Power saving	N/A	21.2793%

B. Branching reduction

Modern CPUs all have branch prediction module which intends to accelerate the execution speed. With the benefit of branch prediction scheme, the modern CPU also suffers from mis-prediction penalty when the CPU predict the branch wrongly. The mis-prediction penalty varies with different design, but it usually takes much longer to recover from mis-prediction state.

C. Loop unrolling and combining

One easiest way to reduce the branching instruction is loop unrolling. Most loop need to judge the condition in every cycle. The condition judgment involves branching instructions. The idea of loop unrolling is removing the condition judgment part. In our testing, the testing program calls a function 5 times. We use implement it in two ways: with and without loop. The results show that the loop unrolling method has 14.0975% power saving.

Another way is we can combine two small loops together. In this manner, the share the same condition judgment statement. this way can also reduce the branch prediction. In our testing, this way can save power as much as 35.1535%.

D. Hardware preferables

In all CPUs, there is at least one ALU which commits the arithmetic operation. Usually it is a fixed size. When we are writing program in high level, it is a good habit to define the operands as a data type that fits the ALU size if we are aware of the power consumption. For example, in 32 bit machine, the 'int' type is 32-bit long in most compiler. It is preferable data type. In our testing, we define a function to operate on 'int', 'short', 'char' types respectively, the results shows that the function with 'int' type operand has 15.944% power reduction. Comparison with 0 is more efficient than with a non-0. Most CPU has a designated instruction to commit comparing an operand with 0. So it is more efficient to do comparison with 0.

In our testing, we execute 9 iteration loops in two styles:
a) for(i=0; i<10; i++) {...}
b) for(i=10; i>0; i--) {...}
The results show that the example in b) has 12.6985% power saving.

Conclusion

This paper briefly summarized techniques for power reduction at system level and general state-of-the-art system-level low power techniques are evaluated. This paper provides common coding tips to software engineers who are considering ultra-low-power design at early stage of the design abstraction hierarchy.

References

[1] Kaushik Roy , Mark C. Johnson, Software design for low power, Low power design in deep submicron electronics, Kluwer Academic Publishers, Norwell, MA, 1997.

[2] Harris, E.P., Depp S.W., Pence W,E., Kirkpatrick S. et al, "Technology directions for portable computers" Proceedings of the IEEE, 83(4), 636-657.

[3] DeFreef E., Gatthoor F. Deman H. "Memory access coalescing: Atechnique for elimilating redundant memory access" ACM SIGPLAN Notices, 26(6).

[4] John L. Hennessy, David A. Patterson. "Computer Architecture: A Quantitative Approach"3rd edition. Morgan Kaufmann. 2003.

[5] Mike Tien-Chien Lee, Vivek Tiwari, Sharad Malik, Masahiro Fujita. "Power Analysis and MinimizationTechniques for Embedded DSP Software", IEEE Trans. on Very Large Scale Integration Systems, Vol. 5, No. 1, March 1997.

[6] Vivek Tiwari, Sharad Malik, Andrew Wolfe, Mike Tien-Chien Lee. "Instruction Level Power Analysis and Optimization of Software" Proceedings of International Conference on VLSI Design. p326-328, January 1996.

[7] David Brooks, Vivek Tiwari, Margaret Martonosi. "Wattch: A Framework for Architectural-Level Power Analysis and Optimizations," Proceedings, International Symposium on Computer Architecture. p83-94, Jun 2000.

[8] Doug Burger, Todd M. Austin. "The SimpleScalar Tool Set, Version 2.0", University of Wisconsin-Madison Computer Science Department Technical Report #1342, June, 1997.

A novel depth estimation method for uncalibrated stereo images

Maziar Loghman	Amin Zarshenas	Kwang-Hoon Chung	Yunsik Lee	Joohee Kim
Illinois Institute of Technology 3301 S. Dearborn St. Chicago, U.S.A. mloghma1@hawk.iit.edu	Illinois Institute of Technology 3301 S. Dearborn St. Chicago, U.S.A. mzarshen@hawk.iit.edu	Sane Systems 439 Dongahn-Gu, Anyang-So Gyunki-do, South Korea jkh@sanesys.com	Korea Electronics Tech. Inst. Bundanga-gu, Sungnam-si Gyungki-do, South Korea leeys@keti.re.kr	Illinois Institute of Technology 3301 S. Dearborn St. Chicago, U.S.A. joohee@ece.iit.edu

Abstract

In this paper, a depth map estimation algorithm that performs stereo matching without explicit image rectification has been proposed. In the proposed method, the fundamental matrix is estimated by using Random Sample Consensus and the 8-point algorithm. Then, the epilolar line equation obtained by the projective mapping is derived and the search for point correspondence is performed along the epilolar line. Simulation results show that the proposed method produces accurate depth maps for uncalibrated stereo images with reduced computational complexity.

Keywords- Stereo vision; depth estimation; image rectification; fundamental matrix

Introduction

Three-dimensional (3-D) geometry information is crucial for various applications, such as 3-D reconstruction and stereo-based object detection, and extensive research on dense stereo correspondence has been performed over the past decades. A survey study on dense two-frame stereo correspondence is available in [1]. Most stereo matching algorithms make assumptions about camera calibration and epipolar geometry. In these approaches, given a pair of stereo images, image rectification is performed so that pairs of conjugate epipolar lines become collinear and parallel to the *x*-axis of image. The major advantage of rectification is that point correspondence becomes much simpler because search is performed along the horizontal lines of the rectified images [2]. However, image rectification is computationally expensive and sometimes causes undesirable distortions. Recently, several depth estimation methods for uncalibrated stereo images have been proposed [3]. The major benefit of theses approached is reduced computational complexity.

In this paper, we propose a depth map estimation algorithm that performs stereo correspondence without explicit image rectification.

Proposed algorithm

In the proposed architecture, we assume that we are given a pair of non-rectified images and the proposed method computes disparities without explicit image rectification. The Radom Sample Consensus (RANSAC) algorithm [4] is used to calculate the fundamental matrix and the Census transform is used as the matching criterion to obtain disparities The proposed technique consists of three steps: estimating fundamental matrix, updating epipolar search lines. and computing disparities.

A. Estimating fundamental matrix

The fundamental matrix is the algebraic representation of epipolar geometry. Given a pair of stereo images, to each point in one image, there exists a corresponding epipolar line in the other image. Any point in the second image matching the point in the first image must line on the epipolar line and a projective mapping from a point in one image to its corresponding epipolar line in the other image is represented by the fundamental matrix. For robust estimation of the fundamental matrix, we use RANSAC to remove the effect of outliers and apply the 8-point algorithm [5]. To improve the accuracy in estimating the fundamental matrix, we choose the matching points that have s larger Euclidean distance This modification can increase the probability that the coplanar matching points which lie on the same object are not chosen to estimate the fundamental matrix.

A. Updating epipolar search lines

Based on the relation between the epipolar line and the fundamental matrix, the line equation is computed. In the proposed method, the epipolar line is used as the search line. We bypass the rectification step which is time consuming and in many cases causes shearing and resampling effect. While calculating the line equation for each pixel, an implementation optimization is applied which can reduce the computational complexity in an M × N image from 9MN multiplications plus 6MN additions to only 3(M+N) additions. The epipolar line equation l_r can be written as:

$$l_r : c_1 x + c_2 y + c_3 = 0, \quad (1)$$

where c_1, c_2 and c_3 are the line equation coefficients for the pixel (x,y). The right side image epipolar line coefficients vector can be written as:

$$\begin{bmatrix} c_1 \\ c_2 \\ c_3 \end{bmatrix} = \begin{bmatrix} f_{11} \\ f_{21} \\ f_{31} \end{bmatrix} \times x_l + \begin{bmatrix} f_{12} \\ f_{22} \\ f_{32} \end{bmatrix} \times y_l + \begin{bmatrix} f_{13} \\ f_{23} \\ f_{33} \end{bmatrix}, \quad (2)$$

where x_l and y_l refer to the column and row coordinate of the left side image point, respectively. Starting with (0,0) coordinate in the left side image, the right image epipolar line coefficients vector can be initialized by the third column vector of the fundamental matrix. To reduce the computational complexity significantly, we only consider an addition of the second column vector of the fundamental matrix while switching from the corresponding epipolar lines of two consecutive points. Stepping from one row to the next one, we only apply a single vector addition of the fundamental matrix first column vector and the epipolar line coefficients.

B. Extracting the matching pixels

For a pixel in the left side image, the start point of searching strategy is the projection of the pixel in the same coordinate in the right side image onto the epipolar line. This point has the least distance to the reference pixel compared to the other points on the line. The maximum disparity range is defined by the user and varies based on the image resolution. The search direction is on the epipolar line. The matching metric used for cost calculation is the non-parametric census transform [6] due to its robustness to illumination changes. Census transform maps a block surrounding the reference pixel into a bit string. The cost function is calculated by finding the Hamming distance of the obtained bit streams. The final matching cost function is used for optimization. The optimum disparity is the value which minimizes the cost function.

Experimental results

We demonstrate the efficiency of our depth estimation and refinement algorithms by using the Middlebury dataset [7]. The workstation runs the Windows 7 operating system with Intel Xeon Quad-Core processor and 8 GB RAM. Four different image pairs from the Middlebury dataset [7] are chosen for the evaluation. In the case where the dataset are rectified images, we have used a transformation to make non-rectified images. Table I indicates the error statistic of percentage of bad pixels with respect to the provided ground truth depth map.

Table II shows the processing time by running the algorithms

TABLE I
PERCENTAGE OF BAD PIXELS FOR THE MIDDLEBURRY DATASET[7]

Stereo image	Proposed method	Hierarchical [3]
Tsukuba	2.64 %	5.52 %
Cones	7.64 %	11.7 %
Teddy	12.8 %	13.3 %
Venus	0.84 %	1.65 %

TABLE II
COMPUTATIONAL TIME COMPLEXITY

Stereo image	Time (ms) - proposed method	Time(ms) - Hierarchical [3]
Tsukuba	59.7	63.5
Cones	71.3	74.3
Teddy	71.9	76.1
Venus	63.4	60.2

using C programming on CPU. The performance of the proposed algorithm is compared with the Hierarchical method [3].

Conclusion

In this paper, an efficient depth estimation algorithm is proposed that finds the stereo correspondences by searching along the epipolar lines. The algorithm skips the rectification process which makes it less computationally complex. Simulation results indicate that the proposed methods achieves the goals by improving the quality of the estimated depth maps and reducing the computational time complexity.

Acknowledgment

This work was supported by the Technology Development Program for Commercializing System Semiconductor funded By the Ministry of Trade, industry & Energy (MOTIE, Korea). (No. 10041126, Title: International Collaborative R&BD Project for System Semiconductor).

References

[1] D. Schartsein, and R. Szeleiski, "A taxonomy and evaluation of dense two frame stereo correspondence algorithm," in *International Journal of Computer Vision*, vol. 47, no. 1, pp. 7-42, May 2002.

[2] Y. S. Kang, and Y. S. Ho, "An efficient image rectification method for parallel multi-camera arrangement,", in *IEEE transaction on Consumer Electronics*, vol. 57, no. 3, pp. 1041-1048, Aug. 2011.

[3] L. Nalpantidis, A. Amanatiadis, G. Sirakoulis, N. Kyriakoulis, and A. Gasteratos, "Dense disparity estimation using a hierarchical matching technique from uncalibrated stereo vision," in Proc. Intl. Workshop on Imaging Syst. Tech., May 2009.

[4] M. A. Fischler, R. C. Bolles, "Random sample consensus: a paradigm for model fitting with applications to image analysis and automated cartography" in *Communications of the ACM*, vol 24 (6), pp 381-395, Jun 1981.

[5] Richard I. Hartley,"In Defense of the Eight-Point Algorithm", *IEEE Transaction on Pattern Recognition and Machine Intelligence*, vol. 19 (6), pp 580–593, 1997.

[6] R. Zabih1 and J. Woodfill, "Non-parametric Local Transforms for Computing Visual Correspondence" in *Proceedings of European Conference on Computer Vision (ECCV)*, Stockholm, Sweden, pp 151-158, May 1994

[7] D. Scharstein, and R. Szeliski, "High-accuracy stereo depth maps using structured light," in *Proc. of IEEE Conf. on Computer Vision and Pattern Recognition (CVPR)*, vol. 1, pp. 195-202, Jun. 2003.

978-1-4799-5128-4/14 $31.00 © 2014 IEEE

Ring Projection Transforms by Using CUDA Implementation for Recognizing Objects in Automation System

Sang-hyeob Song, Seong-muk Kang, Hyeong-jun Cho, and Jun-dong Cho

College of Information and Communication Engineering, Sungkyunkwan University
Suwon, Korea
songsang7@skku.edu, jdcho@skku.edu

Abstract

This paper presents a Ring Projection Transforms (RPT) for recognizing object in robot manufacturing automation system. Our proposed method is based on CUDA (Compute Unified Device Architecture), which is a kind of GPGPU (General-Purpose computing on Graphics Processing Units). Compared with the conventional method without utilizing CUDA, our method achieves better performance in terms of processing time. The experimental validation indicates that our proposed method is faster about 35 times and reliable for recognizing object in real-time automation system.

Keywords-Vision; Recognition; Ring Projection Transforms; RPT; CUDA; GPGPU; Template Matching

I. Introduction

Nowadays, the mobile device automation facilities replace human's role especially in manufacturing of small product such as digital camera modules. In factory automation, robot arms replace human manual assembly. To assemble small-sized parts in manufacturing camera modules, pattern recognition is essential to align the parts to PCB mounts, and should guarantee both reliability and speed. However, image processing normally demands much time, so satisfying both reliability and speed is a challenging task.

In order to improve speed of processing for image pattern recognition, this paper suggests an effective way of using GPU on vision algorithm to attain a desirable solution.

II. Related Work

A. Vision Algorithm

In vision algorithm, *Ring Projection Transforms (RPT)* [1] is a kind of template matching to recognize static object on scene. The RPT can be defined at Fig. 1 as (1):

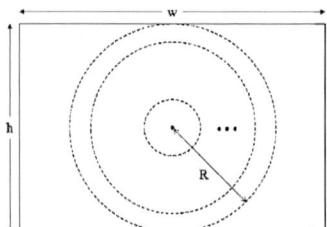

Fig. 1 Template image projected to ring with size (w x h)

$$\vec{T}(r) = \frac{1}{2\pi r} \int_0^{2\pi} I\left(\frac{w}{2} + r\cos\theta, \frac{h}{2} + r\sin\theta\right) d\theta$$
$$(0 < r \leq R) \quad (1)$$

At the Equation (1), T(r) means vector at the r and I(x, y) is the value of pixel at (x, y) on the template image. The value R is limited to half of min(w, h).

After calculating T, we should match it at every points on the scene. The Fig. 2 shows the matching process.

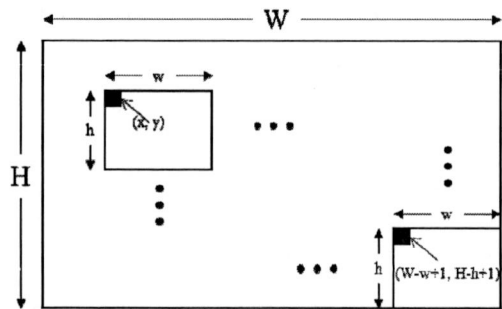

Fig. 2 Matching process on the scene with size (W x H) by template with size (w x h)

This method as shown in Fig. 2 is called a block (or areal) - based matching. In case of RPT, O(x, y) which is output at (x, y) can be calculated as below:

$$O(x, y) = \text{NCC}(\vec{T}, \vec{S}_{(x,y)}) = \frac{Co\text{-}variance(\vec{T}, \vec{S}_{(x,y)})}{(R-1)\sigma(\vec{T})\sigma(\vec{S}_{(x,y)})} \quad (2)$$

$$\overrightarrow{S_{(x,y)}}(r) = \frac{1}{2\pi r} \int_0^{2\pi} I'\left(x + \frac{w}{2} + r\cos\theta, y + \frac{h}{2} + r\sin\theta\right) d\theta$$

$$\begin{pmatrix} 0 \leq x \leq W-w+1 \\ 0 \leq y \leq H-h+1 \\ 0 < r \leq R \end{pmatrix} \quad (3)$$

At the equation (2), NCC [2] means *Normalized Cross-Correlation* and is also called *Coefficient of Correlation*. The parameter σ(X) is standard deviation of X. The value of NCC is known as barometer of relationship between vectors. Therefore, if a certain point has the highest value of NCC, it should be the location of object that we want to find. However, computing (1) ~ (3) demands too much computation time for deriving a result, thus it is hard to adapt for them into automation system.

B. GPGPU

Traditionally, only CPU has been in charge of computing for general purposes. In recent years, GPU venders have proposed using GPU not only for rendering objects but for general computing. Unlike CPU, GPU includes a lot of processing units, becoming appropriate to parallel calculation. This usage is called *GPGPU (General-Purpose computing on Graphics Processing Units)*. Especially, *CUDA (Compute Unified Device Architecture)* [3] is one of GPGPU provided from Nvidia™. The CUDA produces high computing power with logically hierarchical structure with grid, block and thread.

III. Proposed Method

A. Strategy of Using Memory

CUDA provides various kinds of memory to user such as shared memory, constant memory, global memory and texture memory [4]. Each type of memory has own unique characteristic. For example, shared memory has the widest bandwidth, the fastest clock speed and the smallest capacity since this memory is one of on-chip memory.

In the proposed method, we would use constant memory, texture memory and global memory. Constant memory is a kind of global memory that is read-only and cashed to on-chip memory. Furthermore, the constant memory spreads data to 16 threads (half-warp) like broad casting when we access the constant memory. Therefore, this memory has advantage if all thread read same data. To maximize this advantage, we would use look-up table for accessing from whole threads.

B. Thread Allocation

In this algorithm, we would get (W-w+1)*(H-h+1) number of values as a result. Though CUDA can allocate threads with hierarchical structure, we make it 1-dimension vector because both values W-w+1 and H-h+1 cannot be guaranteed being multiple of 32. The number 32 means unit size of warp in GPU. Therefore, to make occupancy-rate higher, using 1-dimension vector is better than others. Finally, we allocate threads and blocks as below:

$$
\begin{cases}
\text{Threads / block} = 1024 \ (= \text{Maximum number to allocate}) \\
\text{Blocks / grid} = \frac{(W-w+1)(H-h+1)+Threads\ per\ block-1}{Threads\ per\ block}
\end{cases} \quad (5)
$$

This method to allocate threads and blocks is general used way to treat like 1-D vector.

IV. Experimental Result

The main purpose is to improve speed of RPT algorithm. Therefore, we measure time for calculating. To be compared with non-CUDA method, we also run the same algorithm on CPU without GPGPU. The experiments were conducted on a PC with running Windows 7 64bit, 16 GB RAM, Intel Core i5-4670 3.4GHz processor and GeForce GTX 770. The graphics card has 2GB RAM on board. The test image set contains a scene image with size 640x480 and template image with size 99x99.

The result for RPT is given in Fig. 3. Applying both RPT using CUDA and without using CUDA show that the template

image is well found on the scene image.

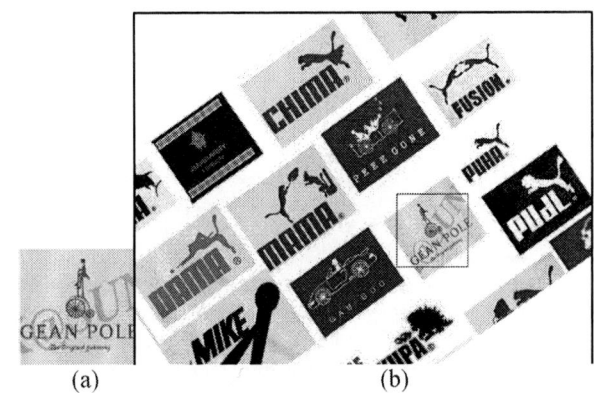

(a)   (b)

Fig. 3 (a) Template image (b) Scene image with applying RPT

TABLE I summarizes the processing time and indicates that CUDA implemented version is faster than the other about 35 times.

TABLE I
TIME CONSUMPTION FOR PROCESS

Without CUDA	With CUDA	Degree of improvement
2645.00[ms]	75.37[ms]	x35.1

V. Conclusion

Our proposed method provides high computing power and makes this algorithm available option in automation system. Therefore, our proposed method can be a suitable solution to shorten processing time for recognition. Furthermore, we also can apply our method not only to recognition but also to other computing problems.

Acknowledgment

The research was partially supported by MOTIE (Ministry of Trade, Industry and Energy) Foundation of The World-Class300 Project: Development of automated manufacturing robot system technology integrating with the 6 DOF robot mechanisms and the S/W platform for assembling mobile IT products. (10043213)

This research was supported by Basic Science Research Program through the National Research Foundation of Korea (NRF) funded by the Ministry of Education (NRF-2013R1A1A2058942)

References

[1] Kim, Hae Yong, and Sidnei Alves De Araújo. "Grayscale template-matching invariant to rotation, scale, translation, brightness and contrast." *Advances in Image and Video Technology*. Springer Berlin Heidelberg, 2007. 100-113.

[2] R. Gonzales and R. Woods, Digital Image Processing. Pearson Prentice Hall, third ed., 2008

[3] Sanders, Jason, and Edward Kandrot. *CUDA by example: an introduction to general-purpose GPU programming*. Addison-Wesley Professional, 2010.

[4] Cuda, C. "Programming guide." *NVIDIA Corporation, July* (2012).

A High Performance Low Power Implementation Scheme for FSM

Shuai Li, Ken Choi

VLSI Design and Automation Laboratory

Department of Electrical and Computer Engineering

Illinois Institute of Technology

3301 S Dearborn St., Chicago, IL 60616, USA

sli97@hawk.iit.edu, kchoi@ece.iit.edu

Abstract

Finite state machine (FSM) takes an important part in digital logic system. FSMs partition is one of the effective methods in regards to low power technique. Most of time only one of sub-FSMs need to be clocked, consequently power is saved. In this paper, we propose a high performance algorithm based on state transitions probability and low complexity control logic to implement the partitioned FSMs. The cost from one sub-FSM to other sub-FSMs could be minimum, and the transitions probability in one sub-FSM should be maximum. Based on the low complexity of control logic, we further give an optimized hardware architecture for the partitioned FSM model. Our proposed scheme has been implemented by using tsmc 45 nm technology library. Experimental results show that an average power reduction of 59% has been obtained for a set of standard FSM benchmark circuits.

Keywords-Low power; finite state machine (FSM); partition model; hardware structure

I. Introduction

Power consumption is a critical consideration in the performance of VLSI design. There are many different methods utilized in power reduction [1]. Partitioned finite state machine (FSM) is a useful technique widely used in low power design.

Power reduction is computed using (1):

$$P = P_{dynamic} + P_{short\_circuit} + P_{leakage} \qquad (1)$$

The three terms indicate dynamic power, short circuit power and leakage power respectively. To address the low power design problem, a wide range of FSM partition algorithms and architectures have been developed [2-6].

In [2], the author proposed a transformed states transition graph (STG) in decomposed FSM. However, the extra transformed STG led to more area dissipation and control logic. In [3], the author proposed an asynchronous model for FSM partition, however, the control logic takes too much power consumption and the asynchronous sequential circuit is complex.

In this paper, we utilize a high efficient partition algorithm to divide the FSM to two or several sub-FSMs, we optimized the control logic by using the state transition table, and the synchronous structure we proposed is much easier to implement than the asynchronous structure.

The rest of the paper is organized as follows. Section II describes the proposed FSM partition algorithm. In section III, we propose the high efficient control logic and its corresponding structure. The experimental results of our proposed method are given in section IV. Finally, section V concludes this paper.

II. The proposed partition algorithm

A. Probability prediction

For a n-state FSM, the transition probability from state S_i to S_j is

$$P_{ij} = P\{X_n = j \mid X_{n-1} = i\} \qquad (2)$$

We define each bit of the input signal has the same probability, e.g. p(0)=p(1)=0.5, so we can get the FSM states transition probability matrix P.

If the Markov chain is irreducible and aperiodic, there is a unique stationary $\pi = \{\pi_j, j \geq 0\}$, we have:

$$\sum \pi_j = 1, \pi_j > 0 \qquad (3)$$

$$\lim_{n \to \infty} P_{ij}^{(n)} = \pi_j \qquad (4)$$

$$\sum_j \pi_j P_{ij} = \pi_j \qquad (5)$$

Through (3) to (5), we got the dcistribution π of a FSM.

$$\pi = \lim_{n \to \infty} P^{(n)} = [P_1 \quad P_2 \quad \cdots \quad P_{n-1} \quad P_n] \qquad (6)$$

P_i denotes the steady probability of state S_i.

Finally, we got the total transition probability between S_i and S_j as shown in (7):

$$P_{i,j} = P_{ij} \cdot \pi_i + P_{ji} \cdot \pi_j \qquad (7)$$

B. Partition algorithm

Since we got $P_{i,j}$, The partition flow is shown in Fig.1.

Algorithm: partition the monolithic FSM into n sub-FSMs

Input: $P_{i,j}$ and its corresponding states S_i and S_j;

Output: sub-FSMs M_1, M_2, $\cdots$, M_n

Begin

    Sorting $P_{i,j}$ in decreasing order and marking them as P_1, P_2, ...,

P_m, (here m>n), thus the state pair (S_i, S_j) also sorted in decreasing order.

for (h=0;h<m;h=h+n) // h is the variable of state pairs

    for (k=1;k<=n;k=k+1) // k is the variable of sub-FSM

        if state pair S_i and S_j are never appeared before

            then { $P_{k+h} \rightarrow (S_i, S_j)\} \in M_k$;

        else if S_i (or S_j) already existed in sub-FSM M_{sub}

            then S_j (or S_i) should also merged into the sub-FSM M_{sub};

        else if both of S_i and S_j already existed in sub-FSM M_{sub};

            then skip;

        end if

    endfor

endfor

End

Figure 1. FSM partition algorithm

Table 1. Comparison of the results between monolithic FSM and proposed method

Bench mark	states	transitions	Monolithic FSM		Conventional method				Proposed method			
			Power (μW)	Area (μm^2)	Power (μW)	Area (μm^2)	Power Red.	Area Incr.c	Power (μW)	Area (μm^2)	Power Red.	Area Incr.
dk27	7	14	17.4	127	9.03	129	48%	1.6%	8.21	131	52.8%	3.1%
dk512	15	30	27.08	224.3	19.67	335.5	27.4%	49.6%	16.32	359.48	39.7%	60.3%
bbtas	6	24	13.22	122.01	5.80	155.33	56.1%	27.3%	2.743	136.10	79.2%	11.5%
lion	4	11	7.37	74.15	3.431	108.9	53.4%	46.9%	1.445	99.02	80%	33.5%
dk17	8	32	28.44	212.84	14.80	257.18	48.0%	20.8%	12.46	255.30	56.2%	19.9%
dk16	27	108	56.95	656.08	32.10	954.56	43.6%	45.5%	30.79	978.96	45.9%	49.2%
avg	-	-	-	-	-	-	46.1%	32.0%	-	-	59.0%	29.6%

III. Control logic and hardware structure

We take MCNC [7] benchmark "bbtas" as an example to illustrate our control logic as shown in Fig.2.

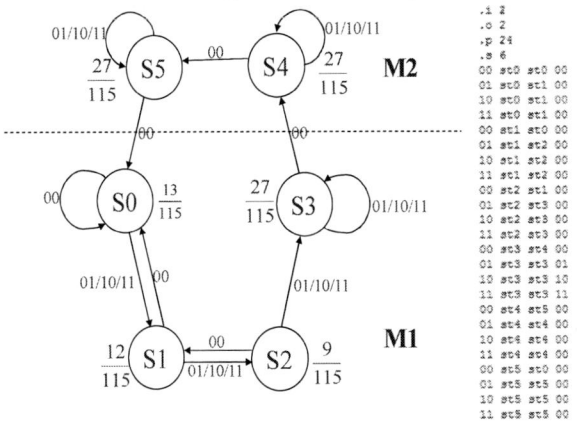

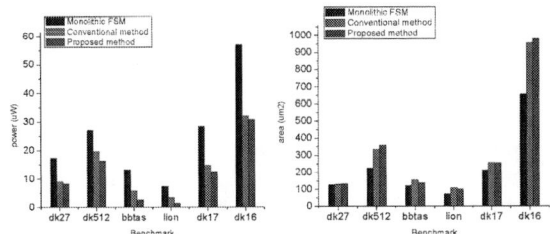

Figure 4. Comparison among monolithic FSM, conventional method and proposed method

As shown in table 1 and Fig.4, compared with orginal monolithic FSM and conventional method, the proposed method has three advantages:

1) Power dissipation is around 59% lower than monolithic.

2) The overall area is smaller than the conventional method.

3) The implementation is less complex.

V. Conclusion

We proposed a low power implementation model for partitioned FSMs and its corresponding structures. The average power reduction is 59% compared with the monolithic FSM, and the area is only increased by 29.6%. The results demonstrate that the proposed method is superior in terms of power dissipation to the anchor.

Figure 2. State transition graph of "bbtas"

We utilize the algorithm in section II to divide the FSM into two groups: $M_1:\{S0,S1,S2,S3\}$ and $M_2:\{S4,S5\}$. Through transition state table we can get the control logic. The proposed hardware structure as shown in Fig.3.

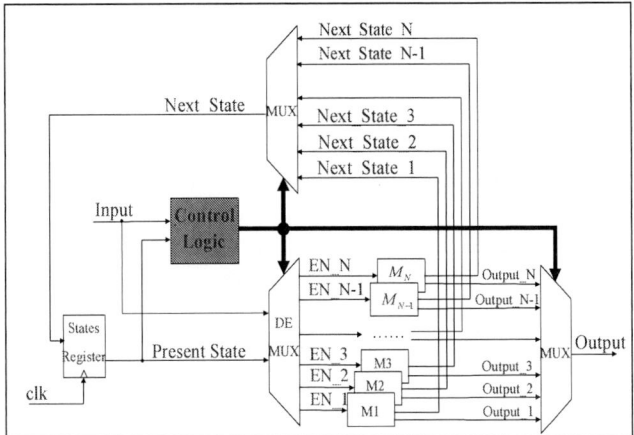

Figure 3. Optimized FSM hardware structure

The implementation model includes: State registers, control logic, sub-FSMs, MUX and DeMUX.

IV. Experimental results

References

[1] M. Pedram, et al, "Low-power RT-level synthesis techniques: a tutorial," Computer and Digital Techniques, IEE Proceedings, Vol 152, Issue 3, Page(s): 333-343, 6 May 2005.

[2] Cao C, et al, "Synthesis tool for low-power finite state machines with mixed synchronous/asynchronous state memory," Computer and Digital Techniques, IEE Proceedings, Vol 153, Issue 4, Page(s): 243-248, 3 July 2006.

[3] Oelmann B, et al, "Asynchronous control of low-power gated-clock finite-state-machines," Electronics, Circuits and Systems, Proceedings of ICECS, Page(s): 915-918, 1999.

[4] Sue-hong Chow, et al, "Low power realization of finite state machine – a decomposition approach," ACM transactions on Automation of Electronic System, vol. 1, No. 3, July 1996, Pages 315-340.

[5] Monteiro, et al, "Finite state machine decomposition for low power," Design Automation Conference, 1998.

[6] Luca Benini, et al, "Automatic Synthesis of Low-Power Gated-Clock Finite-State Machines," IEEE transactions on computer-aided design on integrated circuits and systems, vol. 15, No. 6, June 1996.

[7] Saeyang Yang, "Logic synthesis and optimization benchmark user guide version 3.0," 1991.

Survey on Security techniques for AMI Metering System

SungJin Kim, HyunSoo Chng, Taeshik Shon

Dept. Computer Engineering, Ajou University
206, World cup-ro, Yeongtong-gu
Suwon-si, Gyeonggi-do, Korea
+821031021376, +82312191621, ksjskyblue@ajou.ac.kr

Abstract

AMI(Advanced Metering Infrastructure) handles metering data. Since it transmits private data, security of network is essential. In this paper, DLMS/COSEM and ANSI C12.22, the most popular smart metering standards, was surveyed for their security features. And security features of Data Link Layer protocols that can go under the mentioned standard's protocol was surveyed. And based on recommendation in RFC 6142, appropriate security feature of Application layer was mapped for each Layer 2 protocols.

Smart Metering; Advanced Metering Infrastructre; ANSI C12.22; DLMS/COSEM; Security

Introduction

Contrast to the past, today's grids, that are called Smart Grid, supports bi-directional communication between entities. In this network, AMI takes important role in remote metering data collection and payment request of the Smart Grid System. Since AMI handles customer's private information, it must guarantee security in its communication. But AMI devices demand optimized security in order to provide secured service with lowest cost.

This paper briefly describes AMI, Metering Standards, and security elements they supports in the upper layer defined in DLMS/COSEM and ANSI C12.22. And then it handles security elements that are supported by Data Link Layer protocols that can be used by DLMS/COSEM and ANSI C12.22. This could help with setting up appropriate security for Smart Metering Systems depending on the low level protocol.

Advanced Metering Infrastructure

AMI consists of Smart Meter, DCU(Data Collection Unit), and AMI Head-End. Such devices demands low cost for their service. AMI provides various functionality, but remote metering function is considered most important. This functionality can cause many problem if it doesn't support security means, such as overpayment and underpayment of electricity charge by changing metering data or manipulating DR(Demand Response) domains. So optimization of security features is essential in their communication.

Metering Protocol

Most famous standards for communication between Smart Meter and AMI Server are DLMS/COSEM and ANSI C12.22. Both standards considered security and adopted encryption and authentication techniques. This Section briefly handle what DLMS/COSEM and ANSI C12.22 are and what security techniques they provied.

A. DLMS/COSEM

DLMS/COSEM is a communication standard designed for European smart metering environment by IEC TC13 WG 14. This standard specifies only application layer, so it is possible to select lower layer protocols. It can be used over TCP/IP protocol or Data Link Layer Protocols such as HDLC(High-Level Data Link Control), PPP(Point-to-Point Protocol), Ethernet, etc[1]. For security features, DLMS/COSEM provides application level authenticaiton and encryption. Following is the algorithms and methods that DLMS/COSEM supports for authentication and encryption[2].

```
Hash              : SHA-256, SHA-384
Encryption        : AES 128, AES 256
Digital Signature : ECDH 256, ECDH 384
(ECDH : Elliptic Curve Diffe-Hellman)
```

B. ANSI C12.22

ANSI C12.22 is a communication standard designed for American smart metering environment[3]. It's very similar to DLMS/COSEM in many aspects. This protocol Uses ANSI C12.19, EPSEM(Extended Protocol Specification for Electric Metering), ACSE(Association Control Service Element) to construct messages and transmit them over various Data Link Layer protocols. ANSI C12.22 can transmit data message over various Data Link Layer Environment including general Ethernet, PPP, 3G, PLC(Power Line Communication), and etc. In ANSI C12.22, security features as authentication and encryption algorithms and methods are defined[4]. Following is the algorithms and methods for authentication and encryption that ANSI C12.22 provides.

```
Hash              : MD5
Encryption        : AES 128
Supports PKI(Public Key Infrastructure)
(Particular algorithm was not specified)
```

Security Features for Usable Data Link Protocol

With the mentioned Standards, Data Link Layer protocol can be selected for certain environment. Among these low layer protocols, some support authentication and encryption. But many of them don't support such security features. This section contains information about security features of Data Link Layer Protocols that can be used by DLMS/COSEM and ANSI C12.22.

- HDLC

HDLC focused on transmission of synchronous data between Point-to-Point. It doesn't support security features for message such as encryption and authentication.

- PPP

PPP also focused on data transmission between Point-to-Point but considered security of the transmitting messages with encryption(RFC1968) and authentication such as PAP/CHAP/EAP[5].

- 3G

3G is communication protocol designed for communication between mobile system. It uses KASUM/SNOW3G/AES to support encryption and authentication.

- Ethernet

Ethernet is the most generalized protocol for communication between devices. Since many of application services for Ethernet support security features, it doesn't guarantee security in this layer.

- IEEE 802.11

IEEE 802.11 is famous for wireless communication in LAN network. This protocol supports encryption and authentication of the message with algorithms and methods defined in IEEE 802.11i standard[6].

Security Recommendation

In ANSI C12.22, application layer security is categorized in 3 security levels with conditions of Encryption and Authentication as following table.

And according to IETF RFC 6142, appropriate security features of application layer should be applied depending on

TABLE I
Provided Security Levels

Security Level	Encryption	Authentication
Level 1	X	X
Level 2	X	O
Level 3	O	O

security provided by Data Link Layer. It is recommended that application should support security that Data Link Layer doesn't support.

If appropriate security level is adopted to the previously mentioned 5 Data Link Layer Protocols, it would be as following table.

TABLE II
Recommended Security Level for Data Link Layer Protocol

Protocol (Layer 2)	Security supported by Data Link Layer	Recommended Level
HDLC	None	3
PPP	Encryption and Authntication	1
3G	Encryption and Authntication	1
Ethernet	None	3
802.11	None	3

With this recommendation, it would be possible to avoid overlap of the security features and reduce the cost of the device activity.

Conclusion

In DLMS/COSEM and ANSI C12.22, security algorithms and methods such as AES 128, ECDH 256, SHA 256, etc are used to provide security of the messages. However AMI environment is formed with low power embedded devices, therefore it is important to optimize security features with low cost.

In this paper, security features supported by Application Layer level protocols defined in DLMS/COSEM and ANSI C12.22 was surveyed. Then security features of Data Link Layer protocols that can be used by DLMS/COSEM and ANSI C12.22 was surveyed. And with the recommendation of to IETF RFC 6142, appropriate security level for each Data Link Layer protocols was mapped. With the result of this paper, it would be possible to set up Application Layer level security features without overlapping features of Data Link Layer protocol. Also it could help in research on quantifying and accounting vulnerability of smart meters.

References

[1] UA.DLMS, "DLMS/COSEM Architecture and Protocols." Green Book , 2009.
[2] L.Weith, "DLMS/COSEM Protocol Security Evaluation."
[3] A. Moise and J. Brodkin, "ANSI C12. 22, IEEE 1703, and MC12. 22 Transport Over IP." 2011.
[4] J.Wang and V.C. Leung, "A survey of technical requirements and consumer application standards for IP-based smart grid AMI network."Information Networking (ICOIN), 2011 International Conference on. IEEE, 2011.
[5] W.Simpson, "The Point-to-Point Protocol", RFC 1661, Jul 1994.
[6] J.C.Chen, M.C. Jiang, and Y.W.Liu. "Wireless LAN security and IEEE 802.11 i." Wireless Communications, IEEE 12.1, pp. 27-36, 2005.

A Digital Lock Detector for a Dual Loop PLL

Chang-Hyun Bae[1,2] and Changsik Yoo[2]

[1] DRAM Design, Samsung Electronics, Hwaseong, Korea
[2] Department of Electronics and Computer Engineering, Hanyang University, Seoul, Korea
chbae@hanyang.ac.kr, csyoo@hanyang.ac.kr

Abstract

A digital lock detector capable of detecting frequency variations occurred during the operation of a digital filter of the detector is proposed. In this paper, the digital lock detector is applied to a dual loop PLL to give the frequency lock information to another PLL loop. The proposed lock detector with the PLL fabricated in a 0.13-μm CMOS process occupies 0.17-mm^2 and consumes 35-mW from a 1.2-V power supply. The measured frequency offset between two VCOs is 0.16-%.

Keywords-Lock Detector; PLL; CMOS

Introduction

In order to increase PLL locking speed and decrease phase noise, a dual loop structure is used in PLL and a lock detector is widely used to acquire frequency lock state in this structure [1], [2].

In [3], [4], a digital lock detector which is composed of two flip flops and delay elements is presented. This detector has advantage of an immediate response, however, the output of the lock detector can be sensitively responded to little voltage variations because it does not have any filter.

Another digital lock detector which the frequency of the feedback clock is compared with that of the reference clock as every 2^N-1 cycles is presented in [5]. The detector is a fully digital lock detector but frequency variation cannot be detected during the filter operation by N-bit counters.

In this paper, a fully digital lock detector capable of the detection of frequency variations during the N-bit counter operation is presented. The operation of the proposed detector is verified by applying to a dual loop PLL.

Circuit description

Fig. 1 shows the proposed lock detector which is composed of an N-bit counter enable block, a Lock counter enable block and a Lock counter. An N-bit counter enable block consists of delay elements and D-flipflops. This block is operated by comparing the phase of CLK_{REF} with the phase of CLK_{FB}. The outputs of two D-flipflops are connected to NOR gate. N-bit counters count the high going edge of CLK_{REF} and that of CLK_{FB} respectively while *EN* is logical "high". The two outputs of N-bit counters are compared and a Lock counter counts the high going edge of CLK_{REF} until the output of the counter gets the pre-determined value.

Fig. 2 shows the timing diagram of a conventional lock detector with N-bit counters (a) and the proposed lock detector with N-bit Counter Enable block (b). In the conventional

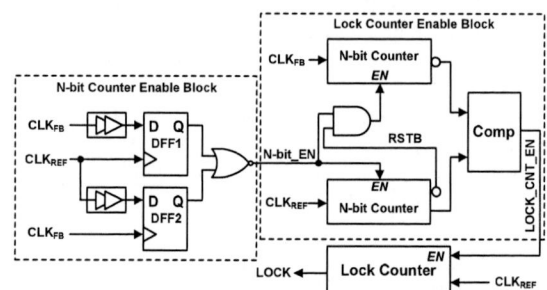

Fig. 1. The proposed lock detector.

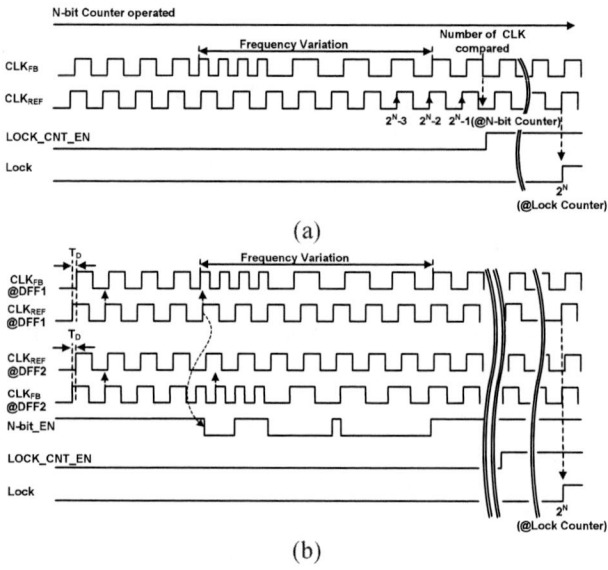

Fig. 2. Timing diagram of a conventional lock detector with N-bit counters (a) and the proposed lock detector (b).

detector, the frequency variation which is occurred during the operation of the counter cannot be detected as shown in Fig. 2 (a). In Fig. 2 (b), the delayed CLK_{FB} is sampled by CLK_{REF} in DFF1 and the output of DFF1 is logical "low" when the phase of CLK_{FB} is locked to the phase of CLK_{REF}. If the variation is occurred, *N-bit_EN* cannot be logical "high" due to the phase offset and N-bit Counter is not operated. The DFF2 is needed to quickly detect another (increase or decrease) direction as the frequency variation of CLK_{FB}. The output of DFF2 is also logical "low" at the lock state. When *N-bit_EN* is logical "high", each N-bit counter counts CLK_{REF} and CLK_{FB} respectively until each counter gets its maximum value. After every 2^N-1 cycles, the outputs of two counters are compared and the counters are reset. If the compared results are same, Lock counter is enabled and counts high going edge of CLK_{REF}. If it gets its pre-determined maximum value, then the *LOCK*

978-1-4799-5128-4/14 $31.00 © 2014 IEEE   194

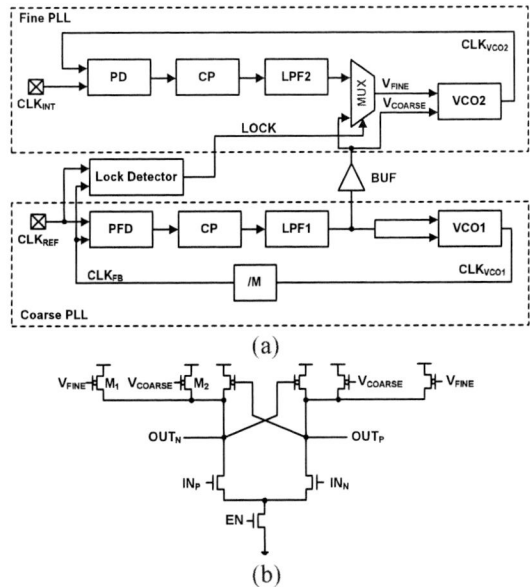

(a)

(b)

Fig. 3. Block and circuit diagram (a) dual loop PLL architecture with the proposed lock detector, (b) unit cell of VCO .

signal is issued as logical "high". This means that the PLL is in lock state.

Although there are many applications using a lock detector, the proposed detector is used to give the lock information of a Coarse PLL to a Fine PLL in our work. Fig. 3 (a). shows a dual loop PLL including the proposed digital lock detector and Fig.3 (b) shows the unit cell of a VCO. In order to enable the coarse and fine lock, the ratio of the size of M1 and M2 is 1:10. The test PLL is composed of a Coarse PLL, a Fine PLL, a full-range unit gain buffer and a Lock Detector. Each VCO of two PLLs has the same structure but K_{VCO} is different each other. K_{VCO} of the Coarse PLL is larger than that of the Fine PLL so that the PLL can lock with fast tracking speed and low jitter. For two PLL loops to operate stably, the bandwidth of the Fine PLL is determined to 1/10 of that of the Coarse PLL. The Coarse PLL path makes optimum frequency and the control voltage is supplied to the VCO of the Fine PLL so that the frequency of the Fine PLL can match the frequency of the internal clock CLK_{INT}. The control voltage generated in the Coarse PLL passes through a full-range unity gain buffer to prevent from delivering drastic voltage variation occurred in the Coarse PLL path. Initially two VCOs have the same control volage from the coarse PLL until the frequency of VCOs is the same as the frequency of CLK_{REF}. If the Lock detector confirms lock state, the connection of the input V_{FINE} of VCO2 is switched to the output of LPF2 and then the Fine PLL is operated to tracking to the phase of CLK_{INT}.

Measurement Results

The prototype digital lock detector with the PLL has been implemented in a 0.13-μm CMOS logic process. The microphotograph of the test chip and detailed layout are shown in Fig. 4. The test chip occupies an active area of 0.17-mm².

Fig. 5 shows the clocks of VCO1, VCO2 and CLK_{REF} respectively. CLK_{REF} whose frequency is 15.625-MHz is used as a triggering clock and CLK_{INT} is 375-MHz. The frequency offset of VCO1 and VCO2 is 0.16-%.

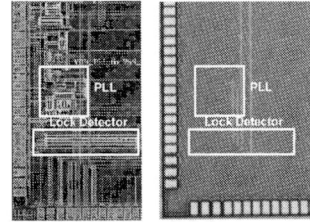

Fig. 4. Layout and chip photo.

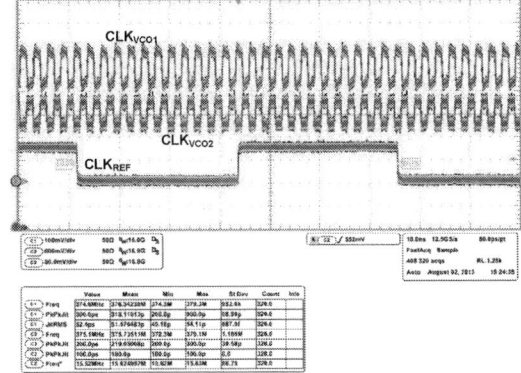

Fig. 5. Measured clock signal triggered by CLK_{REF}.

Conclusions

A digital lock detector for application in a dual loop PLL is presented. It is possible to inform more accurate lock state becasuse this full digital lock detector is capable of filtering out the frequency variations during the clock counting operation.

Acknowledgement

This work is sponsored by IT R&D program MKE/KEIT. [No.10035202, Large Scale hyper-MLC SSD Technology Development]. The CAD tools were provided by IC Design Education Center (IDEC), KAIST, Korea.

References

[1] K. H. Cheng, W. B. Yang, and C. M. Ying, "A dual-slope phase frequency detector and charge pump architecture to achieve fast locking of phase-locked loop," *IEEE Trans. Circuit and System II*, vol. 50, pp. 892-896, Nov 2003.

[2] Ming-ta Hsieh and Gerald E. Sobelman, "Architectures for Multi-Gigabit wire-linked clock and data recovery," *Circuits and Systems Magazine, IEEE*, vol. 8, issue 4, pp. 45-57, Fourth Quarter 2008.

[3] Shao-Ku Kao, Bo-Jiun Chen, and Shen-Iuan Liu, "A 62.5-625-MHz anti-Reset all-digital delay-locked loop", *Circuits and Systems II: Express Briefs, IEEE Transactions* vol 54, issue 7, pp.566-570, July 2007.

[4] Yue-Fang Kuo, Ro-Min Weng, and Chuan-Yu Liu, "A fast locking PLL with phase error detector", *Electron Devices and Solid-State Circuits, 2005 IEEE Conference*, pp. 423-426, Dec. 2005.

[5] V. Melikyan, A. Hovsepyan, M. Ishkhanyan and T. Hakobyan, "Digital lock detector for PLL," *Design & Test Symposium (EWDTS), 2008 East-West*, pp. 141-142, Oct 2008.

978-1-4799-5128-4/14 $31.00 © 2014 IEEE

Automatic Current Range Detection for Biochemical Sensor

Wei-Jun Liu, Ming-Han Yu, Bin-Da Liu, and Chia-Ling Wei

Department of Electrical Engineering, National Cheng Kung University,
One University Road, Tainan 70101, Taiwan
+886-6-2098202, bruce1223776@gmail.com, moonplume@hotmail.com, bdliu@mail.ncku.edu.tw, and
clwei@mail.ncku.edu.tw

Abstract

In this paper, a current sensing mechanism for biochemical sensor is presented. The sensing circuit consists of current conveyor, current mirror, current range detection circuit, and current to digital converter. The chip can be integrated in a potentiostat system. This work is implemented in TSMC 0.18-μm CMOS process, and the area is 0.5625 mm^2. The measured current ranges from 20 μA to 1200 μA with average linear correlation (R-squared) of 0.9986. The power consumption of the proposed circuit is 1.41 mW from 1.8 V power supply.

Keywords-Biochemical sensor, potentiostat, current mirror, current range detection circuit.

Introduction

In recent years, with the improvement in technology, medical instruments has ground-breaking development. The public has gradually raised their requirement for health care condition. Therefore, the demand for user-friendly medical monitoring systems is increasing. The core architecture of this kind of medical device is the biomedical sensor which has several features such as high specificity, high sensitivity, and instant output response. By changing different sorts of sensor interfaces, various measuring purposes can be achieved. Users can know about their physical conditions faster and more conveniently than before with the help of medical devices enhancing the probability of early treatment and prevention of diseases. For a general biomedical sensor, the concentration of analyte in the test solution is the basis of disease diagnoses. Thus, a potentiostat consists of three electrode (working, reference, and counter electrode) is applied to detect the concentration [1]. By placing the electrode of the potentiostat into the test solution, chemical reaction caused by potential change will produce current flowing through the proposed circuit. The current value is proportional to the concentration of analyte. Hence, a current sensing circuit is required.

Circuit Implementation

The architecture of the proposed circuit shown in Fig. 1 consists of a current conveyor, a current mirror, a current range detection circuit, and a current to digital converter. The function works will be described in the following sections.

This work was supported by the Ministry of Science and Technology, Republic of China, under Grant #MOST 103-2220-E-006 -007.

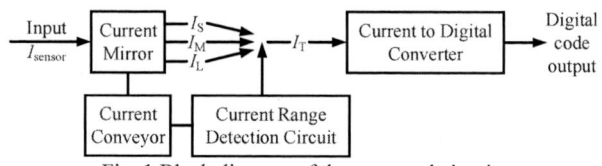

Fig. 1 Block diagram of the proposed circuit.

A. Current Conveyor

When measuring the current flowing through the working electrode of the potentiostat, the current conveyor [2] is used to maintain the potential of the working electrode to be stable.

B. Current Mirror

When measuring the input current, the linearity of the output current data is the most important item and must be taken into consideration. Therefore, the first step is to reduce the input current to an appropriate current range by a current mirror. To meet the design specification, in which the input current I_{sensor} is from 20 μA to 1200 μA, a three ratio reduction structure is presented. The current value is normalized to the same range from 0.25 μA to 1 μA with better linearity. The current flows through the three current paths are denoted as I_S, I_M, and I_L, which are the mirrored current of the original input current I_{sensor}. The current ratios of I_S, I_M, and I_L to I_{sensor} are designed as 1/80, 1/320, and 1/1200 respectively.

C. Current Range Detection Circuit

The structure of the current range detection circuit is shown in Fig. 2. Similar to the structure of current mirror, this circuit is also attached to the current conveyor. First, the sensor current I_{sensor} is mirrored by the p-type MOS transistors so that the current flowing through the resistor R_1 and R_2 are identical to I_{sensor}. Therefore, the voltages of V_{R1} and V_{R2} are determined by the product of the sensor current I_{sensor} and the resistance value of R_1 and R_2 respectively. The resistance of R_1 is 2 kΩ, and that of R_2 is 8 kΩ, V_{ref} is set to be 0.64 V. *Out1* and *Out2* are the signal outputs indicating which current range is selected. Switch1 and Switch2 are transmission gates with same structure but the latter one is placed inversely. If the control signal is low, Switch1 turns on, vice versa. The function of Switch2 is inverted. If I_{sensor} is less than 80 μA, *Out1* and *Out2* remain low, only the current path I_S flows through. If I_{sensor} is in the range from 80 μA to 320 μA, *Out1* remains low and *Out2* changes to high, only the current path I_M flows through. If I_{sensor} is larger than 320 μA, both *Out1* and

978-1-4799-5128-4/14 $31.00 © 2014 IEEE

Out2 are changed to high, only the current path I_L flows through. The value of I_T equals to one of the three currents each time.

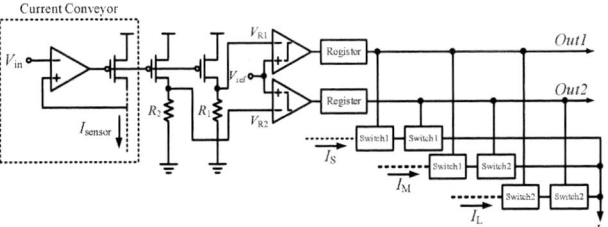

Fig. 2 Structure of the current range detection circuit.

D. Current to Digital Converter

This ciucuit consists of a current to timc converter, a counter, and a register, and its function is to generate digital output codes.

Experimental Results

The circuit has been fabricated in TSMC 0.18-μm CMOS process. Fig. 3 shows the micrograph of the IC. The chip area is 0.75×0.75 mm^2. The relationship between the input current to corresponding reciprocal time values obtained from digital output codes are shown in Figs. 4 to 6. The trend lines and theoretical lines of the output data are depicted in these figures. In addition, the *R*-squared values are also shown in the figures. Table I shows the comparison with other works of current sensing architecture.

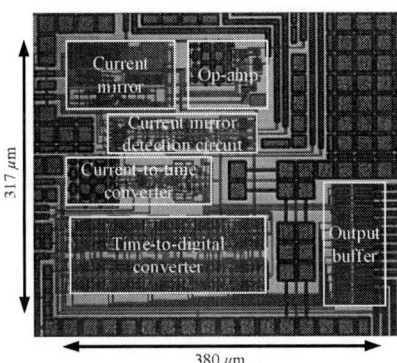

Fig. 3 Micrograph of the chip.

Conclusion

In this paper, a current range detection circuit for electrochemical sensing was presented. By using a DC current source generator to conduct the measurement, a comparison chart of input current value to output code can be established. This study with small area and micro-ampere sensing range is available for different potentiostat systems which can be integrated into multifunction biomedical devices. The measured current range achieves from 20 μA to 1200 μA and has good linearity with average *R*-squared value of 0.9986 while consuming 1.41 mW from 1.8 V power supply.

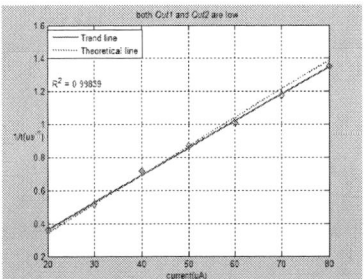

Fig. 4 Linear relationship between both *Out1* and *Out2* at low level.

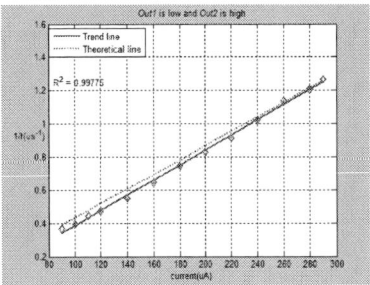

Fig. 5 Linear relationship between *Out1* at low level and *Out2* at high level.

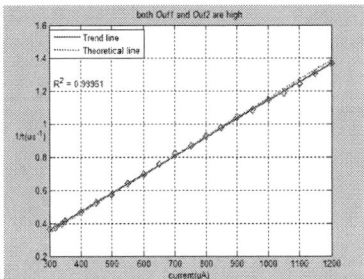

Fig. 6 Linear relationship between both *Out1* and *Out2* at high level.

TABLE I
Comparison of Performance

Specifications	This work	[3]	[4]	[5]
Process (μm)	0.18	0.5	0.5	0.35
Supply voltage (V)	1.8	5	3	5, 3.3
Clock frequency (Hz)	100M	1M	100k	100k
Input current range	20 μA~1200 μA	1 pA~ 1 μA	100 fA~ 1 μA	0~12 μA
Power consumption (mW)	1.41	0.5	0.45~1.2	5
Core area (mm^2)	0.5625	1.21	9	1

References

[1] P. M. Levine, P. Gong, R. Levicky, and K. L. Shepard, "Active CMOS sensor array for electrochemical biomolecular detection," *IEEE J. Solid-State Circuits*, vol. 43, pp. 1859-1871, July 2008.

[2] H. S. Narula and J. G. Harris, "A time-based VLSI potentiostat for ion current measurement," *IEEE Sensors J.*, vol. 6, pp. 239-247, Apr. 2006

[3] A. Bandyopadhyay, G. Mulliken, G. Cauwenberghs, and N. Thakor, "VLSI potentiostat array for distributed electrochemical neural recording," in *Proc. IEEE ISCAS*, May 2002, pp. 740-743.

[4] M. Stancacevic, K. Murami, A. Rege, G. Cauwenberghs, and N. Thakor, "VLSI potentiostat array with oversampling gain modulation for wide-range neurotransmitter sensing," *IEEE Trans. Biomed. Circuits Syst.*, vol. 1, pp. 63-72, March 2007.

[5] C. Y. Huang, "Design of a voltammetry potentiostat for biochemical sensors," *Analog Integr. Circ. Sig. Process*, vol. 67, pp. 375-381, June 2011

A Compact Comparator Block for Laser Radar Front-end Receiver

Jongsun An, Joo-Young Choi, Eun-Gyu Lee and Choul-Young Kim

Department of electronics engineering, Chungnam National University
99 Daehak-ro, Yuseong-gu, 305-764
Daejeon, Korea
E-mail : jsan@cnu.ac.kr

Abstract

This paper describes a comparator block fabricated in 0.18um CMOS process. The comparator block composed of transimpedance amplifier(TIA) as a preamplifier and comparator with hysteresis. The regulated cascode(RGC) structure TIA is used as a preamplifier. The measured results provide 79dBΩ transimpedance gain, 148MHz -3dB bandwidth for a 0.5pF photodiode capacitance and 6.3mW power consumption for 1.8V supply voltage. The comparator has been integrated in a core size of only 30 X 35um^2 . The proposed comparator detect -33dBm optical power from 9A/W avalanche photo diode(APD) responsivity and has 13.65cm of walkerror.

Keywords-Laser Radar; RGC; TIA; Comparator; Hyesterisis; Laser Radar; front-end receiver;

Introduction

The laser radar systems using short laser pulse and time-of-flight (TOF) method have a great advantage of resolution compared with sonar or microwave. Therefore, laser radar systems have been deployed in various applications such as autonomous vehicle, remote sensing, robot vision and surface mapping for buildings and scenes. [1]

In this paper, comparator block is described as laser radar front-end receiver for real-time 3D images is to use a focal plane array (FPA) of the photodetectors. [2]

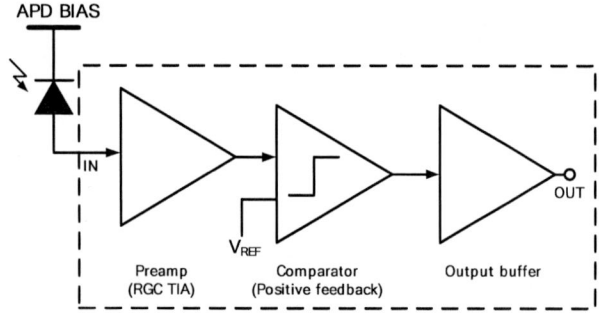

Fig. 1 Comparator block diagram for laser radar receiver

Design of Unit Comprator Block

A. RGC Transimpedance Amplifier

Fig. 2 shows the schematic diagram of the RGC circuit. The RGC circuit has the characteristics of high output impedance and wide output voltage range. The photocurrent is amplified to be a voltage at the drain of M1. The M2 and R3 stage reduces input impedance as a local feedback. [3]

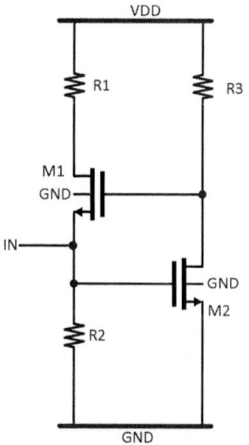

Fig. 2 Schematic diagram of the RGC TIA

Equation 1 shows the input impedance of RGC TIA where $(1+g_{m2}R_3)$ is the voltage gain of the local feedback and the input impedance is reduced by voltage gain

$$Zin(0) \cong \frac{1}{g_{m1}(1+g_{m2}R_3)} \qquad (1)$$

The input impedance of RGC is smaller than that of Common gate(CG) input. [4]

B. Comparator

Fig. 3 shows the schematic diagram of comparator with hyesterisis. The comparator composed of decision stage using positive feedback, self-biased differential amplifier and output buffer. Hysteresis characteristics have a immunity of noise in decision stage. [5]

978-1-4799-5128-4/14 $31.00 © 2014 IEEE   198

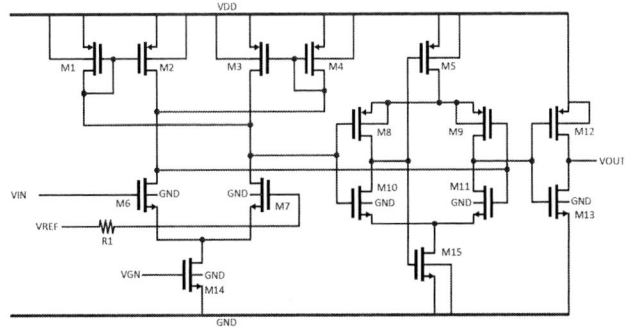

Fig. 3 Schematic diagram of the Comparator

C. Unit Pixel

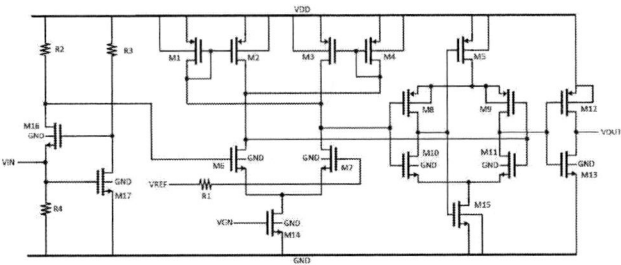

Fig. 4 Comparator unit pixel schematic

A optical current from a APD is amplified in the TIA stage, and then comparator convert analog voltage signal to digital voltage signal with a distance information

Measurement result of Comparator

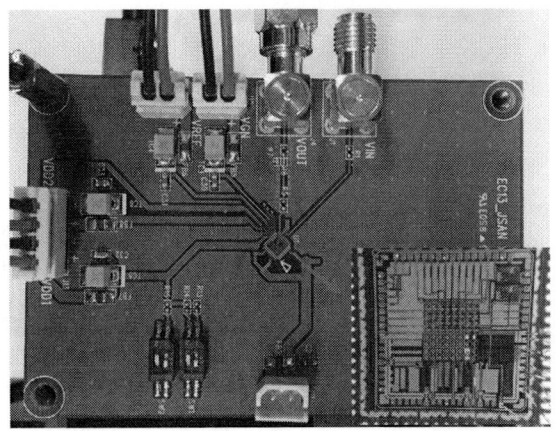

Fig. 4 PCB & Chip

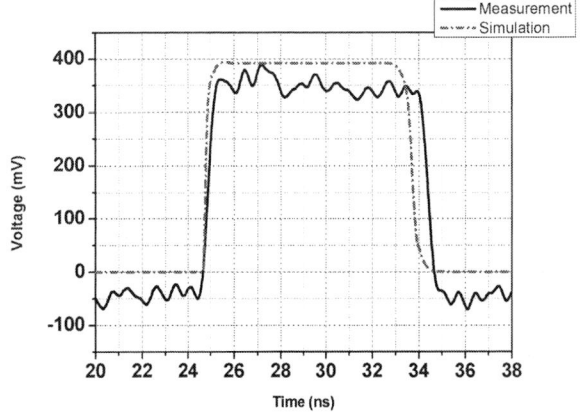

Fig. 5 Comparator output pulse measurement

A test chips were measured on a FR-4 PCB. Table 1 shows the comparator measurement result.

TABLE I
SUMMARY OF MEASUREMENT RESULTS

Process	0.18um CMOS
Minimum detectable optical power	-33dBm
Walkerror	13.65cm
Power consumption	6.3mW
Core size	30 X 35um^2

Conclusion

The comparator with RGC TIA as a preamplifier provides 79dBΩ transimpedance gain, 148MHz -3dB bandwidth for a 0.5pF photodiode capacitance, 6.3mW power consumption for 1.8V supply voltage and -33dBm minimum detectable optical power in a core size of only 30 X 35um^2.

In this work, the proposed comparator has a advantage of high gain, high resolution and small size. Consequently, it can be applied to laser radar front-end receiver.

References

[1] R. Stettner, H. Bailey, and S. Silverman, "Three dimensional Flash LADAR focal planes and time dependent imaging," *Int. J. High Speed Electron, Syst.*, 2008, 18, (02), pp. 401-406

[2] S. Thrun, M. Montemerlo, H. Dahlkamp, D. Stavens, A. Aron, J. Diebel, P. Fong, J. Gale, M. Halpenny, G. Hoffmann, K. Lau, C. Oakley, M. Palatucci, V. Pratt and P. Stang, "Stanley: the robot that won the DARPA grand challenge," *J. Field Robot.*, 2006, 23, (9), pp. 661-692

[3] S.M. Park and H.J. YOO, "1.25-Gb/s Regulated Cascode CMOS Transimpedance Amplifier For Gigabit Ethernet Applications," *IEEE J. of Solid-State Circuits,* vol. 39, no. 1, pp. 112-121, Jan. 2004

[4] S.M. Park, "CMOS Transimpedance Amplifiers for Gigabit Ethernet Applications," Journal Title, ICCT(Publisher Name), vol. 43, no. 4, pp. 16-22, 2006.

[5] Phillip E. Allen, "CMOS Analog Circuit Design," Oxford, 2002

978-1-4799-5128-4/14 $31.00 © 2014 IEEE

A 6bit 550Ms/s Small Area Low Power Successive Approximation ADC

Zhou Peng[1], Chenxi Han[2], Dongmei Li[2], Zhihua Wang[1]

Tsinghua National Laboratory for Information Science and Technology
[1]Institution of Microelectronics, Tsinghua University, Beijing 100084, China
[2]Dept. of Electronic Engineering, Tsinghua University, Beijing 100084,China
pz_tsinghua@foxmail.com, aasdehcx@163.com

Abstract

This paper presents a new capacitor array architecture to achieve a 6 bit 550Ms/s energy-efficient SAR with 65nm CMOS, which also takes up smaller area than traditional SAR. The bypass logic is a key feature to speed up the SA algorithm. Dynamic logic is used in the critical path to accelerate the speed. The whole circuit is supplied with 1.2V voltage. Simulation results show that the SAR achieves ENOB of 5.72, power consumption of 3.12mW with sampling rate at 550Ms/s, input frequency at Nyquist frequency.

Keywords-SAR ADC; capacitor array; bypass logic

Introduction

Analog to digital converters(ADCs) have become one of the most important components in electronics devices. The performance of ADC is constantly driven by the new applications which needs higher speeds and resolutions. In high speed ADC area, the Flash and pipeline type ADC is not power and area efficient, thus it is not suitable in energy limited applications. However, for SAR ADC it's better when used in energy efficient devices. The proposed SAR-type ADC in this paper can achieve very high speed with no time-interleaving and no flash-SAR hybrid structure (like 2 bit/cycle[1] or two alternative comparator[2]). In addition, the SAR in this design is easily used in Time-interleaving (TI) ADC. Because of its high speed, the number of channels can be reduced, which saves the area and reduces the complexity of TI.

SAR Architecture and Curcuit Design

A. SAR Architecture

Fig.1 shows the proposed SAR architecture, mainly including three parts: DAC, comparator, SAR logic. The logic is asynchronous, before the start of the conversion, the input signal is sampled onto all the capacitors. The external clock will initiate the first comparison, after it is done, the RDY token signal is generated to produce the reset signal. The next bit cycles are very similar with the first, the difference is that the regeneration of the comparator is triggered by the previous rising edge of the reset signal not the external clock.

B. DAC capacitor array design

In SAR design, the DAC capacitor arrays take up most area

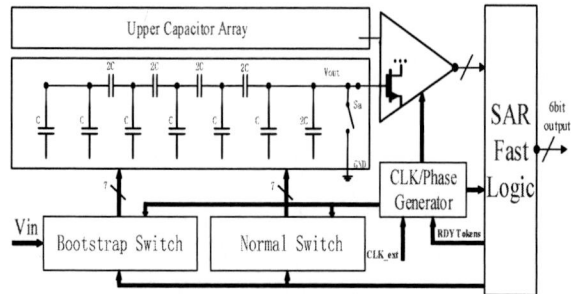

Fig. 1 The proposed SAR architecture.

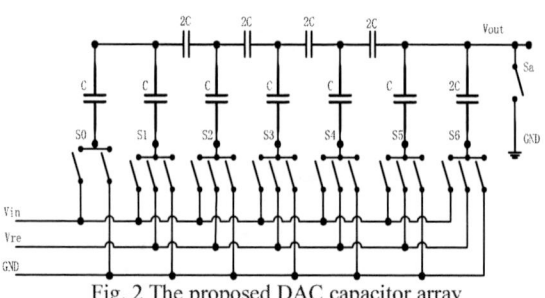

Fig. 2 The proposed DAC capacitor array

TABLE I
THE COMPARISON BETWEEN DIFFERENT DAC
ARCHITECTURES

Architecture	BWCA	TBWCA	C-2CCA	This work
Area cost	128c	32c	34c	32c
Input capacitive load	64c	8.875c	2c	4c

of the whole circuit. The proper choose of capacitor array is quite important to reduce the area of the circuit. There are several candidates for DAC capacitor array, traditional binary-weighted capacitor array (BWCA), two (multi)-stage binary-weighted capacitor array (TBWCA), C-2C capacitor array (C-2CCA). In this design, we propose a different architecture which is based on C-2C DAC, as is shown in Fig. 2. The comparison between different architectures in area cost and capacitive load for 6 bit ADC design is shown in TABLE I (assume c stands for the unit capacitor and It's a differential circuit).The architecture we proposed takes up very small area, the input capacitive load is a little larger than that of C-2C based capacitor array, which is also good for recudcing charge sharing effect between the capacitive load and parasitics of comparator.

C. Comparator design

The dynamic comparator shown in Fig. 3 is a good choice in this design[4]. The comparator includes two stages, a preamplifier and a sense amplifier. We add two cross-coupled transistors to reduce the kick back noise. The loads of the preamplifier is a pair of two PMOS transistors, the proper size of them can improved the speed of the comparator, Decreasing the size of the PMOS load is good for speed, but it will also reduce the gain of the amplifier, which is bad for the offset and noise of the second stage. In addition, the small load will lead to more mismatch.

C. Bypass SAR logic

A whole clock cycle includes sampling period and six bit decision cycles, the sampling period is determined by RC constant of sampling switch and capacitor array. In this part, we mainly discuss the speed of the six bit decision process (SAR logic), because in this high speed SAR ADC, SAR logic takes up much more time than sampling period. For every bit decision cycle, it includes compare time, DAC settling and SAR logic delay. Once the comparator determine its result, the more quicker the result is utilized to DAC settling, the more shorter time the bit cycle will take. Here we propose a bypass logic which can speed this process, as is shown in Fig. 4. After the comparator completes the comparing, a ready signal is generated, which is used as the clock of the MUX to switch the CMP_result to right channel for DAC settling, the critical path after the compare only includes one D-flip-flop, and this flip-flop is optimized in dynamic logic to accelerate the speed.

Simulation Results

The proposed high speed SAR is designed in TSMC 1P9M 65nm CMOS technology. Fig.5 shows the simulation result at Nyquist frequency. TABLE II compare the result with the current state of the art.

Acknowledgment

This research was supported by the National High Technology Research and Development Program of China (Grant No. 2012AA012301) and the State Natural Sciences Foundation Project of China (Grant No.61171001)

References

[1] Y.-C. Lien, "A 4.5-mW 8-b 750-Ms/s 2-b/step Asynchronous Subranged SAR ADC in 28-nm CMOS Technology," *Symp. On VLSI Curcuits*, pp.88-89, June 2012.

[2] L. Kull, T. Toifl, M. Schmatz, et al., "A 3.1mW 8b 1.2GS/s Single-Channel Asynchronous SAR ADC with AlternateComparators for Enhanced Speed in 32nm Digital SOI CMOS," *ISSCC Dig. Tech. Papers*, pp.468-469, Feb.2013.

[3] V. Tripathi, B. Murmann, "An 8-bit 450-Ms/s Single-Bit/Cycle SAR ADC in 65-nm CMOS," *ESSCIRC*, pp.117-120,Sept.2013.

[4] C.-H. Chan, Yan Zhu, U-Fat Chio, et al., "A Reconfigurable Low-Noise Dybamic Comparator with Offset Calibration in 90nm

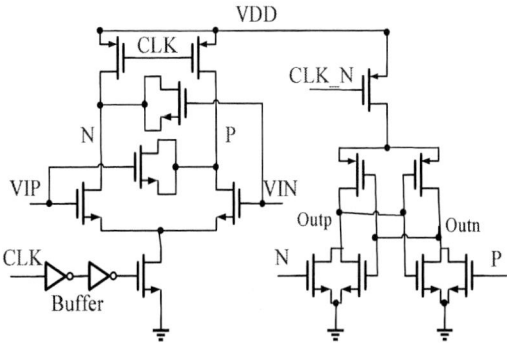

Fig. 3 Schematic of the proposed dynamic comparator.

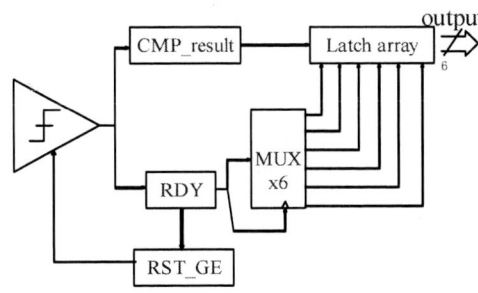

Fig. 4 Block diagrams of the proposed bypass SAR logic.

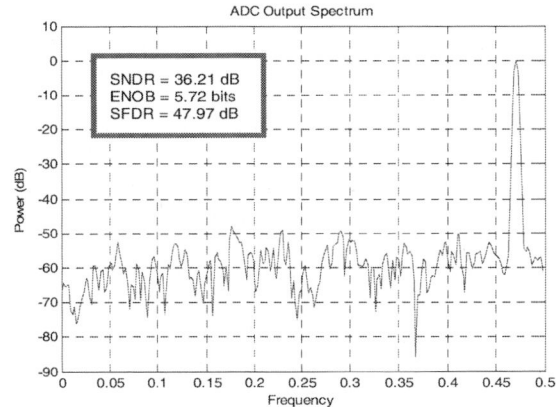

Fig. 5 ADC output FFT(1024pts) at 550Ms/s.

TABLE II
PERFORMANCE COMPARISON

	[1]	[2]	[3]	This work
Architecture	2b/cycle	1b/cycle (2 cmp)	1b/cycle (1cmp)	1b/cycle (1cmp)
Technology [nm]	28	32	65	65
Resolution	8	8	8	6
Fs[MS/s]	750	1200	450	550
SNDR[dB]	43.3	37.8	47.3	36.2
FOM [fJ/conv-step]	50.2	32.9	76	88.6

CMOS," *IEEE ASSCC*, pp.233-236, Nov.2011.

A 10-bit 20-MS/s Asynchronous SAR ADC with Controllable Analog Input Voltage Range and Meta-stability Detection Circuit

Sang-Min Park, Yeon-Ho Jeong, Dong-Gil Jeong, Seung-Wuk Baek, Yu-Jeong Hwang, Pil-Ho Lee, and Young-Chan Jang

Department of Electronic Engineering, Kumoh National Institute of Technology
61, Daehak-ro, Gumi-si, Gyeongsangbuk-do, Korea E-mail: ycjang@kumoh.ac.kr

Abstract

A 10-bit 20-MS/s asynchronous SAR ADC, which has a controllable analog input voltage range and a meta-stability detection circuit, is proposed. The proposed SAR ADC with the area of 0.095 mm² is implemented using a 130-nm CMOS process with 1.2-V supply. The measured peak ENOBs for the full rail-to-rail ±1.2V (peak-to-peak) differential sinusoidal input signal is 9.56 bits. The FoM achieves 41 fJ/conversion-step.

Keywords- asynchronous successive approximation register; analog-to-digital converter; half rail-to-rail analog input range

Introduction

Recently, there has been an increase in use of an ADC with an 8-bit to 12-bit resolution and a sampling rate of tens of MHz in wireless communication systems. Given these specifications, a successive approximation register (SAR) ADC is the most suitable architecture because it has minimal analog circuits, resulting in a small area and low power consumption. Generally, a SAR ADC with an analog input voltage range of the full rail-to-rail does not use an additional reference voltage by using a supply voltage as a reference voltage for a capacitor DAC [1][2]. To change the analog input voltage range of a SAR ADC, the additional reference voltages are required for a capacitor DAC. The prior literature [3] proposed the reference-free SAR ADC with a scalable analog input voltage range. This scheme equivalently amplified by two the analog input signal using additional sampling capacitor (C_S) and a 4-input preamplifier. However, the relative capacitance mismatch between the output node of a DAC and an additional sampling node using C_S may cause a gain error and a linearity error. In this paper, an SAR ADC, which can select an analog input range of the full rail-to-rail or the half rail-to-rail, is proposed and it does not use an additional reference voltage but changes the capacitor for a most significant bit of a DAC for an analog input voltage range of the half rail-to-rail.

Circuit and Implementation of proposed SAR ADC

A. Architecture of SAR ADC

Fig. 1 shows the block and timing diagram of the proposed 10-bit 20MS/s asynchronous SAR ADC. It consists of a DAC using the V_{CM}-based switching scheme [3], a dynamic comparator with a meta-stability detection circuit, and a SAR

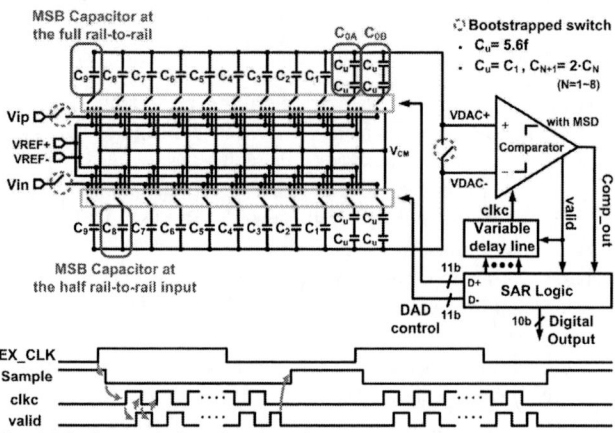

Fig. 1 Block and timing diagram of proposed SAR ADC

control logic. An analog input voltage range of the proposed SAR ADC is selected to be the full rail-to-rail and the half rail-to-rail by changing the capacitor for a most significant bit of a DAC. At the rising edge of *EX_CLK*, the sampling process is completed and the conversion process for a 10-bit resolution is started. The conversion process of ten times is asynchronously performed by the SAR logic. After this conversion process, the SAR ADC samples an analog input signals until the next rising edge of *EX_CLK*. According to the operation sequence of the proposed SAR ADC, its performance is independent the duty cycle ratio of *EX_CLK*.

B. DAC for proposed SAR ADC

The DAC designed in this SAR ADC uses the V_{CM}-based switching scheme to reduce total capacitance of the DAC. To implement a SAR ADC with analog input voltage ranges of the full rail-to-rail and the half rail-to-rail, the capacitor for a most significant bit of a DAC is controlled without the additional reference voltages according to an analog input voltage range. For the operation of a SAR ADC with the full rail-to-rail analog input voltage range, the capacitors of from C_9 to C_1 in the DAC are switched for a 10-bit resolution according to Eq. (1). In this case, C_{0A} and C_{0B} are only operated as a sampling capacitor.

$$
VDACM - VDACP = (Vip - Vin) \pm \{\frac{1}{2}(Vrefp - Vrefm) +
$$
$$
\dots + \frac{1}{2^{n-1}}(Vrefp - Vrefm)\} \tag{1}
$$

The operation of a SAR ADC with the half rail-to-rail analog input voltage range should be performed by Eq. (2).

The capacitors of from C_8 to C_{0A} in the DAC are switched for a 10-bit resolution. To reduce the reference voltage in half, C_9 and C_{0B} are operated as a sampling capacitor.

$$VDACM - VDACP = (Vip - Vin) \pm \frac{1}{2}\{\frac{1}{2}(Vrefp - Vrefm) +$$
$$... + \frac{1}{2^{n-1}}(Vrefp - Vrefm)\}$$
(2)

C. Comparator with Meta-stability Detector Circuit

An asynchronous SAR ADC requires the circuit to detect a meta-stability of a comparator. Fig. 2 shows the block diagram of comparator with meta-stability detector circuit. When the original comparator operates at a meta-stable state, the signal *valid*, which informs the completion of the comparison operation of a comparator, is generated by the delay time of three replica comparators. According to Hspice simulation, the comparison delay time of the original comparator is similar to the delay time of three replica comparators when the input voltage difference of the original comparator is about 0.1 LSB.

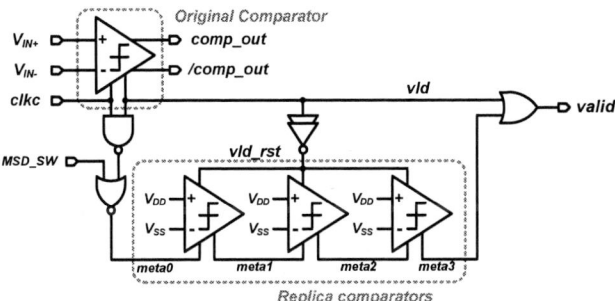

Fig. 2 Block diagram of comparator with meta-stability detector circuit

Measurement Results

The proposed SAR ADC with a sampling rate of 20 MS/s was implemented using a 130nm CMOS process with a 1.2-V supply. Fig. 3 shows the micrograph and layout of the fabricated SAR ADC. The total active area is 0.095 mm^2. The performances of the implemented SAR ADC were evaluated at the analog input voltage ranges of the half rail-to-rail ($\pm0.6V_{pp}$) and the full rail-to-rail ($\pm1.2V_{pp}$). The measured DNL and INL are +0.65/-0.66 LSB and +0.56/-0.89 LSB, respectively, in the case of the full rail-to-rail The frequency spectrum of the SAR ADC output is shown in Fig. 4 when the sampling rate is 20 MS/s and the input signal is the full rail-to-rail $\pm1.2V$ (peak-to-peak) differential sinusoidal with a frequency of 9.97 MHz. The measured SNDR and SFDR are 59.31dB and 70.15dB, respectively. Fig. 5 shows the measured ENOBs according to the frequency increase of the analog input signal at two analog input voltage ranges. The total power consumptions of the SAR ADC for two analog input voltage ranges are 0.61mW and 0.62mW, respectively. Table I

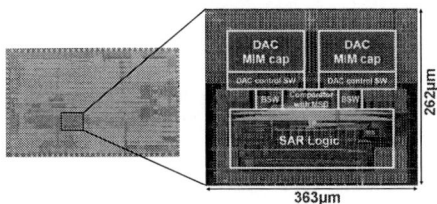

Fig. 3 Micrograph and layout of implemented SAR ADC

summarized the performance and comparison of the proposed SAR ADC. This comparison was achieved by selecting the SAR ADCs designed using a comparable CMOS process of 90 nm or 130 nm.

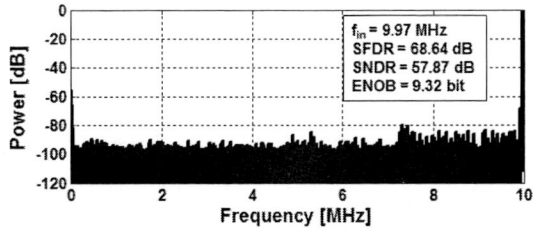

Fig. 4 Measured frequency spectrum @ full rail-to-rail analog input voltage range

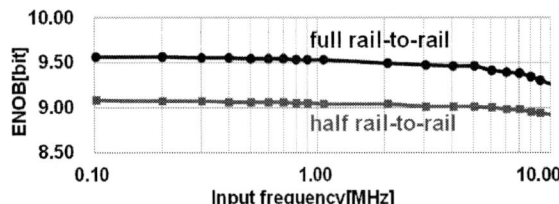

Fig. 5 ENOB versus input frequency at half rail-to-rail and full rail-to-rail analog input voltage ranges

Table I. Performance summary and comparison

Specifications	[1]	[2]	[3]	This work	
Process [nm]	130	130	90	130	
Supply [v]	1.2	1.2	1.2	1.2	
Resolution [bit]	10	10	10	10	
Sample rate [MS/s]	50	40	100	20	
Area [mm^2]	0.075	0.32	0.18	0.095	
Input range[V]	±1.2	±1.2	±0.6	±0.6	±1.2
ENOB [bit]	8.48	8.35	9.1	9.08@0.1MHz 8.94@Nyq	9.56@0.1MHz 9.32@Nyq
INL [LSB]	+2.2 /-2.09	+0.9 /-1.55	+0.86 /-0.78	+0.56 /-0.89	+0.22 /-0.47
DNL [LSB]	+0.88 /-1	+0.72 /-0.78	+0.79 /-0.78	+0.65 /-0.66	+0.34 /-0.32
Power [mW]	0.92	0.55	3	0.61	0.62
FoM [fJ/c-s]	52	42	55	56	41

Acknowledgment

This research was supported by ETRI, IDEC, and the Basic Science Research Program through the National Research Foundation of Korea (NRF) funded by the Ministry of Education (2013R1A1A4A01012914).

References

[1] C.-C. Liu, S.-J. Chang, G.-Y. Huang, and Y.-Z. Lin, "A 0.92mW 10-bit 50-MS/s SAR ADC in 0.13μm CMOS process," *Symp. VLSI Circuits Dig. Tech Papers*, pp. 236-237, June 2009.

[2] S.-H. Cho, C.-K. Lee, J.-K. Kwon, and S.-T. Ryu, "A 550-μW 10-b 40-MS/s SAR ADC with Multistep Addition-Only Digital Error Correction," *IEEE J. Solid-State Circuits*, vol. 46, no. 8, pp. 1881-1892, August 2011.

[3] Y. Zhu, C.-H. Chan, U-F. Chio, S.-W. Sin, S.-P. U, R.P. Martins, and F. Maloberti, "A 10-bit 100-MS/s reference-free SAR ADC in 90 nm CMOS," *IEEE J. Solid-State Circuits*, vol. 45, no. 6, pp. 1111-1121, June 2010.

A 24-mW 60-GHz OOK RF transceiver for 3-Gbps data communication

[1)]Hui Dong Lee, [1)]Tae Young Kang, [2)]Ki Chan Eun, [1)]Moon-Sik Lee, and [1)]Bonghyuk Park

[1)]Electronics and Telecommunications Research Institute
218 Gajeong-ro, Yuseong-gu, Daejeon, South Korea
[2)]KORF Incorporated
Phone: +82-42-860-6238, Fax: +82-42-860-6732, and E-mail: leehd@etri.re.kr

Abstract

This paper presents the design and experimental results of a low-power 60-GHz OOK transceiver. For low-power operation, a simple OOK structure is utilized. A switching modulator in the transmitter consumes less power due to no power consumption at the low-state of data signal. In addition, an LO generater in the receiver is not required. The transceiver has been implemented in a 65-nm CMOS technology. The transmitter and the receiver consume 23.6-mW and 24-mW from a 1.0 V supply, respectively. The transmitter has an output power of -5 dBm, on/off isolation of 34.5 dB, and tuning range of 54 to 61.5 GHz. The receiver has a conversion gain of 22 dB and a bandwidth over 3 GHz.

Keywords-CMOS; millimeter wave; MMIC; on-off keying (OOK); 60-GHz; transceiver; VCO;

Introduction

Recently, the demand for the wireless communications using the millimeter wave has been rapidly increasing due to the bandwidth limitations of conventional communications [1-5]. A simple modulation scheme, such as OOK or ASK, for gigabit data communication using the unlicensed 60-GHz band has been utilized [3-5]. It is more important to have a low-power performance in implementation of the 60-GHz system due to the feature of the wireless communication. However, recently published 60-GHz systems have consumed a relatively large power [2-4]. This paper presents the design and measured results of a low-power 60-GHz OOK RF transceiver.

OOK Transceiver Design

Fig. 1 shows the block diagram of 60-GHz RF OOK transceiver. Because of the millimeter wave and low-power operation, a simple structure is suitable. The RF transmitter (Tx) consists of OOK modulator and 60-GHz VCO (30-GHz VCO + doubler). And the RF receiver (Rx) consists of LNA and OOK demodulator (detector + baseband amplifier (BBA)).

A. Transmitter Design
Fig. 2(a) shows the detailed schematic of RF Tx. To generate a stable 60-GHz signal, we have designed the 30-GHz cross-coupled LC-VCO with n-type MOSFETs. And then, a single-ended LO signal of 60-GHz is achieved throughout the frequency doubler with a push-push structure by utilizing the second harmonic components. Next, the operation of the modulator is described. The generated LO signal and the

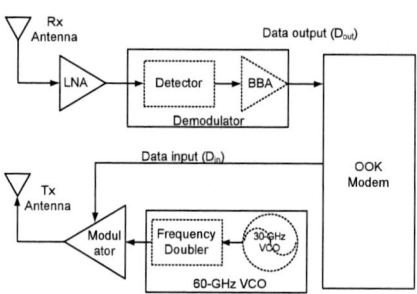

Fig. 1. Block diagram of 60-GHz OOK transceiver

high-speed data signal (D_{in}) are inputs of the OOK modulator. When D_{in} is in high state, LO signal is transmitted to the output. In the other case, the transistor of the current source, M_5 in Fig. 2(a), is turned off and no DC current flows. Therefore, this switching OOK modulation can reduce DC power consumption. The differential signal is converted into the single-ended OOK output signal by the on-chip transformer to deliver it to Tx antenna. The Tx outputs the OOK RF signal including an LO signal. Because the Rx can obtain the data signal by self-mixing the RF signal, it is possible to configure the OOK Tx and Rx with the simplified structure.

B. Receiver Design
Fig. 2(b) shows the detailed circuit of RF Rx. The received signal via Rx antenna is amplified by LNA. We have designed the 2-stage common-source LNA using a current-reused structure for high-gain and low-power operation. The demodulator is composed of a detector and a baseband amplifier. Specially, the gain-boosting technique, M_5 and M_6 in Fig. 2(b), is applied to the detector to increase the gain and operation speed [5]. The LO signal component in the amplified signal by LNA can be removed at the detector by using a square-law detection method and the data signal is extracted and amplified by the baseband amplifier. This detection method is referred to as a non-coherent detection technique and LO signal is not required. Thus, we have implemented the low-power OOK Rx based on square-law detection. The BBA is based on 3-stage differential amplifiers and designed to have a gain of 26 dB and a bandwidth of 4 GHz. All inductors, transformer, and interconnection lines have been implemented as the microstrip line using the metal9 as the signal line and the metal1 as the ground plane. Precise EM analysis has been increased the design accuracy.

Experimental Results

The proposed OOK RF transceiver has been implemented in a 65-nm CMOS technology. The core chip size of the Tx is 270 x 340 μm². And, those of the LNA and demodulator are 280 x

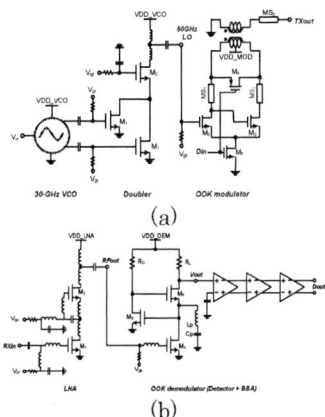

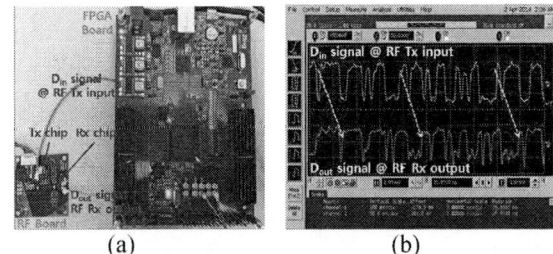

(a)

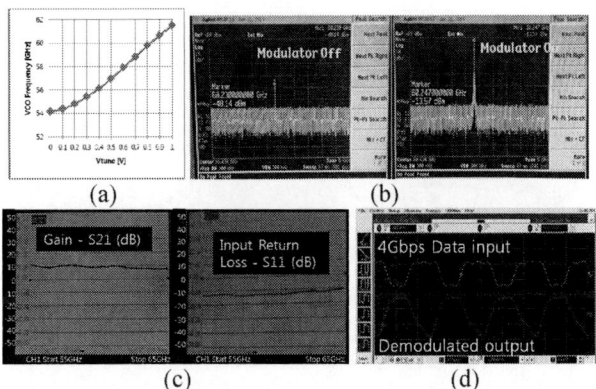

(a)   (b)

Fig. 4. Data communication experiment; (a) Test setup and (b) input and output waveforms of OOK RF transceiver.

Fig. 2. Schematic of the 60-GHz OOK transceiver; (a) transmitter and (b) receiver.

(a)   (b)

(c)   (d)

Fig. 3. Measured transmitter and receiver results; (a) 60-GHz VCO curve, (b) Modulator on/off characteristics, (c) S-parameter results of LNA, and (d) Demodulator results.

220 μm^2 and 140 x 200 μm^2, respectively. The measurement results are given below.

Fig. 3(a) shows the generated frequency curve of 30-GHz VCO and frueqney doubler. The wider frequency tuning range of 54.1 to 61.5 GHz and the maximum output power of -10 dBm are achieved while power consumption of 30-GHz VCO is 5.6 mW and that of the doubler is 10.7 mW. The measured power includes cable and probe losses of about 9 dB. Fig. 3(b) represents the output power levels of the modulator at low-state and high-state of the data signal, respectively. And, the on/off isolation is higher than 34 dB. The reason is that pseudo- differential structure and load switch M_4 of the modulator in Fig. 2(a) increase the on-off isolation. The average power consumption is about 7.3 mW. The output power is -14 dBm at 50 Ω terminal. A data signal more than 5-Gbps may be operated on this OOK Tx. The measured results of the Rx are shown in Fig. 3(c) and 3(d). The current-reused LNA has a gain of 9 dB and a wide bandwidth of 55 to 65 GHz while power consumption is around 6 mW from a 1 V supply. The output levels of demodulator are measured as 150, 140, and 100 mV_{pp} at the input data speeds of 2, 4, and 6 Gbps at the single-ended 50Ω terminal. The conversion gain of the demodulator is calculated as 13 dB while the DC power consumption is about 18 mW.

Fig. 4(a) shows the BER test-setup. The left board mounted on RF Tx and Rx chips is for the RF transceiver and the right board is the FPGA board for 3 Gbps OOK modem. The wireless transmission via antennas and air is replaced by the lossy transmission line on FR-4 board for simple verification. The attenuation of the transmission line implemented with the length of 1.2 cm on FR-4 boad is simulated as about 20 dB by the EM analysis. Fig. 4(b) represents the input and output waveforms. Throught the waveform results, we can confirm that the high-speed data of 3 Gbps is restored after delay time of 2-ns at the RF Rx output node. Additionally, a PRBS pattern with a length of 2^7-1 is applied to RF Tx input and then we have analyzed the restored data signal. The BER is measured as less than 10^{-12} for 2^7-1 PRBS pattern.

Conclusion

A 60-GHz OOK RF transceiver has been implemented in a 65-nm CMOS technology. The simple OOK transceiver with the switching modulated transmitter and the gain-boosting receiver using square-law detection method is applied for the low-power consumption. Thus, the transmitter and the receiver consume low-power of 23.6 mW and 24 mW, respectively. Prior to the interworking test with antennas, we have confirmed that high-speed communication of 3-Gbps is well operated by wired connection between transmitter and receiver. The experimental results indicate the proposed OOK RF transceiver is suitable to low-power 60-GHz wireless communications.

Acknowledgment

This research was supported by the ICT R&D program of MSIP/IITP. [14-000-04-001, Development of 5G Mobile Commnuication Technologies for Hyper-connected Smart Services]

References

[1] W.-H. Chen, *et al.*, "A 6-Gb/s wireless inter-chip data link using 43-GHz transceivers and bond-wire antennas," *IEEE J. Solid-State Circuits*, vol. 44, no. 10, pp. 2711-2721, Oct. 2009.

[2] Y. Tanaka, *et al.*, "A versatile multi-modality serial link," *IEEE Int. Solid-State Circuits Conf. Tech. Dig.*, Feb. 2012, pp. 332-333.

[3] E. Juntunen, *et al.*, "A 60-GHz 38-pJ/bit 3.5-Gb/s 90-nm CMOS OOK digital radio," *IEEE Trans. Microw. Theory Techn.*, vol. 58, no. 2, pp. 348-355, Feb. 2010.

[4] J. Lee, Y. Chen, and Y. Huang, "A low-power low-cost fully-integrated 60-GHz transceiver system with OOK modulation and on-board antenna sssembly," *IEEE J. Solid-State Circuits*, vol. 45, no. 2, pp. 264-275, Feb. 2010.

[5] C. W. Byeon, J. J. Lee, K. C. Eun, and C. S. Park, "A 60 GHz 5 Gb/s gain-boosting OOK demodulator," *IEEE Microw. Wireless Compon. Lett.*, vol. 21, no. 2, pp. 101-103, Feb. 2011.

Small size 433 MHz On-off Keying Transceiver IC for Sensor Application

Jeong-Ho Park, Han-Won Cho, Ji-Hoon Suh, and Hyung-Joun Yoo

The Department of Electrical Engineering and the Mobile Sensor and IT Convergence Center
Korea Advanced Institute of Science and Technology
373-1 Guseong-dong, Daejeon, Republic of Korea
pjh0656@kaist.ac.kr, myjhw@kaist.ac.kr, sj3995@kaist.ac.kr, hjyoo53@kaist.ac.kr

Abstract

In order to make wireless sensor nodes, low power consumption and low cost RF transceiver are required. In this work, a 433 MHz On-Off keying transceiver is designed and simulated with Dong-bu 0.11 μm CMOS process. For high data rate and power efficiency, digital pulse shaping is adopted. For small size and low power, the receiver is designed using the digital pulse-shaped signal for demodulation. The chip size is 180×180 μm^2 without pads.

Keywords-OOK; transceiver; pulse shaping; self-sampling;

Introduction

Wireless sensor nodes are recently used in many applications. Especially, environmental monitoring system and health care system use wireless sensor nodes [1, 2]. In order to make the wireless sensor nodes, low power consumption and small size RF transceiver design are required. Many kinds of on-off keying (OOK) transceivers have been studied. In the transmitter, the efficiency increases using pulse shaping method [3]. In the receiver, the envelope detector is used for low power consumption [4]. However, the size of the receiver that uses the envelope detector is large because of the capacitor.

In this work, small size a low power OOK 433 MHz RF transceiver is proposed. In order to reduce the spurious emission and increase power spectral efficiency, the digital pulse shaping method is adopted for the transmitter. For small size and low power consumption, the demodulator in the receiver is designed without the capacitors and the oscillator. The demodulator uses a digital pulse-shaped signal which is adopted instead of the capacitors and the oscillator.

Proposed Architecture

A. Transmitter

A transceiver adopting OOK modulation can be easily implemented. However, OOK signal covers a wide bandwidth. The digital pulse shaping can reduce the signal bandwidth [3]. Therefore, the higher data rate can be obtained through the digital pulse shaping even in the same bandwidth. As a result, spectrum efficiency increases. In this work, the digital baseband signal is processed by a 9-point digital moving

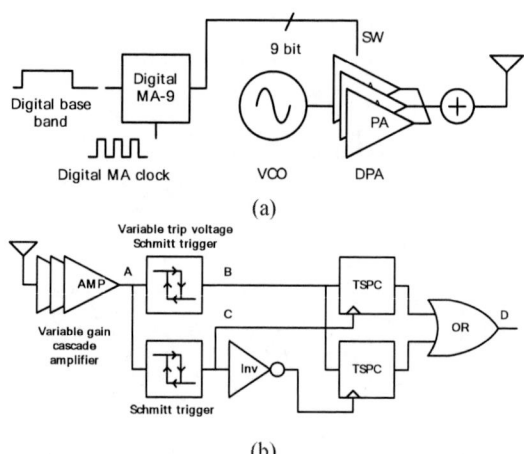

Fig. 1 The proposed OOK (a) transmitter and (b) receiver

average (MA) in order to implement pulse shaping. Fig. 1 (a) shows the transmitter architecture. The OOK transmitter contains a digital MA block, a voltage controlled oscillator (VCO) and digitally controlled power amplifier (DPA). The digital MA block is made up of 9 shift registers. The shift registers constitute 9 control bits in order to switch the DPA. DPA is a 9-bit switching power amplifier as shown by [5]. DPA can control the amplitude by 10 levels. Ring type VCO is used to reduce the power consumption and size.

B. Receiver

Fig. 1 (b) shows the receiver architecture. The receiver has no oscillator in order to reduce the power consumption and chip size. The receiver consists of a cascade amplifier, Schmitt triggers, true single phase clock (TSPC) D flip flops, and logic gates. The cascade amplifier is made of inverter-based amplifiers that can control the gain by adjusting the feedback resistors. Two types of Schmitt triggers are key blocks of the receiver. A variable lower trip voltage (VLTV) Schmitt trigger makes the pulses from the received signal. The VLTV Schmitt trigger can handle input signals with various amplitudes. A normal Schmitt trigger extracts the clock signal from the received signal. Using the pulse signal and the extracted clock, the OOK signal can be demodulated by TSPC. Two TSPCs (rising edge and falling edge) are used to resolve the sampling timing problem.

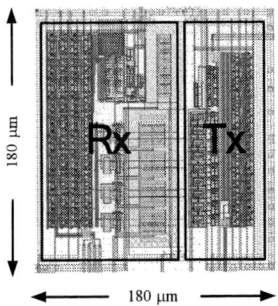

Fig. 3 Chip layout

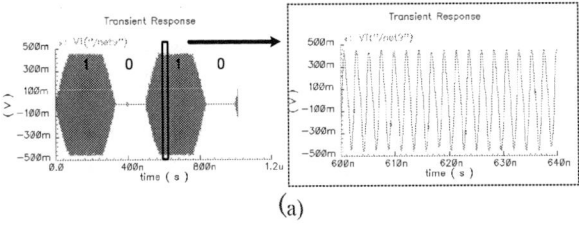

(a)

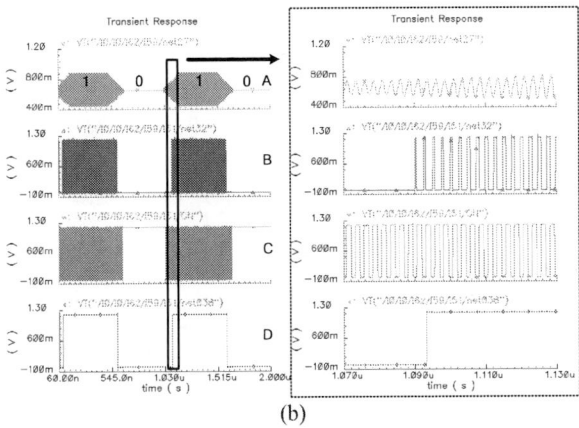

(b)

Fig. 4 Simulation results of (a) the transmitter and (b) the receiver

Simulation Results

Fig. 3 shows the chip layout of the OOK transceiver. Dong-bu 0.11 μm CMOS process is used for layout and simulation.

The simulation results of the transmitter are shown in Fig. 4 (a). The result is observed at the output of the antenna. The frequency of baseband signal is 2 MHz. The frequency of the digital MA clock is 60 MHz. The result shows that the signal amplitude can be controlled by 10 levels. The amplitude step voltage varies from 41.4 mV to 65.5 mV. As the amplitude of the output signal increases, the controlled step voltage decreases due to the impedance mismatch. However, the effect on the signal spectrum is not significant.

Fig. 4 (b) shows simulation results of the receiver. We assume the ideal transmitter signal for the receiver chain test. The frequency of the baseband signal is 2 MHz. The power of digital pulse-shaped input signal is -50 dBm. The simulation results after TSPC show the demodulated OOK signal.

Conclusion

Table I shows the performance summary of the proposed OOK transceiver. The proposed OOK transceiver with a low power consumption and small size can be used for various wireless sensor applications that require low cost and low power. The test chip will be measured after fabrication.

Acknowledgment

This work was supported by IC Design Education Center (IDEC).

References

[1] M. Akhondi, A. Talevski, S. Carlsen, and S. Petersen, "Applications of Wireless Sensor Networks in the Oil, Gas and Resources Industries," *24th IEEE International Conference on Advanced Information Networking and Applications (AINA)*, 2010, pp. 941-948.

[2] Ze Zhao and Li Cui, "EasiMed: A remote health care solution," *27th Annual International Conference of the IEEE-EMBS*, 2005, pp.2145-2148.

[3] X. Huang, P. Harpe, X. Wang, G. Dolmans, and H. de Groot, "A 0dBm 10Mbps 2.4GHz ultra-low power ASK/OOK transmitter with digital pulse-shaping," *Proc. of IEEE Radio Frequency Integrated Circuits (RFIC) Symposium*, 2010, pp.263-266.

[4] C. Chou, L. Liu, and C. Wu, "A MedRadio-Band Low-Energy-Per-Bit 4-Mbps CMOS OOK Receiver for Implantable Medical Devices," *35th Annual International Conference of the IEEE-EMBS*, 2013, pp.5171-5174.

[5] R. B. Staszewski, J.Wallberg, S. Rezeq. C.-M. Hung, O. Eliezer, S. Vemulapalli, C. Fernando, K. Maggio, R. Staszewski, N. Barton, M.-C. Lee, P. Cruise, M. Entezari, K. Muhammad, and D. Leipold, "All-digital PLL and transmitter for mobile phones," *IEEE Journal of Solid-State Circuits (JSSC)*, vol. 40, pp. 2469-2482, Dec. 2005.

TABLE I
PERPORMACE SUMMARY

	Parameter (unit)	[4]	This work
Tx	Carrier Freq. (MHz)	401~406	150~500
	Output power (dBm)	-15.55	5
	Data rate (Mbps)	4	2
	Power (mW)	0.27	3.36
	Energy per bit (nJ/bit)	0.067	1.68
	Size (mm^2)	-	0.011
Rx	Sensitivity (dBm)	-45.67	-50
	Data rate (Mbps)	4	2
	Power (mW)	0.27	1.5
	Energy per bit (nJ/bit)	0.07	0.750
	Size (mm^2)	-	0.021
	Technology	0.18 μm	0.11 μm
	Total size (mm^2)	0.57(pad)	0.032

A 24 GHz PMOS Body Voltage Controlled Oscillator with Transformer Coupled Varactor

Jae-Hoon Song[1], Byung-Sung Kim[2], Sangwook Nam[1]

[1] Institute of New Media & Communication, Seoul National University, Seoul, Korea
[2] School of Information and Communication Engineering, Sungkyunkwan University, Suwon, Korea
doritos43@ael.snu.ac.kr

Abstract

This paper includes a 24 GHz VCO design in 110 nm CMOS using transformer coupled varactor and body voltage control technique. This VCO maximizes frequency tunung range by using transformer coupled varactor and PMOS body voltage control. The measured results of the proposed VCO shows 24 GHz center frequency with 6.7 % frequency tuning range. The output frequency curve has wide linear tuning region. The phase noise of the VCO is -102 dBc/Hz at 1 MHz offset frequency. DC power consumption of the VCO core is 9 mW at 1.2V V_{DD}.

Keywords-component; VCO, Body Voltage Control, CMOS, Transformer Coupled Varactor

I. Introduction

With the recent increase in the demand for low-cost high-integration systems, CMOS has become an attractive solution. In case of CMOS communication systems, it has been great concern to achieve wide linear tuning range and good phase noise of voltage controlled oscillator (VCO) because the VCO mainly affects the sensitivity of fast-tuning systems such as radar sensors [1]. Conventional varactor-tuned CMOS VCOs suffer from nonlinear tuning range owing to inherent non-linearity of the varactor capacitance [2]. In this paper, by using transformer coupled varactor [3] and body voltage tuning [4], a VCO with wide linear frequency tuning range is proposed.

II. Circuit Description

The proposed 24 GHz VCO circuit is shown in Fig. 1. The cross-coupled core is realized by PMOS because PMOS parasitic capacitor makes rising output frequency curve when body voltage increases from 0V to V_{DD}. Also, PMOS has inherently low flicker-noise property compared to the NMOS counterpart. Width of PMOS core is 69 μm. A 3-D stacked center-tapped 1:1 transformer which size is 98×115 μm^2 is designed. The primary and the secondary inductors of the transformer are connected to the VCO core and the varactor pair, respectively. An advantage of utilizing the transformer coupling is extending the tuning range and the linear frequency tuning region of the VCO without generating an additional burden on the entire system such as the use of negative control

voltage or higher control voltage than the system V_{DD}. It is noted that to maximize the linear tuning range of the VCO, it is necessary to set the transition point of the varactor capacitance curve in the middle of the available tuning voltage range, usually $V_{DD}/2$ [5]. Thus, the center tap of the secondary inductor is biased by $V_{DD}/2$ generated by resistor dividing.

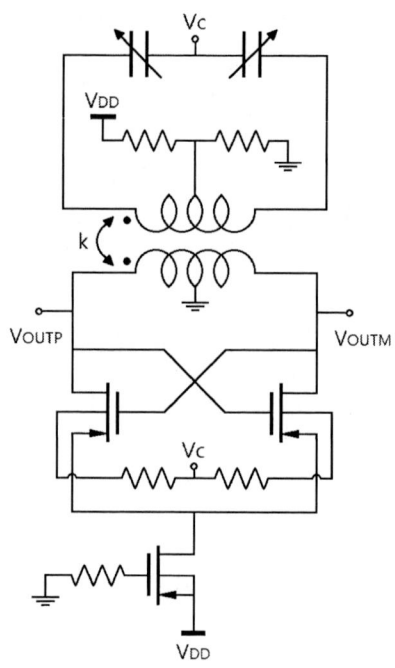

Fig. 1. Proposed VCO schematic

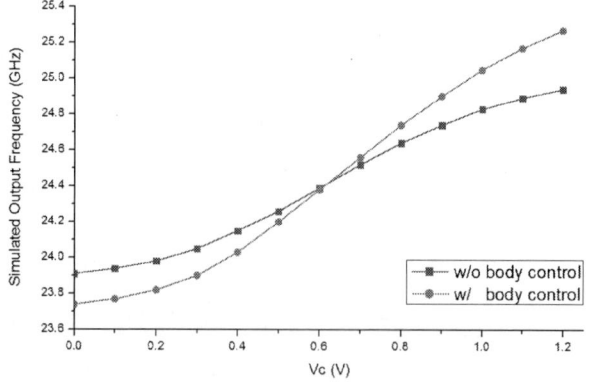

Fig. 2. Simulated output frequency curve.

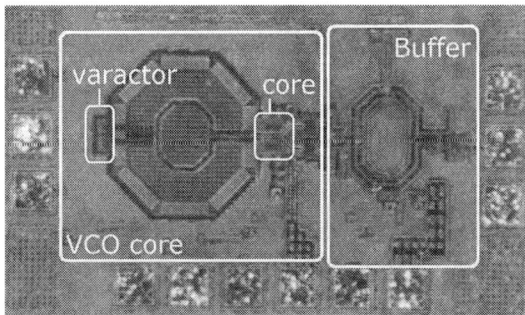

Fig. 3. Chip micro photograph

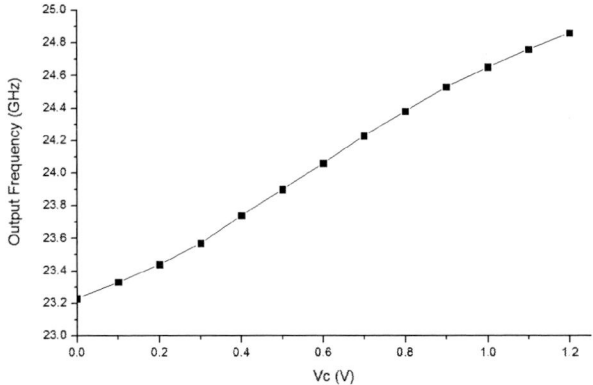

Fig. 4. Measured output frequency curve.

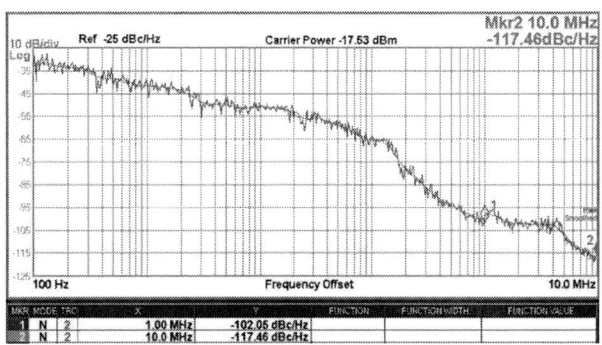

Fig. 5. Measured phase noise.

TABLE I
Performance of the Proposed VCO

Technology	110 nm RF CMOS
Center Frequency	24 GHz
Tuning Range	1.6 GHz (6.7%)
Phase Noise	-102 dBc/Hz @ 1MHz offset
Core DC power	9 mW
Full-Chip Size	$840 \times 470 \ \mu m^2$
FOM$_T$	-176.75

Fig. 2 represents simulated output frequency curve of VCOs with transformer coupled varactor. The simulated frequency tuning range of the proposed VCO (with body control) and without body control is 1.53 GHz and 1.03 GHz, respectively. By comparison between the two cases, it can be proved that the body voltage control technique extends frequency tuning range as much as 50%.

III. Measurement Results

The proposed 24 GHz VCO with transformer coupled varactor and body voltage control is fabricated in shirink 110-nm RF CMOS process. Fig. 3 shows a chip micro photograph. The chip size is $840 \times 470 \ \mu m^2$, including output buffer and test pads. The measured output frequency is shown in Fig. 4. The center frequency and the tuning range is 24 GHz and 1.6 GHz, respectively. The frequency curve has wide linear region at the center of the curve. The measured phase noise shows a value of -102 dBc/Hz at 1 MHz offset frequency [Fig. 5]. The core DC power consumption is 9 mW at 1.2V V_{DD}. FOM$_T$ of the proposed VCO is -176.75 from Eq. (1). The measured results are arranged in Table I.

$$FOM_T = L\{\Delta f\} + 10\log((\Delta f / f_{osc})^2 \cdot P_{dc}) - 20\log(FTR/10) \quad (1)$$

IV. Conclusion

A 24-GHz PMOS body voltage controlled oscillator with transformer coupled varactor is presented. The wide linear tuning range is achieved through simple and easily realizable techniques such as transformer coupled varactor and PMOS

body voltage control.

Acknowledgement

This work was supported by the National Research Foundation of Korea (NRF) grant funded by the Ministry of Science, ICT & Future Planning (MSIP) (No. 2009-0083495).

References

[1] J. D. Park and W. J. Kim ,"An Efficient Method of Eliminating the Range Ambiguity for a Low-Cost FMCW Radar Using VCO Tuning Characteristics," IEEE Trans. Microwave Theory and Techniques., vol. 54, no. 10, pp. 3623–3629, Oct. 2006.

[2] J. W. M. Rogers, J. A. Macedo, C. Plett, "The Effect of Varactor Nonlinearity on the Phase Noise of Completely Integrated VCOs," IEEE Journal of Solid-State Circuits., vol. 35, no. 9, pp. 1360–1367, Aug. 2002.

[3] J. H. Song, B.S Kim, S. Nam, "A 24 GHz Transformer Coupled CMOS VCO for a Wide Linear Tuning Range," IEICE Transactions on Electronics., Vol.E96-C, No.10, pp.1348- 1350, Oct. 2013.

[4] L.Geynet, E. De Foucauld, P. Vincent and G. Jacquemod, "Fully-Integrated Multi-Standard VCOs with switched LC tank and Power Controlled by Body Voltage in 130nm CMOS/SOI," IEEE Radio Frequency Integrated Circuits (RFIC) Symposium., Jun. 2006.

[5] S. Levantino, et al., "Frequency dependence on bias current in 5 GHz CMOS VCOs: impact on tuning range and flicker noise upconversion," IEEE Journal of Solid-State Circuits., vol. 37, no. 8, pp. 1003–1011, Aug. 2002.

A Spread Spectrum Clock Generator Using Discontinuous Modulation Technique for Reduction of Time Interval Errors

Taiming Piao, Jae-Kyung Wee, Inchae Song, Boo-Gyoun Kim

School of Electronic Engineering
Soongsil University, Seoul, Korea
ptm@ssu.ac.kr, wjk@ssu.ac.kr, isong@ssu.ac.kr, bgkim@e.ssu.ac.kr

Abstract

In this paper, we propose a novel discontinuous spread spectrum clock generator with a low maximum time interval error (MTIE) and low electromagnetic interference (EMI). The proposed circuitry was fabricated with 0.35μm CMOS process and operated with 3.3V supply voltage at the average center frequency of 100MHz. The measured results showed the MTIE of 11.59ns with the EMI reduction of 14.57dBm.

Keywords-component;Electromagnetic Interference, Spread Spectrum Clock, Discontinuous Modulation, Time Interval Error

Introduction

Nowadays, faster operating speed of automotive electronic units brings about such serious problems as functional failures due to electromagnetic interference at high frequency [1-2]. An effective solution to diminish EMI is to modulate system clock frequency, which is known as a spread spectrum clock (SSC). This method reduces EMI by spreading energy of discrete frequency harmonics over a wide range of frequencies. However, asynchronous serial data protocols in vehicle networks require a strict time interval error (TIE) range because receivers in asynchronous CAN circuitry may detect wrong data when the TIE is out of range [2].

Several methods have been reported to find efficient solutions considering trade-offs between the TIE and EMI reduction. One method solving the TIE issue is to reduce a modulation index. While this approach reduces the MTIE, it degrades the EMI reduction [3]. Another method reported in [2, 4] can reduce the MTIE without sacrifice of EMI performance. However, its implementation needs complex and large additional circuitry. In this paper, we propose novel SSC technique which satisfies requirement on the MTIE and reduces EMI effectively in asynchronous vehicle networks.

Design Descriptions

A. EMI Reduction and MTIE

When a conventional triangular modulation profile is adopted for SSC, the reduction of EMI peaks can be estimated by Equation (1) [5]. In Equation (1), f_c, A_{mo} and F_{mo} are center spread frequency, modulation depth, and modulation frequency, respectively. In the center-spread method, the TIE is varied from 0 to MTIE within one modulation cycle. The MTIE level is also given in Equation (1) as a function of A_{mo} and F_{mo} [2, 5].

$$\text{EMI Peak Reduction} = 10 \times \log\left[\frac{A_{mo} \times f_c}{F_{mo}}\right], \quad \text{MTIE} = \frac{A_{mo}}{8 \times F_{mo}} \quad (1)$$

$$\text{EMI Peak Reduction} = 10 \times \log[8 \times \text{MTIE} \times f_c] \quad (2)$$

Equation (2) shows EMI peak reduction expressed with MTIE and f_c. In general, the higher MTIE gives rise to the higher EMI peak reduction as shown in Equation (2). However, requirement on the MTIE limits the use of SSC for asynchronous interfaces. Therefore, a new method to effectively reduce MTIE becomes indispensable in automotive networks requiring low EMI generation.

B. Proposed Discontinuous Modulation Technique

Fig. 1 shows concepts of the conventional SSC and the proposed discontinuous spread spectrum clock (DSSC). Fig. 1 also shows comparison between theoretical TIE plots. The proposed method is based on up-spread spectrums while the conventional SSC uses the center-spread spectrums with respect to the center frequency (Non-SSC frequency). In the conventional method, the modulation index (A_{mo}/F_{mo}) should be properly reduced to comply with the allowed MTIE. On the other hand, the proposed method can control TIE within the allowed MTIE by resetting TIE with selection of its half-frequency clock whenever the TIE runs over one clock cycle. Therefore, the proposed modulation depth A_{mo_p} can be larger than A_{mo_c} of the conventional SSC. The proposed DSSC technique with high modulation depth can reduce EMI effectively while maintaining the targeted MTIE.

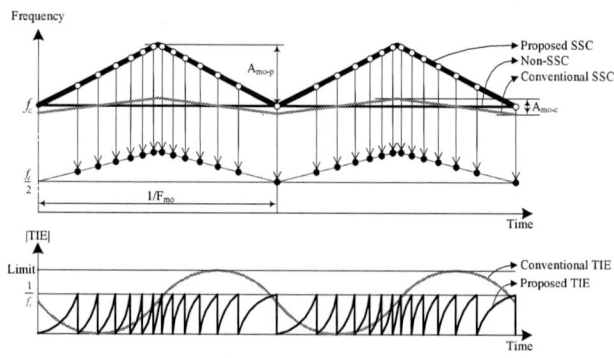

Fig. 1 Comparison between the conventional SSC and the proposed DSSC.

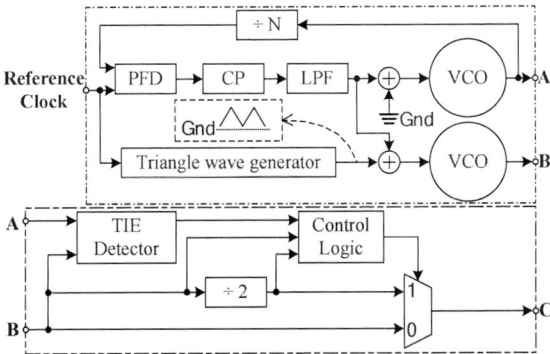

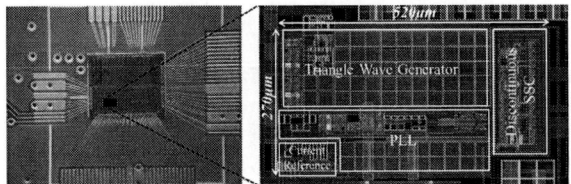

Fig. 2 Block diagram of the proposed discontinuous modulation SSC (Signal A: Non-SSC, Signal B: Up-SSC, Signal C: DSSC).

Fig. 3 Test chip photograph and layout.

C. Circuit Descriptions

Fig. 2 shows the block diagram of the proposed discontinuous modulation SSC which consists of a PLL (Non-SSC) block, an up-spread spectrum clock (Up-SSC) block, a TIE detector, and a tiny control logic block. The Non-SSC clock is generated at the node A and the Up-SSC is simultaneously generated at the node B. In the next stage, the TIE detector compares the signal B with the signal A. It checks out whether the accumulated TIE is within one clock cycle. The detected signal is transferred into the Control Logic. The Control Logic and MUX select Up-SSC clock or its half-frequency clock generated through the 1/2 divider. If the accumulative TIE reaches one clock cycle, the TIE is reset to zero by selection of the half-frequency clock. The repeating reset sequence can always keep the MTIE within one clock cycle. The photograph of the test module and the chip layout are shown in Fig. 3. The designed chip size is 520 μm x 270 μm. The chip was fabricated with 0.35μm CMOS process and operated with 3.3V supply voltage.

Measurement Results

The fabricated chip was tested with the modulation frequency (F_{mo}) of 50kHz and the modulation depth (A_{mo}) of 7% with respect to the center frequency (Non-SSC frequency) of 100MHz. Fig. 4 shows the measured clocks of Up-SSC (B), Non-SSC (A), and DSSC (C). It shows that the measured DSSC (C) resets the TIE to be zero whenever the TIE of Up-SSC is accumulated over one period of Non-SSC clock. Fig. 5 shows the measured EMIs of the Non-SSC and the DSSC. The EMI characteristic of the DSSC was reduced by 14.57dBm. Fig. 6 shows comparison between the simulated TIE of the conventional SSC and the measured TIE of the DSSC under the same EMI performance. The MTIE of the proposed DSSC is improved from 175ns to 11.59ns; i.e. by MTIE reduction of 163.4ns.

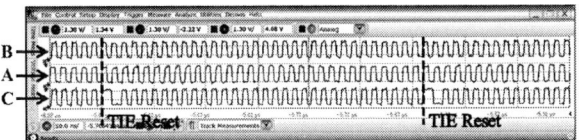

Fig. 4 Measured discontinuous modulation SSC (Signal A: Non-SSC, Signal B: Up-SSC, Signal C: DSSC).

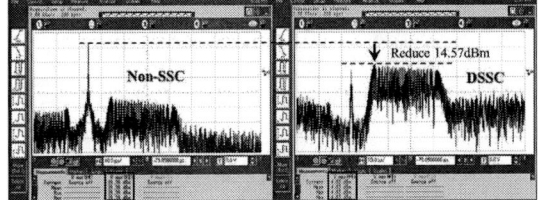

Fig. 5 Measured spectrums of the output signals (DSSC and Non-SSC).

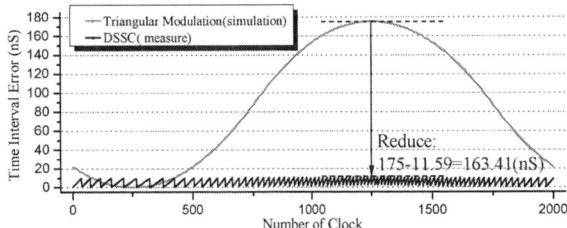

Fig. 6 Measured time interval error.

Conclusion

The proposed SSC with discontinuous modulation technique can reduce EMI peaks while maintaining very low TIEs compared with the conventional SSC. Therefore, this scheme is very useful to apply to asynchronous interfaces where a strict MTIE and relatively low EMI are required. The measured results show that the EMI level is reduced by 14.57dBm compared with the Non-SSC and the MTIE is 11.6nS at the target frequency of 100MHz.

Acknowledgment

This research was supported by a grant "High efficient heat rejection and supply technologies for zero emission vehicle (10035530)" from the Korea Evaluation Institute of Industrial Technology.

References

[1] H. Chang, et al., "A spread-spectrum clock generator with triangular modulation," IEEE Journal of Solid-State Circuits, vol. 38, no. 4, pp. 673-676, Apr. 2003.

[2] T. Steinecke, "Low-jitter frequency-modulated PLL", Asia-Pacific Symposium on Electromagnetic Compatibility, pp. 329-332, May 2012.

[3] J. Zhou and W. Dehaene, "A synchronization-free spread spectrum clock generation technique for automotive applications," IEEE Trans. Electromagn. Comp., vol. 53, no. 1, pp. 169–177, Feb. 2011.

[4] N. Da Dalt, et al., "An all-digital PLL using random modulation for SSC generation in 65nm CMOS," Solid-State Circuits Conference Digest of Technical Papers (ISSCC), pp. 252-253, Feb. 2013.

[5] Y. Komatsu, et al., "Bi-directional AC coupled interface with adaptive spread spectrum clock generator", Proceedings of IEEE Asian Solid-State Circuits Conference, pp. 71-74, Nov. 2007.

978-1-4799-5128-4/14 $31.00 © 2014 IEEE

K-Critical Path Search Based Multi corner Multi mode Static Timing Analysis

Deokkeun Oh, Eunsuk Park, Myeoungwoo Jin and Juho Kim

Department of Computer Science and Engineering, Sogang University
35 Baekbeom-ro, Mapo-gu
Seoul, Republic of Korea
{ pci123, qwerpes, jinmw9168, jhkim } @ sogang.ac.kr

Abstract

As CMOS technologies are scaled down into the nanometer range, effects of process variation have increased. Finding a critical path is important for performance and optimization of circuit. In this paper, we proposed efficient K-critical path detection in Multi Corner Multi Mode Static Timing Analysis(MCMM STA). We analyzed the characteristic of circuit and used MCMM timing model for considering various corners and modes. Using this timing model and pruning method, we can find more efficiently K-critical paths than traditional STA.

Keywords-process variation; critical path; static timing analysis; pruuning method; MCMM

I. Introduction

With the rapid development of electronics device, time to market is becoming important. As CMOS technologies are scaled down into the nanometer range, the process variations appear on device parameters that affect circuit performance such as oxide thicknesses (Tox), effective channel length and width in semiconductor production process stages [1]. And this has led to an increase in the number of corners required to cover the process parameter space. However static timing analysis such as traditional corner-based approach does not handle the process variations. Also, performing static timing analysis (STA) over all corners can be time consuming [2]. Finding a best/worst corner on circuit is becoming difficult. So to efficient analyze the performance of circuits is very significant. Circuit performance is affected by its critical path delay, it determines operation frequency. Therefore, efficiently extracting K-critical path is very important [3].

In this paper, we presented K-critical path search method using MCMM static timing model. So, we constructed the cell library for transistor characteristic and MCMM STA applied to existing K-critical path search algorithm.

The rest of the paper is organized as follows. Section 2 describes MCMM timing model used in this paper. In section 3, we explain K-critical path search method using MCMM STA as the proposed method and show overall flow. Experiment results are represented in section 4. Finally, we give our concluding remarks in section 5.

II. Multi Corner Multi Mode Static Timing Analysis

The timing information of a circuit is modeled by a timing graph $G = (V,E)$, where vertices, $v \in V$, correspond to pins in the circuit, and directed edges, $e \in E$, correspond to pin to-pin delays in cells or interconnect. The primary inputs are vertices with no incoming edges. All vertices with no outgoing edges are primary outputs but there may also be primary outputs with outgoing edges. A complete path is a sequence of edges, connecting a primary input to a primary output [4]. Gate delays are described by affine functions of process and operational parameter variations as shown in (1).

$$Delay_{gate} = a_0 + a_1 P_1 + a_2 P_2 + \cdots a_n P_n \qquad (1)$$

where a_0 is the nominal delay computed at the nominal values of the parameters, $a_{1 \sim n}$ is the sensitivity of $P_{1 \sim n}$, P_1 is normalized process parameter value and the sensitivity can be written as:

$$a_i = \frac{Delay @ P_{i,MAX} - Delay @ P_{i,MIN}}{P_{i,MAX} - P_{,MIN}} , i = 1,2,3,\cdots \qquad (2)$$

where $P_{i,MAX}$ is maximal value of P_i, $P_{i,MIN}$ is minimal value of P_i, To obtain the maximal delay, we set 1 to all the process parameter with positive sensitivity and -1 to all the process parameter with negative sensitivity. Thus smallest/largest delay is obtained by setting each parameter to one of its extreme values [5]. Using this model, we can verify timing at all corners in a single time run [6]. By computing the delay deviations of multiple process corners. MCMM STA can ensure the worst case or best case delays and thus act as a conservative but efficient variation-aware timing analysis approach.

III. K-Critical Path Search

The pseudo code of proposed method is shown in Fig. 1. First, proposed method take a circuit, constant K as inputs and extract K-critical paths as outputs. Second, given circuit as input converts to timing graph. This is an acyclic directed graph. Third, for the purpose of easy handling, we modify the timing graph by adding a source node s and a sink node t. A dummy edge is added from node s to each of the starting

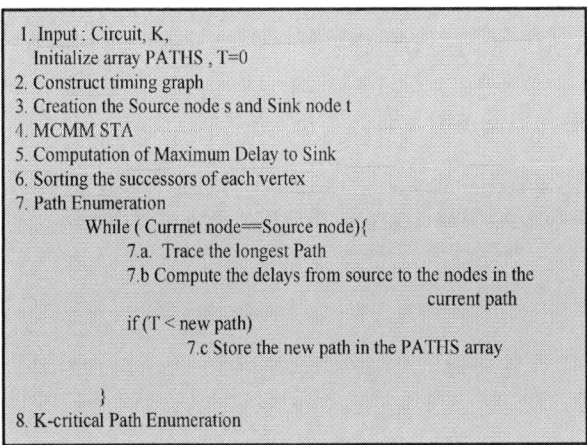

```
1. Input : Circuit, K,
   Initialize array PATHS , T=0
2. Construct timing graph
3. Creation the Source node s and Sink node t
4. MCMM STA
5. Computation of Maximum Delay to Sink
6. Sorting the successors of each vertex
7. Path Enumeration
       While ( Currnet node==Source node){
           7.a  Trace the longest Path
           7.b Compute the delays from source to the nodes in the
                                        current path
           if (T < new path)
                7.c Store the new path in the PATHS array

       }
8. K-critical Path Enumeration
```

Fig.1 Psuedo code of proposed method

vertices and from each of the ending vertices to the node t. the weight of a dummy egde is 0 [3]. Next, we compute the delay from each vertex to sink node t and apply to MCMM timing model, Also, rank edge for each vertex and enumerate the path. In path enumeration phase, path pruning method is very significant for reducing runtime cost. Thus, this paper uses a modified depth-first search with pruning. Fig. 2. represent path

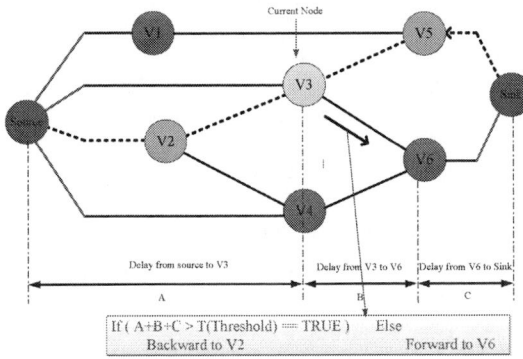

Fig 2 Path prunning

pruning during critical path search. To reduce search space, it is required to set the proper threshold(n% of critical path delay). So, proposed method prune out as significant amount of the search space. Finally, we can obtain K-critical path at multiple corners. If V is the number of gate, E is the number of connection in a circuit, time complexity of the proposed algorithm is $O(V^2E)$.

IV. Experiment Results

We have implemented our K-critical path search method using C/C++ and have tested it on the ISCAS85 benchmark suite. The Predictive Technology Model(PTM) was used for the SPICE model parameter. For process variations, we considered threshold voltage, oxide thickness, channel length, and width. Fig. 3 represented that accuracy of proposed method is similar to HSPICE and Table I shows run time comparison with traditional method.

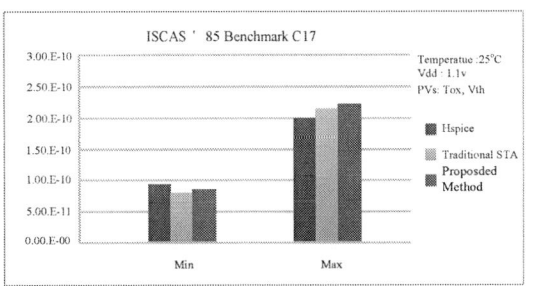

Fig. 3 ISCAS85 Benchmark c17

Table I.
Run time comparison with traditional method

ISCAS'85 CIRCUIT	Critical Path Search run time (sec) , K=100		
	Traditional Method(DFS)	Proposed Method	Improvement (%)
C432	6.415	4.193	53
C499	5.234	3.466	51
C880	1.909	1.317	45
C1355	5.532	3.592	54
C1908	15.433	10.428	48
C2670	10.638	7.439	43

V. Conclusion

This paper has presented K-critical path search method using multi corner multi mode static timing analysis. As technology is scaling, the number of process parameters is growing. Run time complexity is exponential. Thus, this paper proposed method that extracts K-critical path and built methodology to efficiently analyze circuit. We construct a cell library for timing analysis and efficiently analyze circuit performance at all corners. This can reduce critical path search time.

References

[1] L. Xie and A. Davoodi, "Bound-based statistically-critical path extraction under process variations," *IEEE Trans. Computer-Aided Design Integrated Circuits and Systems,* vol. 30, no. 1, pp. 59–71, Jan. 2011.

[2] S. Onaissi and F. N. Najm, "A linear-time approach for static timing analysis covering all process corners," *IEEE Transactions on, Computer-Aided Design of Integrated Circuits and Systems,* vol. 27, no. 7, pp. 1291-1304, Jul. 2008.

[3] K. S.H.C. Yen, D.C. Du, and S. Ghanta, "Efficient Algorithms for extracting the K most critical paths in timing analysis," *26th ACM/IEEE Design Automation Conference,* pp. 649-654, June. 1989.

[4] L.G. e Silva, L.M. Silveira, and J.R. Phillips, "Efficient computation of the worst-delay corner," *Proc.Design Automation and Test in Europe (DATE),* pp. 1–6, 2007.

[5] C. Visweswariah, K. Ravindran, K. Kalafala, S. Walker, S. Narayan, D. Beece, J. Piaget, N. Venkateswaran, and J. Hemmett, "First-order incremental block-based statistical timing analysis," *IEEE Trans. Computer-Aided Design of Integrated Circuits and Systems,* vol. 25, no. 10, pp. 2170–2180, Oct. 2006.

[6] Heloue, K.R, Onaissi, S, Najm, F.N, "Efficient Block-Based Parameterized Timing Analysis Covering All Potentially Critical Paths," *Computer-Aided Design of Integrated Circuits and Systems, IEEE Transactions on,* vol. 31, no. 4, pp. 472-484, April. 2012.

A Comparison-Free Sorting Algorithm

Saleh Abdel-hafeez[1] and Ann Gordon-Ross[2]

[1]Jordan University of Science and Technology
IRBID 22110, Jordan, sabdel@just.edu.jo
[2]University of Florida, FL. 36211, USA, ann@ece.ufl.edu
Also with the NSF Center for High-Performance Reconfigurable Computing (CHREC) at the University of Florida

Abstract

We propose a novel sorting algorithm that sorts data elements without data comparison operations—a comparison-free sort. Our hardware-based sorting algorithm leverages Hamming memory, which is an SRAM-based memory structure that stores the data elements based on the elements' Hamming maximum order representations. The data elements are also stored in a serial shift buffer in binary representation, and a simple matrix multiplication between this buffer and the Hamming memory produces the outputted sorted elements in $2N$ clock cycles for N data elements.

Keywords- Hardware-based sorting, comparison-free sorting, SOC design, Hamming memory, Hamming maximum order representation, 8T-CELL memory.

I. Introduction and Motivation

Prior research in sorting algorithms must consider the complexity of efficiently sorting data elements while maximizing the capabilities of the available computing resources, thus making efficient hardware realization challenging [1][2]. Sorting algorithms iteratively move data between comparison units and memory, requiring wide, high-speed data buses, complex control logic, and numerous shift, swap, comparison, etc. operations [3][4], thus requiring special design considerations for scalability to big data and specialization for certain data-type particulars.

We propose a new sorting algorithm that leverages the data elements' binary and Hamming weight representations to sort

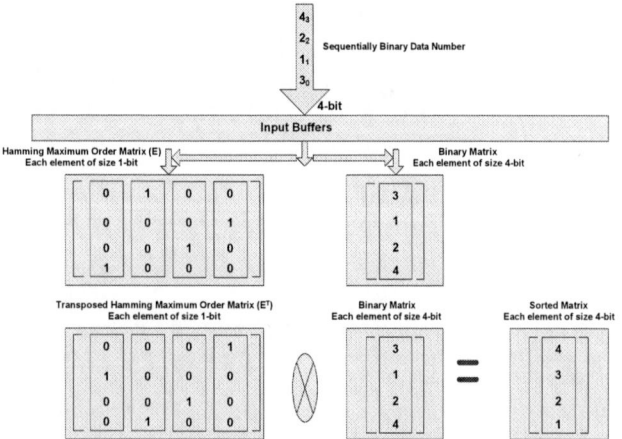

Fig. 1 Sorting example using matrix multiplication operations considering a 4-bit data input bus.

the data elements without comparison operations. A simple matrix multiplication (ANDING) operation outputs the sorted data elements, and the associated hardware structure alleviates the iterative movement of data elements between the memory and processing units. Our sorting algorithm's complexity is on the order of $O(N)$, which makes our sorting method suitable for a wide range of sorting applications, and is competitive with state of the art sorting methods.

II. Comparison-free Sorting Algorithm

The sorting algorithm's input is an m-bit bus carrying the data element's binary representation, which enables sorting $N=2^m$ data elements where each element has a Hamming representation of size $K=N$ for a lossless representation. For example, 5 has a binary representation of "101", and could have several Hamming representations, such as "10101011", "11100011", "00111110", etc. (i.e., covering all possible maximum order representations). However, our binary-to-Hamming converter deterministically converts 5 to "00011111", with a Hamming maximum order representation of "00010000". This Hamming maximum order representation ensures that different elements are orthogonal with respect to each other when projected to a R^n linear space.

Our sorting algorithm operates in two sequential phases: the write phase and the read phase. During the write phase, the data elements are sequentially inputted, converted to the element's Hamming representation, and stored into an SRAM-based memory [5] with a counter-based decoder address [6] in Hamming maximum order representation. We refer to this memory as a Hamming memory due to our Hamming representation storage methodology. The data elements are interpreted as a two-dimensional (2D) Hamming matrix E of size NxK where every element of the Hamming memory/matrix is of size 1-bit. In parallel, the element's binary representation is also sequentially stored in a serial shift buffer of registers, creating a one-demensional (1D) binary matrix B of size Nx1, where each register is of size m-bit. Since there are N data elements, the write phase requires N clock cycles.

The read phase effectively sorts and outputs the data elements using a matrix multiplication (ANDING) operation, rather than comparison operations, as in prior work. The matrix multiplication multiplies the transposed 2D Hamming representation matrix E^T (i.e., the transpose of E) with the 1D binary matrix B. This multiplication essentially enables a read from the associated binary matrix B's register that is aligned with a '1' in the read column of the Hamming matrix E. The result is the sorted matrix $S=S^T$xB, where S is of size Kx1-bit

978-1-4799-5128-4/14 $31.00 © 2014 IEEE   214

```
1.  Input: integer Element[0 : n − 1]
2.  Output: integer Sorted[0 : n − 1]
3.  Hamming memory: Boolean H[0 : n − 1][ 0 : n − 1] initialize to zero
4.  while i < n-1 do
5.      H[i][Element[i]-1] ←1
6.  endwhile
7.  k ← 0
8.  while j >= 0 do
9.      while i < n-1 do
10.         then if H[i][j] = 1
11.             then Sorted[k] ← Element[0: n-1]
12.                 k ← k+1
13.         endif
14.     endwhile
15. endwhile
```

Fig. 2 Pseudo code for our sorting algorithm assuming a uniprocessor system with no threading.

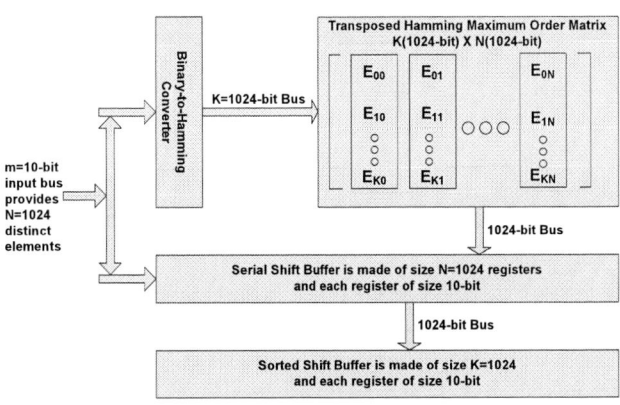

Fig. 3 Block diagram of our sorting algorithm's hardware data path using Hamming memory.

and contains the sorted data elements that are shifted into a sorted shift buffer, and outputted after the read phase completes.

Duplicated data elements are represented using the same vector space, such that the corresponding Hamming matrix column has multiple '1' values. These multiple '1's enable multiple registers in the binary matrix and these registers store duplicate data elements. Therefore, our sorting algorithm counts the number of '1's in the Hamming representation matrix column using simple control logic, and sends the repeated register value to the sorted shift buffer.

Fig. 1 illustrates a sorting example for four 4-bit data elements {3,1,2,4}, which generates the sorted matrix (sorted shift buffer) S = {1,2,3,4}. Fig. 2 shows the pseudo code for our sorting algorithm, assuming a single-threaded uniprocessor system (future work will extend this to multi-threaded multiprocessor systems).

III. Comparison-free Sorting Hardware Data Path and Functional Details

Fig. 3 depicts a block diagram for our sorting algorithm's data path assuming a sample $m=10$-bit input bus, which sorts $N=2^m=1024$ distinct data elements. The binary-to-Hamming converter generates the Hamming maximum order representation by converting the m-bit binary representation to the N-bit Hamming representation using a simple one-hot decoding unit, which directly connects to the SRAM-based 8T-Cell [5][6] Hamming memory's input bus. During the write phase, the data elements are stored serially into the serial shift buffer, and in parallel, the data elements are converted to their Hamming representation and stored in the Hamming memory in row order. After all elements are written to the Hamming memory and the serial shift buffer, the read phase processes the elements from the Hamming memory in column order. This transpose during the read phase facilitates the matrix multiplication (ANDING) operation $E^T xB$, where E^T is of size KxN and B is of size $Nx1$, and produces the sorted output.

IV. Conclusions

We presented a novel comparison-free sorting algorithm and associated hardware implementation. To the best of our knowledge, our design is the first to simultaneously leverage the data elements' Hamming weight and binary representations to sort the data elements without any comparison operations, and using only a simple matrix multiplication (ANDING) operation. Since comparison operations are eliminated, a key contribution of our sorting method is the elimination of the associated comparison units' high power dissipation. Additionally, our sorting algorithm is data-type/ordering/duplication independent, and can sort N data elements in $2N$ clock cycles—on the order O(N)— using simple control logic. We implemented and evaluated our sorting algorithm for a sample sorting of $N=1024$ data elements using 90 nm TSMC technology at 1V and an 8T-cell memory. Results verified that the sorting requires $2N=2048$ clock cycles at an operating frequency of 0.25 GHz.

V. Acknowledgments

This work was supported in part by the National Science Foundation (CNS-0953447). Any opinions, findings, and conclusions or recommendations expressed in this material are those of the author(s) and do not necessarily reflect the views of the National Science Foundation.

VI. References

[1] Enzo Mumolo, Gabriele Capello, and Massimiliano Nolich, VHDL Design of a Scalable VLSI Sorting Device Based on Pipelined Computation, Journal of Computing and Information Technology, Vol. 12, pp. 1-14, 2004.

[2] A. A. Colavita, A Cicuttin, F. Fratnik, and G. Capello, SORTCHIP: A VLSI Implementation of a Hardware Algorithm for Continuous data Sorting, IEEE Journal of Solid-State Circuits, Vol. 38, No. , pp. 1076-1079, June 2003.

[3] Li Xiao, Xiaodong Zhang, Stefan A. Kubricht, Improving Memory Performance of Sorting Algorithms, ACM Journal on Experimental Algorithmics, Vol. 5, 1-21, 2000.

[4] L. M. Busse, M. H. Chehreghani, J. M. Buhmann, The Information Content in Sorting Algorithms, IEEE International Symposium on Information Theory Proceedings (ISIT), pp. 2746-2750, 2012.

[5] Saleh Abdel-Hafeez and Anas Matalkah, "CMOS Eight-Transistor Memory Cell for Low-Dynamic-Power High-speed Embedded SRAMS," Journal of Circuits, Systems and Computers, Vol. 17, No. 5, pp. 845-863,Oct. 2008.

[6] Saleh Abdel-Hafeez and Ann Gordon-Ross, "A Digital CMOS Parallel Counter Architecture Based on State Look-Ahead logic", Journal of IEEE Transactions on Very Large Scale Integration (VLSI) Systems, Vol. 19, Issue 6, pp. 1023-1034, May 23, 2011.

A Carry Look-ahead Adder Designed by Reversible Logic

Junchao Wang, Ken Choi
Department of Electrical and Computer Engineering,
Illinois Institute of Techonoly,
Chicago, Illinois State, USA
jwang181@hawk.iit.edu, kchoi@ece.iit.edu

Abstract: **In recent years, a lot of attentions have been attracted by the reversible logic due to the characteristic of zero energy dessipation. Reversible logic can be applied in fields such as low power CMOS circuits, quantum computation and DNA computing. In this paper, the author proposed a 16 bit carry look-ahead adder is constructed by four 4 digits groups based on the theory of reversible logic, which has the advantages of theroretical zero power dissipation and high efficiency.**

Keywords: *Reversible logic, Carry select adder, Quantum gates, Quantum Cost*

I. INTRODUCTION

With the increasing importance of energy dissipation in integrated circuits, the reversible logic implementations gain in prominence as a way to reduce power since irreversible computing is one of the most significant factors to generate energy dissipation. By the Landauer's principle, each bit of information lost will generate $kT\ln2$ joules of heat energy, where T stands for the absolute temperature at which computation is performed and k is Boltzmann's constant[1], so how to avoid information lost is an efficient way to decline the energy dissipation in digital integrated circuits.

Reversible logic circuits implement bijective Boolean functions[2], where the output vector is a permutation of all the input combinations. Therefore, input vectors can be always uniquely reconstructed from the outputs, which means no information lost occurs resulting in very low power dissipation, so using reversible logic to design ICs casues wide attention.

A non-reversible specification has typically a different number of inputs and outputs, that is, $f: B^n \rightarrow B^m$. To guarantee a bijective mapping, the number of inputs and outputs must be equal, so the additional inputs or outputs must be often used to generate a reversible embedding of an irreversible function f_r. Such additional inputs are called ancilla bits, while extra outputs are referred to as garbage bits (g).

Every reversible circuitis is constructed by certain basic reversible gates. However, the reversible gates differ in architecture and quantum cost. There are three typical reversible logic gates illustated in the Fig.1. The gates with one or two inputs are called NOT gate and Controlled Not (CNOT) gate, respectively. Gates with three or more inputs are called TOFOLLI gates. Similar to traditional digital circuits, some sets of reversible logic gates, such as the set

of Controlled NOT gate and TOFFOLI gate, are functionally complete sets. Hence, all the reversible logic functions can be realized by Controlled NOT gates and TOFFOLI gates.

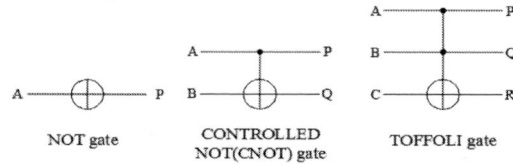

NOT gate   CONTROLLED NOT(CNOT) gate   TOFFOLI gate

Fig. 1 Three Typical Basic Reversible Gate

II. Proposed 16-bit Carry Look-ahead Adder Designed by Reversible Logic

Addition is one of the most fundamental operation for any kind of digital systems, and Ripple Carry Adder (RCA) is the foundation of digital adders although it has certain obvious defects. Carry Look-ahead Adder (CLA) is a kind of optimization to conventional RCA, which overcomes the defects of RCA such as low computing efficiency and long delay, so CLA becomes one type of wide-used adders. In this paper, we present a new design of 16-bit CLA which can both enhance the computing efficiency and decline the amount of energy dissipation based on reversible logic theory. Fig.3 illustrates the total structure of the proposed reversible CLA.

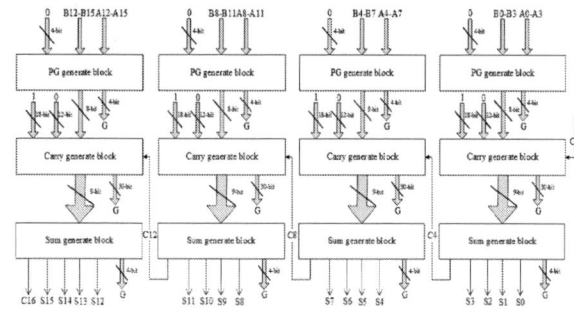

Fig. 3 Three Typical Basic Reversible Gate

The structure of proposed reversible CLA is identical as the classic electrical circuit. However, we had to rebuild all the parts by quantum gates so that the circuit has the characteristic of reversible logic. As mentioned before, proposed 16-bit CLA is constructed by four 4 digits groups

978-1-4799-5128-4/14 $31.00 © 2014 IEEE   216

which means the four 4-bit CLA are identical and we just need to cascade the carry-out of each stage to the carry-in of next stage. Therefore, what we mainly concern about is how to rebuild every part of 4-bit CLA,which is PG generate block, Carry generate block and Sum generate block, by quantum gates.

Fig.4 demonstrated the internal quantum circuit of PG generrate block which utilized 4 CNOT gates and 4 TOFFOLI gates. The quantum cost of this block is 24. The functions of this block are $P_i = A_i \oplus B_i$ and $G_i = A_i \cdot B_i$. Fig.5 showed the internal quantum circuit of Carry generate block, it has 10 CNOT gates and 20 TOFFOLI gates. The quantum cost of this Carry generate block is $10*1+20*5=110$. The function of this block is to generate the carry out $C_i - C_{i+4}$ of current stage by $P_i - P_{i+3}$ and $G_i - G_{i+3}$. Fig.6 illustrated the quantum circuit of the Sum generate block. It utilizes 4 CNOT gates, so the quantum cost is $4*1=4$. The function of this block is $S_i = C_i \oplus P_i$. In all the figures above, "g" stands for garbage line.

By cascading those three parts, the 4-bit CLA designed by reversible logic had been finished as Fig.6 shows. It has 18 CNOT gates and 24 TOFFOLI gates which means the quantum cost of this 4 bit reversible CLA is $18*1+24*5=138$. For the 16-bit reversible CLA, it utilizes 72 CNOT gates and 96 TOFFOLI gates, so the quantum cost is 552. It has 136 ancilla bits and 152 garbage lines as shown in Fig.3.

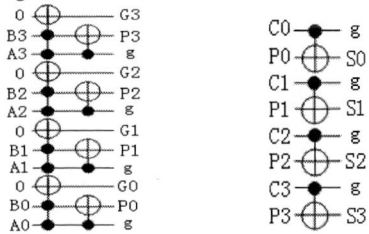

Fig.4 Quantum circuit of PG generate block Fig.6 Quantum circuit of Sum generate block

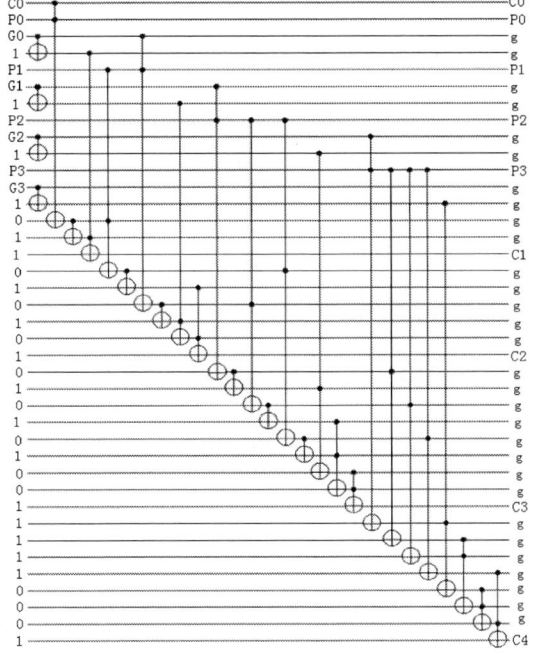

Fig.5 Quantum circuit of Carry generate block

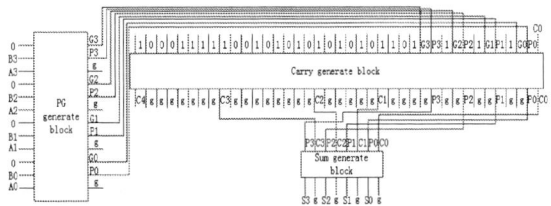

Fig.6 Structure of 4-bit Reversible CLA

III. Conclusion

In this paper, we propose a new reversible design of 16-bit CLA that consists of 72 CNOT and 96 TOFFOLI gates with total quantum cost of 552. This 16-bit Carry Look-ahead Adder designed by reversible logic has advantages of short delay and low power dissipation, so it can be applied to quantum computing, wireless sensors and other areas require very low power dissipation very well.

Reference

[1]Langauer R., Irreversibility and heat generation in the computing process. IBM Jr. Research and Development, 5, 183-191, 1961

[2] Bruce J W, Thronton M A, Shivakumaraiah L, Kokate PS, Li X, "Efficient adder circuit based on a conservative logic gate", IEEE Computer Society Annual Symposium on VLSI, 2002.

[3]Yu Pang, Shaoquan Wang, Zhilong He, Jinzhao Lin, Sayeeda Sultana, Katarzyna Radecka, "Positive Davio-based Synthesis Algorithm for Reversible Logic". IEEE 29th International Conference on Computer Design (ICCD), 2011.

[4]Claudio Moraga and Fatima Zhora Hadjam, "On double gates for reversible computing circuits", Preprint Proceeding International Workshop on Boolean Problems., Freiberg, Germany, Sep 19.21 2012.

[5]Yu Pang, Junchao Wang and Shaoquan Wang, "A 16-bit carry skip adder designed by reversible logic", 5th International Conference on Biomedical Engineering and Informatics (BMEI), 16-18 Oct. 2012

[6]Yu Pang, Junchao Wang and Yang Xia, "A BCD Priority Encoder Designed by Reversible Logic", 2012 International Conference of Wavelet Active Media Technology and Information Processing

Dynamic Stability Estimation for Latch-Type Voltage Sense Amplifier

Woong Choi[1], Jongsun Park[2] and Gyuseong Kang[1]

[1] University of Minnesota, 200 Union Street SE, Minneapolis, MN 55455, USA, {choiw, kangg}@umn.edu
[2] Korea University, Seoul 136-713, Korea, jongsun@korea.ac.kr

Abstract

This paper presents a dynamic stability estimation method for latch-type voltage sense amplifier (LV-S/A) which is widely used in several type of memory. In order to overcome the large computational burden of conventional Monte-carlo simulation, predetermined coordinate based mean-shift most probable failure point (PMS-MPFP) method is proposed. Compared to mean-shift MPFP (MS-MPFP) [2], proposed PMS-MPFP needs approximately 2 less iteration number for 1E-10 bit error bound in wide range of simulation conditions.

Keywords-latch, sense amplifier, dynamic stability, mpfp

Introduction

Electronics industry has continuously demends low power system-on-chip (SoCs) in many portable applications. Supply voltage scaling has been considered one of the most promising techniques for recuding power consumption, and it is widely applied to recent integrated circuits. However, in low voltage operation, the impact of device mismatch and electrical noise is becoming significant, especially symmetrical circuits such as SRAM bit-cell and latch-based circuits. For this reason, accurate stability estimation of symmetrical circuits is essential to ensure operational functionality.

Previously, comprehensive analysis on LV-S/A is performed in [1]. However, the dynamic stability estimation is roughly addressed by conventional Monte-calro simulation with 1000 samples. For 95% estimation confidence level, at least 1E+12 samples are required in conventional Monte-carlo simulation [2]. In order to achieve fast simulation time and high estimation accuracy, we propose a PMS-MPFP which is based on predetermined coordinate and efficient sampling strategy. Compared to mean-shift MPFP (MS-MPFP) [2], the PMS-MPFP requires 2 less iteration number for 1E-10 bit error bound in wide range of simulation conditions. By using PMS-MPFP, dynamic stability of LV-S/A will be drawn in this paper.

Predetermined Coordinate based Mean-shift MPFP

A. Basic of Most Probable Failure Point (MPFP)

When threshold voltage (V_{TH}) variations become larger than certain boundary, the operational failures start to apprear. On this boundary, there is a MPFP, which has shortest distance from means of V_{TH} distribuitions. Since the long distance means low probability and short distance means high probability, the failures are most frequently occurred in MPFP. As a result, the operational failure estimation can be performed by finding the MPFP. The detailed explanation on MPFP can be

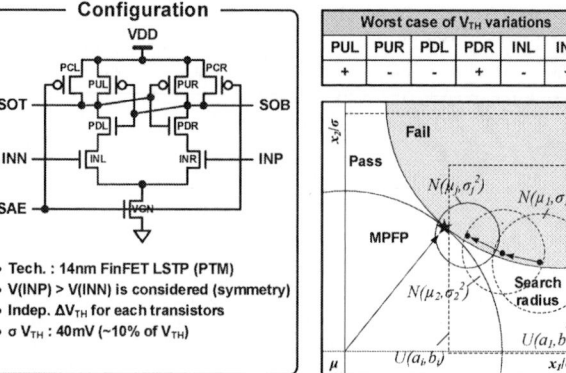

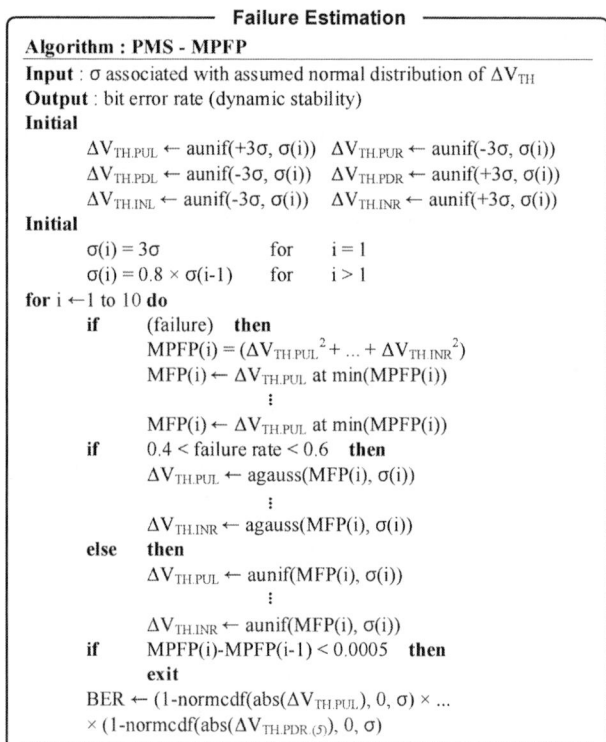

Fig. 1. Framework for dynamic stability estimation of LV-S/A.

found in [2].

B. Predetermined Coordinate and Searching Algorithm

Fig.1 shows the framework for LV-S/A including LV-S/A configuration and PMS–MPFP algorithm. To simplify calculating failure probability, it is assumed that threshold variations (ΔV_{TH}) of each transistor are independently laid on normal distributions. In addition, precharge PMOS (PCL, PCR) and footer NMOS (VGN) are not taken into account to our MPFP

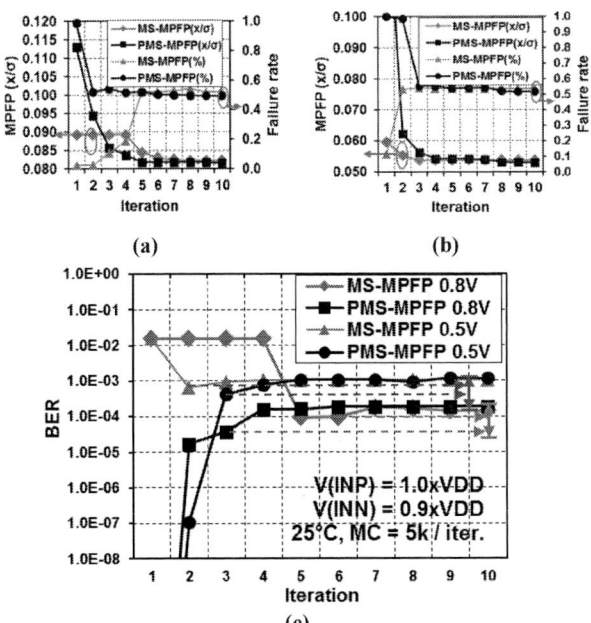

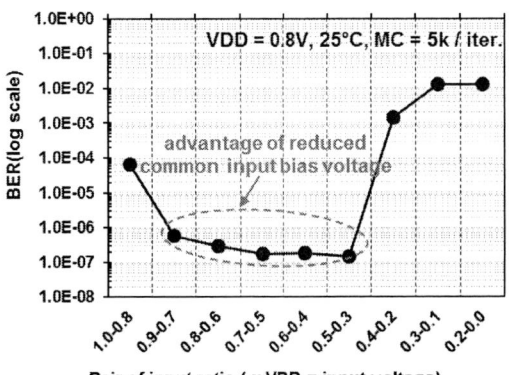

Fig. 3. Impact of input common bias voltage level on BER.

Fig. 2. Computational overhead and accuracy comparison between MS-MPFP and PMS-MPFP. (a) at VDD=0.8V. (b) at VDD=0.5V. (c) bit error rate (BER).

simuluation since the impact of those transistors on the dynamic stability of LV-S/A is negligible. Considering large capacitance of memory bit-line, both S/A input (INP, INN) can be regarded as static [1]. Based on these assumptions, MPFP will be searched by following algorithm. When input voltage (INP) is higher than INN, sense amplifier output (SOT) is generated as logic '1'. In this case, worst case of V_{TH} variations can be predetermined as decribed in Fig. 1. Contrary to MS-MPFP, the proposed PMS-MPFP have initial searching area aiming to eliminate inefficient simulation. Depending on operating conditions, MPFPs are widely spread in predetermined coordinate. As a result, mean based searching algorithm used in MS-MPFP induces wasting iterations, which are the large burden to find pass/fail (PF) boundary. To cope with widely spread MPFP in several simulation conditions, PMS-MPFP adapts uniformly distributed samples in initially searching area since this samples are more extensively distributed than normally distributed samples while both type of samples have equivalent mean and standard deviation. On the other hand, the normally distributed samples are utilized to take advantage of its mean-concentrated property when center of search radius is sufficiently closer to the BF boundary.

Simulation Results

Fig. 2 shows the comparison results between MS-MPFP and PMS-MPFP. During comparison for simulation overhead, 1E-10 bit error bound (at the last 10th iteration) is considered as reference accuracy. When supply voltage is set to 0.8V, the MPFP is far away from means of V_{TH} distributions than 0.5V. As a result, the initial searching algorithm of MS-MPFP require more iterations number to extend the search radius closer to the PF boundary. Meanwhile, during worse operating condition (VDD=0.5), the PF boundary is closer than good operating condition (VDD= 0.8V) so that the MS-MPFP shows fast convergence behavior as drawn in Fig. 2(b). Contrary to the

MS-MPFP, the proposed PMS-MPFP can handle wide range of simulation conditions without MPFP placement dependency. By using the predetermined coordinate and uniformly distributed samples, the PF boundary can be easily found in PMS-MPFP. In addition to this advantage, estimation accuracy is also higher than MS-MPFP since the search radius of PMS-MPFP is continuously diminished by iterations. This searching scheme selects more relevance samples to the MPFP and removes seed dependency in SPICE based simulation. As shown in Fig.2 (b), the result of MS-MPFP is stucked after 4th iteration due to the fixed seed of Monte-calro simulation. When we consider the MPFP searching time based on 1E-10 bit error bound, the required iteration number of PMS-MPFP is 2 less than that of MS-MPFP. Based on the PMS-MPFP, the dynamic stability of LV-S/A is performed in Fig. 3. As early verified in [1], reduced common input voltage level (0.8~0.6 of VDD) is helpful for improving dynamic stability of LV-S/A.

Conclusion

To reduce simulation overhead while maintaing the estimation accuracy in wide range of simulation condition, PMS-MPFP dynamic stability estimation method for LV-S/A is proposed. By using the predetermined coordinate and sample combining technique based on uniformly distributed samples and normally distributed samples, the required simulation time for accurate stability estimation can be signficantly reduced in proposed PMS-MPFP. We also examine the advantage of reduced common input bias voltage which improves the dynamic stability of LV-S/A.

Acknowledgment

This research was supported by National Research Foundation of Korea Grant (NRF-2012R1A2A2A01012471) and IDEC.

References

[1] B. Wicht, T. Nirschl, and D. Schmitt-Landsiedel, "Yield and speed optimization of a latch-type voltage sense amplifier," *IEEE J. Solid-State Circuits*, vol. 39, no. 7, pp. 1148–1158, Jul. 2004.

[2] T. Kida, Y. Tsukamoto, and Y. Kihara, "Optimization of importance sampling Monte Carlo using consecutive mean-shift method and its application to SRAM dynamic stability analysis," in *Proc. ISQED*, Mar. 2012, pp. 572-579.

CAN Data Reduction Using Three Segement Signal Decomposition

Yujing Wu[1], Jin-Gyun Chung[1], Yinan Xu[2]

[1]Division of Electronic Engineering, Chonbuk National University, Jeonju, South Korea
[2]Division of Electronics & Communication Engineering, Yanbian University, Yanji, China
yjwu@jbnu.ac.kr, jgchung@jbnu.ac.kr, ynxu@ybu.edu.cn

Abstract

Controller area network (CAN) was designed for multiplexing communication between electronic control units (ECUs) in vehicles and many high-level industrial control applications. It is desirable to reduce the CAN frame length, since the time duration for the data transmission is proportional to the frame length. In this paper, we present a CAN message compression method to reduce the CAN frame length using three segment signal decomposition. Experimental results indicate that CAN transmission data can be compressed by up to 91.9% with the proposed method.

Keywords-CAN; signal decomposition; DLC

Introduction

The CAN systems have been successfully applied in many fields due to its high reliability and cost efficiency. As the number of ECUs or sensors connected to the CAN bus increases, so does the bus load. When a CAN bus is overloaded, it is not easy to transmit low priority CAN messages due to the increased waiting time. In addition, the error probability of the data transmission also increases.

To reduce the bus load, only the message differences between the current and the preceding CAN messages can be transmitted based on the observation that CAN data do not change rapidly [1].

In this paper, we present a CAN message compression method using three segment signal decomposition. In addition, to indicate compressed signals in a data field, we use Data Length Code (DLC) to avoid the use of additional header bits in the data field as opposed to conventional compression methods [2-3].

Exisiting CAN Message Compression Methods

In the boundary of fifteen compression (BFC) method, if the current value of a CAN signal has changed within the maximum compression range of ±15, then the CAN signal can be compressed using the boundary of fifteen compression algorithm.

By using the DLC in CAN data frame format, enhanced data reduction (EDR) algorithm eliminates the difficulties for the identification of compressed messages such as the use of the reserved bit, the use of the dedicated message IDs or the use of additional bit in the data field. In addition, EDR algorithm uses a method of managing signals of shorter lengths (i.e., <5 bits) by combining them into groups that are handled as single signals.

Proposed Compression Method

The proposed message compression procedures are summarized as follows:

1. The rate of change of each signal in a CAN data field (e.g., Table I) is estimated by computing the difference values between successive data fields using actual CAN data. Regardless of the original signal assignment, a data field is assumed to be composed of 3 signals. (or, 3 segements) The length of each segment can be determined such that the number of the required bits of the compressed signal is minimized. In general, best results are obtained when the numbers of the bits of each segment are the same. In addition, signals are arranged such that slowly changing signals are placed in the most significant parts and frequently changing CAN signals are placed in the least significant parts of a two dimensional map (refer to Table II).

2. For each signal, compute difference values of signals between current and the preceding frames (refer to Table III). After computing the difference values, each difference value is represented using a modified sign-magnitude number. That is, as oppose to the conventional sign-magnitude number, the sign is denoted by the least significant bit. The length of each modified sign-magnitude number is one larger than the length of each segment.

3. Generate a compression data map (refer to Table IV) by placing the non-zero difference values together, starting from the first row.

4. Beginning from the leftmost column, delete a column if every element in the column is zero. The region from the column with the first non-zero element to the last column is selected as the compression area (the shaded area in Table IV).

5. Beginning with the data bits in the last column of the compression area, the data bits are rearranged into two-demensional memory with 8 columns (refer to Table V).

6. Since the number of the CAN signals in a data field is 3, three DLC bits (DLC[2:0]) are used to avoid the use of additional header bits in the data field. If the difference value of the first segment is zero, DLC[2] bit is set to 0. Otherwise, DLC[2] bit is set to 1. For second and third

segments, repeat the same procedure with DLC[1] and [0] bits.

By comparing Tables II and V, it can be seen that we achieve a data compression rate of 66.7% since 2 bytes are required instead of the original 6 bytes.

Performance Analysis

In this section, the performance of the proposed method is compared with those of the EDR algorithm and the BFC algorithm. The CAN signals used for the simulation are EMS1(Engine Management System 1), EMS2 and EMS4 signals.

Tables VI and VII compares the compression efficiencies of the three methods obtained through MATLAB simulations of actual CAN signals under normal driving and sudden stop /acceleration conditions. It can be seen from Tables VI and VII that we can obtain additional compression efficiency of 5-36% with the proposed method, compared with other CAN data compression methods.

Fig. 1 shows the test environments including the CAN Pro analyzer. By using a Cortex M3 embedded test board, 64-bit EMS CAN data compression can be performed within 0.16 ms; consequently, the proposed algorithm can be used in automobile applications without any problem since EMS CAN signals are usually transmitted every 10 ms in automobiles.

Conclusions

In this paper, we present a CAN message compression

TABLE I EMS4 CAN SIGNAL (ID: 545)

Sginals	Signal Description	Bit Len.	Bit Add.
Free	Free	1	0
LMIL	Lamp "check engine for OBD"	1	1
IMST	Status "immobilizer"	1	2
AMP	Atmospheric pressure	5	3
FCO	Fule comsumption	16	8
VB	Battery voltage	8	24
TQIA	Acual engine torque	16	32

TABLE II THREE SEGEMENT SIGNAL DECOMPOSITION OF EMS4 CAN SIGNAL

Segments	Bit Address	
	bit 15 ~ bit 8	bit 7 ~ bit 0
S1 (16 bits)	TQIA[15:8]	Free[0] ~ AMP[4:0]
S2 (16 bits)	FCO[7:0]	FCO[15:8]
S3 (16 bits)	TQIA[7:0]	VB[7:0]

TABLE III EXAMPLE OF DIFFERENCE VALUES

	S1	S2	S3
Previous Frame	665605	983080	15360
Current Frame	665600	983091	15360
Difference value	-5 (17 bits)	11 (17 bits)	0

TABLE IV COMPRESSION DATA MAP

Segments	bit 16 ~ bit 5	bit 4	bit 3	bit 2	bit 1	bit 0
S1	0...0	0	1	0	1	1
S2	0...0	1	0	1	1	0

TABLE V MEMORY MAP OF TABLE IV

	bit 7	bit 6	bit 5	bit 4	bit 3	bit 2	bit 1	bit 0
byte 0	S2[3]	S1[3]	S2[2]	S1[2]	S2[1]	S1[1]	S2[0]	S1[0]
byte 1	0	0	0	0	0	0	S2[4]	S1[4]

TABLE VI COMPARISON OF THE NUMBER OF COMPRESSED DATA BITS (COMPRESSION EFFICIENCIES) UNDER NORMAL DRIVING CONDITIONS

Comp. methods	EMS1	EMS2	EMS4
Original	3,020,160 (0%)	3,020,112 (0%)	3,020,112 (0%)
EDR	1,370,496 (54.62%)	1,069,776 (64.58%)	1,527,824 (49.41%)
BFC	1,321,760 (56.24%)	1,012,376 (66.49%)	1,789,864 (40.74%)
Proposed	724,376 (76.02%)	243,376 (91.94%)	1,100192 (63.57%)

TABLE VII COMPARISON OF THE NUMBER OF COMPRESSED DATA BITS (COMPRESSION EFFICIENCIES) UNDER SUDDEN STOP / ACCELERATION CONDITIONS

Comp. methods	EMS1	EMS2	EMS4
Original	195,264 (0%)	195,120 (0%)	195,168 (0%)
EDR	96,464 (50.60%)	72,016 (63.09%)	76,688 (60.71%)
BFC	92,040 (52.86%)	65,808 (66.27%)	79,152 (59.44%)
Proposed	25,264 (87.05%)	54,992 (71.84%)	41,048 (78.97%)

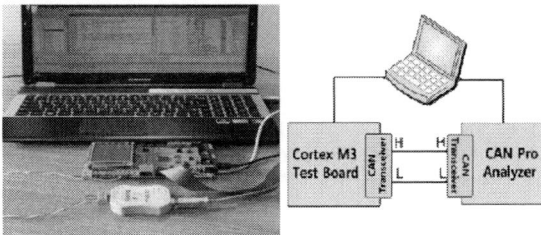

Fig. 1 Test Environments

method using three segement signal decomposition. By the proposed method, it is not needed to predict the maximum value of the difference in successive CAN messages as opposed to EDR and BFC algorithms. By use of DLC, the use of additional header bits in the data field is avoided. It is shown that the CAN transmission data is further reduced up to 3-36% by the proposed method, compared with conventional methods.

Acknowledgment

This work was supported by the IT R&D program of MOTIE/KEIT. [10044092, Development of Core IPs of OFDM PHY and RF Transceiver for 60GHz Wireless LAN/PAN in application of 7Gbps Wireless Multimedia Services]

References

[1] Lawrenz, W., 1997. CAN System Engineering: From Theory to Practical Applications.

[2] Kelkar, S., kamal, r., "Boundary of fifteen compression algorithm for controller area network based automotive applications," *Circuits, Systems, Communication and In-formation Technology Applications (CSCITA)*, pp.162-167, April, 2014.

[3] Miucic, R., Mahmud, S., "An enhanced data-reduction algorithm for event-triggered networks," *IEEE Transactions on vehicular Technology*, vol. 58, no.6, pp.2663-2678, July, 2009.

Underbody Component Monitoring System of Railway Vehicles using the Infra-red Thermal Images

Min-Soo Kim, Seh-Chan Oh, Geun-Yep Kim and Seok-Jin Kwon

Metropolitan Transit System Research Division,
Korea Railroad Research Institute
360-1 Woram-dong, Uiwang-si, Gyeonggi-do, 437-757, KOREA
+82-31-460-5205 and ms_kim@krri.re.kr

Abstract

This paper is on the study of the status monitoring system using infra-red camera for the key underbody components of railway vehicles (traction motors and reducers). Image processing and fault diagnosis algorithms that enable the efficient detection of abnormality of the key underbody components of railway vehicles are important factors for safe railway operations. For the field tests, underbody images of the KTX trains have been collected as they enter the depot. By calculating the position of the components with the reference to the maximum temperature of the targets, the abnormal overheat is detected using template matching technique. Moreover, advanced methods have been added to the conventional template matching process for the fast image processing using high speed camera such as minimum temperature setting for the region, pattern size designation, and discarding repeated images. And in order to improve the detection rate, the templates with different suitability in the same frame are compared using AND calculation.

Keywords-Infra-red Thermal Image, Underbody Component, Railway Vehicle

Introduction

The underbody components of railway vehicles are composed of traction motors, axles, wheels and bearings and the failures of these components occur due to numerous causes. Minor defects may not cause troubles in early stages; however, they can eventually cause big accidents due to the enlargement of the defect area after repeated operation, especially for the driving system of the vehicles. Therefore, in order to prevent the accidents due to the minor defects, it is necessary to detect the defects in early stages using nondestructive diagnosis technologies without disassembly.

Infra-red thermal image inspection is a nondestructive inspection technology that examines the status of an object by measuring the temperature based on the infra-red emission. Infra-red thermal image inspection technology is applicable to any types and status of objects and it is widely used in many fields including mechanical, electrical, chemical, medical, and manufacturing processes. For railway systems, infra-red thermal image inspection has been used for status diagnosis for some railway vehicle components. Infra-red thermal image method captures the thermal image of an object based on its infra-red radiant energy using infra-red thermal image camera. The images can be analyzed using image processing techniques to estimate the abnormal temperature distributions or changes. Internal failures can be detected by measuring the temperature difference between normal components and faulty components since they show different heat conductivity. Moreover, using high speed infra-red thermal image cameras, large areas can be measured in short period of time[1]-[4].

This paper suggests a novel detection algorithm for abnormal overheat using infra-red camera as a nondestructive diagnosis technology for traction motors and reducers, the main underbody component of KTX(Korea Train eXpress) vehicles and urban metro trains.

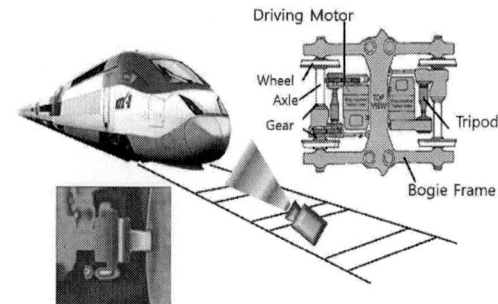

Fig. 1 Inspection System for the underbody components of railway vehicle using the infra-red camera

Underbody Component Monitoring System

A. Overheat Detection System using Infra-red Thermal Images

Detection system for overheat of the underbody components of railway vehicles is composed of infra-red thermal image camera device, train detection device, and abnormal overheat detection algorithm. Overheat monitoring system has been developed to detect the abnormal overheat of the underbody components of railway vehicles such as traction motors and reducers. Infra-red camera installed on the railroad track captures the images of the underbody of railway vehicles once the trigger signal is received from train detection device. Then the monitoring system has been designed to run the detection algorithm for abnormal overheat. Moreover, the monitoring system is designed so that it can be applied to various underbody components of different railway vehicles including urban metro trains, regional trains, and high speed trains by implementing the functions such as region selection among underbody components by the users, template setting that selects the region of interest in the entire image, and selection of image processing algorithm that detects overheat.

B. Overheat Detection Algorithm using Infra-red Thermal Images

Figure 2 shows a result of monitoring system that detects abnormal overheat of the traction motor of KTX train.

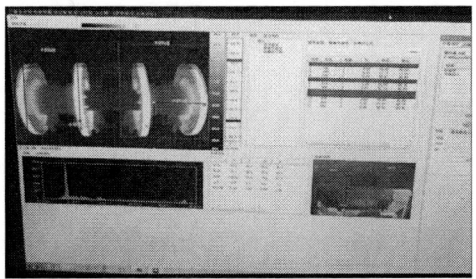

Fig. 2 Real-time detection progress about the braking disc

The KTX train set is composed of 20 cars and 3 driving bogies are located in the front and the back, respectively. In this study, the traction motor of the front driving bogie has been selected as the target component and the suggested detection algorithm has been verified on this component. Figure shows the detected traction motor in the front and the maximum temperature of the detected region.

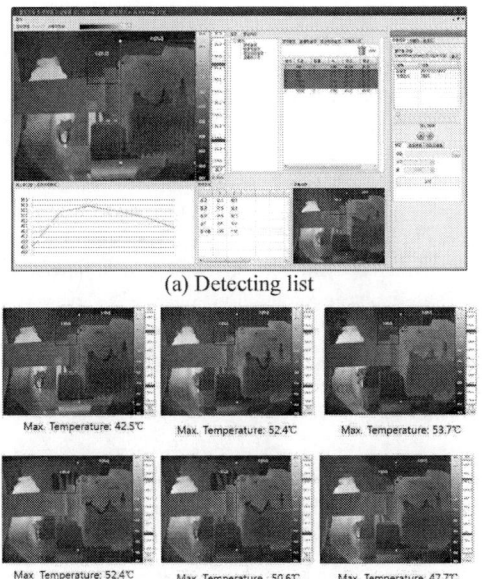

(a) Detecting list

Max. Temperature: 42.5℃ Max. Temperature: 52.4℃ Max. Temperature: 53.7℃

Max. Temperature: 52.4℃ Max. Temperature: 50.6℃ Max. Temperature: 47.7℃

(b) Maximum temperature of driving motor
Fig. 3 Detecting list and image processing algorithm

The algorithm that has been implemented to monitor the target, the traction motor in the front, is the template matching technique. As shown in Equation (1), the template matching technique utilizes normalized square difference method between the original image and the template. The abnormal overheat is detected by analyzing the frame that has the minimum value among the consecutive images and it can be calculated with the following equation:

$$R(x,y) = \frac{\Sigma_{x',y'}\ (T(x',y') - I(x+x', y+y'))^2}{\sqrt{\Sigma_{x',y'}\ T(x',y')^2 \Sigma_{x',y'} (x+x', y+y')^2}}$$
(1)

where the original image is denoted by I(x,y) and the template

by T(x,y).

Figure 3 shows the list of detected images for the axle and the traction motor with the matched templates. In order to process the large number of images acquired by the high speed camera, a number of functions have been added to the procedure such as setting the temperature of the region, setting the pattern size, and discarding repeated images. First, "temperature setting of the region" function compares each frame with the minimum temperature of a certain region and discards the frames that have the temperature lower than the minimum temperature. Second, "pattern size setting" function checks the size of the binary image in the template matching process and the comparison is preceded only when the binary image is larger than the pattern image. Third, "discarding repeated image" function increases the efficiency of the algorithm by ignoring the repeated patterns in the following images after detection is made in the previous images. This function also eliminates the repeated detection of the same item at the same location that is present in the images from high speed cameras. The components of the front traction motor of traction bogie that have the temperature of 50℃ or higher (axle #2, 3, 4, 5) are highlighted in red in the list of Figure 3.

Conclusion

Image processing and diagnosis algorithms that help monitoring the status of main underbody components of railway vehicles are an important factor for safe railway operation. And the abnormal conditions can be efficiently detected by using thermal images from infra-red thermal image system. In this study, overheat detection algorithm using infra-red thermal camera for the key underbody components of railway vehicles including traction motors and reducers. The experiments have been carried out for the KTX trains and the proposed template matching algorithm has been applied to the underbody components to detect overheat by calculating the positions of the components and observing the maximum temperature of the targets.

References

[1] Michael Vollmer, Klaus-Peter Möllmann, *Infra-red Thermal Imaging*, WILEY-VCH, pp. 189-190, 2010.

[2] D. Almond, G. Busse, E. Grinzato, J. C. Krapez, X. Maldague, V. Vavilov, and W. Peng, "Infra-red thermographic detection and characterization of impact damages in carbon fiber plastics: round robin test result," *Proceedings of Quantitative Infra-red Thermography '98*, Lodz, Poland, 7-10 September 1998

[3] X. Y. Han, L. D. Favro, Z. Ouyang, R. L. Thomas, "Thermosonics: Detecting cracks and adhesion defects using ultrasonic excitation and infra-red imaging," *Journal of Adhesion*, Vol. 76, No. 2, pp. 151-162, 2001.

[4] Michael Vollmer, Klaus-Peter Möllmann, *Infra-red Thermal Imaging*, WILEY-VCH, pp. 73-77, 2010.

Implementation of Efficient SHA-256 Hash Algorithm for Secure Vehicle Communication using FPGA

Chanbok Jeong and Youngmin Kim

School of Electrical and Computer Engineering, Ulsan National Institute of Science and Technology (UNIST)
UNIST-gil 50, Ulsan 689-798, Republic of Korea
+82-52-217-2182, {jcbcom, youngmin}@unist.ac.kr

Abstract

In this paper, we implement the SHA-256 FPGA hardware module for the security protocol of the IEEE 1609.2 vehicle communication (VC). VC requires high-throughput and low-latency hardware architectures. For fast and efficient design, we exploit parallel structures for preprocessing and hash computation in SHA-256. The proposed design is implemented in Vertex-5 and verified for correct hash operation. As a result, we can achieve up to 179.08 MHz with 2796 slices.

Keywords-SHA-256, FPGA, Security for Vehicle Communication, Hash Algorithm, IEEE 1609.2, FIPS-180-3

Introduction

Recently, vehicle communication (VC) has become the most important electrical function for intelligent vehicles and systems. IEEE 1609.2 Wireless Access in Vehicular Environments (WAVE) standard method is used for secure VC. To implement WAVE protocol, hash module is required to process both Elliptic Curve Digital Signature Algorithm (ECDSA) and Elliptic Curve Integrated Encryption Scheme (ECIES) security algorithms. In addition, the hash algorithm is vital and widely used in many cryptography processes [1].

In [2], authors have implemented the hash function in SHA-256 module for HMAC (Hash-based Message Authentication Code). To improve the performance, they use four pipeline stages for hash computation of four different input messages, and a carry save adder. In [3], a compact FPGA processor for the SHA-256 algorithm is implemented without preprocessing unit. To optimize the SHA-256 hash function, they have proposed several techniques, such as minimization of the critical path, reducing of the memory access by using data reuse, and a specific 4-input arithmetic logic unit (ALU).

In this paper, we implement an entire SHA-256 algorithm using parallel architecture of the preprocessing and the hash computation, a proposed Latest Message Scheduler (LMS) block, and Xilinx adder IP for WAVE protocol to support fast and efficient cryptography using FPGA. In contrast to other research, we implement the preprocessing function because we use the SHA-256 module for embedded environment such as vehicle communication. By aforementioned efficient methods, we can achieve SHA-256 FPGA hardware with 179.08 MHz speed.

SHA-256 Hash Function

The SHA-256 is an one-way hash function that has been published as FIPS 180-3 by the National Institute of Standards and Technology (NIST) and included in SHA-2 family. Input data of SHA-256 is processed by preprocessing and hash computation. The preprocessing conducts padding a message first, then parses the padded message into N x 512-bit blocks and sets initialization values. After that, hash computation follows to generate message schedule blocks using the padded message and a series of the hash value using the message schedule blocks. Finally, the message digest, which is the output of the SHA-256, is generated as the output. The size of the message digest of SHA-256 is 256 bits. The message digest always contains unique values depending on the input message [4].

In SHA-256, all operations are processed based on 32-bit unit. Thus, all arithmetic operations are implemented based on 2^{32}. The SHA-256 has six logical functions (i.e., Ch, Maj, $\sum_0^{\{256\}}(x)$, $\sum_1^{\{256\}}(x)$, $\sigma_0^{\{256\}}(x)$, and $\sigma_1^{\{256\}}(x)$), eight working variables (a, b, c, d, e, f, g, and h), and two temporary variables (T_1 and T_2). These logical functions, working variables, and temporary variables are combined and mixed together to process the hash computation stage.

Hardware Implementation

To construct a fast and efficient SHA-256, we apply a parallel architecture for the preprocessing block and the hash computation block in the proposed SHA-256 architecture as shown in Fig. 1. In our design, the preprocessing block for new message and hash computation block are pipelined for higher throughput. At the preprocessing block, the input data (Input_Text) is parsed in the 512 bits padded blocks. The last block consists of the end of message bits, one bit '1' value, sequence of '0' bits, and the 64-bit value of the length of message. After that, the block scheduler prepares the message schedule blocks (W_t). The W_t are used to calculate the eight working variables in the Compute Memory block and the two temporary variables (T_1 and T_2). The W_t from 0 to 15 are equal to the corresponding padded blocks stored in separate 16 registers, but the W_t from 16 to 63 require additional calculations using the previous W_t in a single 32 bits register. To calculate the W_t from 16 to 63, we propose the LMS module for reducing total number of registers. The LMS stores current padded message and schedules to issue a proper message for each hash computation round. At the final, in 63th round,

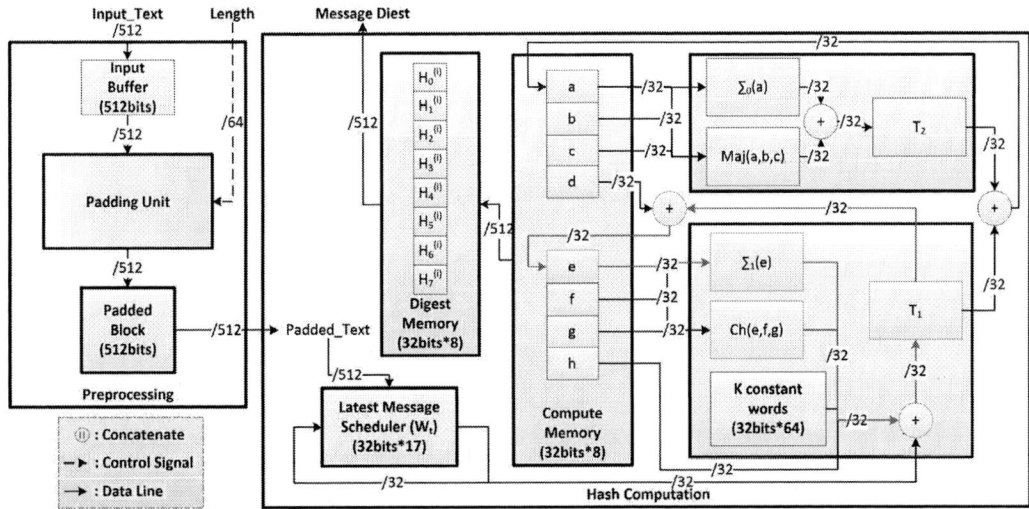

Fig. 1. The proposed parallel structure of the SHA-256.

TABLE I
FPGA Implementation result and comparison with previous works

Device	FPGA slice	Clock Freq. (MHz)	Implementation of Preprocessing
Virtex-5 [2]	1885	169.00	Software
Virtex-5 [3]	139	64.45	No
Virtex-5 [this work]	2796	179.08	Yes (Hardware)

Fig. 2. The result of timing simulation showing input text and output hased message ($o\_message\_digest$).

eight working variables are added with previous hash values to generate the message digest of the input message. In Fig. 1, the critical path of the hash computation block is drawn in red line. As shown, the path has data dependency to calculate the variable itself. And many 32-bit add operations affect the performance of the path. To reduce the delay of this critical path and improve the overall performance, we exploit the adder and subtracter IP logics provided by Xilinx ISE tool [5].

Result

The proposed architecture is implemented in the Virtex-5 and tested in various cases. The FPGA hardware implementation is summarized and compared with other designs in Table I. As shown, we can achieve the best clock speed (i.e., up to 179.08 MHz) with a hardware preprocessing unit by using of 2796 slices in the FPGA hardware. Fig. 2 is timing simulation result of the synthesized SHA-256 FPGA hardware. For the timing simulation, the NIST SHA-256 test vector is used [6].

Conclusion

We have successfully implanted the SHA-256 module in Virtex-5 FPGA chip for use in the security protocol of the IEEE 1609.2 VC. To meet the performance requirement of the VC systems, several techniques for fast and efficient design of the preprocessing and the hash function are proposed and exploited. As a result, the implemented hardware operates correctly in a 179.08 MHz system clock by using of 2796 slices.

Acknowledgments

This research was financially supported by the "R&D Infrastructure for Green Electric Vehicle (RE-EV)" through the Ministry of Trade Industry & Energy (MOTIE) and Korea Institute for Advancement of Technology (KIAT).

References

[1] IEEE Std 1609.2-2013. (April, 2013.) IEEE Standard for WAVE–Security Services for Applications and Management Messages. [Online]. Available : http://standards.ieee.org/findstds/standard/1609.2-2013.html

[2] H. E. Michail, *et al.*, "On the Exploitation of a High-Throughput SHA-256 FPGA Design for HMAC," *ACM Trans. on Reconfigurable Tech. and Sys.*, vol. 5 no. 1, pp. 1–28, 2012.

[3] R. Garcïaa, *et al.*, "A compact FPGA-based processor for the Secure Hash Algorithm SHA-256," *Computers & Electrical Engineering*, vol. 40, no. 1, pp. 194–202, 2014.

[4] NIST FIPS PUB 180-3. (October, 2008), "Secure Hash Standard (SHS)," [Online]. Available : http://csrc.nist.gov/publications/fips/fips180-3/fips180-3_final.pdf

[5] Xilinx LogiCORE IP Adder/Subtracter v11.0. (March, 2011), [Online]. Available : http://www.xilinx.com/support/documentation/ip_documentation/addsub_ds214.pdf

[6] SHA-256. (October, 2007), [Online]. Available : http://csrc.nist.gov/groups/ST/toolkit/documents/Examples/SHA256.pdf

978-1-4799-5128-4/14 $31.00 © 2014 IEEE

An Online Test and Debug Methodology for Automotive Image Processing System

[1]Hyunggoy Oh, Inhyuk Choi, Taewoo Han, Won Jung, [2]Byungin Moon, [3]Sungho Kang

Department of Electrical and Electrionic Engineering
Yonsei University
Seoul, Korea
Email: [1]{kyob508, ihchoi, twhan, dino12h}@soc.yonsei.ac.kr, [2]bihmoon@knu.ac.kr, [3]shkang@yonsei.ac.kr

Abstract

In the digital system where safety is a major issue, the reliability issue has been more important. However, as the circuit design has been more complicated, the number of some errors which escaped from the pre-silicon verification has been increased and the undetected errors have a bad influence upon reliability. To solve this problem, an online test and debug methodology for the automotive image processing system is proposed in this paper. Experimental results show the proposed methodology has high system reliability and provides the concurrent operation with a negligible test time and a small hardware overhead compared to the previous works.

Keywords - online test, online debug, automotive image processing system.

Introduction

In the semiconductor industry, the test issues of a digital system are more challenging due to the complexity in the integrated circuits. One of the challenges is some errors can't be detected in the pre-silicon verification [1]. Unlike most image processing systems, these unrevealed errors adversely cause serious problems to the environment recognition, position recognition and autonomous driving technology in the automotive image processing system, which are directly connected to people's lives. Therefore, in the automotive image processing system where safety is a major issue, the methodology which detects these unrevealed errors during the normal system operation is strongly required to increase the system reliability. However, the previous methodologies of the online test [2] and the online debug [3] do not provide the concurrent operation and require lots of hardware and test time overheads to increase the system reliability. To solve this problem, an online test and debug methodology for the automotive image processing system is proposed.

Proposed Methodology

Figure 1 describes the proposed online test and debug module for the automotive image processing system. The online test is composed of the structural and functional tests to verify the whole system. In addition, the online debug verifies each module to increase the system reliability. Since the online test and debug are performed during the proper time intervals, they do not interfere with the system operations. The

characteristic of the proposed methodology is described in Table 1.

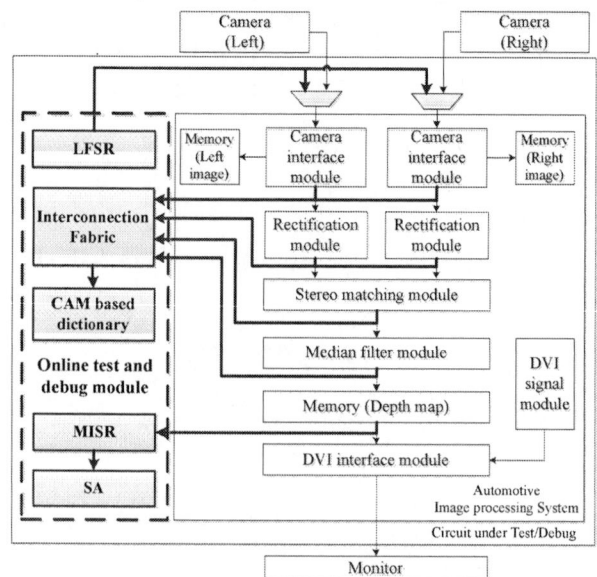

Fig 1. The proposed online test and debug module for the automotive image processing system

TABLE I
GOAL AND OPERATION TIME FOR ONLINE TEST AND DEBUG MODE

Test / Debug mode		Goal	Operation time
Online test mode	S_mode	Structural test	Preoperation time
	F_mode	Functional test	Blanking interval time
Online debug mode		Debug for modules	Normal system operation

A. Online test

The online test module consists of a linear-feedback-shift register (LFSR), a multiple-input signature register (MISR) and a signature analyzer (SA). In the S_mode, the structural test is performed before the system is operated in order to guarantee the correctness of the system. Using DFT circuitry used in the offline test, the hardware overhead of structural test is minimized. In addition, using the self-test feature of DFT circuitry, the extra I/O interface due to the structural test is minimized. After the system starts, the functional test is performed during the blanking interval time, which is the idle time between each frame occurred by the characteristic of the camera interface. Since the blanking interval time is typically enough to add one pixel line [4], it can be used for the functional test. In order to reduce the test overhead, the characteristics of the stereo cameras used in the image processing system are used in the F_mode. Since stereo camera input data is totally random, the pseudo-random patterns can be used for the test. In addition, since there are duplicated modules such as Camera interface module and Rectification module, the same test patterns are applied to the modules and compared the outputs each other. The hardware overhead for the functional test is drastically reduced due to pseudo random pattern generation using the LFSR and output comparison of the same modules.

B. Online debug

To increase the system reliability, each module is verified by the online debug. To debug the real time data of each module, the online debug is performed during the normal system operation. The online debug module consists of the interconnection fabric and a content-addressable memory (CAM) based dictionary architecture. The interconnection fabric is composed of XOR-MUX cells and used to select the proper signals to debug sequentially [1]. Through the interconnection fabric, the traced input and output data signals of each module are transmitted to the CAM based dictionary to verify the functional operations of each module. And then the traced signal data is compared to the golden data stored in the CAM based dictionary in a single clock cycle using dedicated comparison circuitry [5]. However, the dictionary method typically requires a lot of data volume when all the debug cases are considered. Instead of considering all the cases, some major debug cases are only considered to reduce the hardware overhead. The major debug cases are determined by designers' knowledge based on the module functionality.

Experimental Results

To apply the methodology to the real case, the automotive image processing system for context awareness is used, which provides a 752x480 sized frame with 30 fps. To get the 95% fault coverage and 100% functional coverage, a 28bit sized LFSR is used to generate the test patterns. The several debug cases used in the experiment are shown in Table 2. Table 3 shows the performance comparison of the previous approaches and the proposed methodology. The test time of the proposed methodology is less than that of the online debug since the only major debug cases are verified. Comparing to the online test, the test time overhead related to the functional test is negligible since the functional test is performed during the blanking interval time. Using the proposed methodology, the required functional test time is just 88 frames. The hardware overhead of the proposed methodology is larger than that of the online test because the additional dictionary hardware is added to perform the online debug process. However, comparing to the online debug, it is almost the same due to the heavy reuse of the offline BIST and utilization of the characteristic of the system. The system reliability of the proposed methodology is much higher than the previous works since the whole system is tested structurally and functionally with verifying the major debug cases. Furthermore, the proposed methodology provides the concurrent operation of the online test and debug.

TABLE II
DEBUG CASES AND SOLUTIONS OF EACH MODULE

Module	Debug case		Solution
Rectification module	Valid signals for each module		Check the left frame valid signal and the address data
Stereo matching module	Disparity range	16	Check whether the fifth output data is 0
		32	Check wheter the fifth output data is 1
Median filter module	Proposterous value		Check the output data of the previous three inputs

TABLE III
PERFORMANCE COMPARISON

	Online test [2]	Online debug [3]	Proposed
Test time	Medium	High	Medium
Hardware	Low	Medium	Medium
Concurrency	X	X	O
System reliability	Medium	Low	Very high

Conclusion

The proposed methodology shows how the online test and debug verify the automotive image processing system. The online test consists of the structural and functional test for the whole system, which are performed during the preoperation time and the blanking interval time. To increase the system reliability, the online debug verifies the major debug cases during the normal system operation. Experimental results show that the proposed methodology has high system reliability and provides the concurrent operation with a negligible test time and small hardware overhead compared to the previous works.

Acknowledgment

This research was supported by the MSIP(Ministry of Science, ICT & Future Planning), Korea, under the C-ITRC(Convergence Information Technology Research Center) support program (NIPA-2014-H0401-14-1004) supervised by the NIPA(National IT Industry Promotion Agency).

References

[1] Xiao Liu and Qiang Xu, "On Efficient Silicon Debug with Flexible Trace Interconnection Fabric," IEEE International Test Conference (ITC), pp.1-9, 2012.

[2] R.S Oliveira, J. Semiao, and et al, "On-line Bist for Performance Failure Prediction under Aging Effects in Automotive Safety-Critical Application," Latin American Test Workshop (LATW), pp.1-6, 2011

[3] Zhang Peng, Fan Xiaoya, and Huang Xiaoping, "An on-chip debugging method based on bus access," IEEE International Conference on Signal Processing, Communication and Computing (ICSPCC), pp.1-5, 2013.

[4] Vasquez, Maximino, and et al, "Techniques for Aligning Frame Data," U.S. Patent Application 12/655, 410, 2009.

[5] Pagiamtzis, Kostas and Ali Sheikholeslami,"Content-Addressable Memory (CAM) Circuits and Architectures: A tutorial and survey," IEEE Journal of Solid-State Circuits, pp.712-727, 2006.

978-1-4799-5128-4/14 $31.00 © 2014 IEEE

Analysis of Structural Variation and Threshold Voltage Modulation in 10-nm Double Gate-All-Around (DGAA) Transistor

Myunghwan Ryu and Youngmin Kim

School of Electrical and Computer Engineering, Ulsan National Institute of Science and Technology (UNIST)
UNIST-gil 50, Ulsan 689-798, Republic of Korea
Phone : +82-52-217-2182, Fax : +82-52-217-2109 {arshall, youngmin}@unist.ac.kr

Abstract

Increasing short channel effects (SCE) interrupt the further technology scaling in the CMOS transistors. Beyond 10 nm technology node, the gate-all-around (GAA) FET is considered as a promising solution for continuing the Moore's law. In this paper, we report the analysis of the double gate-all-around (DGAA) FET in terms of structural variations and the effect of the threshold voltage modulation by independently controlled inner gate. The impact of inner gate thickness and gate oxide thickness variations on the electrical characteristic of the DGAA FET are investigated. In addition, we propose the inner gate utilization to modulate the threshold voltage of the transistor for providing more design options.

Keywords - Gate-All-Around (GAA), DGAA, SGAA, IGAA, low-power, oxide thickness variation

Introduction

The scaling limits of CMOS technology make hard to follows the Moore's law, which requires novel device structures to increase gate controllability and suppress the short channel effects (SCEs) [1]. In ultra-scaled devices (i.e., 45-nm beyond), SCEs, such as subthreshold swing (SS) degradation, source/drain leakage current problems, larger drain-induced barrier lowering (DIBL), threshold voltage (V_{th}) roll-off and V_{th} mismatch caused by random dopant fluctuations (RDF), limits further scaling. And the increased SCEs occur mainly due to the reduced gate controllability. The industry and academia has been proposed a number of next-generation transistor candidates [2]. Among the several candidates, multi-gate device topologies are considered as a leading technology for further scaling. Recently, the 22 nm technology has adopted FinFET which is an example of the multi-gate technology and is expected to work for channel length down to 10 nm [2]. Beyond 10 nm, the gate-all-around (GAA) FET is considered as the most promising multi-gate technology. The GAA FET provides theoretically perfect electrostatic control on the channel, which enables further reduction in transistor size while maintaing low leakage currents and makes it highly attractive for low power applications [3]. The DGAA FET can be classified either shorted-GAA (SGAA) or independent-GAA (IGAA) according to the existence of the different inner gate bias. The circuit designers cannot change the V_{th} of device because V_{th} is set by channel doping (traditional way) or appropriate gate

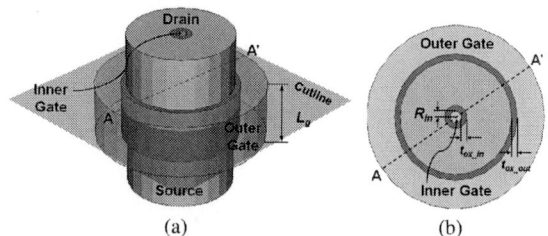

Fig. 1. DGAA FET structure used in device simulations. (a) bird's eye view and (b) cross-sectional view.

workfunction. But, we can change the threshold voltage of the transistor by controlling the inner gate separately. By using of the IGAA scheme in circuit design, designer have various V_{th} of device, which provides more design options. In this study, the inner gate thickness and oxide thickness variations both for inner and outer gate are analyzed in 3D TCAD simulations and the effect of the IGAA is investigated.

Device Structure and Simulation

The 3D TCAD structure used in this study is shown in Fig. 1. The nominal physical parameters used in the device simulations are summarized in Table I. The channel length (L_g) is 10 nm and both inner and outer gate oxide thickness (t_{ox}) are set 1 nm. The radius of silicon channel is 10 nm. The channel region is lightly doped with $1 \times 10^{16} cm^{-3}$ (Boron) to reduce the RDF effects and avoid the mobility degradation and source/drain is doped with 2×10^{19} (Arsenic) for N-type FET. A high-k metal-gate (HKMG) process, with tungsten and HfO2 as gate electrode and gate dielectric, respectively, is used to construct the gate region. The supply voltage is set to be 0.75V according to ITRS roadmap [2]. TCAD Sentaurus Device [4] is used to perform device simulations.

Results and Discussion

A. Impact of Inner Gate Thickness (R_{in})

Simulation results of the on and off current by the inner gate thickness (R_{in}) variation are shown in Fig. 2. The outer gate radius is fixed thus increased R_{in} reduces the channel thickness. The currents are normalized by the average of inner and outer circumference of the channel. As shown, both I_{on} and I_{off} decrease as the R_{in} increases. The on-current has linear dependency on the inner gate thickness because increased inner gate thickness reduces the channel region for current flow.

978-1-4799-5128-4/14 $31.00 © 2014 IEEE   228

TABLE I
Physical parameters used in the device simulation

Parameters	Value
Channel length (L_g)	10 nm
Inner gate oxide thickness (t_{ox_in})	1 nm
Outer gate oxide thickness (t_{ox_out})	1 nm
Radius of silicon channel	10 nm
Channel doping concentration	$1 \times 10^{16} cm^{-3}$
Source/Drain concentration	$2 \times 10^{19} cm^{-3}$
Supply voltage (V_{DD})	0.75 V
Gate material	Tungsten
Gate oxide material	HfO_2

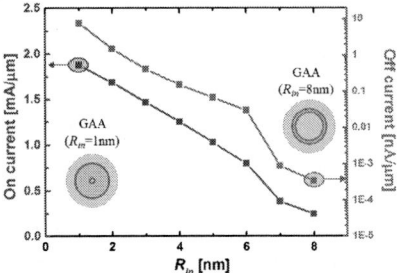

Fig. 2. Impact of the inner gate thickness variation on the on- and off-current. The blue region is the channel where current flows and two green regions are two gates.

However, there is exponential decrease in the leakage current by the increased inner gate thickness due to the better gate controllability by the narrower channel region when R_{in} increases. For example, when R_{in} becomes 4 nm from 1 nm, only 30% on current (per unit width) reduction occurs but approximately two orders of magnitude leakage current can be reduced.

B. Impact of Gate Oxide Thickness (t_{ox})

There are two type of gate oxide, t_{ox_in} and t_{ox_out}, since there exists an additional inner gate terminal in DGAA devices. The on- and off-current results for different inner and outer gate oxide thickness variations are shown in Fig. 3. At first, both I_{on} and I_{off} decrease as the inner gate oxide thickness increases from 1 nm to 3 nm because increased oxide thickness reduces the channel thickness. Second, I_{on} becomes smaller for thicker outer gate oxide due to reduction of inversion capacitance but the leakage can be further reduced by the shallower outer gate oxide thickness due to the better control of the channel. Therefore, the minimum gate oxide thickness both for inner gate and outer gate provides the best leakage current property with the highest driving current.

C. Impact of Inner Gate Bias in IGAA

The gate voltage is applied to make a conductive channel between the source and drain for current flow. In this paper, V_{th} is set by the metal-semiconductor work function difference. When inner gate and outer gate terminal are tied together and biased equally, they are involved together to induce a conductive channel. However, when inner gate is biased differently from the outer gate, metal-semiconductor workfunc-

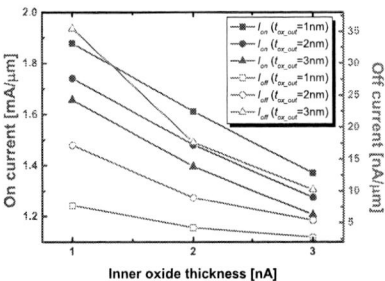

Fig. 3. Impact of thickness variation of inner and outer gate oxide.

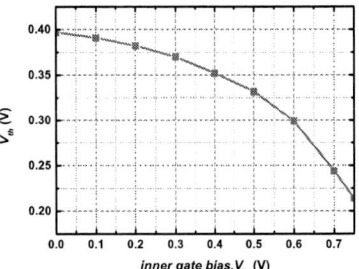

Fig. 4. Threshold voltage (V_{th}) modulation by introducing of the additional inner gate bias (V_{in}).

tion will differ between in the vicinity of inner gate and other gate region. In other words, the V_{th} modulation can happen by separating two gates in IGAA. Fig. 4 represents the threshold voltage for different inner gate bias (V_{in}) in the IGAA FET. As shown, the V_{th} reduces from 0.397 V to 0.214 V when V_{in} varies from 0 V to 0.75 V. The dynamic threshold voltage change in the IGAA transistors can provide another options for the circuit designers in terms of performance and power optimization according to the requirements of the integrated circuits.

Conclusion

In this paper, we investigate the impact of the inner gate thickness and gate oxide thickness variations on the electrical characteristic of the DGAA FET through 3-D TCAD device simulations. For SGAA FET, the on-current has linear dependency on the inner gate thickness, whereas off-current has exponential dependency as the inner gate gets thicker. The oxide thickness simulations reveal that the minimum inner and outer gate oxide thickness results in the best GAA FET in the performance and leakage. In addition, we confirm the feasibilty of the inner gate utilization to modulate the threshold voltage of the IGAA FET which provides more design options for the circuit designers.

References

[1] Y. Taur and T. Ning, "Modern VLSI Devices", Cambridge, U.K.: *Cambridge Univ. Press*, 2009.

[2] ITRS, 2013 edition, [online] Available at http://www.itrs.net/

[3] H. M. Fahad and M. M. Hussain. "Are nanotube architectures more advantageous than nanowire architectures for field effect transistors?", Scientific reports, 2. 2012.

[4] Sentaurus, ver. H-2013-03, [online], http://www.synopsys.com/

978-1-4799-5128-4/14 $31.00 © 2014 IEEE

Performance and Leakage Optimization with 22 nm Bi-level FinFET

Jaemin Lee and Youngmin Kim

School of Electrical and Computer Engineering, Ulsan National Institute of Science and Technology (UNIST)

Unist-gil 50, Ulsan 689-798, Republic of Korea

jm3430202, and youngmin@unist.ac.kr

Abstract

In this paper, we analyze on- and off-current characteristic of the 22-nm Bi-level FinFET in which two different fin widths are formed. From the 3D Technology CAD (TCAD) simulations, we find out that the narrower the fin width the lower the leakage current. However, narrower fin width results in the reduced driving current for the triple-gate FinFET structure. We propose the optimal shape parameters of Bi-level FinFET such as the fin width and height of the narrower fin for providing better leakage current while keeping the required driving current of the nominal FinFET. Simulation results show that up to 4% speed up with 33% leakage current reduction by the optimal Bi-level FinFET compared to the nominal one.

Keywords-FinFET, Bi-level, on-current, off-current, optimization, 3D TCAD

Introduction

Transistor scaling has resulted in astounding increases in the transistor density and performance, leading to greater functionality at higher speeds for single chips. However, as the chip size continues to shrink, the leakage and short-channel effects can render silicon-based transistors less reliable [1]. As a solution to the scalability and leakage issues, a multi-gate transistor, such as a fin-type field effect transistor (FinFET) has been emerged as a promising substitute for bulk CMOS at the 32-nm node and beyond [2].

FinFETs have been proposed and investigated as strong candidates for sub-32-nm technology CMOS devices because of their excellent immunity to the short channel effect (SCE), low leakage current, high driving controllability, and high output resistance [3]. A Bi-level FinFET (BL-FinFET) structure where the fin regions consist of two different level is proposed in [4]. They compare between the BL-FinFET and the conventional FinFET (C-FinFET) in various electrical properties and have suggested that BL-FinFET has many benefits over C-FinFET.

In this work, we investigate the Bi-level FinFET in 20 nm technology node by using of 3D TCAD simulations [5] and propose the optimum fin shape in the BL-FinFET for performance (on-current) and leakage (off-current) characteristics.

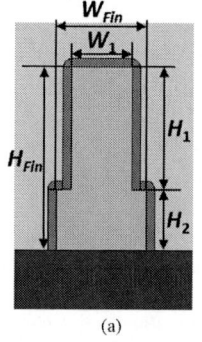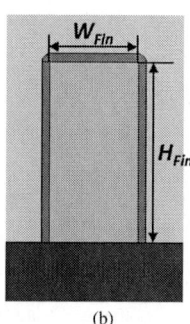

Fig. 1. Cross-sectional images of the 3D TCAD structure with parameters in (a) BL-FinFET and (b) C-FinFET.

TABLE I

Parameters of Bi-Level and Conventional FinFET 3D TCAD simulations.

Device Parameters	Values
Gate length (L_g)	20 nm
Fin width (W_{Fin})	15 nm
Fin height (H_{Fin})	30 nm
Fin_1 width (W_1)	2 nm $\sim$ 13 nm
Fin_1 height (H_1)	2 nm $\sim$ 28 nm
Channel doping concentration (N_{ch})	$2.0 * 10^{15} cm^{-3}$
Source/Drain doping concentration (N_{sd})	$2.5 * 10^{19} cm^{-3}$
Gate oxide thickness (t_{ox})	1.4 nm
Supply voltage	0.9 V

Simulation setup

The cross-sectional view of the BL-FinFET and C-FinFET in the 3D TCAD simulation are shown in Fig. 1(a) and (b), respectively. As suggested in Fig. 1(a), the fin region consists of two different fin widths and heights in BL-FinFET. The first fin (Fin_1) has the narrower fin width at the upper region, and the other one (Fin_2) is located at the bottom of the fin. TCAD simulation parameters are also depicted in Fig. 1. The W_1 and H_1 are the width and the height of Fin_1, and W_{Fin} and H_2 are the width and the height of Fin_2. In the Fig. 2(b), W_{Fin} and H_{Fin} are the nominal fin width and height of C-FinFET. In this study, the width of the bottom fin (Fin_2) is fixed and the $H_{Fin} = H_1 + H_2$. The nominal FinFET structure is calibrated to match the I-V characteristics of the 22-nm PTM FinFET model [6, 7]. The parameters are summarized in the Table I.

978-1-4799-5128-4/14 $31.00 © 2014 IEEE

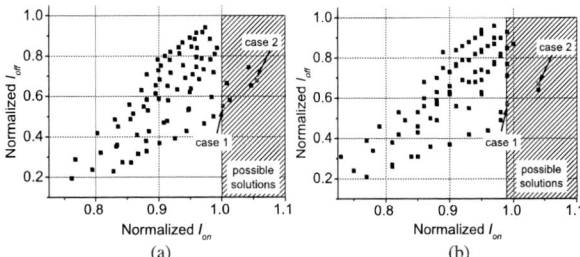

Fig. 2. I_{on} and I_{off} values of the BL-FinFET structures for different W_1 and H_1. (a) NMOS and (b) PMOS. The currents are normalized to the C-FinFET. Two red dots indicate the selected combinations.

When determining the optimum values of W_1 and H_1, we sweep the fin height and width of the Fin_1 with the other values fixed, and measure the on-current (I_{on}) and off-current (I_{off}) of the BL-FinFET and compare with C-FinFET results.

Optimal Bi-level FinFETs

Fig. 2 shows the simulation results (normalized) of BL-FinFETs of all the W_1 and H_1 combinations. As shown, the leakage current in BL-FinFET is smaller than that of the C-FinFET for all cases due to the narrower fin thickness and the BL-FinFETs provide less driving currents due to the narrower channel area by the Fin_1 in most cases. However, there are several cases for higher on-current (i.e., larger than 1.0) by BL-FinFET. The leakage current reduction in the BL-FinFET can be explained by the improved channel controllability in the smaller fin thickness [8]. We select two cases from Fig. 2 for the circuit-level analysis; First, the case 1, W_1 is 12 nm and H_2/H_1 is 0.25 (H_1 and H_2 are 24 nm and 6 nm, respectively) for NMOS and H_2/H_1 is 0.36 (H_1 and H_2 are 22 nm and 8 nm, respectively) for PMOS, provides similar I_{on} with large reduction in the I_{off}. Second, the case 2, W_1 is 13nm and H_2/H_1 is 0.36 (H_1 and H_2 are 22 nm and 8 nm, respectively) for both NMOS and PMOS, results in the maximum I_{on}.

Fig. 3 shows on/off current density in NMOS of the BL-FinFET (case 2) and C-FinFET. As shown in (a) and (b), the current flows at the surface of the fin for 'on' cases of the narrow fin FinFET [9]. The top surface region of the BL-FinFET is narrower than that of the C-FinFET but higher current density can be achieved by the narrower Fin_1 at the two sidewalls in the BL-FinFET than C-FinFET. In addition, as shown in (c) and (d), the leakage current flows at the farthest region from the gates in the FinFET. Therefore, less I_{off} can be achieved by the BL-FinFET due to the improved channel controllability by the narrower Fin_1.

Mixed-mode TCAD simulations using an inverter implemented by the optimal N/PMOS BL-FinFETs are conducted to analyze the dynamic property of the proposed BL-FinFET shape [5]. The propagation delay and leakage comparison are summarized in the Table II. As shown in Table II, the case 2 BL-FinFET results in up to 4% speed up while providing 33% leakage current reduction than C-FinFET. Moreover, we can reduce about a half of the leakage current by the case 1 structure with minimal delay overhead (e.g., 1%) in the optimal BL-

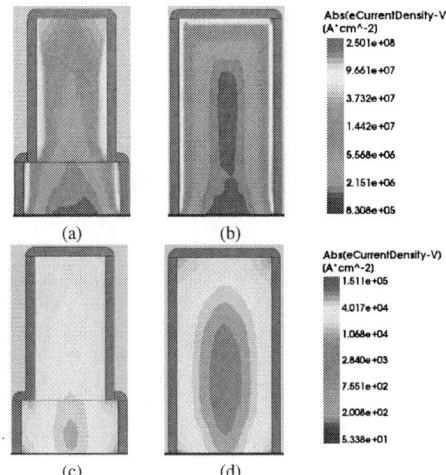

Fig. 3. On-current density of (a) BL-FinFET (b) C-FinFET, Off-current density of (c) BL-FinFET (d) C-FinFET.

TABLE II
Mixed mode TCAD simulation results of an inverter logic.

	propagation delay		leakage current	
	(ps)	(normalized)	(nA/μm)	(normalized)
C-FinFET	50.52	1.00	163.05	1.00
case 1	51.05	1.01	91.30	0.56
case 2	48.65	0.96	109.86	0.67

FinFET inverter.

Conclusion

In this paper we analyze the BL-FinFET and propose the optimal shape of the Bi-level structure. By comparing the on- and off-current, we suggest two different cases in the BL-FinFET. The mixed-mode simulation indicates that approximately 4% speed up becomes possible with significant leakage reduction by the optimized BL-FinFET.

References

[1] Y. X. Huang, W. C. Lee, C. Kuo, and D. Hisamoto, "Sub 50-nm FinFET: PMOS," *Proceedings of IEDM*, pp.67-70, 1990.

[2] ITRS 2012, [online], http://public.itrs.net.

[3] T. -J. King, "FinFETs for Nanoscale CMOS Digital Integrated Circuits," *Proceedings of ICCAD*, pp. 207–210, 2005.

[4] M. Mehrad and A. A. Orouji, "A new nanoscale and high temperature field effect transistor: Bi level FinFET," *Physica E: Low-dimensional Sys. & Nano.*, vol. 44, no. 3, pp. 654–658, 2011.

[5] Sentaurus, ver. H-2013-03, [online], http://www.synopsys.com/

[6] 22nm FinFET PTM models, [online], http://ptm.asu.edu/

[7] HSPICE, ver. H-2013-03, [online], http://www.synopsys.com/

[8] C. -L. Lin, *et al.*, "Effects of Fin Width on Device Performance and Reliability of Double-Gate n-Type FinFETs," *IEEE TED*, vol. 60, pp. 3639–3644, 2013.

[9] N. K. Jha, and D. Chen, "Nanoelectronic Circuit Design," Springer Press, 2011.

Implicating Logic Functions with Memristors

Pravin Mane*, Nishil Talati**, Ameya Riswadkar, Ramesh Raghu, C. K. Ramesha

Department of Electrical and Electronics Engineering
BITS Pilani K.K. Birla Goa Campus, Goa- 403726, India
Contact Number: +91-9657711842, Fax Number: 0832-2557030
E-mail: pravinmane@goa.bits-pilani.ac.in*, nishil.talatiwork@gmail.com**

Abstract

Conventional FPGAs with SRAM cells, in addition to low density, low speed and higher power consumption, are facing the problems in terms of low retention time due to increase in leakage current as MOS size shrinking continuously due to progress in technology. With the invention and fabrication of memristor as a suitable candidate of 1-bit non-volatile memory, hybrid CMOS-memristor based 3D FPGA architectures have been proposed in literature. In this paper, we propose implication-NOR logic gate based FPGA architecture for implementation of logic functions.

Keywords- Memristor, FPGA, ASIC, Microprocessor, CMOL, FPNI, Material Implication

I. Introduction

Due to the ineluctable advantages and disadvantages of the methods employed for computing (i.e. FPGA, ASIC and Microprocessor based designs) in the modern age, reducing the performance gap between them has become an interesting arena for the researchers [1, 2]. Memristor based design is one of the possible solutions to improve this performance gap. Memrisor is the fourth fundamental passive circuit element, theoretically first discovered by Leon Chua in 1971 [3] and came into the physical existence in 2008 as a result of the research carried out at HP Labs [4].

Hybrid CMOS-Memristor circuits fully compatible in terms of the materials and processing technologies employed to fabricate it. The most straightforward application of described hybrid circuits is embedded and stand-alone memories with their simple matrix structure. In this architecture, memristor will be used to store single bit of data and CMOS is used for coding, decoding, line driving, sensing, and input/output functions [5].

II. Background

A. CMOL and FPNI architectures

Hybrid CMOL (CMOS/Molecular hybrid) architecture of Strukov and Lihkarev is one of the probable ways to extend Moor's Law beyond the limits of transistor scaling. Density, simplicity, and clean separation of configuration and data communication are some of the appreciable merits of CMOL architectures.

Field Programmable Nanowire Interconnect (FPNI) is an another generalized hybrid architecture, which trades off some of the speed, density and defect-tolerance of CMOL in exchange for easier fabrication, lower power dissipation, and the greater freedom in the section of nanodevices in the crossbar junctions [6]. In FPNI architecture, logic is performed at CMOS layer and routing operation is implemented through the nanowires, reducing the static power dissipation significantly .

B. Characteristics of a Memristor

Some of the interesting properties of a memristor include resistance switching, pinched hysteresis loop etc. To evaluate them, the circuit shown in Figure 1 (left) has been designed and the current is varied as shwn in Figure 1 (right). The simulation results for this (voltage, current, and resistance variation) can be shown in the figure 1 (right).

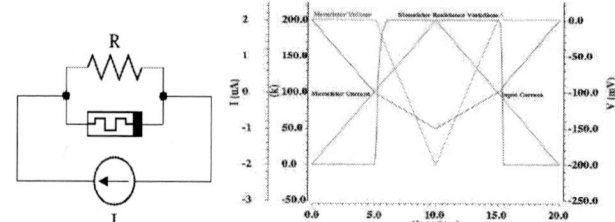

Fig. 1: Memristor Characteristocs Evaluation Simulation Circuit (left), Simulation Results (right).

C. Material Implication

Material Implication is the most natural operation implemented using a memristor. The circuit diagram are shown in figure 2. Here, the inputs and outputs are in terms of the resistances of the memristors. After applying inputs, V_{COND} and V_{SET} are applied across them. The result is achieved on the memristor Q shown in the figure. Note that here, $V_{SET} \square V_{COND}$ is always followed.

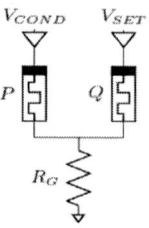

Fig. 2: Implication using Memristors.

III. Generalized Architecture

In this paper, we propose a novel CMOS-Memristor based generalized architecture for logic function implementation. Using Material Implication property, NOR gate has been designed, which is the smallest block of this architecture.

A. Implication Based NOR Block

The important observation in Implication is the inequality between V_{COND} and V_{SET}. The only transition possible for Q is from *OFF* to *ON* state. This observation has been utilized to design NOR gate (figure 3). There are three phases involved to evaluate this logic. Write phase includes applicaiton of *WRITE* signal which applies the inputs to the memristors. Evaluate phase includes application of *EVAL* signal, which applies V_{COND} and V_{SET} as shown in figure. Read phase applies *READ* signal, which converts the resistance of Evaluation Memristor (M_{Eval}) into voltage. The symbolic representation for the further interpretation of *n*-input NOR is also shown in Figure 3 (right).

B. Comprehensive Description of the Architecture

Figure 4 shows the detailed architecture to implement 3-input function. *A,B,C* are the data lines. *EN* the enable control signal. *LEN* is the level enable signal and *W,E,R* are write, evaluate, and read control signals respectively. LEVEL1 consists 7 NOR blocks and a 7-input priority circuit. LEVEL2

978-1-4799-5128-4/14 $31.00 © 2014 IEEE   232

consists of *n*-input NOR blocks, having inputs depending on the function being implemented. The number of input combinations that produce a different output than the outputs with all inputs logic high is the number of input memristors of the corresponding NOR block.

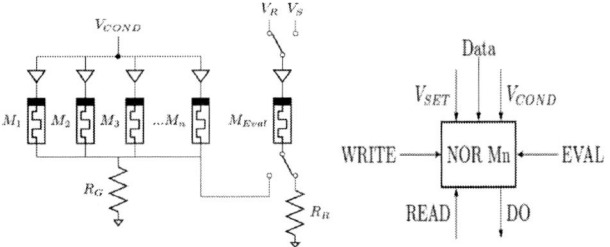

Fig. 3: NOR gate using Material Implication circuitry (left), symbolic reprentation (right).

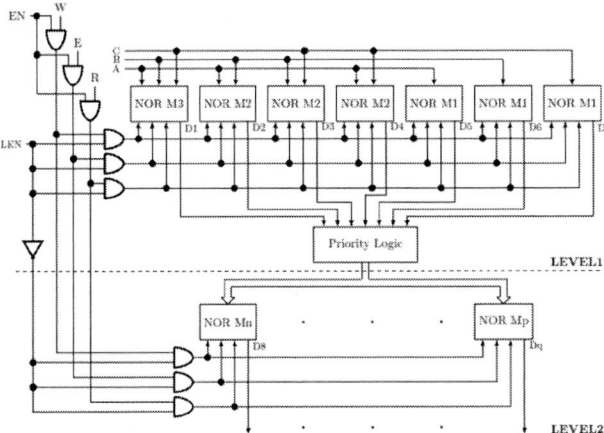

Fig. 4: 3-input logic block architecture.

There is a restriction in terms of fan-in of this architecture. The maximum number of inputs (n) can be

$$n \leq R_{ON} / [(V_{COND} \cdot R_G / [V_{SET} - i_{ON} \cdot R_{OFF}]) - R_G] \quad (1)$$

To remove this restriction on fan-in, the bigger architecture has been designed using the above architecture (shown as LB3) as shown in the figure 5. In this, the function is divided according to the fan-in. Here, it is assumed to be 3. Thus, I_{1-3} are given to LB3 and rest of the inputs are used to select any such block.

C. Performance Evaluation

The detailed analysis of the architecture has been carried out and compared it the CMOL FPGA architecture considering some benchmark circuits produced by UCB. This comparison has been given in Table 1.

IV. Conclusion

In this paper, we have introduced a novel Implication based architecture to implement any logic function. It is based on the hybrid CMOS-Memristor circuit.

V. Acknowledgment

This work is supported by DST-FIST grant for the department of Electrical & Electronics Engineering, BITS Pilani, K.K. Birla Goa Campus, Goa by Government of India.

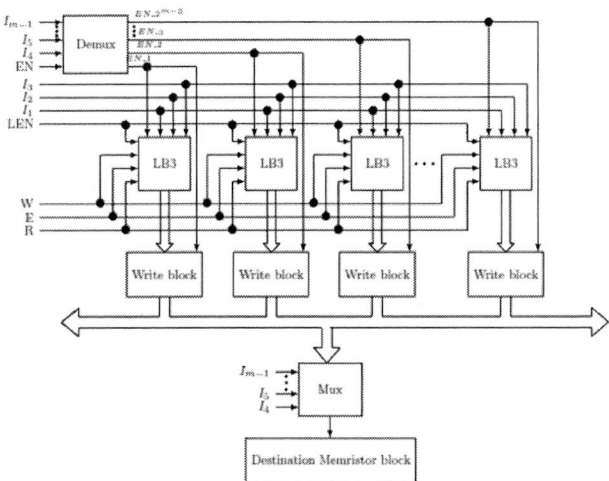

Fig. 5: Architecture to implement any function without any restriction on fan-in.

Table 1: Analysis of the Architecture

Benchmark Circuits	CMOL		Proposed Architecture	
	Tran-sistors	Mem-ristors	Tran-sistor	Mem-ristor
ALU4	22459	8157	3169	91
APEX2	21916	9907	1484	46
APEX4	4178	6257	6876	190
BIGKEY	6339	8835	64840	1738
CLMA	383123	70403	10583	289
EXP5P	3034	4885	21704	586
EX1010	63879	33298	3843	109
MISEX	15061	7598	5191	145
PDC	12336	35552	6140	379
S298	52088	13321	2495	73
SEQ	24499	8897	12268	334
SPLA	132792	29251	15975	443

References

[1] I. Kuon and J. Rose, Quantifying and exploring the gap between FPGAs and ASICs. Springer, 2009.

[2] G. Cuniberti, G. Fagas, and K. Richter(Eds.), Introducing Molecular Electronics. Springer, 2005.

[3] L. O. Chua, "Memristor - the missing circuit element," IEEE Transactions on Circuit Theory, vol. CT-18, no. 5, pp. 507–519, September 1971.

[4] D. B. Strukov, G. S. Snider, D. R. Stewart, and R. S. Williams, "The missing memristor found," Nature, vol. 453, pp. 80–83, May 2008.

[5] D. B. Strukov and K. K. Likharev, "Defect-tolerant architectures for nanoelectronic crossbar memories," Journal of Nanoscience and Nanotechnology, vol. 7, pp. 151–167, 2007.

[6] G. S. Snider and R. S. Williams, "Nano/CMOS architectures using a field-programmable nanowire interconnect," Nanotech- nology, vol. 18, pp. 1–11, January 2007.

978-1-4799-5128-4/14 $31.00 © 2014 IEEE

SEU-Tolerant Active Body-Bias Inverter

Youngkyu Jang, Ik-Joon Chang, Jinsang Kim, and Seungjoo Lee*

Department of Electronics and Radio Engineering, Kyung Hee University, Rep. of Korea
* Department of Electronics Engineering, Hyejeon University, Rep. of Korea
jskim27@khu.ac.kr

Abstract— **PD-SOI (Partial Depleted Silicon On Insulator) process is a good candidate technology for space system designs, since it features excellent insulation to the silicon substrate compared to the conventional bulk process. However, the radioactive particles from the low earth orbit can causes the single event upset (SEU) or the charge collection in a circuit node, leading to a logical error. We propose SEU-tolerant CMOS logic inverter using a novel active body-bias scheme.**

I. Introduction

The radioactive particles can ionize a material and form a transient conductive track that can lead to single-event-upset (SEU) which upsets the voltage level. Especially, the electron easily can be swept by the node current because the electron is lighter than charges [1]. Every transistor is isolated in the SOI technology. Therefore, the SOI technology provides better performance at radiation environment. There are two types of SOI structures based on the thickness of the top silicon layer: fully depleted transistors (FD) and partially depleted (PD) transistors [2]. PD structures use thicker silicon films and exhibit a floating body isolating the front device from the buried oxide. The floating body is coupled to the MOSFET transistor through the gate capacitance and source and drain diode junctions. Our research has been focused on the circuit development using PD-SOI structure.

If a radioactive particle pass throw drain and body, the reverse biased p-n junction between body and drain would be collapsed in a very short time and it allows SEU. The best way to prevent the SEU is to physically isolate nodes without potential difference. The floating body without the body tie is helpful to prevent SEU. In order to deal with both SEU and the side effects of SOI devices simultaneously, we propose a novel active body-bias scheme for the logic designs of space applications. The conventional active body-bias circuits [3][4] use auxiliary transistors to implement the bias circuitry and to increase the Vdd limit of the body for faster operation. However, the existing active body biasing circuits does not incorporate SEU protection scheme and hence are vulnerable to the logic error from the radioactive particles.

II. SEU-Tolerant Inverter Using Active Body-Bias

The conventional active body-bias circuits control the body potential using system assisted enable signal [3] and output controlled scheme [4]. However, [3] and [4] do not divide the transition state and the static state for the body bias, and the body potential is controlled during the static state. Therefore, they are not robust to SEU, since a conductive path can be formed during the static state in case radioactive particles hit on a circuit node.

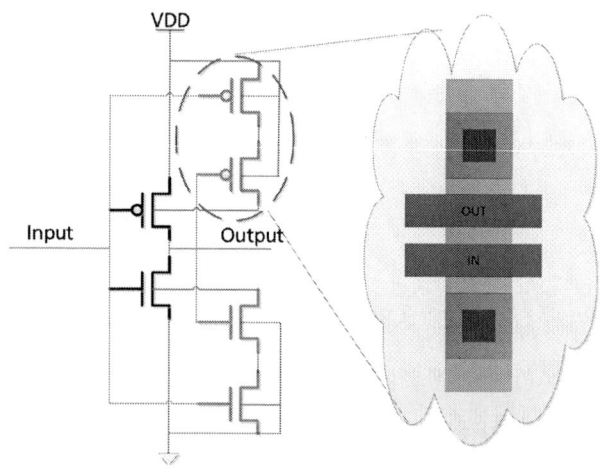

Fig. 1. Inverter using the proposed active body-bias scheme

The key idea of the proposed SEU-tolerant inverter is that the body-bias is on during the transition state and is off during the static state. The inverter using the proposed active body-bias method is shown in Fig. 1. The proposed body-bias method uses both input and output data simultaneously and the auxiliary transistors for the body biasing are in series. The size of the additional body-bias transistors is much smaller than that of the main transistors, since the functionality of the body-bias transistors are only to control the body potential and they can easily stacked without contacts for layout design. The body-bias is off during the static state and the bodies of the nMOSFET and the pMOSFET keep the floating state.

Since the proposed active body-bias method is bias-on and bias-off during transition state and static state, respectively,

978-1-4799-5128-4/14 $31.00 © 2014 IEEE   234

we can easily expand the proposed design concept to other logic gates.

The proposed inverter requires two additional small bias transistors per main transistor. As shown in Fig. 2, since one bias transistor can share with other transistor's body-biasing, the number of bias transistor can be reduced as the number of inputs of gate increases. The proposed NAND and NOR gates require 6 transistors for active body-biasing. [9] shows that for an inverter circuit, 6 additional body bias transistors are required.

III. SIMULATION RESULTS

The proposed active body-bias inverter was simulated using 0.13 um BSIMSOI 4.3 MOSFET model. An inverter with fan-out4 (FO4) load is tested using twice smaller bias circuit transistors.

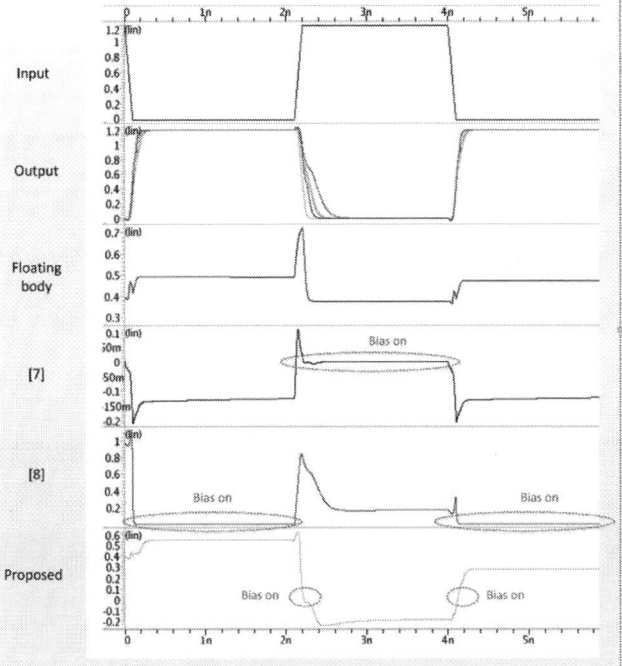

Fig. 3. nMOS body potential of inverters

In the proposed circuit the body-bias is on only during the transition state. In Fig. 3, we can see the nMOSFET body is about 0V only during the transition state. After the transition state, it returns to the floating state, leading to more robust to SEU. Also, we verified the functionality of other logic gates.

IV. CONCLUSION

In this paper, we proposed a SEU-tolerant active body-bias inverter using PD-SOI technology. The proposed active body-bias inverter control the body-bias only during the transition state and the bodies remain at the floating state during the static state. We showed that the proposed inverter has better performance in terms of Q critical capacitance and the history effect. As a future research, we will focus on the optimization of the area overhead for the auxiliary body-bias circuit and application studies for system level implementations.

ACKNOWLEDGMENT

This work was supported by Global Surveillance Research Center (GSRC) program funded by the Defense Acquisition Program Administration (DAPA) and Agency for Defense Development (ADD) and the national research foundation of Korea (NRF) under grant 2012R1A2A2A01011.

REFERENCES

[1] Petersen, E.L., "Predictions and observations of SEU rates in space," IEEE Transactions on Nuclear Science, Volume: 44 , Issue: 6, Dec-1997, 2174 – 2187

[2] Sarajlic, M, Ramovic, R," Analytical Modeling of the Triggering Drain Voltage at the Onset of the Kink Effect for PD SOI NMOS," Microelectronics, 2006 25th International Conference on p325-p328

[3] Casu, M.R., "Comparative analysis of PD-SOI active body-bias circuits," 2000 IEEE International SOI Conference, 2000, 94 – 95

[4] Joonho Gil, "A high speed and low power SOI inverter using active body-bias," 1998 International Symposium on Low Power Electronics and Design, Aug-1998, 59 – 63

Design of Hysteretic Buck Converter with a Low Output Ripple Voltage and a Fixed Switching Frequency in CCM

Tae-Jin Jeong, Woo-Seong Kang, Ji-San Choi, and Goang-Seob Yoon
College of Electronic Engineering
INHA University, Korea
specizy@naver.com, ksyoon@inha.ac.kr

Abstract—**An efficient fast response hysteretic buck converter suitable for mobile application is proposed. The problems of large output ripple and difficulty in using of small power inductor that conventional hysteretic converter has is improved by adding PLL (Phase Locked Loop) structure. The circuits are implemented by using Dongbu 0.35um BCDMOS process. The settling time of charge pump output is less than 6us and output voltage overshoot is 21mV when load current changes from 50mA to 500mA. The efficiency is 90~95% under 50~500mA load condition and output voltage is 1.2mV with 5.5mV ripple voltage.**

Keywords—hysteretic buck converter, ramp generator, delay time control circuit, switching frequency

I. INTRODUCTION

It has been found in the literature[1~2] that the conventional hysteretic buck converter inherently showed a faster transient response with load current change situation than different kind of feedback methods such as PWM voltage/current mode control and has simple circuit architecture because of no compensation. However, they suffered from not only EMI problem because of unpredictable switching frequency, but also large output ripple voltage because of wide hysteresis window range. The inherent switching frequency of the conventional hysteretic converter becomes too low to employ small power inductor, and to make output ripple small. The proposed circuit employed a ramp generator, CCM/DCM select circuit and delay time control circuit with a PLL structure to resolve the problems conventional hysteretic converter has. Section II describes the proposed architecture and design of the main block. The simulation results are presented to compare the performances of the proposed circuit with those of the conventional circuits in Section III. The conclusions are drawn in Section IV.

II. THE PROPOSED ARCHITECTURE

The overall architecture proposed in this section is presented in Fig. 1. It consists of the conventional hysteretic buck converter with ramp generator, delay control circuit and CCM/DCM select circuit. Ramp generator is connected with the input node (V_{FB}) of the comparator to add ramp signal to output ripple voltage. Delay control circuit is connected between the output node of hysteresis comparator and dead

time control to adjust the switching frequency to reference clock frequency. CCM/DCM select circuit is composed of CCM/DCM detector and 2 to 1 multiplexer. CCM/DCM detector receives the output signal ($V_{OFFTIME}$) of ZCD (Zero Current Detector) and drives the multiplexer to separate each conduction mode.

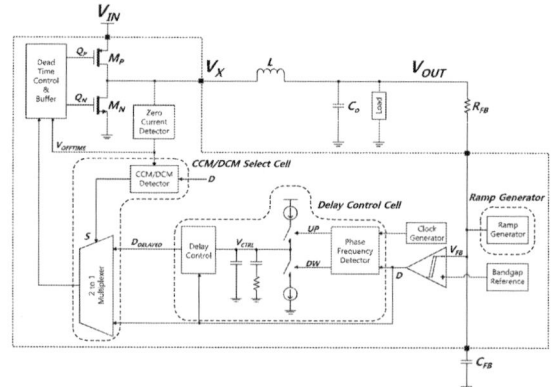

Fig. 1 Proposed hysteretic buck converter block diagram

The ramp generator is presented in Fig. 2 (a). The multiplexer controlled by pMOS drive signal (Q_P) generates a square waveform at node V_S. The low pass filter composed of R_{ADD} and C_{FB} allows the square waveform to be converted to the ramp signal. The amplitude of the ramp signal is usually greater than that of output ripple voltage. This results in the switching frequency more dependent on the magnitude of ramp signal than output ripple voltage. Also parasitic values of the output filter such as ESR have less effect on the output voltage ripple and switching frequency. [3] The switching period is described as (1).

$$T_S = \frac{V_{IN} \cdot C_{FB} \cdot V_{HYST} \cdot R_{ADD}}{V_{OUT} \cdot (V_{IN} - V_{OUT})} + T_{\nabla 1} \cdot \frac{V_{IN}}{V_{OUT}} + T_{\nabla 2} \cdot \frac{V_{IN}}{V_{IN} - V_{OUT}} \qquad (1)$$

,where V_{IN} and V_{OUT} are the input and output voltage of the buck converter, V_{HYST} is the window voltage of the hysteresis comparator, and T_{DEL1} and T_{DEL2} are the internal delay times of the switching on time and off time, respectively.

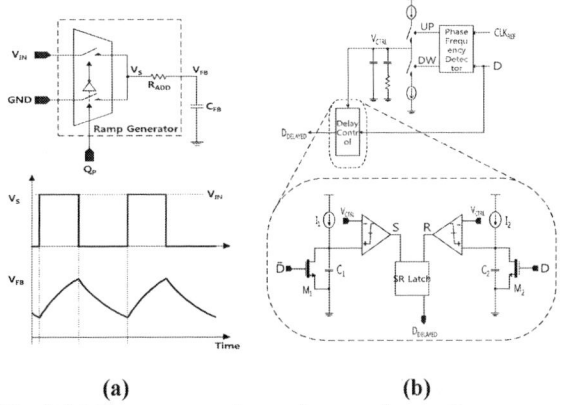

(a)   **(b)**

Fig. 2 (a) Ramp generator and general waveforms sample for ramp generator, (b) Delay time control block diagram

The delay time control circuit consists of PFD, charge pump and loop pass filter followed by the delay control cell show Fig. 2 (b). The digital output of the hysteresis comparator D controls the switches M1 and M2 to charge capacitors C1 and C2 with current I1 and I2. The delay times T_{DEL1} and T_{DEL2} can be determined by (2).

$$T_{DEL1} = \frac{C1}{I1} \cdot V_{CTRL}, \; T_{DEL2} = \frac{C2}{I2} \cdot V_{CTRL} \qquad (2)$$

The PFD and charge pump detect the phase and frequency between the reference clock (CLK_{REF}) and the switching signal D and adjust V_{CTRL}. If the switching signal runs slower than CLK_{REF}, the PFD will control the low-side charge pump to discharge V_{CTRL}. Because delay is directly proportional to V_{CTRL}, decreasing V_{CTRL} increases the frequency. This mechanism will make the switching frequency same with reference clock. [4] The above fixed switching frequency can't be applied to the DCM, resulting in less switching loss in DCM.

III. SIMULATION RESULTS

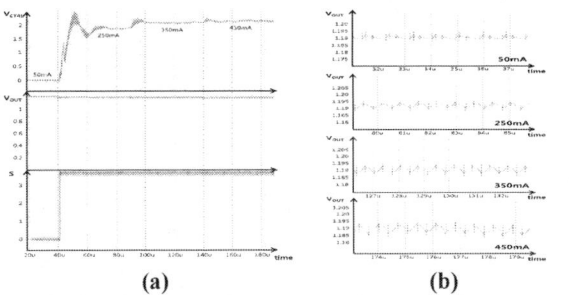

(a)   **(b)**

Fig. 3 Simulated waveform : (a) output of charge pump (V_{CTRL}), output voltage (V_{OUT}) and output of CCM/DCM detector (S), (b) output voltage in 50mA, 250mA, 350mA and 450mA load condition

As the load current varies from 50mA to 450mA, the simulated waveform of the V_{CTRL}, V_{OUT} and output of CCM/DCM detector (S) are presented in Fig. 3. When the

load current is more than 250mA, the S is logic '1' and the switching frequency is fixed as 1.42MHz. When the load current is 50mA, the S is logic '0' and the switching frequency is lower than 1.42MHz.

Table. 1 Performance Comparison

	[5]	[6]	This work
Freq(light load)	50kHz	670kHz	1.08MHz
Freq(heavy load)	1.38MHz	1.2MHz	1.42MHz
Inductor	10uH	4.7uH	1.2uH
Capacitor	20uF	4.7uF	15uF
V_{IN}/V_{OUT}	2.7~4.2V/2V	2.5~3.6V/1.5V	3.7V/1.2V
Ripple voltage	5.73mV	4.63mV	5.5mV
Load range	50~500mA	20~600mA	50~500mA
Transient response/overshoot	8.26us 79mV	x	5.5us 21mV
Efficiency	89~95%	88.7~96.4%	90~95%

IV. CONCLUSIONS

This paper describes the proposed hysteretic buck converter for mobile applications. Methods and circuits for achieving low output ripple and fixed frequency in CCM are presented. The proposed circuit employs the ramp generator to increase the switching frequency and delay time control block composed with PLL structure to fix the switching frequency of 1.42MHz. The simulation results and table 1 show that the output ripple, overshoot and efficiency of the proposed circuit with load current ranging from 50mA to 450mA are 5.5mV, 21mV and 90~95% respective

ACKNOWLEDGMENT

This research was supported by the MSIP (Ministry of Science, ICT and Future Planning), Korea, under the ITRC (Information Technology Research Center) support program (NIPA-2014-(H0301-14-1008)) supervised by the NIPA (National IT Industry Promotion Agency).

REFERENCES

[1] R Miftakhutdinov-TL Unitrode "An Analytical Coomparison of Alternative Control Techniques for Powering Next-Generation Microprocessors." Power Supply Design Seminar. pp26-30 2001.

[2] L.K.Wong and T.K.Man. "Steady State Analysis of Hysteretic Control Buck Converters" 13th International Power Electronicsand Motion Control Conference. pp400-403. 2008

[3] B.C. Bao, J. Yang, J.P. Xu, X. Zhang and G.H. Zhou. "Effect of output capacitor ESR on dynamic performance of voltage-mode hysteretic controlled buck converter" 26th Electronics Letters. Vol.49 No.20 pp1293-1294. September 2013.

[4] Chung-Hsien Tso, student Member, IEEE, and Jiin-Chuan Wu, Member, IEEE. "A Ripple Control Buck Regulator With Fixed Output Frequency." IEEE Power Electonics Letters. Vol.1 No.3 pp61-62. September 2003.

[5] C.J. Chuang and H.P. Chou. "An Efficient Fast Response Hysteretic Buck Converter with Adaptive Synthetic Ripple Modulator." 8th International Conference Power Electronics-ECCE Asia. pp625-627. May 30-June 3 2011.

[6] Kwang-ho Kim. and Bai-Sun Kong. "Adaptive Frequency Controlled Ultra Fast Hysteretic Buck Converter for Portable Devices." SoC Design conference(ISOCC). pp7-8. 2012.

978-1-4799-5128-4/14 $31.00 © 2014 IEEE   237

A Buck DC-DC Converter Using Automatic PFM/PWM Mode Change for High-Efficiency Li-Ion Battery Charger

Thanh Tien Ha, Do-Young Chung, Dasom Park, Hyun-Sik Lee, and Jong-Wook Lee

School of Electronics and Information
Kyung Hee University, Suwon, 446-701, Korea, E-mail: jwlee@khu.ac.kr

Abstract

We present a high efficiency buck DC-DC converter for switch-mode Lithium-ion (Li-ion) battery charger. To achieve a high efficiency over a wide load range, the converter enters a power saving pulse-frequency modulation (PFM) mode at light load currents. And the proposed buck converter features an automatic mode change between pulse-width modulation (PWM) and PFM depending on the load current level. The chip was fabricated in a 0.18 μm 1-poly 6-metal standard CMOS process in a 1.28×1.19 mm^2. The maximum efficiency is 93%.

Keywords - DC-DC switching converter; Li-ion charger; PFM/PWM techique;

Introduction

Recently, there has been a rapid growth of mobile devices (laptop, digital camera, household robot, mobile phone…), and the portable battery-operated devices are becoming more popular. For these devices, there are high demand for low-cost, small-size, light weight, high efficiency, and long battery run-time. For the mobile device power conversion, DC-DC converters are widely used. Achieving a high efficiency over a wide load current range is required [1]. For mobile device charging, Li-ion battery is widely used for its high capacity and desirable output current characteristics. And there are active researches on a high efficiency Li-ion battery charger [2], [3]. To achieve a high efficiency, in this paper, we present a buck DC-DC converter which features an automatic mode change between pulse-width modulation (PWM) and pulse-frequency modulation (PFM) depending on the load current level.

Design

Fig. 1 shows simplified block diagram of the proposed DC-DC converter. In PWM mode, the control loop senses the output voltage V_{OUT}, and subtracts this from a reference voltage V_{REF}, to generate an error signal V_C using an operational transconductance amplifier (OTA). The amplified error signal is compared to a fixed frequency carrier waveform, usually triangular or saw tooth in shape, voltage ramp V_{RAMP}. The comparison between V_C and V_{RAMP} results in pulse signal V_{PULSE}, which inputs to a SR latch. Thus the SR latch adjusts the duty cycle of PWM signal in a logic block. The logic block includes dead time control so that it can avoid "shoot-through" of both power transistors. The driver block provides output

pulses DRV_P and DRV_N to power transistors MP and MN.

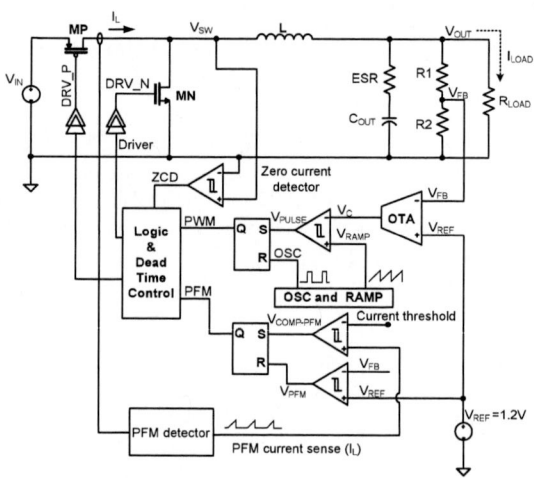

Fig. 1. Simplified block diagram of the proposed DC-DC converter.

In PFM mode, the current I_L of the high-side PMOS MP (it is also the inductor current) is detected by PFM detector. The I_L is compared with current threshold I_{COMP}. When I_L decreases below I_{COMP}, the converter is switched to PFM mode in which the power stage operates intermittently based on load demand. During PFM mode, switching activity at power stage is reduced. Thus, the switching loss is decreased and power transistors run with minimum quiescent current to achieve a high efficiency. When V_{OUT} drops below V_{REF} = 1.2 V, the output voltage V_{PFM} of the comparator resets the SR latch. Thus, DRV_P signal turns on PMOS MP and I_L ramps up. The current comparator detects the point when I_L reaches the preset peak current. And the output signal $V_{COMP-PFM}$ of comparator sets the SR latch, which turns off MP while turning on MN. This result in a ramping down I_L achieving voltage regulation. During I_L ramps down, reverse current is passed through the body diode of MN. The reverse current is detected by a zero current detector (ZCD). The output of the ZCD turns off MN to prevent drawing current from output capacitor back to ground. The next cycle starts when the output voltage V_{OUT} drops below V_{REF} again.

The schematic of mode control block is shown in Fig. 2. The combined mode change logic is implemented by using a SR latch, a D flip-flop, a PFM detector, a MUX, and comparators. Fig. 3 shows the waveform at the boundary between PWM and PFM modes. When I_{LOAD} decreases, the inductor current I_L ramps down below current threshold I_{COMP}, then, the output

978-1-4799-5128-4/14 $31.00 © 2014 IEEE

signal of current comparator $V_{COMP\text{-}PFM}$ detects the change. The mode change signal from PWM to PFM mode (W2F) is determined at the boundary between PWM and PFM mode. When the converter starts to operate in PFM mode, W2F signal goes high and the output signal MODE of SR latch changes to high, selecting PFM mode control signal.

Fig. 4 shows the waveform in case when the operating mode is changed from PFM to PWM mode. While the converter in the PFM mode, if the load current I_{LOAD} increase, then PFM mode cannot drive current I_L through the power transistors. In that case, V_{OUT} falls down and V_{FB} become smaller than $V_{REF} =$ 1.2 V (See Fig. 2). This results that the boundary signal (F2W) is switched to high level and resets SR latch. This sets the MODE signal go to low, then the converter changes mode from PFM to PWM mode.

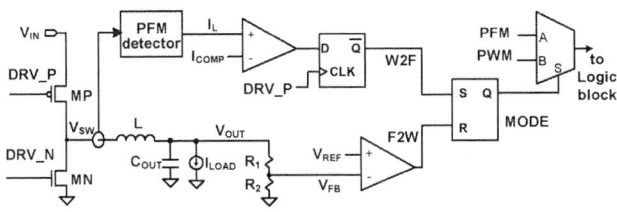

Fig. 2. Schematic of mode control block.

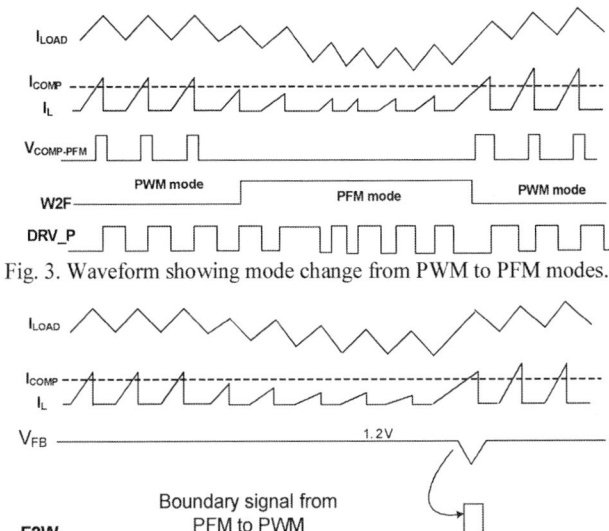

Fig. 3. Waveform showing mode change from PWM to PFM modes.

Fig. 4. Waveform showing mode change from PFM to PWM modes.

Result

Fig. 5 shows the layout of the proposed buck converter using a 0.18 μm 1-poly 6-metal standard CMOS process. The total area of the chip including the I/O pad is 1.28×1.19 mm² where the power transistors occupy 394×304 μm². Fig. 6 shows transient response for positive load dump of the buck converter. Fig. 7 shows correct operation of PFM / PWM mode by detecting the load current change. In PFM mode, we observe a damped harmonic oscillation at switching node V_{SW}. Efficiency of the designed buck converter as a function of load current is obtained and the maximum efficiency is 93%.

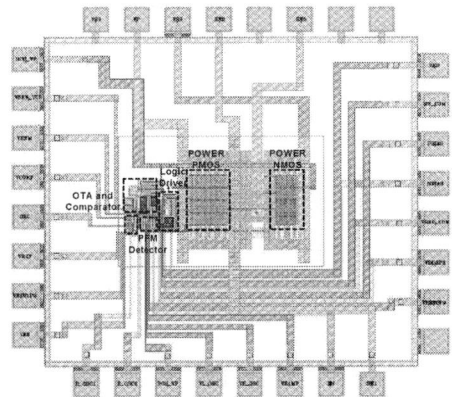

Fig. 5. Layout of the proposed buck DC-DC converter.

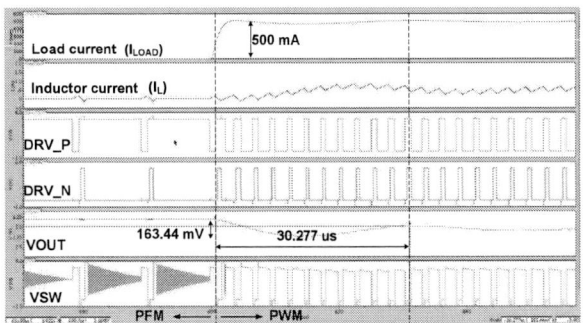

Fig. 6. Transient response under positive load dump.

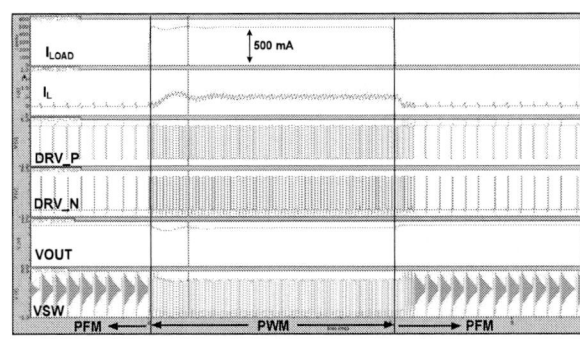

Fig. 7. Transient response under PFM / PWM mode change.

Acknowledgment

This work was supported by the Industrial Strategic Technology Development Program (no. 10048344) funded By the Ministry of Trade, Industry & Energy, Korea. The chip fabrication and CAD tools were made available through the IC Design Education Center (IDEC), Korea.

References

[1] L. F Shi and W. G Jia, "Mode-selectable high-efficiency low-quiescent-current synchronous buck DC–DC converter," *IEEE Trans. Ind. Electron*, vol. 61, no. 5, pp. 2278-2285, May. 2014

[2] R. L. R. Chen, S. L. Wu, D. T. Shieh, and T. R. Chen, "Sinusoidal-ripple current charging strategy and optimal charging frequency study for Li-ion batteries," *IEEE Trans. Ind. Electron.*, vol. 60, no. 1, pp. 88–97, Jan. 2013.

[3] J. M. Liu, P. Y. Wang, and T. K. Kuo, "A current-mode DC–DC converter with efficiency-optimized frequency control and reconfigurable compensation," *IEEE Trans. Power Electron.*, vol. 27, no. 2, pp. 869–880, Feb. 2012.

Design of a Sub-1-V Low Dropout Regulator
for Supply-Regulated Active-Loop-Filter VCOs

Young-Min Jang*, Ji-Geun Kim*, Jae-Hyun Cho*, Bruce C. Kim**, Sang-Bock Cho***

*·***Department of Electronics Engineering
University of Ulsan
Ulsan, Republic of Korea
**Department of Electrical Engineering
City University of New York, CUNY
New York, USA
min-s2@nate.com, bkim91us@gmail.com, sbcho@ulsan.ac.kr

Abstract

Full-swing voltage-controlled oscillators (VCOs) are very sensitive to supply noises and require supply regulators with increased supply voltages, which may entail more power consumption.

Further, in a phase-locked loop (PLL) with a conventional second-order passive-loop filter (PLF), PLL output frequency range is restricted by the compliance range of the charge pump output voltage (V_{CP}) even though supply-regulated VCOs (SR-VCO) can provide a wider frequency range.

In order to solve these problems, V_{CP} be separated from the V_{CTRL} using the low dropout regulator (LDO). This paper proposes a low dropout regulator of 0.83V with bandgap references (BGR) of 0.62V output voltage in supply-regulated active-loop filter VCOs (SR-ALF-VCO).

Keywords-Internet of things (IoT), RF/PMIC, low dropout regulator (LDO), voltage-controlled oscillator (VCO), active-loop filter (ALF)

Introduction

The Internet of the next generation is being expanded to IoT[1] (Internet of Things) for M2M (machine to machine) and V2V (vehicle to vehicle) communication using mobile devices. Therefore, not only existing RF/PMIC core components but also appropriate research and development are required.

The PLL and VCO[2, 3] are a control system that generates an output signal related to the phase of an input signal in chip-to-chip interconnect. In particular, characteristics related to phase noise and frequency stability of the local oscillator will have a significant impact on the data error characteristics of the entire system.

In this paper, we propose a sub-1-V LDO that has references independent of the supply as well as well-defined behavior regarding the temperature of supply-noise-independent VCO.

Supply-Regulated VCO

As shown in Fig. 1(a), PLLs based on SR-VCO operate such that the V_{REG} tracks V_{CTRL} and drives the supply-controlled oscillator (SCO). Fig. 1(b) shows SR-VCO with a conventional PLF. Because this PLF does not include active

devices, it has the advantage of degrading flicker noises and PLL phase noise. However, the PLL output frequency range is restricted by the compliance range of the V_{CP} even though SR-VCO can provide a wider frequency range.

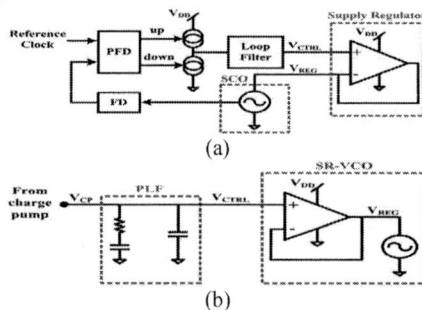

Fig. 1 Conventional block diagram of (a) PLL based on SR-VCO (b) second-order PLF + SR-VCO.

The LDO in SR-ALF-VCO shown in Fig. 2 can be a solution for the problem of PLF + SR-VCO. The V_{CP} can be separated from the V_{CTRL} using the LDO. Therefore, V_{CP} and V_{REG} are fixed to each input. Also, the SR-ALF-VCO makes it easier to design because it does not need a wide V_{CP}.

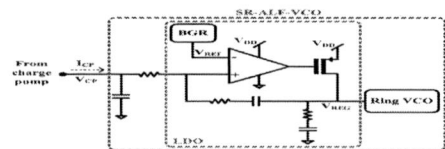

Fig. 2 Proposed low dropout regulator for SR-ALF-VCO.

The LDO is intended for insensitive Ring VCO for supply noise using LDO output voltage standards.

Low Dropout Regulator

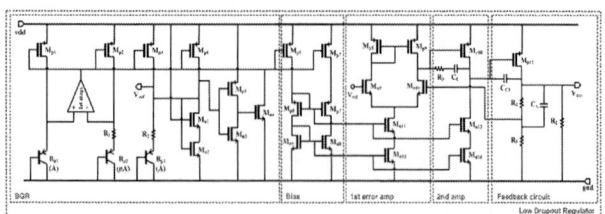

Fig. 3 Schematic of low dropout regulator.

As shown in Fig. 3, the LDO circuit consists of BGR, bias, 1st error amp, 2nd amp, and the feedback circuit. The output voltage of BGR and LDO can be shown in (1).

$$V_{ref} = V_{BE3} + \frac{R_2}{R_1} V_T \ln n \; ; \; V_T = kT/q \quad , \quad V_{od} = \left(\frac{R_4}{R_5} + 1 \right) \times V_{ref} \qquad (1)$$

The error amp with references and feedback input detects changes in output. The pass transistor (M_{p11}) is used as a voltage-controlled current source of input in order to have a constant DC value. Moreover, in order to prevent the transistors from falling to the linear region, the current source uses cascode current mirror topology so that the transistors are more stably saturated.

Simulation Results

Our concept is implemented using the cadence tool and 0.11um dong-bu mobile CMOS technology. The results of the simulation show the dropout voltage, line regulation, load regulation, temperature variation, and ripple rejection. Because other factors are not necessary, they are outside the scope of this paper.

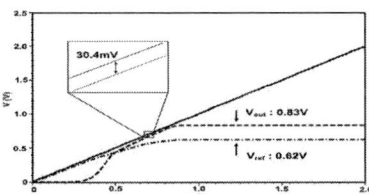

Fig. 4 Simulation result of dropout voltage (Iout:10mA).

Fig. 4 shows dropout voltage of LDO and BGR when output current is 10mA. Figs. 5 and 6 show corner simulation of line/load regulation. Line regulation value can be shown by (2).

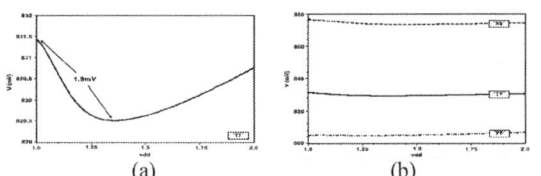

(a)   (b)

Fig. 5 Simulation result of (a) line regulation "TT" (b) line regulation "FF" "TT" "SS" corner.

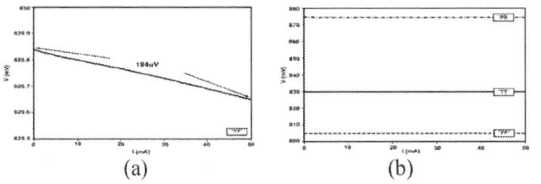

(a)   (b)

Fig. 6 Simulation result of (a) load regulation "TT" (b) load regulation "FF" "TT" "SS" corner.

$$Line = \frac{\Delta V_{od}}{\Delta V_{in}} \times \frac{100}{V_{od}} = \frac{1.9mV}{2 - 1V} \times \frac{100}{0.83V} = 0.223\%/V \qquad (2)$$

Further, if the output currents change from 1mA to 50mA, load regulation changes by 184uV. Fig. 7 shows the temperature coefficient with 1.9[ppm/° C].

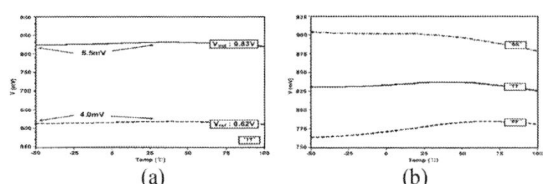

(a)   (b)

Fig. 7 Simulation result of (a) temperature variation "TT" (b) temperature variation "FF" "TT" "SS" corner.

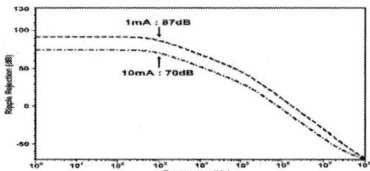

Fig. 8 Simulation result of ripple rejection (Iout:1mA, 10mA)

Fig. 8 presents ripple rejection, while Table I arranges each parameter value. Fig. 9 shows the final layout with an area of 230×230 um^2.

Table I Summary of simulation results

	Min(1mA)	Typ	Max(10mA)	Units
Dropout Voltage	3.2	-	30.4	mV
Line Regulation		0.223	-	%/V
Load Regulation	-	0.184	-	mV
Ripple Rejection	87	-	70	dB

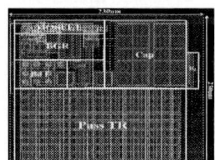

Fig. 9 Final layout of LDO

Conclusion

In this paper, we designed a sub-1-V LDO of 0.83V with BGR of 0.62V output voltage in a supply-regulated active-loop filter VCO. The line regulator of the proposed circuit is 0.223%/V, the load regulation is 184uV, the dropout voltage is 3.2mV and 30.4mV respectively and ripple rejection is 87dB. The temperature coefficient is 1.9 ppm/° C. The chip area is 230×230 um^2.

Acknowledgment

This work was supported by the Industrial Core Technology Development Program (10049009, Development of Main IPs for IoT and Image-Based Security Low-Power SoC) funded by the Ministry of Trade, Industry & Energy. In addition, we appreciate partial support by Semiconductor Research Corporation through TxACE.

References

[1] Wikipedia, "http://en.wikipedia.org/wiki/Internet_of_Things,"
[2] Kwang-Chun Choi, Sung-Geun Kim, Seung-Woo Lee, Bhum-Cheol Lee, and Woo-Young Choi, "A 990uW 1.6-GHz PLL Based on a Novel Supply-Regulated Active Loop Filter VCO," Journal of latex class files, Vol6. No.1, pp. 1-5, Jan 2007.
[3] Behzad Razavi " Design of Analog CMOS Integrated Circuit" McGraw-Hill International Edition 2001, pp. 377~403, 2001.

978-1-4799-5128-4/14 $31.00 © 2014 IEEE

Design of a Digital Controller for an LED lamp driver IC

Kilsoo Seo[1], Wonkyeong Park[2], Jusung Park[3]

[1]Power Semiconductor Research Center, KERI, Changwon 642-120, Korea
[2]Department of Circuit Design, Silicon Mitus, Korea
[3]Department of Electronic engineering, Pusan National University, Korea
[1]+82-55-280-1532, 1590 and ksseo@keri.re.kr
[3] juspark@pusan.ac.kr

Abstract

This paper presents a digital controller with a new dimming method for LED driver core circuits. The designed circuit enables the LED driver operate without a bulky capacitor and voltage dividing resistors. The proposed digital controller circuit was designed and fabricated using the 1 μm-650 V DMOS process. The results obtained from the simulation and measurement of the chip indicate that the inductor current waveform is similar to the input voltage to gain a high power factor, because the inductor current is synchronized to the zero voltage signal as expected.

Keywords - LED driver, digital sine-signal, zero voltage detection, dimming, soft-starter

Introduction

In the past several years, LED lights have emerged as replacements for incandescent light bulbs, cold cathode fluorescent lamps (CCFLs), and high intensity discharge (HID) lamps in various fields. This trend is attributed to the fact that LED lights have high light efficiencies, long operating lifetimes, small size, and are environmentally friendly [1,2,3]. Technological advancement in recent years enables a high-brightness LED to be used as a replacement for an incandescent bulb. However, unlike other competing lighting solutions, LEDs require driver circuits with considerable complexity when directly driven by commercial AC supplies. The asymmetric I-V characteristic of LED devices as diodes makes the load current profile considerably different from that of the AC offline voltage, which results in significant degradation in the power factor. In order to comply with standards such as IEC61000-3-2 Class C [4] that mandate higher than 85% power factors, most existing LED modules employ power-factor correction (PFC) circuits that function in the continuous-conduction mode(CCM)[6, 7]. These conventional CCM converters are quite expensive as they typically consist of many external devices such as bulky electrolytic capacitors, voltage dividing resistors, and power MOSFETs. The short life time of bulky capacitors results in a reduction in the LED module's lifetime [5]. In addition, CCM PFC controllers contain resistive dividers that are used to extract the sine-wave phase reference from the AC line. These resistive dividers increase power consumption and cause the peak current to vary according to the amplitude of the AC line voltage; also, it is hard to integrate these resistors on an IC.

In this study, we have developed a simple DCL that operates in the discontinuous-conduction mode (DCM) with a high PF for LED lighting applications. The DCM approach not only eliminates the shortcomings of conventional converters as mentioned before but also adds a new simple step-dimming method. The DCL can be easily implemented in a single chip without external components.

Proposed digital controller for LED driver

The proposed controller uses a sine-wave reference signal, which is in phase with the AC line voltage causing the LED current to also be in phase with the AC line voltage. As a result, the PF becomes very close to unity and the LED current shows a low THD, as outlined in [4].

Fig. 1(a) shows the proposed, simplified block diagram of the DCL. The proposed DCL is composed of three main components: a zero voltage detector (ZVD), a digital sine-signal generator (DSG), and dimming control. The operation of the DCL is described as follows. The ZVD circuit generates the ZVD signal, using the transmission gate synchronized with the clock that is used in the inductor current signal generator, comparator, and reference voltage for the ZVD. The ZVD signal is generated by comparing the Vdrain generated by the inductor current with a reference voltage for the ZVD. The ZVD detects when the Vdrain is close to 0, and it generates a signal synchronized with the input voltage. By determining the start and end of the DSG signal relative to the ZVD signal, it is possible to create a waveform similar to the input voltage. Furthermore, the DSG design is accomplished without using bulky capacitors or voltage dividing resistors.

A dimming system is beneficial for all lighting systems, not only for energy saving purposes but also for electricity demand reduction, visual comfort for the room occupants, and better productivity at the work place. Adjustment of the brightness is realized via regulating the inductor current. The duty ratio is determined by comparing Vsense signal and the DSG signal. The brightness is adjusted by adjusting the magnitude of the DSG signal. In this paper, a new dimming control method with 4 levels is presented. By using an external switch, an adjustment request signal will be generated and applied to the dimming circuit. This changes the reference signal as well as the DSG signal amplitude. Further, the soft-start function is added to the dimming circuit to prevent the inrush current from damaging the power MOSFET and LED string when the external switch is turned on.

Fig. 1(b) shows a circuit for generating a reference voltage. The reference voltage is the most important voltage for determining the magnitude of the DSG signal, as used in the brightness adjustment and soft-starter functions. The soft-starter is composed of D flip-flops (D-FFs) that are clocked by the ZVDb signal.

978-1-4799-5128-4/14 $31.00 © 2014 IEEE

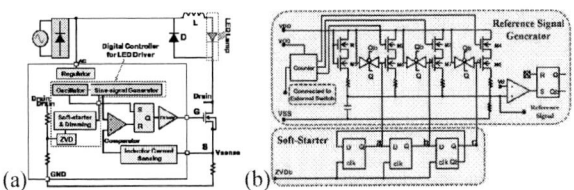

Fig 1. Block diagram of an LED driver and schematic of the reference signal and soft-start sub-block

The a, b, and c outputs become high in turn as the clock advances turning on M2, M3 and M4, respectively. Therefore, the reference voltage increases gradually. The reference voltage is set to be higher than the VB to ensure that the FFs of the soft-starter will not work anymore in the RS-FF mode. The brightness can be adjusted in four steps when the counter is turned on/off in the order M6 ~ M8 by the external switch. The counter works in the following sequence: 111,011,001,000,100,110,111. It is possible to expand the number of dimming levels if required by using a counter with a higher resolution.

Chip fabrication and measurements

As shown in Fig. 2(a), the ZVD signal is created by detecting when the input voltage is low. The ZVD signal determines the start and end of the DSG causing the DSG signal to be synchronized to the input voltage that makes the LED current waveform is close to a sinusoid synchronized with the input AC voltage as expected as shown in Fig. 2(a). This indicates that the input impedance is nearly constant and independent of the input voltage. As a result, the PF becomes very close to unity and the LED current has a low THD.

When the external switch operates, it changes the output states of the counter to change the reference signal amplitude as shown in Fig. 2(a). Next, it has been found that the magnitude of the inductor current according to the reference voltages changes as shown in Fig. 2(a). These results confirmed the function of the proposed dimming method as expected. Also, the reference signal changes according to the soft-starter's function to prevent damage to the power MOSFET and LED string when the LED driver is turned on.

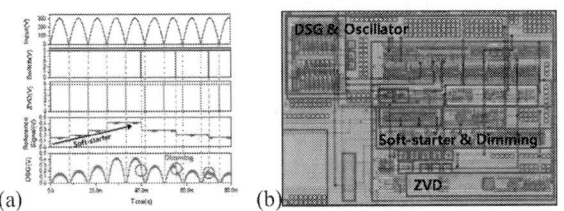

Fig. 2. Simulated results of the proposed DCL and photography of chip

The circuit was simulated and fabricated using the 1 μm-650V DMOS process that provides double-diffused MOS transistors with a breakdown voltage of 600V and a diode. Fig. 2(b) shows an image of the photomicrograph of the fabricated chip that occupies an area of $1.11 \times 1.78 mm^2$.

Fig. 3 shows the measured waveforms of the fabricated chip. As shown in Fig. 3(a), the oscillator generates a frequency of 120 kHz; Fig. 3(b) shows the generated ZVD signal, which is synchronized and matched with the input voltage. Fig. 3(c) shows the DSG according to the reference signal, which can be

changed by an external switch for dimming purpose; from these figures, it can be seen that the DSG signal is similar to the input voltage. These results confirm the design functions and the high PF that proposed DCL can be gained. Table I summarizes the measured data of the proposed DCL from the fabricated chip.

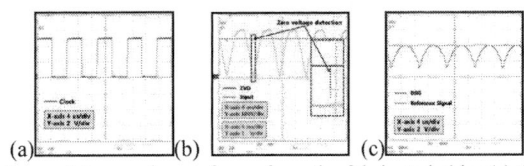

Fig. 3. Measured waveforms from the fabricated chip. (a) Clock signal. (b) ZVD signal. (c)) Reference and DSG signal

Table 1. Performance of fabricated chip

Subject		Value	Unit
IC Process		1 μm-650V DMOS process	-
Input Voltage		220 Vrms @ 60 Hz	-
Frequency of Oscillator		120	kHz
Number of Bits of DSG		6	Bit
Number of Bits of Soft-start		4	Bit
Magnitude of DSG	Max.	440	mV
	Min.	160	mV
Number of brightness steps		4	Step

Conclusions

In this paper, we propose a new digital controller that has a high PF and involves the use of a new dimming method for an LED driver. The proposed controller was designed and fabricated using the 1-μm-650V DMOS process. Through simulation and experimental results, we conclude that the proposed controller has advantages over the conventional one in the point of PF and small number of components. And also the proposed circuit is suitable for a low-cost IC implementation that can mitigate the high costs of the currently solutions.

Acknowledgment

This research was supported by the project "AC direct driving LED IC with dimming control" of KERI(Korea Electrotechnology Research Institute), Korea.

References

[1] M. Arik, J. Petroski, and S. Weaver: Proc. IEEE 8th Intersoc. Conf. Thermal Thermomech. Phenomena Electron. Syst. (2002) 113.

[2] M. R. Krames, O. B. Shchekin, R. Mueller-Mach, G. O. Mueller, L. Zhou, G. Harbers, and M. G. Craford: J. Display Technol. 3 [2] (2007) 160.

[3] S. Y. Hui, S. N. Li, X. H. Tao, W. Chen, and W.M.Ng: IEEE Trans. Power Electron. 25 [10] (2010) 2665.

[4] International Standard: IEC 61000-3-2 (2001) http://webstore.iec.ch.

[5] B. Wang, X. Ruan, K. Yao, and M. Xu: IEEE Trans. on Power Electronics 25 [3] (2010) 592.

[6] R. L. Lin, S. Y. Liu, C. C. Lee, and Y. C. Chang: IEEE Transactions on Industry Applications 49 [4] (2013) 1854.

[7] M. Ali, M. Orabi, M. E. Ahmed and A. El-Aroudi: IEEE Int. Conf. on Power and Energy (PECon) (2010) 501.

A Low Power High Resolution Digital PWM with Process and Temperature Calibrations for Digital Controlled DC-DC Converters

Jing Lu[1], Ho Joon Lee[1], Kyung Ki Kim[2], Yong-Bin Kim[1]

[1]Department of Electrical and Computer Engineering, Northeastern University
Boston, MA, USA
[2]Department of Electronic Engineering, Daegu University
Gyeongsan, South Korea
{jinglu, hjlee}@ece.neu.edu, kkkim@daegu.ac.kr, ybk@ece.neu.edu

Abstract

In this paper, a 12-bit high resolution, power and area efficiency hybrid digital pulse width modulator (DPWM) with process and temperature (PT) calibration is proposed for digital controlled DC-DC converters. The complete circuits design has been verified under different corners of CMOS 0.18um process technology node.

Keywords-DC-DC converter, PT calibration, DPWM, ADC

Introduction

Digital control implemented in switching power converter is receiving increasing attention [1]-[5]. It offers additional advantages in system energy management such as improved flexibility and increased functionality compared to an analog design. Available DSP or microcontrollers, together with ADC (Analog-to-Digital Converter) and DAC (Digital-to-Analog Converter) can do the job of digital control. However, this approach has disadvantages such as significant cost, area, and power overhead. Therefore, a customized low power digital controller DC-DC controller is becoming more popular and realistic solution for SoC. With the detailed structure analysis of digital controlled DC-DC converter, one can tell that its major disadvantage is that the performance is limited by the realistic resolution of ADC and DPWM. Meanwhile, the resolution of DPWM should always be higher than that of the ADC to prevent an unstable state called limit-cycle [2]. Therefore a high resolution DPWM is always demanded in such systems. Reference [3] runs simulation on the output precision versus the resolution of ADC and the resolution of DPWM. The result shows that the optimum precision is reached when the DPWM is 12-bit and the ADC is approximately 11-bit. Beyond this resolution, the output ripple becomes the limiting factor of the output accuracy.

This paper presents an effective design of a 12-bit hybrid DPWM that incorporates a low 6-bit differential tapped delay line ring-mux (DTC) with process and temperature calibration and a high-resolution 6-bit counter-comparator DTC.

Hybrid Differential DPWM

The conventional hybrid DPWM is analyzed in Ref. [1] and [2]. A tapped delay line and a digital counter together with

comparator are used to build the structure. This structure smartly combines the existing small size counter-comparator DPWM and low power tapped delay-line ring-mux DPWM. Ref. [3] improves the conventional structure by employing differential delay line cells to further reduce the area. Ref. [3] utilizes the delay-line ring oscillator frequency as the clock of the counter so that a high frequency clock generation circuit is saved. However, the structure proposed in [3] suffers two major problems. One is that its delay cell is an analog differential amplifier with common feedback loop, which causes its delay cell to be much larger than the conventional digital delay cell. The other one is that the delay line is usually highly dependent on the process and temperature, which results that the switching frequency will vary as the process and temperature change. To overcome these problems, a novel 12-bit DPWM with process and temperature calibration is proposed in this paper as shown in Fig. 1. Eight 1X cells and three 4X cells are connected into a ring oscillator frame. The low 3-bit [L0-L7] and middle 3-bit [M0-M7] are tapped out as denoted in Fig. 1.

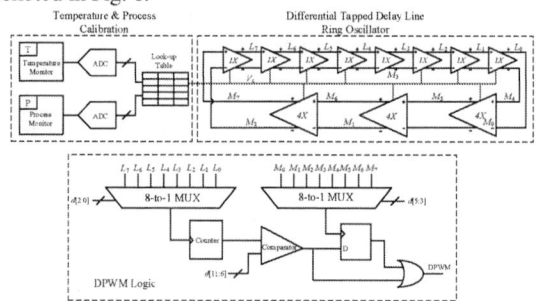

Fig. 1 Proposed 12-bit hybrid DPWM with process and temperature calibration.

The relationship of the low 6-bit signals is shown in Fig. 2(a), where T_d is the oscillation period of the delay line. The rising edge sequence is from L7 to L0 and M0 to M7. For the final DPWM pulse, the Low 3-bit determines the starting edge of the pulse and the high six bits [H6-H11] determines how many T_d is following, and the middle three bits determine the falling edge of the pulse. The total period is $2^6 \times T_d$ as illustrated in Fig. 2(b). The differential tapped delay line ring oscillator is composed of 8 1X delay cells and three 4X delay cells. A control voltage called V_c can adjust all the cells' delay time. The control voltage is generated by the PT calibration circuit, which works as following. First, the PT monitor circuits keep monitoring the process and temperature variations and output an analog output. The output is converted to digital code by two 2-bits flash ADC. Combining the information of process and temperature variations, an appropriate voltage is selected through a loop-up table.

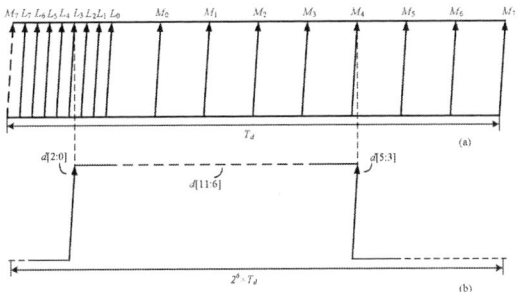

Fig. 2 Time chart of the proposed DPWM.

Circuit Implementation

The detailed structure of 1X cell in Fig. 1 is revealed in Fig. 3 (a), which are to ensure the opposite phase of the differential output. The delay element, as shown in Fig. 3 (b), is constructed by a voltage controlled inverter and a gain boost inverter, which provides sharper transient edges, and a full digital output swing for the delay element.

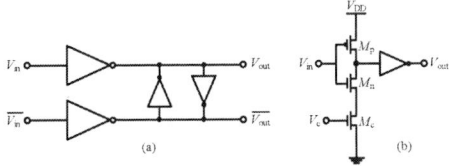

Fig. 3 (a) Differential delay cell (b) Voltage controlled delay element.

For temperature monitor, it is necessary to design an effective circuit is process insensitive and linear to temperature. The simulation results are summarized in Table I. The output voltage differs a lot as temperature changing but only with little process variation. The 2-bit flash ADC described in next sub-section will encode three cases of temperature output.

TABLE I
TEMPERATURE MONITOR CIRCUIT SIMULATION RESULTS

Temp	Vout (ff)	Vout (tt)	Vout(ss)	T_1T_0
-40°C	769 mV	775 mV	780 mV	11
25°C	693 mV	700 mV	706 mV	10
125°C	575 mV	585 mV	593 mV	01

The proposed process monitor circuit is illustrated in Fig. 4. It is composed of start-up circuit, PTAT current generation circuit, and a process dependent output.

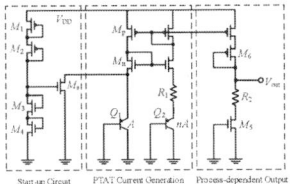

Fig. 4 Process monitor circuit.

The simulation results are summarized in Table II. The Vout changes about 100mV at different process corner, but only varies less than 15mV across the whole temperature range.

TABLE II
PROCESS MONITOR CIRCUIT SIMULATION RESULTS

Corner	Vout (-40°C)	Vout (25°C)	Vout (125°C)	P_1P_0
ss	800 mV	802 mV	815 mV	11
tt	704 mV	700 mV	705 mV	10
ff	625 mV	619 mV	621 mV	01

B. Look-up Table

Both the process and the temperature will affect the delay time at the same time. Table III lists all possible combination of PT variations. A look-up table circuit is designed to select different output according to the PT output code.

TABLE III
ALL THE POSSIBLE COMBINATION OF PT VARIATIONS

Process Corner			Temperature			Control Voltage
ss	tt	ff	-40°C	25°C	125°C	
11			11			620 mV
11				10		610 mV
11					01	602 mV
	10		11			510 mV
	10			10		500 mV
	10				01	480 mV
		01	11			400 mV
		01		10		385 mV
		01			01	370 mV

Conclusion

This paper proposed a power and area efficient architecture of 12-bit hybrid DPWM with process and temperature calibration. It is constructed by 6-bit differential tapped delay line ring-mux DTC and 6-bit counter-comparator DTC. To overcome the common problems of process and temperature variation of most delay cells, a process and temperature monitoring and self-calibration circuits are proposed and analyzed in this paper along with he simulation results.

References

[1] B. J. Patella, A. Prodic, A. Zirger, and D. Maksimovic, "High frequency digital PWM controller IC for DC/DC converters," IEEE Trans. on Power Electronic, Vol. 18, Issue 1, Part 2, pp: 438-446, 2003.

[2] A. V. Peterchev, S. R. Sanders, "Quantization resolution and limit cycling in digitally controlled PWM converters," IEEE Trans. on Power Electronic, Vol. 18, Issue 1, Part: 2, pp. 208-215, 2007.

[3] H. C. Foong, Y. Zheng, Y. K. Tan, and M. T. Tan, "Fast-transient integrated digital DC-DC converter with predictive and feedforward control," IEEE Trans. on Circuits and Systems I, Vol. 59, Issue 7, pp. 1567-1576. 2012.

[4] A. Syed, E. Ahmed, D. Maksimovic, E. Alarcon, "Digital pulse width modulator architectures," IEEE Annual Power Electronics Specialists Conference, Vol. 6, pp: 4689-4695, 2014.

[5] A. V. Peterchev, J. Xiao, S. R. Sanders, "Architecture and IC implementation of a digital VRM controller," IEEE Trans. on Power Electronics, Vol. 18, Issue: 1, Part: 2, pp. 356-364, 2003.

An Output-Capacitorless Low Dropout Regulator without Resistors

Jie MEI, Hao ZHANG, Tsutomu YOSHIHARA

Graduate School of Information, Production and Systems, Waseda University, Fukuoka, Japan. 808-0135

Email: meiwaseda@moegi.waseda.jp

Abstract—An all-MOSFET low-power low-dropout regulator is designed in CMOS technology, featuring low sensitivity with respect to input voltage and temperature. Supply voltage can be as low as 800mV. An error amplify (EA) with an embedded voltage reference (EVR) is employed and a buffer is used to improve the load transient. The circuit is simulated in 0.18 μm CMOS technology. Simulated results verify that the proposed LDO is stable for a capacitive load from 0 to 10 pF and with load capability of 50 mA. The maximum overshoot and undershoot under a 0.8 V supply are less than 90 mV for full load current changes within 1 μs edge time, and the recovery time is less than 1.5 μs. The temperature coefficient (TC) is 37.8 ppm/°C ranging from -25 °C to 100 °C.

Index Terms—low voltage, embedded voltage reference, temperature coefficient, Output-Capacitorless low-dropout regulator.

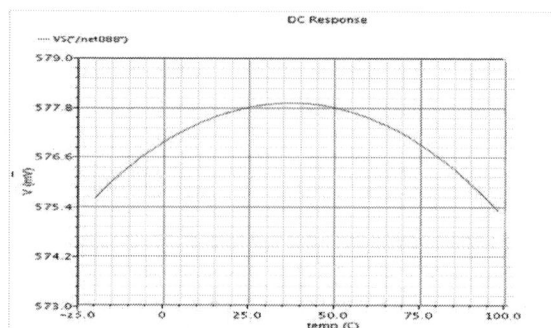

Fig. 2. Temperature coefficient

I. INTRODUCTION

In recent years output-capacitorless low-dropout voltage regulators (LDOs) have been widely used in the recent power management units due to its good regulation and fast transient response. However, the sub-1 V voltage reference is difficult to design with conventional LDO because that the offset voltage in the error-amplify seriously affects the value of V_{ref}, and output voltage is very high [1]. A latest art is proposed in [2] to solve this problem. While there is a trade-off between power dissipation and chip area due to the use of off-resistor. To solve this problem, a pure-MOSFET structure is proposed in this paper.

II. PROPOSED OUTPUT-CAPACITORLESS LDO REGULATOR

Fig. 1.shows the simplified schematic of the proposed regulator, including the proposed EA with EVR, the buffer, and power MOSFET. All the MOSEFTs, excluding M_{N1}, M_{N5}, M_{N8-9} and M_{PS}, operate in the sub-threshold region. The MOSEFTs of M_{N6-9} and M_{P5-7} generate the reference current with a value of $V_T ln(N)/R_{MN8}$ [3], where V_T is the thermal voltage. V_T is equal to kT/q, where k is Boltzmann's constant and T is the absolute temperature. N is equal to S_{MN7}/S_{MN6}. S is the aspect ratio of the MOSFET, and M_{P5} and M_{P6} have the same aspect ratio. M_{N9} and M_{P7} provide a gate voltage for M_{N8} working in linear region. Moreover, as Fig 1 shows, the N-type current mirror pairs (M_{N5} and M_{N1}, M_{N5} and M_{N6}, M_{N6} and M_{N7},) and the P-type current mirror pair (M_{P3} and M_{P4},) cause the current

I_{D1} of M_{N1} to be K times the current I_{D8} of M_{N8}. With M_{N1} in saturation region, the output voltage can be expressed as,

$$V_{OUT} = V_{GS_M N1} = V_{TH} + \sqrt{2KV_T ln(N)/\mu_n C_{ox} S_{MN1}}$$
(1)

where V_{GS-MN1} is the gate-source voltage of M_{N1} and V_{TH} is the threshold voltage. V_{TH} has a complementary to absolute temperature (CTAT) coefficient because it dominates the TC, whereas another part of V_{OUT} has a proportional to absolute temperature (PTAT) coefficient. The error-amplify, the buffer and power MOSFET consist a 3-stage amplifier with a feedback-loop path. The buffer is consisted of M_{P1-2} and M_{N2-3}. Due to the EA with EVR, a sub-1V voltage can be adopted. The buffer is used to increase the gain and improve the stability. The values of the compensation capacitors C_1 and C_2 are 2 pF and 1.2 pF respectively.

III. SIMULATION RESULTS

The circuit is simulated in 0.18 μm CMOS process. The simulated result of temperature coefficient is shown in Fig. 2 and the TC is 37.8 ppm/°C. The simulated loop gains of the proposed LDO at different I_{LOAD} are shown in Fig. 3. The low-frequency loop gains are 85 dB and 75 dB at light load and heavy load. In all conditions, the phase margins are more than 60° and bandwidths are above 6 MHz. The sufficient loop gain and wide bandwidth show that the proposed LDO has good voltage regulation and high closed-loop stability.

The proposed LDO is stable when the input supply voltage ranges from 0.8 V to 1.8 V with capacitive load of no more than 10 pF. Fig. 4 shows the simulated load transient response

978-1-4799-5128-4/14 $31.00 © 2014 IEEE

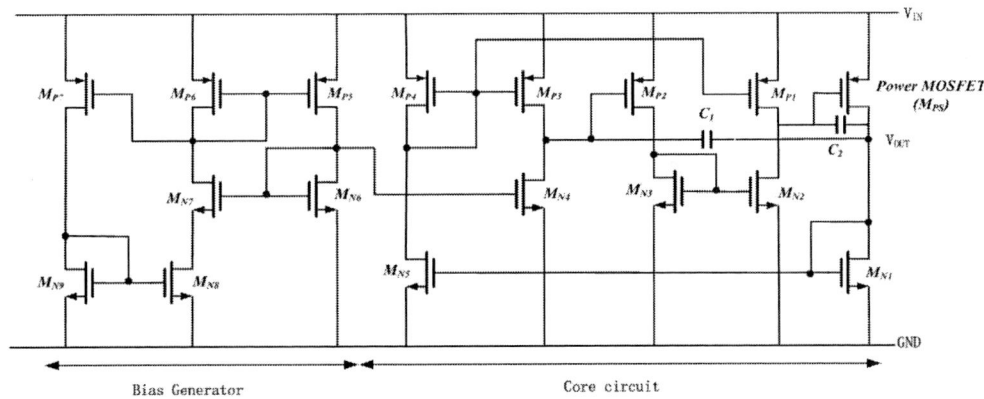

Fig. 1. Schematic of the proposed LDO regulator with EVR

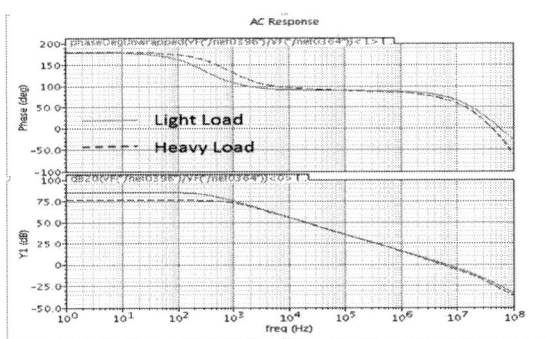

Fig. 3. Frequency response

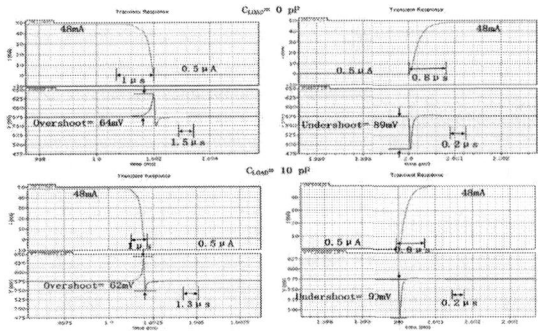

Fig. 4. Simulated load transient response

not need a large output capacitor, and it achieves the best line regulation among other designs.

TABLE I
PERFORMANCE COMPARISON WITH RECENT PUBLISHED LDOs

	This work	[1]	[2]	[4]	unit
Process	0.18–μm	0.35–μm	21nm	0.09–μm	.
Supply Voltage	0.8~1.8	3~5	0.65~0.9	1.2	V
Supply Current	380	170	5	6000	μA
Output Voltage	576	2800	600	900	mV
Output Capacitor	<10	N/A	<100	N/A	pF
TC (ppm/°C)	37.8	N/A	30	N/A	ppm/°C
Load Range	0~50	0~50	0~10	0~50	mA
Load Regulation	0.3	0.02	0.5	1.8	mV/mA
Line Regulation	3	3.3	16	N/A	mV/V

IV. CONCLUSION

In this paper, a sub-1 V all-MOSFET LDO is presented. Simulated result verifies that the proposed LDO is stable for a capacitive load from 0 to 10 pF and with load capability ranging from 0 to 50 mA. The features of the LDO regulator are ultra-fast load transient responses, high loop gain, and small on-chip compensation capacitance. Moreover, it maintains high stability without load capacitor. The input voltage can be reduced to 0.8 V using EA with EVR that reduces energy consumption and the temperature coefficient is less than 40.0 ppm/°C.

when V_{IN} is 0.8 V, V_{OUT} is 576 mV and sub-1 μs edge time without load capacitor and with 10 pF load capacitor respectively. As shown in this figure, when the output current changes from 0.5 μA to 48 mA, the overshoot of V_{OUT} without load capacitor is about 64 mV, while it is about 62 mV when with 10 pF output capacitor, and the undershoot of the two is about 90 mV. The output voltage variation is approximately 0.3 mV. Performance comparison between the proposed LDO and some selected recently-published LDOs is shown in TABLE I. The proposed LDO can be operated at sub-1 V supply voltage. Moreover, the proposed LDO does

REFERENCES

[1] H. C. Yang, et al., "High-PSR-bandwidth capacitor-free LDO regulator with 50 μA minimized load current requirement for achieving high efficiency at light loads," *WSEAS Trans. Circuits Syst.*, vol. 7, no. 5, pp. 428İC437, May 2008.

[2] Wei-Chung Chen, et al., "17.10 0.65V-input-voltage 0.6V-output-voltage 30ppm/°C low-dropout regulator with embedded voltage reference for low-power biomedical systems," *ISSCC*, vol., no., pp.304,305, 9-13 Feb. 2014

[3] Oguey, H.J., Aebischer, D., "CMOS current reference without resistance," *Solid-State Circuits, IEEE Journal of*, vol. 34, no. 5, pp. 670 – 674, may 1999.

[4] Hazucha, P., et al., "Area-efficient linear regulator with ultra-fast load regulation," *Solid-State Circuits, IEEE Journal of*, vol.40, no.4, pp.933,940, April 2005

Fully Digitalized Switch-mode LED Driver without ADC

Jae-Hyoun Park[*], Hyung-Do Yoon

Photonic Convergence Research Center
Korea Electronics Technology Institute
Seongnam, Korea
jhpark@keti.re.kr

Abstract

A digitalized switch-mode LED driver is implemented using a 0.35μm high voltage process, which featured by no use of any ADCs. The comparison result between the output current and the desired current causes the counter to increase or decrease its binary value which determines the duty ratio of the PWM signal. This PWM signal controls switching devices. A delay routine added in the counter prevents oscillation of the counter value owing to a high speed clock.

Keywords-component; LED driver, digitalized, without ADC

Introduction

As is generally acknowledged, a key technology of an LED driver is a DC-DC converter which has been based on an analog control scheme [1], [2]. The analog control systems should require too much complex hardware circuit even though these systems have two principal merits such as a low cost design and a wide system bandwidth [3]. Several research activities about digitalized DC-DC converters have been successfully reported [3], [4]. These digitalized systems have several merits of good noise immunity, insensitivity to process variation and ambient conditions, more flexible design to other processes, and good compatibility [3], [4]. Furthermore, more improved and intricate control algorithms such as a flexible control with sudden transition in various loads and an interface with sensors can be achieved using digital technologies [4]. For the most part, however, an analog-to-digital converter (ADC) has been necessary to realize a digitalized DC-DC converter, and hence, the system has a quite large amount analog block [3], [4]. To get the essential merits of digitalized DC-DC converters, those systems should be implemented without ADCs.

Recently, LED system lighting or emotional lighting has been in the limelight, which includes various functions such as cable and/or radio communications interfaced with multiple sensors. This LED lighting needs more complex controllable LED drivers, which is enough reason for necessity of a digitalized LED driver. Additionally, for an LED driver, a higher switching frequency makes possible that smaller inductors and capacitors are used, though more switching losses and poor efficiency [5]. These smaller capacitors, especially, ceramic capacitors have much longer life than aluminum electrolytic capacitors [5].

In this paper, a novel fully digitalized switch-mode LED driver is presented, which makes a feature of realization without ADC and 1MHz switching frequency. The proposed driver uses a comparator, a digital counter with a delay routine and a PWM converter, which is implemented using a 0.35μm 2poly-4metal high voltage process.

Overall System Configuration

The proposed fully digitalized LED driver schematic and flowchart are described in Fig. 1 and Fig. 2, respectively. This driver forms four-channel buck converters with the external passive devices such as inductors and capacitors, as shown in Fig. 1. To get the desired LED current, the output voltage of the buck converter should be equal to the sum of the forward voltage of serially connected LEDs, neglecting the voltage drop at the current regulator circuit. To do this, firstly, the output current should be compared to the desired current using the comparator. The desired current is set by the external biasing resistor R_{SET} connected with the current regulator. Depending on the comparison result, the counter should be determined to increase or to decrease its binary value. Then a PWM signal can be generated, which its pulse width is equal to the product of the binary value and the clock period. This PWM signal controls the switching devices and the desired output current or voltage can be obtained. In this procedure, the counters and the comparators play the role of ADCs in a general digitalized DC-DC converter.

For the proposed driver, a high speed clock should be required to achieve a high switching frequency. The wanted clock frequency can be easily obtained by the ring oscillator with the external charging resistor R_{CHG}. The counter value, however, will be oscillated because the output voltage cannot immediately response to this high frequency PWM signal. This error will be expressed by the jitters of the PWM signal. To settle this error, a delay routine is added to the counter.

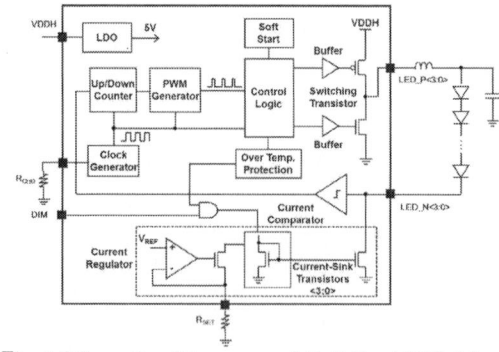

Fig. 1 Schematic of the proposed digitalized LED driver.

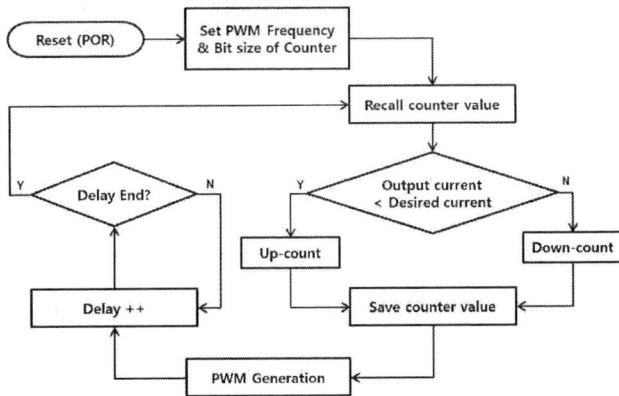

Fig. 2 Flowchart of the proposed fully digitalized LED driver.

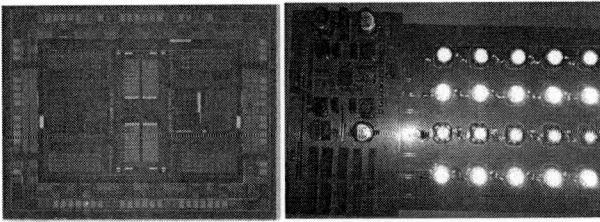

Fig. 3 Photographs of the fabricated digital-controlled LED driver and its driving result using the test board.

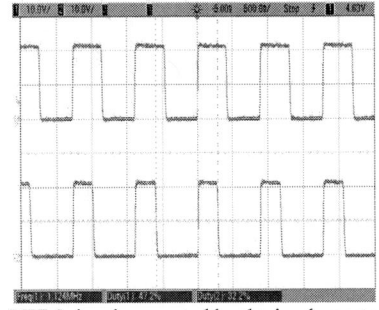

Fig. 4 1MHz PWM signal generated by the implemented LED driver.

Results and Discussion

The proposed digitalized LED driver is realized through a 0.35μm 2poly-4metal 40V process and is packaged using a 40-pin quad flat no-lead (QFN-40L) package. Fig. 3 shows photographs of the fabricated driver and its driving result using the test board which have a four-string LED load. In this test board, in order to get the PWM frequency of 1MHz, a 64MHz clock is generated. The PWM frequency can be generated by the internal clock divider and is inversely proportional to the bit size of the counter. The bit size of the counter is set to 6-bit, which can be calculated by the binary logarithm of the ratio of above two frequency values. In addition, this bit size is closely connected with the resolution and fluctuation of the output voltage of the proposed digitalized LED driver. Hence, the correlation between the two frequency values and the bit size should be carefully considered.

Fig. 4 is the waveforms of 1MHz PWM signals controlled by this digitalized LED driver. Different duty ratios of these signals correspond to different string voltages caused by using different types of LEDs for each string, such as blue-LEDs, red-LEDs, and its combinations. The measured results accord well with the theoretical results which can be obtained by dividing the output voltage by the supply voltage $VDDH$.

The waveforms of the PWM signal jitters as a function of delay time are shown in Fig. 5. Delay time is programmable and is varied from 0 to $64T_{CLK}$, where T_{CLK} is a clock period. As delay time is increased, the jitters are reduced. This means that system stability is improved in a digital domain. At $32T_{CLK}$ delay, this driver has few jitters except a 1-bit digital error.

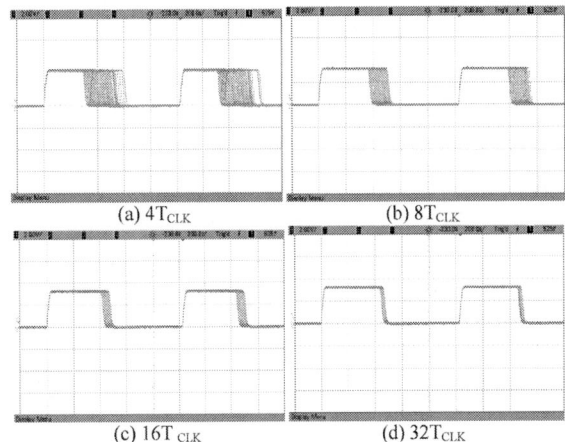

Fig. 5. The PWM signal jitters as a function of delay time.

Acknowledgment

This work was supported by LED System Lighting R&D program of KEIT [10042947].

Conclusions

In this paper, without any ADC used, a novel digitalized switch-mode LED driver is presented. This proposed driver can generate a 1MHz PWM signal without the jitters caused by a high speed clock, which will make possible that a ceramic capacitor is used. This driver also works well with a multi-string LED load, even though LEDs at each string have different forward voltage. Consequently, this digitalized LED driver will be effectively applied to LED system lighting or emotional lighting, which essentially requires a digital control part for cable and/or radio communications interfaced with multiple sensors.

References

[1] Y. Hu and M. M. Jovanović, "LED driver with self-adaptive drive voltage," *IEEE Trans. Power Electron.*, vol. 23, no. 6, pp. 3116–3125, Nov. 2008.

[2] S. Y. Tseng, S. C. Lin, and H. C. Lin, "LED backlight power system with auto-tuning regulation voltage for LCD panels," *Proc. IEEE Appl. Power Electron. Conf.*, pp. 551–557, Feb. 2008.

[3] L. Guo, "Implementation of Digital PID Controllers for DC-DC Converters using Digital Signal Processors," *Proc. IEEE Electro /Information Technology Conf.*, pp. 306–311, May 2007.

[4] T. Jackum, G. Maderbacher, and R. Riederer, "A Digitally Controlled Linear Voltage Regulator in a 65nm CMOS Process," *Proc. IEEE Int'l Conf. on Electronics, Circuits, and Systems*, pp. 984–987, Dec. 2010

[5] S. Winder, *Power Supplies for LED Driving*. Elsevier: Burlington MA, 2008, pp. 182–190.

IC design of DMPPT controller for Cascade and MCCU topology of Photovoltaic Systems

K.H. Cho[1], S.M. Sohn[1], J.G. Jeong[1], S.I. Lim[2], I.S. Cho[2,] J.Y. Kim[1]

[1]RTS Energy, Inc.
R&D Tower #1007 Nurikkum Square
396 World Cup Buk-ro, Mapo-gu
Seoul, 121-795 Korea
E-mail: rts.energy@gmail.com

[2]Dept. of Eletronics Engineering
Seokyeong University
Seoul, Korea
E-mail: silim@skuniv.ac.kr

Abstract

This paper presents IC design of Distributed Maximum Power Point Tracking (DMPPT) controller for Cascade and Mismatched Current Compensation Unit (MCCU) for photovoltaic systems. The Cascade part can handle the conventional DMPPT algorithm and the MCCU part can handle the differential power processing algorithm. This solution can mitigate mismatches in photovoltaic(PV) power system, while introducing no insertion loss and reducing the system cost compared to a microprocessor-based solution. A 0.18um 1-poly 4-metal 5V process was selected for the IC development. MPPT peak efficiency of over 99% is achieved.

Key words - DMPPT, IC, PV module, Cascade, MCCU

Introduction

A DMPPT is a power processing architecture that reduce mismatch-related losses between the PV panels. Various DMPPT algorithms have been proposed in academia. [1]- [5] Easier way to realize those DMPPT algorithms is to use a microprocessor but it is too expensive and the problem with reliability. This paper suggest IC design to replace the microprocessor. The designed IC is a mixed signal chip in a sense that it is composed of digital logics and analog to digital converter (ADC) and protection circuits. The external DC-DC converter used a non-inverting Buck-Boost converter.

Fig. 1 is a IC functional block diagrams.

The Architecture of the Proposed Design

The IC is composed of digital parts (registers, spi, MPPT tracker, PI controller, PWM generator, soft start) and analog parts (ADC, OVP, UVLO, etc). In this paper, suggested IC design saves system cost and gives better reliability with fewer components tran microprocessor. To make this approach a feasible one, a thorough simulation and verification of functions and performance by FPGA have been done. PV voltage and current are connected into ADC ports of the IC and these two signals, after AD conversion, are averaged out to kill noise, then fed into PI controller and MPPT where only PV voltage is used in PI controller.

The PI controller puts out its output so that PV voltage follows the reference voltage handed by MPPT. The output of PI controller is fed into PWM signal gererator. The MPPT tries to find PV MPP by calculating power difference between the present and previous sampling times.

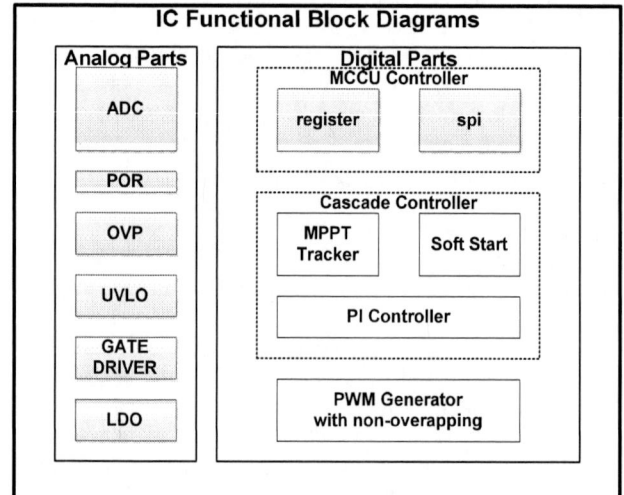

Fig. 1 IC Functional Block Diagrams

A. ADC

The ADC is designed 12 bit SAR architecture. Sampling frequency is 14MHz so 12 bit word conversion is done at 875 kHz. Only one ADC is used to read voltage and current by multiplexing of two inputs. ENOB is set as 11.152bit at worst case, SNDR is 68.89dB, and SFDR is 78.8dB.

B. OVP

The OVP circuits can detect abnormal state and control PWM to prevent the system failure. When the PV voltage goes over the pre-set voltage, then over voltage protect (OVP) circuit detect it, and makes the PWM signals to be low.

C. UVLO

The UVLO circuits monitors the PV voltage, control PWM. When the PV voltage goes down the pre-set voltage, then UVLO circuit detect it, and makes the PWM signals to be low.

D. GATE DRIVER

The GATE DRIVER dives the signal, generated by PWM generator, to the external power FET.

E. PWM

The PWM generates two pairs of complementary PWM pulses. Each pair is dedicated to Buck and Boost switches respectively. To handle dead time issue between high-side and low-side FET switches, generates non-overlap PWM signal between high-side and low-side.

F. MCCU Controller

The MCCU controller supports the MCCU structure which handles the differential power processing. The MCCU structure has each pair of adjacent PV panels in the series string, and a bi-directional dc-dc converter associated with them.

G. Cascade Controller

The cascade controller supports the MPPT function of the cascade structure, in which the output of each PV module is an input to its DC-DC converter and DC-DC converter out is connected in series to form a string voltage.

Simulation and Implementation

A process of 0.18μm 1-poly 4-metal 5V was selected for the IC implementation. The layout of the IC is depicted in Fig. 2. Chip area is 2420x4940μm^2.

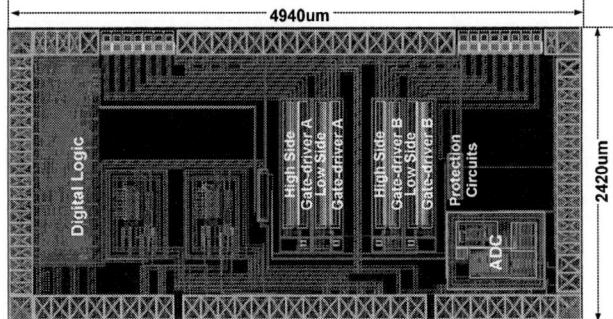

Fig. 2 IC Layout

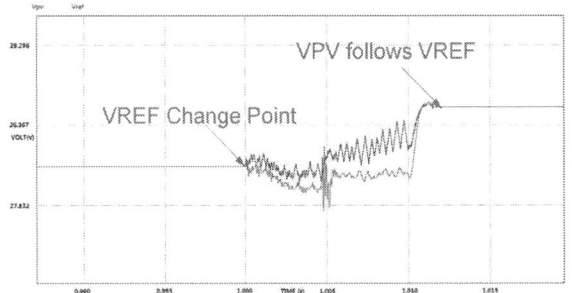

Fig. 3 Simulation result (Vref and Vpv)

Simulation result is shown in Fig. 3 where it is clearly

shown that the PV voltage follows well the reference voltage. The same tracking function is shown in Fig. 4, where real-time testing is done using a Xilinx Spartan 6 FPGA. Table 1 is a list of important system parameters including DC-DC converter and IC.

SYSTEM PARAMETERS
TABLE I

	Value
Input voltage	15V~40V
Solar panel profile	SM240PA8, Pmax=240W
Main clock	54MHz
PWM frequency	100KHz
P&O frequency	50Hz
PI controller	100KHz

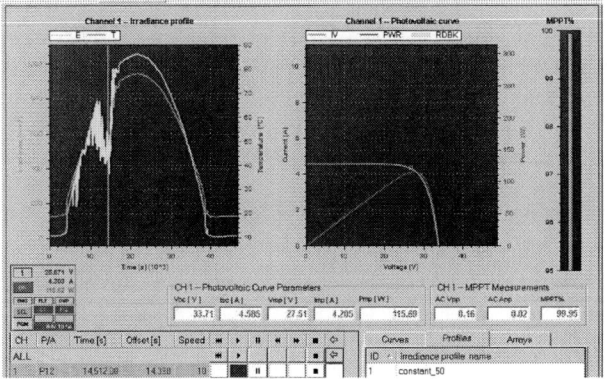

Fig. 4 FPGA real-time test results

Acknowledgment

This research was supported by Korean government fund: Investment-linked business components and materials technology development program and by the Industrial Core technology development program(10049009) managed by Ministry of Trade, Industry and Energy.

References

[1] Pradeep S. Shenoy and Katherine A.Kim, etc "Differential Power Processing for Increased Energy Production and Reliability of Photovoltaic Systems", IEEE Transaction on Power Electronics, Vol. 28, pp. 2968-2979,No. 6, June 2013

[2] G.R. Walker and Jordan C Pierce, "Photovoltaic DC-DC Module Integrated Converter for Novel Cascaded and Bypass Grid Connection Topologies – Design and Optimisation," IEEE PESC, 2006, pp. 1-7

[3] Kim, K.A. and Philip T. Krein, etc " Photovoltaic Differential Power Coverter Trade-off s as a Consequence of Panel Variation", Control and Modeling for Power Electronics(COMPEL), 2012 IEEE 13th Workshop, June 2012, pp.1-7

[4] Blumenfeld, A, " Enhanced differential power processor for PV systems:Resonant switched-capacitor gyrator converter with local MPPT",Applied Power Electronics Conference and Exposition (APEC), 2014 twenty-ninth Annual IEEE, pp.2972-2979, March 2014

[5] Nicola Femia, Giovanni Petrone, Giovanni Spagnuolo, and Massimo Vitelli, "Optimization of Purturb and Observe Maximum Power Point Tracking Method", IEEE Transaction on Power Electronics, vol. 20, no. 4, pp. 963-973, Jul. 2005

Histogram Color Correction For Multi-View Video Cording

Chan-Su Park, Hi-Seok Kim, Hyeong-Woo Cha

dept. electronics engineering
Cheongju University
Cheongju, KOREA
slsipark@cju.ac.kr, khs8391@cju.ac.kr, hwcha@cju.ac.kr

Abstract

In this paper, a histogram clustering threshold based on color correction method is proposed for multi-view video coding. To find the maximum matching regions for the reference-view frame and the target view-frame, SAD is performed. By using an iterative histogram clustering threshold, a minimum histogram difference between the reference-view frame and the target-view frame is computed to correct the target-view frame. Experimental results have shown that the proposed color correction method outperforms the other histogram matching method on the aspect of the PSNR and computation time.

Keywords-component; formatting; Multi-view video coding; color correction; histogram matching

Introduction

To improve the performance of Multi-view Video Coding (MVC), appropriate luminance and color correction scheme is demanded to calibrate significant discrepancies . Fecker et al. [1] proposed the usage of histogram matching (HM) for the calibration of luminance and chrominance variations of different views based on the cumulative histograms. The advantage of this algorithm is that it has low computational complexity. However, it is difficult to handle occlusions between views. Chen et al. [2] proposed a histogram-offset-based color correction method for benefitting multi-view video coding. Disparity estimation in the rank-transformed space is exploited to find the maximum matching regions. But, it has more computation time required if window size of rank transform is very large. This paper proposed a histogram clustering threshold based on color correction algorithm for multi-view video coding. by using an iterative threshold, a minimum histogram difference is computed and then used to correct the target-view.

The rest of paper is organized as follows. Section2 describes the proposed histogram-clustering threshold for color correction scheme. Section 3 shows some experimental results to illustrate the improved performances of the proposed algorithm. Conclusion is drawn in the final section

Computation of histogram clustering threshold for color correction

Histogram-based clustering method is applied to each histogram bin of the luminance component and that of the chrominance component. In this paper, we only calibrate the histogram bins which have high occurrences according to a histogram clustering threshold. For example, Take the luminance calibration for an example, denoted by $H_{r,y}[n]$ and $H_{r,y}[n]$ ($0 \leq n \leq 255$) are the luminance histogram regions of the reference-view frame fr and the target-view frame ft, respectively. Based on the histograms, an iterative threshold is generated as follows.

1) Let $t_0 = $ (Width+Height)/256 be the initial threshold, where Width and Height are the horizontal and vertical size on the target-view frame, respectively.

2) Separate the histogram into two classes, increasing histogram occurrences Ucnt and decreasing histogram occurrences Lcnt according to the current threshold value t_i In this step, two groups of luminance bins are generated and denoted by A and B, where A consists of histogram luminance bins with occurrences $\leq ti$, and B contains rest.

Let $H_{t,y}[i]$ be a histogram luminance bin on the target view image.if $H_{t,y}[i] \leq T_i$ then Lcnt is increased as *Lcnt*+ 1, otherwise *Ucnt* + 1.

3) If ($H_{t,y}[i] < t_i \leq H_{t,y}[i+1]$) ,then Ucnt is regard as a candidate histogram bin. Compute a maximum histogram occurrence Ucnt among the several candidate histogram bins for a given histogram constraint repeatedly. Otherwise, compute a maximum Lcnt. The entire threshold process stops within a constraint ($t_{i+1} - t_i$) <ε and i < 255). Otherwise, let i =i+1 and go to Step 2).

With the generated iterative threshold t_i for the target-view frame denoted as dot line in Fig.1, if the maximum occurrence *Ucnt* and *Lcnt* can be computed at a histogram luminance bin t, then threshold Tt is determined to a histogram clustering threshold (HCT) as shown in Fig.1. In multi-view video sequences, the reference view and the target view might vary drastically under different illumination conditions, leading to inaccurate disparity estimation. The histogram matching algorithm can be used to adapt a distorted target image to a reference image. In previous work[3], The best match is indicated by the minimum distance or maximum correlation between a reference image and a target image, and thus the sum of absolute difference(SAD) is performed in this paper.SAD detects the best matching point to calculate the minimum histogram difference or maximum correlation by searching all the histogram bin; it is defined as follows:

978-1-4799-5128-4/14 $31.00 © 2014 IEEE

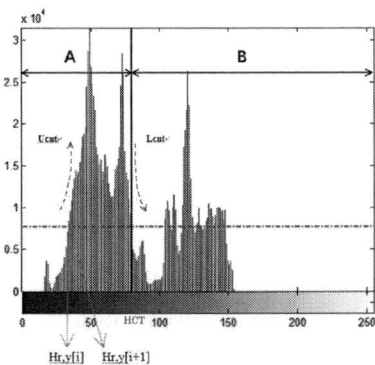

Fig.1 Histogram clustering threshold with maximum occurrence

$$\arg\min \sum_{n=0}^{255} \left| H_{r,y}(n) - H_{t,y}(n+\Delta n) \right| \qquad (1)$$

Where $H_{r,y}(n)$ and $H_{t,y}(n+\Delta n)$ are the luminance histo gram bins of the reference-view frame fr and the target- view frame ft, respectively. Based on the histogram clustering threshold, $H_{t,y}(n+\Delta n)$ can be changed to use the generated histogram clustering threshold.

$$K_A = \arg\min \sum_{n=0}^{HCP-1} \left| H_{r,y}(n) - H_{t,y}(n+n_{kA}) \right| \qquad (2)$$

$$K_B = \arg\min \sum_{n=HCP}^{255} \left| H_{r,y}(n) - H_{t,y}(n+n_{kB}) \right| \qquad (3)$$

Where K_A and K_B are the minimum values of the histogram

bins with occurrences according to the generated histogram clustering threshold of the target-view frame respectively.

With the generated K_A and K_B ,The corrected image fc is

obtained by adding K_A and K_B to the distorted target- view

frame extracted by:

$$f_c[m,n] = f_r[m,n] + K_A \qquad (4)$$

where m and n are the coordinate of a pixel in a frame, Hr (m,n)< HCT. Otherwise, it is obtained by:

$$f_c[m,n] = f_r[m,n] + K_B, \text{Where } H_r(m,n) > HCT \qquad (5)$$

Experimental results

In this section, the color correction is performed frame by frame within each view. For example, Identification of high occurrences histogram bins of stereo-view "baby2" is shown in Fig.1. Its clustering threshold of the target-view frame is determined to 81 by using our proposed approach. The corrected image is much similar to the reference image as shown in Fig.2.Experimental results are obtained by applying our clustering threshold algorithm on two multi-view (with 8 views) video sequences, "ballroom"(with a frame size of 640×480) ,"Exit", and four stereo-view(Middelbury datasets). These results are compared with the ones using HM method in [1][2]. Table I shows average PSNR performance compared to the previous HM methods. The PSNR of our proposed approach is much similar to the method suggested by Fecker.Table II shows computation time compared to the previous ones. These results show competitive value of PSNR and less computation time than that of the method suggested

by Chen in[2].That is, average of total clock number computation for a proposed algorithm is less than twenty four times of that for a conventional one[2].

Fig. 2 Test of stereo –view image Baby2

TABLE I. COMPARISON AVERAGE PSNR OF PROPOSED APPROACH AND OTHER APPROACH.

Image	Image Size	Reference	Chen	Fecker	Proposed
Are	1282x1110	28.804103	28.804103	30.313519	30.617407
Art	1390x1110	28.565059	28.565059	29.432861	29.449379
Baby2	1240x1110	29.654289	29.654289	33.389012	32.993796
Computr	1330x1110	29.306039	29.306039	30.246099	31.683449
Balroom	640x480	32.089695	x	32.097518	32.397693
Exit	640x480	31.806480	x	32.048123	31.961818

TABLE II. COMPARISON COMPUTATION TIME OF

Image	Chen	Fecker	Proposed
Are	95750000	3984375	3890625
Art	111546875	4062500	4140625
Baby2	90265625	3875000	3796875
Computr	102453125	4015625	4609375

PROPOSED APPROACH AND OTHER APPROACH

Conclusion

In this paper, we proposed an efficient histogram clustering threshold based on color correction algorithm for improving the coding efficiency of multi-view video sequences and stereo-view. To find the maximum matching regions for the reference-view frame and the target view-frame, SAD is performed. By using an iterative histogram clustering threshold , a minimum histogram difference between the reference-view frame and the target-view frame is computed to correct the target-view frame. Experimental results have shown that the proposed method outperforms the other method on the aspect of the PSNR and computation time.

References

[1]U. Fecker, M. Barkowsky, and A. Kaup, "Histogram-Based Prefiltering for Luminance and Chrominance Compensation of Multiview Video", *IEEE Trans. Circuits and System for Vodeo Tech.*, Vol. 18, no. 9, pp.1258-1267, September,2008

[2]Y. Chen, K. Ma and C. Cai, "Histogram-Offset-Based Color Correction for Multi-View Video Doding," *Proceedings of 2010 IEEE 17th International Conference on Image Processing*, pp. 977-980, Singapore, September 2010.

[3]M. Mori and K. Kashino, "Fast Template Matching Based on Normalized Cross Correlation Using Adaptive Block Partitioning and Initial Threshold Estimation" *IEEE International Symposium on Multimedia* , pp.196-203 , 2010

978-1-4799-5128-4/14 $31.00 © 2014 IEEE

On FHD 300MHz@60fps, Intra/Inter CU Mode Decision Hardware Architecture for the Hypernova H.265 Encoder

Sukho Lee, Hyunmi Kim, Kyungjin Byun, and Nakwoong Eum

Electronics and Telecommunications Research Institute

218 Gajeong-ro Yuseong-gu, Daejeon South Korea

{shlee99, chaos0218, kjbyun, and nweum}@etri.re.kr

Abstract

H.265 (HEVC) is the latest joint video coding standard with ITU-T SG16 WP and ISO/IEC JTC1/SC29/WG11. Its coding efficiency is about two times compared to H.264. However the burden of coding unit (CU) mode decision with rate distortion optimization (RDO) is too costly to implement it with hardware. The key idea of this paper is a novel mode decision architecture to reduce the HW complexity of RDO that is the most effective on an encoder's performance without a noticeable PSNR loss. To shrink the size the Hypernova H.265 encoder uses simplified RDO blocks and shares the transform resources. Its operating clock frequency is 300MHz@60fps on FHD image and BD-BR increase is negligible at 6.02% on hardware aspect. The estimated gate count of its is around 1M.

Keywords-H.265, High Efficiency Video Coding (HEVC), Coding Unit (CU), Rate Distortion Optimization (RDO).

Introduction

In H.265, pictures are divided into a sequence of coding tree units (CTUs). The CTU concept is broadly analogous to that of the macroblock in previous standards such as AVC. The CTU further partitioned into multiple CU to adapt to various local characteristics. A quadtree denoted as the coding tree is used to partition the CTU into multiple CUs. The maximum allowed size of the luma block in a CTU is specified to be *64x64* in main profile. The coding unit (CU) is a square region, represented as the leaf node of a quadtree partitioning of the CTU, which shares the same prediction mode: intra, inter or skipped. The quadtree partitioning structure allows recursive splitting into four equally sized nodes, starting from the CTU as illustrated in Fig.1.

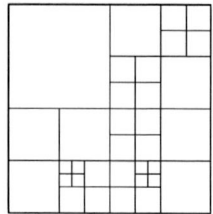

Fig. 1 Example of coding tree structure

The prediction unit (PU) is a region, defined by partitioning the CU, on which the same prediction is applied. In general, the PU is not restricted to being square in shape, in order to facilitate partitioning which matches the boundaries of real objects in the picture. Each CU contains one, two or four PUs depending on the partition mode. The *PART_2Nx2N* and *PART_NxN* partition modes are used for an intra-coded CU.

The partition mode *PART_NxN* is allowed only when the corresponding CU size is equal to the minimum CU size [1], [2]. To estimate an accurate CU mode that there is *MODE_INTRA/INTER* and *PartMode*, encoder have to do RDO operation. But it is a time consuming process and makes hardware fat. In order to solve this drawback, Firstly, Hypernova chooses the minimum number of intra mode which happens a lot by a probability distribution and uses the limited CU size into RDO. The generalized RDO calculates distortion with sum of square error (SSE) through transform, quantization, scaling and inverse quantization but the simplified RDO in Hypernova gets distortion without inverse quantization. Also the estimation of bitrate does with quantization coefficients [3], [4]. Secondly, integer motion estimation (IME) decides the CU size ahead before fine motion estimation (FME) so RDO calculates only a *2Nx2N* merge SKIP and a size decided inter coded CU. Normally intra prediction hardware has two forward transforms (DCT). One is for reconstruction for a reference pixel generation, the other one is for calculation of distortion in RDO. However Hypernova shares with a DCT in RDO as well as in reconstruction by pipelined time scheduling. This results in 50% saving for the forward transform resources.

Rough Mode Decision for Intra/Inter CU Candidates

The Hypernova's largest coding unit (LCU) size is *32x32*. In I-slice, intra prediction has four candidates about *16x16* CU size and two about *8x8* CU. Four intra coded candidates for *16* CU have planar, DC, horizontal, and vertical direction each. These prediction samples go directly into RDO module to get a minimum rate distortion (RD) cost. In case of *8* CU, intra prediction pre-decides two modes among the designated twelve modes by calculating a minimum cost of sum of absolute difference (SAD) or Hadamard transformed SAD (SATD). These two modes also enter into RDO. The final intra CU mode decision chooses one of six modes through RDO. P-slice has not only intra coded blocks but also inter ones, so the number of CU mode decision is more than I-slice because P-slice supports from *32x32* to *8x8* inter coded CU, merge SKIP, and intra coded CU that there are *16x16* and *8x8* size. To cut down hardware complexity, the proposed architecture reduces the number of case for intra modes. Differently to I-slice, P-slice has each one intra coded candidate for 16x16 and 8x8 size respectively. Finally, P-slice decides the best mode with two intra ones, the pre-determined FME and merge SKIP that IME already decided a CU size.

A Hardware Architecture for CU Mode Decision

CU mode decision consists of three stages that there are intra,

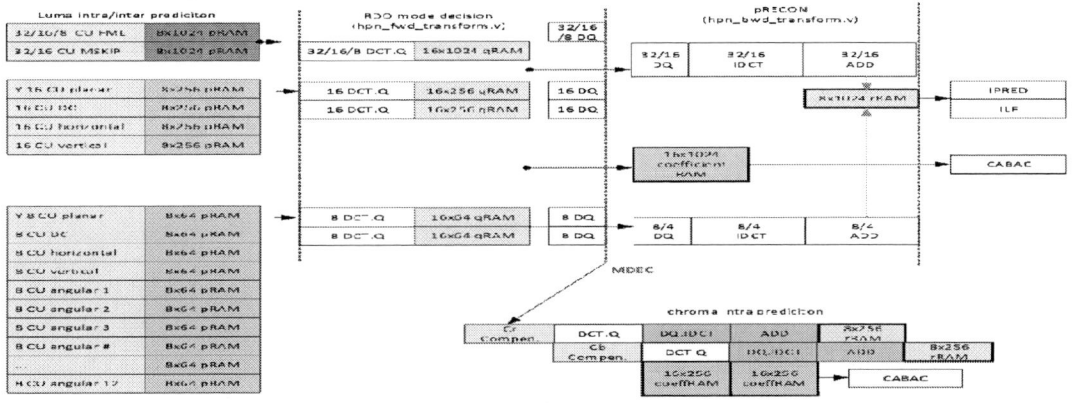

Fig. 2 The hardware architecture for CU mode decision in Hypernova H.265 encoder

inter prediction, mode decision with RDO and reconstruction. Hypernova operates on 300MHz@60fps for FHD encoding and the processing cycles per a LCU is about 2,000. Intra prediction has to do all including prediction, RDO and reconstruction within one pipeline slot because the reconstructed boundary samples of adjacent blocks are used as a reference data. The proposed architecture has five independent RDO blocks for mode decision as shown in Fig. 2. The largest one is a size integrated DCT and quantization block for the inter CU and merge SKIP from *32x32* to *8x8*. The others are two for *16x16* intra CU and two for *8x8* intra CU. RDO calculates distortion with De-quantization and estimates bitrates with quantization coefficients. Finally, it adds two results and make a decision to find out a best CU mode.

A. 16x16/8x8 Intra CU decision in I-slice

In I-slice, intra prediction has all predicted samples in predicted RAMs (pRAMs). Four *16x16* CU size's predicted samples go directly into two RDO blocks with time sharing. In case of *8x8* CU, two predicted samples through rough mode decision (RMD) with SA(T)D enters into RDO. Mode decision block computes a minimum cost and finally decides the best CU mode. A *32x32* LCU can be just split with four *16x16* CUs or a *16x16* CU with four *8x8* CUs if four *8x8* RD-cost summing value is smaller than a *16x16* one. RDO block stores all modes' transformed quantization coefficients to quantization RAMs (qRAMs) in order to reuse them on scaling (de-quantization) in reconstruction after the mode decision. This DCT shared architecture reduces the hardware resources in an entire encoder architecture. Reconstruction restores reference image to reconstruction RAM (rRAM) with addition an intra-predicted sample and a retrieved residual sample via scaling and inverse DCT (IDCT). The Hypernova has only two size integrated inverse transform blocks because a CU size chosen in mode decision makes a decision of the IDCT size.

B. 32x3216x16/8x8 Inter/MergeSKIP CU decision in P-slice

Unlikely to I-slice, inter prediction in P-slice has one candidate mode each for *16x16* and *8x8* CU respectively. The rough mode decision block of inter prediction selects these candidates with SAD. If IME pre-decides an inter coded CU size, one fine motion compensated sample which it is one from *32x32* to *8x8* size and another merge SKIP sample of *32x32* or

16x16 become two inter coded candidates to decide the best mode. Totally four CU size's predicted samples including intra predicted ones go into RDO block. The sub-block of mode decision with a size integrated transform calculates RD-cost value for motion compensation (MC) of a FME as well as a merge SKIP sequentially. Finally the mode decision block decides one of the best intra/inter CU part mode by RDO calculation.

Conclusions

As the prediction modes of H.265 increase, an accurate mode decision with RDO is more important than H.264 because it impacts directly on the performance of an encoder. But the hardware computational complexity and size to implement it with a SoC increases rapidly high. To lighten this hardware load, Hypernova adopts the CU level mode decision for an intra/inter prediction with the RMD and simplified RDO. The processing cycle to deal with a *32x32* LCU for mode decision is about 2,000 and the transform cycle to do RDO operation occupies over 50% of it. The Hypernova operates on 300MHz@60fps for FHD encoding and the increase of BD-BR is 6.02% on average compared to the mode decision method in HM-13.0. The Supernova H.265 decoder has developed and Hypernova is being exploited toward a H.265 codec SoC.

Acknowledgment

This research is supported by the IT R&D program of MOTIE/KEIT. [10039214, Video Codec SoC for Ultra High Definition]

References

[1] JCT-VC, "High Efficiency Video Coding (HEVC) Test Model 13 (HM 13) Encoder Description," *JCTVC-O1002*, Oct. 2013.

[2] I.-K. Kim, J. Min, T. Lee, W.-J. Han, and J. H. Park, "Block partitioning structure in the HEVC standard," *IEEE Trans. Circuits Sys. Video Technol.*, vol. 22, no.12, pp. 1697-1706, Dec. 2012.

[3] L. Liu and X. Zhung, "Cabac based bit estimation for fast H.264 RD optimization decision," *In Proc. IEEE Int. Conf. on Consumer Communication and Networking*, pp. 1-5, Jan. 2009.

[4] L. M. Po and K. Guo, "Transform-domain fast sum of the squared difference computation for H.264/AVC rate-distortion optimization," *IEEE Trans. Circuits Sys. Video Technol.*, vol. 17, no.6, pp. 765-773, Jun. 2007.

Real-time HDR Multi Exposure Fusion Hardware Design

Junwon Moon, and Jaeseok Kim
Department of Electrical and Electronic Engineering,
Yonsei University, Seoul 120-749, Korea
Email: {mjw5554, jaekim}@yonsei.ac.kr

Abstract—**In a high dynamic range (HDR) circumstance, it's hard to describe whole information of the scene as common sensor's dynamic range is not sufficient to cover whole range. In this paper, we describe hardware design which enables HDR multi exposure fusion technique pursuing to be processed on real-time, for high definition (HD) images at 60fps . The proposed design can operate with Virtex5 XC5VLX20T device.**

I. INTRODUCTION

Charge coupled device (CCD) sensor and complementary metal-oxide semiconductor (CMOS) sensor are widely used in industry, however, both of them has a smaller range capacity than a range which real circumstance has. To overcome this weakness, many researchers try to describe high dynamic range (HDR) scene with multi exposure images and represents those whole information in common display devices.

Nowadays, HDR imaging techniques are applied in autonomous industry or camera security. In many applications, they encounter HDR circumstances, such as the situation that a car passes through the tunnel. In that case, without HDR imaging technique, it is hard to detect lane or vehicles.

Two approaches have been studied for HDR processing, one is HDR tone mapping, and the other is multi exposure fusion. With common low dynamic range (LDR) images, HDR tone-mapping technique needs reconstruction of HDR radiance-map step and tone-mapping step, on the other hand, exposure fusion technique can obtain the result without reconsturcting HDR radiance-map [?] [?] [?].

In this paper, we propose a HDR multi exposure fusion hardware design which can operates on real time for high definition (HD) images at 60fps. As HDR imaging is not main application, design aims at extremely low complexity and low computation time. The algorithm uses are designed for two different exposed images captured in common devices, and it requires only simple arithmetic computation.

II. SYSTEM MODEL

Fig. 1 shows that proposed system model consists of three main steps. Two different exposed images are stored in two buffers, and each pixel data width is 8-bit for each R, G, B channel. Each block requires 256×8 ROM table and simple arithematic calculation only, such as adder, bit reversal and shift operation.

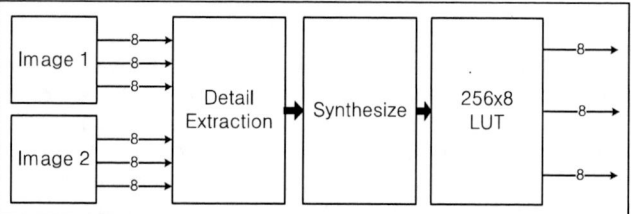

Fig. 1. Block diagram for proposed HDR exposure fusion design

A. Detail Extraction

In the dynamic range scene, one image cannot represent scene's detail for every location. As a first step, it needs to find and estimate which image represents the most information for each location. In the proposed scheme, scene detail is measured as follows:

$$\mathbf{D} = \begin{cases} 1, & \text{if } |\mathbf{G_L} - \mathbf{B_L}| + |\mathbf{G_L} - \mathbf{R_L}| > \\ & \quad |\mathbf{G_S} - \mathbf{B_S}| + |\mathbf{G_S} - \mathbf{R_S}| \\ 0, & \text{otherwise} \end{cases} \quad (1)$$

where $\mathbf{D}$ is the decision variable that shows which image has more information at each location. $\mathbf{G_L}, \mathbf{B_L}, \mathbf{R_L}$ ($\mathbf{G_S}, \mathbf{B_S}, \mathbf{R_S}$) is long (short) image's pixel intensity for each color channel.

B. Synthesis

After detail extraction step is completed, two images are synthesized into proper result without any artifacts, such as blocking effect or halos. In the proposed scheme, result of HDR information image is calculated for each R, G, B color channel as follows:

$$\tilde{\mathbf{G}} = (\mathbf{G_L} + \mathbf{G_S})/\mathbf{2} \quad (2)$$

$$\tilde{\mathbf{R}} = \begin{cases} \tilde{\mathbf{G}} - \mathbf{G_L} + \mathbf{R_L}, & \text{if } \mathbf{D} == 1 \\ \tilde{\mathbf{G}} - \mathbf{G_S} + \mathbf{R_S}, & \text{otherwise} \end{cases} \quad (3)$$

$$\tilde{\mathbf{B}} = \begin{cases} \tilde{\mathbf{G}} - \mathbf{G_L} + \mathbf{B_L}, & \text{if } \mathbf{D} == 1 \\ \tilde{\mathbf{G}} - \mathbf{G_S} + \mathbf{B_S}, & \text{otherwise} \end{cases} \quad (4)$$

where $\tilde{\mathbf{R}}, \tilde{\mathbf{G}}, \tilde{\mathbf{B}}$ are result pixel intensity for each color channel. Note that each pixel is calculated with only corresponding

978-1-4799-5128-4/14 $31.00 © 2014 IEEE

pixel values of long image and short image. This fact makes proposed design can be performed without line memory.

C. Post-processing

Post-processing is necessary to make results better when the scene is extremely bright or extremely dark. Therefore, we implemented simple gamma curve by using look-up-table(LUT) for each channel.

III. EXPERIMENT RESULT

In this paper, we aim to achieve real-time processing design for HD images at 60fps. As a pixel data commonly takes four clock cycles to be transmitted, path delay should be less then 4.52ns.

A. Synthesis result

The proposed design is synthesized with Virtex5 XC5VLX20T for the Xilinx FPGA. As a design requires only simple ALU operations, synthesis result shows that path delay is 3.168ns. Synthesis report is shown in Table. 1. Remarkable thing is that this design does not require any multiplier.

TABLE I
HDL SYNTHESIS REPORT

	Size	Number
ROMs	256 × 8	3
Adder/Subtractor	8-bit / 9-bit	21
Registers	Flip Flop	118
Comparator	9-bit	1

B. Simulation result

The example images for our experiment is from http://www.easyHDR.com, which is one of famous HDR research site. Many samples are here, and we chose the image which has a extremely high dynamic range. Two input images and the synthesized image are shown in Fig. 2.

(a) Long exposed image   (b) Short exposed image

(c) Synthesized image

Fig. 2.   Two different exposed images and a synthesized image

Fig 2.(a) and Fig 2.(b) are input images and Fig 2.(c) is output of exposure fusion image. Long exposed image represents indoor well, but outdoor is over-exposed. Conversely, short exposed image represents outdoor well, but indoor is under-exposed. Synthesized image represents the information both outdoor and indoor. Result shows that synthesized image represents much wide range than a single image.

IV. CONCLUSION

In a high dynamic range circumstance, common camera sensor cannot capture satisfied image. This problem is issued in camera industry or smart car industry for object detection. In those field, they require a HDR imaging technique which can be processed on real-time for high resolution image.

In this paper, we proposed HDR exposure fusion technique pursuing to be processed on real-time for HD images at 60fps. Although many algorithms have been studied, most of them are hard to be processed on real-time. The proposed algorithm is not for a high quality HDR image but fast to apply to other main application such as object detection.

As the experiment results, the algorithm discussed here can achieve the goal operation clock, and require a few components with Virtex5 XC5VLX20T for the Xilinx FPGA. Although simulation result seems unfavourable to see at over-exposed range, but we expect that it's helpful for detecting.

ACKNOWLEDGMENT

This work was supported by the World Class 300 project R&D program of KEIT [10046974, Development of a 640X480 pixels 1,000fps Vision Sensor and a ±0.5μm resolution 100μs Displacement Sensor for Non-contact Measurement and Inspection], and also supported by the IT R&D program of MOTIE/KEIT.[10035389, Research on high speed and low power wireless communication SoC for high resolution video information mining], and CAD tools were supported by IDEC.

REFERENCES

[1] Debevec, Paul. "Rendering synthetic objects into real scenes: Bridging traditional and image-based graphics with global illumination and high dynamic range photography." *ACM SIGGRAPH 2008 classes*. ACM, 2008.

[2] Park, Tae Jang, and In Kyu Park. "High dynamic range image acquisition using multiple images with different apertures." *Optical Engineering* 51.12 (2012): 127002-127002.

[3] Gu, Bo, et al. "Gradient field multi-exposure images fusion for high dynamic range image visualization." *Journal of Visual Communication and Image Representation* 23.4 (2012): 604-610.

Enhanced Tone Mapping of High Dynamic Range to Correspond to Illumination Changes

Young-Min Jang*, Jae-Hyun Cho*, Ji-Geun Kim*, Bruce C. Kim**, Sang-Bock Cho***

*,***Department of Electronics Engineering
University of Ulsan
Ulsan, Republic of Korea
**Department of Electrical Engineering
City University of New York, CUNY
New York, USA
min-s2@nate.com, bkim91us@gmail.com, sbcho@ulsan.ac.kr

Abstract

In recent years, the study of IoT (Internet of things) has been expanded to M2M (machine to machine) and V2V (vehicle to vehicle) communication using image information. Accurate image sensing is required for this kind of technology.

This paper describes a normalized numeric image descriptor used to assess the luminance and contrast of the image. The proposed image descriptor uses each pixel's data as a weighted value of the probability density function. It is further defined by normalization for objective representation. The proposed numeric image descriptor can be used in the adaptive gamma process because it suggests an objective basis of the gamma value selection. In particular, it is possible to use tone mapping in illumination changes and low illumination because of its high dynamic range.

Keywords-Internet of things (IoT), image processing, high dynamic range, gamma correction, illumination

Introduction

The IoT (Internet of Things)[1] industry is experiencing huge growth as technology expands to connect devices and machines in a wide range of industries. M2M (machine to machine) and V2V (vehicle to vehicle) communication is representative technology for IoT. This technology requires accurate circumstance detection of its environment.

A typical camera recognizes only the general surrounding environment. But it is difficult for a camera to recognize the images in a situation of illumination change and low illumination.

To solve this problem, in this paper, we propose a normalized numeric image descriptor used to assess the luminance and contrast of the image. The proposed image numeric descriptor can be used in the adaptive gamma[3, 4] process because it suggests an objective basis of the gamma value selection.

The proposed method offers the following advantages:
1) Consistent normalization technique irrespective of image resolution and data representation bits
2) Suitable method for real-time application

3) Dynamic range[2] compression using adjustable variable factors
4) Possibility of tone mapping with illumination change and low illumination because of high dynamic range
5) Suitable algorithm for security images

Proposed Method

A. Normalization of Luminance

Normalization of luminance (*NoL*) is weighted representing the luminance of the image, where r_k is the k-th gray level and n_k is the number of pixels with a k-th gray level. Because luminance depends on the image resolution, (1) is necessary for normalization. As shown in (2), *NoL* value can be normalized by dividing the image resolution from *LD*.

$$LD = \sum_{k=0}^{L-1} r_k \cdot n_k \tag{1}$$

$$NoL = \frac{1}{LN} \sum_{k=0}^{L-1} r_k \cdot n_k \, , \, 0 \leq NoL \leq 1 \tag{2}$$

B. Normalization of Contrast

Contrast information may be represented by the difference between the maximum and minimum values of the pixels in the image. As shown in (3), c_l is the contrast of the l-th block, and b_l is the number of blocks of c_l contrast value. In the same way, the value of *NoC* in (4) can be obtained through the normalization of (3).

$$CD = \sum_{l=0}^{B-1} c_l \cdot b_l \tag{3}$$

$$NoC = \frac{1}{L} \sum_{l=0}^{B-1} c_l \cdot b_l \, , \, 0 \leq NoC \leq 1 \tag{4}$$

C. HDR Tone Mapping

The normalized equation (5) of image quality concerning luminance and contrast is presented through an arrangement of (2) and (4).

$$HDRTM = (NoL, NoC) \tag{5}$$

Therefore, it is possible to provide objective information about an image as well as removing the ambiguity of conventional methods.

Experiment Results

In this paper, images were captured using the CMOS image sensor. We implemented this system using the environment of Visual Studio 2010 and OpenCV 2.2, Intel (R) Core(TM)2 Quad Q8400 CPU 2.66GHz.

Fig. 1 Left: original images Right: corrected images. (clip 1: situation of low illumination)

Fig. 2 Left: original images Right: corrected images. (clip 2: situation of low illumination)

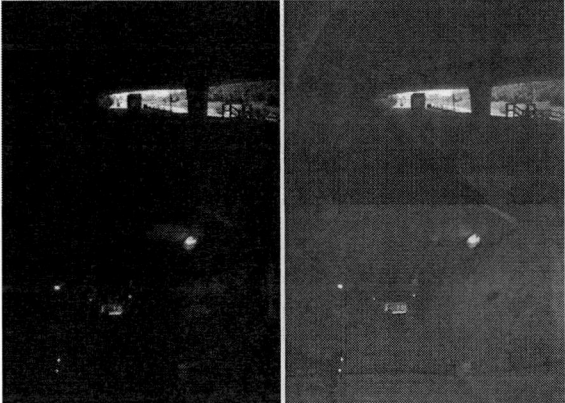

Fig. 3 Left: original images Right: corrected images. (clip 3: situation of illumination change)

Figs. 1 and 2 show the corrected results with low illumination at night. Fig. 3 shows the corrected results for a vehicle passing through a tunnel and under a bridge.

Table I Average values

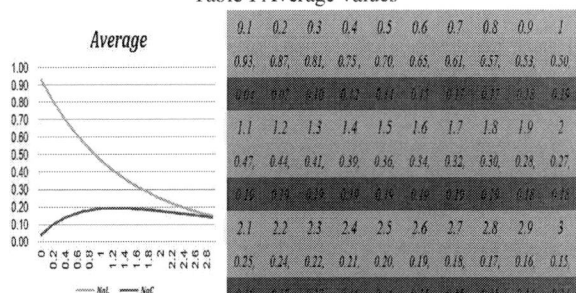

0.1	0.2	0.3	0.4	0.5	0.6	0.7	0.8	0.9	1
0.93	0.87	0.81	0.75	0.70	0.65	0.61	0.57	0.53	0.50
1.1	1.2	1.3	1.4	1.5	1.6	1.7	1.8	1.9	2
0.47	0.44	0.41	0.39	0.36	0.34	0.32	0.30	0.28	0.27
2.1	2.2	2.3	2.4	2.5	2.6	2.7	2.8	2.9	3
0.25	0.24	0.22	0.21	0.20	0.19	0.18	0.17	0.16	0.15

As shown in Table 1, each result is arranged by average result value. If the NoL value is larger than 1, the image is made a little darker using the greater gamma value. Conversely, if the NoC value is smaller than 1, the image is made a little brighter using the smaller gamma value.

Conclusion

M2M (machine to machine) and V2V (vehicle to vehicle) communication is representative technology for IoT and requires accurate image detection of the surrounding environment.

This paper describes a normalized numeric image descriptor used to assess the luminance and contrast of the image. This method covers a wide variety of images including the high dynamic range. It shows consistent normalization irrespective of image resolution and expression of data bits. This descriptor can be used as a real-time application because it is easy to calculate. In particular, tone mapping becomes possible with illumination changes and low illumination because of the high dynamic range.

Therefore, the proposed algorithm can be applied to security images for CCTV, vehicle and mobile cameras to create correct images of the surrounding environment.

Acknowledgment

This work was supported by the Industrial Core Technology Development Program (10049009, Development of Main IPs for IoT and Image-Based Security Low-Power SoC) funded by the Ministry of Trade, Industry & Energy.

References

[1] Wikipedia, http://en.wikipedia.org/wiki/Internet_of_Things,

[2] MH. Kim and LW. MacDonald, "RENDERING HIGH DYNAMIC RANGE IMAGES," EVA 2006 London Conference, pp. 22.1-22.11, 2006

[3] G. Ramponi, Adaptive contrast improvement for still images and video frames," IEEE- EURASIP Workshop on Nonlinear Signal and Image Processing, NSIP-07, Bucharest, Romania,. 10-12, Sept 2007.

[4] A. Restrepo (Palacios) and G. Ramponi, "Word Descriptors of Image Quality Based on Local Dispersion-versus-Location Distributions," 16th European Signal Processing Conference 2008, pp. 25-29, Aug 2008.

Enhanced Test Zone Search Motion Estimation Algorithm for HEVC

Nidhi Parmar and Myung Hoon Sunwoo, *Fellow, IEEE*

School of Electrical and Computer Engineering, Ajou University
San 5, Wonchon-Dong, Yongtong-Gu
Suwon, 443-749 Korea
E-mail: sunwoo@ajou.ac.kr, ndz86@ajou.ac.kr

Abstract

Recent video coding standards like H.264/MPEG and HEVC adopted the Test Zone search (TZS) algorithm due to its excellent performance in reducing ME (motion estimation) time while maintaining comparable RD performance. This paper propose enhanced TZS ME algorithm with reference to its implementation in HEVC reference software. Experimental results show that the proposed change in TZS can save total encoding time and ME time upto 14.5% and 22.29% respectively while maintaining similar PSNR and RD performance compared to the original one.

Keywords - Motion Estimation; TZ search; HEVC

Introduction

High Efficiency Video Coding (HEVC)/H.265 has been jointly developed by video coding standardization project of the ITU-T Video Coding Experts Group (ITU-T Q.6/SG 16) and ISO/IEC Moving Picture Expert Group [1]. The emerging HEVC standard aims to provide a doubling in coding efficiency with respect to the H.264/AVC high profile while delivering the same video quality at half the bit rate [1]. ME is essential process in video coding standards since it can effectively eliminate temporal redundancy of image sequences. The HEVC test model (HM encoder) provides two kinds of ME algorithms: Full search (FS) and Test Zone Search (TZS). Although the FS is the best algorithm in terms of the quality of the predicted image and the simplicity of the algorithm, it is computationally intensive. On the other hand, TZS is faster than FS, however, the computational complexity is still large and thus, many improvements have been carried out [2]-[4] in order to improve the coding efficiency. However, the computation is still large and thus, more optimization steps need to be carried out. This paper proposes improvements to the TZS algorithm with reference to its implementation in HM14.0.

Analysis of TZ search algorithm

The TZ search is the combination of diamond search (DS) and raster search. The first step is motion vector (MV) prediction. The minimum among all predictors is selected as a starting search point for further steps. The next step is coarse search with initial grid search in either diamond or square search with power of 2. In this step, the minimum distortion point is taken as the center for further search and the minimum distance is stored in variable called 'uiBestDistance'. In third step, if the uiBestDistance is greater than 'iRaster', then it performs raster search. Finally, it performs refinement search with either raster or star refinement.

The straight forward way to diminish the computational complexity of ME is to reduce the number of search points. Moreover, it has been concluded from the experimental results that, MVs have higher density in center-biased region and the MVD (motion vector distribution) has the effect of gravity [5]. That means compared to the vertical MVP (motion vector probability) down below the center point is holding a significant probability. Thus, to cover the high MVP area while maintaining the similar distortion error and to reduce the ME complexity, this paper proposes pentagon pattern instead of DS in initial grid search step of TZS algorithm as depicted in Fig. 1.

Proposed Improvements

This Section deals with the proposed TZ search improvement. For the prediction stage this algorithm uses the similar predictors to that of the HM14.0. Moreover, as discussed in the previous section, the initial grid search used in the HM reference encoder use either diamond or square search patterns. Each pattern use 8 points per grid, which results in high computational complexity. However, instead of searching all the points in each grid, only if the most probable points are considered, then the large savings in encoding time can be achieved. Inspired from the center-biased characteristics and the effect of gravity of MVs [5], this paper proposes the pentagon pattern to replace the conventional diamond pattern in the initial grid search. Theoretically, the proposed pattern can reduce the number of search points up to 32% per each search window compared to the original DS. For instance as depicted in Fig. 1, the search window of 16x16 requires 3 grids with 5 points per each grid. Thus, including small diamond search with 4 points, it requires 19 (5x3+4) points. On the other hand, DS requires 28 (8x3+4) points for the same size search

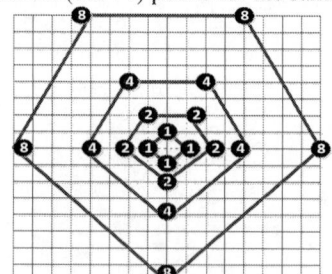

Fig. 1 Pentagon Pattern

978-1-4799-5128-4/14 $31.00 © 2014 IEEE

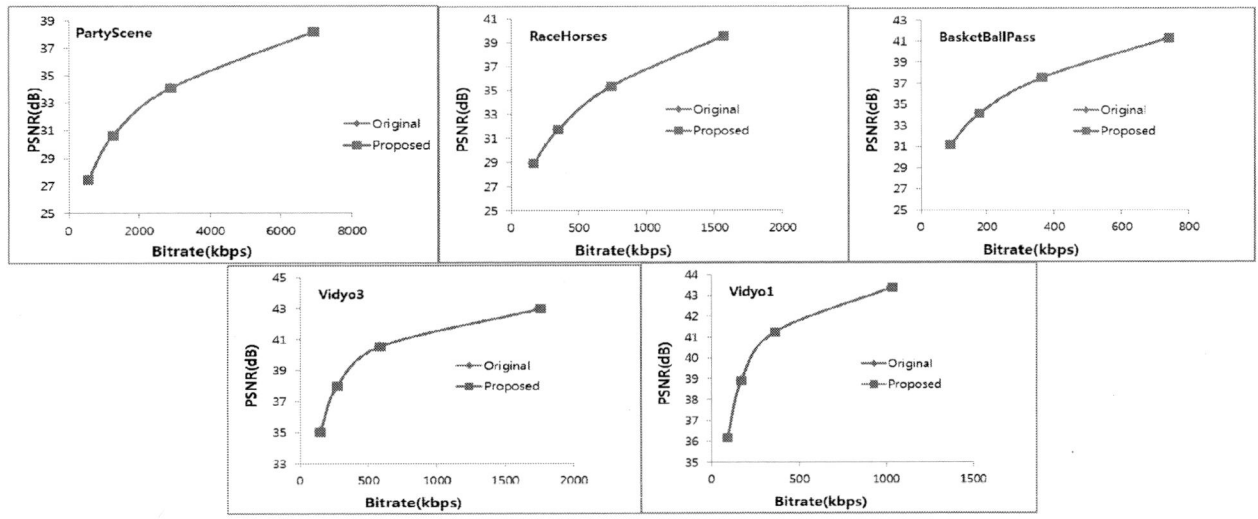

Fig. 2. RD performance for different Video Sequences with QP 22, 27, 32, 37

window. Moreover, for the refinement stage, this algorithm uses the pentagon grids with reduced stride lengths in every loop.

TABLE I
Performance comparisons between HM14.0 and proposed idea

Sequence	QP	HM14.0 (dB)	Proposed (dB)	ΔPSNR-Y (dB)	ΔME Time (%)
PartyScene	32	30.6278	30.6239	-0.0039	21.91
	27	34.1003	34.1002	-0.0001	20.76
Racehorses	32	31.7314	31.7312	-0.0002	18.08
	27	35.3447	35.3429	-0.0018	21.37
Basketball pass	32	34.1106	34.1017	-0.0089	25.94
	27	37.5279	37.5189	-0.009	22.91
Vidyo3	32	37.985	37.9871	-0.0021	20.35
	27	40.5524	40.5515	-0.0009	25.92
Vidyo1	32	38.9061	38.8967	-0.0094	22.41
	27	41.2478	41.2442	-0.0036	23.30
Average				-0.00399	22.295

Simulation Results

The experiments validate the proposed improved TZS algorithm. For simulation, we used reference software HM14.0, running on Ubuntu 14.10 platform with Intel core i3 @ 2.53 GHz CPU and 6GB RAM. The number of frames is taken 100 for each standard sequence with different resolutions. The search range and maximum CU partition depth is 64 and 4, respectively. Table I shows the performance comparisons between HM14.0 and proposed idea. As it can be seen about 22.29% gain has been achieved in terms of the ME time. Moreover, Table II shows the overall increase in BD-BR is only 0.358. Fig.2 shows the RD performance between the proposed idea and original DS for the different sequences, taken at QPs 22, 27, 32, 37. It shows that the RD performance remains the same compared to the original one while reducing 14.5% of the encoding time.

Conclusion

This paper proposed improvements in the TZS algorithm with reference to the HM14.0. The simulation results show that the proposed algorithm can reduce the total encoding time and total ME time by 14.5% and 22.29% with almost the same RD performance compared to the TZS that is implemented in HEVC reference software.

TABLE II
Bit-rate and Encoding time comparison between HM14.0 and proposed idea

Sequence	BD-BR (%)	ΔEncoding time (%)
PartyScene	0.29	15.70
Racehorses	0.55	12.74
Basketball pass	0.58	15.09
Vidyo3	0.13	17.1
Vidyo1	0.24	11.9
Average	0.358	14.506

Acknowledgment

This work was partly supported in part by the IT R&D program of MOTIE/KEIT (10044092, Development of Core IPs of OFDM PHY and RF Transceiver for 60GHz Wireless LAN/PAN in Application of 7Gbps Wireless Multimedia Services) and in part by the Mid-Career Researcher Program through an NRF grant funded by the MSIP (2014R1A2A2A01002952).

References

[1] G. J. Sullivan, J. R. Ohm, W.-J. Han, T. Wiegand, "Overview of the high efficiency video coding (HEVC) standard," *IEEE Trans. Circuits Syst. Video Technol.*, vol. 22, no. 12, pp.1649-1668, Dec. 2012.

[2] N.Purnachand, L.Nero Alves, A.Navarro, "Improvements to TZ search motion estimation algorithm for multiview video coding," *IEEE IWSSIP*, Vienna, Apr. 2012.

[3] N.Purnachand, L.Nero Alves, A.Navarro, "Fast Motion Estimation Algorithm for HEVC," *IEEE ICCE*, Berlin, 2012

[4] Zhaoqing Pan, Yun Zhang, "Early Termination for TZSearch in HEVC Motion Estiamtion ," *IEEE ICASSP*, Vancouver BC, 2013.

[5] Chi-Wai Lam, Lai-Man Po and Chun Ho Cheung, "A Novel Kite-Cross-Diamond Search Algorithm for Fast Block Matching Motion Estimation ," *IEEE ISCAS*, 2004.

Efficient Hardware Architecture for Real-time Semi-Global Matching

Seongbo Sim, Kyoungwon Min[*], Seonyoung Lee[*], Haengson Son[*], Jongtae Kim

Department of Electrical and Computer Engineering
Sungkyunkwan University
[*]Department of SoC Platform Research Center
[*]Korea Electronics Technology Institute
Suwon and [*]Seongnam, Republic of Korea
{sungbo180, jtkim}@skku.edu, [*]{drleesy, hsson, minkw}@keti.re.kr

Abstract

In this paper we propose an efficient hardware architecture for real-time SGM (Semi-Global Matching). SGM has a robust characteristic than previous local stereo matching algorithms. But SGM requires high computational loads and extremely high memory bandwidth to store intermediate results. To overcome these problems, we have maximized data parallelism by adopting systolic array and pipelining. Also we have maximized internal memory recycling efficiency to minimize memory bandwidth. With this method, our architecture not only processes 32 frame of VGA disparity images per second at 100MHz operating frequency but also do not requires external memory to store intermediate data. Our architecture was implemented using Verilog HDL. Our circuit is composed of 529,200 logic gates and 2,030,784 bits internal memory. Disparity map of SGM circuit has been verified using the Middlebury test images and the average error rate is 6.22%.

Keywords-stereo matching; SGM; hardware architecture

Introduction

Stereo matching is a traditional method to acquire 3-D information from image pair. Stereo matching has many advantages over other 3-D sensing methods in terms of safety, operating range, and reliability[1]. SGM is one of the various stereo matching algorithm. The advantage of SGM is to obtain stereo matching result especially at object boundaries, robustness against illumination changes. On the other hand, SGM algorithm requires high computational loads and extremely high memory bandwidth to store intermediate results[2].

In this paper, we propose an efficient hardware architecture for real-time SGM. We have maximized data parallelism by adopting systolic array and pipelining.

Real-Time Hardware Architecture

In this section, we explain our proposed efficient hardware architecture for the real-time SGM. Our architecture is composed of 'Matching Cost Calculator', 'Pathwise Cost Aggregator' and 'Cost Buffer', as shown Fig. 1. 'Matching Cost Calculator' module has 'Edge Extractor', 'BT calculator' and 'Cost Operator'. This module generates initial cost data for

an image. 'Cost Buffer' module contains initial cost data and transfers to 'Pathwise Cost Aggregator' module. This module calculates the minimum cost using the global cost function of SGM equation for each path.

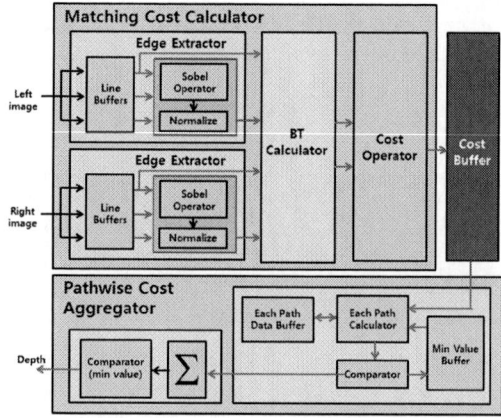

Fig. 1. Overall architecture of the propsed real-time SGM

Sobel and BT(Birchfiled Tomasi) algorithm[3] are designed as a pipeline structure by inputting pixel data into each clock shown as Fig. 2. This structure is available for real time operation due to recycling of memory and arithmetic circuit.

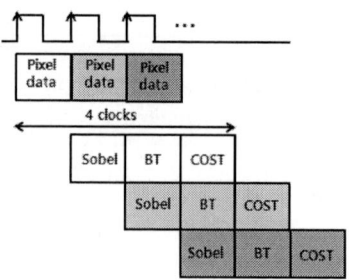

Fig. 2. Pipeline structure for the Maching Cost Calculator.

Figure 3 shows the proposed systolic array structure. The systolic array based structure for the SGM is highly scalable and allows adjustment of the optimal point for a given application, in terms of resource usage, throughput and latency. Pathwise cost aggregator has many repeating parts of same calculations. Both path and disparity calculations are performed simultaneously by systolic array structure using the repeating calculation characteristics. Disparity calculation starts after completing four path calculation for cost data of pixel. Most part of memory and arithmetic circuit are recycled by systolic array, and as the result, the sizes of memory and arithmetic circuit are decreased. To reduce the bandwidth requirements of the algorithm, we use only four directions (0°, 45°, 90° and 180°) for cost aggregation because the quality of the computed disparity map is marginally effected. This idea was also evaluated by Hermann et. al[4].

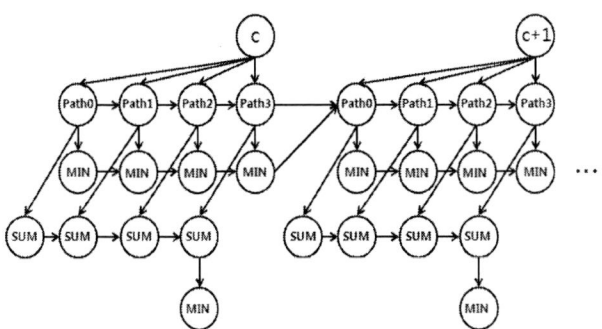

Fig. 3. Systolic array structure of pathwise cost aggregation.

Figure 4 shows the proposed pipeline timing architecture with 128 disparity range. Proposed architecture calculates 16 matching costs value per one clock at the matching cost stage. Also calculated matching cost is directly used on pathwise cost aggregation stage at the next clock cycle. This structure is applied to the whole image, and real time is possible as reducing the waiting time for calculating the required data.

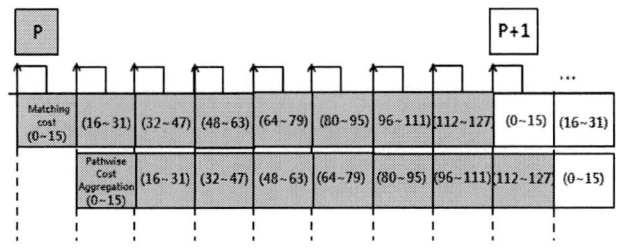

Fig. 4. Proposed pipeline architecture.

Performance and Result

Table I shows the size of the required memory and arithmetic circuit of the proposed hardware architecture when the disparity range is 128 and image size is 640x480. Our architecture was implemented using Verilog HDL. The synthesized circuit size for our architecture is 529,200 logic gates and 2,030,784 bits internal memory. Previous researches used internal and external memories. However our paper contains the internal memory and has the memory size which is 1.14% lower than [2].

Table II shows the error rate based on Middlebury test images. The error rate represents the difference between depth results of our architecture and the ground truth values provided by Middlebury. Fig. 5 shows disparity map results of the proposed hardware architecture. Average error rate of the proposed architecture is 6.22%, and 0.43% less than [2].

TABLE I
SIZE ANALYSIS FOR PROPOSED ARCHITECTURE

Number of Processing Operator per Clock		16	32	64	128
Operation Time (frames per second)		32	49	65	266
Memory (bits)	Matching Cost Calculator	38,080			
	Pathwise Cost Aggregator	1,992,704			
Operator (gates)	Matching Cost Calculaor	292,400	394,800	586,800	958,800
	Pathwise Cost Aggregator	236,800	473,600	947,200	1,894,400
Total	Memory (bits)	2,030,784			
	Operator (gates)	529,200	868,400	1,534,000	2,852,400

TABLE II
PERCENTAGES OF ERRONEOUS DISPARITES

No. of paths	Cones	Teddy	Tsukuba	Venus	Average
4 paths [2]	9.5%	13.3%	6.8%	4.1%	8.43%
8 paths [2]	8.4%	11.4%	4.1%	2.7%	6.65%
Proposed	4.93%	10.1%	6.04%	3.82%	6.22%

Fig. 5. Results for the Cones image set : (left) Cones left image, (middle) ground truth image (right) disparity map on the proposed architecture.

Acknowledgment

This work was supported by the Knowledge & Economy Technology Innovation program (10045774) funded by the Ministry of Trade, Industry & Energy (MOTIE), Korea and the Information and Communication R&D program (10041656) funded by the Ministry of Science, ICT and Future Planning.

References

[1] J. I. Woodfill, G. Gordon, and R. Buck, "Tyzx DeepSea high speed streo vision system," in Proc. IEEE Comput. Soc. Workshop Real-Time 3-D Sensor Use Conf. Compt. Vision Pattern Recog., Washington D.C., 2004, pp. 41-46.

[2] C. Banz, S. Hesselbarth, H. Flatt, H. Blume, and P. Pirsh. "Real-time stereo vision using semi-global matching disparity estimation: Architecture and FPGA-implementation," in intl. IEEE Conference. Embedded Computer Systems, 2010, pp. 93-101.

[3] S. Birchfiled, C. Tomasi, "Depth discontinuities by Pixel to Pixel Stereo", International Journal of Computer Vision 35(3), 1999, pp. 269-293.

[4] S. Hermann, S. Morales, and R. Klette, "Half-resolution semiglobal stereo matching," in Proc. IEEE IV Symp., Baden-Baden, Germany, 2011, pp. 201-206.

Gram-Schmidt tailed High-throughput QR Decomposition Architecture for MIMO detector

Dongyeob Shin[1], Ji-Hwan Yoon[1], Jongsun Park[1], and Woong Choi[2]

[1] Korea University, Seongbuk-Gu, Seoul, 136-701, Republic of Korea, {shindy919, improma, jongsun}@korea.ac.kr
[2] University of Minnesota, 200 Union Street SE, Minneapolis, MN 55455, USA, choiw@umn.edu

Abstract

This paper proposes a high-speed and low latency QR decomposition (QRD) architecture. In the proposed QRD architecture, two QRD algorihms, the CORDIC-based algorithm and Gram-Schmidt, is effectively combined together to reduce hardware latency. The experimental results show that the proposed architecture implemented with Samsung 0.13 um technology achieves 16.7 % speed-up over the conventional Givens rotation-based architecture with only 5.9% area overhead for 4 × 4 matrix decomposition.

Keywords — QR decomposition (QRD); Givens rotation; Gram-schmidt; multiple-input–multiple-output (MIMO); very large scale integration (VLSI).

Introduction

Due to high spectral efficiency with link reliability, multiple input multiple output (MIMO) technology has become a key element in many wireless standards such as IEEE 802.11n/, .16e/m and LTE etc [1]-[3]. In the hardware implementation of MIMO system, one of the most challenging issues is to develop high-throughput low complexity QR decomposition (QRD) which decomposes the estimated channel matrix into an unitary matrix Q and an upper triangular matrix R. Among three general QRD schemes in literature, which are Givens rotation [4], Householder reflection [5], and Gram-Schmidt [6], Givens rotation and Householder-based QRD generally show lower hardware complexity when implemented using Coordinate rotation digital computer (CORDIC) module [4]-[5]. However, in the CORDIC-based QRD architecture, though computational complexity for last 2 columns processing is much smaller than that of the other columns, the latency for processing the last 2 column is still same with others. Thus, reducing the latency of the last 2 comlumns processing is necessary for high-throughput QRD design. To resolve this latency issue, we proposed a novel hybrid QRD architecture, where the CORDIC-based algorithm is effciently combined with Gram-Schmidt to reduce the latency of the last 2 columns processing.

Proposed QRD Approach

A. The Previous CORDIC-based QRD Approaches

Generally, the CORDIC-based QRD approaches shows low computational complexity since CORDIC, its submodule, consist of simple addition and shift operations. However, since the CORDIC is based on the iterative computation, only a few

previous works are proposed to reduce the number of CORDIC cycles [4]-[5]. Eventhough the previous approaches reduce the number of CORDIC cycles, at least 3 CORDIC cycles are still required for the computation of last 2 columns as shown in Fig. 1. Since last 2 columns have only one element (h_{43}) to perform a zero insertion, the computational costs are much smaller compared to that of the other columns. Considering the computational complexity, 3 CORDIC cycles comsumed in last 2 columns are too large. If we arbitrarily choose only CORDIC-based QRD approaches, 3 CORDIC cycles which are needed to compute zero insertion for last 2 columns cannot be reduced. To reduce the clock latency for the computation of last 2 columns, efficient Gram-schmidt method is presented in the next sub-section.

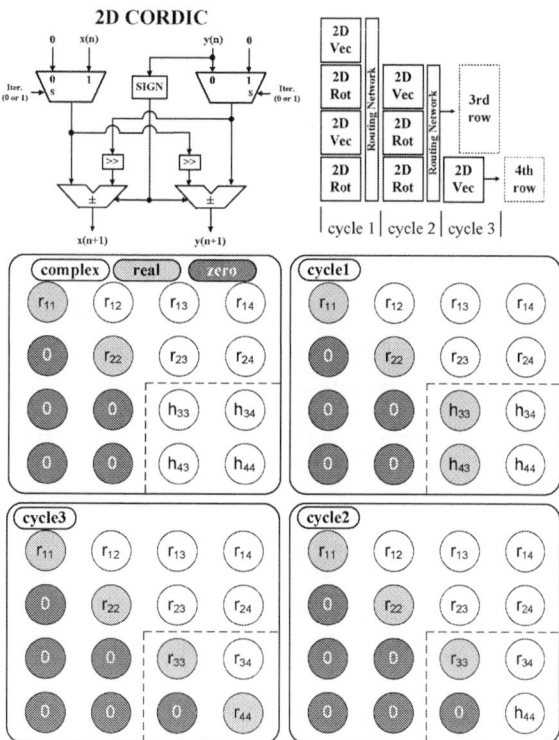

Fig. 1 Decomposition steps by CORDIC-based architecture

B. Efficient Gram-Schmidt QRD Architecture

Gram-Schmidt method has not been frequently used as CORDIC-based Givens Rotation or Householder method for QR decomposition in MIMO systems since it needs a large number of complex hardware modules like multiplication, division and square roots for QRD of 4×4 or larger dimension matrix. However, for implementing smaller 2×2 complex matrix, Gram-Schmidt algorithm can be efficiently used for low latency and area due to exponentially decrease of the number of computations, which involve operations like (2), (3) or (4) for finding the magnitude of vectors and the projection of vectors to other vectors, following the matrix dimension reduction [7]. In the proposed QRD architecture, since Gram-Schmit is exploited only in the last 2×2 matrix, latency of the whole 4×4 matrix QRD process can be reduced without incurring large hardware overhead. Gram-Schmidt algorithm based non-iterative 2×2 matrix process is shown in (1)-(5), and the low cost hardware architecture is illustrated in Fig. 2. Due to non-iterative computations from (2) to (5), the number of clock cycles can be significantly reduced.

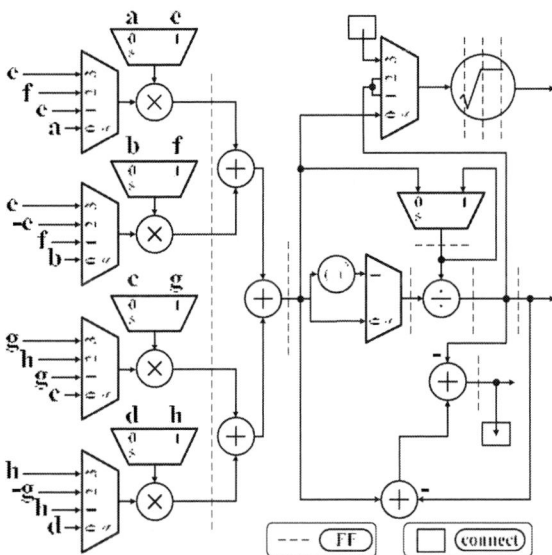

Fig. 2 Proposed 2x2 complex Gram-Scmidt architecture

$$H_{2x2} = \begin{bmatrix} a + bi & e + fi \\ c + di & g + hi \end{bmatrix} \quad (1)$$

$$r_{33} = \sqrt{a^2 + b^2 + c^2 + d^2} \quad (2)$$

$$Re\{r_{34}\} = (ae + bf + cg + dh)/r_{33} \quad (3)$$

$$Im\{r_{34}\} = (af - be + ch - dg)/r_{33} \quad (4)$$

$$r_{44} = \sqrt{e^2 + f^2 + g^2 + h^2 - Re\{r_{34}\}^2 - Im\{r_{34}\}^2} \quad (5)$$

Experimental Results

The poroposed hybrid QRD arachitecture and conventional QRD [4]-[5] are implemented using Samsung 0.13 um technology, and Table I shows the numerical results. As shown in Fig. 2, decomposing last 2x2 matrix, which equals to find the answers of equations (2) to (5) is finished in 1 processing(CORDIC) cycle illustrated in Fig. 1. As a result,

TABLE I

Comparisons with the conventional QR decomposition architectures

	[5]	This work
Process tech (µm)	0.13	0.13
Latency (Clock cycle)	56	40
Max. operating frequency	215MHz	179MHz
Gate count (2nnd)	83.7K	88.6K
Throughput (MQRD/s)	3.84	4.48

the rest of computations, 2×1 CORDIC cycle (8 clock cycle here to compare with conventional) are reduced compared to the conventional work. At the expense of the number of cycles (latency) reduction, the proposed QRD shows 5.9% area overhead. Throughput of the proposed work are improved by 16.7%.

Conclusion

In this paper, a high-speed QRD architecture was presented based on the proposed Gram-Schmidt approach. The proposed work enables further reduction of processing cycles(CORDIC cycles) which could not diminished by CORDIC-based architectures. The proposed QRD design was implemented using Samsung 0.13 um technology, and the proposed architecture achieves 16.7% increase in the throughput with 5.9% area overhead.

Acknowledgment

This research was supported by National Research Foundation of Korea Grant (NRF-2012R1A2A2A01012471) and IDEC.

References

[1] Wireless LAN medium access control (MAC) and physical layer (PHY)specifications, IEEE 802.11 Std., Aug. 1999.

[2] K. Etemad, "Overview of mobile WiMAX technology and evolution," *IEEE Commun. Mag.*, vol. 46, no. 10, pp. 31-40, Oct. 2008.

[3] H. Ekstrom, A. Furuskar, J. Karlsson, M. Meyer, S. Parkvall, J. Torsner,and M. Wahlqvist, "Technical solutions for the 3G long-term evolution," *IEEE Commun. Mag.*, vol. 44, no. 3, pp. 38-45, Mar. 2006.

[4] Min-Woo Lee, Ji-Hwan Yoon, and Jongsun Park, "High-Speed Tournament Givens Rotation-Based QR Decomposition Architecture for MIMO Receiver," in *Circuit and Systems (ISCAS), 2012 IEEE international Symposium on*, May. 2012.

[5] Kurniawan. I. H., Ji-Hwan Yoon, and Jongsun Park, "Multidimensional Householder based high-speed QR decomposition architecture for MIMO receivers," in *Circuit and Systems (ISCAS), 2013 IEEE international Symposium on*, May 2013.

[6] Chitranjan K. Singh, Sushma Honnavara Prasad, Poras T. Balsara, "VLSI Architecture for Matrix Inversion using Modified Gram-Schmidt based QR Decomposition," *VLSI Design*, 2007. *Held jointly with 6th International Conference on Embedded Systems., 20th International Conference on*, Jan. 2007.

[7] G. H. Golub and C. F. V. Loan, *Matrix Computations*, 3rd ed. Baltimore, MD: John Hopkins Univ. Press, 1996.

Efficient Min-Max Nonbinary LDPC Decoding on GPU

Huyen Pham Thi, Sabooh Ajaz, and Hanho Lee

Dept. of Information and Communication Engineering,
Inha University, 402-751, Incheon, Korea
E-mail: hhlee@inha.ac.kr

Abstract

This paper presents an novel modified Min-Max algorithm (MMMA) and an efficient implementation of an nonbinary LDPC (NB-LDPC) decoder on a graphics processing unit (GPU) to achieve both great flexibility and scalability. The MMMA for check node processing removes the multiplications over Galois-field in merger step and significantly reduces the decoding latency. The proposed MMMA provides a better BER performance than previous algorithm. The experimental results show that the GPU-based implementation of the proposed NB-LDPC decoder provides higher throughput and the coding gain under low 10^{-8} BER comparted to CPU-based implementation.

Keywords- GPU; nonbinary LDPC; Min-Max, decoding; CUDA

Introduction

Matthew *et al.* [1] showed that NB-LDPC codes give a significant performance improvement when the code lengths are short and moderate. However, the decoding algorithms for NB-LDPC codes require complex computations and a large memory [2-4]. The demand for new codes and novel low-complexity decoding algorithm for NB-LDPC codes requires a huge amount of extensive simulations. Due to the high complexity of NB-LDPC decoding algorithms, the simulation time on CPU is extremely slow in higher order Galois-filed $GF(q)$ fields.

A GPU can provide massively parallel computation threads with a many core architecture, which can accelerate the simulations of the NB-LDPC decoding. However, the implementation of NB-LDPC codes on GPU is still very challenging. In this paper, the novel modified Min-Max algorithm and an efficient implementation of a parallel-block layered NB-LDPC decoder based on a GPU is presented. The MMMA for check node processing removes the multiplications over Galois-field in merger step and significantly reduces the decoding latency.

Proposed Modified Min-Max Algorithm

In this section, a new modified Min-Max algorithm is provided which removes the multiplications with nonzero elements of **H** matrix in merger step. A block-layered decoding algorithm is proposed in Algorithm 1. The Min-Max decoding, which is implemented by a well-known forward-backward algorithm (FBA), is applied in check node process [3].

Algorithm 1:
Initialization:
$L_n(a) = \ln(\Pr(c_n = s_n|\text{ channel})/(\Pr(c_n = a \mid \text{channel}));$
$L_n^{1,0}(a) = L_n(a);\ R_{mn}^{0,1}(a) = 0;$
Iterations:
For ($k=1; k <= I_{max}; k++$)
  For ($l=1; l<= L; l++$)
    For ($m=0; m < q-1; m++$)
      $Step1: \tilde{L}_{nm}^{k,l}(a) = L_n^{k,l-1}(a) - R_{mn}^{k-1,l}(a);$

      $\tilde{L}_{nm}^{k,l} = \min_{a \in GF(q)} \left(\tilde{L}_{nm}^{k,l}(a) \right)$

      $L_{nm}^{k,l}(a) = \tilde{L}_{nm}^{k,l}(a) - \tilde{L}_{nm}^{k,l};$

      $Step2: R_{mn}^{k,l}(a) = \min_{(a,n')_{n' \in N(m)}^{a \in \gamma_{mn}}(a)} \left(\max_{n' \in N(m\setminus\{n\})} \left(L_{n'm}^{k,l}(a_{n'}) \right) \right)$

      $Step3: L_n^{k,l}(a) = \tilde{L}_{nm}^{k,l}(a) + R_{mn}^{k,l}(a);$
    End for
  End for
Decision:
$\tilde{c}_n = \arg\min \left(L_n^{k,l}(a) \right);$
End for

Forward metrics, $(F_i)_{i=0,dc-2}$ and backward metrics $(B_i)_{i=1,dc-1}$ respectively are calculated sequentially with a conditional equation as follows:

$$a' + \alpha^{h_{v_i c}} a'' = a \qquad (1)$$

where $h_{vi,c}$ is the nonzero element in **H** matrix.

In the merger step, the merger messages are C2V messages, which are updated for the posteriori messages. When the check node degree is equal to d_c that is after finishing the forward and backward processing, two vectors of merger processing are found such as $M_{\theta v_0}(a)$ and $M_{\theta v_{d_c-1}}(a)$. It is remarked that $M_{\theta v_0}(a)$ is equal to $B_1(a)$ and $M_{\theta v_{d_c-1}}(a)$ is equal to $F_{d_c-2}(a)$. The FB messages, which are already multiplied with nonzero elements of **H** matrix in FB processing in the equation (1), are used directly to generate other merger vectors. Therefore, a new conditional equation (2) and a computing merger metrics (4) are proposed for the merger step in the decoding algorithm as follows:

Conditional equation for merger step:
$$a' + a'' = a \qquad (2)$$

Merger metrics:
$$M_{\theta v_0}(a) = B_1(a);\ M_{\theta v_{d_c-1}}(a) = F_{d_c-2}(a) \qquad (3)$$

$$M_{\theta v_k}(a) = \min_{\substack{a',a'' \in GF(q) \\ a' + a'' = a}} \left(\max \left(F_{k-1}(a'), B_{k+1}(a'') \right) \right) \qquad (4)$$

978-1-4799-5128-4/14 $31.00 © 2014 IEEE

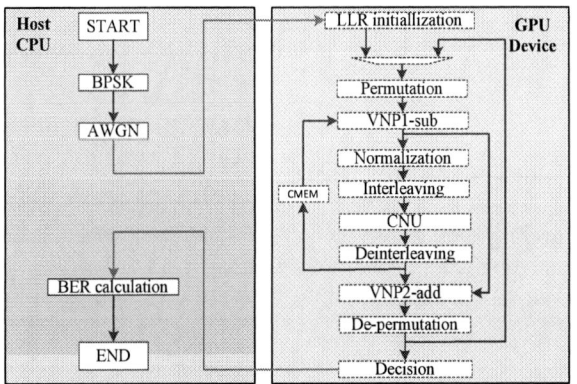

Fig. 1. Data flow of parallel block-layered NB-LDPC decoding on CPU and GPU platforms.

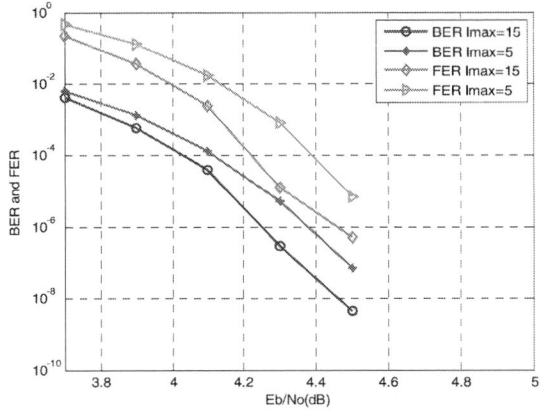

Fig. 2. BERs and FERs of a (744,651) NB-LDPC code over GF(2^5) using GPU.

Block-Layered NB-LDPC Decoder on GPU

NVIDIA GPUs are powerful arithmetic engines capable of running thousands of lightweight threads in parallel following a Single Instruction Multiple Thread approach. Furthermore, the NB-LDPC decoding algorithm satisfies a high computation to memory access ratio.

Fig. 1 shows data flow of layered decoder, which implements the decoding on the CPU and GPU platforms. Each of module in GPU device corresponds to one CUDA kernel. The host CPU transfers the data to/from the GPU device. Most of computations are implemented on GPU and all the intermediate messages are stored in the device memory to restrict data transfer between host and device.

Experimental Results

The experimental setup to evaluate the performance of the proposed NB-LDPC decoder consists of NVIDIA GTX650Ti GPU with 768 CUDA cores, 1024 MB of GDDR5 device memory, and an Intel(R) Core (TM) i7-4770 CPU with 16 GB RAM running at 3.4 GHz.

The simulation results are implemented over an AWGN channel with binary phase-shift-keying (BPSK) modulation. The bit error rate (BER) and frame error rate (FER) performances under the number of different iterations are shown in Fig. 2. Experiment results show that the proposed

TABLE I
EXPERIMENTAL RESULTS

	Proposed		[4]
Code	(744,651)		(744,651)
No. iterations	15	15	15
Program	C++	CUDA C	C++
Run time (ms)	646	43.07	-
Coding gain at 10^{-5} BER	4.13	4.13	4.3
Coding gain at 10^{-8} BER	-	4.43	-

NB-LDPC decoder with 15 iterations can obtain 4.3-dB coding gain at 10^{-5} FER and 4.13-dB coding gain at 10^{-5} BER, which is approximately 0.17 dB higher coding gain at 10^{-5} BER compared to Fang Cai's result for the same code length over $GF(2^5)$ [4]. Moreover, the GPU accelerate to achieve the coding gain under low 10^{-8} BER within hours, instead of weeks of computation.

Table I shows the experiment results using CPU and GPU. The execution times were obtained with CPU timers. In decoding on GPU process, the running time achieves 43.07ms. The GPU-based implementation using CUDA C program provides 15 times higher throughput than CPU-based implementation using C++ program with the long code (744, 651).

Conclusion

This paper presents a novel MMMA and an efficient implementation of a parallel block-layered NB-LDPC decoder based on a GPU. Due to its inherently massive parallelism, a NB-LDPC decoder is more suitable for a GPU implementation than for binary LDPC codes. The proposed MMMA provides a better BER performance than previous algorithm. The experimental results show that the GPU-based implementation of the proposed NB-LDPC decoder provides higher throughput and the coding gain under low 10^{-8} BER compared to CPU-based implementation.

Acknowledgment

This work was supported by the IT R&D program of MOTIE/KEIT [10044092] and by the MSIP, Korea, under the ITRC support program (NIPA-2014-H0301-14-1042) supervised by the NIPA.

References

[1] C. D. Matthew, and D. MacKay, "Low density parity check codes over GF(q)" IEEE Communications Letters, vol. 2, no. 6, pp. 165-167, Jun. 1998.
[2] B. Zhou, et al, "Construction of nonbinary Quasic-cyclic LDPC codes by arrays and array dispersions," IEEE Trans. on Communications, vol. 57, no. 6, pp. 1652-1662, Jun. 2009.
[3] V. Savin, "Min-Max decoding for nonbinary LDPC codes," In Proc. IEEE. Int. Symp. Inf. Theory, Toronto Canada, pp. 960-964, Jul. 2008.
[4] X. Zhang, and F. Cai, "Efficient partial-parallel decoder architecture for quasi-cyclic nonbinary LDPC codes," IEEE Transactions on Circuits and Systems I, vol. 58, no. 2, pp. 402-414, Feb. 2011.

Area-Efficient FFT Processors for OFDM Systems

Sung Kyung Shin and Myung Hoon Sunwoo, *Fellow, IEEE*

School of Electrical and Computer Engineering Ajou University
San 5, Woncheon-Dong, Yeongtong-Gu
Suwon 442-749, Korea
ssk314@ajou.ac.kr, sunwoo@ajou.ac.kr

Abstract

This paper proposes an area-efficient 128-point fast Fourier transform (FFT) processor for OFDM systems. The modified decomposition method and complex constant multiplier can reduce the number of twiddle factor multipliers and its ROMs. Compared with the existing FFT processor, the proposed processor can reduce the number of multipliers by 30% and the size of ROM by 62% respectively, without increase of delay elements.

Keywords - FFT; OFDM; pipeline; mixed-radix

Introduction

Multiple input multiple output orthogonal frequency division multiplexing (MIMO-OFDM) systems are powerful in enhancing communication capacity without bandwidth increase. As multiple spatial streams increase, the hardware complexities in MIMO-OFDM systems also considerably increase. It is important to design a MIMO-OFDM transceiver with minimal hardware cost and power dissipation. A fast Fourier transform/inverse fast Fourier transform (FFT) processor shows large hardware complexity in a MIMO-OFDM transceiver [1]. Hence, it is challenging to design the FFT processor as efficiently as possible. There are various FFT architectures such as memory, pipelined and array architectures. To compute k spatial streams, the most area-efficient approach is using a radix-k multipath delay commutator (RkMDC) which is one of the pipelined architectures [2].

This paper proposes a design of an area-efficient 128 FFT processor for OFDM systems. To deal with eight spatial streams and efficiently support 128-points, the proposed FFT architecture combines R8MDC, R4MDC and R2 single-path delay feedback (SDF) architectures. In addition, to reduce the implementation area, the proposed FFT processor employs complex constant multipliers.

FFT decomposition method

The discrete Fourier transform (DFT) of length N is expressed as follows:

$$X(k) = \sum_{n=0}^{N-1} x(n) W_N^{nk}, \qquad k = 0, 1, \cdots, N-1 \quad (1)$$

The subscript, n is the time index, and k is the frequency index. W_N^{nk} denotes, $e^{-j2\pi nk/N}$, the Nth primitive root of unity with its exponent being evaluated modulo N. To compute 128-point FFT, applying a 4-dimensional linear index map using radix-8, radix-4, and radix-2 algorithms, as follows:

$$n = \langle 16n_1 + 4n_2 + 2n_3 + n_4 \rangle_{N=128},$$
$$k = \langle k_1 + 8k_2 + 32k_3 + 64k_4 \rangle_{N=128}, \qquad (2)$$
$$n_1, k_1 = 0, 1, \dots, 7; \quad n_2, k_2 = 0, 1, 2, 3; \quad n_3, k_3 = 0, 1.$$

Using (2), (1) can be rewritten as

$$X(k_1 + 8k_2 + 32k_3 + 64k_4)$$
$$= \sum_{n=0}^{127} x(n) W_{128}^{(k_1+8k_2+32k_3+64k_4)(16n_1+4n_2+2n_3+n_4)},$$
$$= \sum_{n_4}^{1} \left[\sum_{n_3}^{1} \left[\sum_{n_2=0}^{3} \left\{ \sum_{n_1}^{7} x(16n_1 + 4n_2 + 2n_3 + n_4) \overbrace{W_8^{n_1k_1} W_{32}^{n_2k_1}}^{stage1} \right\} \right. \right.$$
$$\underbrace{(-j)^{n_2k_2} W_{128}^{(n_3k_1+8n_3k_1)}}_{stage2} \left. \overbrace{(-1)^{n_3k_3}(-j)^{n_4k_3}}^{stage3} \right] \underbrace{(-1)^{n_4k_4}}_{stage4} \right] \quad (3)$$

Equation (3) implies that the twiddle factor multiplications are required at stage 1, stage 2, and stage 3. However, the twiddle factor multiplications at stage 3 are the trivial multiplications $-j$ which can be implemented with no extra hardware [3]. Thus, the proposed FFT processor using the decomposition method in (3) can support the 128-points without twiddle factor multiplications at stage 3.

Proposed FFT Processor

A. Delay elements

In [1], the FFT processor employed dual-port RAMs which are used in the input and output memory units for 8×8 MIMO systems. Because the delay elements becomes no longer small as decreasing complex multipliers, for area-efficient approach, using RAMs is more effective instead of using the delay elements for input and output reordering [1]. In addition, the FFT processor in [1] minimized the delay elements for

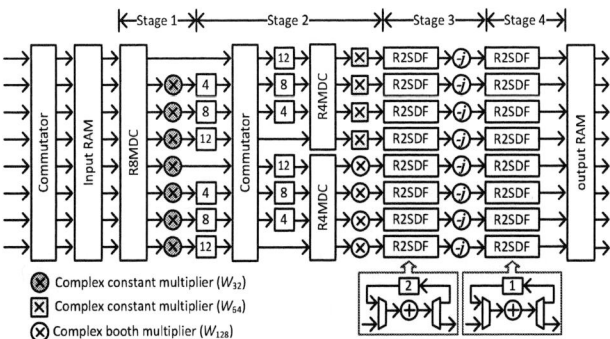

Fig. 1 Block diagram of the proposed FFT processor

commutator. As a result, the number of required delay elements is 120. The proposed FFT processor shown in Fig. 1 applies the input and output RAM architectures and input RAM scheduling scheme in [1]. As shown in Fig. 1, the proposed FFT processor applies the modified decomposition in (4) without increasing the number of required delay elements compared to that of [1].

B. Complex Constant multiplier

In FFT, the twiddle factor multiplier consisted of a ROM and a complex multiplier occupies more area than the other mathematical operators. However, if the twiddle factor requires a small number of coefficients, then the complex constant multiplier can be used for the twiddle factor multiplications. The twiddle factors, W_8, W_{16}, W_{32}, and W_{64} can be implemented in the complex constant multipliers [4], [5]. The constant complex multiplier requires lower hardware cost than that of complex multiplier. In addition, the complex constant multipliers do not need ROMs.

As shown in (4), the proposed FFT processor requires the twiddle factor multiplier to compute W_{128}^i at stage 2 where i is 0 to 127. In the proposed FFT processor, the commutator at stage 2 changes the output data sequence order of stage 1. The changes are explained as follows:

- The data sequence is ordered for R4MDC operation.
- if i of W_{128}^i is even, pass on the first R4MDC at stage 2.
- if i is odd, pass on the second R4MDC at stage 2.

Because the W_{128}^i denotes the $e^{-j2\pi i/128}$, The twiddle factor multipliers for W_{128} at the first R4MDC can be replaced by W_{64} complex constant multiplier, as shown in Fig.1.

C. Comparisons

Table II shows the required normalized area for total twiddle factor multiplication, ROM size and the number of delay elements. Compared with the existing FFT processor [1], the hardware requirement for the twiddle factor multiplication is only 70%. In addition, the complex constant multiplier can be implemented without the twiddle factor ROMs. Thus, the proposed FFT processor can reduce 62% of the ROM size compare with that of [1].

TABLE II.
HARDWARE COMPLEXITY COMPARISION OF FFT PROCESSOR

Architecture	[1]	proposed
No. of Delay element	120	120
No. of CBM[a]	14	4
No. of CCM[b] (W_{32}) (NA[c]=0.46 [4])	-	7
No. of CCM[b] (W_{64}) (NA[c]=0.65 [5])	-	4
Total NA[c] of Complex multiplier	14	9.82
Twiddle factor ROM size (Word length :10bits)	3,360 bits	1,280 bits

[a]CBM denotes the complex booth multiplier.
[b]CCM denotes the complex constant multiplier.
[c]NA denotes the normalized area.

Conclusion

The 128-point FFT processor for OFDM systems is proposed in this paper. The number of twiddle factor multipliers and twiddle factor ROMs are reduced using the modified decomposition method and the complex constant multipliers. This idea can be used for the design of the FFT processor for eight spatial streams in wireless OFDM systems which require lower power and area efficiency.

Acknowledgment

This work was partly supported in part by the IT R&D program of MOTIE/KEIT (10044092, Development of Core IPs of OFDM PHY and RF Transceiver for 60GHz Wireless LAN/PAN in Application of 7Gbps Wireless Multimedia Services) and in part by the Mid-Career Researcher Program through an NRF grant funded by the MSIP (2014R1A2A2A01002952).

References

[1] Yoshizawa, Shingo and Yoshikazu Miyanaga, "Design of Area-and Power-Efficient Pipeline FFT Processors for 8×8 MIMO-OFDM Systems," *IEICE trans. Fundamentals of Electronics, Communications and Computer Sciences.*, vol. 95, no. 2, pp. 550-558, Feb. 2012.

[2] T. Snasaloni, A. Perez-Pascual, V. Torres and J. Valls, "Efficient pipeline FFT processors for WLAN MIMO-OFDM systems," *Election let.*, vol. 41, no. 19, pp. 1043-1044, Sept. 2005.

[3] T. Widhe, J. Melander and L. Wanhammar, "Design of efficient radix-8 butterfly PEs for VLSI," in *Proc. IEEE International Symposium on Circuits and Systems (ISCAS)*, 1997, pp. 9-12.

[4] Cho, T. and Lee, H., "A High-speed low-complexity modified radix-25FFT processor for high rate WPAN applications," *IEEE Trans. Very Large Scale Integr. Syst.*, vol. 21, no. 1, pp. 187-191, May 2013.

[5] Fang-Li Yuan, Yi-Hsien Lin, Chih-Feng Wu, Muh-Tian Shiue and Chorng-Kuang Wang, "A 256-Point Dataflow Scheduling 2×2 MIMO FFT/IFFT Processor for IEEE 802.16 WMAN," in *proc. IEEE Asian Solid-State Circuits Conference (A-SSCC)*, pp. 309-312, Nov. 2008.

Implementation of Multi-Standard Digital Radio Single-Chip

Se-Ho Park, Kyung-Taek Lee

Korea Electronics Technology Institute
11, World cup buk-ro 54-gil, Mapo-gu,
Seoul, Korea
(Tel) +82-2-6388-6620, (Fax) +82-2-6388-6629, sehopark@keti.re.kr

Abstract

Digital radio standards can greatly enhance the quality and efficiency of radio broadcasting compared to analog AM/FM radio. Digital radio standard adoption is taking place in many countries around the world. The adoption process has been rather slow due to the different local frequency and broadcasting policies. Each country is trying to select a standard that best suits their interests. Socio-economic impact factors are also taken into account when choosing the digital transition period. As a result, the development of digital radio receiver chipsets has been rather stagnant over the past years, with relatively slow progress. In this paper, the design and implementation of a single chip that can receive and process multi-standard digital radio broadcasts is introduced. The standards covered include Digital Radio Mondiale (DRM/DRM+), Digital Audio Broadcasting (DAB/DAB+), and HD Radio.

Keywords- Digital Radio chipset; DAB; DRM; HD Radio;

Introduction

Digital radio is the next-generation radio technology that will be replacing analog AM and FM radios following the digital transition. Digital Radio Mondiale (DRM), Digital Audio Broadcasting (DAB), and HD Radio are digital radio standards currently used throughout the globe. These technologies have several functional similarities and common factors but efforts for standards interoperability have been relatively slow. As a consequence, consumers have to deal with incompatibility issues when accessing Digital Radio services. The problem is even more complicated for tightly clustered regions like Europe where different digital radio standards are being simultaneously proposed, developed, tested and deployed.

Implementation

A. Baseband Demodulator Block

Figure 1 shows the high level architecture of the Baseband Demodulator that takes care of the data decoding for DRM, DRM+, DAB, DAB+, and HD Radio. It consists of a Baseband Decoder and Baseband Processor. The Baseband Decoder block extracts the digital data from the RF signal received for each of the different standards. The Baseband Processor

controls the Baseband Decoder and extracts the audio signal from the digital data. The Baseband Decoder block consists of 100% hardware logic. The Baseband Processor includes a DSP Core which enables software changes and updates.

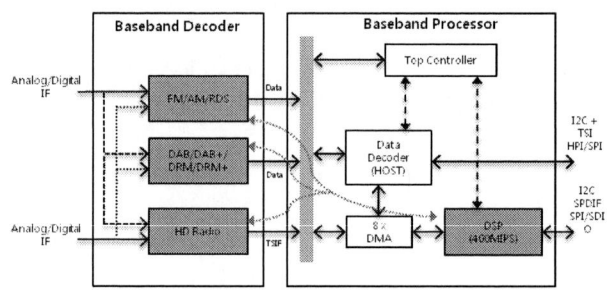

Fig. 1 Baseband Demodulator architecture

B. RF Block

The RF block implemented in this paper is capable of analog AM and FM radio reception. This was a design consideration and decision to enable the use of both analog and digital broadcasting technologies for service compatibility during digital transition periods. Four LNAs are implemented by the RF Tuner Chip, so that amplification of the signal from different frequency bands ranging from a few kilohertz to 1.6 GHz is possible. In order to eliminate the noise and distortion present in the analog signal, the digital signal is processed in the IF band.

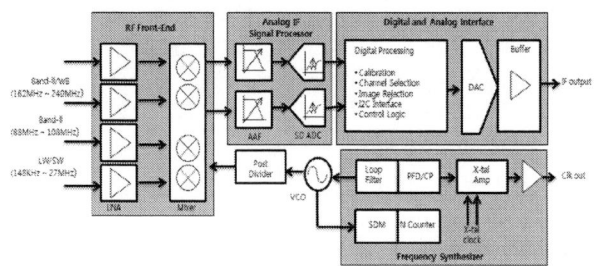

Fig. 2 Overall architecture of the RF Tuner Module

The RF front-end noise figure was designed with target performance of under 4 dB with an additional 1 dB margin. This enables the reception of analog signals from different frequency bands. The 1 dB margin was added to take performance degradation during chip fabrication into account.

978-1-4799-5128-4/14 $31.00 © 2014 IEEE

The design margin for RF Tuner chip's gain (P1dB) and linearity performance (IIP3) was also secured.

C. BER Performance

Figure 3 shows the BER measurements of the implemented Baseband Demodulator chip for digital radio. The BER measurements show the decoding performance of the Baseband Demodulator chip. The C/N ratio measured was 6.9 dB when receiving DAB/DAB+ and 20.1 dB when receiving DRM/DRM+.

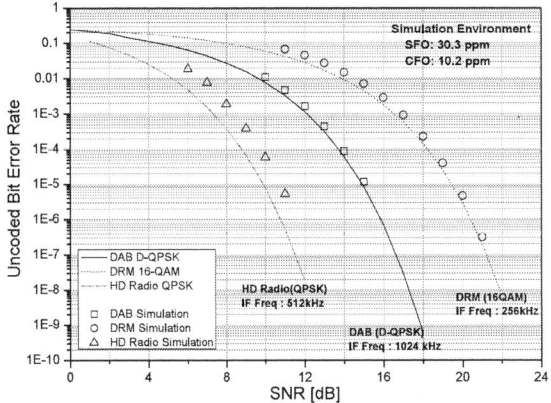

Fig. 3. Performance of the worldwide baseband decoder for Digital Radio standards

D. RF+Demodulator Single Chip

The multi-standard digital radio receiver is composed of RF module, baseband decoder, and DSP Core. Figure 4 depicts an architecture of the designed single chip. RF block supports bandwidths for AM/FM, DAB and DRM alike where the received signals are then converted to 12bit digital values and delivered to a baseband block. HW core for baseband block is capable of decoding the received signals from the RF block. Decoding support for processing analog AM and FM is additionally integrated with DAB/DRM/HR Radio signal processing in the baseband decoder. DSP Core is a 24bit processor which controls various interface devices including USB devices and partly supports the baseband decoder.

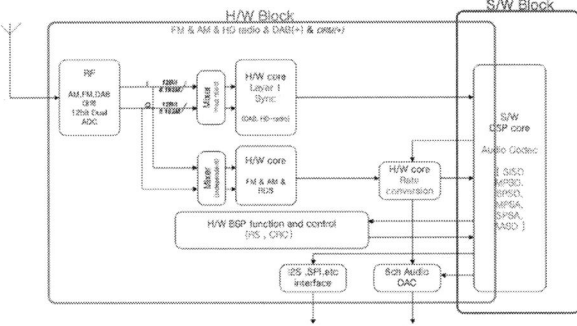

Fig. 4. Overall architecture of Multi-standard Digital Radio Single-Chip

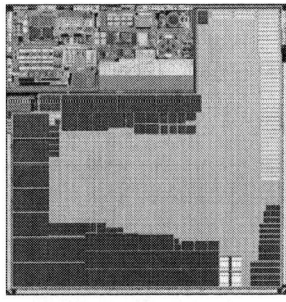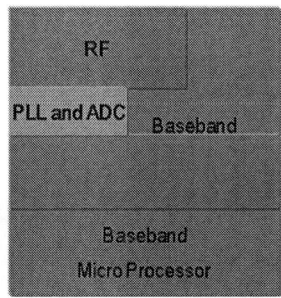

Fig. 5. Layout of Multi-standard Digital Radio Single-Chip

Figure 5 shows a layout of the implemented single chip with both RF block and Demodulator block included for receiving multi-standard digital radio signals. To minimize noises, RF block is strategically positioned into the upper area of H/W Block where as DSP Core is positioned to the bottom. The implemented single chip is fabricated in 65nm CMOS with a size of 8mm by 8mm.

Conclusions

In this paper, a single chip for receiving multi-standard Digital Radio signals (DAB/DAB+, DRM30/DRM+, HD Radio) is introduced. The implemented single chip is deliberately designed to share various hardware blocks used in different standards for minimizing power consumption which makes it an ideal use case for power-constrained mobile settings.

Acknowledgment

This work was supported by the IT R&D program of MSIP/IITP. [10039196, Smart Platform Development for Integrating Worldwide Radio Technology to Smart Devices]

References

[1] ETSI Standard. Digital Radio Mondiale (DRM); System Specification. ETSI ES 201 980, V3.1.1, 2009-08

[2] "Radio Broadcasting Systems; Digital Audio Broadcasting (DAB) to mobile, portable and fixed receivers," ETSI EN 300 401 v1.4.1, Jan, 2006

[3] Seong-Jun Kim, Kyung-Won Park, Kyung-Taek Lee, and Hyung-Jin Choi, "Digital Tuner Implementation Using FM Tuner for DRM Plus Receivers," IEEE International Conference on Consumer Electronics , 2012

[4] Wang, K.P. "RF front-end ICs for digital radio broadcasting DRM and DAB," International Conference of EDSSC. 2009, pp. 39-43.

Combined Channel Estimation

Kyuhoon Lee, Ik-Joon Chang, Jinsang Kim, and Seungjoo Lee*

Department of Electronics and Radio Engineering, Kyung Hee University, Rep. of Korea
* Department of Electronics Engineering, Hyejeon University, Rep. of Korea
jskim27@khu.ac.kr

Abstract— This paper proposes a fast channel estimation (CE) scheme for the OFDM based IEEE 802.11ad wireless local area network (WLAN). The LTF (LTF) -based CE provides better BER (bit error rate) performance at slow fading channel, whereas the pilot-based CE approach is good at fast fading channel. Hence, it is desirable to use both CE methods for higher performance efficiently. In the proposed CE scheme, two CE methods are adaptively selected using the channel quality indicator (CQI) for fast and high performance CE. The proposed CE scheme can increase the CE processing speed with negligibly performance degradation.

I. INTRODUCTION

Multi-gigabit wireless communications has been an emerging next generation WLAN technique. A WLAN IEEE 802.11ad standard uses 7GHz unlicensed bandwidth at 60GHz band [1]. Therefore, IEEE 802.11ad channel suffers from severe inter-symbol interference (ISI). To compensate the data distortion due to the ISI, channel estimation and equalization [2] should be done with better BER performance and higher speed for Gbps throughput. In [3]-[4], the CE performance was improved by jointly performing CE and other algorithms in a receiver. In [5]-[6], the CE performance was improved by adjusting the preamble information. Those approaches focus on the CE performance improvement. In [7], the CE computational complexity is reduced by removing redundant operations for high speed.

We propose a new CE algorithm that can be applied to the 802.11ad WLAN. Due to the inter symbol interference (ISI) caused by the wide bandwidth of IEEE 802.11ad standard, data can be easily distorted during data transmission. For fast and high performance CE, we propose a channel quality indicator (CQI) to select a CE scheme among the two CEs. If the CQI shows that a channel is slow fading, LTF-based CE is used whereas the pilot-based CE is selected since it is robust to ISI for fast fading channel. If a channel is in between slow and fast fading, an efficient scheme combining both LTF and pilot signals is developed.

II. PROPOSED FAST CHANNEL ESTIMATION

In this paper, we propose a scheme to select better CE method in terms of speed and performance. The proposed CE scheme is controlled by the CQI, ρ_c which predicts channel environment as follows:

$$\rho_c = \frac{\sum_{k=0}^{n}(h_{LTF}(k) - \mu_{h_{LTF}})(\delta(k) - \mu_d)}{\left|\sum_{k=0}^{n}(h_{LTF}(k) - \mu_{h_{LTF}})\right|\left|\sum_{k=0}^{n}(\delta(k) - \mu_d)\right|}, \quad (1)$$

where $\mu_{h_{LTF}}$ and μ_d are the mean of $h_{LTF}(k)$ and $\delta(k)$, respectively. If ρ_c is closer to 1, $h_{LTF}(n)$ is closer to the ideal channel and the channel can be estimated as slow fading. If ρ_c is closer to 0, the channel is fast fading and time-varying. In order to find threshold values of ρ_c for classifying the channel state, we did experiments at the various channel environments. In static LOS channel environment, the CQI value is from 0.99 to 1. At NLOS channel environments, the CQI values have wider range from 0.58 to 0.95 as shown in Table 1. For this experiment, we used the TGad channel model [1] and generated various NLOS channels by changing the seed values which initialize the values of scrambler at the channel model.

Table 1. CQI vs proposed CE method

CQI	predicted channel	proposed CE
$\rho_c > 0.95$	slow fading	LTF-based CE
$0.8 \leq \rho_c \leq 0.95$	in between	Combined LTF-based CE and pilot-based CE
$\rho_c < 0.8$	fast fading	pilot-based CE

From the above analysis, in case $0.8 \leq \rho_c \leq 0.95$, it is better to combine both CE methods for high speed. The conventional combined CE algorithm [10] uses an LTF-based CE method as an initial estimate and updates CIR using pilot-based CE result. The proposed combined CE method performs LTF-based CE and pilot-based CE at only the pilot subcarriers (16 subcarriers in IEEE 802.11ad) simultaneously and then computes the frequency response ratios at the pilot subcarriers.

III. SIMULATION RESULTS AND ANALYSIS

A. Simulation Results

We used the TGad channel model providing LOS and NLOS environments. For simplicity, QPSK is used as the modulation. The FFT length is 512 with the cyclic prefix (CP) length of 128 at the bandwidth of 2.64GHz. No channel coding is used and data packet size is 643,072bit.

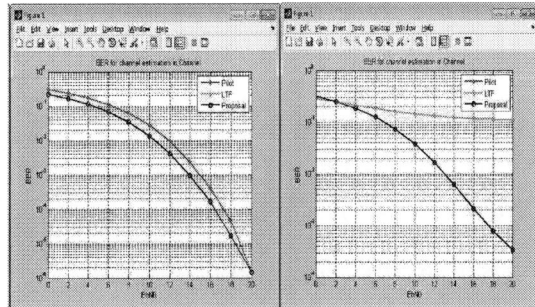

Fig.3. BER Performance in case $\rho_c = 0.9805, 0.7054$

The CE performance results at different channel environments are shown in Fig. 3. The pilot-based CE needs many operations due to the interpolation technique to predict CIRs at non-pilot subcarriers. We found that the interpolation module of the pilot-based CE takes most of area in the pilot-based CE module. Generally for high performance CE, both CE modules are embedded at the IEEE 802.11ad receiver. Compared to the pilot-based CE, LTF-based CE module takes 11% area.

Table 2 Performance vs processing time

ρ_c	loss(dB)	relative processing time (%)
0.946	0.1	76.25
0.8874	1.1	61.25
0.9127	0.15	46.875
0.9146	1.3	85.625
0.8348	0	100
0.9402	0.7	78.125

Table 2 shows the performance degradation and the processing time in case of the combined CE method when $0.8 \leq \rho_c \leq 0.95$. The performance degradation and processing time are relatively compared with those of the pilot-based CE: in case $\rho_c = 0.8348$, the proposed combined CE uses only the pilot-based CE. Therefore, there is no performance loss and the reduction of speed (100%). When $\rho_c = 0.946$, the proposed combined CE utilizes the faster LTF-based CE and the the reduction of the processing time is about 23.75%.

IV. CONCLUSION

We proposed a novel channel quality indicator utilizing the characteristics of the training sequences of the LTF. According to the value of the CQI, the proposed scheme selects a proper CE method among three types of CEs, namely LTF-based and pilot-based and the combined based CEs. Simulation results show that the proposed CE scheme can improve the processing speed up to 53% with maximum 1.5dB loss.

ACKNOWLEDGMENT

This work was partially supported by the National Research Foundation of Korea (NRF) grant funded by the Korea government (MSIP) (No. 2012R1A2A2A01011842) and the IT R&D program of MOTIE/KEIT [7Gbps modem IC design, 10044092]

REFERENCES

[1] http://www.ieee802.org/11/Reports/tgad_update.htm
[2] Singh,H.; Gill,S.S. "Approaches to Channel Equalization", 2012 Second International Conference on ACTT, pp. 172-175
[3] Zhiwei Lin; Xiaoming Peng; Chin, F. "Joint Carrier Frequency Offset and Channel Estimation for OFDM based Gigabit Wireless Communication System with Low Precision ADC", 2011 IEEE Vehicular Technology Conference (VTC Fall), Sept. 2011, pp. 1-5
[4] Jyh-Hau Chen; Yumin Lee, "Joint Synchronization, Channel Length Estimation, and Channel Estimation for the maximum Likelihood Sequence Estimation for High Speed Wireless Communication", 2002 IEEE 56th Vehicular Technology Conference, 2002. Proceedings. VTC 2002-Fall. pp. 1535-1539 vol.3
[5] Su Hu; Gang Wu; Gang Yang; Shaoqian Li; Bo Gao, "Effectiveness of Preamble Based Channel Estimation for OFDM/OQAM System", Networks Security, Wireless Communications and Trusted Computing, 2009. NSWCTC '09. International Conference on, April 2009, pp.34-37 vol.1
[6] hi Ying; Zhang Xiaolin; Shen Liran, "An improved channel estimation method of TD-SCDMA system", International Conference on Intelligent Computation Technology and Automation, March, 2011, pp. 439-442
[7] Haghighi, S.J.; Primak, S.; Xianbin Wang, "Effects of Side Information on Complexity Reduction in Superimposed Pilot Channel Estimation in OFDM Systems", IEEE 71st Vehicular Technology Conference, May, 2010, pp.1-5

Ultra-low Power 12T Dual Port SRAM for Hardware Accelerators

Bo Wang[1,2], Jun Zhou[2], and Tony T. Kim[1]

[1]VIRTUS, School of Electrical and Electronic Engineering, Nanyang Technological University, Singapore
[2]Institute of Microelectronics, A*STAR, Singapore
Email: thkim@ntu.edu.sg

Abstract

Dual-port SRAMs improve the performance of various hardware accelerators. This paper presents a low voltage 12T dual-port SRAM for biomedical hardware accelerators. The proposed dual-port SRAM cell -decreases the disturbance of the common-row-access mode for improving the worst case stability issue and realizing ultra-low voltage operation. In addition, hierarchical bitlines and a virtual ground technique are employed to further lower the power and minimum operating voltage and power consumption. A 16 Kb 12T dual-port SRAM was fabricated in a 65nm CMOS process technology and showed successful dual-port SRAM operation down to 0.4 V in the common-row-access mode.

Keywords: ultra-low power; dual-port SRAM

Introduction

Contemporary computing platforms with big data enable unprecedented interaction between human and computational resources. The ubiquitous computing necessitates high computing power with multi-core processing units and multi-port SRAMs. Dual-port SRAMs are also highly demanded and embedded even by energy-constraint ICs, whose low power dissipation mainly attributes to ultra-low voltage operation [1]-[5]. Hardware accelerators have been widely employed in various emerging ultra-low voltage computing circuits such as wireless sensor nodes, biomedical SoCs, etc. Multi-port memories operating at near-/sub-threshold regime are highly required in various hardware accelerators for achieving higher energy efficiency

Conventional 8T dual-port (DP) SRAM cells are derived from standard 6T single-port (SP) SRAM cells. They inherit the weakness of the 6T SRAM cells at low voltage operation like poor cell stability, reduced read-ability and write-ability, which impedes voltage scaling. The above issues are exacerbated because the 8T DP SRAM cell has 2 more access transistors leading to larger disturbance, especially in the common-row-access mode. This limits the minimum operating voltage (V_{min}) that is generally higher than that of the 6T SP SRAM cell. Various circuits have been proposed to either improve read-/write-ability of 8T DP cell or reduce read-write disturb under the common-row-access circumstance. A wordline-voltage-adjustment system is utilized in [1] to improve read and write against PVT variation. A priority row decoder with bitline shifter is proposed to circumvent the common access mode [2] for enhancing cell stability. It is obvious that the 8T DP SRAM cell has to employ assisting circuits to accommodate challenges from aggressive voltage and technology scaling.

Proposed 12T Dual-port SRAM

A. 12T Dual-port SRAM cell

Dual-port SRAMs boost computation performance and throughput by doubling the number of simultaneous memory access. Fig. 1 depicts the conventional 8T DP SRAM cell. Port A and B are accessed by exclusive address and operation instruction. Each port consists of their corresponding wordline (WL) and a pair of bitlines (BL and /BL). In the common-row-access mode, the selected cell is inevitably disturbed through the second activated WL. Thus, the width of the 2 NMOS drive transistors has to be further expanded (e.g. ×2.7) to maintain cell stability, which is much larger than that in the 6T SP SRAM cell.

Fig. 2 portrays the proposed 12T DP SRAM cell. The proposed cell decouples read paths from write path in each port. RWLA and RWLB control the access to read paths while WWLA and WWLB activate data writing. The conditional discharging of the read bitlines (RBLA and RBLB) is manipulated by VGND employed to suppress leakage. During read, the voltage level of RBLB represents the opposite value of node Q, hence it is connected to a global bitline via a PMOS transistor for inversion. By separating the read paths from the write paths, the amount of read-write disturbance is significantly relaxed and various design metrics such as

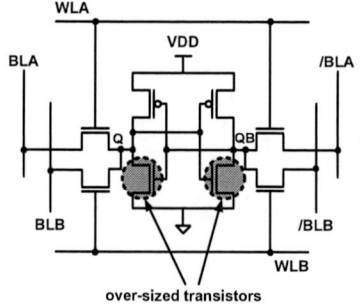

Figure 1. Schematic of conventional 8T dual-port SRAM cell.

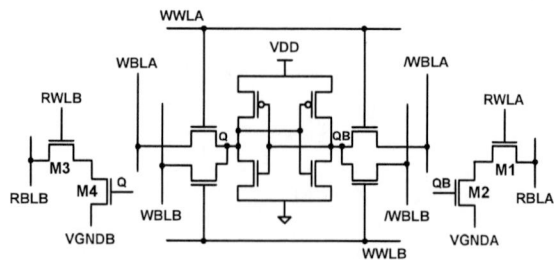

Figure 2. Schematic of the proposed 12T dual-port SRAM cell.

978-1-4799-5128-4/14 $31.00 © 2014 IEEE

Test Chip Implementation and Measurement

The microphotograph of the proposed DP SRAM is illustrated in Fig. 5. The proposed DP SRAM was fabricated in 65nm CMOS technology. Note that each port has dedicated peripheral circuits such as control logic, decoders, read-out circuits, I/Os, etc. The 16Kb array is configured by 256 rows × 64 columns. The test chip occupies the area of 398 μm × 385 μm. Table 1 summarizes the performance of the proposed dual-port SRAM.

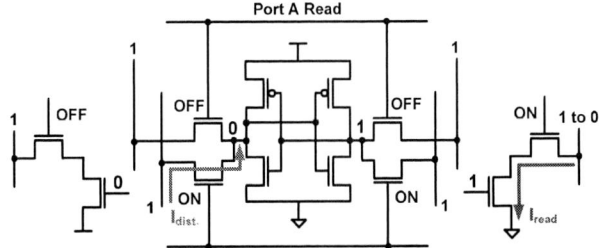

Figure 3. Worst case read disturbance in the proposed DP SRAM cell. Note that the cell is accessed for read through port A while port B is under the half-selected condition.

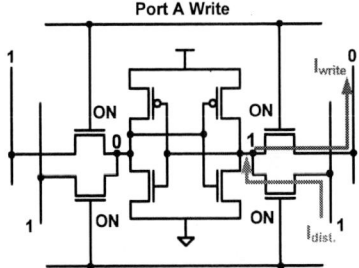

Figure 4. Worst case write disturbance in the proposed DP SRAM cell. Note that the cell is accessed for write through port A while port B is under the half-selected condition.

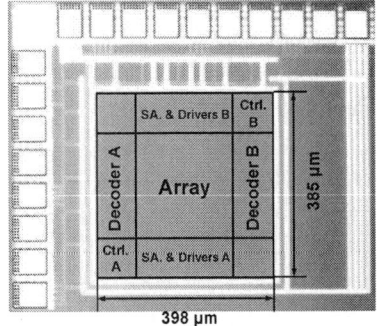

Figure 5. Microphotograph of the proposed DP SRAM.

TABLE I. SUMMARY OF TEST CHIP

Technology	65nm CMOS
Chip Size	439.9 x 437.2 μm²
Cell Size	3.82 x 0.72 μm² (logic design rule)
VDD min.	0.4 V @ 256 rows
Read Access Time	580 ns @ 0.4 V
Leakage Current	63 μA @ 1.2V

stability and read/write margins are improved. The proposed cell eliminates the necessity of over-sizing of the pull down devices while still maintaining performance. Therefore, similar layout area (e.g. ×1) can be achieved in the SRAM cell design.

B. Read Disturb Suppression

When the 8T DP cell [5] is replaced by the proposed 12T cell, the read-disturb is ameliorated in two ways. First, since the read and the write are separated, the proposed DP SRAM cell undergoes the same amount of the disturbance in the conventional 6T SRAM cell (Fig. 3). This improves the read SNM (SNM_1) of the proposed DP SRAM cell compared to the conventional 8T DP SRAM cell. Second, when two simultaneous read operations are executed in the same row, they are interference-free due to the decoupled read ports. The read SNM (SNM_2) in this condition is equal to the hold SNM.

C. Write Disturb Suppression

The worst case write disturbance in the proposed DP cell is depicted in Fig. 4. The worst write disturbance occurs when a cell is accessed for write operation through a port and the other port is in the half-selected condition. In the conventional 8T DP cell, a simultaneous operation from the other port brings in additional disturbance which prevents the node '1' from quickly being discharged and could result in write failure [4]. In the proposed DP cell, the probability of the worst case write disturbance is diminished since the worst case disturbance is generated only when two write operations are executed in one selected row. If a write operation is executed through port A and a read operation is executed through port B, RWLB is enabled instead of WWLB, where no disturb current flows from port B.

Acknowledgement

This work was supported by MediaTek University MPW program.

Conclusions

Dual-port SRAMs are substantial in various applications. This paper explains an ultra-low power near-threshold dual-port SRAM. Decoupled SRAM structures were proposed for improving read and write disturbance..

References

[1] M. Yabuuchi, et al., "20nm High-Density Single-Port and Dual-Port SRAMs with Wordline-Voltage-Adjustment System for Read/Write Assists," Inter. Solid-State Circuits Conf. (ISSCC) Dig. Tech. Papers, 2014, pp. 234-236.

[2] K. Nii, et al., "Synchronous Ultra-High-Density 2RW Dual-Port 8T-SRAM with Circumvention of Simultaneous Common-Row-Access," IEEE J. Solid-State Circuits, vol. 44, no. 3, pp. 977–986, Mar. 2009.

[3] Y. Ishii, et al., "A 28-nm Dual-Port SRAM Macro with Active Bitline Equalizing Circuitry against Write Disturb Issue," in Symp. VLSI Circuit (VLSIC) Dig. Tech. Papers, 2010, pp. 99-100.

[4] L. Chang, et al., "A 5.3GHz 8T-SRAM With Operation Down to 0.41 V in 65 nm CMOS," in Symp. VLSI Circuits (VLSIC) Dig. Tech. Papers, 2007, pp. 250-253.

[5] Y. Ishii, et al., "A 28 nm Dual-Port SRAM Macro with Screening Circuitry Against Write-Read Disturb Failure Issues," in Proc. Asian Solid-State Circuits Conf. (A-SSCC), 2010, pp. 1-4.

Understanding DDR4 in Pursuit of In-DRAM ECC

Sanghyuk Kwon, Young Hoon Son, and Jung Ho Ahn

Department of Transdisciplinary Studies, Seoul National University
Seoul, Republic of Korea
{kkwon114, yhson96, gajh}@snu.ac.kr

Abstract

Continuous DRAM scaling exacerbates problems caused by faulty cells, which lead to lower yields and more frequent errors. In-DRAM ECC is regarded as one of the solutions to overcome these issues. The latest DDR4 SDRAM specification includes new features to further improve reliability, such as an ALERT_n pad, which can be used to report errors detected by a SECDED code in DRAM. This paper identifies the possibilities and challenges of implementing In-DRAM ECC on DDR4 SDRAM devices.

Keywords-DDR4; ECC; SECDED; overhead characteristcs; In-DRAM ECC

Introduction

DRAM devices have exploited steady improvement of process technology over decades. Modern DDR4 SDRAM specification results in a normalized operating voltage reduction of 11% and up to twice peak bandwidth compared to the low voltage DDR3 specification. However, the aspect ratio of DRAM cell capacitors has increased to maintain the cell capacitance on a smaller footprint, which worsens process variation of the DRAM cells. Therefore, without additional treatments, DRAM scaling leads to lower yields. Populating spare rows and columns is a most commonly used technique to alleviate the yield drop. However, as the bit-error-rate (BER) of DRAM devices is increasing, the effectiveness of providing spare rows and columns becomes limited due to area and delay overheads [2].

DDR4 SDRAM supports CRC (Cyclic Redundancy Check) for data writes, Per DRAM Addressability, gear down mode, and CA (Command and Address) parity compared to DDR3 SDRAM that supports only ECC (Error Checking and Correction) at rank granularity for better RAS (reliability, availability, and serviceability). These new features incur latency overhead for parity processing around 4~5 clock cycles and read time (tAA) overhead around up to 2ns [1]. However, the CA parity and CRC features in DDR4 devices target transient errors during data transfers. None of the new features are for permanent or transient faults on the cells and peripheral circuitry of DDR4 devices. Therefore, with the continuous progress in DRAM scaling, providing an In-DRAM ECC feature, which detects and corrects errors within DRAM devices, gains more interest to cope with ever increasing BER. This paper examines the possibilities and challenges of implementing In-DRAM ECC on DDR4 SDRAM devices. We demonstrate that the overhead of supporting In-DRAM ECC depends on the length of a code word and is more prominent on ×4 DDR4 devices.

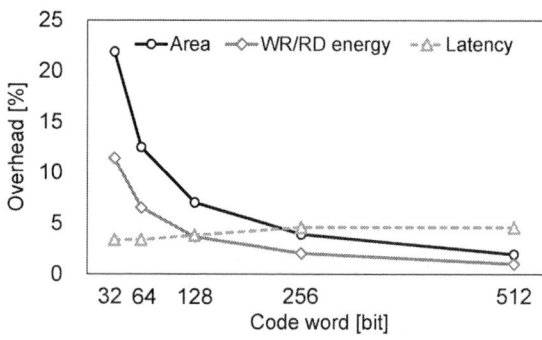

Fig. 1 Overhead of providing a SECDED feature with various codeword sizes in DRAM area, read/write energy, and latency

Exploitable features of DDR4 for In-DRAM ECC

A DDR4 SDRAM device utilizes an ALERT_n pad to report errors detected by CA parity and CRC during data writes for better reliability. The pad can also be used for In-DRAM ECC when it provides single-bit error correction and double-bit error detection (SECDED) within a group of data and parity bits composing a codeword. The data bits of a codeword consist of one or more bursts, each being a unit of data transfer. Single-bit errors can be corrected without any aid outside of a DRAM device, but double-bit errors must be reported because they are not correctable. The ALERT_n pad obviates the need of assigning an additional pad in the existing DDR4 specification.

The overhead of providing SECDED in a DRAM device

Providing SECDED incurs area, energy, and latency overhead. We assumed a 2.4Gbps 8Gb DDR4 device as a reference [1] and modified CACTI-D [6]. Figure 1 shows the overhead associated with the codeword size. When we increase the codeword size, the required parity bits are increased at a slower rate, reducing the portion of parity bits within a codeword and hence the area overhead due to parity bits, which is tightly coupled to DRAM area. As the size of a codeword is increased to 32, 64, and 128 bits (data only), the area overhead due to parity bits is reduced and saturated to 21.9%, 12.5%, and 7.0%. Also, as the codeword size excluding parity bits is increased to 32, 64, and 128 bits, the energy overhead within a DRAM bank due to parity bit extraction and codeword checking is reduced to 11.4%, 6.5%, and 3.7% compared to the baseline without parity.

Figure 2 illustrates In-DRAM ECC operations for reads and writes. We first assume that a codeword consists of one data burst and its corresponding parity bits. On a write, parity bits to support SECDED must be generated and then the pair of the data burst and the parity bits are written to the row buffer of a

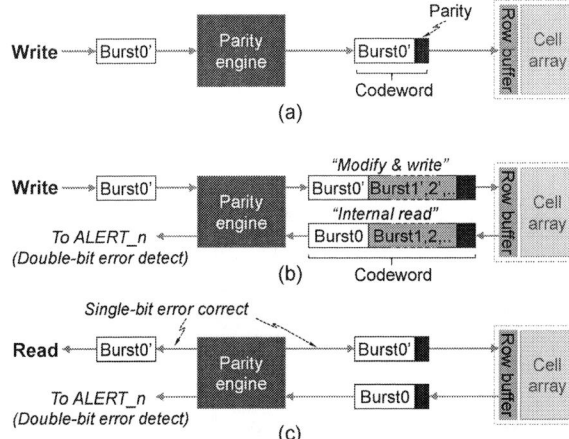

Fig. 2 In-DRAM ECC operations on a write when a codeword includes (a) one data burst and (b) multiple data bursts, (c) on a read

target bank (Figure 2(a)). In this case, an encoder for generating the parity bits is needed, which leads to an additional write latency for parity bit generation. On a read, the data and parity bits of a codeword are read first and then its integrity is tested by a parity checker (Figure 2(c)). If the checker detects a correctable error, it is corrected before leaving the device. In this case, parity bits are re-generated and the new codeword must be written to the row buffer. If In-DRAM ECC targets both transient and permanent faults, its implementation must perform this read-modify-write operation for a normal read, incurring a latency overhead, unless it supports variable latency read operations. If the checker detects a double-bit error, which is not correctable, it reports this through the ALERT_n pad. A data burst consists of 32, 64, and 128 bits on ×4, ×8, and ×16 DDR4 devices. In these cases, the parity bit sizes are 7, 8, and 9 bits, leading to an area overhead due to parity bits of 21.8%, 12.5%, and 6.3% (Figure 3). Their energy overheads are 11.4%, 6.5%, and 3.7% for writes to update additional parity bits.

The parity checking and generation logic parts for SECDED induce latency overheads on reads and writes because they are in the critical paths. Accordingly, write-to-read time (tWTR) and read time (tAA) are increased. However, their gate-level delays are not much sensitive to the codeword size [5]. Therefore, the latency overhead of checking and generating parity bits is less than 5% of a write or read operation with a transfer rate of 2.4Gbps per data pad.

The area overhead of In-DRAM ECC is apparently a burden to DRAM manufacturers. The area overhead is inversely proportional to the size of a data burst, which constitutes a codeword. One way to mitigate the area overhead due to parity bits is to make multiple data bursts constitute a codeword. Figure 3 shows the area overhead due to parity bits of ×4, ×8, and ×16 DDR4 devices when we increase β, the number of data bursts per codeword. With more bursts per codeword, the area overhead is reduced. The impact of increasing β on ×4 DDR4 devices is greater than that on ×8 and ×16 devices.

However, making multiple data bursts be a part of a codeword requires changes on normal writes. To perform a write, the parity bits and data to be overwritten should be read first and combined with the new data to update the parity bits of the codeword. Therefore, a write essentially becomes a read-modify-write and write-to-write time and write-to-read time to the same open row of a bank increase significantly. In summary,

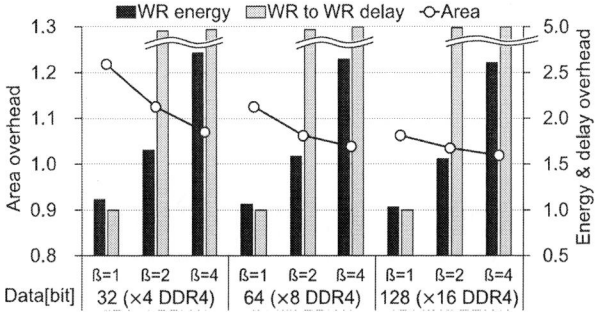

Fig. 3 Overhead of increasing the codeword size (β = the number of data bursts per codeword)

increasing the length of a codeword reduces the area overhead due to additional parity bits, but substantially exacerbates the latency penalty on DRAM writes. There is a proposal [3] to alleviate this latency overhead by provisioning more row buffers [4]. However, it still increases the write energy noticeably due to additional data to be read and processed. For example, when the codeword size of a ×4 device is increased by 2 and 4 times, the energy overhead is increased by 50% or more.

Conclusion

DDR4 SDRAM targets higher capacity and data transfer rates under lower voltage, and enhances reliability using write data CRC and command and address parity. However, DRAM scaling exacerbates process variation, leading to more frequent transient and permanent faults. Current DDR4 specification does not support In-DRAM ECC, but it is considered as a way to improve reliability further. We identified that the SECDED code for In-DRAM ECC has various degrees of impacts on area, energy, and latency of the devices depending on the codeword size, while the overhead is more prominent on ×4 DDR4 devices even with the technique to concatenate multiple bursts within a codeword. Therefore, we expect future studies to further alleviate the overheads of implementing In-DRAM ECC.

Acknowledgment

This work was supported by IDEC (EDA Tool) and the Future Semiconductor Device Technology Development Program (10044735) funded By MOTIE and KSRC.

References

[1] JEDEC, "DDR4 SDRAM Standard," [Online]. Available: http://www.jedec.org/standards-documents/docs/jesd79-4.

[2] S. O et al., "CIDR: A Cache Inspired Area-Efficient DRAM Resilience Architecture against Permanent Faults," Computer Architecture Letters, 2014, doi:10.1109/LCA.2014.2324894

[3] U. Kang et al., "Co-Architecting Controllers and DRAM to Enhance DRAM Process Scaling," In the Memory Forum with ISCA, 2014

[4] Y. Kim et al., "A Case for Exploiting Subarray-Level Parallelism (SALP) in DRAM," in ISCA, 2012

[5] S. Cha et al., "Efficient Implementation of Single Error Correction and Double Error Detection Code with Check Bit Precomputation for Memories," JSTS Vol. 12 No. 4, 2012

[6] S. Thoziyoor et al., "A Comprehensive Memory Modeling Tool and Its Application to the Design and Analysis of Future Memory Hierarchies," in ISCA, 2008.

Ethernet, SD Card and USB Linux Driver Porting on Aldebaran SoC System

Chan Kim, KyungJin Byun, NakWoong Eum

Electronics and Telecommunications Research Institute
218 Gajeong-ro, Yuseong-gu, Daejon, South Korea
{ckim,kjbyun,nweum}@etri.re.kr

Abstract

A partial new design and porting of linux device drivers for a commercial Ethernet chip, a proprietary SD card host controller, and open-source USB host controller for an SoC system called Aldebaran is described. Aldebaran is an SoC including a SPARCv8 based dual issue 13 stage pipeline processor core designed by ETRI. How the drivers work in relation to the kernel core is also explained with the porting procedures. The newly written SD controller driver breaks a single command for a scatterlist into many commands for single buffer to fake the kernel as if the host is processing scatterlist DMA.

The three peripherals and the drivers proved to work properly after some bug fixes. For driver porting with initial hardware and software bugs, the engineer needs to know many areas of the kernel like memory management and scheduling as well as each related kernel subsystem and this paper hopes to give a basic idea about the linux driver for those who are not familiar with the topic.

Keywords: Linux, device driver, Ethernet, SD card, USB, console

Introduction

SoC design involves many peripheral devices and the verification of the new devices usually takes a long time. It is difficult to find the bug when the operation of the device driver or the relevant kernel subsystem operation is not well understood beforehand. In this paper, three device driver portings to an SoC called Aldebaran is described. Aldebaran is an SoC including a SPARCv8 based dual issue 13 stage pipeline processor core designed by ETRI.

This paper tries to describe the driver architecture and the interaction of it with the kernel and how the driver is ported in a plain language so that those not acquainted with this area can easily understand what it is like.

Ethernet Driver Porting

Aldebaran System uses LAN9220 chip for Ethernet interface. By adding SMSC911X driver to the kernel configuration, the initialization routine is added to the init_call list. The source was configured to use 32 bit read/write because the SRAM interface the chip is attached to converts 32 bit access into two 16 bit accesses for the Ethernet chip range. MAC registers are accessed through indirect register access and PHY registers (MII registers) are again indirectly accessed using MAC registers. To send Ethernet frame, the data should be written to TX FIFO with two control words prepended and received frame is also read from the RX FIFO prepended with two control words. For each packet, a status word is inserted in the TX or RX Status FIFO.

The driver is connected to the network core which handles the main protocols including TCP/IP and has some functions registered in net_device_operations which are called by the network core. For packet transmission, ndo_start_xmit functions is used but for packet reception the registered NAPI poll function is called for burst packet read. In linux, the interrupt handler checks the receive FIFO status, and if data has been received, it disables the Rx interrupt and schedules this polling. The NAPI polling function repeatedly reads the rx packets until the given number of packets are read or the receive FIFO is empty. After processing, it enables the RX interrupt. This way frequent interrupt switching is avoided.

SD Card Driver Porting

Aldebaran System uses a proprietary SD/MMC host IP. Fig 1 shows the block diagram of the SD/MMC host controller. Basic SD card interface protocol will not be explained here.

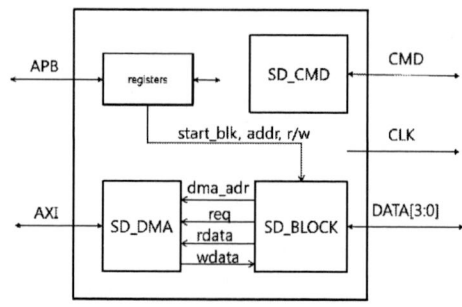

Fig. 1 SD/MMC host controller

For MMC/SD operation, linux kernel uses a data structure called mrq(mmc request). The structure contains information used to communicate with the SD card. Basically it contains the command, optional SD card start block address for the transfer and optional kernel buffer address for the data read and write(when data transfer is associated). Every command is passed to the driver using mrq and completed with response.

The Aldebaran MMC/SD module is added to the configuration together with relevant SCSI disk and file system support. The initialization routine initializes the tasklet for interrupt processing, then calls the slot initialization function which allocates and adds mmc_host to the system(as device). This includes starting the mmc_host, which again schedules the delayed work called mmc_rescan function which starts all the SD card initialization(detecting the SD card and linking it with SCSI driver and file system). The mmc_rescan schedules itself at the end for processing every second . This routine is also called initially by the card detection work(task).

For a single SD card command, many interrupts occur but most time consuming processing is deferred to the tasklet which is a dynamically registered lowest priority softirq defined in linux. Fig. 2 shows how the CMD with scatterlist buffer is handled by the tasklet using many CMDs for single buffer to get around the hardware limitation of not being able to process scatterlist in the DMA. State changes according to interrupts.

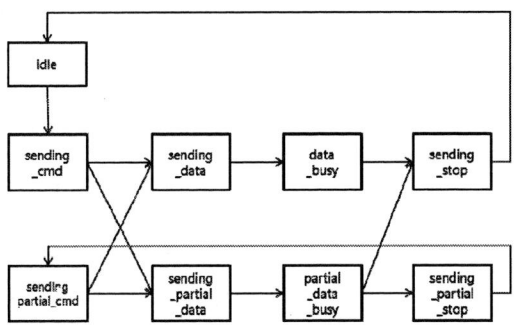

Fig. 2 SD card processing

USB Host Driver Porting

Fig.3 shows the block diagram of the USB1.1 host controller adopted from opencore.

In linux, the data transfers between the kernel and the USB host driver is made through the URB(USB request block). The kernel sends URB to the driver and the driver delivers the data between the kernel buffer and the USB device.

The driver provides several functions to the kernel that are registered to the hcd driver using function pointers. The enqueue function links the URB to the endpoint it is assigned to and optionally schedules the endpoint transfer and triggers the transfer start. In data write case, kernel data is written to the controller FIFO before the transfer but in read case, the controller FIFO data is read in the interrupt routine. When the request type is for the hub itself, the kernel handles it and partly calls the hub control functions for the hardware dependent cases. A registered hub control function is used to control the hub and ports and using separate setup data is linked to the URB for this purpose which is composed of type, value, index and length. If the URB is not for the hub, it is enqueued to the endpoint at the host.

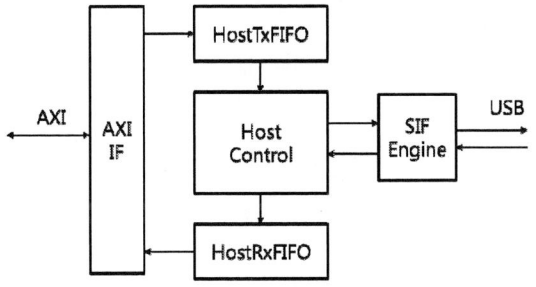

Fig.3 Opencore USB1.1 block diagram

The interrupt routine is called by the hardware interrupt but when the host controller is allocated, a timer is initialized and

every timeout, a registered root hub poll function is called where the interrupt routine is called. This status poll function enqueues the status to the URB so that it is processed by periodic interrupt URB processing.

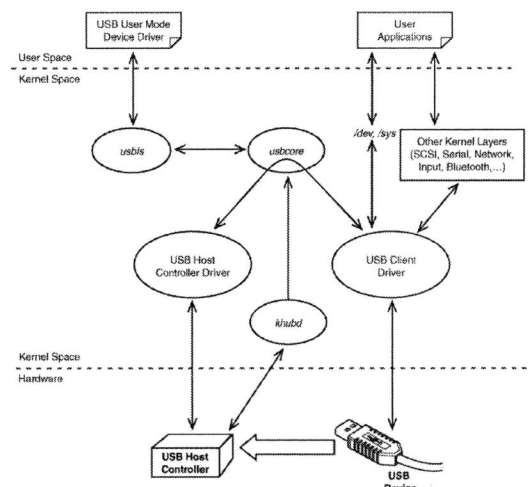

Fig. 4 Linux USB subsystem (Fig. from [2])

The USB subsystem is connected to various other upper subsystems as shown in Fig. 4. In our case, we tested USB mass storage and USB keyboard. The keyboard is connected to virtual console through serial input and tty device. The virtual console also uses frame buffer graphic console using LCD. When configured with frame buffer console, we have 3 virtual consoles on the LCD and the real console on serial port.

Conclusion

After configuration and parameters settings, and fixing some hardware and software bugs, the three drivers proved to work properly. When existing open-source driver is available, just checking read, write and interrupt could make the driver work. But when you don't have a matching driver or hardware has bugs, you need debugging and it requires knowledge about the device and driver itself and also related subsystems and kernel internals. For example the authors had to look into many areas of the kernel including memory management(for frame buffer DMA problem), network, file system(for SD card reading), tty, console, frame buffer, busybox etc. The drivers are being used to build processor emulation system based on QEMU to expedite software development with hardware modeling.

Acknowledgment

This work was supported by the IT R&D program of MSIP/KEIT [10042395, Development of SW emulation and rapid prototyping technology for high-performance SoC based on multicore]

References

[1] linux source v3.3 (http://www.kernel.org)
[2] S. Venkateswaran, *Essential Linux Device Drivers*, Prentice Hall, 2008.

Supporting Linux on Virtual Platform based on Versatile Express Board

Changmin Shin, Yongjoo Kim

Embedded Software Research Division
Electronics and Telecommunications Research Institute
Daejeon, Korea
cmshin@etri.re.kr, y.kim@etri.re.kr

Abstract

In general, when a embedded system is to be developed, the device drivers, operating system, and the application softwares are developed after the hardware of a embedded system is developed. If the hardware and software are developed at the same time, the development time and cost savings of the embedded system are possible.

This paper defines and designs the structure of a SW-SoC Convergence Platform that is a virtual platform for supporting the concurrent development of the hardware and software.

The hardware of the SW-SoC Convergence Platform is implemented using virtual prototypes for ARM Versatile Express Board that is widely used as an embedded reference system. The operating systems (i.e., Ubuntu Desktop V12.11, Ubuntu Nano V13.08) were ported to the virtual platform.

Accordingly, using the operating system ported to the SW-SoC Convergence Platform, the actual application software 's development is possible before the real hardware is developed.

Keywords-Virtual Platform; Intellectual Property; Transaction Level Model; SoC

I. INTRODUCTION

A virtual prototype is a fast functional software model of a system allowing for early software development before the real hardware is available. Virtual Prototypes are created using the industry-standard SystemC language and TLM 2.0 interoperability standard [1, 2].

Virtual Platform is composed of virtual prototypes of hardware IPs, the device driver for the virtual prototypes and the application software.

The EDA (Electronics Design Automation) tool for the establishment of a virtual platform includes Synopsys, Inc.'s Virtualizer and Imperas Software Ltd.'s OVP (Open Virtual Platform), and so on. Platform Architect and Virtualizer of Synopsys, Inc. support the graphical user interface for use of the library provided with the TLM model. OVP of Imperas Software Ltd. supports the composition of a fast instruction-level virtual platform.

The rest of this paper is organized as follows. Section 2 describes the details on how to comprise a virtual platform constructed in this paper. Section 3 is the implementation results of the constructed virtual platform. Section 4 presents conclusion in this paper.

II. SW-SOC CONVERGENCE PLATFORM

A. Definition of SW-SoC Convergence Platform

The SW-SoC Convergence Platform of this paper is a virtual platform that is composed of bootloader, linux kernel, file system and virtual prototypes of hardware IPs.

Using the SW-SoC Convergence Platform, the simultaneous development of hardware and software developers is possible. So the SW-SoC Convergence platform improves the flexibility of the product development. And it supports the configuration using virtual models designed by developers. Also it supports the simulation and verification for minimizing the failure of development.

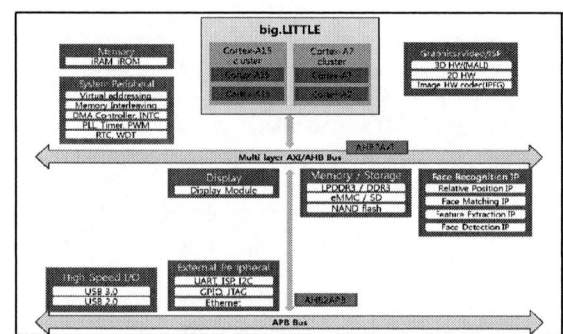

Figure 1. IP Constitution in SW-SoC Convergence Platform

B. Constitution of SW-SoC Convergence Platform

The proposed SW-SoC Convergence Platform is divided into four main blocks.

There is the operating system block supporting a cluster-based multi-core. The virtual prototype block is composed of a cluster-based multi-core, internal/external memory, AMBA bus, display, and system peripherals. The Face Recognition block is integrated into the virtual platform as TLM IP models. The external connection block is composed of camera IP and Wi-Fi IP. The constructed virtual platform can use the hardware resources of the external system that the virtual platform operates onto. The camera IP and Wi-Fi IP applied to the virtual platform are for use of the hardware of the external system.

Figure 2 shows the process of building the virtual platform using the elements of the blocks described above

978-1-4799-5128-4/14 $31.00 © 2014 IEEE

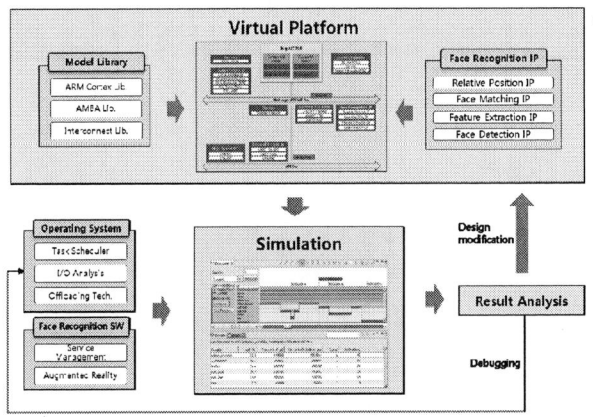

Figure 2. Development Process of Virtual Platform

III. IMPLEMENTATION RESULTS

A. SW-SoC Convergence Platform

Virtual Platform of this paper is based on virtual prototypes of ARM Versatile Express product[3]. ARM Versatile Express product is composed of a Motherboard, various CoreTile boards and LogicTile boards.

Virtualizer tool of Synopsys Inc. supports virtual prototypes of ARM Versatile Express product. Therefore we use Virtualizer tool of Synopsys Inc. to make and simulate virtual prototypes of ARM Versatile Express product.

The CoreTiles are ARM CPU subsystems boards(e.g. Cortex-A15 MPCore, Cortex big.LITTLE). And a user of ARM Versatile Express product can select one of these various CoreTiles. The LogicTiles are FPGA based prototyping boards and consists of USB2, USB3, Ethernet, UART and MobileStorage that is a MMC.

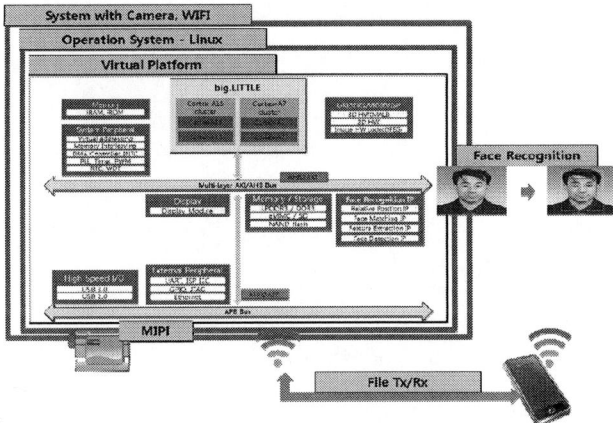

Figure 3. SW-SoC Convergence Platform

B. Supporting Ubuntu Desktop on Virtual Platform

Linux operating systems (i.e., Ubuntu Desktop V12.11, Ubuntu Nano V13.08) are ported on Virtual Platform based on ARM Versatile Express The CPU core of the constructed Virtual Platform is ARM big.LITTLE core. And the internal memory size of the Virtual Platform is 256MB. The size of the MMC external memory is 512MB. MMC(Multi-Media Card) is adopted as a mobile storage. The size of MMC is 512MB.

But the size was modified to support 4GB in size than 512MB. Then, after creating the file system of the Ubuntu Desktop and Nano, the file systems were loaded into MMC.

Figure 6 shows the simulation of Ubuntu Desktop ported on Virtual Platform based on ARM Versatile Express.

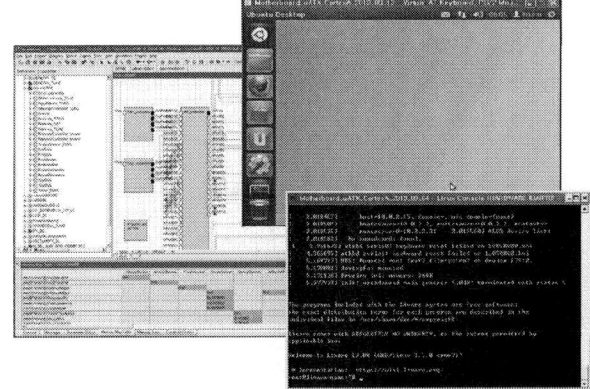

Figure 4. Simulation of Ubuntu Desktop on Virtual Platform

Linaro is open source software for ARM SoCs[4]. Ubuntu Desktop 12.11 in Linaro was adopted as the operating system for virtual prototypes of hardware IPs. Ubuntu Desktop 12.11 has many libraries for supporting a development of multimedia software.

We make a file system of Ubuntu Desktop 12.11 and boot up the operating system on the virtual prototypes of ARM Versatile Express product. And we run application software for the execution of mpeg file. Though the booting time of Ubuntu Desktop 12.11 is about one hour, the running speed of mpeg is sufficient for a development of multimedia software.

IV. CONCLUSIONS

To reduce the development period and cost of embedded systems, the virtual platform that supports the simultaneous development of hardware and software is defined and constructed.

The proposed virtual platform has been built using the ARM Versatile Express Board's virtual prototypes. Ubuntu Desktop V12.11 and Ubuntu Nano V13.08 that is a operating system has been ported to the virtual platform. Therefore, it is possible to develop software before the development of hardware using the constructed virtual platform in this paper.

Acknowledgment

This research was supported by the Technology Innovation Program (No. 10044313) funded by Ministry of Science, ICT and Future Planning of Korea.

References

[1] Synopsys, Inc., CoStart for VDK User's Guide, Version H-2013.06, June 2013

[2] Synopsys, Inc., Virtual Prototype Software Analysis User's Guide, Version H-2013.06, June 2013

[3] ARM Ltd. website. [Online]. Available: http://www.arm.com/

[4] Linaro website. [Online]. Available: http://www.linaro.org/

Development of a Virtual Platform for IP and Firmware Verification

Changmin Shin, Yongjoo Kim

Embedded Software Research Division
Electronics and Telecommunications Research Institute
Daejeon, Korea
cmshin@etri.re.kr, y.kim@etri.re.kr

Abstract

The proposed virtual platform in this paper is composed of the basic IP models and has an integrated simulation functionality. The basic TLM models used to build a virtual platform are ARM9 processor core, memory, AMBA bus, displays, and system peripherals. And these TLM models are connected to each other.

The virtual platform is built by porting the IP TLM model for the IP verification on the previous virtual platform. And the firmware for IP operation is developed. The operation and verification of the IP model are performed by simulating the newly formed virtual platform. In this paper, the verification of UART, Camera IP, and face recognition IP was performed by using the integrated simulation of the constructed virtual platform.

With the integration simulation of virtual platform developed in this paper, it is possible to develop both the IP and the firmware simultaneously. Therefore, the development period is shortened.

Keywords-Intellectual Property; Transaction Level Model; Register Transfer Level; Cycle Level Model

I. INTRODUCTION

The virtual platform consists of a SoC hardware platform modeling in SystemC and an operating system or firmware ported to the virtual platform [1]. As the virtual platform is applied to the SoC development, the rapid SoC development is possible through the optimization reusing the existing IP models

In the early development of embedded system, the virtual system is configured to model a hardware level. It is possible to shorten the development period by providing the virtual platform for software developers to develop the hardware and the software simultaneously.

The virtual platform is made up of various types of models, such as RTL(Register Transfer Level), CLM(Cycle Level Model), TLM(Transaction Level Model), provided to the hardware and the software developers.

RTL model is the model of the signal level. And CLM is the model of the data transmission level. In CLM, the data sent to the IP port of the hardware is operated to the cycle level.

TLM is a model for the data transmission in the transaction. Because TLM has the high level of abstraction and no concept of time, TLM supports a fast simulation. In this paper, the virtual platform used the TLM models to support the fast simulation.

The purpose of this paper is to provide the Virtual Platform for supporting both the IP verification and the software development before the real hardware platform is developed. Virtual Platform is composed of bootloader, firmware and virtual prototypes of hardware IPs.

II. VIRTUAL PLATFORM FOR IP VERIFICATION

As shown in Figure 1, a variety of TLM models(ARM9 processor core, memory, bus, camera interface, UART, interrupt controller, DMA controller, clock, reset) was applied to the virtual platform using Virtualizer tool of Synopsys [2, 3]. And the integrated simulation environment for verifying an IP was built.

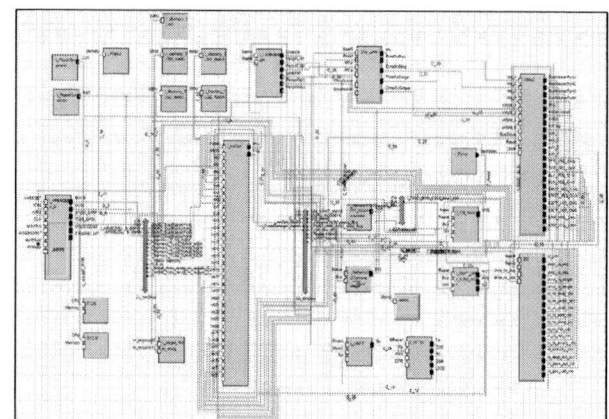

Figure 1. IP Constitution in Virtual Platform

UART, camera IP and face recognition IP were ported on the built virtual platform in order for the operation and verification of IPs. And the firmware for the IPs was developed.

An image data is required by the face recognition IP. The image data is received through the camera equipment in the real system that the virtual platform is installed. The images are transmitted to the firmware by the camera interface IP of the virtual platform.

The integration simulation process of the virtual platform is shown in Figure 2.

The firmware and the test software are mounted on the constructed virtual platform. And the operation and verification for the IP model and test software are performed by the integration simulation of the virtual platform. Further,

978-1-4799-5128-4/14 $31.00 © 2014 IEEE

the modification of the virtual platform configuration, IP models, and test software is made by using a simulation result.

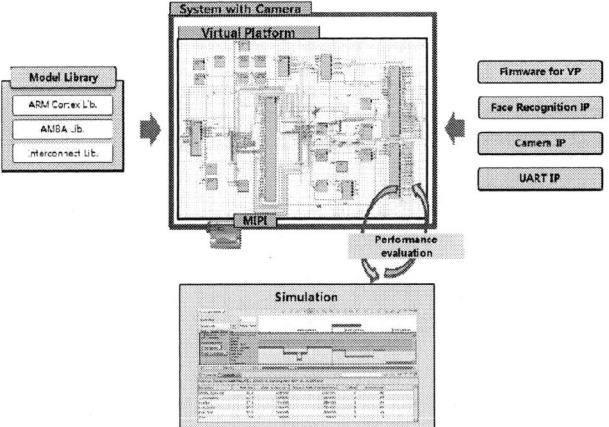

Figure 2. Integration Process of Virtual Platform

III. IMPLEMENTATION RESULTS

The TLM models applied to build the virtual platform for IP verification are ARM9 processor core, memory, bus, interrupt controller, DMA controller, clock, and reset [4]. In this paper, we simulated and verified UART IP, camera interface IP, camera IP, image data interface IP, and face recognition IP using the virtual platform.

The integrated simulation of UART IP and camera interface IP ported on the constructed virtual platform is shown in Figure 3

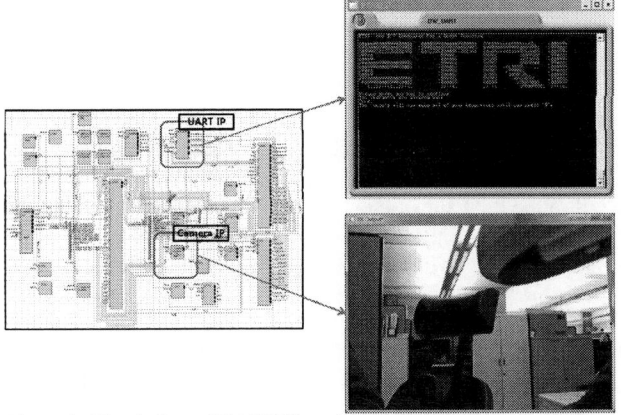

Figure 3. Simulation of UART IP and Camera IP in Virtual Platform

We ported UART IP on the virtual platform and developed the firmware for UART IP verification. UART IP firmware shows a message written on the panel window. And input/output via the keyboard is possible.

The camera interface IP, camera IP, and image data interface IP were ported on the virtual platform so that the camera application software can be used. The virtual platform receives the image from the physical camera in the system that the virtual platform is installed. The image is screened through the display IP of the virtual platform. The camera firmware was developed for the verification of the camera IP. The camera firmware tests the operation of the camera IP and displays the image in the panel of the virtual platform.

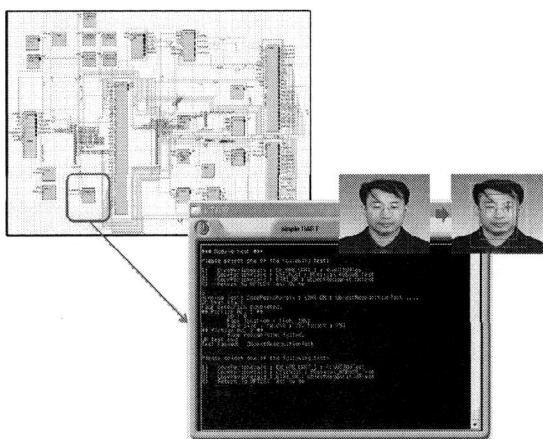

Figure 4. Simulation of Face Recognition in Virtual Platform

For the verification of the face recognition IP, the face recognition IP was ported on the virtual platform. And the firmware for the face recognition IP was developed.

The input face image of the face recognition IP is compared with the images that is registered in advance. If the input image is the same as the registered image, the identity and feature information of the face image are displayed in the panel window.

IV. CONCLUSIONS

In this paper, the virtual platform is composed of TLM models that are ARM9 processor core, internal/external memory, AMBA bus, display, and system peripherals. To verify UART, Camera IP, and face recognition IP, they were ported to the virtual platform. And the firmware for testing them was developed.

Using the virtual platform, the simultaneous development of IP and firmware was possible. And the performance verification of the virtual platform was made.

Acknowledgment

This research was supported by the Technology Innovation Program (No. 10044313) funded by Ministry of Science, ICT and Future Planning of Korea.

References

[1] W. Muller, M. Becker, A. Elfeky, and A. DiPasquale, "Virtual prototyping of Cyber-Physical Systems," Design Automation Conference (ASP-DAC), pp. 219-226, 2012.

[2] Synopsys, Inc., Virtual Prototype Software Analysis User's Guide, Version H-2013.06, June 2013

[3] Synopsys, Inc., VDK Family for ARM Cortex v7 Processors User's Guide, Version H-2013.06, June 2013

[4] ARM Ltd. website. [Online]. Available: http://www.arm.com/

Design of CMOS Dual Antifuse OTP Memory Based On Gate Oxide

Seung-Youl Kim[1], Je-Hoon Lee[2], Tae-Yang Kim[2], Younggap You[1]

[1]College of Electriacal and Computer Eng.
Chungbuk Nat'l Univ.
Cheongju-City, Korea
kimsy@hbt.cbnu.ac.kr

[2]Div. of Electronics, Information and Communication Eng.
Kangwon Nat'l Univ.
Samcheock-City, Korea
jehoon.lee@kangwon.ac.kr

Abstract

This paper presents a one-time-programmable ROM based on CMOS dual antifuses achieving high reliability and performance. The CMOS antifuse exploits the gate oxide layer of a typical CMOS process and thereby does not need an additional mask step for antifuse implementation. The proposed antifuse circuit comprises two antifuse nMOS transistors and two pMOS access transistors. Each antifuse cell employs differential architecture. An antifuse cell compares two differential outputs and determines its final output. This dual antifuse cell shows strong reliability against soft-breakdown failure. This high reliability makes the proposed one-time-programmable ROM can be applicable to high reliability such as encryption systems.

Keywords - Antifuse; ROM; OTP; Sense amp.

Introduction

The OTP(one-time-programmable) ROM can be used to embedded processor, system on chip, circuit trimming, chip ID, memory repairing, CMOS image sensor, display driver ICs, RFID tag and smart card. Additional mask steps are not necessary for OTP ROM designs comprising antifuses based on the gate oxide layer of the standard CMOS process, and thereby reduces fabrication cost of OTP ROM products[1-2]. Conventional OTP memories with antifuses use two or three transistors. Their OTP ROM cell has a sense amplifier taking a single antifuse cell output, and their output can be weak when the antifuse cell suffers soft breakdown[3]. The proposed CMOS dual antifuse cell employs differential cell architecture yielding complimentary outputs, and thereby exhibits strong behavior even under soft breakdown state.

Proposed CMOS Dual Antifuse Cell

The proposed dual antifuse cell can be implemented using the gate oxide layer of the standard CMOS 0.18um process. Figure 1 illustrates its equivalent circuit structure of the proposed dual antifuse cell. As shown in Figure 1a, the dual antifuse cell comprises two antifuse nMOS (Tox: 36Å, 1.8V) transistors of thin gate oxide and two pMOS(Tox: 121Å, 5V) transistors of thick gate oxide. Figure 1b is its equivalent circuit. The antifuses of Figure 1a can be represented with capacitance and resistance values reflecting states before and after programming, respectively, as shown in Figure 1b [4]. The antifuse state before programming can be represented by an equivalent MOS capacitor. The resistance value of the programmed antifuse ranges from 10^1 to 10^4 ohm by breakdown, and can reach 10^6 ohm for soft-breakdown[3]. The proposed circuit can sense the programmed value even under the soft-breakdown state resulting in high reliability.

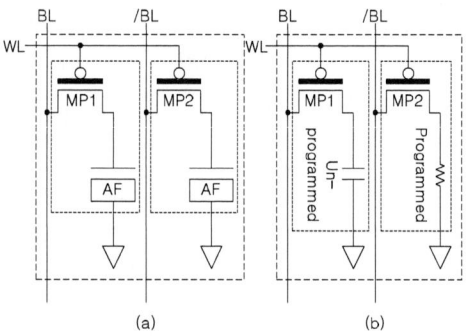

Fig. 1 Dual antifuse cell(a) and equivalent circuit(b)

The proposed circuit has two operation modes: the programming mode and the reading mode. Figure 2 shows the detail of the dual antifuse cell OTP ROM. The programming mode is used only once since this device is one-time programmable. The high voltage, VPP, is to breakdown the antifuse nMOS. It is approximately 7V. The thin gate oxide nMOS of the standard CMOS 0.18um process breaks down at this voltage level[4]. VPP is sent to BL(Bit Line) and WL(Word Line) in the programming mode. The differential structure of BL and /BL takes complementary signals. BL or /BL has high voltage, and the corresponding nMOS structure is programmed while the other stays unprogrammed when the pMOS transistor turns on. Figure 1b illustrates after this programming operation completes.

The reading mode begins with the precharge step, where the VDD of 1.8V is applied to both BL and /BL by selecting WL and turning on pMOS. The voltage level of the BL at the unprogrammed side maintains VDD since it functions as a capacitor. The other bit line discharges through pMOS since the equivalent resistance represents the programmed antifuse. The sense amplifier detects the resultant voltage differences between the two bit lines, and yield reliable output.

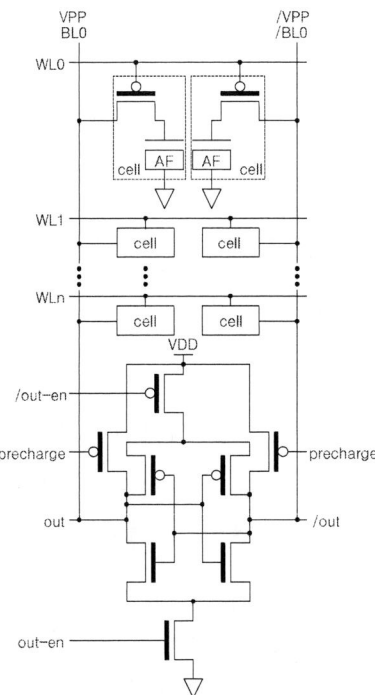

Fig. 2 Dual antifuse cell OTP ROM

Simulation Results

The operation of the dual structure has been verified through simulation. Figure 3 shows the operation of the unprogrammed antifuse and the programmed antifuse using the dual antifuse cell and its equivalent circuit of Figure 1. The simulation shows the circuit behavior by increasing the programmed antifuse resistance from 1MΩ to 10MΩ. The voltage difference becomes 0.75V at 10MΩ.

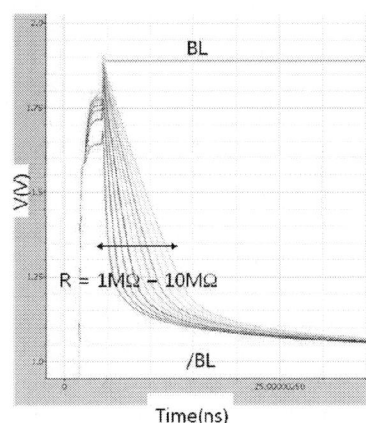

Fig. 3 Dual antifuse cell simulation results

The OTP operation has been evaluated based on this structure. Simulation result of Fig. 4 uses the OTP ROM circuit of Figure 2. The resistance value of the programmed antifuse assumes 1MΩ for simulation. The simulation begins with precharging of BL and /BL by activating Wl and turning on pMOS in turn. The out-en signal activates nMOS, and then the differential sense amplifier compares voltage levels of BL and /BL. It takes less than 20ns for the reading operation.

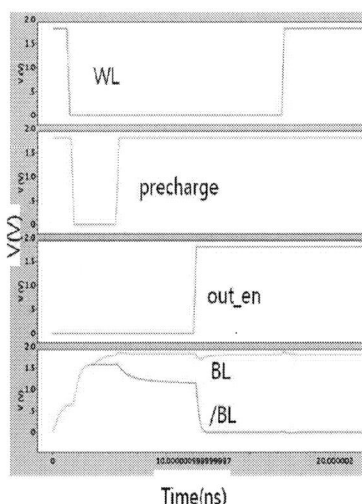

Fig. 4 Dual antifuse cell OTP ROM simulation results

Conclusion

The proposed dual antifuse OTP ROM design is based on the standard CMOS 0.18um process without additional mask steps. The circuit includes 4 transistors where two of them are used to build antifuses creating differential architecture. This structure is strong with respect to soft-breakdown. The voltage difference between BL and /BL is sufficiently big since the resistance of the programmed antifuse is 10MΩ even under the soft-breakdown state. It takes less than 20ns for the reading operation. The dual antifuse OTP ROM can be used for applications requiring high speed and reliable operations.

Acknowledgment

This research was financially supported by the Ministry of Education (MOE) and National Research Foundation of Korea (NRF) through the Human Resource Training Project for Regional Innovation (2012H1B8A2026055). This research was supported by Basic Science Research Program through the National Research Foundation of Korea (NRF) funded by the Ministry of Science, ICT & Future Planning (2011-0013219).

References

[1] M. Deloge et al., "Application of a Tddb model to the optimization of the programming voltage and dimensions of antifuse bitcells," *IEEE Electron Devices Letters*, vol. 32, pp.1041-1043, 2011.

[2] N. Mathur et al., "One time programming device yield study based on anti-fuse gate oxide breakdown on P-type and N-type substrates," *Integrated Reliability Workshop Final Report*, 2005 IEEE International, Oct. 17-20, 2005, pp.111-113

[3] M. Depas, T. Nigam, and M. Heyns, "Soft breakdown of ultra-thin gate oxide layers," *IEEE Trans. Electron Devices*, vol. 43, pp.1499-1504, 1996.

[4] N. D. Phan, I. J. Chang, and J.-W. Lee, "A 2-Kb one-time programmable memory for UHF passive RFID tag IC in a standard 0.18um CMOS process," *IEEE Trans. Circuits and Systems*, vol. 60, pp.1810-1822, 2013.

978-1-4799-5128-4/14 $31.00 © 2014 IEEE

Design of Bio-signal Measurement System for Array Sensor and Application

Seungpyo Jung, Youngju Park, Sangman Kim, Jusung Park

Pusan National University
2, Busandaehak-ro 63beon-gil, Geumjeong-gu
Busan, Korea
spyam@pusan.ac.kr

Abstract

This paper introduces a bio-signal measuring system, which is suitable for array type bio-sensors. The system consists of 32-bits RISC processor, RAM, flash memory, UART, video controller, and ADC module. The system shows that the maximum difference of each chanel capacitance without sensor loading is under 1fF. The system is used for liver cancer dection because it has an inteligence about the disease through a neural network .

key words: array sensor, bio-signal measuring, liver cancer

Introduction

Recently, the research on the bio-signal measurement is actively being carried out for accurate disease diagnosis. And also the sensing materials for a specific disease are intensively being discovered. Generally, the more sensing materials are used for disease diagnosis, the more accurate results come out. Therefore array type sensor that has many sensing points is used for bio-signal measurement. Up to now, many different sensing materials are coated on the spot of slide glass to implement array type sensor. These type methods not only requires redundant time on scanning optically the spot points and also have non uniform temperature profile from spot to spot due to temperature increase during optical scanning. To solve these problems, the chemo-mechanical sensor, which transduces the chemical reaction on sensor to displacement of the sensor, generates the capacitance depending on the displacement. We can exclude the optical scanning step by using ADC that changes the displacement to electrical signal.

Disease measuring method with array sensor

The conventional method requires three step as shown in Fig. 1. At the first step, different sensing materials are coated on many spots of glass sheet. Secondly, this glass sheet is reacted with positive and negative clinical specimen. Finally, reacted glass sheet is read by special optical scanner. The scanner reads color on each point and translates it to reaction rate.[1]

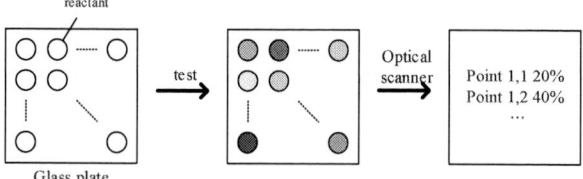

Fig. 1 Conventional method with glass sheet

Since this method uses almost similar sensing materials on the sensing spot, the spots are located very close to each other and have the same environment to extract something different from each sensor. However, environment difference exists from spot to spot point due to brightness, temperature and humidity. The average result of all material represents environmental influences, so normalization step helps to remove environmental effects. However, these methods using glass sheet consumes redundant time like optical scanner reading time. And there is no chance to identify outside effect like short time high temperature environment, unstable humidity from scanner result. And the precise reaction results are difficult to measure by indirect optical method.

In order to overcome such problems, we adopt chemo-mechanical array sensor. The array sensor method replaces the glass sheet and optical scanner with array sensor. The sensing material is coated on sensor and the sensor translates the reaction rate to electrical information as shown in Fig. 2. Direct connection between sensor and reactant has some advantages. Realtime monitoring is one of advantages. In the case of glass sheet which reactant place apart from target point, it is discovered at scanning step. However unconnected sensor is simply detected by connection test step. In the case of sudden environment difference, monitoring log help to find the effect. Thus we design measuring system with array sensor to get precise result of sensor.

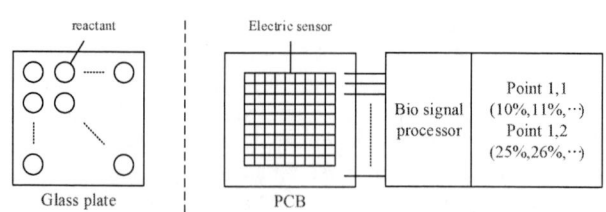

Fig. 2 Array sensor method architecture

Design of the proposed system

978-1-4799-5128-4/14 $31.00 © 2014 IEEE

The measuring system consists of system controller, sensor module, LCD module, communication module, memory block, and ADC module as shown in Fig. 3. The LCD module displays the measuring sequence at first, and then displays the waveforms of each channel. The communication module is used to connect the measuring system and host computer, and the measuring sequence and measured data are directionally transmitted via this module. The memory module, which consists of RAM and flash memory, store the measured data, system control program, and disease diagnosis program.

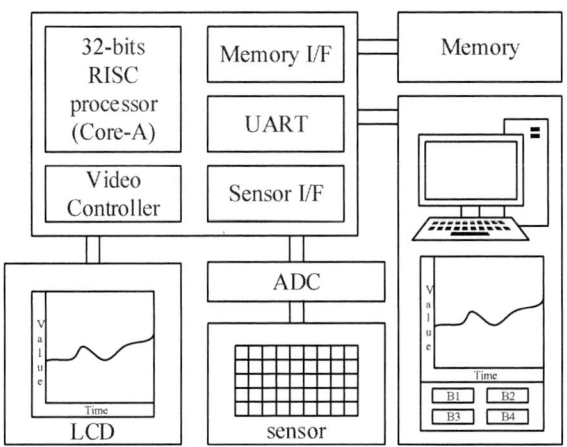

Fig. 3 Architecture of the designed system

The system controller is composed of 32-bits RISC processor(Core-A)[2], memory interface, video controller, UART, programmable interrupt controller, sensor interface and AMBA bus. The video controller access memory through DMA. And Core-A has program memory interface and data memory interface. So the bus is three master, five slave architecture. Since the sensor information is slowly changed than system controller clock speed, the AMBA version 2.0 bus is adopted. Video controller accesses frequently memory and uses burst transmission. Thus the video controller and memory interface are connected to AHB slave port of bus. The UART, programmable interrupt controller and sensor interface are connected to APB slave port of bus.

Implementation of the proposed system

The function simulation is done with basic system which consists of CPU(Core-A), memory interface, UART, AMBA bus and SRAM model. Through the function simulation we confirm Core-A AHB interface and proper operation of the program code which is compiled to monitor the basic operation. We verify the operation of LCD module and sensor module by FPGA module because it takes lot of time to confirm by simulation.

We put all the digital parts including CPU into a FPGA because the CPU is synthesizable core. We carefully design ADC board PCB to make the distance same between array sensor to ADC for the same stray capacitance and make two ground system for signal integrity as shown in Fig.5.

Fig. 4 Photograph of the developed system

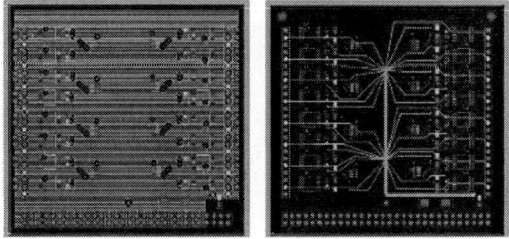

Fig. 5 PCB pattern of ADC board

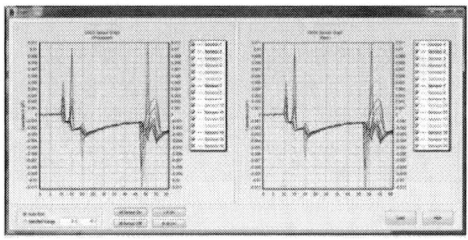

Fig. 6 Photograph of the measured waveform from proto-system

Conclusion

We have designed and implement bio-signal measuring system which is suitable for array sensor. The system detects capacitance variation of array sensor with resolution of sub femto farad and also is used for liver cancer diagnosis. The developed system can be used as POC(Point-of-Care) equipment for many kinds of diseases with proper sensing material on the sensor and pre-learned intelligence in flash memory.

Acknowledgment

This work was supported by Ministry of Science, ICT and Future Planning and IDEC Platform center(IPC)

References

[1] Jae-Hong Eom, Sung-Chun Kim, Byoung-Tak Zhang, "AptaCDSS-E: A classifier ensemble-based clinical decision support system for cardiovascular disease level prediction," *Expert Systems with Applications 34*. pp. 2465-2479, 2008

[2] www.core-a.net

Re-visit Blocking Texture Cache Design for Modern GPU

Jhe-Yu Liou

Chung-Ho Chen

Department of Electrical Engineering, National Cheng Kung University
No.1, University Road
Tainan, Taiwan

elvis@casmail.ee.ncku.edu.tw

chchen@mail.ncku.edu.tw

Abstract

Texture cache plays a significant position in GPU design especially in a limited memory bandwidth environment such as mobile SoC system. In this paper, we evaluate the 6D blocking texture cache design through a sophisticated GPU simulator using DRAM memory model. Our experiment reveals that using a larger block can take advantage of the spatial locality of texel accesses, however, fetching a larger block which requires several burst runs in DRAM access, results in poor memory access efficiency. As a result, the block size used has to match with the DRAM burst length for the best memory access efficiency.

Keywords-component; DRAM model, Texture cache, GPU architecture.

I. Introduction

Mobile graphic processor unit (GPU) is the most bandwidth demander in a hand-held device, especially for those using shared memory approach. The major memory access demand in GPU comes from the texture operations, and as a result, all GPU architectures have used some form of texture caches.

Due to the texture image characteristics, Igehy has proposed to use 6D blocking texture data to dramatically increase cache efficiency[1]. The main idea, shown in Fig.1, is to re-address the texture data, which in the past is linearly placed in the memory like Fig.2(b), into a 3-level and 2-dimension organization and convert the texture coordinate to perfectly fit the general cache hierarchy as tag, entry (index), and block. This method not only exploits texture spatial locality, but also takes the advantage of DRAM burst access because of the ordered sequence of texels in the blocks as depicted in Fig.2(c).

In this paper, we evaluate a 6D blocking texture cache design for modern shader based GPU with detailed DRAM model.

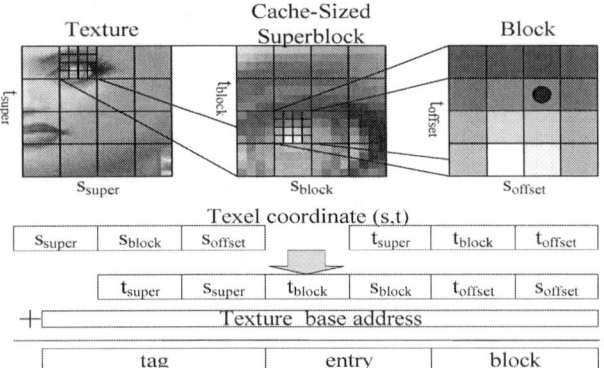

Fig.1 Texture data in 6D blocking organization and how texel address is represented in general cache from the texel coordinate (s,t).

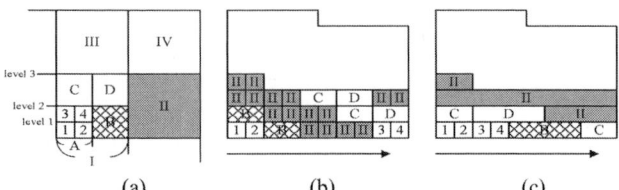

(a)   (b)   (c)

Fig. 2 (a) texture data sequence (b) traditional linear fashion and (c) 6D blocked fashion in memory.

II. Simulation environment

In Hakura's research[2], Igehy chose 16 texels (4x4 texels) per block which worked quite well and used lesser bandwidth than the other configurations. However, the result in Igehy's work has not been based on contemporary DRAM models. Also they use the fixed-function OpenGL test cases unlike modern graphic programming model that bases on the OpenGL Shading language. To render the issues we address above, we develop our own simulation environment which includes a full OpenGL ES 2.0 API implementation and a soft-pipe GPU simulator. To cope with the concept of 6D blocking data organization, we leverage the triangle traversal strategy in Intel larrabee[3] that will test a triangle in 16x16-pixel supertiles. Once a supertile is covered by the triangle, the traversal scheme will subdivide the supertile into four quarter tiles recursively until the 2x2 tile size is reached, and then put these four pixels on the GPU's shader. The GPU simulator has a unified shader which allows 4 threads to be synchronously executed under single instruction multiple thread (SIMT) scenario. The unified shader is also accompanied by a texture unit with the 6D texture cache which can be flexibly configured.

A. Memory model

To precisely model DRAM timing, we create a DDR DRAM simulator based on the following core DRAM parameters, dram clock(CLK), column latency (CL), row to column delay (tRCD), and row precharge time (tRP). The timing scheme is described as below:

1. *Initial access: time = tRP + tRCD + CL*
2. *All following accesses:*
   if this access's desired row differs from previous.
       time += tRP + tRCD + CL
   else
       if this access is a continuous burst access
           time += CLK/2 (Because of double data rate, DDR)
       else
           time += CL

To model a GPU in a mobile SoC chip in which the GPU has no dedicated local memory, we pick low power DDR(LPDDR2) DRAM from Micron instead of graphic DDR(GDDR) DRAM as our target simulated DRAM even though GDDR has larger

TABLE I
Parameters of Micron LPDDR2 800Mhz DRAM. This table hides write latency because all texture operations are read only.

CLK	CL(read)	tRCD	tRP	width	burst length
2.5ns	15 ns	18 ns	18 ns	32 bits	1, 2, 4, 16

memory width and longer burst length. The Micron LPDDR2 DRAM parameters can be found in Table I.

III. Benchmarks

In this paper, three benchmarks are used. They are teapot, StoneFloor, and FourShapes. Each benchmark's characteristic can be found in Table II. The teapot is a simple object but is popular in 3D graphic tutorial with single texture mapping. The StoneFloor achieves the normal mapping effect by utilizing two texture maps as the color map and the normal vector map respectively. The FourShapes demonstrates the parallax occlusion mapping[4] effect by using only two triangles to accomplish lots of image details just like constructed by hundreds or thousands triangles.

TABLE II
Benchmark scenes and their characteristic

Benchmark	Teapot	StoneFloor	FourShapes
Texture image			
Triangles	6400	2	2
Texels / pixel	6.48	10.72	288.15

To analyze the result, all textures are stored in 32 bits per texel even though the last 8 bits are redundant in some textures. Besides, All screen resolution in benchmarks is set in 1024x768 with 2X anisotropic filter enabled.

IV. Performance Analysis

In order to evaluate the relationship between cache organization and memory utilization, we configure the different combination of the cache block-size in 2x2, 4x4, and 8x8, and cache size in 4, 16, and 64KB. Because at most two texture images are used in our benchmarks, the cache is 4-way set association. The maximum burst length in our DRAM model is 16 words. This is obviously harmful to fetching of an 8x8 cache block size because once a cache miss has occurred, the cache needs four burst read accesses to retrieve the data. In contrast, both 2x2 and 4x4 block-size cache can mostly get data in a single burst read access.

Fig. 3 shows the simulation result under various cache block sizes, total cache size, and different benchmarks. Both teapot and StoneFloor show very similar trend. This is because the way they use texture mapping is completely the same except the number of texture images used.

For these two benchmarks, 4x4 block size works quite well in

TABLE III
Cache miss penalty. Note: 8x8i means 8x8 cache block size but with ideal 64 burst length supported in the DRAM model.

block-size	min (ns)	max (ns)	avg (ns)
2x2	18.75	104.5	61.625
4x4	33.75	119.5	76.625
8x8	135	207	171
8x8i	93.75	179.5	136.625

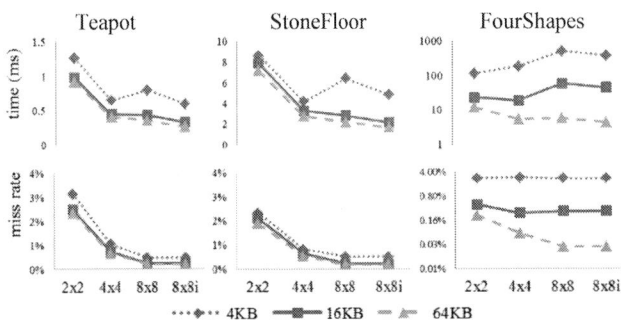

Fig.3 The total memory access time and cache miss rate in different cache parameter and benchmarks. Note: the Y-axis in both FourShape diagrams are drawn in log scale.

all cache sizes. The 8x8 block size has even better performance in 16KB and 64KB. The range of miss penalty can be simply calculated, which is listed in Table III. Considering the miss penalty in Table III, an 8x8 block size needs to decrease the cache miss by about 223% in average compared to 4x4 case to have the same total memory latency. This is the reason why the 8x8 block size in 4KB cache size has higher total memory access time due to the insufficient cache miss drop which, however, can be improved by using the DRAM to have longer burst length for the 8x8 configuration in Table III and Fig.3. Let's look at the FourShapes. The miss rate is about 2% for any block-size under 4KB cache size. Having an average of 288 texel fetches per pixel means that each pixel encounters at least one conflict miss before the next neighbored pixel goes. This destroys the cache spatial locality and leads very poor cache efficiency. Consequently, a large texture cache is required.

V. Conclusion

In this paper, we have developed an OpenGL ES 2.0 GPU simulator which includes a realistic DRAM model and we evaluated the design of 6D blocking texture cache. Our experiment shows that using a larger block can take the advantage of spatial locality, however, fetching a larger block which requires several burst runs in DRAM access, results in poor memory access efficiency. We have observed that the block size used has to match with the DRAM burst length for the best memory access efficiency.

Acknowledgement

This research is supported by the National Science Council of Taiwan under grand NSC 102-3113-P-006 -018.

References

[1] H. Igehy, M. Eldridge, and K. Proudfoot. "Prefetching in a texture cache architecture," In Proc. ACM SIGGRAPH/ EUROGRAPHICS conference on Graphics Hardware, pp. 133-142, 1998

[2] Z. S. Hakura and A. Gupta. "The design and analysis of a cache architecture for texture mapping," In Proc. 24th annual international symposium on Computer architecture (ISCA '97), pp108-120, 1997.

[3] L. Seiler, D. Carmean, E. Sprangle, T. Forsyth, M. Abrash, P. Dubey, et al. "Larrabee: a many-core x86 architecture for visual computing," ACM Trans. Graphics (Proc. SIGGRAPH '08), vol.27, no. 3, Aug. 2008.

[4] N. TATARCHUK. Dynamic parallax occlusion mapping with approximate soft shadows. In: Proceedings of the 2006 symposium on Interactive 3D graphics and games. ACM, 2006. p. 63-69.

Analysis of On Chip Decoupling Capacitor in the Double-Gate FinFETs with PEEC-based Power Delivery Network

Jaemin Lee, Yesung Kang, and Youngmin Kim

School of Electrical and Computer Engineering, UNIST, Ulsan, 689-798, Republic of Korea
{jm3430202,yeskang,youngmin}@unist.ac.kr

Abstract

As the technology node has scaled down below 32 nm, the supply voltage has decreased to 1 V and the current demands for active devices have increased. Therefore, the supply noise due to the IR drop in the power delivery network (PDN) has become a critical problem for robust circuit operation. Huge decoupling capacitors are introduced to overcome the supply voltage fluctuations. In this study, we investigate a 32-nm double-gate FinFET device for a decoupling capacitor in the PDN. The circuit designers can independently control both the gates in the double-gate FinFET. We compare the supply and ground noise reduction in the conventional planar CMOS and in various FinFET structures in a PEEC-based practical PDN and propose the best decoupling capacitor design strategy for double-gate FinFETs. The simulation results show that we can achieve an increased reduction in the supply voltage noise up to 50% by shorting the front gate and back gate together.

Keywords-component; FinFET, decap, IR drop

Introduction

Recently, the demand for a high-speed, high-performance integrated circuit operation with a lower supply voltage and current has resulted in a significant amount of supply voltage fluctuations [1, 2, 3]. A practical approach to reduce the supply noise is to insert on-chip decoupling capacitors between the power supply rail and the ground rail. When the on-chip decoupling capacitors (decap) are placed close to the current load, they act as a low-pass filter to the power line with a wire resistance. Therefore, we can effectively reduce the dynamic IR drop related to the power supply noise with the introduction of on-chip decap.

As the power delivery network (PDN) has become complicated and the wire resistance has increased owing to the interconnect scaling, the supply voltage fluctuations due to the IR drop have become a significant problem in the PDN design. A partial electrical equivalent circuit (PEEC) method [4] has been introduced for accurate extraction of RCL components of wires.

Transistor scaling has resulted in astounding increases in the transistor density and performance, leading to greater functionality at higher speeds for single chips. However, as the chip size continues to shrink, the leakage and short-channel effects can render silicon-based transistors less reliable. As a solution to the scalability and leakage issues, a multi-gate transistor,

TABLE I
Metal interconnect parameters used in the PDN of Fig. 2

Metal	Width (nm)	Height (nm)	Pitch (μm)	Length (μm)	number of tracks
M1	32	58	-	1	-
M3	100	61	5	400	43
M6	300	720	15	200	29

such as a fin-type field-effect transistor (FinFET) is emerging as a promising substitute for bulk CMOS at the 32-nm node and beyond [5]. The additional back gate of the FinFET provides the circuit designers many options and abundant design space. The two gates of the FinFET can be either shorted for achieving a higher performance or independently controlled for a lower leakage or reduced transistor count [6, 7].

In this paper, we compare the performances of the on-chip decoupling capacitors used in the 2D conventional planar transistors and 3D multi-gate transistors (i.e., FinFETs) and investigate the decoupling capacitor design strategy in a double-gate FinFET under different gate bias conditions. We use the PEEC method to extract RLC components from a practical PDN having four VDD and four GND pads. And dynamic transient analysis with $40\times$ size clock buffers is performed to simulate the IR-drop-related voltage fluctuations and evaluate the double-gate FinFET decoupling capacitor.

Simulation Setup

FinFETs, which are one of example of the multi-gate transistors, can be operated in either a shorted gate (SG) mode or

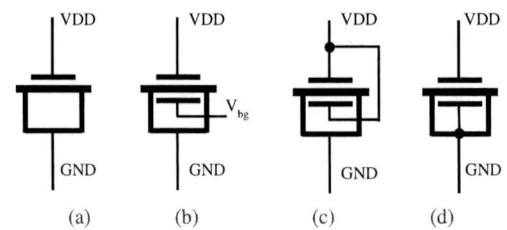

Fig. 1. CMOS and FinFET decap strategy. (a) conventional CMOS moscap, (b) FinFET_1: back-gate-biased FinFET, (c) FinFET_2: VDD-connected back-gate FinFET, and (d) FinFET_3: GND-connected back-gate FinFET.

978-1-4799-5128-4/14 $31.00 © 2014 IEEE

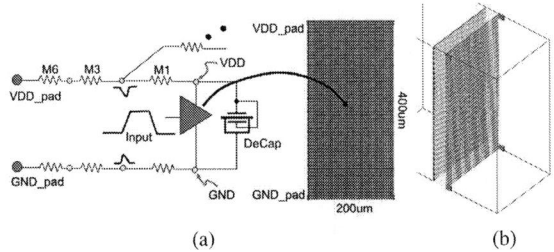

(a)   (b)

Fig. 2. (a) Power delivery network (PDN) structure with clock buffers and FinFET_2 decap used in the dynamic simulation. (b) is the 3D bird's eye view of the PEEC-based PDN.

TABLE II
IR drop analysis results (VDD and GND) of various types of decap

nominal voltage (V)	without decap VDD drop(V)	CMOS decap VDD drop(V)	CMOS decap Reduc. (%)	FinFET_2 decap VDD drop(V)	FinFET_2 decap Reduc. (%)	FinFET_3 decap VDD drop(V)	FinFET_3 decap Reduc. (%)
0.7	0.066	0.061	7.7	0.052	21.5	0.061	7.1
0.9	0.098	0.094	4.1	0.049	49.9	0.068	30.5
1.1	0.141	0.133	5.7	0.088	38.1	0.116	18.3

nominal voltage (V)	without decap GND bump(V)	CMOS decap GND bump(V)	CMOS decap Reduc. (%)	FinFET_2 decap GND bump(V)	FinFET_2 decap Reduc. (%)	FinFET_3 decap GND bump(V)	FinFET_3 decap Reduc. (%)
0.7	0.061	0.059	3.3	0.009	85.7	0.009	84.8
0.9	0.092	0.087	5.4	0.011	88.4	0.012	86.6
1.1	0.135	0.126	6.7	0.019	86.0	0.021	84.8

an independent-gate (IG) mode by separating both the front gate and the back gate [6]. In this study, we analyze the decap behavior for several modes of the FinFET. One CMOS NMOS capacitor (moscap) and three FinFET-based moscaps are shown in Fig. 1. As shown in Fig. 1, (a) is the conventional CMOS moscap, (b) is the back-gate-biased FinFET decap, (c) is the VDD-connected SG-mode FinFET, and (d) is the GND-connected SG-mode FinFET. We use the practical PDN scenario in which there are four power supply pads (VDD_pad) and four ground pads (GND_pad). Further, VDD and GND pass along the M6 layer through M3 and finally to M1 to supply current to the dynamically operating gates. We use 40× size clock buffers with 100-ps slew and 1-GHz clock in 32-nm Predictive Technology Model (PTM) [8]. For all cases, a 600 μm size decap is attached between power supply and ground.

The PDN structure with clock buffers that is used for transient analysis is shown in Fig. 2. and the metal interconnect parameters [1] used in the PDN are summarized in the Table I. Power runs through M6, M3, and finally M1 to supply required current. There are four VDD pads and four GND pads in total at each corner. Interconnect length is 200 μm (M6) in horizontal and 400 μm (M3) in vertical, respectively. We insert total 29 tracks of M6 interconnect runs horizontally and 43 tracks of M3 vertically to construct the PDN. A program implemented with C++ and standard template library (STL) is used to extract PEEC-based RLC components of the PDN structure. Fig. 2(b) shows the 3D visualization of the PEEC elements of the PDN structure showing all the wire segments and vias. Ports for the VDD and GND are shown in the red at each corner and the center.

Results

The dynamic IR drop analysis results are summarized in Table II. As shown in the table, we can effectively reduce the voltage droop and ground bump by introducing a decoupling capacitor. Further, as expected from the previous analysis, connecting the back gate of the FinFET to the power supply (i.e., FinFET_2) provides the best result in minimizing the voltage droop and ground bump. Table II summarizes the normalized reduction of the IR drop by adding same-size decap in the CMOS and FinFET. In the case of FinFET_2, the back gate is connected to the VDD; and in the case of FinFET_3, the back gate is connected to the ground. It can be observed that Fin-

FET_2 exhibits the best results in minimizing the VDD droop by up to 50% and the GND bump by up to 88% at a nominal voltage (i.e., 0.9 V). Further, the FinFET_2 structure provides approximately 50% better IR drop reduction than the CMOS with the same-size decap (i.e., 94 mV droop with CMOS decap vs. 49 mV droop with FinFET_2 decap). The FinFET_2 is known as a single-gate FinFET in which both the front and back are tied together. And this structure provides better manufacturability and results in smaller cell size compare to the independent-gate FinFET (e.g., FinFET_1 or FinFET_3) [6].

Conclusion

In this study, we investigate the decoupling capacitor design strategy for a 32-nm double-gate FinFET with PEEC-based PDN. The dynamic IR drop analysis shows that we can achieve the best results in both voltage droop and ground bump reduction by connecting the back gate of the FinFET to the VDD pin. We can obtain up to 50% VDD droop reduction and 88% ground bump reduction by applying the proposed FinFET decap structures in 32-nm node.

Acknowledgement

The EDA tools were partially supported by IDEC at KAIST.

References

[1] ITRS 2012, *http://public.itrs.net/Links/2012ITRS/Home2012.htm*

[2] A. V. Mezhiba, and E. G. Friedman, "Scaling trends of on-chip power distribution noise," *IEEE TVLSI*, vol. 12, no. 4, pp. 386–394, 2004.

[3] M. Popovich *et al.*, "Efficient Distributed On-Chip Decoupling Capacitors for Nanoscale ICs," *IEEE TVLSI*, vol. 16, no. 12, pp. 1717–1721, 2008.

[4] A. E. Ruehli, "Equivalent circuit models for three dimensional multiconductor systems," *IEEE TMTT*, vol. MTT-22, pp. 216–221, 1974.

[5] D. Hisamoto *et al.*, "FinFet – a self-aligned double-gate MOSFET scalable to 20-nm," *IEEE TED.*, vol. 47, no. 12, pp. 2320–2325, 2000.

[6] M. C. Want, "Independent-Gate FinFET circuit design methodology," *IAENG International JCS*, vol. 37, no. 1, 2010.

[7] F. He *et al.*, "FinFET: From compact modeling to circuit performance," *IEEE International Conf. of EDSSC*, pp. 1–6, 2010.

[8] Y. Cao, PTM models, [online] Available: *http://ptm.asu.edu.*

978-1-4799-5128-4/14 $31.00 © 2014 IEEE

Memory efficient data structure for graph representation of DSPF netlist

Jinwook Kim[1], Hae-seong Park[1], and Young Hwan Kim[1,2]

Dept.of Electrical Engineering[1], Dept. of Creative IT Engineering[2]
Cheongam-ro 77, Nam-gu
Pohang-si, Gyeongbuk, Republic of Korea
{detest[1], seastar13[1], and youngk[1,2]}@postech.ac.kr

Abstract

This paper presents a memory efficient data structure for graph representation of a circuit described in detailed standard parasitic format (DSPF) netlist. The proposed data structure is designed for full-chip DSPF netlists of recent VLSI designs, which contain several hundred millions of elements. Experimental results using an industrial full-chip DSPF netlist of 25Gbytes, the implemented DSPF parser for the proposed data structure showed its ability to handle a full-chip DSPF netlist of a recent VLSI design.

Keywords-component; DSPF, Data Structure, Computer aided design, parser

Introduction

The detailed standard parasitic format (DSPF) is a commonly used file format to describe VLSI designs in transistor-level including parasitic elements such as parasitic resistance and capacitance. As recent VLSI designs contain several hundred millions of transistors, DSPF netlists also contain several hundred millions of transistors, nodes, and parasitic elements. Thus, not only processing circuits described in DSPF is difficult, but also even parsing DSPF netlists is very difficult.

A filesize of a DSPF netlist for a recent VLSI design usually exceeds several tens of Gbytes. Thus, the memory efficient data structure is necessary to read a DSPF netlist and construct an appropriate data structure to avoid swapping of 2nd-level memory storage such as DRAM to 3rd-level memory storage such as hard-disk.

This paper proposes a memory efficient data structure for graph representation of a circuit described in a DSPF netlist for a recent VLSI design. Although the proposed data structure does not consist of innovative data structure, this paper provides a way to build an appropriate data structure by

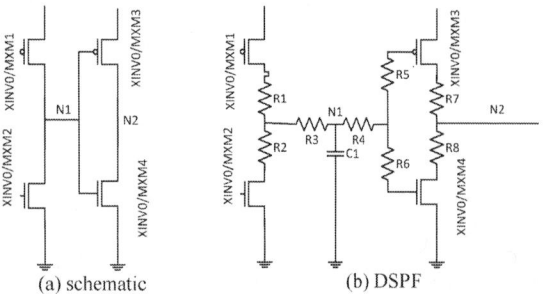

(a) schematic   (b) DSPF

Fig. 1 Example of Net (Node N1 in (a) becomes a Net N1 in (b)).

composing well-known data structures and the proposed data structure might be a good guidline to users who want to implement a DSPF parser and do something on a graph represented circuit such as graph traversal, partitioning, and so on.

The rest of this paper is organized as follows. Section 2 details the proposed data structure and Section 3 presents the experimental results and discussion briefly. Finally, Section 4 concludes this paper.

Proposed data structure

The proposed data structure is designed to describe a circuit structure in a DSPF netlist as a graph. A DSPF netlist describes a circuit structure by using several special keywords which are regarded as comments in SPICE netlist format. 1) *|NET kyword describes a net consisting of R and C elements. A net is a node, a connection between terminals of transistors, in the schematic or transistor level before parasitic components extraction (Fig. 1). After parasitic components extraction, a node in the schematic level is translated to RC-tree which consists of several nodes. 2) *|P describes a pin that represents an external pin. 3) *|I describes a pin (terminal) of instance connected to a net, and 4) *|S describes a node which is introduced during the parasitic components extraction.

Followings are defined classes to describe a circuit structure in DSPF netlist, and skeleton of these classes is also available in our website (http://soc.postech.ac.kr/DSPF.h, until 2017).

A. Classes for DSPF components

Following classes are defined to describe DSPF components. 1) Node is defined to describe a node in DSPF netlist, and it has a name, a pointer to connected pin of instance, a pointer to parent net, and index value as member variables. 2) Net is defined to describe a net stated with *|NET, and it has a name, a list of contained pins defined with *|P statements, a list of connected pins defined with *|I statements, and a list of child nodes. 3) Instance is defined to describe transistors, diodes, and instance from sub-circuit macro, and it has a name of instance, a list of its child pins, and a type as member variables, and 4) Pin is defined to describe pins of class Instance, and it has a name of pin, a pointer to its parent instance, a pointer to its related node, and a type of pin as member variables.

A global node list is stored all nodes in a circuit, and each node has a pointer to its parent net. A net also contains a list of indices of its child nodes and a list of related pins, and its related instances are reachable through pins in the list. An instance also has a list of pins and its related nets are also reachable through pins in the list. Pin is a defined to connect a node with an instance. Overall connectivity and graph representation for Fig. 1 (b) are shown in Fig. 2 and 3, respectively.

B. Underlying data structure

The proposed data structure employs the dynamic array and hash table for storing basic circuit elements, such as node, net, and instances. In addition, sparse matrix is used to store R,C elements.

1) A special type of dynamic array is used to store lists of nodes, nets, and instances. Because the total number of nodes is not known before completely reading of DSPF netlist, dynamic storage, whose size is not fixed, is necessary. Linked list can be one of good choice, but it contains additional pointer, 8-bytes for 64-bit machine, and it is not suitable for random accessing.

A used dynamic array is designed not to change the addresses of already stored elements after resizing of array. Because defined data types such as node, instance, pin, and net are referencing each other with pointers, the memory address of each object should be maintained.

To maintain the memory addresses of already stored elements, the used dynamic array is designed as a kind of two-dimensional array. The designed dynamic array is defined with an array of pointers (symbol ** in C/C++ language). The designed dynamic array allocates a one-dimensional array with given length and stores its first address in array of pointers. When the allocated one-dimensional array is full, the designed array allocates new one-dimensional array with given length and stores its first address in array of pointers. During this procedure, the addresses of already allocated one-dimensional arrays are not changed. The i-th element of array can be accessed at $[i \; / \; N][i - i \; / \; N]$ for a given unit length of

TABLE I
RESULTS FOR BENCHMARK CIRCUIT

Runtime (s)	89 mins
Memory usage	Approx. 27 Gbytes
Throughput	4.8 Gbytes / sec

one-dimensional array N.

2) Hash table is used to efficiently find an instance/net/node. A designed hash table finds an array index of the requested element in the global list of elements. 3) Compressed sparse-row format is used to store the connectivity of resistors and their values. **G** and **C** matrices in the modified nodal analysis [1] are stored in CSR format to reduce the memory space to store RC elements. RC elements are hidden from the viewpoint of both Net and Instance, but there are still reachable through node indices and **G**, **C** matrices, and traversing is also possible through **G**, **C** matrices.

Results

We implemented a DSPF parser for the proposed data structure in C++ language and with Lex and Yacc. The implemented DSPF parser was tested by using an industrial full-chip DSPF netlist, whose filesize was 25Gbytes. The Synopsys Finesim [2] spent about 5.9 hours to read and build an appropriate data structure for circuit-level simulation.

Table I shows the results of the implemented DSPF parser. The implemented DSPF parser spent only about 1.5 hours to read and build the proposed data structure, and it was remarkably faster than the commercial tool, Synopsys Finesim. Clearly, this comparison is not fair and this result do not indicate that the implemented parser is faster than the commercial tool, because the commercial tool might perform additional processing for simulation. Nevertheless, this result indicates that the implemented DSPF parser is efficient and practical enough.

Conclusion

This paper presents a memory efficient data structure for graph representation of a circuit described in a DSPF netlist. The proposed data structure is designed to reduce the required memory space as far as possible. Through the experiments for an industrial full-chip DSPF netlist, the implemented DSPF parser for the proposed data structure showed its practicalness; the implemented DSPF parser spent only about 1.5 hours for parsing, whereas the commercial tool spent about 5.9 hours for parsing.

Acknowledgment

This research was supported by SK Hynix semiconductor, co., Ltd and the MSIP(Ministry of Science, ICT and Future Planning), Korea, under the "IT Consilience Creative Program" (NIPA-2014-H0201-14-1001) supervised by the NIPA(National IT Industry Promotion Agency).

References

[1] C. Ho, AE Ruehli, and PA Brennan, "The modified nodal approach to network analysis," *IEEE Transactions on Circuits and Systems*, vol. 22, no. 6, pp. 504-509, Jun. 1975

[2] Synopsys Finesim, [online] Available: http://www.synopsys.com/Tools/Verification/AMSVerification/Pages/finesim-ds.aspx (Accessed: Jul. 31, 2014)

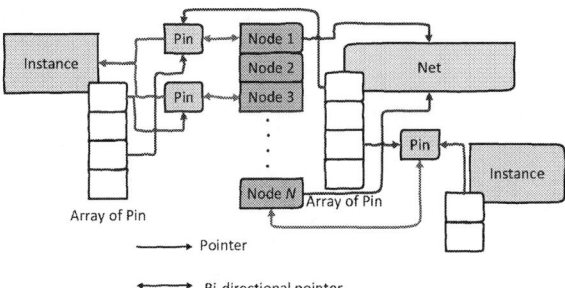

Fig. 2 Connectivity of the proposed data structure

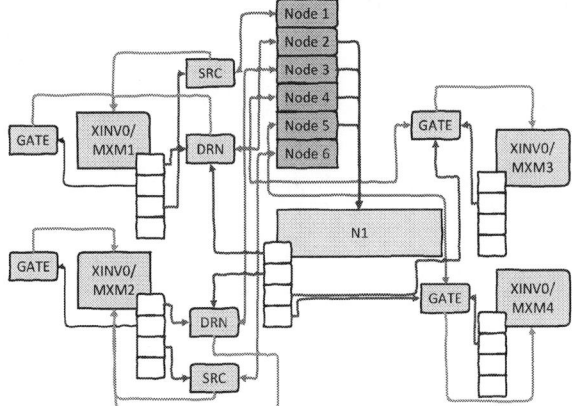

Fig. 2 An example of the proposed data structure for Fig. 1 (b)

Study of Sensor Fault Detection Based on Modified SVM Algorithm

Zhijia Yu, Yinan Xu[*] and Yonghu Ma

Division of Electronics &Communication Engineering
Yanbian University
Yanji, China
{2013050243, ynxu[*], yhma}@ybu.edu.cn

Yujing Wu and Jin-Gyun Chung

Division of Electronic & Information Engineering
Chonbuk National University
Jeonju, South Korea
{yjwu, jgchung}@jbnu.ac.kr

Abstract

Sensor is an essential device in automation control system, and the reliable sensor signal has the massive benefits of the normal operation of whole system. This paper proposed an improved support vector machine (SVM) which is modified by clone select algorithm (CSA) for sensor fault detection. CSA - SVM observer is established by monitoring sensor in real-time, so as to find sensor fault in time. Simulation results show that the algorithm can be effective for fault detection.

Keywords-Fault detection; Support Vector Machine; Clonal Selection

Introduction

Sensor is a very important approach and methods which is used to perceive, test and obtain the information. Therefore, fault detection technologies of the sensor are particularly important and necessary. SVM algorithm is a new machine learning algorithm based on statistical learning theory which has very strong capability to handle the complicated nonlinear system [1]. However, the performance of SVM depends on the problems of parameters selecting. Therefore, according to the problem of the sensor fault detection for nonlinear systems, a fault detection model of SVM based on clonal selection is proposed and simulation results are given.

CSA-SVM Algorithm

SVM is a powerful learning method machine which is proposed by V. Vapnik [2], SVM is mainly used in classifying and regressing problems. The main idea of SVM is constructing a classification function to classify samples for the training set. Assuming that, the training set is $T = \{(x_1, y_1), (x_2, y_2), \cdots, (x_l, y_l)\} \in (X \times Y)^l$. Where, X denotes the input space, Y denotes the output space. By introducing the transformation of $x \mapsto \varphi(x)$ from the input space X to another higher dimensional Hilbert space H, the original input space X training sets are translated into new training set in the Hilbert space H.

$$\bar{T} = \{(\varphi(x_1), y_1), (\varphi(x_2), y_2), (\varphi(x_i), y_i)\} \tag{1}$$

It makes the Hilbert space H linear separable. The Hilbert space H means the feature space. In order to rigidly divide the training set $\bar{T}$, the hyperplane $(\omega \cdot \varphi(x)) + b = 0$ is obtained through the space H. By introducing the Lagrange multiplier and according to the duality theorem Wolfe [3], convert the problem to its dual problem.

$$\max_{\alpha} \quad -\frac{1}{2} \sum_{i=1}^{l} \sum_{j=1}^{l} y_i y_j \alpha_i \alpha_j K(x_i \cdot x_j) + \sum_{i=1}^{l} \alpha_i \tag{2}$$

$$s.t. \quad \sum_{i=1}^{l} y_i \alpha_i = 0$$

$$0 \le \alpha_i \le C, i = 1, \cdots, l$$

In order to improve the robustness of the support vector regression, the insensitive loss function ε has been introduced as follows.

$$L_\varepsilon(f(x), y_i) = \begin{cases} |f(x) - y_i| - \varepsilon & |f(x) - y_i| \ge \varepsilon \\ 0 & el\ se \end{cases} \tag{3}$$

For a given training set T, the appropriate precision parameter $\varepsilon > 0$, penalty parameter $C > 0$ and suitable kernel function $K(x, x')$ have been chosen. Structuring and solving optimized problems as follows.

$$\min_{\alpha^{(*)} \in R^{2l}} \frac{1}{2} \sum_{i,j=1}^{l} (\alpha_i^* - \alpha_i)(\alpha_j^* - \alpha_j) K(x_i, x_j) +$$

$$\varepsilon \sum_{i=1}^{l} (\alpha_i^* - \alpha_i) - \sum_{i=1}^{l} y_i (\alpha_i^* - \alpha_i) \tag{4}$$

$$s.t. \quad \sum_{i=1}^{l} (\alpha_i^* - \alpha_i) = 0$$

$$0 \le \alpha_i, \quad \alpha_i^* \le \frac{C}{l}, \quad i = 1, \cdots, l$$

So to construct the decision function is as follows.

978-1-4799-5128-4/14 $31.00 © 2014 IEEE

$$y = \sum_{i=1}^{l} (\alpha_i^* - \alpha_i) K(x_i, x_j) + \overline{b} \qquad (5)$$

When the distribution area of two kinds of sample points in the input space is seriously overlapped, it can make the distribution area of the sample points which are reflected to each class of feature space more concentrated by selecting the proper kernel function and parameters. However, the basic approach to improve the performance of SVM is to select proper parameters.

CSA can effectively overcome problem such as premature convergence which is suitable for processing optimization problem [4]. The optimization process of the SVM parameter optimization problem can be with the immune system to identify antigens and the process of antibodies evolution.

Simulation Setup and Results

Common sensor faults mainly include stuck fault, constant gain fault and constant bias fault. Through the residual errors which is between the prediction model outputs and the actual system outputs, the sensor fault can be detected.

In order to verify the validity of sensors fault detection model proposed in this study, simulation of sensor faults are carried out by using MATLAB and the simulation results are shown in Fig.1-Fig.3.

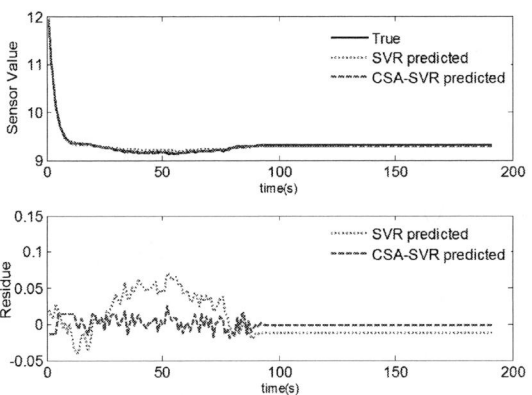

Fig. 1 Sensor stuck fault signal detection and estimation.

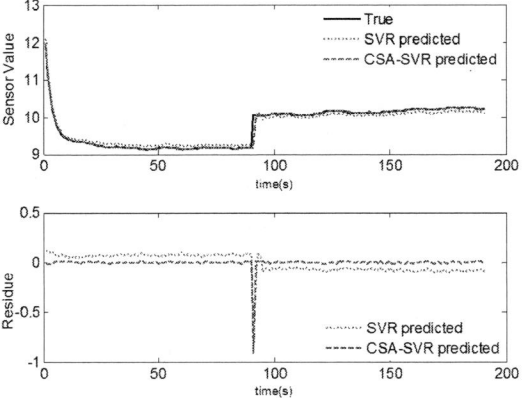

Fig. 2 Sensor constant gain fault signal detection and estimation.

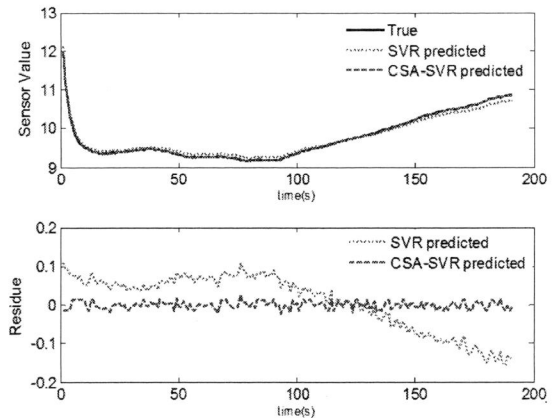

Fig. 3 Sensor constant bias fault signal detection and estimation.

It can be seen from the above results, the SVM model which parameter optimizes by CSA has the higher detection accuracy and stability than the traditional model.

Conclusion

This article researches on sensor fault detection are based on SVM. At the same time, the SVM parameter optimization model has been proposed which is aimed at a big problem of the predicted accuracy of SVM and the stability of the fault detection. Simulated result of experiment indicates that the proposed CSA-SVM fault detection model can accurately and stability detect the sensor fault which laid a solid foundation for the sensor fault-tolerant control technology.

Acknowledgment

This research was supported by the National Natural Science Foundation of China (61361003) and the open fund of Jilin University State Key Laboratory of Automotive Simulation and Control (20121120).

References

[1] WU Jian, Zhao Yang, HE Rui, "Fault detection and diagnosis of EMB sensor system based on SVR," Engineering and Technology Journal of Jilin University, vol. 43, pp. 1178-1183, May 2013.

[2] VAPNIK V, The nature of statistical learning theory, New York, Springer-Verlag, 1995.

[3] Zhi-Quan Luo, Shuzhong Zhang, " On extensions of the Frank-Wolfe theorems," Computational Optimization and Applications, vol. 13, pp. 87-110, April 1999.

[4] Liu Haisong, Wu Jiechang, Chen Guojiu, "Analog circuit fault diagnosis based on SVM optimized by CSA," Journal of Electronic Measurement and Instrument, vol. 24, pp. 1132-1136, Oct 2010.

An Optimal Power Methodology through Constrained Register Sharing

Liu Wan, Chi-Ho Lin, Su-Yeon Song

School of Computer, Semyung University
xibeimy@gmail.com, ich410@semyung.ac.kr, davinchy00@hanmail.net

Abstract

In this paper, we present a new optimal power methodology through constrained register sharing that can be used to optimize the power consumption in the synthesized data path. This paper is the first attempt to the high-level synthesis of constrained scheduling which computes for given number of clock domains. First, the constraints are substituted by subgraphs, and then the number of subgraphs is minimized by using the inclusion and overlap relation efficiently. Also, we show that the algorithm through constrained register sharing which lead to an increase in the number of register. As a results, the proposed algorithm can reduce power in order to share registers and interconnections connected to functional units, as much as possible. The effectiveness of the proposed algorithm has been proven by the experiment with the benchmark examples.

Keywords-Power Consumpution, Clock Period, Register Sharing, Synthesis

I. Introduction

In the past, the most work on high-level synthesis has focused in techniques for area and performance optimization. In recent year, with the increasing performance and integrity of the VLSI circuit as well as the popularity of portable device, power consumption has emerged as an important issue in the design of electronic system. The increasing demand for battery-operated devices, such as cellular phones and notebook computer, requires special attention to power instead of the speed in which is a typical concern in conventional VLSI design. and it have required digital system design for power dissipation increasing the cost packaging or cooling for the power loss is large circuit. It is insufficient high-level research for supporting low power consumption [1-4]. In generally, there are two factors of power dissipation in CMOS:

1) Static dissipation mainly due to leakage current.
2) Dynamic dissipation due to switching activity and short-circuit current.

In modern high-speed circuit design, the optimal power methodology through constrained register sharing has been widely utilized as a manageable resource to improve the circuit performance. The proposed algorithm works on the scheduled input graph and simultaneously allocates and binds registers, functional units and interconnections in stages by considering interdependency between operations and storage element in each control step, in order to share registers and interconnections that are connected to functional units, as much as possible

The rest of the paper is organized as follows. Section 2 describes the optimal power methodology through constrained register, Section 3 describes experimental result in our proposed methodology, and finally section 4 gives conclusion.

II. A new optimal power methodology through constrained register sharing

In this paper, it will empower the chip designers, for the first time, to optimize power simultaneously with area and speed at the early stages of the chip design. This paper's approach rests on the analysis of the multiple sources of power dissipation on a chip (Fig 1.) and development of domain-specific techniques for in behavioral level power optimization for.

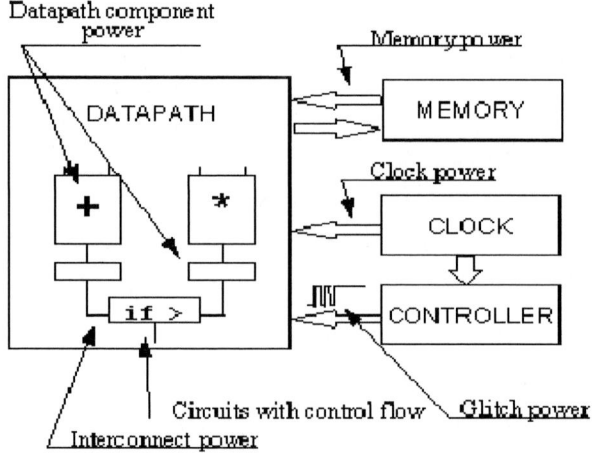

Fig. 1. Sources of total chip power dissipation targeted by power management.

The goal of our work is to introduce an optimal power algorithms used in behavioral level synthesis. Behavioral level synthesis comprises of the sequence of steps by means of which an algorithmic specification is translated into hardware. These steps involve breaking down the algorithm into primitive operations, and associating each operation with the time interval in which it will be executed(called operation scheduling) and the hardware functional block that will executed it (called hardware allocation). Clock-period minimization, throughput constraints, hardware resource constraints and their combination make this a non-trivial optimization problem.

978-1-4799-5128-4/14 $31.00 © 2014 IEEE

Consider the scheduled CDFG shown in Fig. 2 Each operation in the CDFG is annotated with its name (placed inside the circle representing the operation) and the name of the functional unit instance it is mapped to (placed outside the circle representing the operation). Each variable in the CDFG is annotated with its name. Clock cycle boundaries are denoted by dotted lines. The schedule has five control steps, s1,...,s5. Control step s5 is used to hold the output values in the registers and communicate them to the environment that the design interacts with, and to load the input values into their respective registers for the next iteration.

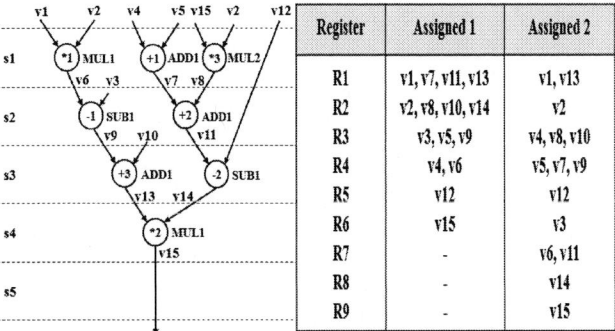

Register	Assigned 1	Assigned 2
R1	v1, v7, v11, v13	v1, v13
R2	v2, v8, v10, v14	v2
R3	v3, v5, v9	v4, v8, v10
R4	v4, v6	v5, v7, v9
R5	v12	v12
R6	v15	v3
R7	-	v6, v11
R8	-	v14
R9	-	v15

Fig. 2. An optimal scheduled result analysis

In order to assess the impact of variable assignment on power consumption, consider two candidate assignments, Assignment 1 and Assignment 2, shown in Fig 2. The architectures obtained using these assignments were subject to logic synthesis optimizations, and placed and routed using a 1.2 micron standard cell library. The transistor-level netlists extracted from the layouts were simulated using a switch level simulator with typical input traces to measure power.

III. Experimental results

In this paper, our synthesis methodology have been implemented in C++ programming language to implement our heuristic algorithm. Also, it have been tested on the Table 1. In this experiment results, we results the calculated power reduction ratio to adopt the HLS benchmark through the result of an optimal data path scheduling technique for low power circuit design.

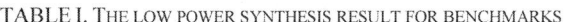

TABLE I. THE LOW POWER SYNTHESIS RESULT FOR BENCHMARKS

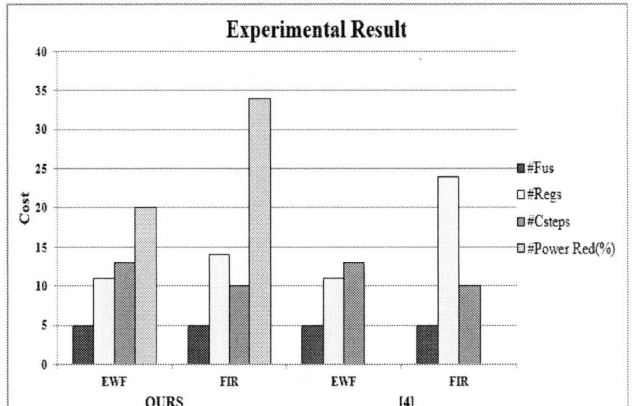

IV. Conclusions

The new behavioral level power optimization methodology through constrained register sharing technique for ASIC design have been performed. The new efficient synthesis methodology for low power design automation has been performed. Unlike most previous work, we also consider the interaction among these tasks in order to better explore the power consumption. We have implemented the algorithm, and presented experimental results to demonstrate its effectiveness. Also, in order to explore this trade-off during behavioral level synthesis, future work will be studied to extend the presented approach to handle accurate power estimation techniques for the clock network, data path, control logic, and registers.

Acknowledgment

This research was supported by the MSIP (Ministry of Science, ICT and Future Planning), Korea, under the Human Resource Development Project for SoC support program (NIPA-2014-H0601-14-1001) supervised by the NIPA (National IT Industry Promotion Agency).

References

[1] A. Chandrakasan, R. Brodersen, "Minimizing Power Consumption in Digital CMOS Ciecuits", *IEEE Proceedings*, vol. 83, no. 4, pp498-523, April 1996.

[2] A. Chandrakasan et al., " Low-Power CMOS Digital Design,", *J. Solid-State Circuits*, vol.27, no.4, pp.473-484, April 1992.

[3] A. Bellaouar and M. I. Elmasry, "Low-Power Digital VLSI Design Circuits and Systems", *Kluwer Academic Publishers*, 1995.

[4] A. Ghosh, "Estimation of Average Switching Activity in Combination and Sequential Circuits", *in Proc. 29th DAC*, pp253-259, June 1992.

[5] P. Landman, "Power Estimation of Behavioral level Synthesis", *in Proc. European DAC*, pp.361-366, Feb 1993.

[6] K.Ravindran, A. Kuehlmann, and E. Sentovich. Multi-domain clock skew scheduling. *In ICCAD*, 2003.

[7] L. Liu, T. Chou, A. Aziz, and D.F. Wong. Zero- skew clock tree construction by simultaneous routing, wire sizing and buffer insertion. *In ISPD, 2000.*ligital VLSI Design

978-1-4799-5128-4/14 $31.00 © 2014 IEEE

Procedures of the Fault Code in the Driverless Mode of the Communication based Train Control System

Min-Soo Kim

Metropolitan Transit System Research Division,
Korea Railroad Research Institute
360-1 Woram-dong, Uiwang-si, Gyeonggi-do, 437-757, KOREA
+82-31-460-5205 and ms_kim@krri.re.kr

Abstract

The Communication Based Train Control(CBTC) system performs a real-time train position report and the movement authority transmission through wireless communications networks between a Wayside Control Center(WCC) and Vehicle Onboard Computers(VOCs). And it has increasing effects of railway capacity by allowing the high-density train control through a high information transmission amount between the WCC and the VOC. Also, it reduces the maintenance cost because it does not use the existing track circuits. This paper deals with the procedures of the fault code including set/release for the emergency braking that occurs during driverless operation in the CBTC system. Also, the CBTC is being developed for application of train operation that driverless, auto, manual, emergency and yard mode operation are possible.

Keywords-Communication Based Train Control; Fault Code; Emergency Brake, Automatic Train Protection

Introduction

Railway signal or Train Control System(TCS) is in charge of ensuring safe railway operation and improving the operational efficiency of railway management by maintaining a safe separation of a preceding train and a following train [1]. Especially urban transit system is a complex system that provides driverless operation by integrating of ground facilities, rolling stocks, railway signals and radio compex networks etc[1]-[3].

Communications Based Train Control(CBTC) system introduced in Korea recently is a railway signal system that based on communications and information technology and it is comprised of subsystems such as Automatic Train Operation(ATS), Automatic Control Protection(ATP), Automatic Train Operation(ATO), Electronic Interlocking (EI) and Radio Complex Networks (RCN) etc. Of these, ATP/ATO is a core subsystem of the signal system, and it covers safe operation of urban rail and automatic operation of trains.

In Europe, the system is incorporating interoperability between equipment by improving the existing line signal system to the CBTC system, for example, the subway in Paris, France. Therefore, reflecting experiences of the Paris subway in improvement of the signal system for urban rail in domestic is possible; it would be helpful in promoting commercialization [3]-[6].

This paper deals with fault code processing and procedure of emergency brake applied by ATS command or vehicle errors which are subsystem of CBTC for urban transit system that driverless operation is possible. This paper is organized as follows. Section 2 and 3 discusses concisely the fault code procedures and experimental results of the CBTC, respectively. The main conclusions are then summarized in section 4.

Procedures of Fault Code

Emergency Brake(EB) code is defined as a subset of fault code sets, and vehicles manage all database. After EB is applied (defiend as 'EB set'), if the cause disappeared (defined as 'EB reset'), onboard ATP(oATP) of the train deletes the corresponding items on the list. Fault code does not need onboard ATP/ATO manufacturer is all same, but ATS shall have all DB that submitted by onboard ATP/ATO manufacturer for the fault code display. It shall be determined by prior consultation as not same fault code assigned for different failure. Also, EB application created by ATS/EI, wayside ATP(wATP), oATP and displayed as fault code, release is possible only 'EB Release Command' of ATS. ATS displays failure data that related to EB from the fault code list.

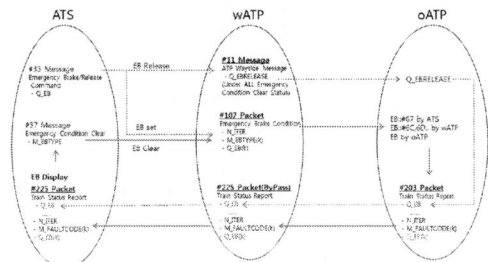

Fig.1 applied and released procedures of fault code(or EB)

A. Fault Code Procedures: EB Set by oATP

① If EB set by error or failure of vehicles, oATP displays EB status by setting EB field of 'Train Status Report' packet, informs error or failure type to wATP through FAULTCODE LIST.

② wATP transmits the 'Train Status Report' packet received from oATP to ATS.

③ ATS identifies EB set on the vehicle through EB field of the 'Train Status Report' packet, identifies EB type through FAULTCODE LIST occurred on the vehicle.

Figure 1 shows ATS, wATP, oATP, delivering relationship of EB and fault code.

B. Fault Code Procedures: EB Release by oATP

① If EB field of the 'Train Status Report' packet is set, ATS identifies EB type through FAULCODE LIST.

② If EB creation cause that set in oATP was deleted, oATP informs to wATP after delete EB item(Fault Code) by ATS between FAULTCODE LIST while maintaining 'EB set' of the 'Train Status Report' packet.

③ wATP transmits the 'Train Status Report' packet received from oATP to ATS.

④ ATS identifies 'EB set' is maintained to the vehicle through EB field of the' Train Status Report' packet.

⑤ ATS transmits 'EB release command' to wATP through the message 'Emergency Brake Release Command'.

⑥ If wATP received 'EB release command' from ATS, change from EBRELEASE field values of wATP message, the message to EBRELEASE('01') and transmit to oATP.

⑦ If oATP received EBRELEASE as '01' values, after release EB, changes its values '10' as EB reset from EB field of 'Train Status Report' packet and transmits to oATP.

Experimental Results

Tests of fault code(or EB) were divided into two cases; EB set by 'overspeed' and EB set by 'loss of movement authority'.

(a) ATS screen before EB and after EB applied by overspeed

Fault Code List: Over-speed (Check propulsion / brakes)

(b) detail screen of ATS (Fault Code is '19')
Fig.2 EB applied by train overspeed

Figure 2 shows 'EB set' by overspeed. By providing temporary speed limit to the operating train, and the train is 'EB set' by overspeed. If the train is operated normally, the train is displayed in green color, and if EB is set to the train, it is changed to pink color. Also, whether temporary speed limit is set as confirm by checking yellow bar under the line. Therefore, it is possible to also check fault code ('19 ') in the detail screen of the ATS.

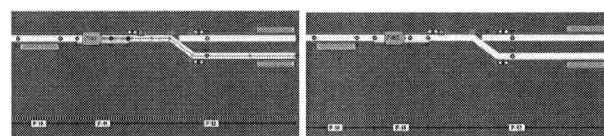

(a) ATS screen before EB and after EB by loss of MA

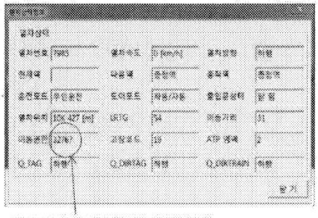

The loss of the Moving Authority(MA)

(b) detail screen of ATS (Moving Authority is 'NULL')
Fig.3 EB applied by loss of the moving authority

Figure 3 shows 'EB set' by 'loss of movement authoority'. It is a test that set the protection area to the train and loss of movement authority ('32767 'or NULL). In order to prevent damage of the train, the operator, in a state of setting the speed limit temporarily in advance, and commanded the protection area. The protection area is possible to confirm by checking the red bar on the line. In addition, confirmation is possible also that the value of the movement authority is given NULL ('32767 ') on the detailed information screen of ATS.

Conclusion

TCS is comprised of automatic control technology, vital software technology, information communication technology, system engineering technology focus on safety etc. CBTC system is a railway signal system that based on communications and information technology and it is comprised of subsystems such as ATS, ATP, ATO, EI etc. ATP/ATO cover safe operation of urban rail and automatic operation of trains.

This paper deals with fault code display and applied and 'EB set' and 'EB release' processing by an error.

References

[1] KRRI, *Train Control System Requirements Specification (SRS)*, Ver.4.0, 2012.

[2] KRRI, *Train Control System Requirements Specification (SRS)*, Ver.4.0, 2012.

[3] *Urban Railway Signal System Standard Specification (proposal)*, Korea Railroad Research Institute, 2007.

[4] *Compendium on ERTMS*, Eurail Press, 2009.

[5] MODURBAN, *Comprehensive operational, functional and performance requirements*, 2009.

[6] IEEE Std 1474.1-2004, *IEEE Standard for Communications Based Train Control (CBTC) Performance and Functional Requirements*, 2004.

[7] IEEE Std 1474.3-2008, *IEEE Recommended Practice for Communications Based Train Control (CBTC) System Design and Functional Allocations*, 2008.

[8] M-S KIM, S-C OH, Y-K YOON and Y-K KIM, "Study on the Procedure of the Emergency Brake in Driverless Mode of the Korean Radio-based Train Control System," *The 12th International Conference on CSECS '13*, Budapest, Hungary, December 10-12, 2013.

Scan Cell Reordering Algorithm for Low Power Consumption during Scan-Based Testing

Wooheon Kang, Hyunyul Lim, and Sungho Kang

Department of Eelectrical & Electronic Engineering, Yonsei University, Seoul, Korea
{sudal, lim8801}@soc.yonsei.ac.kr and shkang@yonsei.ac.kr

Abstract

Power consumption during scan-based testing can be higher than that of normal mode opeartions, which can cause yield loss and degradation of reliability. This paper proposes a scan cell reordering algorithm to reduce the test power consumption during scan-based testing. The proposed algorithm considers both shift-out operations and shift-in operations. A cumulative weighted transition (CWT) is proposed and compared to minimize the test power consumption. Experimental results show that the porposed method greatly reduces the average power during scan testing.

Keywords-component; low power test, scan cell reordering, scan testing

Introduction

Power consumption during testing is important because excessive power consumptions can damage the circuit under test (CUT). Power consumption of integrated circuits (ICs) in scan-based tesing can be significantly higher than that during normal operation [1]. A low correlation between successive test patterns can result in significant switching activities during scan shifting operation.

Power consumption during testing is classified into two types: average power and peak power. The average power is the amount of power consumed by scan shifting operation. In shift mode, a test pattern is shifted into a scan chain one bit by one bit. The peak power is the highest value of the instantaneous power when the test pattern and its test response have opposite logic values. In capture mode, the test response from CUT is loaded into all scans at a time. Excessive test power can damage the CUT, decrease its reliability, or result in yield loss due to IR drop.

To estimate power consumption during the scan shift operation, the number of weighted transitions (WT) is used [2]. The method counts the cumulative switching activity of scan cells during the scan shift-in and shift-out operations. Test patterns with unspecified bits (X-bits) are usually generated during ATPG. To reduce test power, various X-filling techniques are proposed. A widely used method for reducing shift power is adjacent X-filling technique [3]. Because this technique targets only the shift-in operation, it may lead to excessive power during the shift-out operation.

In this paper, a scan cell reordering algorithm is proposed to reduce test power during shift-in and shift-out operations. Scan cells are reordered according to the test patterns and the responses to reduce WT of each test pattern and response.

Proposed Scan Cell Reordering Algorithm

A test pattern and a response to the test pattern are bit streams of 0s, 1s, and X-bits. In order to reduce WT of a test pattern, X-bits in the test pattern are filled with 0s or 1s using X-filling methods. However, this may lead to excessive power during scan shift-out operations because X-bits in the response are not considered during X-filling. Therefore, the new scan cell reordering algorithm is proposed to reduce WT in shift-out operations as well as in shift-in operations after X-filling.

When a test pattern is shifted into a scan chain, a response to the previous test pattern is shifted out of the scan chain concurrently in shift mode. For the point of view of shift-in operations, a transition in the end of the test pattern affects more than a transition in front of the test pattern. When the test pattern is input into the scan chain, the LSB (least significan bit) of the test pattern is input into the scan chain at first and shifted to the end of the scan chain. For this reason, WT can be reduced when the transitions are in the front of the test patterns. Similarly, for the point of view of shift-out operations, a transition in front of the response affects more than a transition in the end of the response. When the response comes out of the scan chain, the MSB (most significant bit) of the response comes out of the scan chain at the finish. For this reason, WT can be reduced when the transitions are in the end of the responses for the point of view of shift-out.

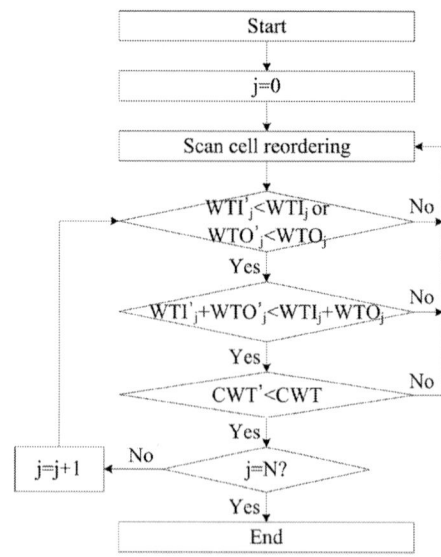

Fig. 1. Flow of the proposed scan cell reordering.

Test patterns /Responses	1	2	3	4	5	6	7	8	9	10	11	12	WTI=124/WTO=97/CWT=221
T_1/R_1	0/1	0/1	1/0	1/0	0/1	0/1	1/0	1/0	0/1	0/1	1/0	1/0	$WTI_1=30/WTO_1=30$
T_2/R_2	0/1	0/1	1/1	1/1	1/1	0/0	1/0	0/1	1/0	0/0	1/0	0/1	$WTI_2=38/WTO_2=18$
T_3/R_3	0/1	0/1	0/1	0/1	1/0	0/1	1/0	0/1	0/1	1/0	1/1	1/0	$WTI_3=22/WTO_3=11$
T_4/R_4	1/1	1/0	1/0	0/1	0/0	1/0	1/0	0/1	0/1	1/0	0/1	0/1	$WTI_4=34/WTO_4=38$

(a)

Test patterns /Responses	1	2	5	6	9	10	3	4	7	8	11	12	
T_1/R_1	0/1	0/1	0/1	0/1	0/1	0/1	1/0	1/0	1/0	1/0	1/0	1/0	$WTI'_1=6/WTO'_1=6/CWT=12$
T_2/R_2	0/1	0/1	1/1	0/0	1/0	0/1	1/1	1/1	1/0	0/1	1/0	0/1	$WTI'_2=16/WTO'_2=16/CWT=44$
T_3/R_3	0/1	0/1	1/0	0/1	0/1	1/0	0/1	0/1	0/1	0/1	1/1	1/0	$WTI'_3=18/WTO'_3=19/CWT=81$
T_4/R_4	1/1	1/0	0/0	1/0	0/1	1/0	1/0	0/1	1/0	0/1	0/1	0/1	$WTI'_4=36/WTO'_4=38/CWT=155$

(b)

Test patterns /Responses	1	2	5	9	6	10	3	7	4	8	11	12	
T_1/R_1	0/1	0/1	0/1	0/1	0/1	0/1	1/0	1/0	1/0	1/0	1/0	1/0	$WTI'_1=6/WTO'_1=6/CWT=12$
T_2/R_2	0/1	0/1	1/1	1/0	0/0	1/0	1/1	1/0	1/1	0/1	1/0	0/1	$WTI'_2=20/WTO'_2=24/CWT=56$
T_3/R_3	0/1	0/1	1/0	0/1	0/1	1/0	0/1	0/1	0/1	0/1	1/1	1/0	$WTI'_3=20/WTO'_3=16/CWT=92$
T_4/R_4	1/1	1/0	0/0	0/1	1/0	1/0	1/0	1/0	0/1	0/1	0/1	0/1	$WTI'_4=14/WTO'_4=32/CWT=138$

(c)

Fig. 2. Scan cell reordering example.
(a) WTs of original scan cells and test patterns/responses. (b) WTs after the scan cell reordering in middle stage. (c) WTs after the scan cell reordering in final stage.

Fig. 1 shows a flow of the proposed scan cell reordering. The notation N is the number of test patterns/responses. The WTI_j and the WTO_j are WTs of the jth test pattern/response. The WTI'_j and the WTO'_j are WTs after the scan cell reordering. The CWT is a cumulative WT from first to the current test patterns/responses, and the CWT' is a cumulative WT after the scan cell reordering. There are three conditions to reduce the WT. After the scan cell reordering, the WTI'_j and WTO'_j are compared to the WTI_j and the WTO_j. If both WTI'_j and WTO'_j are larger than WTI_j and WTO_j, the scan cells are reordered to reduce WT. If one of the WTI'_j and WTO'_j may be larger than original WT, the sum of WTI'_j and WTO'_j is compared to the sum of WTI and WTO. If the scan cell reordering is performed, the WT's are changed. Therefore, CWT' is compared to CWT. If all conditions are satisfied, the next test pattern/response is selected and WT's are compared to the WTs until the last pattern/response. Otherwise, the scan cell reordering is performed and the comparison procedures are repeated until conditions are satisfied.

Fig. 2 shows an example of the scan cell reordering when there are four test patterns/responses and a scan chain consisting of twelve scan cells. Fig. 2(a) shows scan cells and original test patterns and responses after the adjacent filling. For each test pattern/response, the numbers of weighted transitions are calculated. Because the scan cells 1, 2, 5, 6, 9, and 10 have 0s and the scan cells 3, 4, 7, 8, 11, and 12 have 1s in the first test pattern as shown in Fig. 2(b). For the second and third test patterns/responses, scan cells are not reordered since the numbers of weighted transitions satisfy the proposed three conditions. However, the WTI' and the WTO' for the last pattern/response are larger than the WTI and the WTO, the scan cells 9 and 6 are swapped and the scan cells 4 and 7 are swapped as shown in Fig. 2(c). Finally, the total number of weighted transitions is reduced using the proposed scan cell reordering.

Experimental Results

To verify the effectiveness of the proposed technique, the

TABLE I
COMPARISONS OF POWER CONSUMPTION.

Circuits	No. of scan cells	No. of patterns	Adjacnet filling [3]	Proposed WT	Reduction ratio
s9234	228	108	1,478,474	1,145,020	22.55%
s13207	669	90	9,213,072	5,046,707	45.22%
s15850	597	99	8,044,821	5,913,345	26.50%
s38584	1,452	125	52,446,305	47,676,271	9.10%

experiments are performed to estimate power consumption. The ISCAS'89 benchmark circuits were synthesized using the Synopsys Design Vision and the test patterns were generated using the Synopsys TetraMax.

Table I shows the comparison results of power consumption based on the number of weighted transions (WT) using ISCAS'89 benchmark circuits. Since the sacn cell reordering is dependent on the test patterns/responses, the test power reductions may be different according to the CUTs. As presented in Table I, the more number of scan cells, the more WT is generated because the WT depends on the number of scan cells. If a single scan chain is partitioned to multiple scan chains which have the small number of scan cells, the test power reduction can be increaed. The test power is reduced up to 45.22% for s13207 circuit and an average test power reduction is 25.85% for four ISCAS'89 benchmark circuits. Therefore, the proposed algorithm can reduce the test power consumption through reducing the WT.

Conclusion

In this paper, a new scan cell reordering algorithm is proposed to reduce test power consumption during scan-based testing. The proposed algorithm considers shift-out operations as well as shift-in operations because the WT varies according to the scan cell locations. In order to simplify the test power calculation, the cumulative WT is calculated and compared with the previous cumulative WT when the scan cells are reordered. Therefore, the proposed algorithm can greatly reduce test power consumption during scan shift operations.

Acknowledgment

This work was supported by the National Research Foundation of Korea (NRF) grant funded by the Korea government (MEST) (No. 2012R1A2A1A03006255).

References

[1] P. Girard, "Survey of low-power testing of VLSI circuits," *IEEE Des. Test Comp.*, vol. 19, no. 3, pp. 80-90, May/Jun. 2002.

[2] K. Sankaralingam, R. R. Oruganti, and N. A. Touba, "Static compaction techniques to control scan vector power disspation," *Proc. 18th IEEE VLSI Test Symp.*, May 2000, pp. 35-40.

[3] K. M. Butler, J. Saxena, A. Jain, T. Fryars, J. Lewis, and G. Hetherington, "Minimizing power consumption in scan testing: pattern generation and DFT techniques," *Proc. Inter. Test Confer. (ITC)*, Oct. 2004, pp. 355-364.

Fast Allocation of Post-Silicon Tunable Buffers to Mitigate Timing Variation

Hyungjung Seo
shj@snucad.snu.ac.kr

Jeongwoo Heo
jwheo@snucad.snu.ac.kr

Taewhan Kim
tkim@snucad.snu.ac.kr

School of Electrical and Computer Engineering, Seoul National University, Seoul, Korea

Abstract—It is widely accepted that post-silicon tunable (PST) buffer, which can adjust its delay after manufacturing, is an effective design element that can mitigate the timing variation in circuits. However, since the size of PST buffer is not that small, it is very important to minimize the number of PST buffers to be allocated while meeting the timing yield constraint. Recently, two noticeable progresses have been made in the literature: (1) one is devising a formulation of 'timing criticality' which facilitates identifying a set of circuit paths that are more likely to be susceptible to the timing variation; (2) the other is developing a graph based timing representation which enables a fast timing yield computation. In this work, we exploit the two features of (1) and (2) on the PST buffer allocation. Namely, we extract timing critical paths according to (1), rather than relying on a simple rule of thumb, and iteratively allocate PST buffers to resolve the timing criticality based on the fast timing yield computation according to (2), instead of using a very slow Monte-Carlo simulation. Through experiments with benchmark circuits, it is shown that our proposed PST buffer allocation methodology speeds up the PST allocation by 10x∼10000x over a Monte-Carlo simulation based approach. Furthermore, when compared with that produced by a full PST buffer allocation, ours is able to use 89.5% less number of PST buffers on average.

Keywords: Post-silicon tunable (PST) buffer; statistical static timing analysis; yield; timing variation, allocation algorithm.

I. INTRODUCTION

As the process variation seriously increases the uncertainty of circuit timing, many techniques to mitigate the adverse effects of timing variation on circuit timing have been developed. One dominating research direction is developing statistical static timing analysis (SSTA) techniques (e.g., [1], [2]) to model the circuit timing statistically. On the other side, another research direction is towards developing post-silicon tunable techniques, such as body-biasing and voltage scaling, among which exploiting post-silicon tunable buffers (PSTs) has been accepted as one of the promising solutions to the timing variation problem [3].

A PST buffer is a circuit element which can adjust its delay, either faster or slower after the manufacturing stage is completed. Thus, a chip with timing failure can be revitalized by properly tuning the delays of some of PSTs allocated in the circuits. Since the size of a PST buffer is not so small, it is very important to minimize the number PST buffers to be allocated while meeting the timing yield constraint.[1]

Recently, two noticeable progress have been made on PST buffer allocation: (1) one is the work by Tsai *et al.* in [4], in which they devised a cost formulation which enables to identify (so called 'timing critical') flip-flops (FFs), which

are highly likely to demand PST buffers to control the clock signal arrival times to the FFs. Later on, Tsai *et al.* in [5] utilized the cost formulation to iteratively allocate PST buffers under timing yield constraint, in which they used Monte-Carlo simulation in computing the timing yield of each instance of PST buffer allocation; (2) the other is the work by Li *et al.* in [6], in which they transformed the problem of timing yield computation for the instance of the 'full' PST buffer allocation[2] into the problem of computing a probability that each cycle in a graph has a non-positive total (arc) weight. The graph based yield computation enabled a drastic improvement in run time over Monte-Carlo simulation, but the work did not address how the yield computation method can be exploited to minimally allocate PST buffers.

In this work, we exploit the two features of (1) and (2) on the PST buffer allocation. In other words, we extract timing critical paths according to (1), rather than relying on a simple rule of thumb, and iteratively allocates PST buffers to resolve the timing criticality based on the fast timing yield computation according to (2), instead of using a very slow Monte-Carlo simulation.

II. FAST PST BUFFER ALLOCATION

Since our proposed methodology of fast allocation of PST buffers combines the two strengths in [4] and [6], brief reviews on [4] and [6] are covered, followed by summarizing our allocation methodology.

● *Timing critical paths* [4]: Let us examine the setup time constraint:[3]

$$x_i + D_{ij} \leq T + x_j - s_j \qquad (1)$$

where x_i and x_j respectively indicate the clock arrival time to flip-flops FF_i and FF_j, D_{ij} represents the maximum delay from FF_i to FF_j, and s_j is the setup time of FF_j. By [4], slack $\delta(e_{i,j})$ is defined to $\delta_{i,j} = x_j + T - D_{ij} - s_j - x_i$. Then, *normalized slack* λ_{ij} is defined to

$$\lambda_{ij} = \frac{x_j + T - \mu(D_{ij}) - s_j - x_i}{\sigma(D_{ij})}. \qquad (2)$$

The *normalized slack* λ_i for FF_i is then defined to $\lambda_i = \max\{\lambda_{ij} | \forall FF_j\}$. Thus, the smaller the value of λ_i is, the more timing critical the paths connected to FF_i are.

● *Graph based yield computation* [6]: The work in [6] proposed to use constraint graph to compute the timing yield of circuit. The timing relations among flip-flops can be transformed into a constraint graph like that shown in Fig. 1,

---

[1]In this work, timing yield is referred to as the percentage of chips which do not fail due to the timing.

[2]The full PST buffer allocation means to the PST buffer allocation by which the clock time to each FF is exclusively controlled by a distinct PST buffer.

[3]The hold time constraint can be explained similarly.

978-1-4799-5128-4/14 $31.00 © 2014 IEEE

in which the clock arrival time x_i represents the delay of PST buffer B_i, w_{ij} represents the sum of of the maximum delay from FF_i to FF_j and the setup time of FF_j, and r_i is the adjustable delay range of B_i. The work used the fact [7] that a circuit after the insertion of a PST buffer B_i at the clock path connected to each FF_i, $i = 1, 2, \cdots$ is working if there is a value assignment of x_i, $i = 1, 2, \cdots$ that satisfies the setup and hold time constraints, which is equivalent to the constraint that all the loops in the corresponding constraint graph are non-positive.

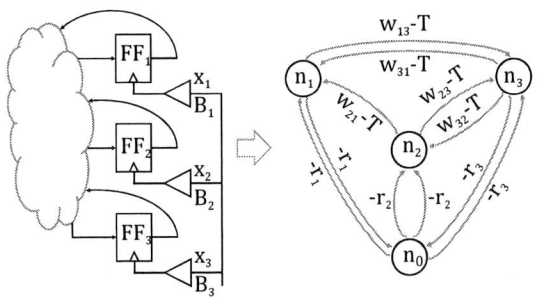

Fig. 1. Constraint graph representing the timing relation between flip-flops: (a) a portion of circuit instance, I, with the allocation of PST buffers B_1, B_2, and B_3 that are used to adjust clock arrival times to flip-flips FF_1, FF_2, and FF_3, respectively; (b) constraint graph $G(I)$.

• *Proposed allocation methodology*: Our iterative PST buffer allocation can be described using flow diagram as below.

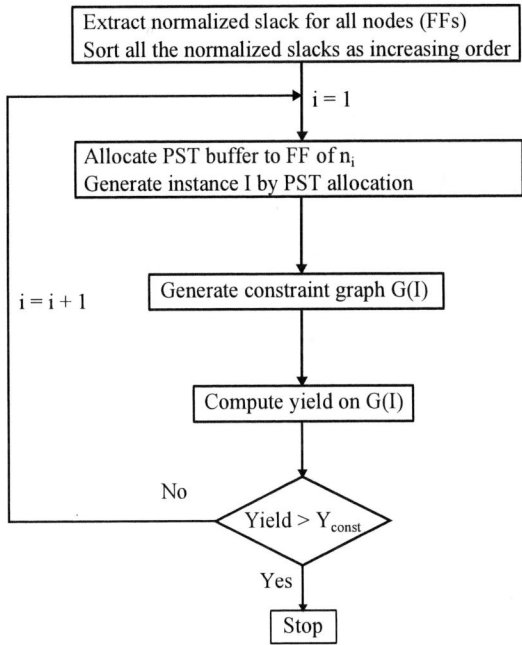

Fig. 2. Flow diagram of our PST buffer allocation.

III. EXPERIMENTAL RESULTS

Our proposed PST buffer allocation method, called PST-alloc, is implemented with Python3 and C++ language on Linux machine with 8 cores of 3.60 GHz Intel i7 CPU and 16 GB memory. We used ISCAS89 benchmark circuits for the experiments. The circuit elements in the benchmark circuits are mapped to 45 nm Nangate Open Cell library [8]. An SSTA engine is implemented, following that in [1] and [6]. Table I shows a comparison of run times used by PST-alloc and the work in [4] with Monte-Carlo simulation. The comparison shows that PST-alloc achieves 10x∼10,000x speed-up. Table II shows a comparison of the numbers of PST buffers allocated by PST-alloc and the 'full' PST buffer allocation. The comparison shows that PST-alloc uses on average 89.5% less number of PST buffers over that by the full allocation at the expense of only 1.6% yield degradation.

TABLE I
RUN TIMES BY [4] + MONTE-CARLO SIMULATION AND PST-alloc

Run time (s)	[4]+Monte	PST-alloc
s386	409.4	0.080
s382	1141.0	0.270
s15850	968.7	26.640
s5378	8160.1	323.060
s13207	10689.4	271.860

TABLE II
THE NUMBERS OF PST BUFFERS ALLOCATED BY FULL-ALLOC+[6] AND PST-alloc

	Full-alloc + [6]		PST-alloc	
	Yield	#PST	Yield	#PST
s386	0.97261	20	0.944625	3
s382	0.99153	30	0.994948	6
s15850	0.99565	235	0.934544	11
s5378	0.93312	247	0.937961	26
s13207	0.98168	482	0.98493	10

IV. CONCLUSION

This work proposed a new PST buffer allocation methodology that combined the merits of two prior works: (1) one is devising a formulation of 'timing criticality' and (2) the other is developing a graph based fast timing yield computation. Through experiments, it was shown that the proposed methodology speeds up the PST allocation by 10x∼10000x over a Monte-Carlo simulation based approach, and compared with that produced by a full PST buffer allocation, it was able to use 89.5% less number of PST buffers on average.

Acknowledgment: This work was supported by the CISS of Global Frontier project by MSIP (CISS 2011-0031863) in Korea, the ITRC program of NIPA by MSIP (NIPA-2013-H0301-13-1011) in Korea, the Seoul R&BD Program (RI130006) in 2014, and the Brain Korea 21 Plus Project in 2014.

REFERENCES

[1] C. Visweswariah, K. Ravindran, K. Kalafala, S. G. Walker, S. Narayan, D. K. Beece, J. Piaget, N. Venkateswaran and J. G. Hemmett, "First-order incremental block-based statistical timing analysis," *IEEE TCAD*, vol.25, no.10, Oct., 2006.

[2] Z. Feng, P. Li and Y. Zhan, "Fast second-order statistical static timing analysis using parameter dimension reduction," *DAC*, 2007.

[3] S. Rusu and S. Tam, "Clock generation and distribution for the first IA-64 microprocessor," *JSSC*, 2000.

[4] J. Tsai, D. Baik, C. C. Chen and K. K. Saluja, "A yield improvement methodology using pre- and post-silicon statistical clock scheduling," *ICCAD*, 2004.

[5] J. Tsai, Z. Lizheng and C. C. Chen, "Statistical timing analysis driven post-silicon-tunable clock-tree synthesis," *ICCAD*, 2005.

[6] B. Li, N. Chen and U. Schlichtmann, "Fast statistical timing analysis for circuits with post-Silicon tunable clock buffers," *ICCAD*, 2011.

[7] T. H. Cormen, C. E. Leiserson and R. L. Rivest, *Introduction to Algorithms*, MIT Press, 1990.

[8] "Nangate 45nm library," http://www.nangate.com/

CURRAN ASSOCIATES INC.
proceedings
.com

9781479951284